And he had in his hand a little book open: and he set his right foot upon the sea, and his left foot on the earth.

REVELATION 10:2

LANDING OPERATIONS

Strategy, Psychology, Tactics, Politics, From Antiquity to 1945

DR. ALFRED VAGTS

Author, A History of Militarism

> **Some people opine that theory would always advise what is most cautious. That is wrong! If theory advises anything, it is in the nature of war that it would counsel what is most decisive; therefore what is boldest.—*Clausewitz*.**

Military Service Publishing Company

Harrisburg, Pa.

Second Printing, January, 1952

Printed in the United States by
THE TELEGRAPH PRESS,
Established 1831
HARRISBURG, PENNSYLVANIA

CONTENTS

PART 1

The Overall Picture

PART 2

Ancient and Medieval Operations

PART 3

17th and 18th Centuries

PART 4

The Age of Steam

INTRODUCTION

Landing operations across the seas are "shockers" among military enterprises, perhaps the greatest among them. This the attacker intends them to be, this the defender fears most. They are apt to upset not only a military status quo, but a psycho-political one as well. Even if military practice and theory have neglected landings for long periods, the memory of peoples along sea shores never forgets them. Hence it is that landings may provide a shock that as in a wounded man may prove more fatal than the loss of blood.

The shock strategy and shock tactics of landings can force war out of its ruts, its "halts in the mud," by providing new openings, new fronts. They are reminders of the fact which military and other conservatives may overlook that war is a political and, potentially at least, a revolutionary act.

Professional militarists of the past two and a half centuries avoided landings that were potentially decisive largely because they shunned cooperation and the inevitable subordination of one service or the other. The battle behind the scenes between the "Easterners," calling for expeditions to the Balkan Peninsula, and the "Westerners" in England and other Allied countries during the First World War, was largely over landings which the most conservative military leaders examined and dismissed with a shoulder-shrugging "Sweet are landings to the non-expert." The same struggle was resumed to a certain extent during the "phony" period of the Second World War when Weygand unsuccessfully proposed a massive landing operation against and beyond Salonika.

Total war has, however, abrogated the old pro- and con inhibitions, against landings and, more broadly speaking, amphibious and tri-elemental war. And today landings achieve, under the impact of necessity and the guidance of unified leadership, a new synthesis of the various forms of violence which military, naval and air force can provide.

PART 1

The Overall Picture

Chapter 1

CONTINENTAL AND LITTORAL WAR

FOR LONG PERIODS war was too continental, too earthbound for landings to be considered either fundamentally necessary or technically feasible, despite the fact that they were most often successful. In Alexander's generalship, although he carried the war across the sea into Asia, landings played no part. On the other hand, it is said by Lord Fisher that to one who was preeminently a land general, Frederick the Great, came a distressing shock when during the Seven Years' War Russians landed on the shores of his Baltic provinces while he was battling far inland. Caesar's versatility and the localities where he had to meet, or seek his enemies made him the leader in amphibious operations of all the generals of classical antiquity.

Wolfe, John Moore, and, more restrictedly, Bonaparte and Wellington knew the technique and understood the strategical importance of landings. For the past two and a half centuries, however, coinciding with the division of war between armies and navies, generals and admirals, opportunities for landings were generally believed unlikely to appear, or be important, or unfeasible. But World War II, a war of sea-borne invasions, has called for and provoked landings again and again, and thereby forces a reconsideration of the history and principles of landing enterprises.

Geographically considered, wars may be classified according to the tendency to wage them on land, or on the sea, or on both, as either continental, naval, or littoral (from *litus*-shore). Littoral war as here understood is the war fought across the shore lines of the .territories of one, or more participants. War is nearly always either littoral, or continental, so much so in fact that probably the majority of sea battles have been fought in sight of land. Except for fairly short phases and for the role played possibly by one participant in a coalition war, or for such a purely naval conflict as the first Anglo-Dutch War (1652-54), there can be no war only on the sea, or only in the air. All war is land based and war inherently demonstrates the strongest possible tendency towards landing operations.

The so-called sea powers have always striven to keep enemies from landing on their own shores, whether they were situated on coastal points of continents favorable for landings, such as Athens, or the United Netherlands; or on islands (Venice and Great Britain); or on separate continents (Carthage and the United States of America). And still, in spite of Mahan, greatest belittler of land warfare and the unity of warfare, there has hardly ever been a sheer sea war, and the chances are slight that there will ever be an exclusively air war, in spite of Douhet's insistence. Martial history shows that final decisions are made on land, and therefore it is reasonable to assume that conclusive results in military enterprises always will be sought upon terra firma, upon which war has a will to land.

MAHAN'S INFLUENCE MISLEADING.

The error of placing wars' decisions with sea power arose in the 17th century, with its early-capitalistic outlook and its economy heavily depending upon foreign trade. Before the 19th century it was possible to damage the interests of influential, or actually governing classes so seriously by sea warfare as to impel them sooner or later to welcome peace. In the 19th and 20th centuries, although the system of industrial production changed so greatly, this belief, now accepted as erroneous, and in large part drawn by Mahan from the "teachings" of 18th-century sea warfare, persisted. If sea power proved its effective potency in World War I, it was for reasons not appreciated by Mahan. It was victorious because sea power could be, and was, used to prolong the war beyond the period of the *blitzkrieg* planned by Schlieffen. It was this prolongation by the Allies that made the blockade effective, partly because the fascination of sea power, as expounded by Mahan and embraced by the Kaiser and his navy and the Reichstag, and not barred by the General Staff, had misled the Germans into misinvesting a certain part of their war power in battleships. If a memorial were to be set up to Mahan in England or America, it should be inscribed in gratitude by those nations: "He taught the Germans the wrong kind of war."

This vain investment on the part of Germany, which it never was able to utilize with decisive effect, enabled the Anglo-Saxon peoples to mobilize their manpower and industry for war and bring them, men and machines together, to the European mainland as the theater of war chosen by them for the decision. Once U-boat warfare, the German way of warding off, or restricting landing, had failed, German military thinkers like Seeckt, who lacked Ludendorff's gambling pertinacity, knew that, considering war technology as then evolved, England and the United States could not yet be restrained, or beaten by Germany's war power.

The error of Douhet in his insistence upon placing the decision with the air force is even more singular. Briefly, this Italian Fascist general expounded the view that concentrated air attack on enemy cities and other centers of population, administration and industry would lead the terrorized population to force their government to sue for peace. Mahan predicted and relied upon the active and passive pressure exerted by the weight of ships on the seas. Douhet theorized that it was equally the mechanical pressure of air ships, neither he, nor Mahan concerning themselves greatly with the thought of how this mechanical pressure would be transmuted into a social-political one. The special mistake of the Fascist authority lay in supposing that the democracies would yield to the pressure of panic so soon as the external mechanical weight was strongly applied, and that the Fascist process, supposedly controlling popular movements, would free Fascist governments from such a menace. Actually, of course, England survived the air blitz, while Mussolini and his Fascist dictatorship fell victims to the silent pressure exerted by a war-weary populace when Italian soil was invaded.

Army-Navy-Air Triangle

In a way, Mahan's theories were the outcome of anxiety on the part of the sea powers to prevent war from becoming total; while Douhet similarly expressed the same concern on the part of the original air powers. Total war had been feared in a measure and become highly suspect by Western society ever since the French Revolution, and since Clausewitz delivered his preachments. It was, so to speak, the fear-dream, the trauma, of that society, and it is not to be wondered that a somewhat neurotic personality like Mahan gave it voice and theory. So-called sea power, working not alone, but in combination with other restraining influences, did keep war within certain bounds for a long time. As Sir Francis Bacon once put it: "He that commandeth the sea is at great liberty, and may take as much, or as little of the war as he will." The air arm of war overreached the long arm of sea power some time after World

War I. And since then there has no longer been any choice in the acceptance, or refusal of war. It is no longer possible to conduct war within a single department, or reach a decision by or in air or sea, without regard to land operations. Actually, the decision lies in the ever changing center of gravity within the army-navy-air force triangle.

Up to the burgeoning into effectivity of aviation, at least, the pure type of continental war was common enough in history, remote, or late. It would be represented by such conflicts as the Indian wars of the United States, or the Russo-Polish War of 1919-20. A land war with a very small naval participation would be the Franco-German War of 1870-71. The vast majority of the medieval wars were continental, although the Hundred Years' War was of necessity littoral, as were the Crusades which the feudal potentates and commanders tried to keep land-bound; hence the arduous and manpower-wasting marches to the Holy Land, when the water routes were so much shorter, safer and better known. The Thirty Years' War, in spite of a few maritime episodes, such as the futile siege of sea-supplied Stralsund by Wallenstein (1583-1634), Austrian generalissimo and admiral, without a fleet, and the invasion by Gustavus Adolphus (King of Sweden, 1611-22) of Germany, was overwhelmingly continental. The series of struggles between England and Spain were, in the nature of things, of littoral type, and only occasionally of the completely maritime type, such as the Armada which was planned to end in a great and triumphantly-conquering landing. World War I was never as littoral, or amphibious as it might have been, considering the opportunities that it offered.

The defensive policy of the sea powers in their struggle against the threat of German domination from the land in World War I had France as a *pied-à-terre* with all necessary harbor and supplying facilities. In a way, France aside from being a military ally, was from 1914 to 1918 one great terminal for the war-making Anglo-Saxon powers. But in World War II, after the disasters of Dunkirk and Greece, when England evacuated the European continent for three or four years, and after Pearl Harbor, the imperative task of landing on hostile shores was one that could not be eluded by the political, military and naval leaders of the United Nations. In a large measure, both Germany and Japan were reasonably immune from the effects of conventional blockading. Therefore, amphibious warfare was forced on the United Nations by the Axis powers, and also by the expectations of the Russian ally. Since the United Nations had made little, or no practical, or theoretical preparation for this kind of warfare, they proceeded experimentally against the German-controlled eastern Atlantic coast, the Pacific islands, North Africa, Sicily, Italy, in preparation for the D-Day of 1944.

War Will Land

Differently put: Total war allowed no more of evasion of amphibious or, better named, tri-elemental war. (Triphibious in analogy to amphibious, from *amphi,* both, and *bios,* life, is clearly an impossible word, even if attempted by such a sublime master of language as Mr. Churchill.) War could no longer exclude landing operations. War was to come back to earth again for all decisions. Even the Axis admirals realized that "combined operations represent the most effective form of naval warfare" (Admiral Dr. Groos. *Seekriegslehren).* They landed from the air, as they did in Franco's Spain, or from the sea, as the Japanese did in China, using for the first time the modern, steel landing craft, planned to be bullet proof, and motor-driven for speed and provided with facilities for rapid disembarkation of men and materiel.

After these landings, World War II, in its initial phases, assumed a predominantly continental character. The campaign of Poland and the "phony" war along the Maginot and Siegfried Lines seemed to continue the tradition of land-locked struggle, even though Britain was involved in it from the outset. With the attacks on Norway, Holland, Belgium and France presenting a nearer threat to Britain, it took on far more of littoral qualities, graphically emphasized by the delimitation between unoccupied and German-occupied France which, with its extension to the Franco-Spanish boundary on the Bay of Biscay, included in the German zone all of France's Atlantic coast line.

At the end of the fighting season of 1940, the German Atlantic Wall extended from Hendaye to Hammerfest, representing the littoral line of *Festung Europa* in the west. A new Continental System, even more extensive than that of Napoleon, had come into being. The Balkan campaign brought a part of the Black Sea littoral and practically all of the Aegean coast line under German control, but the German-Russian campaign again made the war largely continental, as far as concerned the actual fighting engaged in by the largest part of the armed forces.

With the United States in the war, the littoral character was intensified. The cry for a second major front at a time when there were actually far more than two fronts, was perhaps due to the vague general awareness by military men of the pendulum swing, and the public's desire for more action along the shores of the Axis-controlled continents. Although the demand was undoubtedly of sincere origin and justified, there was relatively little popular realization of the technical tasks involved in actual littoral war *on land* and also in the air, where it had never really ceased. It was regarded far too much in the terms of the old-style, pre-aviation landings. Or, in other words, our long continued unfamiliarity with landings and their problems, due to the series of continental wars, had been reflected in a general neglect of the landing problem.

Whether a war is to be conducted along the shore-line, or deep inland, may be determined by various elements; by geography, the fighting materials available to the combatants and also by a less calculable factor, the mental, or temperamental predispositions of the war-leaders. The term "Continentals" used in the American War of Independence pointed to the original tendency on the part of the Americans to keep the war inland. When the British followed them there, instead of sitting down on the coast, the war was lost by the British, according to one of the best 18th century military writers, General Lloyd. One of the few landing operations of the Continentals, the expedition sent to the Penobscot in May, 1779, from Boston, without Washington's concurrence, failed. Instead of making use of the surprise they had achieved and taking the unfinished works by storm, the forces from Boston began regular siege operations and sent for reinforcements. A British naval force destroyed their ships in August and forced them to march home through the wilderness. It was on this occasion that John Moore, later one of Britain's best landing generals, began his military apprenticeship.

The choice between the two types of war is seldom a free one. For one thing, the departmentalization of war, a process whereby not only the conduct of campaigns, but the relative investment in either land, or sea armaments was divided between two or three departments, one each for army, navy and, latterly, air forces, has been one of the determining factors ever since that process began, in the 18th century, if not earlier. Of all human labor and activities, war was the first to be specialized. Nowhere was division of labor carried out so early and in such a far reaching manner.

Specialization Determines Campaigns

In many more respects than people nowadays realize, did war work and war organization, with war as the largest employer of men and the greatest consumer of materials, precede industry in organization. This specialization at last went so far that the instruments created and maintained for war in the end determined the forms of warfare, rather than vice versa. Even where conditions might have called for other wars than plain land, or sea war, the existing specialization for warfare determined whether a campaign was made on land, or on the sea, or, since the accession of air power, in the air. This is another way of saying that there never has been so much amphibious warfare as there might have been. Amphibious warfare seemed to be none of the business either of the army or the navy—the two traditionally separated services.

No military historian, or soldier will maintain that this division has always been a balanced one. War's successful faring has often suffered by this division. Nothing can better demonstrate that than the history of landings. Too often has a land-minded department overlooked the possibility of a coastal invasion, or a naval office failed to carry the war from the sea to the coast of the enemy.

It is fully in keeping with this departmentalization of war that none of the great military theoreticians, such as Clausewitz, Jomini, Mahan, have ever done full justice to the problem of landing. They themselves and their frame of reference could not escape the sharp bi-sectioning of war in their own day. Their thoughts and works on war dealt with either land war, or sea war, with either one, or the other being taken as "decisive," a term so often a synonym for service egotism. Landings, in their outlook, were something distinctly marginal and not truly decisive.

The mere co-existence of the two services led, of course, to numerous arrangements for, and writings about, their cooperation in war. But these for the most part proved merely perfunctory, and whenever undertaken in peace time maneuvers seldom provided true, or useful preparation for war. National defense policy, grand strategy combined operations, of which landings form today the most important, really the culminating, part, were then too often merely harmonizing formulae.

Even when outside authorities could determine the politico-strategical division and assignment of forces for war, there was no assurance of the tactical cooperation necessary to prepare and carry out amphibious enterprises. The Dardanelles and Salonika expeditions were presented to the services from the outside as fundamentally sound ideas. They were subsequently bungled by the services, which is not to say that the civilian war-makers ought not to have known far better than they did the mentalities and professional prejudices of those to whom they entrusted the notion for execution.

Chapter 2

LANDINGS—SCOPE, MEANS, POSSIBILITIES

OF THE RECENT war we can say that none was ever so littoral in character. Even before the geographers had finished the detailed measuring of the total sea coast lines of the globe, the evolution of aviation added immensely to the *litus* of war. Indeed, it introduced a new *litus,* the line of immeasurable length at which air and land meet. While the tremendous length of the earth's coast lines indicates the greatest theater for what was usually called amphibious warfare, the meeting line of earth and sky, and to a certain extent the meeting line of sea and air, indicate the practically ubiquitous littoral character of most modern warfare and the possible stretching of the war fronts to unguessable expansions.

During the 18th century existing roads, however poor, passes, fords and bridges would prescribe to a land invader where he must make his incursions. In the 19th century, and to 1939, army movements were still narrowly circumscribed and tied largely to metalled highways and to railroads, while navies, in spite of the greater measure of self-propulsion and control given them by steam, were mostly linked, or linked themselves, to ports and were dependent upon them and their facilities for ingress and operation. Air ships of all types enjoyed only a limited emancipation from air-ports and hangars. What may be termed, for armies, ships and planes, the roadless era, freedom in movement from existing routes and bases, began in earnest in 1939; freedom from the road coming through the tracked vehicles, from the seaport through the landing craft, from the airport through the plane carrier, the glider and parachute landings. Of course, much of this emancipation is true only in connection with the opening phases of an attack which will have to be sustained from bases and over roads newly won.

Landings, or what corresponds to them, within the range of such operations now can take place nearly everywhere. Air bombardment, after all, is nothing but a transformation of nearly horizontal firing to nearly vertical firing, with the potential odds in favor of the air bombardier. Two bombardiers firing from land and sea respectively are more nearly equal. But in the exchange of fire between air force and land force, gravitation has given the odds to the former. It should be remarked that vertical air bombardment can be changed again into horizontal firing by landing airborne artillery. Air landing is clearly limited, but whether landing from the air is, by broad comparison, more limited than landing from the water, provides one of those problems in warfare that invite narrow and specific generalization beyond that which can as yet be given it.

In all save linguistic definition, landing from the air is feasible on the sea as well as on land. The thought of hostile landing from the air was bound to occur soon after the first ascents and descents of the Montgolfières and other air balloonists in the 1780's. The military thinkers and theorists of the 18th century, and even of the Napoleonic wars, were on the whole too professionally conservative to appreciate the possibilities of the balloon, and so were the majority of those in the 19th. An illustration of this is offered by the experiences of the aeronautical inventor, Count Ferdinand von Zeppelin (1838-1917).

War Usefulness of Aviation

To a revolutionary thinker and politician like Benjamin Franklin, however, the thought of the war-usefulness of the gas-bags came quite naturally when he observed balloon ascensions in Paris in the 1780's. Might not military commanders land troops behind the enemy's lines? In spite of such thought-inspiring maxims as that of Frederick the Great, that two soldiers at the enemy's rear are worth more than ten opposite him in the front, the idea of landing from the air—if we omit the few balloon flights from besieged Paris in 1870-1—did not find acceptance until again revolutionaries like the Bolshevists obtained actual control over an army, although the irritation and apprehension provoked by the involuntary landing of a Zeppelin at Luneville, on French soil, early in this century, showed in what concern the possibility of a hostile landing of such craft was held by the French, and presumably others.

The Russians, accustomed like Franklin to propagandize ideas in the enemy's rear, evolved military mass landings by parachutes, which was further encouraged by the opportunities provided by the vast territories of Russia and the prospective theaters of warfare. What was originally a technical evolution in transportation was wedded to political revolutionary propaganda schemes; the offspring of the union was a piece of military progress. This is one of war's complicated evolutionary tricks.

So recent is this evolution of air landing that the language has hardly kept pace with it. We call combined land and water expeditions, including landings, amphibious, from the Greek *amphi,* meaning both, of both kinds. Since the introduction of aircraft amphibious enterprises may combine land and air, or sea and air, operations. But it is the combination of all three which makes modern war tri-elemental, a term that seems preferable to the usually proposed three-dimensional.[1]

Of all the various operations involved by littoral war, such as blockade, bombardment of hostile shores, mining of coastal waters and overseas transportation, landing is the most important, the most significant, the most complicated, and therefore most elaborate. Indeed, littoral war need not always begin with landing—Sherman marched to the sea instead of from it. But the moment felt by one to be truly momentous is landing from the sea; it may easily be a decisive stroke.

Landing, hostile landing, (in French *descente* or *debarquement hostile,* in German *Landung,*[2] in Italian *sbarco,* and in Latin *copias ex classe in hostium educere)* has been variously defined. An 18th-century war, engineering, artillery and sea lexicon thought it meant "to send troops over seas and disembark them on enemy's shore for the execution of a definite purpose." [3] As a conclusive and illuminating definition this stands up rather better than a French one of more than a century later, according to which hostile debarkation, or maritime descent is "an ingression by sea and the aggressive outrage which a land army offers to the shore of an enemy country." For a technical definition, the personal element here injected is rather too obtrusive. It goes on to say that landing "ordinarily takes place either stealthily and at night, in order to protect oneself from the batteries which fire red-hot shot; or under the protection of vessels which sweep the shore, take care of the guard ships and fight down the small forts, martellos, towers erected to dominate possible landing points. . . . To form for oneself a base of operations should be the next move in a landing. . . . The landings at Quiberon (France, by British, 1759), Abukir (Egypt, by the British, 1801), Helder (Holland, by the British), etc., have testified how very adventurous such enterprises are." [4]

Both definitions are those of army officers, who view landings as made from the land, so to speak; naval officers will see them rather as ex-sea operations. To them a landing operation will be "the putting on land of part, or all the people on board a ship for warlike purposes." As a matter of principle a ship lands only so many people as can be spared, the while keeping aboard enough to man her for action, so that landing and re-embarkation, if the latter become necessary, can take place under her protection. Landing maneuvers in our own time belong among the most common exercises of war ships, and can be planned to involve relatively small exertion on the part of crews since the introduction of power pinnaces. The troops disembarked by one, or several ships are called landing corps, and they always include the marines, or sea infantry if any are among the ship's company. The boat weapons, small guns and rocket batteries serve as field artillery in case of a landing.

"Considering the means of defense and communications at the disposal of most modern states, landings became increasingly more difficult. Preferentially, flat-lying coastal regions, allowing good approach and vision, will be chosen which, however, should not be too close to important places, or strong points. Nowadays (1878) landing will only have chances to succeed when and if the decision on land cannot be reached, or large numbers of troops appear dispensable near the beginning of war; both cases, however, will rarely occur." [5]

It is significant of the thought encouraged by the strategic-political situation of that time, the era of the later Moltke, that this German army-edited encyclopedia of the military sciences which is quoted left the authorship of such an article to a naval officer. It meant in a way that the Great General Staff was disinterested, in fact believed landings were unlikely and impractical, if not downright impossible except, perhaps, in connection with police activities in colonies, or in colonial expeditions. If an army man had written on the topic, in the last years of Moltke, he would have relied on the naval opinion and almost copied the naval man's words.[6]

Discarding the most obvious time-imposed limitations, we might define landing at present, or in the days of pre-aviation warfare, as the invasion of enemy territory by crossing its sea frontier, actual usage of the term being largely restricted to overseas operations, excluding river and lake crossings. However, much of what is to be said of trans-marine landings also holds good for wars involving river conflicts, as on the Yangtse in China during the Taiping rebellion and World War II, and for lake warfare as well. It will be either uncontested, or contested by forces which are land-based or sea-based, or both.

Landings vary according to the intentions of the invaders, who may come as a raiding, a diversionary, or an invasion force, for a hostile foray, or a prolonged occupation. Raiders, prepared for more or less prompt reembarkation, may have been sent for reconnoitering, for a diversion of enemy forces, to weaken morale of the enemy, or strengthen that of the population in enemy-occupied lands, or to destroy an important installation on, or near the seaboard. Invading forces come to stay (the long outmoded metaphor of military language insisting that they "burn their ships behind them") and are prepared to accept battle at once, or eventually.

A landing is either nearly point-like or line-like at the outset, the expansion of point, or line depending on the objective pursued, or the success of the post-landing attack. The use of airborne invasion forces landing further inland, behind the defenders' line, allows an invasion in depth from the very beginning. The water-borne invaders will, if possible, join the inland parachutists who are bound to be without support for a time, and are often faced with annihilation as they try to inflict a maximum of damage upon enemy communications and installation, and to behind-the-front morale.

Air and Sea Landings

Parachutists, in mobility and action, are definitely ruled by the weather, and subject to air currents even more than are ships to water currents and apt to be driven far from their destinations. Whether orientation, the placing of forces where they are wanted, from the air is easier than orientation from the sea would seem to depend much on training and chance.

The common first object of a landing enterprise from the sea is to gain the control of a beachhead, a certain length of coastline—38 miles for the Americans and 37 for the British in the original landing on the shores of Sicily—plus the hinterland of this shore. A country's coastline is usually a thinly occupied frontier, in large part defended by what is called "nature." Only a small part is permanently occupied and fortified with coastal defenses. These seldom can provide an uninterrupted fire curtain over the whole length of the nation's shore. In modern war belligerents have defended large stretches of their coasts with wire entanglements and other field fortifications; where danger of landing seemed more serious martello systems of coastal fortifications have been introduced. England, uncommonly impressed by the strength exhibited by one, or two such towers on Corsica in 1794, in face of English landings, soon after built at points on her coast numerous martello towers of solid masonry with two or three guns. While they were never put to any serious test, afterwards the Germans in World War II embraced a modified martello system for their Westwall along the ocean shores of France and the Low Countries.

Their vicinity with relation to the potential points of embarkation for landing forces and certain features of the enemy coast line, such as proximity of the 5-fathom line to the coast, gently sloping beaches, make certain coasts more invitingly vulnerable than others, which may be rockbound, or their water approach too shallow. Certain other features of coastal formation seemed for a long time to make landing practically impossible until the necessities, or unconventionalities of World War II forced such attempts to be initiated. Pre-1939 conceptions of the unapproachability of certain coasts, like those apparently underlying the theoretical impregnability of the defenses of Singapore and other tropical places, were expressed by a military geographer like Professor-General Haushofer, who added malaria to the obstacles in the way of an invader, although handicapping the defender as well. Haushofer said: "Let someone try an approach at night on a mangrove coast and see what then becomes of troops not absolutely seasoned!" For him the "classical terrain of littoral war geography are coasts with not too rich a precipitation (because otherwise coastal forests, mangrove coasts, etc., become more important for the defense than the mere contact of sea and land)," thereby implying that mangrove coasts seemed closed to war in general and landings in particular.

Further, he thought, such ideal terrain required "coastal mountain chains, not too many rivers—such regions as Phoenicia, Hellas, Liguria, parts of the Black Sea coasts, the Pacific coast of America and, generally speaking, the Pacific type of coast as compared with the Atlantic type and its deeper inlets and cutins; the Malabar coasts, the longitudinal coasts of Burma, the Malay Peninsula." Writing his War Geography in 1932, Haushofer promised a vast future extension of military geography, including landings, to "parts of Australia, southern China, Norway, Alaska, the Russian coastal province in the Far East, etc.—enough fighting space to make us suspect variants," according to the geopolitician, whose old relationship with the Japanese military may help to explain some of the choices of new theaters of war made by him at the time.[7]

Landings form by far the largest and most important part of so-called combined—or as the 18th century termed them, conjunct—operations. In Jomini's system they rank under "mixed operations" (mixing and combining not so much army and naval forces as strategy and tactics), and he groups debarkments on hostile coasts, consequently, with river crossings, retreats, pursuits after a battle and winter quarters. The term "combined operations," wrote Britain's Chief of Combined Operations in 1943, Lord Mountbatten, in demonstrating how little had been done as yet in the theoretical clarification of his own tasks, "is vague and does not convey more than a general meaning; but their scope is definite and precise." And then goes on: "A combined operation is a landing operation in which, owing to actual or expected opposition, it is essential that the fighting services take part together." [8]

All Operations Are Combined

This would seem simply a doer's definition; to its "native hue of resolution" a combined operation is simply a landing operation. But more logically the latter is merely one of various possible combined operations, which would also include feints by a naval force to support movements of the army and *vice versa,* involving the threat of landings, the carrying of troops and their supplies, the mutual lending of fire support during combat, either directed from ship to shore or from shore to ship.[9] This combination was rather rare before the present war when it has become almost commonplace. Comparing this with other wars, it would now astonish the public less than the historian to hear of a submarine cannonading a train on a line running close to the shore, or exchanging gunfire with German tanks.

No longer ago than 1915 or so, it might exasperate an exceptional layman like Lloyd George that "no attempt was ever made by our powerful Navy to turn its great guns on the submarine nests of Flanders." But when he ventured to suggest such an idea, it was turned down peremptorily.[10]

The comparative strength of land-based and ship-based guns, more loosely prejudged than actually ascertained by experiment or experience, falls under a contemplation of the realities of combined operations, although the common prejudice in favor of the far superior strength of land guns has contributed for long periods to delay such unions. Coastal defense occupies a rather ambiguous position in the field of combined operations. This is evidenced by the ever-changing assignment of its tasks either to the army, or the navy personnel in various countries.

Since the addition of air forces to the two traditional services, there are almost no operations left that are not in one way or another combined. Some have been so bold as to declare, as did three of Britain's elder statesmen in the military field, in advocating a consolidation of all her three services into one, that: "Today all operations are combined operations."[11]

In his final press conference in Paris, prior to his visit to the United States in June-July, 1945, General Eisenhower urged the correspondents to study carefully one thing—"the value of integrated tactical power in war." Explaining he said: "When you put sea, ground and air power together, the result you get is not the sum of their separate powers. You multiply their power, rather than add."

By then already that multiplication of power had become evident in the Pacific, where the integration of the three arms had steadily been on the increase. For example, sea power brought the Marines to Iwo, air power protected them and their ground power won the island, so that until Okinawa and other islands closer to Japan proper were later taken, air power could in turn be conserved at Iwo when the B-29's reached the limit of their flying capacity on the long return flights from bombings of Japan. The fast carrier task force ranged off Okinawa as a barrier, while "jeep" (escort) carriers gave close-in air support to the ground forces until land-based planes could use the airstrips, which were the ground forces' first objective.

NOTES, CHAPTER 2

[1] "In fact, the time may be approaching when it will no longer suffice to study strategy from the 'amphibious' point of view, the issue will be decided in the air, as well as by sea and land; strategy must then be decided 'in three dimensions,' and the relationship between time and space in the movements of the different forces will require careful study. Roughly

speaking, aircraft may move about five times as fast as ships, and thirty or forty times as fast as armies." Sir George Aston: *Sea, Land and Air Strategy. A Comparison.* Boston, 1914. 264. Based on lectures delivered at the Staff College at Camberley, 1904-7.

[2] The German term *landen,* to come or bring from water to land, is of Nether German origin and in the 17th century drove out the High German *länden.* This was part of the slow growth of a unified nautical language in the German.

[3] Lt. Col. Rudolf Fasch: *Kriegs-, Ingenieur-, Artillerie-, und See-Lexikon.* Dresden and Leipzig, 1735.

[4] Général Bardin: *Dictionnaire de l'armée de terre.* Paris, 1851. II, 1789.

[5] Kapitän-Leutant von Holleben, Imperial German Navy, in Colonel B. Poten (ed.): *Handwörterbuch der gesamten Militärwissenschaften.* Bielefeld & Leipzig, 1878. VI, 144.

[6] See article *Landung* by Julius Castner, Royal Prussian Artificers Captain: *Militär-Lexikon.* Leipzig, 1882. 233.

[7] *Wehr-Geopolitik.* Berlin, 1932. 78 ff.

[8] *Combined Operations.* The Official Story of the Commandos. N. Y., 1943. v.

[9] UP despatch from London, 8 April 1944.

[10] *War Memoirs* of Lloyd George. III, 83.

[11] *The Times,* 1 October 1943.

Chapter 3

SUPREME COMMAND AND COOPERATION IN COMBINED OPERATIONS

> "If everything went right, they would win a great success, and although Hornblower had never yet heard of a combined operation of war in which everything went right, he could still hope for one."
>
> C. S. Forester: *Captain Horatio Hornblower*. 1938. p. 402.

BEFORE WAR BECAME DEPARTMENTALIZED, Caesar, the Viking sea-kings, William the Bastard, even William III when he set out from Holland to take the crown of England in 1688, were war lords in every field; each was general *cum* admiral. It was after them and full separation that some of the great landing fiascos took place. When Napoleon Bonaparte held, or controlled all the power there was in France, he found the fact that division of war labor had injected into the conduct of war a convenient excuse for his failure to carry out his long and elaborate preparations to invade England. His navy had failed him, or else (in 1805) he could have successfully sailed across the Channel and on England's soil brought perfidious Albion, as the phase went among the French, under his yoke. Or so he wrote in retrospect from St. Helena.

Specialization in the handling of weapons on sea and land was a factor in bringing about practically complete separation of armies and navies in organization and personnel. The American Congress created originally, in 1789, a single Department of War, but in 1798 the ambitions, or the interests working for an independent navy carried, and a Navy Department came into being. After this separation had been perfected in most countries, only a monarch, a chosen chief executive, or an honorary title-holder could by the nature of things at the same time function, actively or passively, as a general and an admiral. The holding of a commission in both army and navy of a country, simultaneously, or in succession, became rather exceptional, as in the celebrated case of Sir Eric Geddes, a war-organizer of unusual abilities, who was forced on both services from the outside; or in the more recent case of Lord Louis Mountbatten who held high rank, if only temporarily, in the British Navy, Army and RAF.

The relations between the two services were at best *schiedlich-friedlich,* peacefully separated, but more commonly highly competitive and inharmonious when both were forced to work closely in union in the same operations. As seen in the light of these experiences, the Marines of the various maritime powers more, or less provided an element of passably cohesive harmony, as soldiers of the admirals, by enabling the latter to undertake combined operations, on a minor scale, such as the 150 landings of the U. S. Marines from 1800 to 1934, without the assistance and concurrence of generals and the army.[1]

The usual, most charitable, although hardly the truest, explanation of failure of army and navy non-cooperation is that neither understood the business of the other. This is the explanation, or excuse given by Mahan for the disastrous conflicts between Admiral Vernon and the commanders of the British land forces thrown together with him in 1741 and 1742 in the attempts against Cartagena, which had thrice before fallen to invaders, and Santiago de Cuba. "The admiral and the general quarreled, as was not uncommon in days when neither had an intelligent comprehension of the other's business." Mahan quotes another naval writer, Captain Marryat, who seems to have had Cartagena in mind when he wrote: "The army thought that the navy might have beaten down the stone ramparts ten feet thick; and the navy wondered why the army had not walked up the same ramparts, which were 30 feet perpendicular." [2]

Problems of Supreme Command

The vexatious problem of who was to hold supreme command over a landing force found a theoretically satisfactory solution, it is true, in the arrangement that during the naval part of an operation the highest ranking naval officer would rate control, which would pass to the highest ranking army officer so soon as landing and land operations began, unless the navy man preferred to have the latter undertaken by sailors and marines, the presence and use of the marines being, in a sense, the product of the determination of the navies concerned to carry out transmarine enterprises exclusively with forces of their own. The problem became even more complicated in coalition warfare, by the addition of national jealousies and personal ambitions. One recalls such occasions as the Boxer campaigns in 1900 when first an English admiral, Sir E. Seymour, commanded landing contingents of altogether 2,000 men representing no less than eight nationalities. A German Fieldmarshal, the so-called *Weltmarschall* Count Waldersee, later took over supreme command, although this seems never to have extended to the navies in Chinese waters.

In World War I the problem was by no means automatically solved. While the German enterprise against Oesel offered rather exemplary arrangements for changing command and the British Zeebrugge attack of 1918 was altogether naval in character, Gallipoli stands out preeminently as a ghastly illustration of the results of failure to cooperate, or form unified strategic conceptions. To put the experiences of the Dardanelles in the language of one of the survivors of the disaster, Sir Ian Hamilton, the military commander of the enterprise in its middle period, who issued his reminiscences during the early days of the vastly different Sicilian expedition of 1943:

"The original plan had called for the British [and French, it might be added] Navy to blast its way unaided through the Dardanelles to Constantinople, but that failed and the army was called in. This was the extent of the combined operation. The navy did support the landings with its guns and ferried troops back and forth, but it was never taken into the confidence of the admiralty the way it is done today." [3]

Only by looking backward to this by no means distant past with its disasters due to absence of cooperation, can we recognize the great progress represented by the arrangements under General Eisenhower, symbol of unity, in the Sicilian campaign. To ensure the success of this "amphibious operation of peculiar complexity" which Winston Churchill foreshadowed on June 8, 1943, after his return from Washington, collective land, air and water forces were brought together as well as individual national forces, including a variety of troops from the United Nations in addition to the two main Allied contingents.

Pearl Harbor Errors

Modern landings call for overcoming specialized leadership; that is, of leadership trained in, and for, only one element. Insofar as landings form part of a siége, the most violent phase, the storming of a fortress in either its outlying parts, or at its very heart, these are apt reminders that combined leadership in war might have occurred first in the art of sieges when the mind of the general had to combine, as on one thread, the multiplicity of material products required for siege work.[4] Such arrangements are imperative when war forces are expected to land, not only for the attacker, but for the defender as well. They were wanting at Pearl Harbor on the receiving end when we were still somewhat banking on profit accruing to us by the conflict and lack of coordination between the Japanese army and navy.

In an age of mass war when specialization, intended to master mass, produces new mass, and when only vast staffs can cope with the variety of inter-force relationships and their common technical and psychological problems; when staffs are in part research agencies constantly submitting choices and alternatives for the decisions of commanders and superiors, the supreme leader may at times, perhaps at most times, appear as a mere symbol, necessary and highly useful as such, but acting little, or rarely. In his over-topping authority the hierarchical lines of command, never to be over-taxed in the higher extensions, will converge. They must be cleared from the confusions of this convergement to permit such decisions, at the same time unavoidable and imposed by contingencies, as Eisenhower's on June 6, 1944, to start the invasion of Normandy in spite of the indecisive and by no means reassuring weather reports on which so much of the success of the cross-Channel enterprise depended.

The mere division of labor between two, or three services is unfortunately never clear enough in landings, or other combined operations to make the appointment of a supreme commander unnecessary, or superfluous. Even with the best will on the part of all concerned such command is generally needed to establish and maintain convergency in both strategical and tactical fields. Without breaking down customary lines of separation, usually identically associated with the shore line, such features, now very customary, as the artillery support of land forces by naval gunfire reaching as far inland as 20 miles[5] would have remained unthinkable.

Harmonization of the two services had to come from the outside, from civilian initiative and impetus. Inter-service conflicts and mutual non-understandings were to be avoided—or so the civilian governors thought—either by strict delimitation of respective functions and authority, or by the institution of a supreme command, the latter to be valid either for the duration of the specific combined operation for which it had been set up, or by arrangements whereby the ranking naval officer commanded the whole force while en route asea and the military commander from the moment when the troops touched ground. Down to this day there is no unified tendency in the direction of either of these two arrangements.

Conflicts in Authority

The French of the Revolution, and Napoleon, who treated his admirals cavalierly, preferred the monolithic arrangement of a supreme commander in combined operations, who was practically always a general. Lazare Hoche, in the Irish expedition of 1796, had "supreme command of both army and navy, and found by bitter experience the delays which incompetence, or ill-affected subordinates can impose, especially in a branch of service of which the commander-in-chief has no particular knowledge."[6] The British at the same time avoided any such arrangements, hence the numerous violent conflicts between admirals, such as Nelson and Hood, and the generals who were joined with them for some common enterprise, as Stuart and Moore in Corsica in 1794.

During the era between 1814 and 1914, in which few large-scale combined operations took place, when it was axiomatic that "sea power affects terrestrial operations as a general rule only in so far as it governs or renders practicable the transports of troops, of warlike stores, or of sustenance of armies across the ocean."[7] the problem of supreme command in combined operations either did not arise, or was evaded rather than settled. There was no unified command at the Dardanelles where, however, the failure could least of all be attributed to its absence, for in this instance the naval-military cooperation on the spot was fairly harmonious.

In the present war, with its tremendous technical as well as political complications, the command of combined operations has been arranged for in various manners. The Germans, and following them the French,[8] chose to adopt the monolithic order, including the combined staffs, even before 1939. The practice of the Allies has varied. While the arrangement of the supreme inter-Allied and inter-service command during the intervention in Norway in 1940, where the military command over English and French troops was in the ineffectual hands of General Ironside, remains obscure, the double difficulty of coalition warfare and combined operations was finally solved by the plan adopted for the combined staffs of the United States and Great Britain and for a supreme commander in various theaters of war. Eisenhower received supreme command in northwest Africa as far east as Tunisia, beyond which Montgomery's began, and in Sicily, Italy and Normandy.

When Eisenhower relinquished the post of supreme commander of the Allied forces in the Mediterranean theater in January, 1944, it fell to General Sir Henry Maitland Wilson of the British Army, with unspecified powers over the naval units in that sea. The commander of the latter, occasionally and officially called "Commander-in-Chief in the Mediterranean," [9] did not reveal that he held that status when after the landings in southern France in August, 1944, he announced that "responsibility for landing and establishing the army on shore is naval and the naval commander has ordered the assault pressed home with relentless force." Whether such indefinite communiqués, with respect to indicating where supreme command rested, were in themselves suggestive of completeness in the plan of unified command might well be pondered.

Command in the Pacific

In a theater of war more their own, although not exclusively, the Pacific, the United States settled the question of supreme command by simply dividing the area into two, between their army and their navy, with General MacArthur in full and overall command of both military and naval forces in the southwest and Admiral Nimitz with similar powers in the rest of that ocean. In an age of highly perfected communication technique the explanation given by President Roosevelt that "by the nature of Pacific geography there can be no single command, for that would be beyond the comprehension of any man" was not completely convincing to everyone, not even General MacArthur himself, who merely went so far as to provide indications that he thought such command feasible, or to others who considered it indispensable.[10] On the other side of the Pacific and Indian oceans, Admiral Mountbatten was made supreme commander with the late Major General Wingate as his "lieutenant" as Churchill described their relationship (in the House of Commons, August 2, 1944).

The often dangerous and fateful conflicts of generals and admirals, or of the two, and now three, services, the manpower and material waste involved in their separation,[11] all of which in earlier times were taken for granted, have in our current days of total warfare led to proposals, or actual measures for the return of unity to war operations by uniting all two, or three, services in one department of, or for, war. This desire or hope is unfortunately harder to materialize at present than before, considering the enormous difficulties offered in grasping the technicalities of warfare, by any one officer, or group of officers. But unified command beyond and, of course below, the one constitutionally arranged for will remain the minimum demand created by total war, or its peculiar feature, the landing operation.

Liaison (from *ligare,* to bind) is the looser, less imperative, more consultative form of inter-service cooperation. An inefficient liaison officer is a fish altogether out of his particular water and the ideal one he who can swim professionally in different waters as required and who understands the other services at least half as well as he knows his own. He must be able to appreciate in minimum time the wishes of one partner, which he must perhaps be sufficiently clever, and if occasion arises diplomatic, to reduce to reasonable and comprehensive terms before he sends them off to his own command. Some of these requests for support may originally appear somewhat unusual to the other partner, or partners, like the request during World War I for fire against targets on land not immediately visible from shipboard.[12] Before and into the days of wireless the transmission of liaison intelligence might be a time-consuming and purpose-defeating business, especially in the tactical field.

While in today's war the technical transmission will require but little time, a proposal from one to the other service which has to make its way through the hierarchical steps of command, up and down, is apt to arrive, or be acted upon too late for all practical purposes. An example of unsatisfactory and slowed-down army-airforce combined operations may serve to illustrate this danger, in lack of an illustration from army-navy cooperation. In the French forces of 1939-40 army commanders controlled, or were supposed to control, an observation squadron and a pursuit group of planes and corps commanders some observation planes. As for the rest, generals of the army were merely authorized to make requests for air bombardment, distant reconnaissance, etc., to their peers in the air force, who might, or might not grant them. The request coming from a division, or brigade general had to travel upward to the general commanding an army and from him to his peer in the air force, a process which in the French army at that time took five hours,[13] but probably less than that many minutes in the German army.

Whether under the influence of such disastrous experiences, or in anticipation of engaging in much amphibious warfare, the U. S. Navy in 1941 began the training of "shore fire control" personnel, using them first on a large scale in the conquest of Sicily. Their main equipment consists of light-weight radio sets. These parties, one assigned to each assault battalion, go ashore with the first waves, advise military commanders what their artillery can do for them and transmit their wishes as well as the necessary artillery data to the ships.[14]

NOTES, CHAPTER 3

[1] See the various manuals for the training of marines, such as *Training Manual for Landing Operations*: By Authority of the Lords Commissioners of the Admiralty. London, 1910, 96 pp.; *The Landing-Force and Small-Arms Instructions,* US Navy, 1915. Revised 1915. Annapolis, 522 pp.; US Navy Department: *Landing-Force Manual,* US Navy, 1918. Revised under the direction of the Navy Department. Annapolis, 1918. VII, 553 pp.; *Landing-Force Manual,* US Navy, 1920. Revised under the Direction of the Navy Department. Washington, Govt. Printing Office, 1921. 750 pp.

[2] Captain A. T. Mahan: *The Influence of Sea Power Upon History, 1660-1783.* Boston, 1897, 13th ed., 261.

[3] *N. Y. Times,* 14 July 1943.

[4] Clausewitz, book II, ch. 2.

[5] See *N. Y. Times,* 20 July 1944, Washington despatch for such an occurrence.

[6] Mahan: *The Influence of Sea Power Upon the French Revolution and Empire, 1793-1812.* Boston, 1898, 10th ed., I, 349

[7] Major C. E. Callwell: *The Effect of Maritime Command on Land Campaigns Since Waterloo.* London, 1897. 5.

[8] About the fatal workings of this arrangement in France see Pertinax: *Les Fossoyeurs.* I, 52-56.

[9] Communiqué from Rome, 17 August 1944.

[10] Representative Maas (R) has repeatedly demanded a "really united force, with one war plan, under one commander" to defeat Japan. *N. Y. Times,* 13 November 1942.

[11] Examples of such wastage are cited in an article by then Senator Harry Truman in *Collier's,* 21 August 1944.

[12] Tschischwitz: *Armee und Marine bei der Eroberung der Baltischen Inseln,* 166-8.

[13] Pertinax: *Les Fossoyeurs,* I, 72.

[14] For details see *Time,* 9 August 1943.

Chapter 4

STRATEGY AND TACTICS

IF, IN ACCORDANCE WITH CLAUSEWITZ, tactics covers the use of armed forces in combat, and strategy involves the turning to account of combats to accomplish the purpose of war, the dividing line between the tactics and strategy of landings must be drawn somewhere near the beginning of the operation that culminates in the actual landing. This defining of such distinctions is by no means easy, or satisfactory in results, considering the combinations involved of army-navy, or army-navy-airforce, the partners which enter the tactical sphere at quite different points of time and space. The military forces may consider the moment of their leaving their ship as the opening of the tactical sphere of their activity, but even so, less than ever does this mean complete severence in cooperation from the other two forces.

An army force on its way to a landing will of itself have very little influence on sea, or air command. Ships and planes carry the soldiers to collectively selected landing beaches and landing fields despite enemy opposition, if offered. Navigation errors, or difficulties at times interfere with prompt, or completely successful arrivals of landing craft on beaches and still more often with airborne forces in approaching the ground. While the military force thus travels, it is almost wholly inactive, this measure of passivity, in comparison with the past, now having increased rather than decreased. It would be unthought of today, as was proposed to the time of World War I, to employ, or get ready for possible action the ship-transported artillery of the landing force while en route to a landing. Naval participation in, or influence on, landing operations may end abruptly with the landing, or may be called and counted upon for days and weeks after D-Day, while air force cooperation will be similarly demanded.

A landing operation presupposes the command of the sea, or the air, or both by the attacker, for all, or most of its duration. Command of the sea, as regards such operations, whatever else it is made to cover, has been defined by the British Admiralty of 1901, as a relation of sea forces that makes it "possible to transport across the waters commanded a large military expedition without risk of serious loss." Command of the air, *mutatis mutandis,* would mean the same as regards passage in and below the upper sphere. Whether both command of the sea and of the air are required as the first conditions of landing operations will depend on various factors of the situation. Generally speaking, a large landing, a genuine invasion, will require both, while a smaller enterprise, such as the conquest of Crete by the Germans, can be performed under air protection alone, if it is adequate.

Conversely, the defender against an invasion will endeavor to maintain at least a "fleet in being" to act on what corresponds to his sea glacis, a concept of naval strategy that was accepted, if not introduced, by the commanders of the English fleet after their defeat at the hands of the French at Beachy Head in 1690. For a time they then shunned further battle and the risk that accompanied it of losing the rest of their fleet, for with its loss "the Kingdom would have lain open to invasion." Under modern conditions a fleet in being, while still of value as a potential harassing force, will be largely eliminated from exerting an adverse influence on landing operations which are sealed off against its interference.

Concerted Strategy Rare

At most times in the past the strategic tie-up of military operations and of naval movements has been taken for granted, the presumption being that some supreme commander, functioning only constitutionally, but not also technically, or a legislative body in control of the purse strings, had laid down a definite and immutable division of war funds and duties between the army and the navy of their own countries, the greater importance in which one service may have been held being expressed by the grant to it of more money than to the other. Actually, the common basis of military and naval action in war, or in the preparation for war has seldom rested on a concerted strategy of the two. One of the proofs of this may on occasion be supplied by the absence of tactical cooperation between the two. In most periods before the present war such tactical cooperation was distinctly rare, and when it was forced upon the two services, usually against the will of both, oftener than not there was a deficient quality of interest and performance.

Appreciative of the poor results from such cooperation, some generals, like Wellington, have distinctly, and almost as a matter of principle, shunned it. The British Navy kept him supplied and naval units and marines were thrown by him into the defense of the lines of Torres Vedras; and naval units would occasionally lend fire support and harrass French flanks resting on the sea, or French convoys moving on roads closely paralleling the coast, or bring help to Spanish cities and towns besieged by the French.[1] But he discouraged the intentions of the War Ministry to send smaller forces by sea into parts of the Peninsula distant from those where he himself operated. He wrote to a Minister in May, 1811:

"No such maritime operations should be undertaken, for unless a very large force was sent, it would scarcely be able to effect a landing and maintain the situation of which it might take possession. Then that large force would be unable to move, or effect any object at all adequate to the expense, or to the expectations which would be formed from its strength, owing to the want of those equipments and supplies in which an army landed from its ships must be deficient. It was in vain to hope for any assistance, even in this way, much less military assistance, from the Spaniards; the first thing they would require uniformly would be money; then arms, ammunition, clothing of all descriptions, provisions, forage, horses, means of transport, and everything which the expedition would have a right to require from them; and, after all, this extraordinary and perverse people would scarcely allow the commander of the expedition to have a voice in the plan of operations, to be followed when the whole should be ready to undertake any, if indeed they ever should be ready."[2]

Wellington's doubts did not spring from any overly selfish desire to reserve command of all forces available in the Peninsula for himself. The actual experience with landing and landing attempts, or plans accessory to Wellington's own campaign were on the whole futile and discouraging,[3] an outcome to which contributed the propensity for improvisation of British officers and politicians, as well as the incorrigible procrastination of the Spaniards in the coalition that was opposing Napoleon. On the whole, coalitions have impeded rather than helped such complicated operations as landings.

Many Opposing Opinions

In the strategic choice of a site for landing not only the services, but also the sentiment of the people, and in a war of coalition, of the partners thereof are apt to be at variance. It is only necessary to recall the public clamor for the opening of a second European front in the recent war. By itself, the temptation to open a new front across, say, the waters that isolate England from the continent, is considerable. Combined operations offer to laymen or politicians great temptations to which such strong ministers as the two Pitts almost habitually fell victim; temptations based upon prospects that the power, or powers in possession of sea supremacy, or nowadays of air supremacy as well, are thus enabled to move freely, and can seemingly hit at will and surprise and discomfort the enemy. Such speculations are, however, more apt to take into account the enemy's weakness than his strength.

While politico-strategical thought may speculate and improvise in connection with landings, all improvisation must end once tactical planning for such operations begins. This was something the Pitts, among others, would never understand. "War is in the province of chance" or accident, but no part of it more than landings. Their extension through two or three elements increases chances of misadventure as inter-service co-operation does with regard to frictions between commanders with results, as a rule, more adverse and complicating than favorable and simplifying.

While strategic advantage in landing operations may be with the attacker, most of the tactical advantages are likely to be on the side of the defender. The latter must be overcome by most minute planning against contingencies based on voluminous and precise data on tides, depth of coastal waters, currents, weather, coastal terrain, enemy strength, location of his forces, and by the training of special landing troops, and by surprise.

Every operation of modern war demands surprise—a statement which seems less of a truism if we compare modern combat conditions with the open battle arrays of earlier times—but none more than landings. It is the sworn duty of any participant to such an undertaking, so far as he shares the secret of its timing and destination, to keep it and not divulge it through treason, or carelessness. One of the great betrayals of secret landing plans is recorded in the timely warning given the French government through Jacobite channels of the preparations for an Anglo-Dutch descent upon France in the summer of 1694. It was an object lesson in the importance of military taciturnity when General Eisenhower, during the preparations for the landing in Normandy, demoted and sent home a high ranking American officer, commander of the U. S. Ninth Air Force Service Command, who early in April, 1944, at a dinner party allowed himself to say in the hearing of numerous guests that the invasion would take place before June 15, and whose indiscretion apparently was denounced to American military authorities by a member of the sex which is presumed to be far less taciturn than the male.

Of the three forms that may be assumed by landings, depending on their purpose, namely the raid, the diversion and the full-scale invasion, the first depends for its success mostly on surprise, as to time and place. It is a hit-and-run affair, tactics being similar to those of infantry shock troops, based on minute reconnaissance; and undertaken with small forces to test enemy strength, as at Dieppe, or to destroy special installations such as U-boat lairs at St. Nazaire.

Object of Diversion

Diversion is intended to mislead the enemy, or tie down his forces, and its success is less dependent on surprise. Its object is to keep the enemy in the dark about one's own intentions and thereby tie down if possible, by use of an inferior force, superior enemy units and keep them away from striking distance of a main invasion. Hence the landing of the U. S. Marines on Choiseul Island during the attack on the main objective, Bougainville Island, and who were subsequently retired.

In a full-scale landing strategic surprise may have to be sacrificed, on account of the large preparations involved, which cannot well be hidden completely from the enemy, from whom, however, the tactical surprise must be kept secret. For example, before the Allied landings in southern France in August, 1944, the Germans knew that an armada clearly presaging a new invasion had passed Gibraltar into the Mediterranean, but they could not be certain of the point chosen for the landing.

Full-scale invasion is characterized by the intention to stay and strive for a decisive victory and must be prepared to force its way into enemy territory despite resistance. Both sea and air supremacy are required for its initial advance and the maintenance of communications. Only in operations of a restricted nature will supremacy in a single element, air, or sea, suffice. In the German conquest of Crete the air-borne invasion was successful while the water-borne attempt was repulsed with heavy losses by the British fleet, which then and in that sea-area preserved a narrow margin of supremacy. In the full dress invasion the attacker races with time to land both men and material before the enemy has gathered his strength for the counterattack.

Even without a complete record of the landing enterprises set down in history, it would seem certain that in the majority surprise did operate, at least in the initial stages. The landing force often found the enemy unprepared and unsuspecting. Sometimes it moved him to retire in face of clear local superiority on the part of the invader (as in Caesar's second expedition to Britain), and postpone battle until a more propitious time. Most landings have taken place without finding immediate opposition.

Occasionally, indeed, the invader was unable to take the enemy unaware; giving away the secret of surprise in landing is tantamount to self-betrayal. This was the result on two fateful occasions: the Athenian expedition to Sicily during the Peloponnesian War (415-13 B.C.), which sealed the doom of Athens by its failure, and the Allied enterprise against the Dardanelles in World War I. As a rule, however, the surprise element has been maintained and has given at least a momentary success to the invading hosts. The Persians landed unopposed at Marathon; most of the many incursions of the Vikings did not find the enemy ready until hours, or days after their arrival on German, French or English shores. William the Bastard landed on the 28th of September 1066 and the Battle of Hastings did not take place until the 14th of October, which was still too early to enable Harold to assemble his maximum strength.

Battles on Coast Lines

Martial history furnishes remarkably few instances of landings being contested on and across the shore line. Matthew Arnold's "Dover Beach" in reality would be an unlikely site for any struggle in which "ignorant armies clash by night." Modern examples of contested landings are furnished by Gallipoli and, in the present war, North Africa and the Pacific. But the classic fight in and out of the water is still Caesar's first attack on Britain, when his two legions fought their way to the shore through the surf, as the Britons advanced to meet them, some in chariots. The Britons had prepared for the coming of the

Romans by accustoming their horses to run and pull chariots through water—forerunners of our amphibious tanks being used by the defense—while the Romans had to seek a footing, get their legionary formations together and rally to their standards, the enemy at the same time contesting every step.

Artillery of one kind, or another enhanced considerably the depth of the battle zone on either side of the coastline. In most landing enterprises, before the parties to it came to blows, they would exchange missiles, largely across the water and the shoreline. Potentially, the artillery superiority, if he had and used cannon, would be necessarily on the side of the defender. This superiority was expressed in formulas like *"Un cannon à terre vaut deux* (or *dix) sur mer."* (A gun on land is worth two or ten on the sea.) This conviction became so prevalent that only after severe experiences in the present war did the lesson finally sink in that practically every immobile target, everything non-moving in modern war, is predestined to be overcome.

Superior tactics, technical advantage, surprise on the part of the attacker go a long way toward establishing his own fire superiority, with air bombardment as the latest method of obtaining it. On the whole and until a fairly recent date, coastal defense was only faced seaward. Landing in depth, through the air, has changed that one-sided range and the units of the martello system, whether that of Napoleonic, or Hitlerian times, must provide a field of fire all around the aiming circle.

War history of antiquity records several occasions when a sea-borne invader had to contend with and overcome the original missile superiority of the defender. Caesar, as we shall see in greater detail further on, used his marine artillery against the javelins of the Britons. After the fall of Carthage, when the Romans became zestful to possess the Balearic Islands, they ran up against the most specialized missile-throwers of the time, lined up along the shore with their slings, but the Roman Consul, Q. Caecilius Metellus, made their shots ineffective by covering the decks of his ships with hides, thus protecting the soldiers until the moment of debarkation. His were the armored landing craft of antiquity. Rushing through the hail of stones and the surf, they drove the natives inland. Rome later was to draw upon those islands for the contingents of slingers for her armies.

Forces in the act of landing usually offer excellent targets to the defenders. In order to avoid detection, or make observation difficult, the invaders will be inclined to prefer night landings, or approaches, often making the landing coincidental with sunrise, or fog, or weather favoring low visibility, which helped the British at the Zeebrugge landing in 1918. Troops employed in night landings, or at times where low visibility will be of definite help, have learned the profit of blackening their faces, or streaking them green and brown when coming to islands with jungle growth, in order to slip ashore as "dark invaders."

SCREENING AND ILLUMINATING

Artificial fogs, either designed to hide the approach, or otherwise hinder the defenders' observation, have been devised by modern chemistry and put at the disposal of navies and armies, who have taken up their use since World War I. Smoke screens as cloaks of invisibility were employed among the older stratagems on a few earlier actions which had much in common with landings from the sea—river crossings. Charles XII of Sweden crossed the Düna on July 18, 1701, "under the protection of an artificial fog which was hiding his movements, before the old Steinau, who commanded the Saxons, had observed him. The Swedes, once carried across, moved instantly into battle order. . . . What presence of mind and what energy to assign to the troops at the very moment of landing a suitable battle terrain!" Thus were the terms in which Frederick the Great, even in his last days, expressed his admiration of the bold and unconventional Swede *(in his Observations about the Military Talents and the Character of Charles XII.* 1786).

To break through the veil of night in which the invader will seek to enfold himself the defenders will employ illuminating devices of various kinds, such as Very lights, flares, star-shells, searchlights. Zeebrugge and St. Nazaire are actions that called forth the greatest display of these illuminants.

Air power has robbed coastlines of most of their original character as military obstacles by over-flying them and dropping parachute, or glider-borne troops and materiel in their rear. This not only has put an end to old-fashioned methods of coastal defense that looked only to the sea and disregarded the rear, but also has given depth to seashore combat from the outset. While much of the airborne troops must in the nature of their tasks be considered "forlorn hopes," one man back of a defended coastline, so long as he lasts, is easily worth ten in front of such a line—more so, even, than in pure land warfare.

Contrary to common opinion, effecting a first landing is not usually the most difficult element in landing operations. So far as this stage of the enterprise is concerned, the history of landings shows that many more have been initially fortunate than have failed. The reaching of the coastline is not the difficult thing, but the maintenance and widening out of the beachhead and its use for victory. In ways, a landing enterprise resembles the action of fever. Gaining the land merely corresponds to the shock of toxic action in fever when it is setting in, while the crisis arrives later, when once in the body the enemy gathers his forces for the real assault upon the vitality of the patient. A large-scale landing thus involves a double race, one to the land and another to gain a field spacious enough to battle upon, to maneuver. The invader must be in strength and well enough provided with supplies simultaneously to cope with the enemy, who in his own race to contain, or crush the offense, will after a few days comprehend the invader's strength and intentions and formulate a counter-plan.

In most cases the attacked party will be unable to man the whole coastline. The so-called Atlantic Wall of the Germans must not be understood as an unbroken fortified line co-extensive with the coastline. It was no shoreline *limes* in the sense of the Roman Wall in Britain, or the great wall of China, but rather blocks of fortifications, hedgehogs along the sea, intended to dominate likely invasion terrain by flanking rather than by merely frontal resistance; or to keep the invader from winning and using desirable permanent harbor installations.

Actual, or prospective, invasion problems which face the defender, the measures taken by him, form the reverse side of the theme under consideration. We notice them only in passing, the field being too wide for extensive comment here, including as it does such broad features as coastal defense, or the absolute and relative areas of land and sea involved and, more recently, air forces, to be contained, or considered by the defending nation; or the more specific question whether the regions invaded, or likely to be so, should be evacuated and the scorched earth policy applied.

Many of the more or less constant problems entailed by such a defense were discussed by a group of Elizabethan statesmen on the occasion of an invasion scare in November, 1596. These "*Opinions Delivered by the Earl of Essex, Lord Burleigh, Lord*

Willoughby, Lord Burroughs, Lord North, Sir William Knollys, Sir Walter Raleigh, and Sir George Carew, on the Alarum of an Invasion from Spain in the Year 1596, and the Measures proper to be taken on that Occasion"—they were reprinted in London during the 1790's—took the form of questions raised by Essex after news had arrived of Spanish concentration of ships and troops.

DEFENSIVE BATTLES HAZARDOUS

Was it likely that the Spanish would come so late in the year, or rather during a more propitious season? Did they intend an invasion, or a mere incursion? If an incursion, although it be impossible to provide against it everywhere, which measures should be taken? If invasion, where would it most probably be attempted? Which measures to stop an inland march of the Spaniard? If he should arrive with a large force of infantry, should the English venture battle and when? What was to be done regarding food supplies which might become available to the enemy?

On the last point Essex proposed that nothing should be left the enemy "to relieve himself in his march, but what he has upon his carriages, by forcing him to march slowly and thereby consume such victuals as he has, etc., for so we shall both win time and waste his numbers which are in this case two infinite advantages." Several of the statesmen were of the opinion that an invader like the Spaniard with his excellent infantry must be "overtopped with horse" and that battle must not be sought at once—as Harold did to his own disadvantage—on the part of the defenders. Consensus was that:

"There is nothing so perilous to a kingdom invaded as to hazard battle when it may be avoided, nor nothing in policy better agreeing with the state of an invader than the contrary, if possibly he may procure it. The reasons are apparent, for that the one in losing the victory (besides the loss of men) may endanger a kingdom, whereas the other in losing a battle loses but men only, which with often skirmishing, sickness, want of clothes and victuals (inconveniences and wants evermore incident to invaders) will in a small time consume of themselves: And therefore in my judgment it is not good at any time to hazard the danger of a battle, unless we be enforced unto it, or that the General in his wisdom find such apparent reason that (unless God fight against us) we must of necessity be victors."

Assembly of the defending forces, who were to be warned and called up by beacons—which occasionally were fired by inadvertence—should take place at distances of at least six or seven miles from the seashore, "because by that means you may have time to put your men in order and consult what is best to be done: Yet did I wish that by all possible means we may give the enemy continual alarums, both at his landings and afterwards, with horsemen in the plains, to the intent they straggle not, and some numbers of shot on horseback, at every strait and place of advantage to win time, whilst our greater forces may be put in better order to defend."

Uniform organization and action which are today so much taken for granted in affairs that are even remotely military, seemed to these counsellors to need special emphasis, hence their insistence "that an uniform order in martial causes be observed throughout England, knowing that disorder breeds confusion, and confusion brings ruin. For establishing thereof her Majesty may be pleased to settle an authority in some such whose valour, virtue and judgment may make him worthy of so great a charge, and whose religious love to her Majesty, and true zeal to his country, may deserve so great a trust."

NOTES, CHAPTER 4

[1] Napier's *History of the War in the Peninsula.* ed. N. Y., 1862. III, 157, 309, etc.

[2] Napier, III, 16.

[3] For examples see Napier, III, 19-20, 134, 157-170, 435 ff.; IV, 95-6, 316, 397-8.

Chapter 5

LOGISTICS AND STATISTICS

"People who have seen, or planned this kind of operation, even over short distances, do not speak glibly about landing great expeditions on a few days' notice, or on all the beaches of Europe at the same time."

President Roosevelt, *Message to the Congress*, 17 September 1943.

PREPARING, CREATING, CARRYING and maintaining armed forces to be used in landing operations lie in the so-called logistical field. This covers the preparation and assembly of troops, ships and supplies; the building and maintenance of bases and camps; provisioning, measures for the comfort of troops and crews, care of wounded and sick. Transportation of troops, which has so much in common, theoretically, with marches and movements on land, might fall into either the logistical, strategical, or tactical sphere, depending largely on how closely it approximates combat.

At most times before the present war landing troops, including sailors, whose help in certain operations extended beyond the shoreline, have been taken from the average run of the forces. Among the ancients the Athenians excelled in commando-style landings. With the adding of marines to the various navies the first troops to be trained for landings, but for many other activities besides, appeared on the martial scene. But they were never very numerous, and for large-scale invasions it was still necessary to employ infantry, artillery, engineers and, rarely, regular cavalry, who had often not seen the sea, nor boarded ships before. Only under the most favorable conditions, when there was time and an intelligent commander at their head, were they given opportunity for drill in embarkation and debarkation. More often even this preparation was denied them.

Only modern specialization has resulted in the training and use of commandos, rangers and other amphibious assault troops. Usually volunteers, they are employed either as raiders, trained in detail for assault on limited objectives, or as assault and advance troops, in case of a large landing which is commonly carried through with troops less specifically taught for amphibious campaigning. These superior bodies have taken the place of the raw soldiers who were used under the tough conditions at the Dardanelles, and by over-age and under-energetic generals. They must be especially trained and equipped and made familiar with all three elements in which they have to fight, or traverse on the way to combat. For artillery support they may have to depend for some time after landing upon ship artillery, or bombers, communication with which may be difficult to establish, or maintain.

Size of Landing Forces

Military writers of about 1900 might have been tempted to evolve a maximum-optimum size of landing forces on the strength of experiences in the preceding century, envisaging a body of 25-35,000 men. But this check of the hardly comparable sizes of landing corps and their original strength at disembarkation, will show the extreme variations from the days of Herodotus, to whom we owe the very dubious numbers of the Persians who landed before Marathon—figures stated some fifty years after the event—or of the Mongolian invasion of Japan in the 13th century, in which 3500 ships are said to have been used, down to the time of the Normandy landings. (See table following)

The numerical opposition met by these and other landing forces on the beach, seen after disembarkation, or later, are not easily established for purposes of comparison. Usually their strength against the beachhead was increased as the day of the first battle between invader and defender approached. For this reason it does not mean very much to learn that in the Sicilian campaign "the German and Italian defenders in all of the island had actually outnumbered the invasion party by one third at the outset."[4] From whatever comparisons are possible, it is clear that the view prevalent even in the earlier years of World War II that the defenders' strength may be reckoned as being as much as five times that of the attackers', is hardly tenable. Even the situation which would seem most to promise the evolving of something like accurate formulae of comparative strengths, the "throwing back of landed forces into the sea," is too atypical and too much under the influence of contingent factors to make possible the establishment of anything like fixed numerical relation in strengths between invader and defender.

	Date	Men	Horses	Guns	Transports
Persians at Marathon	490 B.C.	50,000	....	....	600 triremes
Julius Caesar in Britain.					
First landing	55 B.C.	10,000		....	80
Second landing	54 B.C.	5 legions	2,000 cavalry	....	800
William the Bastard	1066 A.D.	3-4000 mounted men and 8-9000 foot soldiers (Oman), or 60,000 men and upward (*Chronicles*) and more than 3-4000 horses.			696 the minimum of the *Chronicles*
Spanish Armada	1588	30,000		2,630	160
Bonaparte in Egypt	1798	35-38,000	1230	....	....
French in Algiers	1830	35,000			
Interallied landing force in Crimea, Sept. 13[1]	1854	51,000 inf.	1,000 cavalry	128	150
English in Egypt	1882	31,850	....		
German Expeditionary Force in China	1900	21,622	....	861 (?)	60
Sir Ian Hamilton's original force at Dardanelles	1915	75,000	....	....	....
German landing force on Oesel, Oct.	1917	23,000	5,000	54 and 12 trench mortars	19[2]
Interallied landing in Sicily, initial force	1943	160,000	1,400 vehicles	1,800 plus 600 tanks	2,000[3]
Ditto in Normandy, initial assault	1944	1,000,000 in first 28 days	183,500 vehicles		4,000 ships and several thousand smaller craft

Such numbers remain, then, to a very large extent incomparable. To make them more nearly available for satisfactory comparison, such logistic factors as the length of and dangers offered by, the approach; the variety and weight of equipment and supplies would have to be brought in. As President Roosevelt told the Congress in a message on the progress of the war (Sept. 17, 1943), the Allied landing in North Africa in November, 1942 "was in point of numbers the largest military movement over the longest number of miles to landings under fire that history records anywhere."

LAND TRANSPORTATION HEAVY

Ship movements are preceded by those of land transports in the hinterland of the embarkation ports. To illustrate the modern magnitude of combined operations, the London *Economist* has stated that the British railways in moving the North African expeditionary force from camps in the British isles to embarkation points used 440 special troop trains and 60 freight trains for three weeks in order to move 185,000 men, 20,000 vehicles and 220,000 tons of supplies. So far as American statistics for corresponding movements in this country are available, experts consider that they would dwarf the British performance, considering the great distances, etc.; an American armored division alone requires, for example, about 75 special trains of from 28 to 45 cars and 15 Liberty ships.[5]

With landing operations involving movements in, or through, all the four elements, water, air, earth and enemy fire, the items of equipment and supply needed to deal with all of them and their particular hazards are vast in variety as well as in quantity. It has been stated that the preparations for the North African and the Normandy landings included 70,000 different articles, "from tanks to watch springs, and many of these must be provided in millions." They had to be designed, ordered, manufactured, transported ahead of the invasion date, or soon after so far as their shipment was concerned. The materials asked for by Eisenhower for the invasion of Sicily, or at least all his "essential requirements," were furnished to him a month before the opening of the campaign. A load of materials which he had requisitioned on January 26 for the campaign in Tunisia, including 5,000 trucks, 400 dump trucks, 20 locomotives, 40 flat cars and some smaller items, comprising together 22 ship loads, had to be, and was, ready for shipment on, or before February 15.[6]

Armies no longer march on their bellies, as Napoleon is credited with saying. Their impedimenta and requirements, beyond food, have grown constantly in numbers and weight, despite a few attempts at reformatory reduction. Mechanized warfare has in its turn doubled the amount of materiel needed per man in the armies since the World War I. Even after all the logistic streamlining to which landing troops must be subjected to enable them to negotiate a supply bottleneck while being carried by ships and boats and fighting through the beachhead, the weight and variety of their supplies and replacements remain inordinate, and nowhere more so than in the U. S. Army. Of all modern armies, it functions on the most expansive footing, to the accompaniment of the greatest waste and with the most physical comfort for the troops.

No quartermaster of any European army can readily be brought to understand that under combat conditions a pair of shoes for the American soldier lasts only two weeks on the average and clothes not much longer, and that he will need one ton of supplies a month, including ammunition, clothing, food, medical supplies, etc. "Counting in everything, such as trucks, clothes and weapons, it takes ten tons of organic equipment to get one man into the European theater of operations and it takes sixty pounds of supplies per day to keep him there." [7]

Cherbourg Defense Vital

In the first 28 days of the invasion of Normandy 1,000,000 men. 183,000 vehicles and 650,000 tons of supplies were deposited on beachheads in artificial harbors by a force of 2-3,000 craft and some 15,000 men.[8] Such figures will help to explain why it was so important for the Allies after the Normandy landings to obtain so soon as possible the use of continental ports with their facilities and why the Germans stuck tenaciously to the defense of Cherbourg, 300 miles closer to New York than London, or Brest. To these places supplies could be brought directly from the United States without breaking cargo at British ports, whose facilities were severely taxed, thus saving at least one unloading and one loading.

Supplies for landings and operations immediately following them present problems involving packing and wrapping, possibly waterproofing and rustproofing,[9] and above all loading, which must receive special consideration because most of the unloading takes place not alongside regular wharfs and quays, but on open beaches. Equipment and the first supplies at least have therefore to be "combat loaded," that is to say, with due priority for their use. The neglect of the basic principle "First used, last loaded" was exemplified in the packing of materiel for the initial Gallipoli landings and cost the British in the World War I precious days and weeks spent in sending the freighters which had arrived at the Dardanelles away to Alexandria to have them "combat loaded" there before returning to Tenedos.

Considerations of combat loading come to a climax in operational planning and deciding how much of the tonnage assigned to the initial attack and the first, second, or succeeding waves is to be reserved for troops and their weapons and how much for supplies and materiel reserves. Such distributions have to take place in accordance with contingencies like enemy interference and weather conditions, both of which might temporarily isolate the beachhead from rear communications. They will be decided with such factors in mind as the possible greater vulnerability of secondary shore-bound waves while in transit, as compared with lesser threats to firstcomers, profiting by surprise. Later, the enemy's air forces and artillery will have been alerted and may do serious damage to supplies being landed, or already piled up. Aerial counterattacks may come first, preceding those of the enemy ground forces, and will require the early landing and setting up of antiaircraft guns. Once landed, supplies have to be removed from exposed beachlines to depots within the beachhead, the locations of these dumps being selected ahead of the landings and marked early in the operations by markers, lanterns, etc. Shore party commanders and their battalions keep order and direct traffic in these restricted and crowded localities. The space requirements for storage within a beachhead are considerable. They have been put, for an invasion army of 25,000 men, at some 2,000,000 square feet for open and 1,750,000 square feet for covered storage space.[10]

Hospitals and hospital care must be among the first concerns of the landing forces. Heavy losses which are generally expected to result from landing operations seem to have bred fear, or belief that casualties in landings, particularly those undertaken by co-national forces, are apt to be exceedingly high. Losses in ancient historically recorded landings are usually given as so great that, as told in the Gudrun Saga, the blood of the dead and wounded dyed the waters. In one fight where battle line and coast line were identical and forces of the iconoclastic Byzantines had made an attempt against Ravenna, held by the defenders of the Papacy and the holy images, "the populous sea coast poured forth a multitude of boats; the waters of the Po were so deeply infected with blood that during six years the public prejudices abstained from the fish of the river." Thus Gibbon (II, 593) expresses about all that we seem to know of this probably super-sanguinary battle. Actually, few amphibious operations have been as costly as that, nor have they been anything akin to the loss of Pharaoh and his hosts in the Red Sea when "there remained not so much as one of them," while the children of Israel "walked upon dry land in the midst of the sea."

While many landings, in particular those in which small forces engage, have failed, or been achieved only at cost of an inordinately high percentage of lives and ships, especially among those in the first waves, there have been land, as well as sea, actions in which the percentages of casualties were equally high, if not higher. The numbers in category of "missing," however, are apt to loom impressively in casualty reports on landings, indicating that enemy resistance reaching out beyond the shoreline was effective. Probably without checking by the complete record of these particular casualties, one may judge that General Omar N. Bradley, at the time commander of the U. S. Ground Forces in Britain, kept well within the bounds of accuracy when before the great invasion of 1944, he told American infantry officers that contrary to many expectations, "this stuff about tremendous losses is tommyrot. Some of you won't come back, but they will be very few. In the Tunisian campaign we lost only an average of three or four men to 1,000." [11]

Normandy Losses Less

Such optimism was verified subsequently in Normandy. There the Allies during the first three weeks, when they had landed about a million men, suffered casualties just above 40,000, as against British losses in such indecisive battles as the Second Ypres of 120,000, or nearly 400,000 in the Paschendaele offensives of 1917. German losses during the same period were given by the Allies as 70,000; and although one hesitates to accept such estimates, considering the difficulties of ascertaining them on terrain like that of Normandy, the ratio, even if it should not be quite as favorable as seven to four, is still surprisingly encouraging for a landing party.

As combined operations, tying together land, naval and air forces, all or two of them at a time, landings are apt to entail losses due to faults and misunderstandings in the combination. While tragic, these casualties have not nearly approached losses caused, for example, by artillery to its own infantry, in World War I in all combattant armies. There have been a number of misfirings during landings, three or four in the American forces, causing the death of some 440 men altogether. They were due either to naval fire hitting landing infantry, co-national and allied, or to ground artillery mistakenly attacking planes.

During a landing on Parry Island, Eniwetok Atoll, in the Marshall Islands on February 22, 1944, such a misadventure cost 13 men killed and 46 wounded when three LCIs were shelled a few hundred yards off shore by an American destroyer. According to the official version:

"At the time the destroyer was providing fire support to the first landing wave of assault troops approaching Parry Island in landing craft through heavy smoke and dust caused by the preparatory bombardment. The primary source of error was that under the difficult conditions of navigation both destroyer and landing craft were slightly out of scheduled positions, with restricted visibility as a contributing factor. . . . In any landing operation on a hostile shore close fire support is essential to prevent heavy losses during the landing and assault and that this involves a calculated risk must be accepted." [12]

The gravest incident in a landing operation occurred in July, 1943, at the very start of the Sicilian campaign, when 21 or 23 American troop-carrying planes were shot down by American navy and army gunfire and about 400 men killed. A number of British planes were also shot down, as if to illustrate most drastically the added complication of coalition warfare during a landing. The Secretary of War, reluctantly revealing the incident, ascribed it to "a combination of circumstances that may occur in spite of the best laid and most meticulous planning. I know of no more complicated operation in war than an amphibious operation at night involving not only sea and ground personnel but airborne troops as well." It was "one of those terrible hazards that must be taken in any bold, complicated modern maneuver such as we had off the coast of Sicily."[13]

NOTES, CHAPTER 5

[1] The total of the troops disembarked in the Crimea was: 309,268 French, 97,864 English, 21,000 Piedmontese and *circa* 60,000 Turks.

[2] This expedition included also 1400 vehicles, 150 MG., supplies for 30 days, the guns reaching up to 21 cm. caliber.

[3] Participating in the first assaults on the Sicilian beaches before dawn on July 10; in the operations "as a whole" 3266 ships took part.

[4] Major General Walter B. Smith, General Eisenhower's chief-of-staff, quoted in the *N. Y. Times,* 16 August 1943.

[5] Joseph B. Eastman, Director of ODT, quoted in *N. Y. Times,* 24 June 1943.

[6] Interview with Maj. Gen. W. D. Styer, Chief-of-Staff of the Army Service Forces. *N. Y. Times,* 7 August 1943.

[7] Interview with Maj. Gen. W. D. Styer, Chief-of-Staff of the Army Service Forces. *N. Y. Times,* 7 August 1943.

[8] UP despatch from Allied Naval HQ, London, 8 August 1944.

[9] For details see an article "Invasion Vehicles made Waterproof." *N. Y. Times,* 9 June 1944.

[10] Lt. Gen. James Harbord, quoted *N. Y. Times,* 20 May 1944.

[11] AP despatch from somewhere in England, 7 April 1944.

[12] AP despatch from Pearl Harbor and UP despatch from Washington, 2 April 1944.

[13] *N. Y. Herald Tribune,* 24 March 1944.

Chapter 6

THE PREPARATION AND PLANNING OF LANDINGS

THE USUAL MILITARY PRACTICE in overcoming fear and uncertainty among troops in war, and to a certain extent warding off accidents, is by way of special exercises which reproduce the hazards of actual combat. However, everyone familiar with former training is aware that certain enterprises of a highly risky, or involved character, which are bound to occur in war, were rarely given attention in peace time maneuvers; among these were night fighting and landings.

Schlieffen once had the problem of landing in England treated by the Great General Staff, but it does not appear that a single naval officer was present at that *Kriegsspiel*. Pre-1914 landing maneuvers were usually undertaken by navies alone and by the fairly small marine corps at their disposal, and often chiefly provided occasion for jokes at the expense of the "mounted navy." Very few of the landing enterprises in the not remote past were preceded by methodical exercises. If such were attempted, it was more as the result of chance than of plan.

Napoleon caused his troops to practice in the camp of Boulogne, ordering that they be taught to swim and exposing them to considerable hazards, not all of them entirely reasonable, or probable. One of the most successful landings, that of Abercrombie in Egypt in 1801, probably owed much to his being compelled to wait in Cyprus for six weeks, until remounts came. This period was used by Abercrombie to train the troops to embark and disembark with great speed "so that when our fleet of nearly 200 sail anchored in Aboukir Bay, every soldier understood his duty."

The special training which the German expeditionary corps for Oesel received was by no means planned originally. Only when the preparatory minesweeping was prolonged, due to bad weather, for a fortnight, did the commanders find time and occasion to take account of stock and to find out that they had not half enough landing craft, and to train their men in the embarkation port of Libau, putting them through exercises that were to prove of great value. It was not until the present

"War of Landings" that such preparations found true appreciation and the men adequate preparatory coaching. Only today has realization of the need for special training against the unknown, or the little known led to working out interservice collaboration and the most minute reconnoitering and to exercises in which prospective possible difficulties and hardships connected with the enterprises planned are reproduced as closely as possible.

Careful Training Essential

This reproduction of actual conditions is basic in the education of the commandos and in many ways epitomizes shock troop tactics and techniques, including working with models, to and across the seas, familiarising men with novel situations, devising and adopting timetable exactitude for the voyage into the largely unknown. The less specialized formations in present landing operations do not undergo such detailed training, but it would seem to have become axiomatic that all troops intended for assault on hostile shores first must be confronted with some simulacrum of the realities, momentous and fortuitous which they are about to face.

Careful training is particularly necessary in cases where untried troops are to be used. One of the new lessons taught by the present war is that raw troops, American and Canadian, can be dependably and successfully employed in landings. These had come into the war fairly late and as neophytes who had to war not in defensive lines, as the Americans at first did in France in 1917-18, but on the offensive, indeed in extremely offensive landings, as in North Africa and Sicily. For these, however, nothing proved more advantageous than the fact that Italy had entered the war on Germany's side. For this offered them a war-training approach to the Festung Europa without in the first actions running up against the resistance might of that fortress and its best forces.

What is the best composition for the landing corps forms a subject for special ponderation by planners. What percentage should there be of infantry, or marines, of engineers, of artillery, of cavalry—a serious question in days agone, when horse transportation and landing usually proved difficult—of tanks, of communications troops, of parachutists, of medical corpsmen? How should they be distributed among the various waves of an assault? How large should, or must, be the naval forces employed? And how large the air forces, provided one can draw from a plethora of resources? It would hardly seem, at least as measured in relative strength, that the infantry is any longer queen of the beachhead, as it traditionally is of battle, being now the smallest major division of the U. S. Army, less than 20 per cent as against the nearly 50 per cent of the air force.[1] We have not yet seen in connection with this war any break-down of a landing corps, or landing army according to their component parts. But every detailed description of a landing operation focuses attention on the increasing importance in most recent landings of the tasks of engineers and communications troops.

In a manner, surprising the enemy is another way of overcoming one's uncertainty about the latter's strength, intentions and positions. Thus regarded, plans are timetables for resolute martial adventures into the realm of uncertainty, extending to the moment of collision between forces. Planned surprise, with its aleatory factors, must go far in its calculations for upsetting or equalling the strength inherent in the position of the land defender, for it has been estimated even fairly recently that "a suitably disposed force of infantry, machine guns and light artillery if not surprised, can withstand attack from the sea by five to ten times its number." [2] It does seem, however, that the actual experiences of the present war have not altogether borne out this estimate of 1940.

Reckoning With Contingencies

However carefully made, plans can never carry with them full assurance that unforeseen contingencies will not seriously interfere with landings. These may be offered by man—like the crossing of a German unit over the path of a part of the commando raid on Dieppe—or by nature. While land war has become measurably emancipated from restrictions imposed by weather and climate, landings by water and air, of all operations in war, are still most affected by weather. Winter gave the ancient nations along the shores of the North Sea their only respite from the depredations of the Vikings, until the latter moved into winter quarters on islands close to the mainland, like Thanet and Noirmoutier, and from there undertook winter raids. One or more of the various expeditions attempted by Kublai Khan against Japan, the last in 1381, among other mishaps, ran into typhoons and failed, with losses so great that a last attempt to cope with the islanders had to be given up, so strong thereafter was the disinclination of the landbound Mongols.

Closer to our own time, steam power and the internal combustion engine diminished the domination of weather conditions, which still present strong claims to respectful consideration, however, for was it not reported that "large parts of the first troops landing in Sicily became seasick in the swell of the Mediterranean"? It seemed to be accepted at the beginning of the present war that no debarkation in small boats could take place when the wind velocity exceeded 20 miles per hour and that such craft could be beached only if the waves did not exceed those incidental to light surf, or ground swell. How far the evolution of special landing craft has overcome these limitations seems impossible as yet to state definitely. In any case, landings in the Pacific still have to consider and respect the typhoon season, lasting in the western ocean from October to December.

Tides, even although in their basic movements and average rise and fall they may be well known in advance, are apt to be influenced by winds and thus present considerable deviations from the normal in either height, or pace of flood, or ebb, and thereby disturb the plans of invaders. This happened at Tarawa. Only in inland waters like the Black Sea and the Mediterranean do tides not interfere with landing operations such as those in Sicily in 1943, while along the Channel coasts tides of ten to twenty feet, or even forty, are the rule. The tidal race there is strongest along the classical invasion coast from Ostend to Cherbourg, while lessening farther outward to the west and the south.

Meteorological aid is one of the greatest concerns of any landing planner. To a large extent it prescribes the season for the attack. Most often in the past summer has been chosen, and only now has war's still partial declaration of independence from the weather led to landings all around the clock of the year, or the day. But yet, as formerly, the dawn remains the ideal hour for the assault, with the approach managed during pre-dawn darkness.

Napoleon's Weather Problems

The disadvantage that summer in some latitudes brings to operations so dependent upon surprise lies in the higher local degree of exposure to detection. For example, nights are shortened by light for 18 out of 24 hours in the south of England during May and June. Napoleon, who frequently uttered his military maxims more dogmatically and with more seeming conviction than the fact actually warranted, was uncertain about the best season for landing on such shores as those of England, which he and his predecessors contemplated for some 15 years. Originally he believed that "it needed long nights, consequently winter, and that, once April was over, one could not undertake anything" (1798).

One of the many landing projects submitted to him, or to the Directoire proposed Christmas Eve as the most suitable time for invasion of England, as well as the eight festival days that followed. Later, in his camp at Boulogne, Napoleon seemed sure that summer, or specifically August, was best, a change of opinion which does not make it easier to settle the moot question whether Bonaparte actually and seriously ever did contemplate the invasion, and thought it feasible with the forces at his command.

Bad weather has thwarted numbers of landings, launched either close to shore, or from a great distance. These failures include such large expeditions as that of the Spanish Armada of 1588, which was thwarted in part at least by a series of western gales of unusual violence. This and other occasions, when blasts from the west aided against Catholic continental invaders, earned for them among the English the popular term, "Protestant winds." Of far more modest scale, the French expedition into Bantry Bay in Ireland in 1796 failed because the naval commander thought wind and weather too foul to dare debarkation of troops, although they would not have found any British force immediately to oppose them. The worst this storm had done was to separate the military commander, Lazare Hoche, who was the very soul of the enterprise, from the main body of the ships. He might profitably have forced the timid naval commander to attempt the assault, regardless of the weather.

A constant gale postponed the debarkation of Abercrombie's force at Aboukir in 1801 for four days. Luckily, the French on the beach did little, or nothing to make profitable use of this delay. After the landing of troops, the wind blew too heavily for two days to allow unloading of necessary supplies and stores, thus immobilizing the main body of the British army.

The assumptions as to weather, wind and surf conditions prior to the Sicilian landings of July, 1943, as based on long term meteorological observations, were reasonably hopeful that ideal conditions would present themselves. Consequently long range forecasts were equally favorable and short range forecasts were free of alarming features. For four days the invasion fleet on its approach sailed in ideal weather. But early on July 9, twenty-four hours before the scheduled landing time, a gale blew up, making the launching of boats from transports and the push through heavy surf to the landing beaches apparently impossible.

It became necessary to delay debarkation by one hour, which imposed a daylight assault, thereby washing out entirely the element of surprise. What might have been lost in surprise advantage, was in a manner cancelled by the creation of hopes on the defenders' side that the weather barred landing, on that day, at least. Even after the wind abated, the swell and surf remained heavy, more so on the American assault beaches on the southwestern coast than on those of the British on the eastern shore.

EISENHOWER'S DIFFICULT DECISION

The invasion of France in the spring of 1944 was decidedly vexed by weather. "This possible change in weather certainly did hang like a vulture poised in the sky over the heads of the most sanguine" (Churchill in Parliament, Aug. 2, 1944). Generally speaking, the first half of June offered, as weatherwise British merchant sailors told beforehand, the ideal time for operations in French west-coast waters, with its usual accompaniment of the smallest number of gales year by year and favorable tidal conditions. That June was also the worst month for fog and haziness could prove helpful, as well as adverse, to landing operations.[3]

The choice of June 6 for D-Day, 24 hours later than originally intended, was made by the Allied commander-in-chief, swayed by weather reports and prognoses for the next twelve hours which all around were not too favorable. While the forecasts predicted weather unsuitable to bombing operations against shore batteries, it encouraged operations of air-borne troops, profiting by the cloud cover. Wind pressure was such as to produce desirable wave conditions on the beach, although not far offshore.[4]

General Eisenhower made his choice and decided to go ahead after receiving the 4 A.M. forecast on June 5, rather than postpone the whole gigantic enterprise for possibly a month, or at least ten days, or a fortnight. Had he delayed, developments proved that in the two weeks until the next quiet period of Channel weather, as indicated by a study of conditions over long periods, the invasion craft would have been scourged by a gale. In fact, the Channel coasts were struck shortly after the 20th of June by "the worst June gales for forty years."

As it was, the weather, worse on the whole than in the Sicilian landings, did not improve until the 11th when the sea calmed sufficiently to allow a considerable acceleration in the ferrying and unloading of heavy supplies. Even then the full display of Allied air superiority was often prevented. Further away on the European war fronts, the weather, or rather the season, also played a role in the choice of June. By that time in the east what the Russians call "the wayless season," of roads and terrain deep in mud as the result of spring thaws, when communications would be at their worst, would be over, and Russian attacks could be expected to tie down German reserves and keep them away from the newly opened Western front.

Prolonged fair weather is highly desirable when landings have to take place over open beaches and before a harbor has been won on the invaded shore. The moon phase may be a factor in the timing of the landing, depending on whether a bright moonlit, or a moonless night is wanted, in connection with certain tides. A prime combination might be a flood tide setting in not long before sunrise to help carry the first troops to the shore, while later craft more heavily laden with materiel could come in with the full flood. A combination such as this occurs only once, or twice a month, a circumstance of which the enemy could not be unaware.

Considerations involving the most favorable weather, season and hour for landings play an equally important part in determining the landing and movements of paratroops and glider-borne infantry and other formations. Two factors stress themselves most heavily: the risks incurred during the last stages before landing and the orientation of the plane pilots and of the landed troops. If undertaken in daylight, as the Germans have usually preferred to do, casualties are apt to be high, particularly during that passive transit to land when these troops are most in peril from ground fire. Orientation in the direction of the chosen objectives is much easier, however.

At night losses are apt to be much lower, but disorientation greater in the air and on the ground, as measured by the percentage of troops alighting far from their area of destination. The Allies were inclined to favor night operations, as giving the participants a greater chance of survival, often being satisfied with the mere disturbing and dispersive effect of air-borne troops in the enemy's rear. The German attempts were more frequently directed toward concentric effects. The use of gliders of increasingly large capacity enhances vulnerability and casualties, at the same time ensuring a better initial concentration of troops. Some of these considerations prevail in the choice of landing craft. When the fire of the defender becomes too heavy, or there is reason for believing that it will become so, it is preferable to adopt evasive tactics and use smaller vessels.

Pre-Landing Reconnaissance

Maps now present the conditions prevailing on *either* side of the shoreline. In the period when separate military and naval interests were the rule, maps displayed blanks on either the sea, or the land side. Data which go to the making of maps and moonlight and tide tables, etc., are assembled from various sources such as old maps, in the directions accompanying which nature may have made many alterations since they were drawn; or books like *Directions on Navigation Between the Islands and Atolls of the Gilbert Groups;* or the experiences of practical navigators in specific waters. It may be advisable to reconnoiter waters and shores more or less immediately before a landing by fast vessels running as close to the coast as possible, or by special hydro-geographic landing parties debarking at nights to ascertain geophysical shore conditions in chosen places, to determine if and where soft ground would interfere with the movement of vehicles, etc., etc.[5]

The most recent means of reconnoitering a landing strip, observing and noting changes in conditions, by nature, or enemy, provided the latter are not camouflaged as they were to a large extent on Tarawa, is through aerial photography. The eye of the lens penetrates to a certain extent even below the surfaces of waters. While the maps provided for the projected Allied expedition to Norway were militarily worthless,[6] aerial photography, it is said, in the two years of preparation for the

Allied landings in Normandy "produced more original maps of France than that country has made since the days of Julius Caesar." Based on these surveys 125,000,000 maps, with 1 to 25,000 the largest scale, were at the disposal of the American invasion forces.[7]

Aerial mapping can furnish even forces already on the move to their objective with the latest maps, sometimes made after they have set out. It is reported that "in a matter of hours, instead of a minimum of three days expectable for such a mission, aerial map makers of the U. S. Army, Navy and Marine Corps completed composite pictures of the reef-strewn and hitherto uncharted Kula Gulf, delivering the pictures to our task force even as it steamed against the Japanese . . . . At least 20 per cent of the Kula Gulf was unfit for navigation and the existence of submerged reefs was of tremendous concern to ships' captains entering this body of water. We accomplished what had never been done before—doing hydrographic mapping with nothing but photographs in the interests of finding submerged menaces." [8]

Map reading, not to speak of the memorizing of topographic features from maps, does not come easy to every participant of an action in a given locality. Realizing this, the American command has usually arranged to supply relief models of various invasion shores, such as Tarawa or Gela. With the help of these, available on board the invasion transports, all ranks were given opportunity to familiarize themselves with the topographic features of their specific objectives. Still, how unfamiliar everything sometimes remains ashore after a night landing. "True, we had the map of this area burned into our brains," writes a participant of the landing at Gela, "but now at night, in the experiencing instead of the studying, there was nothing that even faintly resembled the markings on that map. Everything was unknowable, not a piece of ground we had ever seen before, not one familiar landmark, not a guide, not a signpost, not a house along the lonely way where we might stop and ask directions. Every tree, every bush, every clump of ground and dune of sand—what hidden surprises did they not hold?—a lurking sniper, a well-emplaced machine gun or a ready, waiting ambush. This alien land seemed filled with all the mystery of life and ever-waiting death. . . ." until "at last, remembering our maps, we knew where we were." [9]

To hamper and disorient an invader, the country threatened with a landing from the sea, or the air will, like England from 1940 on, attempt to remove such helps as the enemy might find immediately useful, such as highway signposts and place names on railway stations, public buildings and the like.

NOTES, CHAPTER 6

[1] According to data in the *N. Y. Times,* 9 January 1944.
[2] Col. C. A. Baehr in *The Field Artillery Journal.* September-October, 1940.
[3] UP despatch from London, 5 May 1944.
[4] *N. Y. Times,* 21 June 1944.
[5] For some details see report from Supreme HQ. AEF, in *N. Y. Times,* 1 June 1944.
[6] Pertinax: *Les Fossoyeurs.* I, 40.
[7] *N. Y. Times,* 7 June 1944.
[8] Lt. Col. E. E. Pollock, USMC, speaking before the American Society of Photogrammetry. AP despatch from Washington, 22 January 1944.
[9] Jack Belden: *Still Time to Die.* N. Y., 1944. 258-60.

Chapter 7

LANDINGS ACROSS FROZEN WATERS

He smote the sledded Polack on the ice.

SHAKESPEARE.

WARLIKE ENTERPRISES over the ice of frozen rivers, ditches, moats, lakes—such as the battle on the ice of Peipus Lake between Alexander Nevski and the Teutonic Order on April 5, 1242, a spring date!—and, much rarer, across winter-bound seas involve uncertainties which nearly equal those over open waters. While on the whole they are physically less hazardous, to many of the participants they may seem even more fraught with peril. Southern warriors, the Romans,[1] or Spanish, usually refraining from war in winter, would dread such operations at least as much northern soldiers might dislike a tropical campaign.

The bloodiest of all winter battles was that of Lund (Dec. 3-4, 1676) between two equally winter-hardened forces. The Swedish army had crossed the Lydde river by night and in fast marches reached the Helgonabacke, a high plateau near Lund. This they scaled on one side, and the Danes on the opposite side. Such clashes were exceptional until the present war, which knows no season, and in which problems of landing, crossing and battling under winter conditions were offered in the Russian-German campaigns and by the Japanese threat to Alaska in the Aleutians.

For the first time progress was made beyond use of the traditional ice-crossing vehicle, the sled drawn by men or animals, when preparations were undertaken against a Japanese landing in Alaska proper. A "winterized" motor vehicle was designed and built that could travel fast over deep snow, or on ice as well as over other difficult terrain, using tank treads of unusual lightness, combined with "fins" instead of wheels. This was the so-called "weasel" of the U. S. Army. It was found that special training in suitable surroundings could mentally and physically prepare troops for the hazards of winter campaigning.

Descriptions of the complexes of mental reactions involved in such enterprises seem to be rare, hence mention of the personal experiences of the writer in crossing the ice of a large inland body of water during World War I possesses a pertinence which merits it a place here. In command of a detached infantry trench mortar company, in February, 1918, he marched his men across a lake in the vicinity of Pskov on his own initiative, after having measured the thickness of the ice and looked up its weight-bearing capacity in the Field Service Regulations. The information in this amply tested book and inquiries among the inhabitants of the region seemed thoroughly reassuring. Crossing the ice appeared to be the only possibility that offered of reaching a certain rendezvous in time, besides sparing the strength of horses and men who already were well jaded by prolonged marching through deep snow drifts. Compared with battling through snow, a march across the lake promised a progress as easy as walking over a parquet floor.

One Commander's Dilemma

The column was strung out as thinly as possible. In spite of all safety precautions, when about one-third of the five miles across the lake had been covered the ice began to quake, its cracking became audible. It was a most unpleasant situation for everyone, and most so for the responsible commander. There was so little one could do and so much one could think of doing, most of which was obviously impracticable. To avert a panic, one had to meet the apprehensions of the men by assuring them, which might have been true, that the sinking of the lake's water level caused the cracking of the ice,[2] and to order them to remain strung out when their preference was to walk closely together. All the while one thought that the heavy vehicles would probably go down first, and that ordering them to keep still further behind the column would probably set the drivers off into a mad rush for safety and thus tax the ice more severely, and recalling to mind again and again the silly German proverb, "Water has no beam," and adding "neither does ice."

But the worst did not occur, although the rest of the march would compare in gloomy apprehension with any Nordic doom myth, the men silent, the commander trying to talk them out of fear and the ice cracking from time to time like an enormous whip. The relief after gaining land was beyond picturing; the snow's firm texture, deep as it was, was hailed as something safe to tread upon. It seemed like a mad mocking of our anxiety when we reached the first Russian village by the lake shore and saw a man, lobster red, from a steam bath, rushing across the snow to plunge into a water hole cut through the thick ice of the lake. Many men were unable to sleep for hours afterwards. In retrospect it seemed lucky that none of us knew at the time of the disastrous winter battle that had taken place nearly centuries ago, fought and lost by the Teutonic Knights whose mission we were presumed to carry on. (The mission of the Knights, a military religious order founded in 1089-91, was to crusade in the east against Moslems, and also Slavs.)

Landings across ice, and in particular that of seas, are naturally limited to northern climates and exceptionally cold winters. Such a winter was that of 1657-58. Sweden was at war with Poland and Denmark and in the late summer and autumn her army was occupied in reducing Jutland, the last fortified place held against her, Fredriksodde, or Fredericia, not falling until late in October. The diplomatic and military situation boded not too well for Sweden. Following the conquest of Fredericia, King Charles X, the grandfather of Charles XII, prepared for a crossing of the Little Belt into Fünen in ships, but soon winter provided a substitute for a transport fleet. By the middle of December a great frost set in. It became so cold that the soldiers had to cut their bread with axes.

After a few weeks it seemed to the Swedes that even the swift currents of the Little Belt, less than a mile wide at the narrowest spot, might freeze over and afford them safe crossing if they waited a little longer, although some of the commanders like Wrangel, of Thirty Years' War fame, were skeptical. Watching the weather and measuring the ice, the Swedes waited until the 30th of January when early in the morning their partly mail-clad force started for Fünen.

Charles chose a route in the direction of Brandsö, a small island midway between Jutland and Fünen, while the Danes had expected him at the narrowest extension of the Belt and had only 4,000 men to meet him where he chose to land. "The Danes," according to a participant from France, "kept up an incessant fire with their artillery to break and weaken the ice. The Swedish army was also much troubled by the canon balls which skimmed along the ice where it was smooth and shot up [ricocheted] where impeded by hillocks of ice and snow." They were strung out in a thin line, the horsemen dismounting where the ice seemed weak, and reforming when in sight of the island shore. Charles, who had crossed in a sled.

mounted his horse and proceeded to the front, his sled and horses soon after disappearing in the waters, as one spot in the icefield gave way beneath.

NIGHT MARCH ACROSS ICE

Two squadrons of cavalry also broke through the ice while attacking the left wing of the Danes, and the King in order to quiet his soldiers' fears rode close to the gaping hole in the ice to prove that it was still firm. The Danes on the beach were soon worsted and all of Fünen occupied. Not only was fortune in the Swedish camp, but a very good engineer as well to make use of the remarkable chances which luck offered, Eric Dahlberg, who thoroughly understood the structural characteristics of ice. When Charles contemplated crossing the much wider Great Belt as well, and thus force Copenhagen to surrender, Dahlberg advised him to avoid the direct route, from Nyborg to Korsör (12 miles), and choose the safer, although more circuitous, route by way of the islands of Langeland-Laaland-Falster, the sustaining power of the ice decreasing in direct proportion with the increase of the distance from one land point to another.

A Swedish council of war thought even this proposal too hazardous, if not outright criminal; but Charles, impressed only momentarily by the consensus of the council, endorsed Dahlberg's suggestion and ordered the crossing of the Great Belt to begin in the night of February 5-6. Cavalry led the march over the snow-covered ice. The temperature must have risen by that time of the year, for it is said that the snow melted under the horses' hoofs and that the infantry following up was forced to wade half an ell [about 22 inches] deep through slush. The march of some 20 miles, from Svendborg on Fünen to Grimstead on Laaland, with Dahlberg at the head of the column, took until three o'clock in the afternoon, and not a man was lost. Two days later Falster was reached and on the 11th Zealand. At the end of the trip Charles was met by Cromwell's minister in Copenhagen whom the Danes had sent as mediator. The king greeted him with a tart, "You here, Mr. Ambassador?" The Englishman answered: "Upon my word, I am even more astonished to meet your Majesty in this place."

This unique feat, coupled with the landing of the Swedes from the west, in the rear of the Danes, and the resulting surprises and shocks to the enemy, led at once to the signing of the preliminaries to peace at Taastrup, February 18, and the final Peace of Roskilde, February 26 by which Denmark surrendered half her territories. Charles had a medal struck to commemorate the "glorious transit of the Baltic Sea," bearing the inscription: *Natura hoc debuit uni.*[3]

The Swedes had no monopoly of victories won across frozen waters. Some years after the death of Charles, the "Nordic Alexander," in 1660, Swedish troops invaded Eastern Prussia from Livonia. Frederick William of Brandenburg, the Great Elector, in a magnificent winter campaign threw them out again, setting forth with his troops from Berlin in January, 1679, crossing the newly frozen Frische Haff and Kurische Haff, two Baltic lagoons, with the infantry on sleighs, pursuing them as far as Riga. With 16,000 men the Swedes had started out on their winter campaign; with only 3,000 did they return.

Russian Daring Paid

With the extreme declensions of degrees of winter temperature around the Baltic Sea and its low salinity, the gulfs of Riga, Finland and Bothnia and the waters around the Aland Isles are frozen over in many winters, allowing traffic upon the ice.

In the war between Sweden and Russia over the possession of Finland in 1808-9, these gulfs were crossed repeatedly. A Russian force under Prince Peter Bagration, by a daring march across the Gulf of Finland in 1808, captured the Alands and in 1809, Barclay de Tolly by a similar march across the Bothnian Gulf surprised and seized Umea. There was not a little hesitation on the part of the Russian troops to take advantage of the conditions produced by an unusually severe winter and it is said that the order of the Czar to do so was only enforced by the presence and activity of Arakcheev, Inspector General and War Minister, even more of a martinet than the average *ancien régime* general, and that he "compelled an unwilling general and a semi-mutinous army to begin a campaign which ended in the conquest of Finland." [4]

After the Germans had taken the Islands of Oesel and Moon in October, 1917, they had to make preparations against a Russian crossing of the sounds between the islands and the mainland which are frozen over during most winters from January to April and can be traversed by wagons. But nothing happened, since the Russians were preoccupied with their revolution. Some of the subsequent fighting in the rebellion took them over frozen rivers and lakes, but only once, it would seem, across a frozen sea. When Wrangel on his retreat to the Crimea in November, 1920, had blocked the Isthmus of Perekop by fortified lines, the Red Russians, not provided with heavy artillery, passed over the Sivash Sea east of Perekop, which was frozen over at the time, and thus penetrated into the Peninsula.

In another Russian war, against Finland in 1940, the capture of Koivisto on February 21, 1940, enabled the Russians to cross the ice of the Gulf of Finland and close in on Viborg.

A few shore-to-ship and ship-to-shore operations over the ice occurred in the various wars of Holland. Water, rather than land, was the element on and through which the Netherlands won their freedom from Spain, then the greatest land power. Even when water became ice and seemed to favor land power, it was turned to their advantage. In the beginning of the armed resistance against Philip II a small fleet of armed vessels belonging to the Province of Holland had been caught in the ice near Amsterdam. Spanish soldiers were sent across the ice to attack them, but the crews had left a moat, so to speak, around their ships by agitating the water about the boats. From this frozen citadel musketeers on skates sallied forth and drove off the Spanish, who left several hundred dead on the ice. "'Twas a thing never heard of before," reported Alva, the Spanish commander, "to see a body of arquebusiers thus skirmishing upon a frozen sea."

From further attacks across the ice the Dutch were saved by a flood and rapid thaw which enabled them to escape with their craft to Enkhuysen, while another hard frost made pursuit by the water impossible. The incident, small in itself, was taken as an encouraging omen by the Dutch. Only after a decline in Holland's military power had set in did frozen seas enable her enemies to strike at her last bulwarks, fortresses and ships, as in January, 1795, when the Dutch fleet, iced-up in the Texel, were captured by French hussars.

NOTES, CHAPTER 7

[1] A battle on river ice was fought by the Romans against the Iazyges, a Sarmatian tribe, in the winter of 172-3 A.C., "on the river" as the historian Cassius Dio writes, "and by this I mean not that any naval battle took place, but that the Romans followed them as they fled over the frozen Danube and fought there as on dry land." The Sarmatians decided to make a last stand on the Danube, a battlefield so unusual that they hoped to disconcert the Romans, who were not accustomed to ice, whereas even

the Iazygian horses were trained to run across the smooth surface. The Romans at first were surprised and hard put to it to maintain a footing. This they gained at last by throwing their shields on the ice. Standing on those they awaited the enemy charge. Many of the latter they pulled down, men and horse alike, and continued fighting with scores of them on the ice, more often lying on the ice than standing up. At last the Romans won, and few of the Iazyges escaped death or slavery. Dio LXXII 7.

[2] As the march took place towards noon, the more likely explanation is that the cracking was caused by the ice expanding like any solid body under the influence of heat.

[3] Robert Nisbet Bain: *Scandinavia.* Cambridge, 1905. *Memoires du Chevalier de Terlong,* in appendix to *Memoirs of Sir Andrew Melville.* Ed. by Sir Ian Hamilton. London, 1918.

[4] Encyclopedia Britannica, XIth ed., II, 316.

Chapter 8

POLITICAL-PSYCHOLOGICAL ASPECT

PURELY MILITARY TREATMENT and exposition can never do full justice to the subject of landings. What is required above all is an inclusion and evaluation of the psychological factors involved, for the effect of a successful landing goes far beyond merely strategic gains and affects the morale of the attacked population to a degree that from some aspects may be regarded as unreasonable. The psychological impact of an overseas expedition, of the bare threat to a nation, or a people of actual, or menaced invasion, is akin to that of an immense *trauma,* to use the language of psychoanalysis, mass fear, or anxiety. It is as though an old "dream wound" of the psyche-politic were re-opened. On the other hand, it is comforting and reassuring for a people to read in a communique that their own army, or navy has repulsed enemy landing attempts. A somewhat typical German communique that helped to build confidence in the security of the Fortress Europe would read: "On the Kuban bridgehead enemy groups which again attempted to land were annihilated" (Oct. 1, 1943).

Water-borne invasions are apt to shatter the calm and complacent sense of security produced by centuries of undisturbed national existence, such as that of the Incas in the 16th century, or of the Chinese and Japanese in the 19th. Nearly always such an invasion has been preceded by apprehensions born of the memory of some primeval experience, or a prophesy. The various threats of invasion directed against the British Isles would bring back to the English nation the memory of the last great and profoundly revolutionizing invasion suffered by them, that of 1066, for that of 1688 of the Dutch army who aided in placing William and Mary on the throne in succession to James II, was considered an invasion by only a small minority, composed of those whose sympathies were with the fugitive Stuart monarch.

The inherited, or historiographically-induced mass memory of past invasions may give rise to panic, when the people are confronted with actual, or imagined dangers to their customary security. Many nations or tribes include in their collective memory a myth of landing, a *Landungssage,* in either a passive, or an active version. The passive is likely to foreshadow the coming of foreign warriors, usually possessed of rumored strange new weapons and distinguished by a different color of skin, whose advent heralds a cataclysm or revolution. Such evokers of fear were the white men from Spain who broke through the bulwarks of Montezuma's Aztec Indian empire, or the Caucasian invaders of Samoa, where the people called them Papalangi, cleavers of the sky, because they broke the long untouched horizon line around the island group with the prows of their ships, which the natives believed bore "sailing gods."

Ancient Fears Deep-rooted

Technical innovations eradicate few, if any, ancient beliefs, or fears. They merely modify them. The airborne invader is even more frightful to those who fear, or have cause to fear, him and his irruption, such as that "sudden and quickly terminated, but terrific rape of Pearl Harbor;" [1] or the completely irrealistic radio-trumpeted "invasion from Mars" which a few months before the Japanese attack agitated a portion of the American public. To calm such terrors which the landing of the "yellow" Japanese on the "white" outposts of Attu and Kiska had inspired in the hearts of many Americans on the Pacific coast, President Roosevelt told them, once the invaders had been cleared out: "You've dreamt of Japanese marching up the streets of Bremerton, or Seattle tomorrow. You have thought that the Chiefs-of-Staff in Washington were not paying enough attention to the threat against Alaska and the coast." Assuming the *pater patriae,* if not the psychoanalyst's role, he went on: "We realized, of course, that such a Japanese threat could become serious if it was unopposed. But we knew also that Japan did not have the naval and air power to carry the threat into effect without greater resources and a longer time to carry it out." Claiming for himself the statesman's foresight he concluded: "Preparation to throw the Japanese from that toehold, had been laid even before the Japs got there . . .[2]

While its mythical character and its adventure-story features have obscured its true historical nucleus, the Greek saga of the Argonauts, a myth having to do with the feats of gold-seeking marines, really rests upon the colonial expansion of the seafaring Greeks to the shores of the Black Sea. Among the tactical features of the Argonaut legend, it may be pointed out, was that which stressed that Jason's expedition was indeed prepared for combined operations including those in the air! For among the Argonauts were the sons of Boreas, the North Wind, who, carrier-based, rose into the air from the ship to slay the hideous Harpies. Other threats from the air were provided by the Stymphalian birds, pre-historical aerial machine gunners, so to speak, who shot arrows made from their own brass feathers.

Retelling the tale of the Argonauts in modern phraseology, we might also say that they found the help from within the enemy's lines without which they could not have succeeded. This "fifth column" of the Argonauts was a woman (as in the case of Ariadne), the sorceress Medea who had been under a shadow before the invaders arrived. Her fate as well as that of Ariadne on Naxos, incidentally, warned the fifth columnists of all time that wage for their activities in support of foreign intruders are not apt to be paid over-punctiliously, or agreeably.

Synthetic War Scares

The fear of sea-borne invasions that lies dormant in the popular mind has often been awakened by interested sea power nations, by persons and groups intent upon putting over armament, or preparedness programs for which no market offered without the stimulus of threat. These invasion scares have been used where public discussion appeared hopeless in whipping up support for big army, or navy plans. While these scares certainly do not form the most reasonable, or desirable element in democratic systems of government, at times statesmen and executives have considered them indispensable to the prudent and necessary shaping of public opinion. As the late Viscount Esher, one of the few pre-1918 British civilian officials who permanently concerned themselves with imperial defense, wrote to Lord Fisher in 1907:

"No state of mind in a nation is more dangerous—or as experience shows, more foolish—than over-confidence. . . . It is the discussions which keep alive popular interest, upon which alone rest the Navy Estimates. A nation which believes itself secure, all history teaches is doomed. . . . An invasion scare is the mill of God which grinds you out a Navy of Dreadnoughts, and keeps the British people war-like in spirit. . . . Invasion may be a bogey. Granted. But it is the most useful one, and without it Sir John Fisher would never have got 'the truth about the Navy' into the heads of his countrymen." [3]

Invasion scares are mixed blessings. Besides "wholesome

fears" as their proponents term them, they may well unduly and perilously excite national hysteria, like that which measurably prevailed in parts of the United States at the start of the war with Spain in 1898. Every eastern seaboard city or town demanded at least one gunboat to protect it from the apprehended depredations of Spanish torpedo boats which, technically speaking, could hardly have navigated to within range of any of these places.

These built-up armament scares may be considered to disguise ancient mass fears in modern form and to be used for modern purposes, political, or otherwise. Before the coming of the Vikings to England, the *Anglo-Saxon Chronicle* noted, about the year 793, long *after* the event: "Dire forewarnings came over the land of the Northumbrians, and miserably terrified the people; these were excessive whirl winds and lightnings; and fiery dragons were seen flying in the air. A great famine soon followed these tokens; and a little after that the ravaging of heathen men lamentably destroyed God's church at Lindisfarne through rapine and slaughter."

Stock Market and Book Reactions

Those seismographs of public moods, reflected upon stock prices and transactions on bourses and stock exchanges—in our time Wall Street and The City, in London—proved particularly susceptible to invasion scares, reacting to them by registering slumps in prices on such occasions as the Stuart restoration attempts in 1715 and 1745. Then the price decline on the London Stock Exchange threatened to upset the economic structure of England to such an extent that military men like General Lloyd later advised the government to contemplate moratoria and other measures in order to prevent these semaphoric sources of public alarums.

A fairly modern form of playing on invasion fears, which cry out loudly enough—but so far in vain—for rigid psychoanalytical treatment, is the novelistic portrayal of invasions by whatever theoretical enemy that at one time or another seems in public view to be the most likely foe. In a way a history of the inception and burgeoning of anti-German feeling in England might be dated from a military man's novel, *The Battle of Dorking,* by General Sir George Tomkyns, published in 1871, when the German armies were hardly yet marched homeward from conquered France. In his book an invasion of England by the victorious German forces was vividly described as impending. Many such novels have been produced since, by military men occasionally, but more often by non-military writers, down to the late Hendrik van Loon. While their contribution to serious literature and sane military thinking, which at the same time neglected the problem of landing operations, has been slight, occasionally they may have proved more wholesome than pernicious.

The failure of an intended, or actual landing attempt will raise the morale of the threatened nation boundlessly. England abandoned its widespread popular conviction of Napoleon's invincibility when he laid aside his invasion bugaboo, which inflicted upon him a serious moral defeat. In July, 1943, when the Allied landing in Sicily was going on apace, a Berlin communiqué announced that the Germans had warded off a Russian attempt at landing in northern Norway, whereupon the Russians denied that the attempt had taken place. This exchange of claim and denial, not uncommon in war communiques, was promptly followed by an impressive concrete revelation of the effect of the landing in Sicily. Before it, the whole Fascist structure in Italy broke down and Fascism suffered everywhere what German military historians call "an inner Jena." It was caused not by a comparatively minor military defeat, but by the shock of the invasion of Fascism's soil by way of Mussolini's *mare nostro,* which was then no longer *nostro,* if indeed it had ever been.

Spies and Other Agents

Public anxiety in countries expectant of an imminent invasion is apt to reach its climax when it contemplates "the dark invader." Behind such anxiety there may, or may not be awareness that on occasions it has been proposed, or supposed that a water-borne invasion may be undertaken *à la japonaise,* without a declaration of war. Not all overseas invaders have been, or are, of regular, embodied military quality, a circumstance which in cases might further heighten public fears. In practically all past times enemies have sent out spies, saboteurs and other agents of espionage, destruction and sedition, landing them in disguise, or in desolate parts of the invaded country from surface ships, submarines, from the air, or across the ice. This latter, and not so "innocent passage" was used by the

German army in the winters from 1914 to 1918 to send and receive its agents working behind Russian lines.

Generally speaking, only a few agents, or perhaps only a single one, will be landed at a time, as was Sir Roger Casement in Ireland, or the group of German saboteurs put ashore in the United States in 1942, both being conveyed by submarines. The Directory of the French Revolution once planned to loose jail-birds *en masse* on English shores to terrorize that nation.

Considered in terms of troop psychology, landings involve a drawn-out process of "going over the top." They mean advancing into the utter dark and unknown, often mentally as well as topographically. Every landing, as far as extensive complexes of feelings among the participants is concerned, takes place on the shore of the equivalent of darkest Africa. This prolonged uncertainty calls for careful psychological handling of the troops by their commanders, high and low. The latter must avoid anything that might add to their men's burden of anxiety and render their already heavy physical load, including cramped quarters, still more damagingly irksome. There is no place for the anxious type of man, the worrier, in commandos, whose directors, it may be supposed, have now evolved psychological screening tests to eliminate them, although they cannot be altogether excluded from the more all-inclusive landing corps.

Men going over the landing top, so to speak, must be spared unnecessary occasion or excuse to worry, or grouse. It was stupid on the part of the leader of one of the several expeditions sent into Ireland by the French Revolutionaries to withhold from his men the three months' advance payment which had been promised them; a near-mutiny forced him to pay the men while in transit. Far more exemplary behavior by a general during a long sea-crossing was shown by Belisarius during the Byzantine enterprise against Carthage and the Vandals in 534 A.D. During the tedious voyage from Constantinople to North Africa, he carefully observed his soldiers, many of whom were inclined to "explore with anxious curiosity the omens of misfortune and success," and by appropriate acts of discipline and ministering in sundry ways to the health of their souls and bodies he kept them in good spirits.

The eyes of the men at the crucial moment will be on their leader. A feature common to many stories of landings from

Caesar onward has to do with the stumbling of the general as he disembarks, an occurrence apt to be at once interpreted by the soldiers as an evil omen. Aware of this, however, the conqueror traditionally is supposed to cry out that he purposely went down on his knees symbolically to grasp and take possession of the soil of the invaded country, as did William of Normandy, with his: "I take and hold thee, England."

Where the approach to the landing shore is of long duration, reading material is the minimum of entertainment that should be provided. It was a sound realization of this that suggested the *Infantry Journal's* booklet *What to do Aboard a Transport.*

Psychology and Landing Men

Various arrangements have been made to give the troops in an amphibious operation something they usually can learn only while en route—a sense of the importance of each man's personal participation in the adventure, by "briefing" them, by offering them for study profile and other maps and models of the landing beaches, by letting the men share in trifling, or exciting incidents of a trip, or enabling the seafaring personnel below decks to sense events and proceedings through the eyes of a qualified observer who speaks to them from the bridge over a ship broadcasting system.[4]

The psychological situation in which the landing men of the rank and file find themselves has as yet not been explored by any deep-reaching process of psychological analysis; nor is promising material with which to work readily available. Some reports would indicate that there is a definite tendency on the part of many of the invading troops to cling to the landing craft, to hesitate to abandon it for the uncertain, possibly insecure, footing of sandy, rocky, or miry beach; not to exchange the ills he has for others he knows not of.

Psychoanalysis might call this a throw-back tendency to cling to the mother womb, caused by unwillingness to face the unknown world of strife. Be that as it may, and some may find in the notion a humorous suggestion which may help them in faring on valorously, the moment of final debarkation is one that calls for high qualities of leadership, particularly on the part of the lower command. It is this sort of leadership which Caesar immortalized in his story of the eagle bearer of the Tenth Legion who was the first to jump overboard into the

English waters and from there admonished the hesitant legionnaires to follow him and the legion standard.

A new uncertainty presents itself to be felt so soon as landing has been achieved. There is nothing immediately available to fall back on, it will seem to those inclined to look backward. The ships which conveyed the expedition retire from the scene, sometimes too quickly, as it seemed to the Marine "expendables" left on the beach at Guadalcanal. A somber hostile land is staring at the invader, as if to sap his feeling of security at standing once more on terra firma. It is to overcome this backward glance, this hesitation, and give the troops a desperate knowledge that they "fight with their backs to the sea," that leaders of landing expeditions have sometimes resorted, literally, to burning their ships behind them. The military metaphor which employs "bridges" is indeed, more common than the actual military practice.

But we have the actual example of Cortes, whose campaigns in Mexico beginning with 1519 offer a rich collection of stratagems useful to the water-borne invader. How far the actual destruction of boats may go is illustrated by measures taken by the landing parties of the "well governed and prosperous voyage of M. James Lancaster," a privateering enterprise against "Fernambuck" (Pernambuco) in Brazil in 1594-5, details of which are supplied later on. As Hakluyt narrates, the admiral of that expedition before landing "gave express order to all who had charge of governing the boats or galleys to run them with such violence against the shore that they should be all cast away without recovery, and not one man to stay in them, whereby our men might have no manner of retreat to trust unto, but only to God and their weapons." The main ships, of course, were saved and used to carry off the precious loot found in the Pernambuco warehouses, over which their very efficient and successful landing operations gave these buccaneers control.

NOTES, CHAPTER 8

[1] Lt. Cdr. E. Rogers Smith, M.D., USN, quoted in the *N. Y. Times*, 12 May 1943.

[2] Radio address at Puget Sound Navy Yard, 12 August 1944.

[3] *Journals and Letters of Viscount Esher.* London, 1934. II, 249-51.

[4] For such an arrangement in the U. S. fleet going to Sicily in 1943 see Lt. John Mason Brown: *To All Hands. An Amphibious Adventure.* Foreword by Rear Admiral Alan G. Kirk. N. Y., 1943.

Chapter 9

INSULARITY AND INVASION

IN HUMAN CONSCIOUSNESS and sub-consciousness the tendency is to regard islands as actual, or mythical localities of innate poetical, religious, political, economic, or military significance, more or less haloed by fervent wish, or hope that they remain inviolate to strife and other harsh manifestations of life. "Wie Träume liegen die Inseln im Nebel auf dem Meer," as a German poet, Theodore Storm (1817-88), in exile remembered the eerie peacefulness of his homeland on the North Sea—"Like dreams the islands are resting in the fog upon the sea."

The word island in many languages connotes a sanctuary, acquires an almost religious ring, as that imparted by William Blake to his Jerusalem, built "in England's green and pleasant land," or the Isle de France—an inland island so to speak, the "heart land" of French nationalism. Such usage of the word has easily led to erroneous, although quite understandable etymologies—Heligoland has been interpreted as Heiligland, or Holy Land, when it actually means Halligenland, the Halligen being small islands close to the Holstein coast.

Great Britain includes at least three Holy Islands, so-termed, one each in England, Scotland and Wales, not to mention such hallowed isles as Iona and Lindisfarne. They are called "holy" because they are "whole," in all Germanic languages, with their politico-psychological etymology of whole holy, hale, was hal, wassail, heil and Heil, hail, Heiland, meaning Savior. The island may not represent entity in human coexistence. "No man is an island, entire of itself; every man is a piece of the continent, a part of the main" (John Donne, 1573-1631). But man, venturing into geopolitical mysticism, would still often see in island form, or seek to see in it, the only possible physical token of separation from the evil in the rest of the world, and particularly from war, thus hoping to achieve something "entire of itself."

In art, although sometimes it is not easy to depict it as such, the island may be presented as an image of repose and final rest, like Boecklin's popular, more than truly great, "Isle of the Dead," which appositively illustrates the lines of the English poet.

> *To die is landing on some silent shore*
> *Where billows never break, nor tempests roar;*
> *Ere well we feel the friendly stroke, 'tis o'er."* [1]

Poe employed the island as a simile of the untouchable:

> *Thou wast all that to me, love,*
> *For which my soul did pine—*
> *A green isle in the sea, love,*
> *A fountain and a shrine.*
>
> (*To One in Paradise*)

In many mythologies islands are represented as localities where the blessed abide in after life. Babylonian cult presents to us a large dark cave as the after-life abode, close to the ocean which surrounds the earth and flows beneath it. Here the dead are assembled for a depressive hereafter of inactivity, gloom and dust; and only a few escape this general fate and are transported by the gods to a paradisiacal island. The Hesperian island, so-called, of the Greeks, which in fact was Italy, placed in the west while the strife along the frontier of Greek civilization took place in the east; the Atlantis of Plato and other writers, the Fortunate Isle of the Romans, the "abode of the blest" [2] in Virgil, Avalon of Welsh mythology, the kingdom of the dead originally and finally, in the Arthurian cycles, the dwelling place of heroes whether Arthur is conveyed after his last battle—all were islands in the west where the favorites of the gods, or of God are borne after death. Always toward the setting sun, to which Tennyson turns the face of Ulysses, with his: "For my purpose holds to sail beyond the sunset and the baths of all the western stars until I die."

Ancient Sacred Islands

Medieval Ireland was often called Island of the Saints, and numerous monasteries, refuges from the world and worldliness and mundane materialism were on islands like Lindisfarne, or Moore's "sweet Innisfallen," in the lakes of Killarney; or as Reichenau in Lake Constance, or Frauenchiemsee, some occasionally replacing devotion to heathen cults centered in such islands as Rügen, where the Germanic Hertha, or Nerthus cult was superseded by a Slavic temple which in turn was destroyed by Christian proselyters from Denmark. "Est in insula Oceani castum nemus," "There is in an island of the ocean a holy grove." [3]

Ceylon historically may be the island most invaded; its records in myth and history, since 543 B.C., begin with invasions. After the conversion of the inhabitants to Buddhism in the 3rd century B.C. it became a kind of holy island studded with Buddhist monasteries, but warlike invaders intruded there as often as before. The Normans were most violently denounced by the monk-annalists and chroniclers of these holy places, many of which were pillaged and burned by these sea-rovers, "*fortissime gens* who have spread themselves over the earth, ever leaving small things to acquire greater, unwilling to serve, but seeking to have everyone in subjection," as a monk of Monte Cassino wrote of them in the 11th century.[4]

Such sanctification of the island character, which found one of its most recent forms in the inclusion of the Monroe Doctrine among the tenets of Christian Science, made invasions diabolic in character and the invader an ally of Satan, a necromancer of the blackest hue. This was the nature attributed to the leader of a French invasion fleet of the 13th century, Eustace the Monk, a "master of pirates" as Matthew Paris called him, and a leader of mercenaries, a combination that became rare as condottieri specialized in, and turned largely to, land warfare. As stories about him insist, Eustace was a religious and national apostate. English-born he served Louis, son of Philip Augustus, whom the English barons had called into their land to assist them against John Lackland and his son.

While carrying reinforcements and stores from Calais, to the barons, lately defeated at Lincoln, he was attacked by the Dover fleet loyal to the British king; he was beaten and himself captured and beheaded. His defeat (August 21, 1217) sealed the fate of the baronial cause and was hence of some historical consequence. Popular imagination soon after his death changed him into a piratical necromancer who could make his own ship invisible—the supreme hope of any invader and the worst fear of those threatened with invasion—until he was detected by one Stephen Crabbe on the English side, who applied white magic to ward off the danger to his fatherland. Being the only person able to see this phantom ship, Crabbe boarded it; his fellow mariners saw him swinging an axe in the air above the waters and beheading Eustace. This feat restored the ship to the vision of all, although the brave and patriotic Crabbe himself was torn to pieces by demons, who thus probably rendered Eustace the last of the services for which he had contracted his soul.[5]

BAPTISM AFTER CONQUEST

This story makes invasion a sin, one not mentioned in the Bible. Islands often were presumed to derive their greatest defensive strength from the power of the gods domesticated there; a strength not always merely defensive. Many islands have become the lairs of pirates like Sardinia, Rhodes at various times, or Rügen in the 12th century when it was the stronghold both of heathen religion and piracy. From Rügen the Wends set out to ravage and lay bare a third, or more of Denmark until warlike Bishop Absalom proceeded against them, conquering in 1168 the main Wendish fortress and temple of the god Svantevit, at Ankona.

After the fall of holy Ankona and the introduction of the Christian God, the belief of the Slavs in the impregnability of their island and in the sacrosanct character of their divinities was deeply shaken. When Absalom with his Danish ships appeared before Garz in southern Rügen, that strongly fortified capital surrendered at the mere sight of the ships. Together with a few housecarls Absalom went ashore. As if paralyzed, the Wend warriors, 6,000 strong, allowed him to pass along their ranks and by the hidden road to the sanctum of the seven-headed god Rügievit. The idol was torn down and destroyed by command of Absalom, who then had all the inhabitants of Garz baptized.[6]

Utopianism is in part secular religion. Consequently many, if not a majority, of the so-called Utopias, descriptions of ideal societies, have been placed by authors ever since Plato on islands, and after the discovery of exotic countries several of these Icarian and other idealizations were actually transferred to those distant regions, whose populations in turn sometimes adopted and applied such attitudes and convictions as were calculated to prosper their defense policies and measures. As a result of this sanctification, mythical, religious and humanitarian, there was woven around islands a political *Gestalt* psychology. They are seen and felt as fixed totalities, framed by nature herself in a perfection unobtainable through all longings for and theorizing about, or conquests of, "natural boundaries" of continental countries.

If and when in its own people's minds the bold and defiant island outline is damaged, or destroyed by invaders, what follows is a shattering, or collapse of something holy and wholesome in the psyche of the assailed. This does not exclude the after-thought that the revolutions, or evolutions started by island-invasions were despite everything regarded as beneficial by many, or possibly most of the invaded populations, from that of Brutus, or Britus in Geoffrey of Monmouth's *Historia Regum Britanniae,* or the great-grandson of Aeneas, who after some Oedipus-complex experiences fled to Britain where he drove out a tribe of giants ruling the island and founded a new Troy on the banks of the Thames, down to William the Third and the Glorious Revolution of 1688. But there still remained among considerable minorities in England a long surviving regret that the invaders succeeded, and not Harold and his Saxon defenders; or perhaps even the Gaels long before, or the later Stuart monarchy. In leaving this theme, we realize that we have merely skirted a problem of the political psyche, yet hardly analyzed.[7]

The strong island-feeling of the British—"continent isolated" as a reporter on weather conditions announced when a fog interfered with Channel communications—has been in a sense transmitted to the American political consciousness and subconsciousness. The Americans as situated in the field of defense, amidst the seas of war, tended to consider their continent as a huge island. The spirit of insularity, prone to regard the sea around island confines as an unbridgeable moat and more important as a medium of insulation rather than of communication, underlies much English and American defensive or other policy, including the Monroe Doctrine which in military terms is above all *Vorfeldsicherung,* safeguarding as it is does the glacis of the United States. It involves also a state of mind with regard to defense which may vary, and often abruptly, from complacency to panic and fright when forced to contemplate the danger of an invasion, to which sea-surrounded societies are more exposed than those of continental countries, the latter, being long inured to uncertainty, to attack from beyond their land borders, and with more numerous invasions in their own past, history and memory. Resenting this threat of violation, if not of rape, to their soils and peoples, the two Anglo-Saxon nations abruptly reacted emotionally to what the Germans and the Japanese menaced at Dunkirk and Pearl Harbor. Overcoming their island-prepossessions, they became the greatest water- and air-borne invaders of all times.

INSULARITY AND SAFETY

The deepest, most ardent hope of insularity and isolationism is to live without wars, or at least keep them by naval power far enough from the home shores to defy attempts at invasion. Yet in historical reality, island societies have not always been able to stay aloof from the competitions and ambition of states and resultant wars. As Sir Walter Raleigh, himself one of the Elizabethan navigators and wideners of the British overseas world, wrote: "Yea the Sea it selfe must be very broad, barren of fish, and void of little islands interjacent. Else will it yield plentiful argument of quarrel to the Kingdomes which it severeth. All this proceeds from desire of having, and such desire from fear of want." [8]

The absolute defensive could be maintained only in a very few places and at some periods, as in Iceland; Japan from 1598 to the opening by American warships to western civilization and trade in the mid-19th century; in the United States during long periods of her history; and finally in theory and caricature, the caricature of 18th-century German *Kleinstaaterei.* In the smallest northwest German principality of Schaumburg-Lippe lies the largest lake of this northwest, the Steinhuder Meer *(mare,* not *lacus),* 31 square kilometers in extent. The governor over this principality and lake from the middle of the 18th century was Count Wilhelm, field marshal of the Portuguese army, one of the finest, most humane representatives of enlightened despotism. Having seen much service abroad and overseas, he laid down his conclusions in six volumes of *Mémoires pour servir à l'art militaire défensive* (1775-77), and in sentences like these:

> "The more perfect the sciences of war are, the more dangerous it is to start wars, the more rarely they will be undertaken, the more the way of conducting them is removed from mere ferocious strangling. . . .
>
> "None but the war of defense is just, every aggression is below the dignity of an honest man. . . .

"One must apply the art of war to avoiding the war or at least to reducing its evils. . . ."

Count Wilhelm further was convinced that defensive war led to conscription, which he consequently introduced among his 20,000 subjects, and was one of the first of his era to do so. To deduce from, illustrate and teach his maxims he made his subjects slave for four years (1761-65) to build a fortress island, or island fortress in the middle of the Steinhuder Lake, surrounded by 17 small islands. On the main island a formidable masonry citadel was constructed, provided with ammunition and rations sufficient for a siege of several years.

Early Study of Landings

In expectation of the siege which never came, Fortress Wilhelmstein was made a war academy (1770) from which Scharnhorst, the leading military reformer in Prussia after Jena, graduated as the Count's greatest pupil. Emphasis in instruction was on engineering and artillery and the cooperation of the diverse arms, the increasing estrangement of which the Count had observed. Together with his pupils he undertook hundreds of experiments with ordnance, powder, mines, including ice mines; and even submarine lamps, it is reported—everything relating to landings, either on the Wilhelmstein, or elsewhere, and to make impossible landings as most unjustifiable acts of aggression.

Not all countries that have sought, or maintained isolation have been situated on islands, or moated with waters. China and pre-Petrine Russia were isolated from the outside world largely by vast land masses. Hitler-Germany, situated in the heart of Europe, tried to induce this synthetically, so to speak. Their governors told the Germans that they were "encircled" by a sea of hostility.

After the temporary Fortress Europe which was to provide more living space and either dispose of, or hold at arm's length all enemies, had again shrunk to the Fortress Germany under the pressure of Allied arms, Germany in the autumn of 1944 was declared "holy soil" by Goebbels' propaganda rodomontade. This proclamation was given some binding force by the Allies' insistence on "unconditional surrender"—not to mention such proposals by irresponsible persons as that all Germans should be sterilized—which in its vague grimness could only mean to

Nazis, or native non-Nazis that nothing they considered their own would be left them; left "heil und ganz," whole and total.

The island-dream, the hope of making a country resemble, as Elizabethan England had seemed, a "Fortress built by Nature for herself against infection and the hand of war" is by no means a thing of the past, as many have thought towards the ends of various wars. The Japanese prolonged fighting for the sanctity of their island empire, originally violated by the black ships of Perry, which precipitated a crash of the whole delicate domestic structure of feudal society and esconomy. "This godly State has never been exposed under a free invasion of enemies," declared Premier General Kuniaki Koiso, on August 24, 1944, "If by one in a million chances our enemies, the United States and Britain, should invade the homeland of imperial Japan, what honor have we, the 100,000,000 people, with which to face our ancestors?" While the revolutionary novelty of the atomic bomb might have helped the Japanese, with their backward-directed minds, to justify themselves for surrendering, thus saving face before their ancestral world, the shock of a terrific, destructive manless invasion completely shattered that people's serenity and sense of insular security.

Isolationism in America is by no means dead, but rather waits for technological justification, for the proof that it is militarily feasible. Russia is building up her own isolation, a process begun before this war was ended by annexing parts of Finland and Poland, Rumania and all of the Baltic States and infiltrating the rest of eastern Europe with an amalgam of Panslavism and Communism, the one to serve where the other will not. In face of this, England must realize that *her* isolation is gone, considering the forms and machines of war that have arrived to scourge the world. Isolation, if possible, can be achieved now and later on the continental scale only.

NOTES, CHAPTER 9

[1] Sir Samuel Garth: The Dispensary. Canto III, 1.225 (1699).

[2] Virgil: *Aeneid.* Book VI, 1.639.

[3] Tacitus: *Germania,* 40.

[4] Aime of Monte Cassino: *Ystoire de li Normant.* Cited Charles Homer Haskins: *The Normans in European History.* Boston, 1915. 13.

[5] Wendelin Forster and Johann Trost: *Wistasse le moine.* Halle, 1891: Henry Lewin Cannon: *The Battle of Sandwich and Eustace the Monk.* Engl. Hist. Review, vol. 27 (1912), 649-70.

[6] Saxo: *Gesta Danorum.* Ed. Holder. Strassburg, 1886. Books X-XV.

[7] During the preparation and waiting for the great invasion of Western Europe in June, 1944, American psychiatrists reported that it had "added greatly to the strain of war-time living. The number of nervous breakdowns has increased markedly," with elderly persons "likely to be the ones most strongly affected." *N. Y. Times*, 17 May 1944.

[8] *A Discourse of the Originall and Fundamentall Cause of Natural, Customary, Arbitrary, Voluntary and Necessary Warre, with the Mystery of Invasive Warre.* London, 1650, no pagination.

PART 2
Ancient and Medieval Operations

Chapter 10

GREEK AMPHIBIOUS WARFARE

THE SIMPLE CONDITIONS OF WARFARE in antiquity reveal, perhaps more clearly than the complicated actions of modern times, many of the basic principles of landing. That is the attraction of a study of early warfare even to those who shun the strictly "applicatory" methods of military history. In a more naive age, classical antiquity was often outspoken about the psychological situation of participants in amphibious enterprises. Far more openly than a Victorian soldier, the antique fighter would express his fear of the risks involved, the danger of sinking or being detected, which might at times seem so much greater than the hope of success in *terra incognita.* And it is very evident that the psychological mastery of a Caesar, or Belisarius could turn the hesitations of waiting troops into confidence, whereas the ineptitude of a Crassus in critical moments led to signal defeat at the Euphrates.

Landings began with political expansion. There is hardly a planned imperial policy that would not ultimately entail amphibious enterprises. This is quite obvious in the case of island-based empires. But even such essentially continental empires as China, or Persia will eventually undertake wars which involve overseas expeditions. The Persians, not originally a seafaring people, under the Achaemenides carried war across the Aegean Sea, provoked by Greek expansion. The Persian force that fought at Marathon (490 B.C.) without doubt numbered far less than the usually accepted figure of 50,000 who met 10,000 Greeks; its landing from ships on a small plain extending from the shore was unopposed. The Greek hoplites won the day against the Persian archers, the majority of whom successfully reembarked, despite the capture of several ships and losses which Herodotus gives as 6400 dead as against 192 of the Greeks.

Immediately after the battle the Persian fleet sailed on to round Cape Sunium with the intention of effecting another landing at Phalerum, the harbor of Athens at the time. With the help of an Athenian fifth column they surprised that place from the sea. Suspicious of a plot of that kind and warned by an intercepted signal from Mount Pentelicus, sent by the traitors, Miltiades brought the victorious army at once back

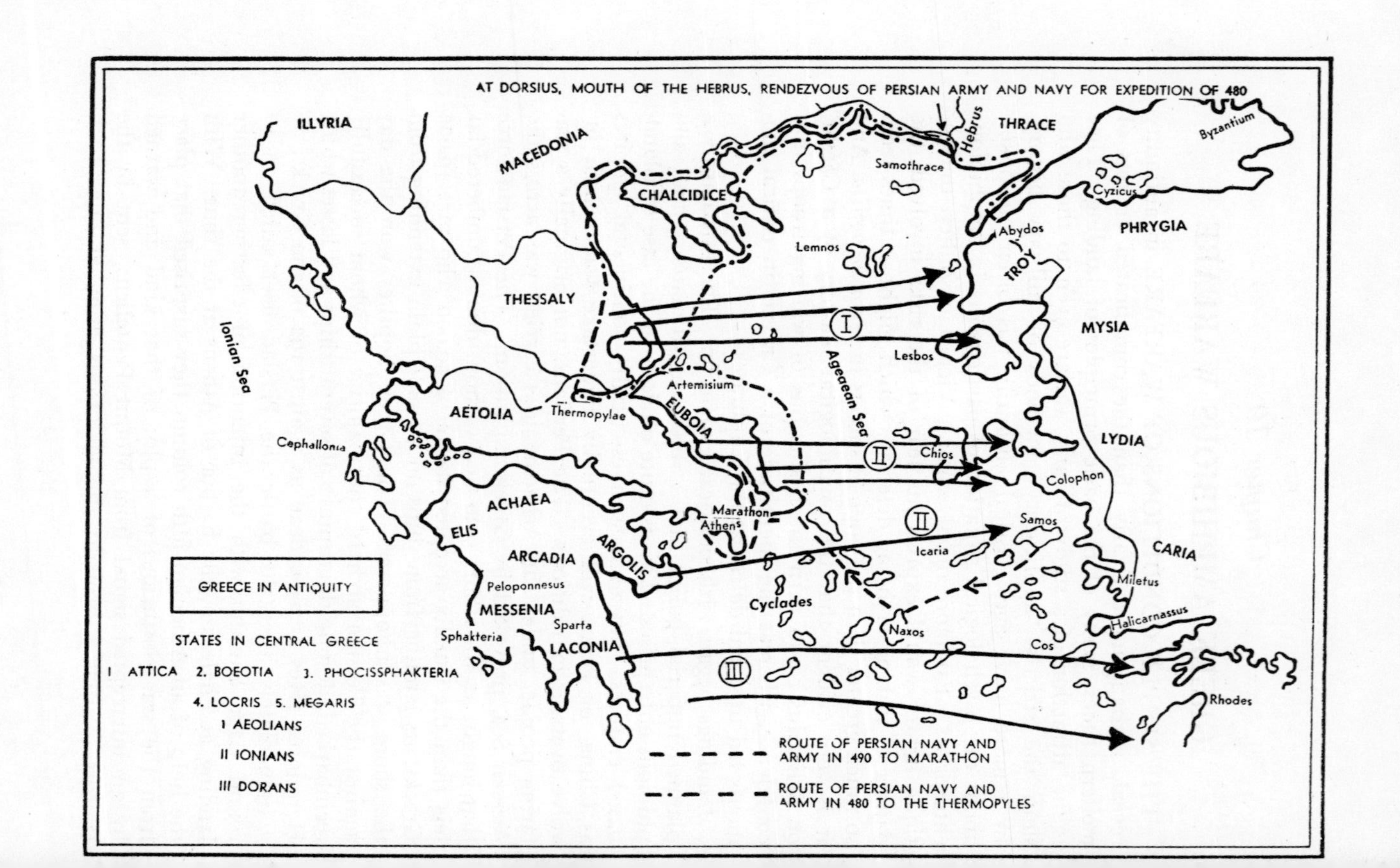
AT DORSIUS, MOUTH OF THE HEBRUS, RENDEZVOUS OF PERSIAN ARMY AND NAVY FOR EXPEDITION OF 480
ILLYRIA
MACEDONIA
THRACE
Hebrus
Byzantium
Samothrace
Cyzicus
CHALCIDICE
Abydos
PHRYGIA
Lemnos
TROY
THESSALY
Ionian Sea
MYSIA
Lesbos
Ageaean Sea
Artemisium
AETOLIA
Thermopylae
EUBOIA
Cephallonia
Chios
LYDIA
Colophon
ACHAEA
Marathon
Athens
Samos
ELIS
ARGOLIS
Icaria
CARIA
ARCADIA
Miletus
GREECE IN ANTIQUITY
Peloponnesus
MESSENIA
Cyclades
Sparta
Halicarnassus
Naxos
Sphakteria
Cos
STATES IN CENTRAL GREECE
LACONIA
I ATTICA 2. BOEOTIA 3. PHOCISSPHAKTERIA
4. LOCRIS 5. MEGARIS
I AEOLIANS
II IONIANS
III DORANS
Rhodes
ROUTE OF PERSIAN NAVY AND ARMY IN 490 TO MARATHON
ROUTE OF PERSIAN NAVY AND ARMY IN 480 TO THE THERMOPYLES

to Athens, thus foiling the second Persian landing. After resting awhile on their oars, they retired from Greek waters to Asia (Herodotus VI, 102-117).

The third expedition of the Persians into Greece, ten years later. found Athens grown into a sea power. For as Herodotus asks: "Of what possible use could the walls across the Isthmus have been if the Persian king had had the mastery of the sea?" According to the historian Thucydides (I 93), it was Themistocles who had set the Athenians on their career of sea power, pointing out the advantages of becoming a naval people; he often advised them, in case they should be hard pressed by land, to go down to the Piraeus and from there defy the world with their navy.

Persian-Greek Conflicts

While most of the ensuing battles between Persians and Greeks—Thermopylae, Salamis, Himera, Plataea—were either sheer land or exclusively sea conflicts, a few amphibious enterprises occurred after the Greeks had taken the offensive. When a Persian general, not trusting the Ionian element in his forces, declined to accept the challenge of a naval battle at Samos and instead pulled his ships ashore at Mycale, a promontory opposite Samos, and dug in there behind a wall and under the protection of a Persian land force, the Greeks landed, stormed the camp and put torch to it (September, 479 B. C.). The Greeks, although in the ascendancy now, were still held back by fears and uncertainties in such a distant land until their leaders removed this psychological barrier by spreading the rumor that their countrymen at home had already beaten the Persians at Plataea, although actually they could as yet hardly have known the outcome of this battle. After that the invaders' hesitations vanished and led by the Spartan Pausanias, the Coalition fleet of Peloponnesians, Athenians and Ionians drove the Persians out of Cyprus and Byzantium. In these and subsequent actions the naval hegemony passed to Athens whose general Cimon defeated the Persians in the land-and-sea battle on the Eurymedon River (466 B. C.). If we may judge by the meager record, it was one of those rare engagements in which both the land and the sea arm proved of equal importance. More usually in such affairs, one arm had been merely the auxiliary.

As this War of the Delian League became entwined with the

First Peloponnesian War, the scene of conflict began to shift rapidly from land to sea and back again. On one occasion, when the Peloponnesians skirted the Corinthian coast, the Athenians drove them into a deserted fort, attacking them on the water and almost at the same time with troops landed from the ships on the shore. In the ensuing disorder and confusion, the Athenians damaged most of the enemy ships on the beach (Thucydides VIII 10). At Cyzicus (April, 410 B. C.) Alcibiades first conquered the Peloponnesians at sea, then immediately putting his men ashore, pursued the enemy and killed their leader Mindanarus, took Cyzicus, made the Athenians master of the Hellespont and thus ensured their grain supply from the Black Sea. Letters were intercepted from remnants of the enemy troops to the Spartan home authorities containing the despairing news: "We are conquered; Mindanarus is dead; the army is dying of hunger; we do not know what to do, or what will become of us." The impression of that land-sea victory on the Spartans was so great that they were genuinely willing to make peace, but the extreme democrats at Athens would have none of it.

The main vehicle of this littoral warfare was the Athenian warship manned by rowers, sailors and marines, the crew of a fast-moving trireme being estimated at 225 men, of whom 174 were rowers, 20 sailors and some 30 marines. According to Plutarch, each of the Attic ships at Salamis had only 18 warriors aboard, 14 hoplites and 4 archers. The extent to which the rowers participated in landing enterprises seemed to have varied, and at times the heavy-armed men might be forced to do their own rowing (Thucydides III 18). The troops required for larger landing enterprises and overseas expeditions were carried in special transports. Horse transports, such as were used by the Persians before them, were rebuilt from old warships by the Athenians,[1] who took cavalry to ravage distant lands (Thucydides VI, 7). The Athenian fleet was commanded by one of the strateges, the *nauarchos* who was originally a land-and-sea general and only much later became an admiral in our modern sense, a specialist.

Early Amphibious Essays

The Peloponnesian War was a first world war in the sense that at one time or another it touched nearly all the world known to the Greeks. Placed on a theater reaching from Sicily

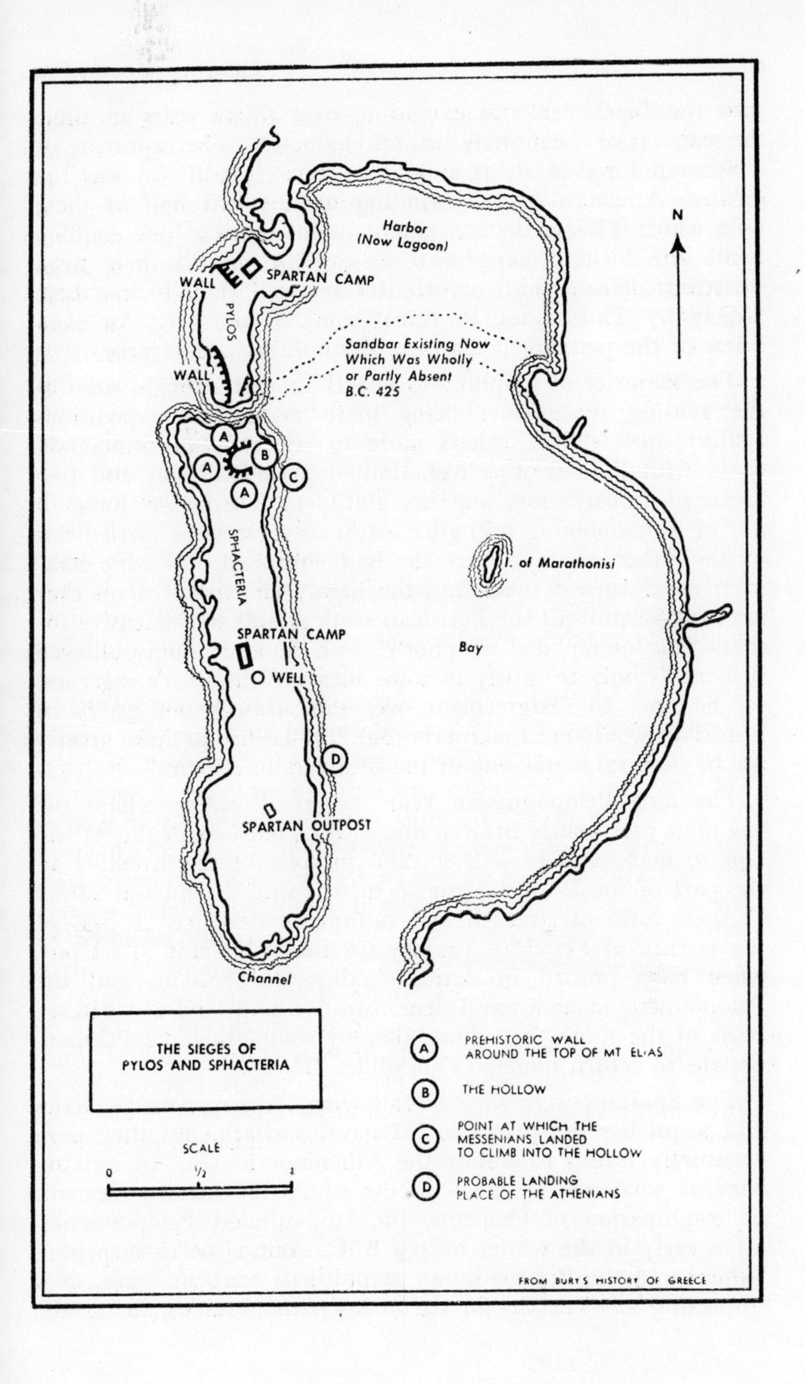
Harbor
(Now Lagoon)
N
WALL
SPARTAN CAMP
PYLOS
Sandbar Existing Now
Which Was Wholly
or Partly Absent
B.C. 425
WALL
A
A
B
C
A
SPHACTERIA
I. of Marathonisi
SPARTAN CAMP
Bay
WELL
D
SPARTAN OUTPOST
Channel
THE SIEGES OF
PYLOS AND SPHACTERIA
A PREHISTORIC WALL
AROUND THE TOP OF MT ELIAS
B THE HOLLOW
C POINT AT WHICH THE
MESSENIANS LANDED
TO CLIMB INTO THE HOLLOW
D PROBABLE LANDING
PLACE OF THE ATHENIANS
SCALE
0
1/2
1
FROM BURY'S HISTORY OF GREECE

into the Black Sea and extending over thirty years in time, the war was of a definitely littoral character. The capturing of a Spartan force in Pylos and Sphakteria (425 B. C.) was the greatest Athenian success, winding up the first half of these wars which Thucydides has taught us to regard as one conflict, while the Sicilian expedition of 415-413 proved their most disastrous defeat. Both occurrences are well enough described, largely by Thucydides, to reveal some of the early, or even a few of the permanent features of amphibious enterprises.

The majority of amphibious efforts in that struggle were of the raiding type, some being mere marauding expeditions against open shores, others more in the style of commando raids, with limited objectives, limited periods of stay and pre-arranged embarkation, and the ability to face enemy forces if not of overwhelming strength.[2] Most of these were carried out by the Athenians who were the best equipped, but who, however, never turned them into the highly effective weapon they might have proved; the Periclean strategy was too defensive for that. It contemplated a "phony" war, and seemingly allowed such raids only to satisfy in some measure the men's eagerness for action. In disagreement over this strategy one group of historians would even maintain that "the failure to make greater use of such raids was one of the blunders of Athens."

The long Peloponnesian War was of all wars perhaps the one most persistently littoral, due to the endeavor of the Athenians to make and keep it so, and in spite of the attempts on the part of the Doric League to turn it into a land war suited to their form of strength and military experience. It was of the essence of Periclean strategy for the Athenians to set out, when hard pressed in Attica, against the Isthmus and the Peloponnese, make a naval demonstration and land at whatever parts of the shore they chose, thereby compelling the Peloponnesians to return home. (Thucydides III, 16.)

The Spartans were slow in following Athenian naval tactics and acquiring the principles of naval warfare, but they were eventually forced to follow the Athenians to sea. In striking back at what was to them on the whole the foreign element, or combination of elements, the land-minded Peloponnesian allies early in the winter of 429 B. C. came close to surprising Athens and the Piraeus by an amphibious *coup de main.* Not suspecting any enemies in the vicinity, the Piraeus was at the time, "as was natural for their decided superiority at sea, left unguarded and open." It was proposed to the crews of a Peloponnesian fleet, just returned to Corinth, that they make use of the opportunity, march overland, each man carrying his oar and cushion, to Nisaea, the port of Megara, where they were to man forty ships which happened to be there and sail to the wide open Piraeus. But once on board, they became afraid of the risk and steered instead to Athens-controlled Salamis in order to raid and lay waste that island, throwing the Athenians into a considerable panic, after they realized what a close escape they had had from an invasion.[3]

Supprise Opportunity Lost

Later, while the Athenians were besieging and at last taking Mytilene in the winter of 428-7, a Peloponnesian fleet, sent out to relieve the island town, tarried on the way, irrupting in Attica and wasting time in other enterprises, being too late, as they learned en route, to keep the Mytileneans from capitulating. It was proposed by one of their officers who had grasped the principles of surprise in littoral warfare, to sail on to Mytilene nevertheless and fall upon the probably very careless Athenians there:

> By sea, indeed, where they have no thought of any enemy attacking them, and where our strength mainly lies, this will be altogether the case; and even their land forces are likely to be dispersed through the houses too carelessly. If then we were to fall upon them suddenly and in the night, I hope that with aid of those in the city (if indeed there be any one left who wishes us well), possession of the place might be gained. And let us not shrink from the danger, but consider that the proverbial 'surprises of war' are nothing else than chances such as this; which if any one should guard against in his own case, and avail himself of them, when he saw them in the case of his enemy, he would be a most successful general. Such was his speech; but he did not persuade the Spartan king Alcidas. (Thucydides III, 30-1.)

The war as fought up to 425 largely along the lines of Periclean strategy, was drawn out without any hopes for a decision, consuming the strength and resources of Athens to no immediately evident purpose, and perhaps more rapidly than it weakened the opposing forces. No plan of the Athenians promised victory in this seventh year of the war, except possibly that of Demosthenes, to whose ideas, regarded as those of an upstart, the rich who were in command would not listen.

Demosthenes at Pylos

It may be remarked that this advocate of the amphibious enterprise was in several respects an outsider. Knowing from his earlier experiences during the campaigns on the western coast of Greece, the conditions prevailing in Messenia which Sparta had conquered and depopulated, leaving some Messenians eager for revolt still in the land, he advocated that the Athenians occupy the excellent harbor of Pylos, the modern Navarino, the best in all the Peninsula, but then unused by Spartan land power. Accompanying an Athenian expedition intended for Kerkyra and Sicily, but holding no command, Demosthenes was able to induce this expeditionary force to fortify the desolate rocky promontory of Pylos, or Coryphasium, which was naturally very strong, particularly towards the land side. Detained there by unfavorable weather, Demosthenes persuaded the lethargic crews to erect fortifications.

After that the main force proceeded to the original destination, leaving him with a small body of men and five ships as a thorn in the side of the Spartans, drawing the rebellious Messenians to his support. The Spartans promptly ceased their seasonal incursions into Attica and drew their forces together around Pylos. Seeing them assembled in strength, Demosthenes sent two of his ships to call the Athenian main fleet from the north, manned the walls of the mountain fortress and posted himself with his best heavy-armed troops and some archers along the shore of the bay where the greatest danger threatened. The Athenians had fortified all good landing sites, probably by palisades, thus producing an artificially steep slope. But some parts had been left untouched on account of the shoals, a weakness of his defenses which the Athenian general thought would invite the Spartans to attack there, even though only a few could make the assault at a time. (Somehow, one is inclined to compare this style of coast defense with D & E submerged wire defenses and mine fields, with D & E lanes, and the timbered beaches commanded by one or the other of Hitler's reinforced concrete bastions within the Westwall.)

After these preparations Demosthenes harangued his men,

in the traditional way often resorted to by an antique author in order to represent for our benefit the thoughts which seemed to underlie a leader's plans. He told them the Spartans would be given no chance to land—always the ultimate aim of the defender of a shore—but would be beaten before they could set foot on land. For what else could be done by a commander who had no appreciable naval power and limited firing range? Cleverly he recalled to the Athenians, and invited them to make the best use of, the most dangerous psychological moment in a landing operation, the one when the landing craft, the *simulacrum* of *terra firma,* must be abandoned by the soldier, when the land warrior becomes the water warrior at least for minutes, and when he does not know how deep the water is or how uneven and treacherous the hidden bottom. In fact, he reminded them of the basic strength of the defense of the land-based force, reversing for the occasion the role of the usual Athenian marine.

> To repulse the enemy we shall find easiest to do while he is on board his ships; but once he has landed, he is on equal terms with us. Nor need you be very much alarmed at their numbers; for great though they may be, they will engage only in small detachments, from the impossibility of bringing to. Besides, they are not an army on land, fighting on equal ground, while superior in numbers, but rather troops on board ships, upon an element where many favorable accidents are required to work with effect. [This would be Thucydides' way of saying that the sea is after all the more accidental element, which it was certainly more in his time than in ours, and that in amphibious enterprises land accident and sea accident are both apt to occur, making this kind of enterprise the most accidental of all.] So that I consider their difficulties a fair equivalent for our (inferior) numbers. And at the same time I call upon you, Athenians that you are, and knowing from experience as you do the nature of a naval descent on a hostile coast, and how impossible it is to drive back an enemy determined to stand his ground, and not retreat for fear of the roaring surf and the terrors of the ships sailing to the shore—I call upon you to stand your own ground, and by resisting them along the very edge of the beach to save both yourselves and the place.

The Spartans and their allies, with 43 ships altogether, thought best to occupy first with 420 of their men the island of Sphakteria which closed the bay of Pylos to the open sea, and thus control the bay with certainty. They then proceeded with their triremes and other vessels to the point of the shore where the small force of Demosthenes was expecting them. But only a few ships could approach at a time. No arrangement for attacking in waves and for relieving the first attackers seems to have been made. At first there was no want of eagerness and mutual exhortation, with Brasidas, the captain of a trireme, as the foremost shouter. Observing the Doric clumsiness on water and water shyness, the unwillingness of the captains and steersmen to risk their ships by running on the rocks, he shouted out that they must not be chary of timbers.

Combatants' Positions Reversed

Having forced his own steersman to run his trireme ashore, he was hotly attacked by the Athenians while stepping on the gangway to land. Others of the Dorians were no more successful that day, or the next when they desisted from further struggle instead of trying to wear down the small Athenian force by continued attacking and sent off ships to obtain timber for systematic siege work. Thucydides does not omit to emphasize the "revolution of fortune," the turn of whose wheel had thus reversed the usual roles of the Spartans, an inland people now trying to land on shores which they considered their own, and of the Athenians, usually the landing experts, who now found themselves in the position of shore defenders.

But their sea power was not far away. After the Spartans had despatched their expedition for lumber, the Athenians answering the call of Demosthenes, arrived with 50 ships of their own and their allies, conquered the Peloponnesian fleet in the bay, and then blockaded Sphakteria where the best, and never very numerous, Spartan manpower was cooped. After a 72 days' siege, topped off by a landing in the rear led by a Messenian guide, the Spartans were forced to make a mass surrender, a thing up to then considered unthinkable for Spartans. Athens could then have obtained a peace more favorable than at any other time of the war, if it had not been for the *hybris* of sea power which possessed the Athenian democracy.

Aside from the choice of Pylos as a strategic moment in this enterprise in which there was not a little improvisation—so far as the war-making state of Athens was concerned, although not on the part of Demosthenes, the originator and local leader, the tactical features of Pylos-Sphakteria far outweighed the strategic ones.[4] Better economy of forces, greater adaptability to the unfamiliar element, better psychological understanding called for on the part of the leader; better knowledge and estimation of the terrain, whether water or land, were all on the Athenian side. No specialization between leaders of the naval and military parts of the enterprise had as yet evolved, although we can observe in Brasidas' endeavors and admonitions to the naval men the impatience of a captain of land forces, a very brave and talented one in his personal case as subsequent history was to show, with the cautious navigators, never more cautious than when approaching land.

Athenians Invade Sicily

Equally fateful, but falling more into the field of the strategy of littoral enterprises, with strategy as "the doctrine of the use of combats for the purpose of war" (Clausewitz), was that other Athenian enterprise towards the west, the Sicilian expedition of 415-13. The dominion of the seas was sufficiently theirs to insure the safety of communications, but never complete enough to prevent the Spartans from giving assistance to the Sicilian enemies of Athens. Persuaded that empire, limitless empire and riches were to be gained by interfering in the quarrels of the Sicilian republics, the Athenians listened to Alcibiades who urged them to discard the cautious nonintervention policy for which Nicias stood. Many were desirous of adventure and others thought "they should both make money at present and gain additional power, from which an unfailing fund for pay would be obtained" (Thucydides VI 24). Voting large marine, land armament and money supplies, the Athenians sent an expeditionary force, one of the largest ever proceeding from Greece, to the conquest of Sicily. This included an unstated number of transport and provision ships besides 134 triremes, 6,500 warriors, a seafaring personnel of 25,460 and a total strength of some 36,000.

The triumvirate of leaders with coordinate powers had different notions about the political and military aspects of the enterprise; the overcautious and superstitious Nicias employed augurs, as did Xenophon on such occasions, to make up his mind for him instead of helping him, as better Greek generals insisted, to make up the minds of his soldiers. Probably the determined Lamachos was the best military leader of the three, the last being the energetic, unscrupulous, erratic Alcibiades over whose comely head a process for sacrilege was impending when he left Athens.

Sicily, we might say, was the Athenians' America, a continent "by herself in the west," although on a much smaller scale, a parallel that could be carried much further in a history of isolationism. Each city-republic held a strip of seacoast, with a hinterland, much like the North American colonies, and there was less war among them than between the states of old Greece and hence comparative backwardness in military institutions. Syracuse, for example, as one of her leaders declared, was much like the earlier Athens, an inland power (Thucydides VII 21); but since Athens under the Persian pressure had become a sea power, why should not Syracuse also take to the sea?

The island had been colonized by Phoenicians, Ionians and Dorians, the Greeks maintaining only slight political connections with the mother cities and being in no way dependent upon them. They were "isolationists," to use modern language, and when the first news of the Athenian projected invasion arrived the panic was great with some, while others professed scepticism about its feasibility. To the former element Hermocrates, leader of the aristocratic party in Syracuse, addressed himself, telling his fellow citizens that few indeed had been the forays either of Greeks or barbarians which had proved successful far from home, an experience, however, which should not lead to unpreparedness in the face of such a threat. For a time, however, the Syracusans paid more attention to Athenagoras, a leader of the populist element among whom the Athenians later found fifth columnists, who argued that the coming and landing of the Athenians was very unlikely, merely a rumor invented by the makers of an invasion scare, a state of mind easily aroused in island people once the feeling of established security gives way to that of general uncertainty and insecurity.

The military plans of the Athenians for Sicily were as hazy as the ambitious notions in general of the people of Athens about that island. Their scouts could not report much more than that the money which their allies in the island had promised to contribute for the pay of soldiers and sailors raised for the project was not actually available. This news caused much dejection among the adventurers. The fleet rendezvoused at Kerkyra, modern Corfu, and then set sail for Italy and hence southward along the coast of Calabria. Most of the coastal towns shut their ports and gates to the Athenians. Not until they got to Rhegium, modern Reggio di Calabria, the southwestern tip of the Peninsula, did the Athenian generals meet to discuss plans. Nicias once more tried to keep the expedition within the limits of its ostensible aims, the support of certain city states in the island. Alcibiades was in favor of coasting along the three sides of Sicily to find who was friend, or enemy and win as many adherents as possible. Lamachos' advice was the best, although he finally sided with the brilliant Alcibiades. He understood better than the others the first condition for the success of amphibious enterprises—surprise, the product according to Clausewitz of secrecy and speed, to which must be added the factor of accident, which is beyond human management.

Catastrophe Before Syracuse

Lamachos urged his colleagues to sail straight to Syracuse, the true military adversary in the island, fight a sea battle under the walls of the city whose inhabitants were quite unprepared and panic-stricken. Unless they did so the psychological effect of the shock of invasion would soon wear off. "For every armament was most terrible at first; but if it spent much time before showing itself, men's spirits revived and felt the more contempt for it on its actual appearance. By suddenly attacking while the enemy was still with terror looking for them, they would obtain the most decided advantage over them and of striking a complete panic into the enemy; by the very sight of their forces—which would never again appear so considerable as at present—by the anticipation of the disaster coming to them and, above all, by the immediate peril of the engagement." Many of the Sicilian cities would be impressed by the defeat of Syracuse and join with Athens.

Precious time was wasted by Alcibiades' diplomatic negotiations and still more uncertainty introduced into the command by his recall to Athens to face criminal proceedings. And nothing much was achieved during the three months of the first summer in Sicily; and only in the following winter were preparations made for an advance against the Syracusans. The latter had regained their courage when no immediate attack was made and now had put their defenses into good order. The time for strategic surprise was over, even if Athenian superiority in war technique might still achieve small successes through tactical surprise. When they landed at last near Syracuse, an assault force occupied a chain of hills only 2,000 paces from the shore, while the Syracusan force intended to cover this line was under arms, but distant, so that they arrived only after a half hour's uphill run, exhausted and disordered, and could easily be beaten off.

The long drawn siege of Syracuse during which the besiegers in turn became the besieged on land as well as in the harbor, need not concern us here. Athenian control of the sea was never complete enough to stop certain support from the Peloponnesian side, including a military leader from Sparta, getting through to the besieged. Their misfortunes led the Athenians to contemplate withdrawal. Pride and fear of public wrath at home delayed them and at last, when withdrawal had been determined upon, superstition stood in the way. The uncertainty produced in many of the troops and in their general Nicias as well by the novel experiences of amphibious warfare induced a paralyzing superstition and the long overdue evacuation was delayed for a more favorable constellation of the stars, which came too late to save the Athenians.

In such a situation the attacker can no longer impose his law upon the enemy; the psychosis of negativism, which good leaders of amphibious enterprises have always known how to combat, progresses too far. Numerous stories have been told about the leaders of fortunate expeditions who stumbled, as they step ashore; lest their followers, including priests, consider this a bad omen, the quick-witted leader insists that he has stooped to claim the land. So, following precedence, William the Conqueror declared: "On the contrary, the very earth invites me to take possession of her. I take and hold thee, England."

At least for the pre-modern landing force the *"Absit Omen!"* should be remembered.

NOTES, CHAPTER 10

[1] Max Jähns: *Handbuch einer Geschichte des Kriegswesens.* Leipzig, 1880.

[2] H. G. Robertson: Commando Raids in the Peloponnesian War. Classical Weekly, 10 January 1944.

[3] Thucydides, II, 25-6, 30, 56, 69; III, 7, 16, 91, 94, 105; IV, 42-5, 53-7.

[4] Tactics being understood in the amphibious enterprise no less than in land warfare for which Clausewitz wrote his definition, as the doctrine of the use of forces in combat.

Chapter 11
CAESAR, AMPHIBIOUS GENERAL

MEDITERRANEAN ANTIQUITY saw much littoral warfare during what Fuller in his *History of Decisive Battles* calls with some, although perhaps not fully sufficient, reason the "thalassic period," lasting until 1453. As a rule the states, or parties carrying it on were either land, or sea powers until the Roman Empire monopolized all power and gave preference to the land. While Alexander remained a master of continental warfare, although in many respects he was heir to Greek traditions, Roman leaders coped repeatedly with the problems of landings. Scipio Africanus defeated Carthage by bringing war through Spain into Africa. Julius Caesar was much concerned with littoral warfare when seeking his victories in Britain, Spain, Africa, Egypt and the Balkan Peninsula.

As the better all-round general of the two, Caesar was at last able to beat Pompey in the civil wars, (beginning 50 B. C.) despite the larger sea power of his adversary. He knew how to utilize at the expense of the enemy not only his superior intelligence service, but the superstitions of his men. While sacrificing to Fortuna at home, the animal dedicated to her altar by Caesar got away from the attendants, fled outside the city and coming to a pond, swam across it. The soothsayers interpreted the occurrence as advice to Caesar not to tarry in Italy, but to cross the water for salvation and victory. Following this prompting to a seaborne invasion, Caesar forced Pompey out of Brundisium, the modern Brindisi and an early Dunkerque, crossed the Adriatic in winter and landed on the Epirotic coast before Pompey's forces had occupied all suitable landing spots.

Two of Pompey's blockading detachments, lying in the harbor of Oricum, in the bay of Avlona, the modern Vallona, and at headquarters on Kerkyra-Corfu, idly observed the action, the one feeling too weak to cope with Caesar's transports and the other not ready to set sail. The blockaders, however, dealt rather severely with the ships sent back by Caesar to fetch reinforcements. He himself proceeded inland at once, making the best use of his surprise arrival, endangering seriously for a time Dyrrachium, the modern Durazzo, the most important trajectory point of the Romans on the Balkan shore of the Adria and then Pompey's main supplying harbor.

But Lady Luck, which will not constantly attend even a Caesar, deserted him for several months during which Pompey regained complete sea control and, given Caesar's intrepidity, might well have beaten the latter on land. Instead, toward the end of the winter a new transport of four legions and 800 horsemen got through under Caesar's lieutenant in Italy, Mark Antony. To show the difficulties confronted by landings in pre-steam days, the winds would not allow Mark Antony's fleet to approach Caesar's camp and forced him to disembark at a point four days' march away, which circumstance happily was on Antony's side for when putting into land he barely escaped Pompey's pursuers who were caught and held back by a suddenly changing wind the moment they appeared off Antony's landing place.

Much strengthened by this junction of his forces, Caesar began to besiege Pompey in and around Dyrrachium and the war became one of positions, calling forth all the Italian diligence in ditching and walling. A fortified line of fifteen miles was thrown around Pompey, the largest field works perhaps constructed during classical antiquity, unless those about Alesia may be considered to have been even larger. The lines of circumvallation and contravallation rested on the sea and when Caesar one night made an attempt on Dyrrachium, in part along the sea coast, he was met not only in front, but in his rear as well by troops of Pompey who had been conveyed along the shore in boats and took him by surprise. In this clash his losses were heavy and he himself was nearly killed. Dyrrachium was by far Caesar's worst military defeat. His bold undertaking to operate on the offensive against a hostile sea power secured by its fleet—amounting to 500 swift ships by the help of which Pompey was able "to touch at many points at once" (Dio Cassius XIL, 52)—had completely failed, as Mommsen sums it up. Only by drawing Pompey away from the strength-giving sea and by eliminating the fleet as a factor of decision, obliging him to carry on the war on equal terms *(De Bello Civ.,* III: 78), could he have obtained the victory won in the land battle of Pharsalus in 48 B.C.

To a widely varying degree every hostile shore is a *terra incognita.* In its topography, the strength on and behind the shore

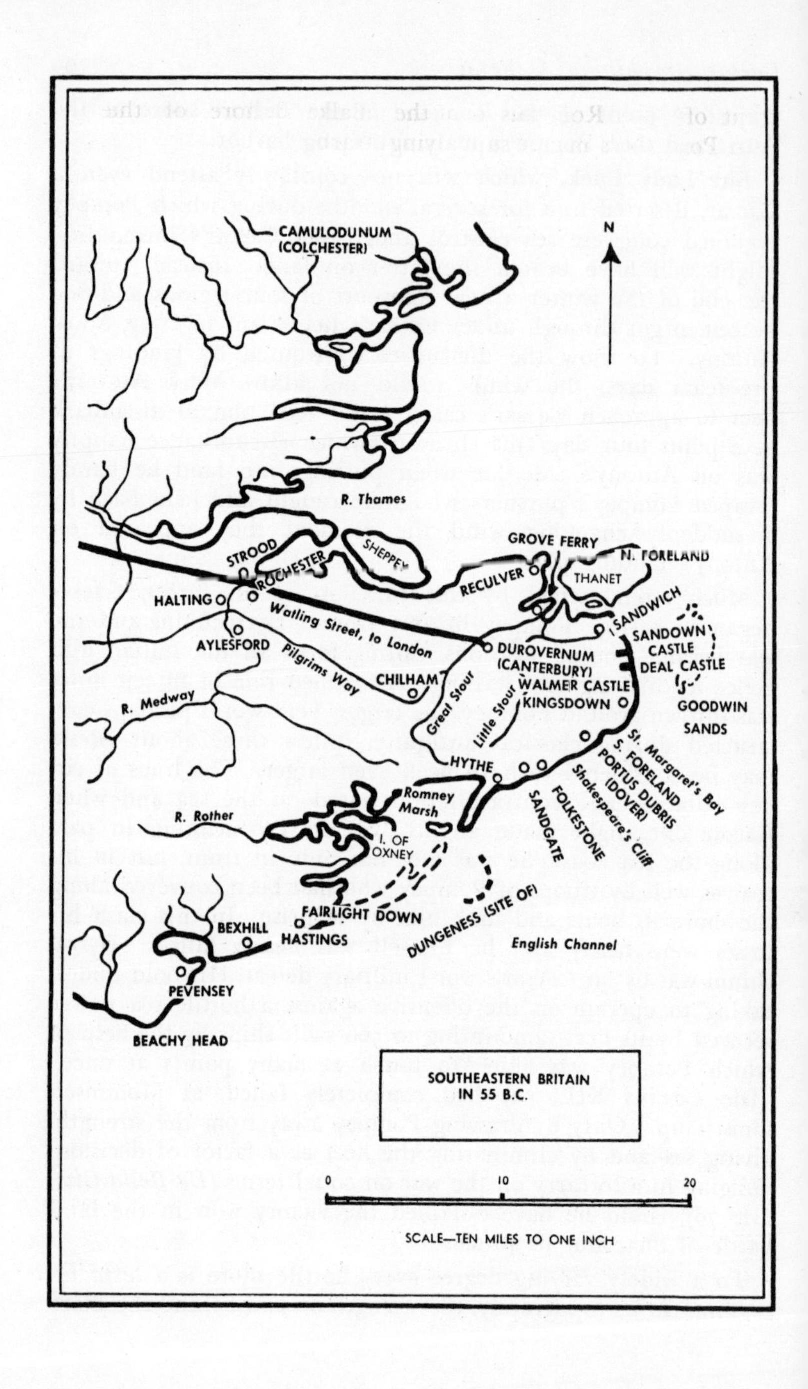

SOUTHEASTERN BRITAIN IN 55 B.C.

line, it may be unknown to the general, and appear threateningly unfamiliar to the rank and file who hesitate before leaving the landing craft which, even though novel to them and frail, appear safe and trustworthy by comparison. An essential part of the preparation before a landing has to consist in the leaders, at least, familiarizing themselves in advance with that hostile shore. Only modern military leadership, however, came to realize that this process should, and must, include the rank and file.

When Caesar contemplated the invasion of Britain, which was not an act of long-pondered statesmanship, but rather one of imperialistic-military improvisation, his knowledge of the island, its coasts, ports, inland waters, the nature of its inhabitants was scanty in the extreme. Merchants who had visited it were called in for council, but could tell him very little about the size of the island, the nation, or nations inhabiting it, their power, experience and versatility in the arts of war, or the capacity of their harbors. To make up for this lack of definite intelligence, Caesar sent out one of his higher officers on a galley to reconnoiter the British coast, a few days before the start of the expedition. After merely a five days' cruise, the latter made his report and it seems evident from Caesar's commentaries that he was far from satisfied with the survey of this officer, who was one of his best.

Boulogne (Portus Itius), situated in the land of the Morini, was chosen as shipbuilding and assembly camp and embarkation point, because the shortest passage to Britain was from thence, and the passage, moreover, "owing to the set of the tidal streams and the prevalence of south-westerly winds was more convenient" (Holmes). It was not until late in the summer of 55 B. C. that the preparations were completed and Caesar felt free to undertake the expedition, realizing that it could not go much beyond a mere reconnaissance. Still, "it would be well worthwhile merely to pay a visit to the island, observe what the people were like and make himself familiar with the features of the country and the landing places." (*De Bello Gallico,* IV: 20.) The expeditionary force was to consist of two legions, both elite troops, the famous Tenth and the scarcely less renowned Seventh; 500 horsemen, slingers from the Balearic Islands, Numidian and Cretan archers, about 10,000 men in all, carried by a small squadron of galleys, scout boats and some 80 transports, none of the latter specially built as landing craft. Archers, slingers and the catapult men with their engines were put aboard the more mobile galleys. Supplies for only a few days were taken.

The barbarians on the other side of the Channel had intelligence of Caesar's movements which was easily as good as his information about them and their land. Some warriors had crossed the straits and fought with their kinsmen against Caesar, who knew and resented "that in nearly all the operations in Gaul our enemies had been reinforced from Britain." (*De Bello Gallico,* IV: 20.) Britain had afforded a refuge for some of the defeated warriors from the continent. Late in the autumn of 56, or early in the spring of 55, the preparations for an invasion were known and some of the Britannic tribes despatched ambassadors to Caesar in Gaul, offering submission and hostages. Well received and hospitably entertained, the emissaries were sent back to their island together with Comius, a Briton by birth and favorite of Caesar's in whose entourage he had been for some time. This fifth columnist was, however, seized by his fellow countrymen, enfuriated at his attachment to a foreigner and an enemy.

Chariots to Resist Romans

British military measures for resistance included training the horses and the drivers of their war chariots to fight in the surf when the Romans quitted their ships. As Caesar describes the war chariots (*De Bello Gallico,* IV: 33) and their uses, they carried a crew of two men, a driver and a warrior, the latter hurling missiles while the vehicle was directed over the field, thereby throwing the enemy ranks into confusion by the terror partly caused by the plunging horses and the clatter of wheels. As soon as they penetrated the hostile ranks, the warriors jumped from the chariots and fought on foot, while the drivers gradually retired from the melee and drew up the chariots where the warriors, if hard pressed, could retreat to them. "Thus they exhibit in action the mobility of cavalry combined with the steadiness of infantry."

Even if Caesar does not tell us so, the British tribes must have arranged a few measures for common defense. As Celts with connection among continental Celts they knew that Romans had command of the Channel after the defeat of the sea power of the Veneti tribe of Gaul in 56 B. C. They secured all landing beaches in the region where Caesar had to be expected, where the Channel was narrowest. More confident than informed, Caesar set out on one of those August days when the Channel can be as smooth as a tablecloth, at one o'clock in the morning and arrived nine hours later at Dover. He found the cliffs lined with armed Britons who were definitely in a superior position. As Holmes, the historian of Caesar's invasions, put it: "It was a maxim of ancient warfare, never disregarded without urgent necessity, to avoid engaging an enemy who had the advantage of higher ground." Hence landing at, or near, Dover was out of the question. While Caesar sailed on up the coast the British chariots and horsemen followed on land, equalling in speed the movement of the Roman ships. This parallel movement perhaps caused the ancients to speculate upon whether the creating of a road system paralleling a coast line should not be a prime consideration in a scheme of coastal defense.

The Roman fleet halted for a council of war, at which Caesar gave his officers their instructions. He weighed anchor at about four in the afternoon and sailed on to the vicinity of modern Deal where a flatter shore seemed to offer a more promising landing. Caesar still found it difficult as he writes, because the Roman ships required a considerable depth of water. His soldiers had their work cut out for them and they were weighed down with heavy armor. Yet they were about to cope with both the waves and the enemy in a site unfamiliar to them, whereas the Britons, standing upon dry land, or wading, or driving their chariots a short distance into the water as far as they knew it to be shallow, and also having freer use of their arms and legs, could boldly cast their darts. They were inured to that kind of combat. This prospect so discouraged the Romans, strangers as they were to such enterprises, that they did not appear to their general to be so cheerful and eager to engage the enemy as in their earlier battles on dry land.

Running Amphibious Combat

Perceiving this, Caesar gave orders that the warships, or galleys (*navis longa,* whereas the *navis oneraria* was a transport), a nimble sort of craft unknown to the enemy, should advance a little ahead of the transports and row along with their broadsides to the shore so that they should force the Britons back from the water by their slings, engines and arrows, "which proved of great value to ourselves." Caesar had mounted part of his artillery on board his ships. Surprised at the shape of the Roman galleys and by their engines, the barbarians slowly yielded ground. In modern terminology, Caesar learned en route that it was advisable to hold the Roman transports beyond reach of the Britons' missiles and to throw his swifter craft broadsides at the land-based enemy during a running amphibious combat. It is not conceivable that very heavy casualties resulted from this exchange of missiles.

At this point the orders to disembark must have been given. "Seeing his comrades still hesitant at venturing into the sea, the eagle bearer of the Tenth Legion after having first invoked the help of the gods, cried out to them that although they might forsake their eagle and suffer it to fall into the enemy's hands, he was resolved to do his duty to the Republic and the general. He then jumped overboard and carried the eagle in the direction of the Britons; the soldiers of his own ship, ashamed not to follow him, encouraged each other and jumped and so did those in the nearby ships, all pressing to the shore." It should be remembered that the *aquilifer* ranked high among the under-officers in the Roman armies, that in combat much of the troop morale depended on him, or was evidenced by him. Possibly Caesar wrote this passage to glorify the eagle bearer and censure, as he did more than once, the hesitancy and incompetence of those young Romans of family, whose birth qualified them for army service, none of whom took the initiative in this landing.

What followed in Caesar's own description of the landing was to become all too common a feature of many future landing operations. The fight was sharply maintained on either side. The Romans, however, were unable to obtain a footing and keep their ranks, or follow the standards to which they belonged. They leaped from their ships and joining the first ensign they encountered, found themselves in great confusion. The Britons, familiar with the shallows, spurred their horses into the water and set upon small groups of Romans, surrounding them with superior numbers, or flanking them. Caesar who had remained on one of his ships to observe the battle better, then had the long boats and smaller vessels manned from the transports and sent them to provide assistance and support where they were required. In this way, the foremost ranks gained footing, followed by the rest of the army, and finally put the enemy to flight. But the Romans were unable to pursue for long, or firmly establish themselves on the island at that time because Caesar had no cavalry, the only thing then lacking to make his success complete. His losses

Caesar Landing In England. . (From Old Engraving.).

must have been high, and probably unnecessarily so, for the difficulties before him evidently had not been well estimated, thought through and adequately planned against.

A Quisling Aided Caesar

Considering the weakness of Caesar's forces and poor preparation, his luck owed him nothing. It did not hold when storms arose, dealing harshly with his ships in the Downs and driving back to Gaul transports which were to bring over cavalry. The Romans had to learn that the tides in the Channel are high, as much as twenty feet, as compared with a rise of at best a few inches off Italian coasts; they flowed particularly high with the full moon. The Romans grew downcast and even panicky; they asked how they were to subsist without grain and how they could return to Gaul in the few ships remaining to them. Courage revived among the Britons, who had been negotiating with Caesar, using the quisling whose life they had spared. In the coincidence of the wreck of the invader's ships and the full moon high tides they found good omen (*De Bello Gallico* I: 50.) and clearly knew that the Romans lacked supplies. They attacked the Roman camp and harassed those seeking to repair some of the ships with timber taken from the twelve worst damaged vessels. Before worse befell, on the day before the equinox Caesar retreated to Gaul.

Whatever Caesar's self-satisfied historiography might claim, the first expedition into Britannia was a very doubtful military success. As a somewhat critical Roman author, Dio Cassius, wrote: Caesar gained nothing for himself, or Rome, except the glory of conducting the first armed force to that unknown land. On this, however, he prided himself greatly and the Romans at home magnified the story. Now that they jumped to the conclusion that the formerly strange Ultima Thule had become accessible and its conquest certain, they celebrated by a twenty days' triumph granted to Caesar for an imperialistic success from which much was expected for the future, rather than for a recent sound military success due to careful planning and forethought. A few were sceptical like Cicero, whose younger brother served in Caesar's army, and who wrote to one of his proteges in Caesar's camp: "I hear there is no gold, or silver in Britain. If so, I advise you to capture a war-chariot and come back in it as soon as you can." Later, in June, he said that in Rome "the result of the British expedition is a source of anxiety. For it is notorious that the approaches to the Island are ramparted by astonishing masses of cliff."

Caesar, ever able to improve his tactics after a failure, or mistake, realized the errors which he had committed during his first Britannic expedition. The second was so much better prepared and conducted that one cannot help feeling that a third one, had such been made, would have become the model landing operation of antiquity. More than ever bent on conquering the island, during the winter of 55-4 he had built an invasion fleet of 800 sail, composed of ships better adapted for landing and transporting than the types used in the Mediterranean. These craft, as Dio Cassius describes them (XL: 1) were halfway between the Roman swift vessels and the native ships of burden, Caesar trying to make them as light as possible and yet entirely seaworthy and keeping them on dry land to avoid damage until the day of departure. When the weather served, he set out with a force increased to five legions, holding three more in reserve at Boulogne, and 2,000 cavalrymen—he had even added animals for transport—and proceeded, more directly than the year before, to the same landing place. The embarkation took place on July 6, after the fleet had been weatherbound for three weeks; sailing near sunset, it arrived about noon the next day.

Fighting in the Hinterland

Although the Britons knew he was coming, he landed unopposed because of the overwhelming strength of his force and also, as Dio remarks, on account of his "approaching the shore at all points at once." He probably had not read Thucydides' report of Pylos in the meantime, but rather had learned from his own experience that *point landing,* as opposed to line, or frontal landing, must be avoided and particularly if the enemy is aware of the invader's approach. *Landing in depth* was as yet technically impossible except along psychological lines which Caesar does not seem to have contemplated this time. Concerted landings as well apparently were not suggested to the Romans.

Yet another thing Caesar had learned since his previous attempt, for after having landed unopposed with his superior force, the Britons retiring with their main forces inland, he left ten cohorts and 300 horsemen to guard the landing place and the ships and at once marched into the country by night to a point twelve miles from the sea where he met the British forces on the Stour river in modern Kent.

In the first engagement, the Britons were routed by the Roman cavalry and retired to a fortress in a wood surrounding a point well strengthened by art and nature. It had probably been built during the intertribal wars of the Britons and the Romans found the paths leading to it blocked by abattis. From this the Britons sent out small raiding and flanking parties that kept the Romans from entering upon the main work until the

Seventh Legion formed a *testudo,* or tortoise, a close formation for assault on fortifications with the shields of the attackers held over their heads, overlapping, or dovetailed, threw up an earthen mound against the Britons' works, carried the place and drove them from the wood.

Caesar's further progress in pursuit of the islanders was stopped for the time by the report that storms had again destroyed a large part of the fleet. Since no division of labor had yet been evolved between general's and admiral's duties, he found it necessary to return to the sea coast, giving orders for the repair of the least damaged ships with timber from those most severely battered. He wrote to Gaul for additional ships and ordered those at hand to be put on dry land and inside a fortified camp, a task involving a prodigious amount of labor such as Italians have nearly always applied to war, and more often in antiquity to better purpose than later in the African wars of Fascism.

Early Scorched Earth Tactics

The Britons from the beginning endeavored to slow up the invaders by evacuation measures, "driving" as it came to be called in Napoleonic days, or the scorched earth policy of the present times. This made it harder for the Romans to live off the country and deprived the mercenary forces, who were accustomed to plunder, of an incentive to vigorous campaigning. While Caesar was repairing his ships the Britons realized that they had to unite their forces to put up an effective resistance against an invader under unified command. Nowhere can disunity have more disastrous consequences than in an invaded country, especially an island. In fact, the first requirement for the defense of isolation (from *insula,* island) should be unity of command. Hence, "the whole management of this war was committed to the care of Cassivelanus by general consent; although he had formerly made war on the rest of his countrymen, yet upon our arrival they all united and concentrated upon him as the fittest personage to direct them at so important a juncture of things," according to Caesar who, like most Roman leaders, always sought for fifth columnists among the barbarians.

When Caesar returned to the field of the last battle he found the Britons greatly strengthened in numbers and leadership. There followed a series of smaller combats in which the Romans did not always have the advantage, nor were the legionary soldiers constantly a match for their enemies. On the other hand, the British at times suffered setbacks. Since these fights and their outcome go beyond littoral war they do not concern us here beyond remarking upon one feature. In their retreat beyond the Thames the Britons planted stakes across the ford, some visible and some submerged.[1]

However, as Mommsen puts it, the Romans marched forward, but did not really get anywhere. At times, after they thought the Britons beaten, their base on the Kentish shore was threatened. The general found no worthwhile laurels and his soldiers no valuable loot. And after some face-saving arrangements with Cassivelanus, Caesar departed altogether from the island, by two evacuation trips, never to return, having shown Britons to the Romans, as Tacitus phrased it, rather than giving them possession of the island.

This is a truer characterization of the two Caesarean expeditions than Caesar's own boast that he had discovered and penetrated into a new world. That world remained largely unknown to the Romans and unpossessed by them until nearly a hundred years later. "Julius, though by a successful engagement he struck terror into the inhabitants and gained possession of the coast, must be regarded as having indicated rather than transmitted the acquisition to posterity." [2]

A ludicrous interlude in the history of landings was provided by the more than half insane Caligula (37-41 A. D.). Invited by the fugitive son of one of the numerous Briton kings to invade the island, which was depicted as rich and easy to conquer by the sheer terror inspired by the name of Gaius Caesar, the Emperor, he assembled a large army on the coast of Belgic Gaul. Here he soon learned, to his great surprise, that on the opposing shores the Britons, informed of his intentions, were gathering under arms to meet his assault. This was enough to make "Little Boots" desist from his enterprise. But to discount accusations of not daring to face the enemy, he sailed along the British coast on a galley and at a safe distance in order to take a look at the British. After this peep-show, he had his soldiers drawn up in battle array, harangued them and gave orders for them to gather specimens of the shells to be found on the shore of the Britanic Ocean. These childish trophies, magnified into symbols of the servitude to Rome of the Britannic Ocean, he sent to the capital with his request for an ovation, a minor form of triumph, which the complacent Senate granted.

Soldiers' Fears of the Unknown

Military and topographical knowledge once acquired was apt to be more easily forgotten in antiquity than today, although even we are not immune to lapses of memory, and what Julius Caesar had learned of Britain was not comprehensively handed down to the later Romans, largely because they lacked maps. We moderns *know* that the ends of the world have been reached, horizontally practically everywhere, if a slight extension vertically may still seem feasible. The ancients *believed* that the world they inhabited ended in the west with the Atlantic Ocean, and that all places and activities beyond the pillars of Hercules and other shore points were out of bounds, to use an appropriate military metaphor. Soldiers of antiquity, from the time of Alexander the Great, often mutinied when commanded to go beyond the last familiar outposts, or too far from home. For them war was localized within a known sphere and the ancients hardly as yet possessed maps, as aids to overcome fear and ignorance of unknow lands and waters. Maps allow and encourage modern invasions, although even they in certain situations, such as landings, will frequently prove insufficient.[3]

Instead of a rational knowledge of distant theaters of war the ancients had preconceptions, often of a fantastic kind. They imagined regions populated by men of inordinate height, or by "the Cannibals that each other eat" and "the Anthropophagi, and men whose heads do grow beneath their shoulders." So when Emperor Claudius in 43 A. D. ordered the praetor Aulus Plautius to invade Britain for the final conquest of this land for Rome, the troops selected for the task mutinied; "they would not make war out of the compass of the known world." But at last, although not until rather late in the season, threats, promises, bonuses brought the troops to assent and Plautius embarked his whole force from three different ports. This was done, as it seems, less for the purpose of convenience in marching than to minimize the chances of failure, particularly in the psychological field. Should one embarkation fail the other might succeed. Some of the men were indeed heartened by favorable winds while the courage of others was heightened by observing a flash of light in the sky starting from the east and proceeding to the west. This probable aurora borealis appearing ahead of the legions loaned itself to interpretation by those in command as an omen favorable to the enterprise.

The Roman rank and file, in their opposition to making war "beyond the line," were obviously not impressed by the official Roman settlement of the war guilt question, as we might call it, or at least the question of the justification for the attack on Britain. "The cause was," according to Cassius Dio (LX 19), "that a certain Bericus, ejected from the island during a revolution, had persuaded Claudius to send a body of troops over." In terms of *Realpolitik,* this meant that the Roman governors were aware of the division among the British and that, in the absence of the unity which had previously gone far to guarantee the island against invasion, the moment was propitious. In addition, their intelligence service had failed the islanders.

Persuaded that the aversion of the legions to making war overseas could not be overcome, they did not expect the Romans, or muster against their coming. Consequently, the Romans found no one to oppose them along the shore. This greatly simplifies the history of that particular landing and allowed the Romans to set ashore their original force of 40,000 men, a well-trained body, aside perhaps from psychological unpreparedness, and able to conquer the various British kings one after the other. Disunity in the face of the enemy sealed the fate of the island tribes, several of whom played quisling. Others fought on, succumbing at last to those wondrous tanks on legs, as they might seem, the elephants brought by the invaders.

NOTES, CHAPTER 11

[1] According to Henry of Huntingdon (Bohn ed. 14), writing not long after 1125, these sharp stakes were planted in the river bank and in the fordable water; the remains were still to be seen in his own day, of the thickness of a man's thigh, "and being shod with lead, remain immovably fixed on the river bed." It is not likely that Henry inspected this ford; he merely copied these sentences from Bede (I, 3), composed some 400 years earlier.

[2] Tacitus, *Agricola,* 13.

[3] It was a distinct and long remembered shock to the writer when during the German invasion of Russia in February, 1918, regions were entered of which cartographic knowledge was scant. Only after the German command had acquired and reprinted the Russian general staff maps, reprints which used Cyrillic letters for a time, could one successfully essay control of the new territories physically, or grasp them mentally.

Sources: In addition to Caesar's Bellum Gallicum, Cassius Dio's *Roman History,* Suetonius (Caesar XXV, Caligula XLIV-XLVI, Claudius XVII) see T. Rice Holmes: *Ancient Britain and the Invasions of Julius Caesar,* 2d ed. Oxford, 1936, and Commander H. D. Warburg: *Caesar's First Expedition to Britain.* English Historical Review, vol. 38 (1923), 226-40.

Chapter 12

LATE ROME AND ITS SEA DEFENSE

IT WAS NOT until comparatively late that the Roman Empire was attacked along its water frontiers. Although it had a very long ellipsoid coastline, as far as the Mediterranean was concerned, whose north-south diameter, the navigable part of the Nile included, equalled a sailing distance of 25 days, the Romans maintained a near-monopoly of land and sea power, the latter being the more complete. This circumstance led them to theorize that defense, both on the sea and so far as their coastlines went, could be neglected with impunity. The empire was first attacked in its land power sector by barbarians. Only rarely was an imperial fleet available to resist them.

If we leave aside the hordes that moved Britain ward along the *litus Saxonium,* a term indicating various coastal regions of the German Ocean inhabited at one time, or another by the motley tribal agglomerations who were called Saxons, the land barbarians were the first and strongest assailants of the Roman Empire. Its *limes* was troubled, fought over and crossed long before the *litus,* which the Romans regarded as more nearly the ideal sea, or river boundary, was imperiled.

Still, the technical all-around military superiority of old Rome was put to more effective use when the defensive struggle came to be waged on and over the seas rather than on land. In this differentiation the personnel equation played a considerable role. The fortifications of the *limes* were rarely put to a full test because when the crucial time came they were seldom occupied by the *couverture,* to use a relevant modern term, in sufficient numbers. The technical superiority of the fleet might also be attributed to that of its seafaring personnel which had remained far more Italian than that of the armies. On the other hand no superior naval leadership provided by the old Roman element became apparent. It was Stilicho, a foreign-born general of western Rome, who equipped fleets in Italian ports and threw one of them across the seas to land on the Isthmus of Corinth and from there roll back the Visigoths under Alaric who had invaded Greece in 396. It was he who despatched another fleet with 5,000 soldiers from Pisa to North Africa in order to suppress the rebellion of Gildo, a Moor, and in 398 win back this indispensable granary of the Roman people in Italy.

Of all the earlier migratory tribes Germanic and otherwise, only the Saxons, which would include the Frisians, and the Vandals were able to build up sea forces. When other migrating and less maritime tribes ventured upon the seas, as did Alaric's Goths in their attempt on Sicily shortly before the King's death in 410, sudden storms which would sink, or scatter their craft dismayed them. "Their courage was daunted by the terror of a new element; and the whole design was defeated by the premature death of Alaric" (Gibbon, I 1129). The Vandals, led across the sea into North Africa from Spain by a Roman official in 429, were transported over the Gibraltar passage by the Iberians who were only too willing to furnish the necessary shipping and see the troublesome Vandals depart. Precursors of the later German *Afrika Korps,* under Genserich, or Geiserich (the Spear-Prince, 427-477), they had moved within a period of only twenty years from the Elbe to the Atlas mountains, and after the occupation of the littoral of North Africa from Tangier to Tripoli they promptly developed into a sea power. North Africa at that time was not only a granary which had been lost to Italy, but the Atlas Mountains were productive of timber and other naval stores. Once more, after an interim of 600 years since its fall, navies went out from Carthage, manned by Nordics allied with dark-skinned Africans, to advance a new bid for the *dominium maris* in the Middle Sea.

Maritime Forays of Vandals

The Vandals conquered Sicily as the earlier Carthaginians had done, sacked numerous towns along the Italian coast and from the mouth of the Tiber marched in 455 to the sack of Rome, which they found undefended. Richly laden with the fruits of systematic rifling and sequestration of the wealth of Rome—the term vandalism, as we understand it, does not at all characterize their treatment of conquered territory, which specialized in looting rather than destruction—the raiders returned to Carthage which they, more unconsciously than consciously, had avenged. The Western Roman Empire received its *coup de grace* by way of Carthage, which then became the base of this early-day Afrika Korps.

Nor was this the last raid of the Vandals against the Roman coast. Every spring a Vandal fleet would set out towards Italy from Carthage, usually with Genserich himself in command, always sailing under sealed orders to prevent intelligence of its purpose from reaching the enemy prematurely. Not until after the departure would the Afro-Nordic sea lord reveal his objectives, and when the pilot asked what course to steer he would answer: "Leave the determination to the winds: they will transport us to the guilty coast whose inhabitants have provoked the divine justice." [1] This war-guilt verdict of the Arian Christian king usually seemed to be interpreted by the winds in such a way that the richest towns would prove to be the guiltiest ones. They were not necessarily always on the sea shore, and the Vandals would take along cavalry to enable them to make inland raids.

Italy began to feel the perilous weight of the inordinate, disproportionate length of her sea coast which was to expose her to sea power again and again. No military vigilance, even if occasionally successful against raiders, no fortifications and, as Gibbon pontificates,[2] only a superior fleet could have protected her from the Vandals. When at last such an Italian fleet was constructed, under the Emperor Majorian (457-461), in him was seen the embodiment of the last rebirth of old Roman strength, of strong personal leadership. He went in disguise as his own ambassador to Carthage in order to inspect the locality before landing at, or near, this "Erbfeind" city, thus himself performing the service of a military or naval attaché. But the counter-intelligence service of Genserich enabled the barbarians, although not a little alarmed at the Romans' approach, to destroy the invasion fleet when it had got as far as Carthagena. The loss of this fleet, the product of three years' labor of organization, was more than the impatience of the Italians and their real leader, Ricimer, could suffer. This general deposed and slew Majorian, the last personification of imperial force in Western Rome, which was now constrained to implore the help of Constantinople, Eastern Rome, heretofore considered as the junior partner in the decadent Romans' hollow pretense of being masters of the world.

To put down the piratical sway of the Vandals, Eastern Rome undertook two long-distance overseas operations, the two largest in antiquity, from Constantinople to Carthage. The first of the two was so costly in blood and treasure, according to Gibbon's calculations, that from it dated the debt economy, or deficit financing of the Eastern Empire—spending in constant anticipation of collection of revenues. The enterprise involved large-space concentric operations. One of the divisions of the Roman forces was formed from the troops of Egypt, the Thebais and Lybias, including Arabs with a train of horses and camels. This part of the expeditionary forces landed on the coast of Tripoli without meeting resistance and surprised and conquered the cities of that Vandal-controlled province.

Gibbon on Overseas Operations

From there the plan was for it to march through the desert and meet the main force, 1113 ships and 100,000 soldiers and mariners, which was also to be joined by an Italian fleet, under the walls of Carthage. Genserich felt seriously threatened and began to negotiate. The Romans under the worthless court general Basiliscus consented to a truce. Here, in comment, we follow Gibbon's very sensible statements on overseas operations and landing; although they seem occasioned and inspired by his knowledge of ancient history, they are easily as much, if not more, influenced by his personal military experiences and provoked by more recent English experiences with combined operations: [3]

> Experience has shown that the success of an invader most commonly depends on the vigor and celerity of his operations. The strength and sharpness of the first impression are blunted by delay; the health and spirits of the troops insensibly languish in a distant climate; the naval and military force, a mighty effort which perhaps can never be repeated, is silently consumed; and every hour that is wasted in negotiations accustoms the enemy to contemplate and examine those hostile terrors which, on their first appearance, he deemed irresistible. . . .
> If Basiliscus had seized the moment of consternation, Carthage must have surrendered and the kingdom of the Vandals was extinguished. . . . Instead, Basiliscus consented to the fatal truce; and his imprudent security seemed to proclaim that he already considered himself as the conqueror of Africa.

During the armistice the wind changed so as to favor Genserich's plan to send out fire-ships against the carelessly assembled Roman fleet which was in large part burned and vanquished. Basiliscus was forced to retire with only about half his fleet and retrace a long, ignominous path to Constantinople, thus re-opening the sea paths for Genserich's maritime rovers who thereby only then obtained for the Vandal empire its largest sway.

To bring down this empire, in spite of the inward decay of that organization, called for the services of the last and perhaps greatest master of amphibious warfare in Rome—Belisarius. His African mission was in large part the outcome of negative Roman experience with coastal fortifications as protection against raids from the sea. As Montesquieu aptly remarked: Justinian's empire was like France in the time of the Norman incursions—never so weak as when every village was fortified.[4] Expressed in terms of technological warfare, this meant that Roman and Romanized societies endeavored to make use of only one of the two kinds of superiority over the barbarians they possessed—land fortification, but nôt ship-building. Greek fire preserved its pseudo-magic character, to be used sparsely rather than as a mass-produced and mass-employed weapon.

The technological means for mass ship-building were still available in several places of Rome, East and West, but not the economic structure; the large over-all economy required to create important sea power, or a large standing army was harder to maintain in late antiquity in consequence of the increasing dearth of precious metals. The military definition, or function of money is after all that of a transformer of a dormant, or floating military potential into a kinetic one, this being true above all in the days before a totally state-directed war economy. The weakness of the monetary economy in late Rome, weaker in the west than the east at most times, combined with war-weariness on the part of the robust elements of original Rome, caused by weakness of the societies that succeeded to the Roman heritage, provided the raiders, Vandals. Saxons and later Saracens and Normans, with opportunities to exercise their barbarian boldness and rapacity unchecked on the shores and even the river banks of more civilized peoples, and for a considerable distance inland.

Victory of Absorption

Indeed, early medieval society many times found hardly any ways, or means to overcome the raiding barbarians except in the Chinese fashion, that is to say, by absorbing, Christianizing and otherwise civilizing them. After that process it was men of barbarian stock, like the Normans in Normandy, England and Sicily, who evolved the state economy and administration which made possible the maintenance of a more modern armed force. Not a small part of military progress has been achieved along the frontiers of civilization and under the shock imparted by foreign barbarians who, when absorbed into a surviving civilized society, often brought an addition of valuable military energy.

Even if it seems totally disassociated from war, early Christianity nevertheless exercised an influence on military spirit and thought. It tended to confirm the passive defensive attitude which was in keeping with the general lethargic character of the socio-economic system of the world that was Rome's heir. The early church did not evolve the theology required for undertaking crusades. But while the western church was not to experience the "rise of the Crusade idea" until about the year 1000,[5] religious bellicose zeal was active much earlier in the eastern than in the western church. In Constantinople it aided governors in carrying the war far over charted, or uncharted seas to enemies who were to many believers close foes only because they were religious opponents. The Athanasian clergy endorsed the expedition against the Arian Vandals, admonishing the Emperor Justinian that "the God of battles will march before your standard and disperse your enemies who are the enemies of His Son."[6] This endorsement made Belisarius' North African campaign the first crusade in the military sense undertaken, it must be remarked, by Christians against Christians, an antagonism which makes such civil war crusades the bloodiest of all. In the end, the eradication of the Vandals proved to be Justinian's great contribution to opening the west of Europe to Mohammedanism.

While dogma made the Vandals close enemies for the clergy, the long distance and the magnitude of the enterprise frightened the administrators of Byzantine finances and other officials. They insisted that it was a 140 days' journey from Constantinople to Carthage and that a whole year would go by before word would come from an army despatched there; and, besides, that eastern Rome was itself perilously exposed to the barbarians on the east and northeast. This was not mere pettifogging in geographic terms, for geographic and logistic ignorance was spreading rapidly; knowledge was sinking below the classic level until sad experiences born of the land marches of the later Crusaders could move a prince of the Church, Aegidius Romanus (1247-1316), to suggest that something like maps would be useful articles for warriors to have. It actually happened, however, that the expedition headed by Belisarius, who knew more about geography and logistics than a prefect of the palace, sailed from Constantinople around June 21, 533 A. D., and fronted Carthage on the 14th of September.

Rome and Carthage at Grips

An expeditionary force of 5,000 horse and 10,000 foot was assembled for this final struggle between Rome and Carthage, unless World War II is counted, for Germany still considered the latter conflict in the light of one between "heroes and traders," *Helden und Händler,*[7] Rome and the "Punic" Anglo-America. Not all the participants under Belisarius were mentally prepared for the enterprise. Troops recalled from the ever-burning Persian frontier "dreaded the sea, the climate, and the arms of an unknown enemy." Some horsemen from far inland, like the Massagetae, or Huns, had to be won over by fraud and deceit to such an unaccustomed voyage. Mariners from the Aegean and adjacent regions, 20,000 in number, manned 500 transports varying in size from 30 to 500 tons and totalling according to Gibbon 100,000 tons, or more than six tons per soldier and nearly three tons per caput of the total expeditionary strength. The convoying fleet was weak, for the type of long ship with several hundred oars once familiar on the Mediterranean had become as extinct as the dodo. Only 92 light brigantines, rowed by youths from the Constantinople waterside, with a mere 20 rowers to each craft, which was but lightly armed against enemy missiles, performed convoy duties.

Belisarius held supreme command both on land and on water. While this double generalship caused Gibbon to remark somewhat cryptically that "the separation of the naval and military professions is at once the effect and cause of the modern improvements in the science of navigation and maritime war," it would seem uncertain whether the historian had in mind and before him the fresh and rather disastrous 18th-century experiences with a too much divided command in such expeditions. They might well have provoked him to comment *ex cathedra historiae* on the solution of the problem of amphibious command. Is it to be given to a military, or naval man, and when and how is it to be divided?

As one of the first large scale Christian overseas enterprises the expedition started off with the blessings of the patriarch of the eastern church. While this may have laid unction to some souls, others "explored with anxious curiosity the omens of misfortune and success." The various stops en route from the Bosporus were used by Belisarius to take aboard more horses, or to stiffen the morale of the soldiery by exemplary, well calculated exertion of the vast authority conferred upon him; to provide new bread after the biscuits issued by a corrupt official had been spoiled, to bring intoxication under control, to obtain the latest intelligence of conditions among the Vandals, to rewater his ships.

Fortune, never more fickle than on the waves, showed herself as once more favoring the Romans. The Vandal king, Gelimer, remained so ignorant of their approach that he was employing a force of 5,000 soldiers and 120 galleys for the conquest of Sardinia, a force outweighing Belisarius' ships in fighting power on the sea. Of this as well as of the baro-psychometric pressure which prevails among brave men on the seas when hostile squadrons are known to be abroad, Belisarius was well aware. He heard his soldiers challenge each other to state the secret fears which they entertained in their hearts during the voyage. They were confident that once again on land they would be quite able to maintain the honor of their arms, but admitted that their courage was insufficient to meet at the same time maritime perils and the barbarians.

These sentiments among the soldiers, engendered by idleness, since their minds, or muscles were apparently not specially employed during the passage, moved Belisarius to land his troops sooner than he had planned in Africa, at Caput Vada, several marches from Carthage, instead of sailing straight into that harbor. After nearly three months in a convoy the seafaring endurance of the Roman soldier seemed indeed near the breaking point.

Vandal Sea Power Destroyed

Now they were returned to earth again. As customary at the end of a marching day of a Roman legion, a fortified camp enclosing a spring was thrown up with ditch and wall, forming a beachhead for the protection of men, horse and stores. In order to win the confidence of the non-Vandal inhabitants, mostly of the Athanasian confession, Belisarius took care that their gardens and houses were spared from looting, whereupon fresh provisions appeared on the markets, thus helping the general to feed his army. Such measures probably did more damage to the Vandals than the first blundering battle fought at the tenth milestone outside Carthage. The city promptly fell to Belisarius. A second battle more stubbornly fought, in which the Vandal force employed had just returned from the conquest of Sardinia, was won by the Byzantine horse, brought all the long way from the Bosporus. This decided the fate of Vandal land and séa power. Once more means offered for the extension of Roman control over the Mediterranean and its southern shore.

Belisarius' subsequent campaigns against the Goths in Italy and Sicily, from whence he began the reconquest, offered fewer demonstrations of his mastership of combined operations which is so apparent from the reports of Procopius, his secretary and historian, who took part in the African campaign. He displayed his ingenious initiative during the siege of Palermo when he saw the need of commanding the place by his missile throwers. Bringing his ships as far into the harbor as possible, he had their boats hoisted with ropes and pulleys to the mast heads, and filled them with archers who from these high positions could rake the walls. As the historian relates, there was no counteraction on the part of the enemy who apparently possessed no archers.

His campaigns against both Vandals and Goths confirmed to Gibbon, who rather outdoes Mahan on this occasion, "the truth of a maxim that the master of the sea will always acquire the dominion of the land," [8] which is certainly not valid for long periods of either the Roman, German, French, Chinese or Russian empires. It would be far more just merely to say that Belisarius knew better than any general for a long time to come when and where to employ military, or naval force or both; to divide the labor that is war and arrange for the division of this labor between the two forces of his day (three in ours) and still apply these labors strictly to one purpose.

Belisarius' competitor and successor, Narses, also well employed amphibious force. When in 552 he started from Salona in Dalmatia with a fleet and marched an army around the head of the Adriatic, he used the fleet to transport soldiers either across the Po and other rivers, or around their estuaries, thus evading the Ostrogothic general, Tejas, who sought to dispute his passage. With the rare exception of the Vandals the barbarians represented mostly land power in one form or the other while Byzantine amphibious power, as long as it existed and was ably utilized, was the most potent factor in keeping them at bay.

NOTES, CHAPTER 12

[1] Gibbon I, 1277.

[2] Ibd. I, 1278.

[3] I, 1285-6. Gibbon has on an earlier occasion given his views about the crossing of a river in the face of an enemy (I, 818-9), views penetrated with a military common sense which must have eased the task of the young Moltke when in 1832-3 he attempted to translate the *Decline and Fall* into German; perhaps the consciousness of having learned from this master work consoled poor Moltke when his publisher went bankrupt and left him largely unpaid. It is interesting to speculate how far Moltke was "influenced" by Gibbon's statements in his views of river crossings and overseas enterprises.

[4] *Considérations sur la grandeur et la décadence des Romains,* ch. XX.

[5] Karl Erdmann: *Die Entstehung des Kreuzzugsgedankens.* 1935.

[6] Gibbon II, 184.

[7] See an article *"Das Karthago des 20. Jahrhunderts"* in Deutsche Wehr, 18 October 1940.

[8] Gibbon II, 299.

Sources: Procopius: *Books About the Wars,* two of which deal with the Vandal War; Gibbon: *The Decline and Fall of the Roman Empire* (Modern Library ed.); Lord Mahon: *Life of Belisarius.* 1829; J. B. Bury: *Later Roman Empire.* 1889.

Chapter 13

SAXON, VIKING AND NORMAN LANDINGS, EARLY MIDDLE AGES

THE TRANSIENT RE-EXPANSION of the Roman Empire reached barely beyond the Alps and never to the northern seas. The cries of the Britons for Roman help against the Picts and Scots remained unanswered. Before the latter tribes had fully realized that the walls built against them were useless without legionaries, they seem to have outflanked them by invading Britain from the sea. The Scots later rarely used this form of attack in their subsequent raids against Britain, perhaps because they could not transport home by water the profits of their yearly cattle lifting.

Making one final effort to succor the Britons, in addition to their stone wall from sea to sea, the Romans built a number of towers along the seacoast to the southward of the wall, because the irruptions of the barbarians were expected there as well (Bede I, 12). Whether these martello-like structures were watch and signal towers, or were also intended as coastal fortifications does not appear in Bede who wrote 300 years after the final Roman evacuation of the island. Without a comprehensive, mobile military organization the Britons remained helpless when attacked by either Picts and Scots, or Saxons. What seafaring abilities survived among them seemed sufficient only to carry those who would not submit to their conquerors, or retire into Wales and Cornwall, across the Channel to the continent where they set up a continental Brittany, or Armorica.

While the descent of the Saxons, or Anglo-Saxons in Britain, beginning in the first half of the 5th century, was fraught with as great consequences as any of the main landings in mythology, or history, it is too poorly documented to supply adequate material to warrant it being given important space in the military account of such enterprises. We cannot say in detail how the Saxons landed, but they seem not to have burned their ships, preferring to establish beachheads around them.

After they settled down, the Saxon invaders became decided isolationists, giving up their martial seamanship. Politics,[1] economy, technology and the Christian religion to which they had been converted, combined to produce and maintain little kingdoms forming political islands within the island. Among them the "bretwaldship" was a coveted hegemonial position, changing from the ruler of one kingdom to another, rather than remaining in a dynasty of over-kings. However, no common danger threatened them for more than 300 years. As in many other places Christianity contributed strongly to isolationist sentiment; the monasteries and other church institutions were in large part aloof from the melee of the intra-Saxon fights and formed islands of peace like Lindisfarne, the Holy Island, which was a real island a short distance off Northumberland, in the North Sea.

Intruding into this pious isolationism, came the dragon prows of the Viking ships. Lindisfarne, whence Christianity had once spread into Northumbria, was one of the first places to be raided and burned by the fierce new invaders (793). "It was not thought possible that they could have made the voyage," as Alcuin, the learned Anglo-Saxon at Charlemagne's court, wrote from home after the sack: "Never in the 350 years that we and our forefathers have dwelt in this fair land, has such a horror appeared in Britain as this that we have just suffered from the heathen."

It was characteristic of the British state of mind at the time that the Vikings were met, on one of their first descents, by a port-reeve "who would fain have known what manner of men they might be." With the hind-sight of the chronicler it is stated that in the fatal year of 793 "dire forewarnings came over the land of the Northumbrians, and miserably terrified the people; these were excessive whirlwinds and lightnings; and fiery dragons were seen flying in the air. A great famine soon followed these tokens; and a little after that . . . the ravaging of heathen men lamentably destroyed God's church at Lindisfarne through rapine and slaughter." *(Anglo-Saxon Chronicle.)*

Comet Caused Panic

As on many occasions when sea-borne invasion threatened, the raids of the Normans gave rise to a panic, one of those "*grandes peurs*" which sweep sections of mankind from time to time. Every curious incident becomes a portent to pious and impious alike. The learned Einhardus (Frankish historian, 770-840), retired from Charlemagne's court to his Seligenstadt (Town of the Blissful), had to explain the comet of 837 to

Emperor Louis the Pious and he wrote, after the raid of the year was over:

> Nearly all the writers of the ancients are of the opinion that the apparition of a new unusual star announces to unfortunate mankind more misfortune and sadness than joy and good tidings. Only the Gospel testifies that the apparition of that new star which the wise Chaldeans saw, promised salvation. . . . The star however which was visible a little while ago, according to the testimony of all who observed it, offered a fearful and sombre sight, big with misfortune. This omen of evil fortune is in my opinion in keeping with our disservices and foreshadows the ill-fortune which we deserve. . . . If only the misery which the Norman fleet of late has brought over part of our country, had atoned for what the apparition of that frightful star has announced! But I fear the affliction which was indicated by that dreadful star will prove still heavier yet, even though all those upon whom the tremendous storm from the Ocean broke have come to feel with their bodies and those of their own relations the hard and bitter punishment. May my Lord, the most pious Emperor, prosper.[2]

While we must not follow too closely the historians writing before and after 1900, many of them eager champions of what then was called sea-power, in their warnings of the dire consequences of its neglect, illustrated by the unhappy state of England and other countries in face of the Viking danger, it is clear enough that the English Saxons were in no way, on land, by sea, or on the sea, prepared to cope with the northern pirates. For they had neglected building up not only sea, but land power, possibly under the influence of Mercia, which may have been a factor in the early British isolationism comparable to that of our midwestern states.

Charlemagne Resists Raiders

It was not astonishing that the Saxons proved so helpless, since even the empire of Charlemagne, with all its might and prestige, was barely able to ward off the raiders of the north and south. During Charles' lifetime, which incidentally included only one "combined operation," a successful siege of Venice on land and sea conducted by his father, King Pepin

(714-768), fleets were built for service against both Moors and Normans. In 807 he sent out an expedition to protect Corsica which defeated the Moorish corsairs, then ravaging the Mediterranean from freshly conquered Spain; but in 809 the corsairs were back in Corsica, plundering a town on Holy Easter Eve and carrying off all its people save a bishop and a few old and sick persons. Charles also tried to form a sea power against the Vikings. As his biographer Einhard (in *Vita Caroli Magni)* relates:

> He built a fleet to make war on the Normans. For this purpose he constructed vessels along those rivers of Gaul and Germany which run into the North Sea. And since the Normans in endless invasions devastated the coasts of Gaul and Germany, he put guard posts into all ports and navigable river mouths. Through these protective measures he thwarted every hostile predatory incursion. The same protective measures he arranged against the Moors . . . along the southern coast of Narbonne and Septimania as well as along the whole coast of Italy down as far as Rome. Thanks to this during his reign neither Italy nor Gaul and Germany suffered any considerable damage; only Civitavecchia in Etruria was taken and devastated by the Moors following treason and in Frisia some islands along the German coast were robbed by the Normans.

As technology decayed in early medieval Europe, the Norsemen remained the best ship-builders, although there is good reason to assume that they purchased their weapons from Germans and other swordsmiths to the southward whom they later visited so unpleasantly, a kind of return to civilization in unpleasant form of its own war products, a danger which the victims have usually seen and repented too late.[3] Their ships were indeed not too strong, for storms sometimes destroyed whole fleets, but they were suitable for short voyages. This raises the question of whether and where the Vikings had naval stations, sites for which the German North Sea islands may have provided. Being of shallow draft, they made excellent landing craft and their size and gear permitted maneuvers along shores, in bays and even inland waters. While they carried one large sail, the form of which they borrowed, with the name for it, from southern peoples—the word sail in the Germanic languages comes from the Latin *sagulum*—these ships depended largely on the labor of oarsmen, all free men and not slaves, as was so often the case in the Mediterranean. The crew were alike sailors and, we might say, marines. The requirements of their voyages, which called for two or three daily shifts of rowers, make it probable that a Viking vessel of twenty oars, the largest during the true era of the Norsemen, would have needed some sixty men. The warriors' shields were hung on the gunwales en route, giving us a forecast of ship armor.

The Viking incursions, extending over all the seas from Russia in the north into the Mediterranean, with the British isles and the states of the Carolingian succession (in France, 751-987) as their preferred objectives, met conditions which could not have been more propitious to their style of warfare. Nowhere was there determination to seek the pirates out in their lairs and "wiks"—the bays from which their name is derived—or in the markets where they disposed of their loot, or to use the type of ship available to meet them in home waters at the end of what must have been exhausting voyages for the invaders. The Vikings' home bases remained as immune as in the days when Tacitus wrote (ca. 98 A. D.) of the Nordic tribes that "promiscuous carrying of arms is not permitted here as it is among the other Germans, but weapons are kept shut up in the charge of a slave who acts as guard. This is because the sea prevents sudden inroads from enemies, and because bands of armed men who have nothing to do often become unruly." *(Germania).*

Tactics of the Vikings

It was rather the exception when, in 851, one of the Anglo-Saxon kings fought the Vikings "on shipboard and slew a great number of the enemy . . . and took nine ships, and put the others to flight" *(Anglo-Saxon Chronicle).* At the same time fortifications as the embodiment of land power were left uncared for, another neglected heritage from Roman days. Conditions were worst in England where the invaders, who surrounded their beachheads and then their winter quarters with stockades, had to demonstrate to the inhabitants the rudiments of fortification, the forgotten art of castramentation, to use the Roman term. Such beachheads made the Norsemen masters of re-embarkation and they were seldom caught off guard by a defending, or offensive land force.

Their strong points, sometimes called *gardr,* a fenced-off piece of terrain, also fort, a name preserved in the Russian gorod, or town (as in Novgorod), were first placed on islands close to the coast, like Lambay in Dublin Bay, or Thanet in Kent; and Sheppey, Walcheren, Rhé or Noirmoutier;[4] or some peninsula.

Later the Vikings selected river islands for this purpose like Oiselles, 10 miles above Rouen, or in 863 an island in the Rhine near Neuss where they would be immune from attacks by foes without shipping. It was considered an epoch-making event when Arnulf (850-99, king of the east Franks) in 891 stormed such a camp. "They were opposed to the bravest people of the Normans, the Danes; it had proved impossible heretofore to take away an entrenched camp from them, or to conquer them inside of it. Hard was the battle, but it did not last long; with God's help the Christians won the victory. The Normans looked for their salvation in flight, but then the river which had protected their rear like a wall, became their fate and death."[5]

Such beachheads began as *pieds-à-terre,* becoming arsenals, then winter quarters into which the Vikings settled for the first time in England in 850; and finally were starting points for the carving out of new principalities, as the invading forces grew in size and the enterprises changed in character from private piracy to folk, or state enterprise, and to what the *Anglo-Saxon Chronicle* in rather helpless admiration calls "the army." From these beachheads the Norsemen would proceed inland, soon trying to find horses from the invaded districts although sometimes bringing mounts with them, as did a force active in England in 893, which was provided with steeds from France. The mounted Vikings did not constitute a heavy cavalry, but merely used horses to increase their mobility as raiders and to transport men and loot when falling back to beachheads and ships. Not until late in their history did the Norsemen become "knightly horse" and even then they continued to rely on their sea-horses, or ships.

Norman Fire Superiority

To ensure unity of command in their "commandos," as one in 1945 is constantly tempted to say, the Norsemen were provided with leaders, the so-called sea-kings. They were at first war-chiefs chosen for the duration of an expedition and for their military qualities by the participants who, like the Saxons, originally represented a diversity of tribes. Their bows and arrows, which were scorned by self-respecting knights in England and on the continent as a plebian war weapon, gave the invaders "fire superiority" over the defender. This went far to balance the latter's shore superiority if the two happened to meet for battle along the coast, a rare occurrence. Other things being equal, the bowmen had as much missile superiority as the range of the arrow is greater than that of the spear.

Considering this fire superiority, it was the height of tactical incompetence for Charles the Bald in 845 to march out against Ragnar Lodbrok who was sailing up the Seine and when approaching him to divide his forces into two detachments, which were to advance on the two banks of the river. Neither of them apparently could harm or stop the Normans with the missiles at their disposal. The latter cut down the two forces successively, hanging many prisoners from the first force on an island in the river in full view of the second detachment which, unnerved by the sight, proved easy to beat.

This was foolishly not turning to profit the one superiority possessed by the invaded, numerical, which occasionally enabled them to overcome the smaller forces of the Vikings. In one of the few Eddic descriptions of their raids that hold value for the military historian it is related that King Eric Bloodaxe sailed first to Wales and harried there and then went on to England, all the people fleeing before him wherever he landed. As a bold warrior with a large force, Eric became so confident that he went far inland where the English and the Danes who had settled in Northumberland caught up with him. "Many of the English fell, but where one fell, three came to his stead from all over the land . . . and by the end of the day Eric Bloodaxe had fallen." [6]

Cohesion among the Viking fighting units was strong, as it is apt to be among pirates who have usually proved excellent cofighters because they must either live, or hang together. As pirates they did not seek battle as often as it was forced on them. They would mix offensive and defensive in such a way that after the original offensive thrust of the invasion and the raid they would act on the defensive on the battle field. When they had to give battle they would form a wall of shields, not the formal *testudo,* the dovetailing of shields practiced by the Roman soldiers, but rather a line of all shield bearing warriors. As such they might on occasion be ridden down by knightly cavalry and in the end be outfought and mastered. Should the Viking line be broken, they might retire to carry on in street fights, among the houses of a nearby town and still gain the day there, in that most stubborn sort of combat. They would fight with the desperation of men separated from their homeland by the vast sea, who can only hope through victory again to see their own hearths.

As little as we can identify the prime force that motivated the great folk wanderings on land, do we know the origin of the southward impulse of the Vikings: probably a mixture of pressure created by overpopulation, zest for adventure, looting propensity, or a reaction against the northward push of Carolingian imperialism and Christian mission, against both of which the Norsemen were warned by Widukind, the duke of the Saxons.[7] In any event, the courage of the Norsemen in the face of the unknown cannot be doubted. "We cannot easily realize how all-embracing that courage was. A trained soldier is often afraid at sea, a trained sailor lost if he has not the protecting sense of his own ship beneath him. The Viking ventured upon unknown waters in ships very ill-fitted for their work. He had all the spirit of adventure of a Drake or Hawkins, all the trained valor of reliance upon his comrades that mark a soldier fighting a militia." [8]

Supernatural Aid for Paris

But such tributes to their bravery must not lead us to believe the Vikings altogether free of fear, of the vast and almost typical dread that seizes upon the man who must land on unfamiliar ground after an ardous voyage. During battle, or looting this fear may have been sublimated into reckless berserk fury, but it did not altogether disappear. While we are nearly devoid of reports of a military character from the Vikings' own side, and must largely depend on Christian chronicles for a history of their raids, an occasional glimpse into their souls, the psyche of the raider, still seems possible. French chroniclers mention one Ragnar in connection with a Norman attack on Paris in 845 which failed because, through the interposition of the local Saint Germain the attackers were suddenly "blacked out," surrounded by darkness—saints more than once are credited with having achieved in war what modern smoke barrages, etc. aim to do—and then slain by the Frankish defenders.

This defeat left a deep impression upon the Norsemen and long lingered in their memory, for Saxo Grammaticus some 350 years later made this defeat occur in hell, which he specifically located in Bjarmaland, now Perm in Russia, hell in Norse mythology being practically identical with Niflheim, the fog-world, or Niflhel, fog-hell, presided over by Hel, goddess of the underworld. Thus the ancient Norsemen sited what has modernly come to be called "the fog of war," in the human psyche rather than in human reason.

Many times populations were either as helpless against invasion or, in the long run, as absorptive, as the old Chinese when their domains were overrun. Others attempted to front the Vikings under new energetic leaders, such as Alfred the Great, or Arnulf of Carinthia, the later emperor of the eastern Franks and last Carolingian monarch. So pressing was the threat from Norman and Slav that the German princes in "the first independent action of the German secular world" as Ranke calls this act, deposed the worthless Charles the Fat, into whose weak hands the empire of Charlemagne had come as an inheritance. His successor was Arnulf, a man of fresh enterprise who conquered the Normans at Louvain in 891 in a strongly fortified camp, placed in a loop of the Dyle, into which he broke with his dismounted knights.

The enthronement of Alfred and Arnulf, their sway over districts larger than customary enabled both to govern under institutions of overlordship like that of the ancient *bretwalda* or king-emperor, which made Alfred "king over the whole English nation except that part which was under the dominion of the Danes" (*Anglo-Saxon Chronicle* ad 902); and indicated the necessity of abrogating at least temporarily the extreme governmental and military decentralization and geographical and political atomization of the west. The burdensome and ineffective military and arms-bearing duties imposed upon freemen were the perfect expression of this anarchy. In face of the common danger from the north there was still no diplomacy, no arrangements for coalition warfare such as would be concerted by diplomatic channels in our own day. And the spiritual centralization of western Europe through Rome provided no counterforce as an asset for war because the church had not yet adopted the practice of blessing, at least, certain wars. At the time of the Norman invasions churchmen would not go beyond uttering certain prayers such as *a furore Normannorum libera nos.* The sums collected regularly from the faithful in England and elsewhere were not used to finance wars against the heathen who threatened the very survival of Christianity.

Alfred's National Navy

Alfred provided himself with a war navy for national service, in which "national navy" he employed mercenaries, either Danish renegades, Frisians, or other foreigners—competent alien warriors who, even today, can so seldom be spared whenever great military reforms are contemplated. His war fleet was composed of long narrow craft embodying ideas in construction gained by experiences with, and observations of, invading squadrons. They were twice as long as the usual type, with sixty oars and more; swifter, steadier and with higher freeboard than the others, though perhaps of shorter cruising range. Whether they had a second bank of oars is uncertain. With these, handled at first "very awkwardly" by the crews, he took the sea against the Danes, who occasionally surrendered, "sorely distressed and wounded before they gave up to him." (*Anglo-Saxon Chronicle* ad 882.)

Alfred's successors carried on this naval construction, with improvements, and his son Edward in 911 could assemble and send out some hundred ships against the Danes in Northumbria in that fairly rare and precarious undertaking—combined operations. So large did this English fleet appear to the Danes that they thought the greater part of the English force must be in the ships and that without exposing themselves to undue peril from it, they could spare enough men to set out on a plundering expedition in safety, a belief in which they found themselves mistaken.

The domestic military reforms of Alfred, as well as those undertaken on the continent, were calculated to repel and discourage invaders and invasions. They went far in the direction of feudalism as a society in which the few were large landholders and leaders of fighting men and the many were serfs dependent on them, the first giving and the latter receiving protection, that being the price of surrendering title to land. Arms-bearing and command of forces became to a large extent exclusive, professional, specialized. The quality of enemies threatening western Europe and the method of their attack contributed powerfully to, and was perhaps the main force in, the evolution of the feudal army, with the mailed warrior as its chief strength and which was definitely more mobile than the Frankish armies, or the English *fyrd* had been. Norse invasion had thus, it may be said, strongly promoted feudalism.

The continental armies contained the rudiments of a cavalry (the *miles,* horseman) as their main reliance, developed to meet and offset the shock-power of fierce and determined Norse attack-tactics. The absence and need of mounted warriors in England led no further than to the evolution of the mailed, but unmounted *thegn,* thane. Whereas while on the continent the horse came into its own in the theater of battle, the English fought on foot and employed horses as pack animals only in transport before and after action, and in approaching, pursuing, or fleeing from the enemy. Mounted men could not only ride down a Viking shield wall; they could also follow and scout out a hostile force raiding inland counties, meet it in open battle if strong enough and harass it on the march, when crossing or fording rivers. It gave to the defenders a large share of the initiative which in the beginning of the Viking invasions had been altogether on the side of the latter. These military-economic consequences rank the landings of the Vikings with other effects upon world history such as were wrought by the Ionians, the Spanish in America, or the western powers in the Orient.

Another long-lasting result of the shocks produced by the Vikings and their contemporaries, the horsed raiders from the east, was the castle and the fortified town, the collections of houses enclosed by wall, or hedge (German *Zaun*). A renaissance, although on a more simple scale, of the art of fortification of the ancients was set in train. The stockaded fortresses, the first artificially strengthened places built in England after the Roman fortified towns, camps and limes, by that time largely decayed, the so-called "burhs," were but faint replicas of the ancient works, which they seem to have imitated less than the stockades which the Vikings threw up about their beachheads. These fortifications were usually placed on elevations near watersides and permanently garrisoned with men who combined military duties with civilian activities, the "burh ware." So great was the obsessive panic-fear produced by invasions, however, that sometimes even formidable castles, however strong, proved useless because the defenders' hearts failed them. Thus when a king of western Francia in 881 constructed a great work for protection against the heathen and the support of the Christians, he could find no one to whom he could confidently entrust the custody of this castle (*Annales Bertiniani,* ad 881).

BRITONS' DEFENSIVE MEASURES

These fortifications were refuges for the people of the neighborhood as well as obstructions in the path of the invaders. Placed along seashores, or river banks, stockades on either side of a stream, connected by a boom or a bridge might suffice to halt the Vikings, or bottle them up on their return. This happened to the Danes who had sailed from the Thames into the Lea and found coming back, that the English had blocked their exit (*Anglo-Saxon Chronicle* ad 876). Along some continental rivers more elaborate works, like bridges fortified at their heads, confronted the Vikings who to then had found no obstacles to raiding far inland and unhindered of course, by artillery on the part of the defenders. Fortified bridges compelled the invaders either to besiege them and the towns of the defenses of which they frequently formed a part, or to portage around them laboriously hauling their ships with them.

Army, navy and fortifications, increased by Alfred's successors, and political unification, which allowed Aethelstan (925-940) to call himself *rex totius Britanniae,* king of all Britain, enabled the English at last to undertake the reconquest of England, of those parts, or Danelaw, held by remnants of long settled Danes who, now immobilized by becoming land lubbers, proved inferior to the forces which assailed them in Wessex. Reconquest often forms the final, occasionally belated, reaction to an invasion. In the case of the 10th-century English it was not permanent for the foreigners came once more and were permanently triumphant under William the Conqueror.

While the army reforms undertaken under the inspiration of the impact of the Nordic invasions had ended the lack of cohesion of national military strength, the defensive debility of a situation wherein all arms-bearing free men theoretically formed, or were supposed to form national armies, but in reality did not, these reforms soon conferred a power on the feudal arms-bearers of the realm which more than balanced that of the king and which made further military reforms impossible for centuries. So long as no solidarity of territorial principalities existed in England, local overlords could be found betraying the country to fresh Danish invasions before and after 1000, or to counsel, as prelates of the Church non-militant seem to have set the example, that tribute be paid to the Danes "on account of the great terror which they caused by the sea-coast" (*Anglo-Saxon Chronicle,* 991). Finding the shores of England once more open to them, the Norsemen under Olaf Tryggveson, Sweyn and Canute landed again and again. The English tried to eradicate the quislings, and not always the right ones, by murdering all the Danes and their domestic supporters (1002), a slaughter of military occupants which other islands have on occasion committed, the Danish blood-letting being akin to the Sicilian Vespers of 1282. (The victims being the French population, the Sicilians acting in reprisal for outrages committed by French troops.)

Although military institutions of sorts were still functioning in England, most of the open battles were lost by the English, either by royal ineptitude, or still prevalent localism which finally went so far that "at the last would not even one shire assist another" (*Anglo-Saxon Chronicle.* 1010); or the faint-heartedness, or treason of the feudal lords. As the *Anglo-Saxon Chronicle* reports conditions for the year 1015 when Canute was ravaging the land with an army: "Then gathered Edric the ealdorman forces, and the etheling Edmund (the king's son) in the north. When they came together, then would the ealdorman betray the etheling, but he was not able: and then they parted without a battle on that account, and gave way to their foes. And Edric the ealdorman then enticed forty ships from the king, and then went over to Canute. And the men of Wessex submitted, and delivered hostages, and horsed the army." In the final battle of Assendune (1016), Edric "betrayed his royal lord and the whole people of the English race." All too often the English found themselves unable to track down the enemy: "If the enemy were east then was the *fyrd* held west, and if they were north, then was our force held south."

English Accept Danish Kings

Despairing of the ineptitude and cowardice of their lords and royalty, and also their bad luck which probably caused the cry of treason to be raised even oftener than it deserved, the English finally accepted Sweyn and Canute for their kings (1013-34), who could draw on their native Denmark to uphold their power in the first uncertain years of their reigns, and later relied on a war fleet manned by paid rowers to secure the coasts of England. Canute ruled wisely and well. The number of these warships was eventually raised to 72, with a rate of pay under Hardicanute of eight marks for each rower. This expenditure was found burdensome enough by the English and that king became highly unpopular, although doubtless he unavoidably incurred that odium by the inescapably costly professionalization of sea service.

Central consolidated power that might have preserved England from invasion by land, or sea was again weak during the reign of Edward the Confessor (1042-66). Britain became wide open hunting ground for foreigners, whom the King after his years in exile at the Norman court came to prefer to have about him, and also to noblemen temporarily banished, or residing abroad, like Harold the later king and his powerful family, who gathered strength on the continent for an armed return. The insularity, or nationalism of the English at last recognized in Harold their most promising representative. "Unwilling that this land should still be exposed to outlandish men, by reason that they themselves destroyed each other," they avoided the battle to decide whether Harold and his family should return from exile, and banished instead most of the French Normans around the King, whom they considered evil counsellors (*Anglo-Saxon Chronicle,* 1052).

Harold did not succeed in establishing a nationality in England whose precondition would have been the subjection of the ever-plotting lords. They, including his own brother Tosti (or Tostig) sought and found foreign support to oppose him. While he organized his own forces on land and the seas fairly efficiently, the English could not bring themselves to organize a cavalry arm before their country lost its primeval English sovereignty. Indeed, in a clash between the Welsh, supporting an outlawed earl returning from an Irish exile, and the English in 1055 at Hereford, the latter "before there was any spear thrown, fled *because they were on horses*" (*Anglo-Saxon Chronicle.* 1055). This boded ill for the English when William of Normandy later crossed the Channel with the knighthood of Europe whom he invited to share in the conquest, and prospective looting, of England.

The Normans were too comfortably domiciled in France at once to follow their Duke in his proposed adventure. The majority of the lords were far from convinced by the flimsy "war guilt" reproaches by William against Harold, and endorsed by the Church; they did not consider them sufficient justification for such a hazardous undertaking. As Henry of Huntingdon represents the story of the political preparation for the invasion, only outright trickery won the approval of the principal men of Normandy, so much did the magnitude of the task appal them. As they were entering the council chamber, William Fitz-Osbert, the Duke William's seneschal, who had a brother in an ecclesiastical post in Sussex, serving the invader, possibly as one of several quislings, "threw himself in their way, representing that the expedition to England was a very serious undertaking, for the English were a most warlike people; and argued vehemently against the very few who were disposed to embark in the project of invading England. The barons, hearing this, were highly delighted, and pledged their faith to him that they would all concur in what he should say. Upon which he presented himself at their head before the Duke, and thus he addressed him: 'I am ready to follow you devotedly with all my people in this expedition.' All the great men of Normandy were thus pledged to what he promised. . . ."

Public anxiety in England was great in 1066; there was seen "a token in the heavens, as no man ever before saw. Some men said that it was cometa, the star which some men called the haired star. . . . And soon after came in Tosti the earl from beyond the sea into the Isle of Wight, with so great a fleet as he might procure; and there they yielded him as well money as food. And King Harold, his brother, gathered so great a ship force, and also a land-force, as no king here in the land had before done; because it was made known to him that William the Bastard would come hither and win this land; all as it afterwards happened."

Harold Victim of a Trick

These celestial portents were reflections and projections of public concern and uncertainty in England. Still more doubt perturbed souls when Pope Alexander II, accepting William's "propaganda" as we might call it, excommunicated Harold and freed all the English from their oath to him, as a perjurer.

There was chicanery on both sides involved in this perjury. William had made unscrupulous use of a visit of Harold to

Normandy to exact from him an oath acknowledging William as heir-apparent to the English crown, upon the death of Edward the Confessor. Had Harold not complied the chances were that he would have been imprisoned, at least; probably killed. William, unknown to Harold, Creasy narrates, "caused all the bones and relics of saints, that were preserved in the Norman monasteries and churches, to be collected in a chest which . . . was covered with a cloth of gold. On the chest of relics, which were thus concealed, was laid a missal." Upon the missal, William insisted, Harold must confirm his oath, which he did, unaware that the chest of relics was beneath. Then, says a Norman chronicler who was present, "the duke made him stand close to the chest and took off the pall that had covered it, and showed Harold upon what holy relics he had sworn; and Harold was sorely alarmed at the sight."

Some of the English came to join William, the one who had the backing of the Church, as the first and oldest Internationale.[9] England was disunited in face of two foreign foes, north and south. Whether these foes had concerted their measures seems uncertain, although the timing of their attacks would lead us to believe it. The first to land was Harold's own brother Tostig with his ally the king of Norway. Harold slew them both, at Stamford Bridge in the north, on the Ouse, pursuing the Norsemen to their ships which only a few of them reached; out of 300 vessels barely 24 got away. Four days later (Sept. 25) William of Normandy and his *conquistadores*—literally and actually those who were jointly questing for new possessions—landed on the Channel coast at Hastings without opposition.

Normans Begin Invasion

Harold had been expecting the Franco-Normans for months and had assembled "a great ship host and land host" to meet them either on land, or water. The English fleet had been on patrol for three months, a full match for the Norman transport since they constituted far more of a fighting navy than William's flotilla, which had been built especially and exclusively for this enterprise, like Napoleon's invasion fleet and later Hitler's. Not until September 8 was it ordered by Harold to proceed to London to revictual and refit before returning to its cruising duties.

Trusting to help from equinoctial gales Harold himself, speeding northward to meet Tostig and the Norwegians, had also dismissed the levy of the *fyrd*, who had indicated a strong desire to gather their crops. But a south wind began to blow, favoring William who had planned for months to come over, holding his fleet ready since August 12 at the Dives estuary, and thence sailing to St. Valery on the left bank of the Somme estuary—about 60 miles from the Sussex coast—where it had been riding at anchor for fifteen days. He had clearly chosen harvest time in England for the crossing in order to simplify the problem of provisioning, always a weakness of the feudal army and in this case made infinitely more complicated by dependence upon sea transportation. It must have been difficult enough to feed the expedition during the tedious waiting period; there were murmurs that William's luck had forsaken him, and the duke had to arrange for religious processions bearing sacred relics to pray for more favorable weather and wind.

On the 28th of September, shortly before midnight, "a great lantern" was hoisted to the mast of William's own ship and all ships were ordered "to bear a light" and follow his craft. Towards nine o'clock in the morning the Norman fleet landed at Pevensey, a Roman settlement with a small harbor, instead of Winchelsea, as intended, where the monks of Fécamp had large holdings. Only two small and overladen ships were lost en route. The landing was totally uncontested and without untoward incidents.

Even then a landing enterprise was felt to be of such potential impact and historical significance that the old writers were tempted to use again and again certain traditional formulae. Conviction that the significance of William's enterprise in their own time was as great as an action of Alexander, or Caesar would incline to make them eager to relate it in classic terms. Our own war reporting today suffers from this tendency toward stereotype narrative, and medieval chroniclers easily yielded to it. According to Wace, the fleet of William was destroyed, or burned after the landing. Others report him as the first, or among the first to land. His foot slipping as he touched earth, he himself, or a soldier nearby, with convenient presence of mind, cried out that this signified the seizing of English soil. Later writers have never wearied in adducing omens and portents turned to good use by the Conqueror, who as "the Bastard," called on fortune for the assistance that legimate birth, then so powerful a personal influence, could not confer upon him. Matthew Paris and, slightly varying, William of Malmesbury, relate that when varlets in their zeal to dress William for battle got some pieces of his armor wrongly placed, he himself cried out: "I accept the augury; it tells me that my duchy will be changed into a kingdom!"

In the history of military amphibious transport the Channel crossing of 1066 is outstanding. It presented in the era of chivalry the first large-scale enterprise of its kind, even if minor in size, whatever the chroniclers may say, when compared with Belisarius' Carthaginian expedition. It constituted a progress beyond the Viking days insofar as the transports, in addition to the soldiers, carried the "tanks" of knighthood, the horse and heavy armor of the nobles. From the process of reduction usually necessary to apply in dealing with military statistics of the Middle Ages—about the lowest figures given by chroniclers for William's expedition is 60,000 men—there would seem reliably to emerge an army estimated variously at from 3,000 to 4,000 mounted knights with their own chargers and 8,000 or 9,000 foot soldiers of various descriptions (Oman I, 158); or at least 5,000 men altogether, of whom 2,000 were knights *(Cambridge Mediaeval History* V, 500). Judging from the kind of transports available, and the recorded number of them, some 700 as seems not unlikely, the lower total of knights would appear to be more accurate.

William Conquers at Hastings

Immediately after landing the Normans proceeded eastward from Pevensey to the somewhat larger town of Hastings where, in proper Viking fashion, they at once constructed a timber castle, or block house as a beachhead. Next they overran and ravaged Kent and Sussex, Harold's home shires. Reports of these depredations reaching him at York in the north on October 1 and 2, stung Harold to quick action. He decided to meet William's army at once, even before his own forces, sorely in need of replacements for the heavy losses suffered at Stamford Bridge, had been fully collected. He marched southward to London at a speed of nearly thirty miles a day. There he might profitably have waited for more levies to strengthen, perhaps double, or triple, his army and adopted a policy of wearing down the invaders by drawing them further inland and forcing them to disperse for foraging.

Instead Harold determined to meet them promptly where they were, and with his small and wearied forces. On October 14 the battle of Hastings, some six miles from the sea, was fought and won by the invading force. It was a victory of amphibiously-transported armored and horsed force over infantry, the former having superiority in projectile-throwing power and being, moreover, the more highly organized, firmly commanded by the hand of one of those nearly absolutistic governors occasionally produced by the feudal era. The defeated side was the more primitive, anarchic force under a leader who could not rely on the cohesion provided by truly national support from a thoroughly alarmed and awakened patriotic people.

Even a skilled and resolute soldier like William could not put complete stop to invasions of England, coming from Scotland, or over the seas from Ireland, Scandinavia, or even Normandy. While these incursions did not seriously endanger the newly established Norman state with its institutions largely novel to England, they demonstrated that the fresh masters were not able for a long time to organize an armed force in keeping with the true requirements of the defense of England, always incomplete without a fleet, even though the building of stone castles was begun almost at once by the Normans.

Less than a year after Hastings followers of the old British regime would come over from Ireland, ravage the English countryside and besiege towns like Bristol. In 1069 a Danish fleet of 240 ships sailed into the Humber, and in cooperation with English foes of the Normans slew hundreds of the latter. William went after their roving bands, but was unable to strike at their fleet which "lay all winter in the Humber where the king could not get at them." So helpless were the Normans in the face of these Danish fleets that William, at the report of the imminence of one such invasion which did not materialize, felt constrained to adopt a scorched earth policy and cause "the country near the sea to be laid waste, that if his enemies landed they might the less readily find any plunder."

When the Danish depredations ended at last, pirates from France and Flanders would frequently land and rob monasteries as the unfortified seats of rural wealth. In their quarrels for dominion, William's successors and the contestants for his crown and possessions, as well as other lands, made numerous crossings of the Channel in either direction. Medieval England was less an island than ever and the Channel more like a river that fails to bar enemy progress.

NOTES, CHAPTER 13

[1] *A History of All the Real and Threatened Invasions of England* (Windsor, 1798, 2d ed.), written with an eye almost too much bent on parallels between that day and earlier events, drew this parallel between the 8th and 9th centuries and the days of the younger Pitt: "The troubles and dissensions in the kingdom of Northumberland gave the Danes frequent opportunities of plundering that coast. Whenever they arrived, they were sure to be joined by the ex-party. The minor faction made no scruple to join with the common enemy, in order to gain the ascendancy; and assisted in the destruction of their country, for the sake of enjoying authority." (p. 33.)

[2] Mon. Germ. Epist. V, no. 40.

[3] In the battle of Hafursfjord (872)

> They (the ships) carried a host of warriors
> With white shields
> And spears from the Westlands (Britain and Ireland)
> And Welsh (French) wrought swords.

History of Harald Harfagr, c. 18 in *Heimskringla Saga.*

[4] Noirmoutier, in the Bay of Bourgneuf, in the Vendée, was originally an abbey like Lindisfarne and pillaged by the Normans similarly. In 1676 the Dutch captured the island which again played a role in the English-supported rebellion of the Vendeans. Rhé also figures in the history of landings. The Normans settled down on it in 843, "fetched from the continent the materials for the building of houses and made arrangements for the winter as if they intended a settlement for eternal times." *Annales Bertiniani* ad 843.

[5] *Annales Fuldenses* ad 891.

[6] Battle at Steinmore, 950. *Heimskringla Saga. The History of Hakon the Good,* c. 4.

[7] *Annales Einhardi* A.D. 782, 809 and 810. A Danish king in 808 determined to protect the frontiers of his kingdom against Saxony by a wall, running from the Baltic to the North Sea shore, with only one gate through which wagons and horsemen could advance and retire—ibd. A.D. 808—the beginning of the later Danewirk.

[8] C. F. Keary: *The Vikings in Western Europe.* 1891. 143.

[9] "The Pope was an international power much more truly than the Emperor. He controlled an organization through which he could exert influence upon every country from within." *Cambridge Mediaeval History* V, 268.

Chapter 14

DESCENTS OF SARACEN PIRATES

SEA POWER IN THE MEDITERRANEAN had become decadent and neglected by the time the Muslim arrived on its shores. The autocrator of that sea still had his seat in Byzantium, as in the days of Belisarius. His position was challenged by the Arabs after they changed from continental warfare to littoral war and by the time Constantine Porphyrogenitus (945-59) assumed that proud title they had already besieged Constantinople several times by land and sea. Sailing up the Hellespont without hindrance, as early as 673, they landed on the European shore of the Marmora Sea a few miles from the capital and began a siege that lasted six years. However, the defense of the older civilization proved superior to the attack, with Greek fire as a valuable weapon in primeval chemical warfare. During the five succeeding winters that the siege endured the Arabs retired to Cyzicus at the entrance of the Dardanelles for the stormy season, much as did the Vikings off the shores of western Europe. There the Arabs assembled and exchanged stores and loot and prepared plans for repeated, but fruitless further attacks, all of which ended in a treaty favorable to the Greeks.

The Muslim siege of 717-8 proved no more successful. After having overcome the Greeks in Asia Minor they crossed the Hellespont at Abydos, where Xerxes once had reviewed his troops before the invasion of Greece, and began to besiege Constantinople from the land side, being soon followed by the fleets of Syria and Egypt which completed the ring of investment around the city. But the fleet was destroyed by Greek fire and the besieging army by a winter too severe for the sons of the desert physically to withstand.

The moral effect of Greek fire, whether set adrift and igniting on the water, or employed by the help of flame throwers, was intensified by its long continued successful use. Such monopoly of a war weapon is always demoralizing. "The Greeks have a fire resembling the lightning from heaven, and when they threw it at us they burned us; for this reason we could not overcome them," as Russians explained the defeat suffered by them outside Constantinople in 941. Greek fire for many years remained an invention that was neither communicated to, nor discovered by, the enemies of the country where it originated. The Occident never knew its composition and the Muslim did not until the end of the 12th century.

The chemical nature of Greek fire is not definitely known. Apparently it contained sulphur and other easily ignitable materials, together with some other substance, like lime, which reacted to water and generated sufficient heat to kindle the other elements.

Early Fifth Columnists

The first important landing of the Muslim in western Europe, that of Tarik at Gibraltar in 711, was effected largely in bottoms of allies whom they had found within the Gothic kingdom of Spain and who hoped to send them back once they had helped them to remove the king. Only rarely have fifth columnists transported their own country's enemies, and probably never have their activities had such fatal consequences. Still, to mark the desperate resolution of the Arabs, or the dire consequences of their arrival, some historians have had them burning the ships that were not theirs behind themselves and make Tarik say before the decisive battle of Xeres (traditionally so-called, actually at La Janda): "The enemy is before you, the sea behind; whither would you fly? Follow your leader, who is resolved either to lose his life, or trample on the prostrate Roman king."

The Moors gained their first foothold on the rock which only much later was to be held by sea power. As peremptory in a military way to the unsuspecting inhabitants as were the Vikings to a certain port official of Britain who inquired their intentions, the Muslims lost no time in defeating the lieutenant sent by Roderic, the last king of the Goths, to arrest and bring in the strangers—a procedure which Bismarck advised following against an army that might dare to land on German soil after 1871. It was an excellent anti-panic formula. But Spain rapidly fell victim to the shock of invasion. In a few months the whole peninsula with the exception of the Asturias had been overrun, for the Goths, due to their isolation in the southwest of Europe, had become strangers to war. And as so often recorded in the case of foreign invasion, "to disarm the Christians, superstition likewise contributed her terrors: and the subtle Arab encouraged the report of dreams, omens, and prophecies, and of the portraits of the destined conquerors of Spain, that were discovered on breaking open an apartment of the royal palace." (Gibbon).

While Constantinople resisted them for another seven hundred years, the Arabs seized a number of islands in the Mediterranean, like Cyprus, which from 644 onwards often changed hands; Crete, 823, and Sicily, 827 and after. These islands were either taken by Muslim pirates, or speedily became bases for piratical forays. Crete was conquered by Arabs coming by sea from Andalusia. When the Arab host had pillaged the island for a second time, the warriors returning to the shore found their own ships in flames, fired by their chief in order to force them to settle down at what came to be known as Candia, originally Chandax, instead of going back to probably overpopulated Spain.

Saracens Ruled Mediterannean

From their fortified camp on the Bay of Suda they soon expanded their domain over all the island. They rebuilt their fleet with the help of timber obtained from the Cretan forests, "the timber of Mount Ida were launched into the main" as Gibbon puts it, and became a thorn in the side of the Greeks for some 140 years.

> Chandax had come to be the immense capital of Saracen pirates over the whole of the Mediterranean, a gigantic den of thieves through which floated the riches of the Levant, the market of Christian slaves, with always new supplies for the purveyors of the harems of the whole Muslim world. Constantly reinforced by adventurers from every corner of Islam, the Arabs of Crete, in this impregnable place, without great risks to themselves, were the Empire's most terrible foes. Every spring, like a monstrous machine for war, Crete spewed forth its countless black-sailed craft, light and wonderfully swift, speeding away in all directions, burning cities, wiping out terror-stricken populations, disappearing with the spoils and the inhabitants of an entire town before the Imperial soldiery, always over-burdened, could arrive.
>
> One must read in the chroniclers of the 9th and 10th centuries the terrifying tale of these adventurers which were endlessly repeated in their frightful monotony. A few hours often sufficed for these admirable pirates who were of in-

comparable agility, audacity, and precision in transforming a flowering city into a smoking solitude. In vain did the detachments of the Imperial fleet cruise constantly through the Dodecanese; they always arrived too late; but a few hours later the bazaars of Chandax would be filled with an immense booty.[1]

Just as Palestine, although the common objective of Christian Crusaders, was saved to the Muslim later on in the Middle Ages by political disunity among the crusading forces, so Europe and Christianity were preserved from complete Muslim conquest through Arabian political disunity, and to a certain extent religious quarrels; dissensions beginning during the reign of Chalif Harun-al-Rashid [764-809]. Potentially, Islam grew ever more dangerous to Europe by building up its own sea power. But since this sea power was largely employed in unpolitical piracy, by small scattered groups who continually pillaged even Italian cities until the opening of the 11th century, it never became as dangerous to the West as Muslim land power.

During one of those several periods of the re-invigoration of Byzantium which we are apt to overlook in Gibbon's inordinately prolonged story of Rome's "decline and fall"—as if any decline could last that long before causing a fall!—Constantinople undertook to reassert its dominion of the known seas, the Mediterranean and the Caspian. Attacking the corsairs, who had been beaten several times on the seas, at their base, the only place where pirates are vulnerable to destruction, several large Byzantine expeditions to Crete had failed. Even if the Greeks had obtained the first condition of success against the pirates, superiority at sea, their "silent pressure of sea power," alone and not translated into military effectiveness, was as inoperative as at other times in war history. If the navy carried the expeditions to Crete safely, the expeditionary forces would fail in the island by neglecting even the most elementary precautions, such as fortifying their base camp after arrival, and could thus be destroyed by corsairs on land although the latter might have been defeated at sea.

Nicephorus Takes to the Sea

At last from Constantinople in 960 appeared a generalissimo who could cope with the Saracens on Crete as well as with the difficulties of a conjoint operation. At his own suggestion Nicephorus Phocas, later Emperor Nicephorus II (963-9), descendant of one of the great landholding families such as the Comnenus, Paleologus, Cantacuzenu, and tried in the endless fighting along the frontiers of Islam, which were "as inconstant as the desert sand," was despatched against them. If he had no special knowledge of sea warfare, he did possess the character, strength and will necessary to hold together soldiers and sailors and apply them to a common purpose. He listened also to technical advice where that seemed required, as in the choice of ships for the expedition. In more than one respect he was another Belisarius.

The fleet entrusted to him consisted of 2,000 or, as some relate, 3,000 war vessels, dromonds, or ships of the line with a crew of 300 men, 230 rowers and 70 marines, and chalands, which were either barges, as some historians think, or scouting ships, according to others; 360 transports and an army of nearly 250,000 men—a number as likely to be exaggerated as those in Occidental military statistics of the age and probably ten times too large.[2] The departure on July 1, 960 from the Golden Horn took place encouraged by optimistic prophecies of soothsayers and with the blessings of the Orthodox Church, whose priests chanted the Hodigitria, the song to the Invincible Mother, Star of the Sea, the song of the multinational forces employed by Byzantium.

Nicephorus proceeded on his 600-mile voyage by cautious stages, halting at a succession of naval stations such as Heraclea, Preconnessus, Tenedos, Mytilene, Chios, Naxos, Nio, Thera, all well provisioned for him in advance. His ships made rendezvous at Phygeles on the Asiatic coast between Chios and Samos. There he was also met by detachments of his scouting ships which had been sent ahead to make a raid on Crete, take prisoners and learn through them of conditions in the island. Hence he knew that the Saracens had heard of the approach of his expedition and that the news had thrown the population into a panic and set the commanders busy providing coastal defenses.

Without losing time, Nicephorus set sail for the northern coast of Crete, using pilots from the island of Carpathos. He found the heights along the coast as thickly held by white-burnoused Arabs on foot and on horseback as Caesar saw the cliffs of Dover lined with Britons. Still, he gave orders to land at once and in face of the enemy. Archers, slingers and shipboard artillery provided the necessary momentary missile-superiority to permit his shallow draft vessels to push close to shore.

Then the Greeks let down the side-hatches of their ships, which were the forerunners of the *huissiers* of the Crusades (from huis, door), and broad gangplanks. Like tanks landing from a modern barge, armored cavalry galloped to the strand, followed by the *gros* of the army, forming at once along the shore with the heavy infantry in front. After this, in the descriptions which are no less conventional by being medieval, the troops wasted their breath singing the Hodigitria while advancing against the Arabs posted on the hills and the hail of their missiles. They were beaten, dislodged and promptly pursued by the Byzantine cavalry.

Early Tri-elemental War

Pursuit by sea, which the oblong form of the island would invite, seems not to have been attempted; but the fleet, divided into squadrons, was put on blockade duty to hold off possible, although unlikely, reinforcements from the seafaring Muslim in Asia, Africa, Sicily and Spain. On the whole, the unity of Muslim power was too much broken to make this a very lively danger. After the final siege of Chandax had begun some emissaries did arrive from Spain, but they convinced themselves speedily that the case of the Cretan Saracens was indeed hopeless and retired. Unrelenting in his mopping up of the island and in his siege of Chandax, Nicephorus applied all the Byzantine arts and techniques of war on the sea, on the ground, underground and through the air, pouring Greek fire on the city walls from mobile high wooden towers which commanded them. The Cretan base of the pirates fell in May, 961.

It would seem fairly futile, although it has been attempted as late as 1941, after a contemporaneous era of combined operations began, to represent this reconquest of Crete as one of the best instances of what Mahan called the "silent pressure of sea power"; or insist that "although the Cretan expedition was a combined naval and military expedition, ... yet the enterprise as a whole was distinctively naval in character and was a tremendous triumph of sea power." Actually, none of the parts of the conjoint operations could have succeeded without the other. "Bi-une they are, inseparable." And Nicephorus was what the Byzantines called a "systrategos," a joint commander, a title they usually reserved for a divine figure, so rare appears to have been this phenomenon of general plus admiral. It was used for Maria Herself, who was Promachus, or foremost champion, Pallas Athena, Tyche and Nike, all in one.

The immense relief which the reconquest of Crete provided Constantinople, (where for once rich trophies and hordes of prisoners and slaves confirmed the truth of bulletins from the front and dispelled the skepticism of the city populace, fed up with over-optimistic official propaganda) netted such a triumph for Nicephorus as Eastern Rome had not granted for a long time, arranged in accordance with the chapter on the "Festival of Thanksgiving for the Renewed Averting of the Muslim Peril" in the *Book of Ceremonies.* Not much later it also won for him the crown of Byzantium. It was as emperor and autocrator of the sea that he boasted to Bishop Luitprand of Cremona, ambassador of the Western emperor Otto I, whose title the Greek would not recognize as valid: "I alone possess sea power"—*Navigantium fortitudo mihi soli est*—"Your master possesses no ships, but I am powerful at sea and might, with my fleet, burn every one of his coastal towns, should I choose." (Liutprand: *Relatio,* c. 11). Calling his bluff, Otto attacked the Byzantine possessions in Italy in 967 and considerably reduced these holdings, after a Greek expedition against Muslim-held Sicily had already failed (964-5).

Decay of Byzantine Sea Power

Before their final fall, wrought by that alliance of mercantile greed and feudal land-grabbing called the Fourth Crusade, the Byzantine Greeks were able to demonstrate by such difficult enterprises as the expedition to Crete how wrong was Gibbon's dictum that "the vices of Byzantine armies were inherent, their victories accidental." Actually, such undertakings show how precisely this statement was the reverse of the truth.[3]

While the claims of Byzantine state theorists to a thalassocratie, reaching as far as the pillars of Hercules, were much reduced by the Muslim in Spain and Africa and the Germans in Italy, who never became heirs to the old Roman dominium of the sea, they did remain masters of the eastern Mediterranean, largely by keeping alive an efficient system of naval administration, drawing for this upon the manpower and other resources of the Asiatic border provinces. Their sea bastions included a string of naval stations such as Dyrrhachium, its marine mainstay being the provincial fleets which, counting about 77 ships when the entire imperial fleet numbered 100, were constantly ready to sail. It was a Byzantine rule to retaliate for an attack on land by the Muslim by a sea attack and vice versa because they reckoned that the strength of these emirates along the frontier was never great enough to enable them to operate on both elements at the same time.

By the middle of the 11th century, however, centralistic tendencies came to prevail in the Greek empire, destroying the ever ready energies and the loyalties of the provincials. Their place was filled by a strange naval mercenarism—that of the Venetians who for a very high price, including commercial concessions, helped in, and often won for the Byzantines, sea battles from the time of Robert Guiscard [Norman adventurer, a conqueror of southern Italy, 1015-1085] on. Venetian greed and sea power at last turned against the Greeks, their old employers, and in the Fourth Crusade achieved the first conquest of Constantinople.

NOTES, CHAPTER 14

[1] Gustave Schlumberger: *Un empereur byzantin au Xe siècle. Nicéphore Phocas.* Paris, 1890. 33-4.

[2] Actually, the force must have been a picked one and hence small, including Tauro-Scythians, Varangians, the Russian Vikings or Vikingized Russians, Chazars, Franks, and Danes. On another occasion it seemed to an unsympathetic visitor from the West that Nicephorus, preparing an expedition into Anatolia, was "not concerned with their quality, but only with their number; and how dangerous a policy this is, he will learn." For the small size of Byzantine armies see *Cambridge Mediaeval History.* IV, 741.

[3] Bury, *Cambridge Mediaeval History* IV, xi.

Sources: *Cambridge Mediaeval History,* vol. IV; Gibbon's *Decline and Fall of the Roman Empire;* Schlumberger; K. Leonardt: Kaiser Nikephoros II. Halle, 1887; Luitprand of Cremona: *Historia Ottonis;* Bertha Diener: *Imperial Byzantium.* Boston, 1938; Arthur MacCartney Shepard: *The Byzantine Reconquest of Crete.* Naval Institute Proceedings, August, 1941; Carl Neumann: *Die byzantinische Marine. Ihre Verfassung und ihr Verfall.* Historische Zeitschrift. Vol. 81, 1 ff.

Chapter 15

FEUDAL WARS AND THE SEAS

"There has never been a knight on the sea."
Sombart: *Der moderne Kapitalismus.*

FEUDALISM SHOWED BUT LITTLE TALENT, or inclination to use the seas and ships in warfare. According to Froissart it was fearful to battle on the water, for there one cannot escape, or flee. The French nobility, as a French tract of 1455 put it, objected to sea service on account of "the hard life one has to live which is not very much in accord with the nobility," [1] and the storms which caused gentlemen an equally uncomfortable and undignified seasickness. The navophobia of the feudal lords contributed to the choice of land routes to the Holy Land, or when they contracted with Italians for sea transportation to insistence upon agreeable fare and quarters, while the common soldiers and pilgrims had to put up with steerage accommodations probably as bad as those on a slave ship.

The greatest medieval sea warriors were the burghers of the Italian, the Hanseatic and the Cinque Ports, the latter with a jurisdiction extending over the south coast of England, from Seaford in Sussex nearly to Margate in Kent, that is to say over the coast that was most exposed to invasion, and whence sorties to the continent could be undertaken. Besides, some French, Flemish and Spanish cities won renown in maritime combat, and the sea power of the Nordic kingdoms was in most respects city-based. Ships and men from cities fought most of the relatively rare sea battles of the Middle Ages. The feudal warrior regarded ships mainly as transportation craft for getting mounted knights and foot soldiery to land battles. When Richard the Lionhearted began a five years war on the continent in 1194, his navy carried the army over to Barfleur but subsequently took no part in the hostilities. The fleet employed on the occasion of an expedition of John Lackland into Ireland in 1210 "was a very large one, yet its only duties seem to have been those of transportation.[2]

Consequently, one may say, most of the crossing and landings were from shore-to-shore, involving a minimum of engineering organization and labor, though there are on record a few landings of the ship-to-shore kind. In the battle of Damme (May

1213) which principally took place inside a harbor, the English used their boats to attack the French fleet, "so that the affair may be regarded as an early cutting out expedition."[3] The landing of Saint Louis at Damietta, as described by Joinville, was distinctly a ship-to-shore operation, effected with the help of the small boats. This sort of operation was never easy to perform, considering the unwieldy bulk of that man-tank, the armored knight, and his horse, whether armored or not. The feudal war makers in addition had little understanding of the special character of sea battle, or sea-to-land battle, trying wherever possible, and even in impossible situations, to transfer and apply the principles and instruments of land warfare to sea combat.

The extreme in this process was perhaps reached when attempts were made to use something which combined in ship-to-land operations the functions of providing morale inspiration with the advantages of a moveable fortress, the *carroccio*. One of the features of the French preparations for a descent on England in 1386 was "the construction of a huge but portable wooden fortress, [Walsingham says: twenty feet high, 3,000 paces long, with towers at intervals, designed to protect the knights after their landing.] But the seventy-two transports conveying it, in sections, to Sluis from Brittany were dispersed by a gale, and some of them were driven into the Thames and taken. The captured sections, set up for a public show near London, seem to have excited ridicule,"[4] as well they might have done. One hesitates usually to call the mediaeval adumbrations of modern battle features "forerunners," although at times that term might be justified, as when Edward III before the battle of Sluys (June 24, 1340), sighting the enemy, sent three knights with their horses to the shore on a reconnaissance. By riding along the beach they were able to ascertain the enemy strength and disposition.[5]

Time and Topographical Factors

The understanding and use of time in war distinguishes mediaeval warfare most strongly from modern combat. Very little comprehension of the factors of time and topography was shown in the days of knighthood, except where imposed by nature, as through the influence of tides. In the year of Crecy, 1346, Edward III sailed from the Isle of Wight on July 11, reached La Hogue on the 12th where he at once landed, while the disembarkation of his troops and unloading of stores took until the 18th. The impracticability of climate would on occasion dictate to the stubborn feudal military mentality. In the second half of the 13th century, for instance, European knights under the influence of experience gained in their overseas enterprises did get around to accepting a certain amount of easening in wearing armor, after the heat had become unbearable for those encased in the heaviest steel equipment,[6] under the Palestinian sun. This might be compared with the American troop equipment for the expedition to Cuba in 1898 which, with flannel shirts etc., was the same as for service in Alaska. The first marked introduction of informality in modern military garb came at the Dardanelles when the Anzacs donned shorts for the sake of coolness.

The Holy Land was an objective that, in the days before the Bagdad Railway, every rational person would have approached by sea. But not the first, and probably the majority of the crusaders.[7] They usually insisted on taking the land route. Still, the most earnest, or most simple minded crusaders, the participants of the Children's Crusade (1212) and of the Shepherds' Crusade of the 1250's were led to believe that the miracle of the Red Sea might be repeated for them, allowing them to march dry-shod through the intervening seas.

In reality the first prerequisite of a landing in Palestine was provided by the rise of the Italian city-states like Venice, Genoa and Pisa which had humbled Saracen sea power in the Mediterranean and had thus given the Christians command of the sea and opened the sea lanes to the Frankish crusaders. These city-states, as well as the Norman state in Apulia and Sicily, had risen to power and wealth in the days of the earlier crusades of their own, by individual or joint naval and overseas enterprises against Muslim seapower, which had sacked Pisa as late as 1011. Gradually they had cleared the waves of these corsairs, freeing the Adriatic and the Tyrrenian (1015-17), as well as the islands of Sicily (1060-91), Sardinia (1016), Corsica and even the Balearic Islands (1013-15). The dominion of the sea in the Western Mediterranean, the Ponent, was in the hands of Christians who in 1064 raided Bona and in 1087 Mahdiyah, the Muslim capital of Tunis, and the chief harbor of the corsairs. Unfortunately, the methods of this renascence of Christian sea power were not—and could not always be, it must be admitted—

repeated in the conflict between Christianity and Islam in the Levant. Here, Islam was land-power.

Feudal mentality lacked grasp and imagination sufficient to enable it to take advantage of such openings as the Italian cities provided. Nor was it in any way gaited for such long distance enterprises as those into and within the Holy Land. Topographical knowledge was scanty; most of what the Romans had known had vanished from man's memory. Knowledge of supply systems was next to nothing. Strict division of labor in war and close observation of specifically assigned duties were either unfamiliar, or irksome to the knightly spirit. Long term planning as opposed to warlike spontaneity, and impetuosity and opportunism was familiar only to such societies as those of the Italian states, or to the rulers of Byzantium. The vanity and inflated ego of individual noblemen gave rise to jealous quarrels, springing either from personal, or national animosities.

Crusaders Without Unified Command

While the constancy of the strategic purpose, the reconquest and defense of the Holy Land, and the unity of will seemed to favor successful coalition warfare, bound up as it was with Christian religion, most of the time the disunity among the crusaders proved far stronger than Christian unity. The clergy admonished the crusaders not to forget the object of their enterprise, yet was itself ignorant of the quality and requisites of warfare which it had finally been persuaded to bless. Hence, no priest of high, or low degree was in a position to head the coalition armies, even though he might have possessed and exercized spiritual authority over them.

Thus, while unity of purpose seemed always firmly established, although not to the inclusion of operational objectives, the unity of command was practically never established. It is true that a man like Godfrey of Bouillon (1058-1100, Duke of Lower Lorraine, leader of the First Crusade) has been portrayed as equally fit to be the light of a monastery, or the leader of an army. Actually he was merely a brave fighter and head of a small contingent of knights, but never a Christian *Weltmarschall,* a supreme commander of an overseas force. Neither a supreme commander, nor his comprehension of supreme necessities could preserve unity, however, given existing conditions. When a representative of the Templars, accustomed to fighting in the Orient, counseled caution, the Count of Artois, brother of Saint Louis, (French King, 1215-1270) charged that the Templars wanted only to prolong the war from which they derived their living. As an English chronicler summed up the results of the Third Crusade and the dissension between Richard Lion Hearted and Philip Augustus of France: "The two kings and peoples did less together than they would have done apart, and each set but light store by the other." [8]

As is apt to occur in wars of coalition, the Crusades produced a nationalism far from wholesome. "Nostris etiam erant importabiles Alemanni," a French chronicler of the crusades reported on the experiences of his countrymen with the Germans, although they found the Greeks equally unbearable.[9] *Divisio* and *propria voluntas,* division and self-will, dominated within the Crusaders' armies until the last. Before Acre, the last Cristian stronghold, in 1291 again fell to the Saracens there was no unity of command in that pent-up place: within its walls were no less than seventeen separate and distinct national quarters, "whence there sprang no small confusion" (Villani), for the Crusaders there fought under separate commands even during the siege, only to fall together before the cohesive might and determination of the infidel. The Templars had the satisfaction of holding out ten days longer than the rest in their section of the city.

Thirteenth Century LST's

The futile hardships of the first two Crusades consumed much of the energies of nobles, knights and men-at-arms long ere the Holy Land was reached by the land route, and brought the more enlightened minds among the feudal lords to realize that a sea-borne expedition was the better way of winning the Holy Land. They learned, too, that the seat of Muslim power was not really in Palestine, but in Egypt, and that they should strike there first. So far as they involved landing enterprises, the Crusades to Egypt in 1218-19 and under Saint Louis in 1249 were futile and muddling. This could not be charged to faults of the builders of the transport fleet and those who supplied it. Some Genoese ships built especially for Saint Louis were equipped with drawbridges and gangplanks for expediting embarkation and debarkation. On the so-called *taridae,* a cross between a sailing ship and a galley, with only one deck, horses were carried on deck and in other boats below deck, with doors at the sterns for entrance and exit. They were medieval forerunners of LST's. One of the largest vessels constructed for Saint Louis had no less than four smaller boats, instead of the usual three, propelled by as many as 52 oars. It carried 100

Landing Ships Wrecked Off Fortified Town. (Courtesy, G. P. Putnam's Sons.)

knights with their horses and attendants or 1,000 and more soldiers.[10]

The Crusade of 1249 against Egypt fell in the wrong season. Setting out from Cyprus late in May, it arrived at the Nile delta when the land there was saturated from the river's annual rise and the increasing heat unbearable. Political conditions were propitious, however, for the Sultan was dying and no certain successor was in prospect. The spirit and leadership of the Saracens were slack. They had a large fleet assembled in one of the mouths of the Nile, but it did not sail out to oppose the approach or the landing of the Franks who later carried the fight to it. They took a number of ships and drove the rest up the Nile.

In face of the Muslim hordes lined up along the shore, raising a noise "with their cymbals and horns fearful to listen to," as it seemed to the men in the French fleet, Saint Louis overruled the cautious war council of his barons and had heart enough to order an immediate landing, even though only a third of his people and ships had yet arrived. To delay, he said, would merely put spirit into the hearts of the foe. Besides there was no port nearby where he could conveniently wait for the rest of his forces. It was settled that the French would land the next day and battle the Saracens should they make a stand.

To judge by the eye-witness description of Joinville, there was neither plan, nor method in the business of landing. He himself was assigned a galley by the king personally in order to carry his own little force and that of a companion to the beach from a great ship which could not get close to shore. Happily, on the return from the council of war, he found that a small ship of his own big enough for eight horses, had just arrived. On the day of disembarkation, the galley assigned to Joinville was not available, whatever the King might have said the day before.

Landing and Fighting

When his party saw that they would have no galley, they let themselves drop pellmell into the ship's boat, so many of them that the boat began to sink and the sailors left it and climbed back into the ship. There was a surplus of twenty knights in the small boat, these had to return to the big ship before the sailors, in three trips, could carry them to Joinville's smaller ship with the horses on board. Here they were luckier than some other knights who essayed to drop into the tenders of the big ships as they cast off, leaving the heavy armored knights to drown.

Before Joinville set out, he forced two of his knights to swear on holy relics to forego an old feud "so that we should not land in company of their enmity." For his own landing craft he selected a small boat sufficient for himself and three of his knights, the two feudists included. As they steered to land, they overtook the barge belonging to the King's own great ship in which was Louis himself. The people in the barge cried out to them that they must land where the King's oriflamme did, which was carried in another vessel ahead of the King's. Joinville did not heed this belated order, but caused his own craft to strike the beach in the face of a great body of Saracens, "at a place where there were full 6,000 men on horseback. So soon as these saw us land, they came towards us, hotly spurring. We, when we saw them coming, fixed the points of our shields into the sand and the handles of our lances in the sand with the points set towards them. But when they were so near that they saw the lances about to enter into their bellies, they turned about and fled."

To the right of this group, a long crossbow shot's distance away, landed the galley with the oriflamme. A Saracen charged them, either because he could not restrain his horse, or in the hope that others would accompany him. He was hacked in pieces. The King himself followed the oriflamme by leaping into the sea fully armed, shield hanging to his neck and helmet on head, the water reaching his armpits, and joined his people who had to restrain him from attacking the Saracens at once, before the French had formed into battle order.

In spite of their numerical superiority the Saracens gave ground. Thrice they sent word by carrier pigeon to the Sultan that the Frankish forces had landed; but receiving no orders in return, he being ill, they thought him dead and abandoned even Damietta to the Franks, to the latter's great astonishment, leaving the draw bridges intact and the city walls in full order. For their cowardice the Sultan had fifty of his officers hanged, but that helped him less than the vacillation of the French, who sat in Damietta for six months while they divided the spoils and waited for the arrival of their companions and a better season for operations, instead of following up their initial success by advancing upon Cairo, where panic prevailed. The Franks became involved in problems of supply and engineering for which they were not in the least prepared, as they extended their campaign beyond Damietta to the river and canal system of the Nile delta.

The most satisfactory landing enterprise during the Crusades formed part of the Fourth, so misscalled, for it never reached nor seriously intended to attack, the Holy Land, or even Egypt. The Venetians told the Crusaders that they could destroy the Turkish power most easily by an assault on Cairo, to which the knightly leaders agreed, but informed "the people at large only that we were bound to go overseas," as Villehardouin puts it. He and the other negotiators of the contract between the Venetians and a group of French crusading magnates kept the objective secret from the majority of those who took the Cross for the enterprise.

The success of this expedition which ended at Constantinople, the Christian capital of the Byzantine Empire, was insured by a combination of canny burgher calculation and gift for organization coupled with knightly martial enterprise and the rapacity of both. The diverse races and social groups as well as military and naval forces in this Crusade were fairly well harmonized under a supreme commander, the blind, aged, but wily Doge, Enrico Dandolo. (1108-1205. He was principally responsible for the diversion of the Crusade from its original object.) Actually, although not formally, heading and guiding the mingled hosts, his diplomacy kept the feudal and burgher elements together, or apart, whichever suited his policy. Another factor in the outcome was the highly practical purpose which he had in view and which he dangled before the Crusaders—the looting of Byzantium.

If the numbers of the Crusaders are once reduced from the fantastic figure of several hundred thousands given by chroniclers to the more likely one of tens of thousands, one sees that the problem of transporting them by sea was quite solvable for the Italian cities, such as Venice, then "the greatest power on the seas." Originally the representatives of a group of French Crusaders had gone to Italy in 1201 to negotiate a contract for transportation to the Levant. Venice promised to build transports for 4,500 horses and 9,000 squires and ships for 4,500 knights and 20,000 sergeants of foot and to furnish food and forage over a period of nine months. The city magnificos also pledged to furnish convoys of fifty armed galleons on condition that, as long as the agreement lasted, Venice was to take half of all conquests in land, or moneys, leaving the rest to the Crusaders. For these services the Cru-

Assault of Zara by Venetians and Crusaders, 1202. (Andrea Vicentino, in Ducal Palace, Venice.)

saders were to pay two marks for each man and four marks for each horse, altogether 85,000 marks, even if the number of participants should fall below 33,500, as it actually did.

A down payment on the contract of 5,000 marks, which their envoys borrowed in Venice, was made by the French contractors, who then tried to attract as many participants as possible. But when the fleet was ready, in the summer of 1202, it was found that shipping space had been provided for at least three times as many men as the French could assemble. While the Venetians had fulfilled all their obligations, the barons had great difficulty in meeting theirs; they were 34,000 marks short, after collecting from every man what he had, because of the many who had preferred to sail from various other ports, and thereby coming to grief, as Villehardouin repeatedly emphasizes, and with good reason.

To obtain repayment for their outlay, the Venetians now proposed that in lieu of their debt the crusaders should help to reduce Zara in Dalmatia, a town recently lost by Venice to the king of Hungary. To this the majority of debtors assented, only the most fervent religious zealots insisting on passage directly to Palestine. Dandolo, the Doge, assumed the Cross himself and accompanied this expedition against Zara, which itself was a Christian city and whose knights also fought under the Cross, as some clerics persisted in reminding the Crusaders. Papal absolution later on was granted for the *fait accompli.*

Commercialism vs. Religious Zeal

Well provided with siege engines and sappers to mine towers and walls, with scaling ladders to be raised from the ships against walls, the Crusaders forced Zara to surrender in November, 1202. It was now too late in the season to proceed further east, although, before setting out, the leaders of the expedition had entered into negotiations with a pretender to the throne of Constantinople who called for help, promising great rewards. The beneficiaries of this projected siege of Christendom's second city were to be the commercially-minded Venetians and the French nobles, avid for lands and loot. The redemption of Christ's sepulchre from the heathen could wait.

As the Doge, a wise and enterprising business man, counseled, the Venetians settled into winter quarters near the harbor of Zara and the Franks close at hand, to await Easter and spring. The separation did not keep the French and the Venetians from coming to blows and smashing each other's skulls and bones in a hardly Christian spirit. Many were killed and at times the host seemed about to disintegrate. Only the cooler, more practical minds of the leaders could control the hotspurs of the rank and file who threatened to spoil their plans, now far advanced, for the conquest of Constantinople, as "affording the

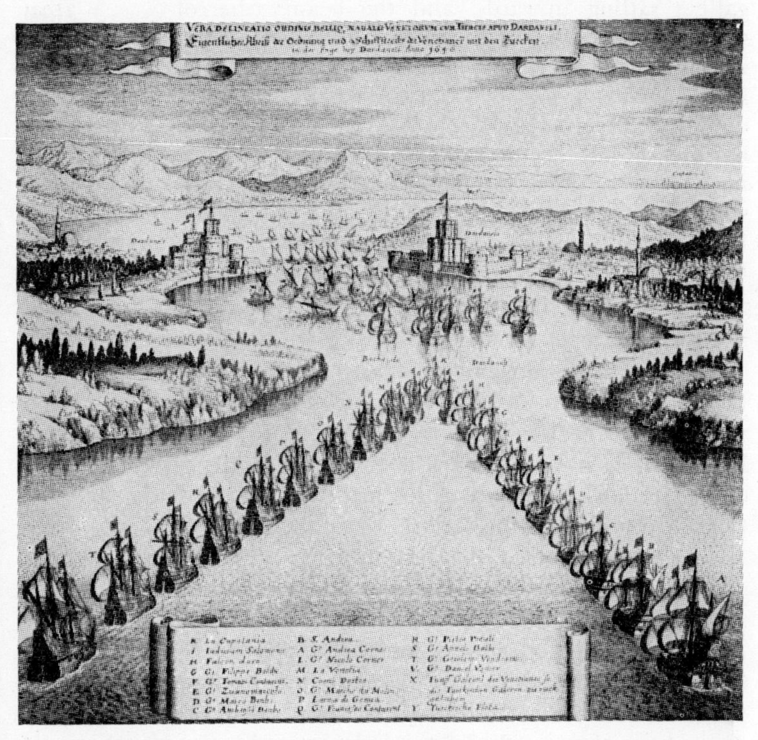

Venetian Fleet in the Dardanelles, to Attack Turks in Constantinople, 1646.

best means by which the land oversea might be recovered." Of this, however, the more pious in heart could not be persuaded. Some of them may have possessed sufficient intelligence to see clearly that the emperor in Constantinople had been marked for deposition chiefly because in the main he had been guilty of favoring Pisa, rather than Venice, in his commercial policies.

In the spring of 1203, leaving Corfu in May, the Venetians and their allies set out for Byzantium, sailed up the Dardanelles and landed without hindrance at Abydos, the narrowest point of the Hellespont where they rendezvoused. Thence they proceeded to St. Stephen's, an abbey three leagues from Constantinople. "Then might you have seen the Straits of St. George, the Bosporus, as it were, in flower with ships and galleys sailing upwards, and the beauty thereof was a great marvel to behold," said Villehardouin, the feudal landlubber, in admiration of the naval splendor. While this sight uplifted knightly hearts, the first view of the walled capital on June 24 depressed them again so much "that no man there was of such hardihood but his flesh trembled; and it was no wonder, for never was so great an enterprise undertaken by any people since the creation of the world." But then "every one looked to his arms, knowing that in time they would have to do business with them."

As one of the few who had been in these waters and lands before, the Doge proposed that the besieging forces be landed on islands in the Marmora Sea where they could be kept under better control and where supplies could be collected, after which the siege was to commence. But the lure of the mainland and its palaces proved too strong for the adventurers, so on the following day the ships made for Chalcedonia, opposite Constantinople on the Asiatic side, the nobles lodging themselves in palaces and the rest in tents. A few days later the ships and the host on land moved further up the Bosporus to Scutari while the Byzantine emperor marched his army out of the city and encamped on the opposite side to keep the Crusaders from "taking land against him by force." The latter proceeded to gather badly needed supplies and were fortunate, or perhaps this too had been calculated by the Venetians, in finding that the grain was just cut in the fields. After some futile negotiations with the Emperor, the host made a demonstration of power by sailing close to the walls of Constantinople and showing their candidate for the throne on shipboard to the Greek people on the city walls. No one dared to begin a rebellion in his favor within the city, however, and the Crusaders realized that they would have to undertake the dreaded task of launching an assault.

To make the necessary arrangements, the nobles assembled "in parliament which was held on horseback, to discuss the order of the battalions." A vanguard and a rearguard and five divisions of the main force were formed and the day fixed when "the host should embark on the ships and transports to take the land by force, and either live or die . . . Then did the bishops and clergy speak to the people, and tell them how they must confess, and make each one his testament, seeing that no one knew what might be the will of God concerning him. And this was done right willingly throughout the host, and very piously." On the appointed day the knights boarded the transports with their horses, both fully armored. The folk of lesser consequence in battle, though perhaps more agile in the manoeuvres of landing, were put on the great ships and the galleys, also armed and made ready for battle.

> The morning was fair a little after the rising of the sun; and the Emperor Alexis stood waiting for them on the other side, with great forces, and everything in order. And the trumpets sound, and every galley takes a transport in tow, so as to reach the other side more readily. None ask who shall go first, but each makes the land as soon as he can. The knights issue from the transports, and leap into the sea up to their waists, fully armed, with helmets laced, and lances in hand; and the good archers, and the good sergeants, and the good crossbowmen, each in his company, land so soon as they touch ground.

This description by Villehardouin, a leading participant, proves how little feudal warriors appreciated the tactical value of surprise and availed themselves of it, even where it is most indispensable, to our way of thinking, in warlike landings; and how much they banked on an impressive demonstration of force and intrepidity. After the Crusaders had landed in Pera, unopposed for the moment, the Greeks, having made "a goodly show of resistance," turned their backs and abandoned the shore as soon as "it came to the lowering of the lances," which here took the place of the bayonet, the sight and menace of which has always put more enemies to rout than its touch.

Thereupon the sailors opened the ports of the transports, which were in a manner mediaeval assault boats, *huissiers,* which let down like a drawbridge and which, sometimes immersed below the loading line, at sea would be caulked, "as when a cask is sunk in water, because, when the ship is on the high seas, all the said door is under water" (Joinville). The landing ramps were lowered and the horses brought forth; the knights

Capture of Constantinople from Turks by Venetians and Crusaders, 1204. (Tintoretto, in Ducal Palace, Venice.)

mounted them and drew up in battle formation. The all important objective in Pera was the tower of Galata where the chain was anchored that ran across the Golden Horn and closed the port of Constantinople. Realizing that they must take the tower and break the chain before they could more closely approach the city itself with ships, part of the assailants settled down before this fort, while the Venetians began to force the boom. The defenders of the tower made a sortie early the next morning, supported by men coming across the Golden Horn in barges, but they were driven off and the Crusaders, rushing after them, obtained possession of the tower and control of the harbor, into which the fleet moved next day. The Venetians proposed to attack Constantinople from there, putting the scaling ladders against the walls from the ships, but the Franks were unwilling to trust such an unsteady base, emphasizing that they were more accustomed to war on land. So the two parties agreed to assault separately, one from the land and the other from the water side, each using its own engines and siege methods.

The Venetians, seeking to command the city walls from the mastheads of their ships, from where they could fire over the heads of their own assault troops, made so vast tumult and noise that it seemed to Villehardouin "as if the very earth and sea were melting together." So much for the impression created by a ship-to-shore and shore-to-ship exchange of missiles even before the days of gunpowder. It would, however, appear that no Greek fire was employed on this occasion, although the Saracens with its help had foiled a closely similiar attack on a fortified place undertaken by the Pisans and others, whose siege engines were all burned up.[11]

Blind Doge Takes Lead

When the Venetian galleys did not dare come close enough to the shore, at one point, the blind Doge cried out to his people to take him on land, or he would punish them with his own hands. "The octogenarian chief, Byzantium's conquering foe," (Byron), had his galley run aground and, shamed by the old man's temerity, his men leapt on land, those from the big ships jumping first into barges and then on the shore, all following the standard of St. Mark which only a moment before had flown in front of the Doge and soon after, by what was hailed a miracle, to which many swore, fluttered from one of the towers of the fortification. After this landing, well timed with the attack of the knights from the land side, to whose succor the Doge had to come before the day was over, Constantinople was taken. The Westerners then and later installed emperors of their own grace who reimbursed everyone for the passage he had paid in Venice.

Neither Greeks, nor Latins were long satisfied with the bargain struck after the conquest which in some respects was highly onerous to the former. War which was declared "lawful and just" by the clergy in the Latin camp, was renewed and the

Fourteenth Century Battle, Between Landing Knights and Garrison of a Coast Castle. (Vienna State Library.)

Latins had to reduce Constantinople a second time. During the second siege, the Greeks made use of a strong wind, but apparently not of Greek fire, and set fire-ships in motion against the pilgrim fleet in the port. That might have resulted disastrously if the Venetians, much to the admiration of Villehardouin, had not known so well how to defend themselves at sea. Jumping into the galleys and small boats, they dragged the fire-ships into the strong currents of the Straits which carried them off. To make the parallel with other landing sagas more complete, the invaders having twice reduced a city heretofore considered impregnable and thus inflicted on the Byzantines an enduring invasion *trauma,* eventually set up an empire of their own, the Latin Empire, lasting until 1261 when a Palaeologue, supported by Genoa, put an end to it.

The Venetian policy in the Levant had colonization as its basis, with civilized Christian populations as the objects of commercial exploitation. It was much the same policy pursued, not very much later, in the north of Europe by the Hanseatic towns, using the same measures in economic and military warfare, like blockade, to which they added boycott against recalcitrant members; and overseas expeditions. The first warlike action of the Hanseatic League was directed against King Eric of Norway for having driven German merchants from Bergen and confiscated their goods; a strict blockade sufficed to bring him to terms. (The Hanseatic League was a commercial association of medieval German towns, the most important of which were Lubeck, Hamburg, Danzig and Bremen.)

After King Waldemar IV Atterdag (Atterdag, tomorrow, a nick name given him probably because of his cunctative inclinations) had taken Wisby from them, a Hanseatic fleet in 1362 sailed to assail and sack Copenhagen. After some successes by Waldemar, 77 Hanse towns, "the hens," as Waldemar called them contemptuously, united in a declaration of war (1367) and went to sea with 21 orlog and 32 smaller vessels, the orlog being a craft of a size to carry artillery and up to one hundred soldiers and twenty horses. A force of 2,000 soldiers was landed wherever the Hanse leaders pleased, plundering Zeeland, Scania and Norway, compelling the Nordic kingdoms to surrender and accept a burdensome peace (1370).

NOTES, CHAPTER 15

[1] J. Huizinga: *Herbst des Mittelalters.* Stuttgart, 1938, 146.
[2] William Laird Clowes: *The Royal Navy.* London, 1897. I, 175, 179.
[3] Ibid. I, 181.
[4] Clowes I, 299.
[5] Clowes I, 257.

[6] For details see Jähns: *Handbuch* 565-6.

[7] Crusades: First, 1066-99; Second, 1147-49; Third, 1190-92; Fourth, 1202-4; Fifth, 1218-21.

[8] *Cambridge Mediaeval History.* V, 311.

[9] Ibid. V, 331.

[10] E. H. Byrne: *Genoese Shipping in the 12th and 13th Centuries.* Cambridge, Mass., 1930; E. Heyck: *Genua und seine Marine im Zeitalter der Kreuzzüge.* Innsbruck, 1886.

[11] *Itinerarium Regis Anglorum Richardi et aliarum in terram Hierosolymorum.*

Sources: *Villehardouin's Chronicle of the Fourth Crusade and the Conquest of Constantinople;* Gibbon, LX.

Chapter 16

MEDIAEVAL ENGLAND—INVADED AND INVADER

WILLIAM THE CONQUEROR'S invasion was called "the last great enterprise of the kind, which during the course of nearly seven hundred years, has fully succeeded in Europe," by the author of a *History of All the Real and Threatened Invasions of England* (2nd ed. 1798. P. 69), a compilation far from complete and which leaves out of consideration the crossing of the Turk to Europe. Actually, England was never so isolated that invasions and raids against her, or by her, were not from time to time attempted, or contemplated.

Some of the landings, usually small, we might term political, like those of Duke Robert of Normandy who in 1088 interfering adversely to his brother, William II, in the latter's domestic troubles, sent a body of troops to the support of his own friends and meanwhile prepared to go over himself with a larger force. "But the English who guarded the sea, attacked the advanced force, and immense numbers of them were either put to the sword or drowned."[1] Robert landed again in 1101 at Portsmouth, after part of the English fleet sent to prevent him from doing so had joined his expedition. He claimed the crown from his younger brother, Henry I, but was bought off before the rebel element among the English baronage could be roused. The perennial growth of such baronial factions was a constant encouragement to most ensuing invasions; that of the Queen-Empress Matilda in the winter of 1139-40, when she nearly succeeded in obtaining the throne; or that of Henry Bolingbroke, later Henry IV, who landed in Yorkshire from exile in July, 1399, while Richard was absent in Ireland; or those of the Earl of Warwick, the Kingmaker, in June, 1460 and in September, 1470. This list, by no means complete, might be extended by adding the landings from Holland at Torbay in November, 1688, of William III who could not wait for a better metereological season since the political ambient in Britain was so favorable, and influential Englishmen had asked him "to bring over an army and secure the infringed liberties of England," jeopardized by James II. Finally, there were the various attempts of the Stuarts.

Since these political landings were implementations of civil war they were especially frequent during the War of the Roses. This strife absorbed the energies of the English at home for nearly fifty years, keeping them away from the continent which became a camp for the opposition party that happened to be exiled at a given moment, and which often was connected by marriage, or political alliances with one of the ruling houses in France, Burgundy, or elsewhere. The main strength of exiles usually lies in their affiliations with elements at home, either attached to them personally, or merely dissatisfied with existing regimes. Their spiritual beachheads were in the hearts and ambitions of Englishmen.

When the exiled Warwick returned in June, 1460, he landed in Kent with a small force that suddenly became a martial York avalanche, sweeping away the Lancastrians within a fortnight. Edward IV, having broken with Warwick who drove him out of England in 1470, came back the next year. With funds provided by his brother-in-law of Burgundy, he had raised a handful of men whom he landed at Ravenspur near Hull on March 14. Discontent again restored the crown to him and gave him London and victories at Barnet and Tewksbury. All these former exiles took great care to insist on and exploit their legitimacy of birth and right to the crown, however spurious or slight such claims might be. There was always much drinking to "the king over the water" among English political partisans

After his initial failure in the autumn of 1483, the Duke of Richmond, later Henry VII, the first Tudor king, whose rights to the succession were sufficiently flimsy, repeated his attempt and landed with a force of exiles and 1,000 mercenaries, borrowed from the regent of France, at Milford Haven on August 7, 1485. Facing him and the many who had joined Henry at Bosworth, August 22, 1485, Richard III was beaten by fifth columnists, by the treason of those who left him even after the battle was joined.

Germans Helped Irish Barons

In the struggle of English factions Scotland and Ireland came to provide in an increasing measure the beachheads for the restoration attempts of English exiles and pretenders. Both Scots and Irish disliked for obvious reasons a strong English government. Irish barons welcomed a Yorkist band of exiles and 2,000 German mercenaries landing among them in May,

1487, and helped to crown a spurious Yorkist heir. Strengthened by some 4-5,000 Irish, these Yorkists a few weeks later landed in Lancashire, hoping for adherents, but few had rallied to them when they were beaten by Edward at Stoke, July 17, 1487.

The Yorkists with another pretender, Perkin Warbeck, repeated the attempt in July, 1495, when 2,000 exiles and mercenaries set out from Flanders; repulsed by shire levies in an effort to land at Deal, they sailed for Ireland where Henry VII had been preparing against them. Frustrated there, they proceeded to Scotland whose king, in the following year, tried in vain to carry the White Rose into England over the border. Warbeck's endeavor to make use of dangerous local discontent by appearing in Devonshire in 1487 proved no more successful, and eventually he was captured and hanged.

No English navy was available to stop the stealthy movements of the exiles, in any case hard to control, or to protect the coasts from forays of French ships which sacked places like Sandwich and Fowey and elsewhere during the Lancastrian regime. And no navy could have halted the infiltration of continental money by which English kings with absolutist tendencies like Edward IV began to emancipate themselves from parliamentary control. By the Treaty of Picquigny (1475) this Edward, having allied himself with Charles the Bold in a war against Louis XI, sold his neutrality for 75,000 gold crowns in cash and an annual pension of 50,000 crowns, a deal duplicated by Henry VII in the Treaty of Etaples (1492) when the English again in what their continental allies called perfidy quitted a continental war. While the English thus shared deeply in the practice and progress of diplomacy, that connected with warfare at the same time largely by-passed them.

There was a great deal of ferrying to and from the wars under the Norman and Plantagenet kings, between England and France and England and Ireland; and later between France and Scotland and Ireland, France's ready allies in the British Isles. Ireland was most often invaded; it felt this so strongly that the sagas about its original peoples, the romantic expression of its invasion trauma, include the deeds of not merely one set of founding warriors, but wave after wave of arrivals from distant shores. Internal dissensions there practically always made impossible the first political condition for resisting invasion, unity under one ruler, or other politico-military head, a consideration which still underlies Irish politics in our days when Republican Catholic Ireland insists on incorporating Ulster, the symbol of all past invasions from which it seeks to be rid and resented the landing of American forces in Northern Ireland in 1942 as endorsing and giving support to the existence of Ulster as an appanage of the Sassenach crown. In military history most landings in Ireland seem devoid of interest (see however Oman I, 407).

Feudal Combined Operations

In only a few of the many invasions of England by Scots and of Scotland by English armies were ships and landings from ships a feature, much less often that one might expect in view of the great length of the coast line of either domain as compared with the short land border. It was an exception for William the Conqueror to lead an army into Scotland in 1072 both by sea and land, or for Edward I to move into Scotland against John Baliol in March, 1296, taking along a fleet of 33 ships to assist him in the siege of Berwick. The technical details of combined operations were not always settled in clear detail, naturally enough, by the feudal makers of war. When on March 30 the commander of the ships, misinterpreting certain moves of the King's land army, sailed into the harbor to participate in the assault, which he thought had begun, he was beaten back by the defenders who took, or burnt a number of his ships. Then Edward, seeing the smoke of his burning vessels, proceeded to the attack, seized the town and slaughtered the whole male population.[2]

The innumerable wars of the English and the Welsh also were conducted over the border and only in the final stages of the conquest did the English make good use of the fact that "Wales could easily be sailed around on the sea side and taken in the rear by landed troops while it was being threatened in front by the main army." And English troops were brought into Wales not only from west coast harbors, but from as far away as the Cinque Ports of the Channel and even Gascony.[3] During the final campaign which ended with the death of Llewellyn, the English while besieging Snowdon castle, constructed a pontoon, or boat bridge across Menai Strait from Anglesey to the mainland. This enterprise was not without its setbacks for early in November, 1282, the Welsh drove the besiegers back to their boat bridge, sinking a number of the craft and drowning many knights and squires and 200 soldiers.[4]

Departmentalization of government and politics, including war, is modern. In consequence of this division of functions the question has been raised whether home or foreign affairs should hold a primary rôle in state affairs. Medieval policies were far more unified. The Church and feudalism were inter- or supranational, loyalties although a much proclaimed virtue were more often honored in the breach than in the performance and wars were consequently to a large extent intra-national, that is to say: many a foreign war was also fought at home and leaders of internal dissensions manifested ignorance of, or lack of respect for the (fairly) modern principle of non-intervention. Once feudal society has been established, these considerations robbed military landings of some of their political shock character.

> For how can tyrants safely govern home,
> When abroad they purchase great alliance?
> (Shakespeare: *King Henry VI)*

Landings in feudal society, which could rely on the help of friends in an invaded country, were far easier to effect than against thoroughly united nations such as we know in modern times, when all or most memberships in so-called internationales appear dormant in wartime. While it is not necessary to generalize, it would seem certain that during that age there were more warriors, supposed to be fighting on one side, who could be bought by the other. William II, Rufus, through intrigues and bribes induced the king of France, allied against him with his brother Duke Robert of Normandy "to abandon the enterprise, and thus the whole army dispersed in a cloud of darkness which money had raised." [5]

Defeat Bred Magna Carta

The Magna Carta, 1215, was a recoil of a continental defeat, that of Bouvines, July, 1214, suffered by the reckless John Lackland who had entered upon a coalition war against Philip Augustus of France. It was a war such as might be schemed, but hardly effectuated by feudal noblemen, constitutionally unable to "be on time," which is the first condition for the success of aggressive coalition warfare. Medieval warfare had few books to consult and certainly, unlike medieval religion, possessed no "book of hours." Bovines in turn was the reaction since the 1190's to old plans of Philip to invade England. He had even negotiated with a later Canute of Denmark to obtain help on the seas in return for recognition of traditional claims to the English throne dating from the great Canute. At one time all necessary, including technical, preparations had been made, the latter including the military aid of conspirators in England, as well as Papal permission to proceed against the then excommunicated John.

In 1213 Philip assembled his invasion fleet of some 1,700 vessels which was to start from Gravelines on the Franco-Flemish border. But when the king himself arrived to head the expedition, the Papal legate forbade him to proceed for John had now decided to listen to reason and subjected himself to the Papal demands. The Count of Flanders, Philip's liegeman but also in John's pay, declined to share in the enterprise, whereupon Philip turned against him, moved his ships into the Zwyn, and the harbor of Damme and Bruges, and took the Flemish cities of Bruges and Ypres. While he was besieging Ghent, the English requisitioned all ships belonging to John's subjects, which were found to be far more numerous than he could use, or at least pay for, and attacked his fleet in the Zwyn, May 30, where eighty of the larger vessels were captured or fired.

A landing was effected the following day when the Count of Flanders openly joined the English. Philip hastened from Bruges, saved Damme, a rich trading place at the time, and drove away the invaders who retired to the island of Walcheren, suffering heavy losses of battle-slain and drowned men. Finding it impossible to keep the rest of his fleet from falling into the hands of his enemies, Philip had it destroyed, abandoned his plans for invading England and made others which culminated in the battle of Bouvines.[6]

In order to defend against John's perfidy their liberties newly won at Runnymede, the king being aided by the Papal threat of excommunication against the observers of the Great Charter, the English barons, as most magnates of the time would have done, looked abroad for help. Their military position in the struggle with John was nearly hopeless, the gentry of England proving once more that they were more apt to advance, or preserve, constitutional liberties than to defend, or enforce, their progress by war. Using foreign mercenaries and foreign leadership, John had gained the upper hand over the constitutionalists, the barons and the city of London. From there, waterways to France being kept open, they negotiated with Philip Augustus, offering to his son Louis the English crown in return for assistance.

Sea Engulfs Army and Treasure

Like those modern putative non-interventionists in the late Spanish civil war, the French at first sent several contingents of knights *sub rosa* to help the barons during the winter of 1215-16. Having eluded a Papal interdict, Louis with a fleet of 1700 vessels late in May, 1216, landed in Kent in considerable force.

The fleet which John, with Dover as his headquarters, had gathered from his subjects in order to protect the invasion coast, had been dispersed by a storm, although later on it did great damage to French ships in French ports. Without its protection John thought it advisable not to face the invaders on the beaches, but to retire inland with his mercenary army, leaving Dover, "the key of England," to be defended by one of his henchmen against a long siege. After this uncontested landing the war dragged on without decision, or events of interest, except a final amphibious enterprise by John who insisted on crossing the Wash from King's Lynn without waiting for low tide, and thereby lost part of his army and all of his baggage train and treasure in the sea.[7]

After the death of John Lackland, who survived this experience by scarcely a week, English nationalism, legitimacy and papal interdict, all great psychological forces, plus a military defeat at the battle of Lincoln, June 19, 1217, made it harder and harder for the invader to maintain his hold. His rearward communications, always precarious, due to the loyalty of the Cinque Ports, were seriously endangered by an English sea victory in the Channel off Sandwich. After that Louis allowed himself to be eased out of the country by papal and other diplomacy and a substantial reparation, 10,000 marks, to indemnify him for invasion costs.

Although both the main participants of the Hundred Years' War and the wars preceding it were adjacent to the same waters, these wars bore much less of a littoral character than might have been expected. For one reason, England had always a *pied-à-terre* on the continent, either in Normandy, Guienne, or elsewhere. That kept the number of contested landings small. Besides, the increased weight of heavy armor made feudal armies less fit than ever to effect landings against opposition, while it is highly dubious whether English archers would have been useful when firing from and across coastal waters in the direction of defenders of the coast.

Henry III, in May, 1230, after requisitioning several hundred ships, landed at St. Malo in the Duchy of Brittany in agreement with the Duke in order to regain Poitou which had been lost to the French. But he achieved nothing in a land of highly divided loyalties, such as Poitou was, above all other medieval territories, and returned home in November, greeted by the resentment of the English who kept him short of supplies, hated all his foreign connections and relatives and cared little for the distant continental fiefs, and who, in the modern sense of the term, were thorough isolationists.

When the king renewed his attack in 1242, landing at Royan on the right bank of the Garonne at its mouth, he had with him only a small force of English, comprising seven earls and 300 knights. If these were intended to be the mainstay against a force largely composed of continentals from Guienne, they proved weak enough, while the leadership of Henry III was worthy of an imbecile. He retired in dismay at the sight of the French camp at Taillebourg, which looked to the English like "a large and populous city," and although only a small force of French crossed the river Charente, despite that the English held the only available bridge, he promptly sued for an armistice which he obtained from St. Louis before the battle had even begun.[8] These vain expeditions quelled the English ambition for overseas enterprise for a century and kept kings like Edward II, ingloriously beaten at Bannockburn, at, or near home, a politico-strategic situation which led Froissart (II, 17) to write: "England is the one country in all the world which is best protected. If it were otherwise, they could not live, and would not know how to, and it behoves the king who is their master that he regulates himself according to them and inclines himself before much of their violences; and if he does the contrary, ill will he fare."

Futile Hundred Years' War

The Hundred Years' War, or more properly wars, brought a few technical-tactical improvements, but as a war of movement towards objectives it proved aimless and almost preposterous militarily. The great battles of Crecy, Poitiers and Agincourt were hardly more than successful attempts of the English to shoot and hew their way out of bad positions into which they had marched themselves while ranging over part of the regions of what is now France. In such a war landings could have no great strategic significance, nor tactical instructiveness. In practically all cases they were uncontested. Warring English factions could often disembark on their own territory in the south of France, or in that of allies such as Normandy and Flanders, and given the state of coastal defenses in general, raids were nearly always and everywhere possible.

The long wars of the Hundred Years were opened by the English a few days after a formal letter of defiance, or a declaration of war had been delivered in Paris—one feature of feudal warfare which the modern Japanese, often boasting of their chivalry, failed to include in their neo-feudal code of warfare—by raiding the island of Cadzand close to the Flemish coast, November 10, 1337. Flemish crossbowmen (arbalestiers) were awaiting them, drawn up along the quais and dykes of Cadzand, and did their best, but the English fleet sailed straight into the harbor, clearing the shore with showers of arrows. The invaders forced their way to shore, the archers a little ahead of the men-at-arms, and shooting smartly. The landing was effected and the ensuing combat on land won, thanks to the fire superiority of the long bow over the crossbow.[9] On November 20 the English force was back in England from a shore-to-shore expedition.

French privateers had already ravaged the south coast of England before the declaration of war, a way of waging the struggle that was at best semi-official. France now in retaliation sent privateers to pillage and burn Southampton and Plymouth, creating such a panic that stakes were driven into the bottom of the Thames to keep them from sailing up to London and that "all sorts of people were roused and approached on horseback as fast as they could," as Froissart described the not very systematic mobilization of coastal defences that, of course, came too late against such raiders.[10] Neither were these the last inroads of the Picard and other privateers-men in England. Their raids of 1359-60 and the stirrings on the part of the Scots, in addition to a threat of famine in the

English camp in France, moved the latter to sign the first peace in the Hundred Years War, that of Bretigny, May, 1360. Command of the sea had been held by the English for twenty years, ever since the sea battle of Sluys (June, 1340), where a French armada intended for the invasion of England had been beaten, giving them thenceforth nearly unbroken dominion of the waters around their isles; but sea power alone neither then, nor later was sufficient to win continental war for Britain.

EDWARD III'S GREAT INVASION

In full control of the Channel, the English staged their first great military invasion of the continent in July, 1346, when King Edward III landed with an army of some 20,000 men near Saint-Vaast de la Hogue in Normandy. This was achieved thanks to the local superiority of the English, their archers and their well-regulated manner of fighting. As if to show that the king followed the long line of landing conquerors with Caesar and William the Bastard, at least in our tradition, Froissart (IV, 386-7) has Edward land first of all and stumble. His knights begged him on that account to return on board ship and not participate in the fight that day. But the king refused, insisting that on the contrary it was a favorable omen, "for the earth desires me and knows that I am her natural master. That is why she has drawn my blood. Forward, then, in the name of God and Saint George and at our enemies!"

Favorably impressed by this answer of the king, the men of his entourage said to one another that he understood very well "how to comfort himself." After some aimless marching the victory of Crecy followed within a month and then, much harder to win, the capture of Calais, the base for troublesome privateers, which was to remain English for two centuries, peopled largely by Englishmen and thus made a safe pied-à-terre and landing port for them much nearer their own coasts than Bordeaux, or Bayonne.

The Hundred Years' War was a spasmodic affair, ceasing and commencing from time to time, always much under the influence of factors like war financing and taxation, internal dissensions both in France and England, epidemic diseases such as the Black Death in England in 1348-50 which took toll of perhaps one-third of English manpower employable in war and work; early nationalism which found its military expression in the campaigns of Joan of Arc who bade the English "to go away, in God's name, to your own countries." Military victory, much less decisive in its influence in this war than the above mentioned factors, since victories as a rule are of less importance the longer a war lasts, veered from side to side with control of the sea. A combined French-Spanish fleet in 1372 destroyed off La Rochelle an English squadron with reinforcements for the south of France, made transport to the Garonne unsafe and several times ravaged the coasts of southern England.

English holdings were for a time reduced to Calais and the coastal strip between Bordeaux and Bayonne, precariously held during a truce from 1375-77, as well as afterwards when a somewhat more lasting peace was concluded in 1395. English dominion of the sea seemed altogether to vanish when the French fleet in 1377 landed in Sussex, burning Rye and Hastings and defeating the local levies. A panic swept England, towns were refortified and the crown jewels pawned to raise funds.

English retaliation took the form of another raid through France, starting and ending in Calais, achieving little against the French who no longer offered resistance in open battle, but retired behind walls of cities and castles. Such an expedition was wasteful and only added to the discontent at home, leading to the peasants' revolt of 1381. Utilizing opportunity offered by English domestic strife, the French on occasions reopened the war, as in 1403, then sending a squadron to burn Plymouth, or providing support to a Welsh guerilla chief. Disturbances in France would force them in turn to abandon foreign war, leaving Calais, Bordeaux and Bayonne still in the hands of the English and permitting them once more to intervene in the fights of the French factions.

Swans A Happy Augury

In August, 1415, King Henry V, having rewon command of the Channel, sailed with a well-equipped, although moderately sized, force on the 8th of the month, aboard a fleet assembled at Southampton and Plymouth.

> After having passed the isle of Wight, swans were seen swimming in the midst of the fleet, which in the opinion of all were said to be happy auspices of the undertaking. On the next day, about the fifth hour after noon, the king entered the mouth of the Seine, and anchored before a

> place about three miles from Harfleur where he proposed landing; and immediately a banner was displayed as a signal for the captains to attend a council; and they having assembled in council, he issued an order throughout the fleet that no one, under pain of death, should land before him, but that the next morning they should be prepared to accompany him. This was done lest the ardor of the English should cause them, without consulting danger, to land before it was proper, disperse in search of plunder, and leave the landing of the King much exposed. And when the following day dawned, the sun shining and the morning beautiful, between the hours of six and seven, the noble knight, Sir John Holland, having been sent by his desire before daybreak, in the stillness of the night, with certain horsemen as scouts to explore the country and place, the King, with the greater part of his army, landed in small vessels, boats and skiffs, and immediately took up a position on the hill nearest Harfleur, having on one side, on the declivity of the valley, a coppice wood towards the River Seine, and on the other enclosed farms and orchards, in order to rest himself and the army until the remainder of the people, the horses, and other necessaries should be brought from the ships.[11]

After the landing had been completed Henry turned at once to the siege of Harfleur which he planned to make a second Calais. In October followed Agincourt—one of those battles into which the English had more or less aimlessly blundered and where the heavily armored French nobility (one-man-and-horse tanks of the era) were bogged down in quite unsuitable terrain and potted at long range by the English archers. In August, 1417, Henry, whose dominion of the sea was never again challenged after a Genoese fleet supporting the French had been beaten a year before, landed again with an army, this time at Trouville, forming with Deauville one of the *plages* of modern France, determined to conquer Normandy rather than reconquer the southern holdings claimed by his house in France.

To the proponents of sea power around 1900 this invasion was "perhaps the first that was attempted on scientific principles," showing by its careful preparation that Henry V "had full understanding of the importance of sea power and of the danger of making any effort of the kind in the face of a potent fleet—

fleet in being, as others called it. Instead of crossing at once, while the enemy was still undefeated, and so running the risk of having to fight an action with his huge convoy of transports in company, he first sent out a squadron to clear the way, and then, so soon as he had learned of the success of the preliminary step, passed unmolested over the path freed for him."

That the king consciously pursued such a plan seems rather more doubtful than his intention of winning a beachhead on the continent as the operational basis for the conquest of all France and his still more romantic dream of a crusade against the Saracens and the rebuilding of the walls of Jerusalem. All came to nought through his early death, and his successors, opposed by burgeoning French nationalism, were unable in purblind stubbornness to appraise effectively either their own strength, or weakness which by 1450 lost them almost the last of their holdings in northern France, where they retained only Calais (until 1588), and the Channel Islands. What remained to them in the south went the same way in 1453. Calais for the time was saved, not by English arms, but by the fact that it was surrounded by neutral Burgundian territory. Gunpowder and siege trains drove the English out of their last strongholds, siege war-craft, as represented by the art and devices of the brothers Bureau, at the time being one of the chief factors of military progress.

Preliminary Steps to Crecy

In a way this attempt on the part of Henry V to organize a vast beachhead on the continent was a confession of belated learning from past errors. Learning from, and organizing for, war were on the whole not characteristic of the workings of the feudal mind in military undertakings. But Henry seems to have been capable of recognizing such mistakes as that of Edward III who, after having landed in France in 1346, neglected to organize an operational base on the continental side of the Channel, preferring to have his fleet serve as a floating base. But when he allowed part of his crafts to return home with sick, wounded and the loot gathered in Normandy, the companies of the rest of the fleet, unable to resist the urge of homesickness, a rather common weakness of the English in those wars, joined them without permission from the king, who was left without a base and temporarily at least without communication lines.

It was this situation which seems to have motivated his march towards the friendly lines of Flanders, across the Seine and the Somme, in the direction of Dunkirk; and apparently more than anything else eventually brought him to the field of Crecy. To obtain at least one beachhead in the north of France the English then moved before Calais, retaining it and Ponthieu, at the mouth of the Somme, in the Treaty of Bretigny.

Although the wars which combined to make up the Hundred Years' conflict were so definitely littoral in character, serious fighting across the shore lines were few and comparatively unimportant.

Chivalry shunned the water as an element foreign to its nature, and knights and men after landing would more or less promptly move, or withdraw, into the interior of the invaded land. The defense put up by seaboard places like Calais, La Rochelle, or Bordeaux on the English side might be termed demonstrations of bourgeois theory and practice of war; and likewise the privateering raids on the English coast. The average knight was neither enough of a sailor, nor engineer, nor infantryman to attempt landings where he could possibly avoid it.

From a study of landings on either side of the Channel one factor stands out distinctly: Nearly all landings, or attempts, whether of a military, or of a political character, took place during seasons of favorable weather. This was due not merely to the fact that medieval war was generally, although not exclusively, restricted to summer, but to the influence of the weather over landings even more than upon other operations. Not even summer landings were uniformly successful. The last war enterprise of the Black Prince, an expedition to France in August, 1372, was foiled by contrary winds which prevented him from landing.

Much more apt to fail were attempts out of season, such as the expedition from the continent of the Duke of Richmond late in October, 1483, whose fleet with exiled Lancastrians and continental mercenaries on board was scattered by a storm that at the same time ruined the campaign prospects for his partisans within England, the swollen waters of the Severn keeping them apart, so that the whole uprising came to nought. One of the very few successful winter expeditions across the sea was that against King Stephen by Duke Henry of Normandy, later King Henry II, in the winter of 1152-53, who achieved his passage in a violent gale, historical sources disagreeing whether he brought a small, or large force. The impression made by his arrival was great and was heightened by the element of surprise, due to the fact that it was accomplished in winter, the unaccustomed season for such excursions.

Surprise Winter Landing

As Henry of Huntingdon, a partisan of the Duke, described the sensation caused by the landing:

> The kingdom was suddenly agitated by the mutterings of rumors, like a quivering bed of reeds swept by the blasts of wind. Reports, as usual, rapidly spreading, disseminated matter of joy and exultation to some, of fear and sorrow to others. But the delight of those who rejoiced at his arrival was somewhat abated by the tidings that he had so few followers, while the apprehensions of their enemies were by the same reports not a little relieved. Both parties were struck at his encountering the dangers of a tempestuous sea in mid-winter; what the one considered intrepidity, the other called rashness. But the brave young prince, of all things disliking delay, collected his adherents, both those he found and those he brought with him, and laid siege to Malmesbury Castle. . . . and presently took it by storm . . .

The *Gesta Stephani* likewise declare that the kingdom was stricken with perturbation at the arrival, rumors going around that the Duke had brought immense sums of money, to the dismay of the English royalist party, "but when it appeared certain that he had brought with him not an army, but a small body of troops; that these had no ready pay, but were to look to the future for their hire; and that he engaged in no brilliant enterprise, but was wasting his time in sloth and negligence, they took courage." Both the *Gesta* and the Archdeacon of Huntingdon agree that the army to which the authors themselves did not adhere was thrown into a panic in the ensuing fights, which would be certain proof that at least one was panicky. The conflict itself was solved by Stephen yielding the throne to Henry.

On the whole, little strategic use was made of the element of surprise in medieval warfare and essays at even tactical surprise were at times held almost in contempt. It might rather be said that often enough medieval troops through lack of co-ordination, planning and mutual exchange of intelligence surprised themselves, or one another instead of the enemy. This happened in the movements which led to the battle of Poitiers, 1356, when the two sides, so to speak, ran into each other. In keeping with this disdainful disregard of surprise, which was perhaps less maintained by the siege engineers who were largely not of knightly rank, the shock exercised by a landing on foreign shores was not systematically exploited. After Edward III had landed in France in 1346, he proceeded on a leisurely plundering expedition through Normandy, instead of pushing at once onward to Paris which was poorly garrisoned at the time. Roads then would have been more open and bridges more intact than four weeks later when he found most of the Seine and Somme bridges broken. This drove him to seek a ford across the Somme, forcing it at low tide, in face of strong enemy opposition, with his cavalry over whose heads the archers are said to have fired during the crossing, thus laying a barrage of projectiles such as a contested landing, or crossing demands.

NOTES, CHAPTER 16

[1] Henry of Huntingdon. Bohn ed. 223.

[2] Clowes: *Royal Navy,* I, 209-10; Hemingford I, 90.

[3] Oman: *Art of War in the Middle Ages* II, 74.

[4] Clowes: *Royal Navy,* I, 204, cit. Knighton, 2464.

[5] Henry of Huntingdon ad 1093. Bohn ed. 223.

[6] Roger of Wendover; Matthew Paris; Rigard: *De Gestis Ph. Aug.*

[7] St. John Hope: *The Loss of King John's Baggage Train in the Wellstream.* Archaeologia, LX (1906), 93-110; A. V. Jenkinson: *The Jewels Lost in the Wash.* History, n.s., VIII (1923), 161-8.

[8] Cambridge Mediaeval History VI, 343; Oman I, 419-420. The two descriptions do not at all tally.

[9] Froissart II, 436; Oman II, 124.

[10] Froissart II, 469 ff., 552.

[11] *History of the Battle of Agincourt,* the narrative of a priest who accompanied the expedition. Cit. by Lt. Gen. Sir Tom Bridges (ed.): *Word from England.* An Anthology. London, 1940, 186-7.

Chapter 17

MEDIAEVAL COASTAL DEFENSE; DEPARTMENTALIZATION OF WAR

IT IS THE PURPOSE OF FORTIFICATIONS to provide protection, both local and supra-local, and obstacle; the emphasis on these purposes, which are not easily disassociated, may vary from place to place and time to time. Those at river crossings may be designed to offer first an impediment to the enemy's crossing and only secondarily protection to the crossing and its hinterland. Coastal fortifications are intended to hinder or prevent hostile landings and invasion at points which nature has made most vulnerable to attack; and next to that protection of the inhabitants in the vicinity and to shipping.

On the whole it might be said that chronologically inland fortification preceded coastal fortification, either because most of the older civilizations had their genesis inland and from there expanded to the sea, as Assur, Babylon, Persia, Athens, Sparta, Rome; or the sea coasts and harbors were largely protected by nature, or because the coast lines were so long and so open that it seemed materially impossible to guard them as strongly as land frontiers. The resolution not to fortify the coasts might also be based on a more or less complete dominion of the seas which they fronted. In such cases the navy provided a mobile line of defense. "The royal navy of England hath ever been its greatest defense and ornament; it is its ancient and natural strength—the floating bulwork of our Island." [1]

Mediaeval coastal defenses owed at least half of their strength to water. A strict blockade was difficult to maintain and siege operations from the water side in addition to those by land appeared almost impossible except for such expert engineers as the Venetians proved themselves to be before Constantinople, whereas the Crusaders failed in their siege of Acre in 1189. After the piratical incursions of the Vikings and the Arabs had come to an end, coastal defenses had been much neglected. The inhabitants, or their rulers relied either on the power of their fleets, or the difficulty of landing troops and siege engines for the assault of coastal towns from the land side, with which garrison forces might effectually interfere. Defense of such places against siege proved often successful, or at least stubborn and drawn-out in the latter part of the Middle Ages as well as later. Topography and the enterprising spirit of the inhabitants (and not merely of the military garrisons) combined to make some coast towns almost impregnable.

Calais, La Rochelle, Mont Saint Michel offered most spirited resistance to the English, and the last named fortress never fell to them in the Hundred Year's War. La Rochelle was held after the massacre of St. Bartholomew by the Protestant defenders for six and a half months, forcing the attackers to raise the siege after losing more than 20,000 men; in 1628 it fell, but not until after a full year's siege directed by Richelieu himself Wallenstein sat before Stralsund in vain for eleven weeks in 1628, despite his oath to take it, "though it were chained to Heaven." The defense of Gibraltar (1779-83) might also be named and that of Kolberg, fruitlessly besieged by the Russians in 1758 and 1760, but taken by them in 1762. In 1806-7 it was one of the very few Prussian places not surrendered to the French, but saved, thanks to the energetic cooperation of the citizenry headed by Nettelbeck, a naval captain, with the regular forces. The list might continue with Sebastopol and Port Arthur.

Coastal Defenses Neglected

During the Hundred Years' War, as in most other epochs, English coastal defense was far more neglected than the French. For a rich French coastal town like Caen, which was taken by the English in 1346, to exist unwalled, was exceptional. At the same time the English relied on dominion of the broad and stormy sea moats around the isles, and Shakespeare was not as anachronistic as in some other references to warfare in his king dramas when he wrote:

> *Let us be back'd with God and with the Seas*
> *which he hath given for fence impregnable,*
> *and with their helps only defend ourselves:*
> *In them and in ourselves our safety lies.*
> (*King Henry VI,* pt. III, act IV, Sc. 1.)

When this control was lost, or seemed threatened, as was the case more than once after Sluys, an invasion scare like that of 1386 would flare up. At that time, a large French force had been assembled from September to December, definitely not the best months for a crossing into England. While it never sailed, panic obsessed the English authorities who, with a fleet stationed in the Channel, called out the shire levies in the south where the towns so far north as Oxford were ordered to repair, or rebuild their walls. Even with a superior fleet it might not always have been possible for the English to ward off an invasion armada, for should their ships be wind-bound at Portsmouth, or Plymouth, those of the French might profit by the same winds and fare westward unhindered, possibly landing on English shores. However, quarrels among princes of the blood in the French camp and a complete want of leadership by the king, or other great personages saved England before she was forced to the test. At such menaces various defense measures would be devised, including the building of a few castles near likely landing points, or the closing of harbors like Dartmouth with chaines, similar to those the Byzantines had used across the Golden Horn.

A new era in the history of coastal defense opened with the introduction of gunpowder and guns and the use of ordnance aboard ships. It would seem clear, moreover, that coastal places were bombarded by ships before ships were made targets from land. The Hanseatic Leaguers who in 1384 tried to come to a monopolistic agreement that those dangerous devices, guns, must not be manufactured by non-member cities, but who failed in this restrictive effort, are said to have used gunpowder in Danish waters about 1354.[2] This would be some quarter of a century later than the first use of gunfire in battle on land. After 1373 reference to guns, gunpowder and shot became frequent in English naval accounts. Between 1370 and 1380 French, Venetian and Spanish fleets began to employ guns in actions. The Spanish off La Rochelle in 1372 used cannon as well as other artillery, some missiles being shot from mast tops, and a severe defeat was inflicted on the relieving English fleet under Pembroke. The use of gunpowder, intermittent at first, is closely connected with the evolution of special war navies, the latter taking the place of armed merchantmen, or all-purpose ships, if one prefers, which heretofore had been hired, or impressed for war. Another important artillery development at sea was brought about by 1500 through the building of war ships that could fire broadsides.

Such artillery enabled the party proceeding against a hostile shore to establish missile superiority in the locale of the landing point, provided it could approach the shore closely enough. A fleet might also, although not too effectually with a firing range of only 400 to 600 yards, fire from the water at coastal towns and perhaps even mediaeval defense towers. It would seem, at least from English experience, that such a threat, rather than actual essays in number, combined with the other threats apprehended in England at the time of Henry VIII that France might use galleys (measurably independent of wind) brought around from the Mediterranean to the Channel as did Caesar, first led to the building of coastal fortifications provided with ordnance firing seaward. Coastal forts of this nature were placed near likely landing shores, or around the entrances to harbors and at estuaries. To such technical considerations were added the consequences of Henry's not very happy continental ventures. As these reasons were expressed by a writer not long after his death:

> King Henrie the eight, having shaken off the intollerable yoke of the Popish tyrannie, & espying that the Emperor was offended, for the divorce of Queen Katharine his wife, & that the French King had coupled the Dolphine his Sonne to the Popes Niece, etc., so that he might more justly suspect them all, then safely trust any one: determined by the aide of God to stand upon his owne gardes & defence, & therefore with all speede, & without sparing any cost, he builded Castles, platfourmes, and block-houses in all needful places of the Realme: And amongst others, fearing lest the ease, & advantage of descending on land at this part, should give occasion & hardiness to the enemies to invade him, he erected (neare together) three fortifications, which might at all times keepe & beate the landing place, that is to say, Sandowne, Dele, & Wamere [Walmer].[3]

These forts were built in considerable number along the southern coast from the Medway to Milford Haven, with the greatest Channel concentration opposite Calais and Boulogne. They were forts, even if they might be called castles, designed for exclusively military purposes, provided with garrisons, at least from time to time. In this respect, coastal fortifications contributed their part to the rise of permanent forces.

The superiority of the land-based gun over ship artillery, not often tested, was in many respects obvious. A ship could never be as well protected as a fort. Guns in a fort, even if they offered a fixed target as compared with the mobile mark of men-of-war, which often anchored for the purpose of bombarding land fortifications, could fire more steadily and with a better knowledge of distances and elevations than a fleet which was under the influence of winds and currents. Moreover, a fort with a low lying silhouette was harder to hit than a tall ship forced to draw near the shore to bring the fort under its guns.

It seems not over bold to say that of all preparations and installations for war, coastal fortifications have least often undergone the acid test of actual combat. Those constructed by Henry VIII experienced their only trial in August, 1545, when a French fleet, in temporary enjoyment of naval superiority in the Channel, had driven an English fleet into the Solent. While the French admiral was eager to follow it there, the experts, his ship captains and pilots, protested, pointing out the superiority of the shore guns over his artillery. The admiral, not an expert himself, followed their timorous play-it-safe advice and stayed out of the narrow waters, returning to France after a few ineffective raids on the Isle of Wight and the English mainland.[4] Whether advice of a more militarily enterprising nature would have resulted in a different decision, we have no means of knowing.

In any case, the incident was one of the several rapidly increasing indications that the departmentalization of war, the division of operations between army and navy, had begun in earnest, which decreased rather than promoted effective combined operations of the two services. It would appear that noncooperation became more and more in evidence as time went on, finally culminating in a dangerous separation in the middle of the 18th century.

The founding and maintenance of special sea infantry, the marines, commencing in England in 1664, was probably as much the outcome of specialization in warfare as of the determination of the navies to bring land fighting units under their own jurisdiction. It is certain that the founding of the Royal Marine Artillery took place in 1804 at the recommendation of Nelson himself, who was far from being a devotee of combined operations, because of his realization that there had been too much friction between the naval officers and the artillery officers detailed for duty aboard men-of-war. In the early period of ship artillery there was a distinct awareness of what might be achieved by cooperation of forces, at least in littoral warfare. Theoretically speaking, a fleet participating in battles close to shore can act as a prolongation of the front; it can be used to outflank the enemy with artillery, if not also with infantry to be landed from the ships, and it can provide artillery fire in the enemy's rear. In such battles sea power and land power are tactically knitted together instead of merely in connection with strategy, the latter term sometimes serving to hide a considerable estrangement between the two services.

Fleet's Services at Pinkie

The potentialities of a fleet under these circumstances, even if not fully understood at the time, were demonstrated in the battle of Pinkie, September 10, 1547. An English force under Lord Protector Somerset had moved into Scotland on land while a fleet of some size had sailed *pari passu* with it, "keeping close touch with the army and reporting daily to receive orders" (Oman), that is to say, supreme command rested with the army. The Scots were met in a position behind the river Esk and across the road from Dunbar to Edinburgh, running inland from the Firth of Forth to a marsh which protected their right wing. In order to seek shelter from the flanking fire of the English fleet, the Scots had strengthened their left, or northern wing by earth works near the beach, provided with a few guns, greatly outnumbered by those of the fleet.

This position forced the English, as they thought, into a frontal attack, to be prepared by the frontal fire of a superior artillery and the flanking fire from the ships that could approach the coast closely enough. Contrary to English expectations the Scots, mistaking the English move to the north as an indication of their intention to retire to Dunbar, undertook the offensive. But while their left wing crossed the Esk near the sea, it suddenly received the fire of the English fleet which must have kept a good lookout, salvos which did great damage, dispersing the Highland archers. This threw the rest of the wing in the way of the center and caused the beginning of the Scottish defeat.[5]

The absence of tactical cooperation between land and sea forces during the 17th and 18th centuries persisted during the period when the rise of gunnery would have made it more feasible than previously. The conclusion to be drawn from this, supported by numerous cases of friction between military and naval leaders, is not therefore to be based on truly technical grounds, but more on the professional rivalries and jealousies of the two forces. It was the fetish and practice of specialization that drove and kept them apart. Only rarely could they be brought into cooperation by a supreme commander, or by a supreme necessity such as arose at the siege of Gibraltar.

On the whole, sea power, already provided with man power of its own for both land combat and landing power by the institution of marines, was more often in a position to render help than the land forces. But only rarely did the two come into tactical contact, as at the close of the battle of La Hougue, May 19 to 23, 1692, or when a French cavalry force, making use of an exceptionally severe winter, crossed the ice and captured the frostbound Dutch fleet in the Texel during the winter of 1794-5. Then a land squadron for once took a naval squadron.

La Hougue was the first of the series of battles, continuing thereafter for more than half a century, in which the English defended the revolution of 1688 against a second Stuart landing and restoration. A force of English exiles with James II at their head and a French military contingent were assembled in the Cotentin in Normandy for the purpose of invading England, and a French fleet, to be composed of a squadron from Toulon and one from Brest, was to be concentrated to cover the Channel crossing. But the Toulon squadron was dispersed by storms and the commander of the Brest fleet who had sailed out to join it, determined to attack the allied Anglo-Dutch fleet in spite of its vast superiority (99 to 45 or 47 ships of the line), trusting that a considerable number of English sea captains, dissatisfied with the new regime, would join the king across the sea and his allies.

Naval Cooperation Increases

In this hope for a maritime fifth column James was disappointed, although there was actually a vast deal of dissatisfaction in the service. On the last day of the battle a number of French ships sought refuge at the anchorage of La Hougue where the invasion troops were waiting to embark. Being mostly of too deep draft to approach the French contingent in the shallow waters, the English and Dutch ships took advantage of the situation to put out in smaller boats and these aided by fire-ships burned a dozen of the French vessels. In vain did the troops try to assist in their defense, cavalry riding out on the shallow beach among the English and Dutch boats, from whence some of the horsemen were pulled off their chargers.

The effective use of the navy in the land battle of Pinkie was not often imitated before the year 1943 when it became almost commonplace, but in Sicily and Italy rather than in North Africa. As far as the 16th century is concerned, Professor Oman points to a minor action which took place in July, 1558, near Gravelines, in the dunes between Calais and Dunkirk where the cooperation of the navy, not arranged beforehand, but rather improvised by the enterprise of an English naval commander who happened to be in the vicinity, seems to have been inspired by the experience of Pinkie. Again this cooperation proved very effective.

But thereafter cooperation between the services waned, not so much in Holland, perhaps, as in England and France; or because of special circumstances land war moved away from the seashore. In any case, such military-naval cooperation, promising of a development as it had seemed, remained rare, and more exceptional, one is tempted to say, than it need have been. Examples of non-cooperation far outnumbered reverse instances as time went on. They were not so evident during the first half century after the close of the Hundred Years' War while Calais was still English and remained so until 1558, when the port was lost in mid-winter through poor garrisoning and provisioning. Under the circumstances it could not have been saved by any navy under sail.

Possession of Calais tempted, or enabled English governments to interfere with military forces in the struggles of the continent which engulfed them even more deeply in the diplomatic field. The military results of diplomacy were the coalition wars suh as Henry VIII embarked upon thrice; once during 1512, when English dismounted cavalry was sent to Biscay on strength of a Spanish promise to provide mounts, which never arrived;

again in 1512-13 (battle of the Spurs or Guinegate), and 1544. In the last war participated the largest English force ever sent into France until Wellington's day, while that of 1513 led to "the furthest irruption" of an English army in France before the 19th century.

In an age of balance-of-power diplomacy the inevitable military consequence of English policies was the often renewed Franco-Scottish alliance. This might involve sending troops, war materials and even a military mission like that from France to Scotland in 1513 which was to bring Scottish tactics to the level of military progress reached on the continent. Practically all the English expeditions to the continent were mismanaged, and raise the question whether failure was due to technical incompetence, or diplomatic duplicity on the part of the coalition partners who used, but hardly profited by, the services of liaison officers.

During the fourth of the wars of Charles V in which Henry VIII supported the emperor in the west and Sultan Soliman the French in the east, Charles and Henry had agreed to meet on the Marne in August in order to join in the grand "Enterprise of Paris," but the emperor never got further than St. Dizier where he settled down for a long siege which lasted until August 18, 1544 and Henry got stuck before Montreuil, having landed with his main force only around July 15. By the middle of September, the king in spite of his large forces had won nothing excepting Boulogne which the English kept in the peace of 1546 and held until 1550 when they sold it back to France, under duress. Both the expeditions of 1512 and 1513 ended in practically complete dissolution of the English forces, in which, among other factors, homesickness figured, a nostalgia which the island character of their homeland seems only to have intensified.

British Military Force Declines

So little of military power was available in England that by 1550 Boulogne had to be sold back to France and the French felt strongly tempted once more to strike at England in retaliation for so many invasions from the island. They entertained close relations with persons in Ireland and still closer ones with Scotland, which might be called upon to invade England at any time convenient to French policy. The French had the plans of fortifications built by Henry VIII in their hands and had recently surveyed the Thames, some of this intelligence being given to them by a Florentine engineer who had served Henry and was dismissed after the king's decease.[6] But then there was Spain as the more important enemy of France, in league with whom the England of "Bloodie" Mary, Henry's successor, had but lately warred against and invaded French territory. It was in this war that England lost Calais, January, 1558, afterwards landing some 6,000 men who were sorely missed at Calais, at Le Conquet in Brittany on July 29.

The landing took place practically without resistance, but soon the Breton militia and the bands of the nobility were mobilized, moving the English, who seem to have had no higher aim than plundering and devastating the region, to reembark. But in spite of the good season their fleet had been unable to keep at the original unprotected anchorage and had moved to the bay of Bertheaume, some six kilometers away, causing a hitch in the movements of the land force whose rearguard was cut to pieces by the Breton militia.

The English troops which Elizabethan diplomacy from time to time sent to the continental wars hardly ever amounted to more than a few thousands, constituting mere token payments on the limited liabilities incurred. By that time the English nation had again become military, although not naval, isolationists. The army was called into existence only in war time and its relations to the navy had become tenuous, as was abundantly demonstrated by the famous "Journey to Portugal" of 1589, undertaken for the purpose of restoring a Portuguese king. Drake, the renowned admiral, and the military commander, Sir John Norris (or Norreys) quarreled from the outset, both being of the choleric type in a choleric age. Drake in fact was to die in 1595 on another expedition to the Spanish Main which ended in disagreement with his co-commander and in disaster.

When he and Norris arrived in Portugal, after spending some time plundering Spanish towns in Galicia, an English force was landed at Peniche, some fifty miles north of the mouth of the Tagus. There was no Spanish force available to offer resistance and the population of Lisbon was ready for an uprising, given some encouragement. They waited for the English to come up the Tagus to the capital and the English in turn tarried for a general uprising of the Portuguese, in the meantime plundering the country for food, since they had left Plymouth with only four days' supplies for some ships. Drake could not be brought to sail up the river from Cascaes where he remained until half of his men and those of the land force had fallen ill and the first Spanish troops had shown their faces. The English expedition had missed a golden opportunity and returned home with nothing accomplished and hardly even a shot fired.

NOTES, CHAPTER 17

[1] Sir William Blackstone: *Commentaries.* Vol. I, bk. I, ch. XIII, 418.

[2] Jähns: *Handbuch der Kriegswissenschaften,* 1265.

[3] William Lambarde: *Perambulations of Kent.* 1576, cit. George Clinch: *English Coast Defences from Roman Times to the Early Years of the 19th Century.* London, 1915. 159-60.

[4] Oman: *Art of War in the 16th Century,* 350-57. It is fairly amusing, if not a plain case of historical relativism, to compare the positive judgment of the military historian like Oman on Henry VIII's coastal fortifications with the negative one of the historian of the royal navy, Clowes (I, 434-5).

[5] Oman, 358-67, with illustration, p. 361 and opp. 366.

[6] Leopold von Ranke: Französische Geschichte. 2d ed. Stuttgart, 1856, I, 128.

Chapter 18

A JAPANESE ARMADA TO KOREA

A DISTANT PARALLEL—distant in space rather than in time, or character—to the Spanish Armada was provided by the Japanese in their enterprise against Korea from 1592 to 1598. Traditionally this has been considered by them a resumption of the earlier, altogether mythological continental expedition of the Japanese Empress Jingo—"appropriately so called" as observed by Lord Curzon—in the early 3rd century. It was also a counter-reflex to the Mongol invasion of Japan in the 13th century.

In character it was a late feudal *conquista,* with Hideyoshi the upstart majordomo playing the rôles of William the Conqueror and a bit of Napoleon into the bargain, reversing the action and plane of these two semi-Frenchmen insofar as his enterprise was island-based and directed against the continent, including among its aims possibly the throne of China. "He so loved action," wrote a later sympathetic daimio, "that he felt aggrieved when his brilliant genius had succeeded in pacifying the land [Japan proper] . . . Most probably his Korean expedition was intended as a safety-valve for the energies of his vassals who might otherwise have lost their heads over their success in the small Empire of Japan and was undertaken in the hope of obtaining lands wherewith to adequately reward their services if fortunes attended his arms so as to bring about the conquest of China." [1]

Much in Hideyoshi's make-up fitted him for such logistics as were involved in a large-scale overseas expedition. For it was said of him that "the stars themselves were not more punctual than his reckonings." But his enterprise against the Asiatic continent resulted in dismal failure, partly because he entrusted leadership in the field to others. His position at home was too doubtful to allow him to cross the sea, a contingency which Napoleon dreaded somewhat when contemplating his descent on England.

Korea must ever disturb the geopoliticians by cogitation as to whether its location makes it a sort of grapnel for China, thrown out in the direction of Japan, to be conquered by the Chinese; or a huge ramp leading from the sea to be utilized by the Japanese in storming the Asiatic continent. It was at least a natural beachhead for the Japanese, and its harbors had previously offered embarkation points for the Mongol fleet which attempted the invasion of Japan. The island of Tsushima, 56 miles from the Korean coast, was the natural halfway station for the crossing. The Japanese invasion army, said to have been 200,000, or even 300,000 strong, numbers which are, of course, far too high considering the small size of the boats at the disposal of the Japanese, with 70,000 first line troops, 87,000 of the second line and the rest as strategic reserve and for replacements, was assembled at Nagoya on Kiushiu Island and the first corps despatched from there on May 24, 1592.

As far as transport and coverage were concerned, Hideyoshi suffered from much the same handicap as William the Conqueror and Napoleon—he did not own a fleet superior in all, or even most, respects to that of his enemy. The transports had to be furnished by the feudatories in somewhat the same manner as taxes and corresponding with these. While this fleet was numerous, it must also have borne a rather motley character and in fighting strength was not up to the Spanish Armada, which was also above all conceived with the idea of transporting troops to insure a military decision to be sought on land.

Hideyoshi's plans were for a first army of three corps and a total strength of some 50,000 men to cross rapidly to Fusan, fan out there into three columns, one to move along the coast to the right, one on the coastal road to the left and the center column northward to the capital, Seoul. The second wave was to follow and more effectually take possession of the regions traversed by the advanced troops, while the third wave was to be carried by sea along the west coast of Korea and join the van. By that time Korea was to have been subdued and the host was to be ready to cross the Yalu River into China. Their rendezvous was to be at Phyöng-yang, on the Taidong river. This was to be the route of the two subsequent Japanese invasions of the Asiatic mainland, that of 1894 against China and of 1904 against Russia.

The plans for this convergent land campaign were almost as post-feudal as Napoleon's, and they succeeded as far as the movements of the first wave were concerned. It crossed the double straits without any interference on the part of the Koreans, took the castle of Fusan on the day after landing, covered the 267 miles from there to Seoul in 19 days, reducing several forts and field fortifications on the way, and fighting one field battle reached the Taidong river by July 15. Meanwhile more corps had landed.

At this stage the time-table of the Japanese came to grief owing to the interference of the Korean fleet with which Hideyoshi seems to have reckoned but little, being himself as much of a land general as Frederick or Napoleon. Japan's isolation and internal fights had never been sufficient to keep her pirates at home; they had harried Chinese and Korean shipping so incessantly that this major aggression finally provoked the Koreans to technological counter-measures. For the most part, the Japanese employed small quick-moving galleys from which they tried to board the more clumsy continental ships and overcome them by superior swordsmanship, much as the Spanish infantry on board the armada had hoped to beat the English. But like the English, the Chinese and Koreans were determined to keep their adversaries at arms' length, or at least from boarding. What the English achieved by superior sailing and gunning, these Asiatics sought to do by introducing, or strengthening the protective armor of their ships.

Decks were covered with strong timber to protect the crew from missiles of all sorts, including shells of weak explosive power, and from boarding parties. In turn the crews would fire with whatever artillery they possessed from behind the primitive fortifications of their ships. Borrowing the idea from the Chinese, the Koreans further strengthened their ships by turning some of them into iron clads of sorts by putting sheet iron roofs over the decks along the sides and equipping them with ram spurs. In some cases the roofs seem to have been studded with chevaux de frise, those "horses of Friesland" which had at about the same period come into use in the Netherlands. The bows are reputed to have been shaped like dragon's heads, spitting the fire of guns from their jaws while the sterns carried a battery of twelve pieces.

While we do not know from the descriptions of these "tortoiseback" ships whether they were the first to anticipate later western developments, just as the initial use of gunpowder is claimed for the east, against the Japanese fleet the ships of the Koreans proved definitely superior. The first attack, made about the time the expeditionary force entered Seoul, found the Japanese fleet at anchor outside Fusan and hit it with devastating effect. Other engagements followed, the last and most disastrous destroying the Japanese fleet which was on the way to Phyöng-

Chinese Raid on Japan. (Bibliotheque Nationale, Paris.).

yang with reinforcements and supplies. With such troubles on their hands and the Chinese interference in favor of the Koreans on the increase, the beginning of the end for Hideyoshi's expedition could be expected.

Some bloody successes were yet achieved over the Chinese and gruesome trophies could be sent home, like the 38,700 pickled and barreled Chinese noses and ears which formed the *mimizuka,* or earmound at Kyoto. But although Hideyoshi, who died in 1598, implored his successors with his last words: "Don't let my soldiers in Korea be made ghosts," *i.e.,* by starving, the enterprise was given up. And despite some face-saving tribute conceded by Korea to the invaders, the shock of failure was great enough to add to the motives for the ensuing 250 years of Japan's isolation.

The protagonists of sea power as conceived since the late

19th century, in their far and wide search for historical precedents that were to help settle for their side a modern technological issue, were eager to point to Hideyoshi's enterprise as an object lesson in sea power, insisting that he "had failed to grasp the first lesson of naval strategy—that command of the sea is an essential prerequisite of successful overseas invasion."[2] But the moral is hardly quite so obvious. With a different war economy as regards supplies and men, a better occupation method, an improved mutual support of Japanese land and naval forces, a more rapid following up of the successes over Chinese incompetence in land warfare, and without the handicap created on the death of Hideyoshi which took the spirit out of the enterprise even though he had never crossed the sea himself, the Japanese might well have carried their invasion of the continent much further.

NOTES, CHAPTER 18

[1] Count Okuma: *Fifty Years of New Japan.* 2nd ed. London, 1910. Vol. I, 38.

[2] Hector C. Bywater: *Sea-Power in the Pacific.* Boston and N. Y., 1921, 130; cf. also Alexander Kiralfy in Earle (ed.): *Makers of Modern Strategy,* 464-7.

Since this chapter was written, an excellent article has been published, dealing generally with early Japanese sea power, including the role played by *Kamikazi,* the "divine wind," in Japan's history, and the invasion of Korea in particular. See *From Jimmu Tenno to Perry; Sea Power in Early Japanese History,* by Arthur J. Marder, American Historical Review, October, 1945.

PART 3

16th-18th Centuries

Chapter 19

SPAIN AS A LANDING POWER

THE FALL OF CONSTANTINOPLE could not fail to affect the dominion of the sea in the Mediterranean. The Arab saying that "God has given the earth to the Musulmans and the sea to the Christians" [1] was perhaps true through the greater part of Muslim history, but at times certain Muslims like Kair-ed-Din, or Barbarossa would arise to be proclaimed Beylerbey of the Sea.

A corsair of Albanian blood, Barbarossa was one of those outsiders, if not renegades, whom the Turks admitted to their armies and navies as a matter of course and who were indispensable for the maintenance of their dominion. He established Turkish rule in North Africa where Spain had acquired Mers-el-Kebir (1505), Oran (1509), Bougie and Tripoli (1510), the latter being handed over to the Knights of St. John by Charles V in 1530, after they had been expelled from Rhodes in 1522. Barbarossa drove the Spanish from Algiers in 1529 and in 1534 seized Tunis, where they had supported local rulers. In the same year he appeared off the Italian coast, destroying shipyards, taking castellos, raiding the countryside and dragging inhabitants into slavery.

Panic was great in southern Italy which had not known such invasion for some four hundred years. Yet another Carthage seemed to have arisen from the ruins and the Holy Roman Emperor Charles V thought it a worthy Christian and necessary Spanish enterprise to dislodge the Turks, in whose attacks his enemy Francis I of France had connived. Francis confided to the Pope that he intended to instigate such forays rather than help to resist them. An armada was prepared by Charles, including some belated Crusaders who found themselves at loose ends, and many more mercenaries, in Spain and Italy. The two columns met off the Sardinian coast near Cagliari whence they sailed to Tunis in June, 1535, landing on the shores of the Gulf of Tunis without encountering opposition.

Goletta, Barbarossa's castle, was besieged both from the harbor and the land. After it had been taken by storm, the Spanish found some of the guns marked with the French lily. Tunis fell likewise, and old Christian prophecies were recalled, of an emperor who would subdue the whole world and on pain of death would decree that every man was to adore the cross, whereupon this ruler would receive the crown from an angel of God in Jerusalem and depart this life.

Charles seemed to be indicated for this role. But the disunion of Christendom more than ever stood in the way of realizing such dreams. Barbarossa hit back at the Spanish in the following year by raiding the Balearic Islands with the remnants of his fleet and Francis I renewed the war against Charles. The Anti-Christ ally of the Most Christian King, Sultan Soliman, sent over from Avlona (Valona) into Apulia 8,000 light horse under the leadership of a pasha and an Italian renegade who wasted the land and carried thousands away into slavery; these bands withdrew when a French invasion planned in Northern Italy did not come off, and Spanish and Venetian squadrons threatened to converge on their communication lines (July-August 1537). Soliman settled down before the Venetian fortress of Corfu for some three weeks, but fearing to be cut off from the mainland by the Christian fleet, retired after some fierce assaults on the place which continued to remain Venetian almost as long as the power of Venice lasted. But about this time Venice lost her hold on the Aegean islands, Scyros, Patmos, Aegina, Paros, Naxos.

Venetian-Spanish Alliance Fails

With Venice and Spain, the two foremost Christian sea powers, engaged together against the Turk, more might have been achieved. But the partners distrusted one another and Christianity once more proved no reliable basis for coalition warfare where diversity of State interests interfered. Andrea Doria, the Emperor's Grand Admiral, holding Barbarossa at bay late in the war season of 1538, took Castel Nuovo from the Turks with a landing force after having battered the place from his ships, an enterprise which as yet did not seem forbidden by formulae about the superiority of land-based guns. When this place, situated within the traditional sphere of influence of the Venetians, was not handed over to them, they retired from the struggle against the Turk for some 30 years and made a peace at Constantinople which saved to them their last island possessions, Zante and Cyprus, Crete and Corfu.

Not until 1541 could Charles think of returning to the corsair-infested African coast where Algiers and Tunis continued to be thorns in the maritime flank of his empire. Too late in that year to be favored by weather he assembled Italian, German and Spanish forces off Algiers, which juncture provided the last successful feature of that Christian enterprise. Algiers would not surrender when summoned on October 25, as only part of the Spanish force had been landed as yet, the commandant replying that he had not only brave men inside his fortress, but a wild sea as an outside ally, of the latter of which Doria, the Pope and others had vainly warned Charles. The next day the bad season began with heavy rain and hail. A three days' storm smashed the ships against one another, or piled them on shore. Provisions and ammunition were lost, or spoiled. Meanwhile Moorish cavalry appeared in the field.

With his cold and unfed troops, panicky by now, Charles was orced back to Cape Matafas, 15 miles away, where the contin-ious bad weather conditions—similar to that which hampered he Allied forces in this region during the winter of 1942-3—onvinced him that he must return to Spain. Although recog-iizing this failure as an "act of God," he would seem to have ssayed to excuse Providence by complaining that if time had een allowed him to complete his landing before the weather roke, Algiers would have fallen. Some in his entourage were old enough to tell him that if the Moorish troops had only een a little stronger, no man of his forces would have escaped.

They might also have added, had they but known it, that he devastating storms had kept at a safe distance the fleet of Barbarossa whom the Sultan had despatched from Constantin-ple at the news of Charles' expedition. The return voyage vhich was made by the Spanish forces at the winds' will, was lso full of peril. In Bugia (Bougie) where Charles paused, olemn processions were held, the Emperor himself participat-ng, to pray for a safe departure from these fatal shores. Not ntil December 1 was Charles back in Cartagena. Of their holdings in North Africa the Spanish lost all except Mers-el-Kebir and Oran.

A small yet important part of the Christian claim to dominion in the Mediterranean was preempted by the Knights of St. John, those sea-going cavaliers of the post-feudal age, long headed by Jehan Pariset de la Valette, who had shared in the siege of Rhodes and in many raids and sea fights and had even for a time been a galley slave until ransomed, an experience as mortifyingly unknightlike as can be imagined. After they had lost Rhodes and became homeless for a time, Charles V, to enable the Order "to use its strength against the pagan enemies of the Christian Commonwealth," conferred on it the islands of Malta and Gozzo and Tripoli in Africa (1530). From these places they offered one of the few constant checks on the depredations of the corsairs.

For only spasmodically would the Latin sea powers unite against this threat, as in the enterprise against Dragut, a succes-sor to Barbarossa, who in certain ways occupied the same re-lationship to the Sultan as the Grand Master of the Knight did to the Emperor. In the attack on his stronghold, Mehdia the Spanish-Italian ships combined with those of the Pope Tuscany and the Maltese knights. An historical reminiscence a quotation from Appian about ancient siegecraft, is said tc have inspired the viceroy of Sicily with the idea of placin siege artillery on a number of galleys immovably anchore together in the harbor and thus smash the town's sea walls Knowledge that these were usually the weakest part of the de-fenses of a coastal fortress might also have encouraged th enterprise.

Turks Besiege Malta

Sultan Soliman II was unwilling to suffer tamely this se back in Africa and sent out a large fleet to clear away th giaours and their most active fighting force, the Maltese knight Malta itself was then found too well fortified and defensivel organized to be taken. The landing force was repulsed b cavalry, but Tripoli was surrendered by the Knights. In 156 the Sultan, intending to crown his vast successes on land b a land-and-sea enterprise, sent out a large force estimated a from 30,000 to 40,000 men against Malta which was held b between 6,000 and 9,000 men, assembled around a nucleus some 500 Knights. The season for the siege, which began i

May, was better chosen than were managed some of the other points involved in the combined operations for this famous investment, the failure of which together with the battle of Lepanto definitely checked Turkish sea power. The Sultan was unfortunate in his timing, for the Maltese knights, headed for eight years by La Valette, had done much to fortify Malta

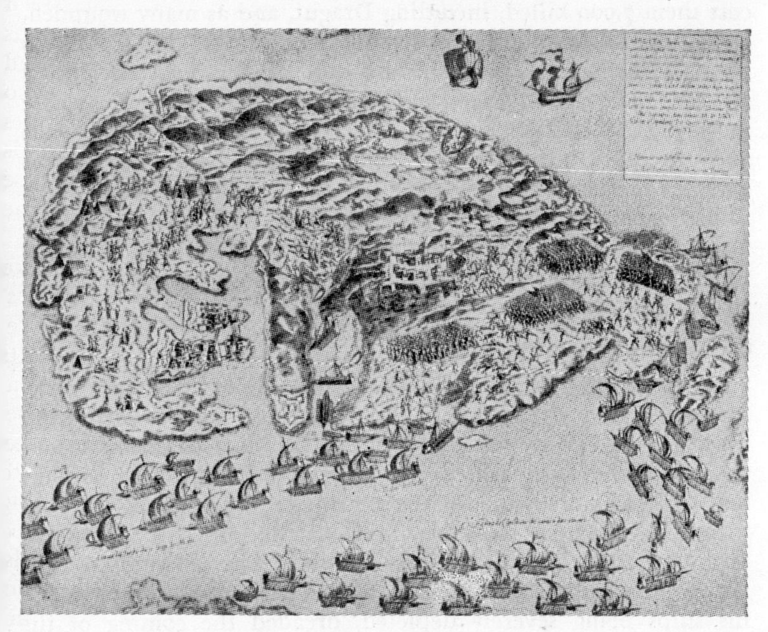

Siege of Malta by Turks, 1565.

against the long awaited assault and their masons were still employed when the Turks arrived.

If the Turks were the foremost military power of the time, their Malta enterprise illustrated the difficulties that had to be surmounted in combined operations. The Sultan, whether for political, or other reasons, had left the question of the supreme command over all his forces in Malta unsettled, putting the land troops under the command of Mustafa Pasha, his vizier, a veteran commander, and the fleet under that of Piali, the capitan-pasha, a young scion of his own house, instructing them to wait for the arrival of Dragut from Africa and settle together with the latter the final plans for the siege. In a council of war, however, the naval commander persuaded the others that the first action must be the assault on St. Elmo, the fort blocking the entrance to Marsamuscetto harbor, in order to obtain there an all-weather safe anchorage for his fleet.

To reduce this fort alone took the Turks, who had landed without immediate opposition, no less than five weeks; it cost them 7,000 killed, including Dragut, and as many wounded, as against losses among the defenders of 130 knights and 1,200 others. During these weeks, until June 24, the Turks neglected all other objects in the island such as Citta Notabile, where most of the civilians were sheltered behind mediaeval walls and towers. Until September 7 the Turks turned their furious energies towards reducing the other Maltese forts, putting them under concentric fire from the land side and from across the Great Harbor, making more than a dozen attempts at storming them, including one undertaken from galley and rowboats against the sea walls and slopes of the forts.

Christians Win Victory

By September the stores of food and ammunition inside the fortress began to run low. And still there was no sign from the Spanish viceroy in Sicily, a tardy if not cowardly leader of a relieving force which 700 impatient knights and soldiers had quit in order to join the besieged. Learning of his ever increasing armada, but not of his vacillations, the Turks determined to deliver one more assault before abandoning the island enterprise, urged by the capitan-pasha who, the crews of his ships being severely depleted, dreaded the coming of the equinox. This final attack took place on September 1 and was repulsed with great losses to the Turks. Prodded into action at last, by his council of war, the viceroy had left Sicily in the meantime, on August 20, with a force of some 10,000 men, a strength nearly equalling that left to the Turks; but the fleet was dispersed and finally got off from Trapani only on September 3, reaching Gozzo on the 5th and disembarking on the following day from six to eight thousand men at the northwest corner of Malta. They set out and soon joined the weak Christian forces holding out in the Citta Notabile, while the fleet sailed around to the harbor to salute and inspire the besieged. Before they had hardly been more than menaced by the armada, the Turks began a hasty re-embarkation, leaving stores and disabled, or unmanned ships behind. They had not gone far when the lack of molestation by the relieving force and reassuring reports about its size, moved Mustafa to land once more in the north of the island. A skirmish with the Spanish showed him that the latter were by far the stronger, and the Turks retired in confusion to their ships, losing another 1,000 men, with disgrace awaiting them on their return to Constantinople.

The siege of Malta, for the attacked a successful maritime Thermopylae, ended the Turkish endeavor to extend their naval dominion to the western Mediterranean where they still maintained mere outposts in the haunts of the corsairs of the Barbary coasts. These pirates, who eventually emancipated themselves from the Porte, were too weak to threaten seriously the holder of sea dominion in the western sector, although thanks to Christian commercial competition and jealousies they were allowed to prey upon shipping for nearly 300 years more. In the eastern sea the Turks drove Venice, which had stood selfishly aside in the earlier sea struggle against the infidels, from Cyprus (1570-1), for the relief of which the Christian sea victory at Lepanto under Don Juan of Austria occurred too late by two months.

Malta, perhaps the strongest sea fortress of the 16th and 17th centuries, was an impressive symbol of Christian unity at sea, although it never saw another siege until the one from the air in 1942. But after Lepanto, this unity was not easily maintained. Even in the year following Lepanto a force of Spaniards and Venetians, much smaller and less cohesive, failed in combined operations around Modon, the site of ancient Pylos and Sphacteria. Into this bay, a narrow harbor with the entrance secured by the fortified town of Navarino and batteries on the opposite headland, the Spanish-Venetian fleet had driven the hastily rebuilt Turkish fleet. Don Juan d'Austria, the Spanish naval commander, declined to face these land guns and try to put his ships into the bay of Modon.

Landing troops, led by Prince Alexander Farnese, later one of the best Spanish commanders in the Netherlands, were repulsed before Navarino by Turkish troops sent overland. Although the Turkish fleet was rapidly disintegrating, with ships leaking, munitions low, epidemics spreading among the crews, the Christians failed to take advantage of this weakness. The season was advanced; it was October now, for politics had postponed the meeting of the squadrons until September, and Don Juan felt he had to abandon the enterprise, allowing the Turk to straggle back to Constantinople.

The Great Armada That Did Not Land

The Mediterranean was the lesser theater for Spanish overseas activities, most of which were directed towards and across the Atlantic. Here, weather conditions and longer distances called for a different fighting equipment in the shape of war fleets under sail. In the Mediterranean the sailing orlog ship had as yet only occasionally appeared and the war fleets were composed almost entirely of galleys propelled by oars. The man-of-war under sail in turn called for battle tactics unlike those directing the galley, for tactics less closely following those of land warfare and definitely more *sui generis.*

The largest single overseas enterprise of Spanish land and sea power, and the greatest failure among landings planned or attempted down to Napoleon's, which it surpassed in the seriousness, not to say fanaticism, of its intent, was that of the abortive and miscalled Invincible Armada of 1588. Planning for this invasion marked an important step in the history of war preparation, shaped as it was by a post-feudal bureaucratic attempt at organization and control on a gigantic scale, but always fatally hampered by the procrastinations of Philip II. The project in its inception went back almost to the time of the Spanish conquest of Portugal (1580), which had given Spain considerably more of an Atlantic sea front. The conqueror of that country, the famous Duque de Alba, was originally chosen to head the Armada against England. Before his death in 1582 he had submitted plans, far more detailed than many others on which wars had been waged in that age, including estimates of the time required for preparation—eight months, the Duke thought—and the necessary ships, military forces and cost.

According to Alba's plans a fleet of sailing ships was to be raised, in large part chartered in the harbors from Italy to the Elbe, consisting of 150 large ships with an aggregate tonnage of 77,250, 40 *urcas,* round bottom freighters, of 200 tons each, and 320 small vessels of altogether 25,500 tons and a fleet of 46 galleys, each with a crew of 80 *hombres de cabo* (noncommissioned officers and soldiers) and 200 rowers. The large ships were to take aboard 20 Italian *fragatas* (a frigate was originally a small oar-propelled vessel) and 20 *falucas,* small coasting vessels with oars and sometimes lateen sails, to be used in the immediate landing operations, together with 200 flat *barcas*, or barges of a type that had been found useful in the Spanish conquest of the Portuguese Azores. This specialization of landing craft also provided an indication, if only a subordinated one, that military thought and practice of the Middle Ages was relegated largely to the past.

Alba's plans also called for 55,000 foot soldiers (28,000 Spaniards, 11,000 Italians and 16,000 Germans), of whom 45,000 would be actually available, allowance being made for 10,000 sick, deserters, etc. Of the 45,000, the landing force would require 35,000, while 10,000 would serve on the vessels. In addition the estimates demanded 1,200 men and the same number of horses for cavalry, 334 field artillerymen, 3,000 engineers *(gestadores)*, 700 men for train personnel and 1,400 mules as draught animals. Altogether, allowing for staffs, etc., Alba's strength was to be some 87,000 men, perhaps the largest military overseas expedition up to then—on paper. The cost was enormous, even after deducting current army and navy expenditure, which would have to be met in any case, and not including the cost of feeding the men.

The whole plan was based on a flourishing shipping and a still solvent economy as well as on the experiences of an able military leader. But Alba, who had lost favor with Philip, died in 1582 and a new leader was chosen in 1583, the Marquis of Santa Cruz, an energetic veteran naval commander who had taken part in the battle of Lepanto and had as Admiral of the Ocean won for Spain the Portuguese island possessions in the Atlantic. He had urged the expedition against England as Spain's most dangerous enemy.

Paper Strength of Armada

His plans were only a little less elaborate than those of Alba, but their carrying out was constantly put off and costs pared down by the procrastinations and vacillations of the King and his constantly recurring financial or political embarrassments. Thus the size of the armada was reduced to a much more modest scale, including in its paper strength only 132 vessels of all sorts, amounting to nearly 60,000 tons with 2,431 guns, carrying 21,621 soldiers and 8,066 sailors; the military personnel comprising 16,973 Spaniards, 2,000 Portuguese, 581 adventureros, or volunteers, from the younger needy nobility, 78 gunners, 180 chaplains (one for every 155), but only 85 surgeons, doctors and their assistants.

Blessed by the Pope and his prelates for the new crusade across the seas, these benedictions serving to launch a quip which began to circulate after 1588, that after all heretics made far better sailors,[2] Philip speculated on receiving armed support from the still strong Catholic element in England, which, however, did not stir in his behalf. The larger part of the armada took to sea from Lisbon on May 29, 1588. It was already behind schedule, which clearly envisaged the conquest of England during the good season, and the delays were still not over. Off Coruna (the Groyne) where the main fleet was to be joined by the ships from Biscay, Castile and Guipuzcoa, it was the target of severe squalls, and another six weeks were spent in repairs and revictualling. Stores had been bad from the outset, due to rascally contractors, and had deteriorated, causing much sickness. However, the weather was not quite as bad as the Admiral reported to King Philip, finding it "the more strange since we are on the business of the Lord and some reason there must be for what has befallen us."

The leader of the armada, as far as it was not directed by the king himself, functioning with true Spanish-Austrian autocracy as director of war from the court, was not a man to cope with these difficulties. The 7th Duke of Medina Sidonia, Santa Cruz having died early in 1588, possessed no military, or naval experience except in connection with the defeat he had suffered at the hands of Drake in the defense of Cadiz, and lacked inborn capacity for leadership. He complained that he was always sick at sea. He was master of one of the greatest fortunes in Europe and Philip may have chosen him as the fittest representative of legitimacy in religion and supremacy in wealth, or in order to demonstrate his sway over even the highest and proudest grandees in Spain, for the Duke accepted the office against his will and solely at the insistence of the King. The case of Medina Sidonia would tend to confirm the thesis of military history and social history, that the owners of great fortunes, such as Nicias, Crassus, Soubise, have seldom proved to be competent and outstanding leaders in war and that for such bold enterprises as overseas expeditions causes would be better served by less well born, or opulent commanders, or even those stigmatized by the bar sinister, like William of Normandy and

Don Juan of Austria. Medina Sidonia was merely cringing in his applications to the King's munificence; an Admiral of the Ocean who hardly even showed personal courage in the combats between the English and the armada.

From the Groyne the armada proceeded on July 12, taking course for Plymouth where the larger part of the English fleet was assembled, while smaller English detachments in conjunction with the Dutch fleet were blockading the Flemish ports in order to prevent the Spanish commander, Prince Alexander Farnese, from joining Medina Sidonia for the descent on England. This possibility had been envisaged by Philip in his sealed orders for the armada which were opened when the English coast was sighted on July 20. They instructed the commander to sail for Dunkirk, avoiding the English fleet, and there to wait until Parma with his army of some 34,000 men would sail in the direction of the Thames and London, under protection of the armada, a conception somewhat resembling Napoleon's last plans for the Channel crossing in 1805. Adhering to these orders, the armada, much encumbered by its transports and harassed by the swifter sailing, better shooting and lighter craft of the British, sailed for Calais, thus missing the opportunity to beat the English main fleet, win control of the Channel, and possibly land on the Isle of Wight.

Beginning of the End

When they cast anchor at Calais on July 26, Medina Sidonia sent messengers to Parma, who replied that he would not be ready for another fortnight, that the landing craft which he had built by royal order could not fight at sea and that in order to enable his troops to leave port the armada must drive off, or destroy the Anglo-Dutch squadron blockading Dunkirk and Nieuport. Before the Spanish could break the blockade, the English sent fire ships among them while they rode at anchor on the night of July 28, to their confusion, fear and loss. Terror was heightened by the memory of many Spanish soldiers of the effect witnessed by them of the infernal engines drifting against Farnese's Scheldt bridge during the Antwerp siege of 1584-5.

The armada scattered, several ships were lost, although not a great number; but in the general discouragement a council of war decided that it was imperative to save as much of the fleet as possible and that they were now too weak to attempt a landing in England. The less timorous proposed to sail for

Hamburg and refit there, but the more pusillanimous majority had no thought beyond fleeing back to Spain, sailing around the north of England since the winds in the Channel remained consistently adverse to the taking of a direct course over the Atlantic. With only 53 ships Medina Sidonia finally arrived in home ports, far more of his vessels falling victim to the exceptionally severe gales of the summer of 1588 than to the guns and fire ships of the English.

Great historical events such as the failure of the armada are not always of an equally great military import, or significance. The inherent weaknesses of the armada did far more to ensure its collapse than the offensive action of the English fleet. Nevertheless the latter did more to ward off the Spanish invasion threat than could probably have been effected by land forces, had the Spanish won beachheads in England. A possibly deservedly low opinion was held of the quality of Elizabethan defense organization. Some of the 16th century Italian diplomats in England who in the absence of military attaches might be said to have exercised that function, thought the military strength of England so low that 10,000 to 12,000 Spanish, or Italian infantry, plus 2,000 cavalry, would suffice to restore Catholicism as the state religion beyond the Channel. In their opinion, which to a certain degree was the product of wishful thinking, "if a foreign prince were only able to land an army on English soil, and win the first battle, he would encounter no further difficulties since, owing to the inability of the people to endure fatigue, and their nature, they would no longer oppose him." True, as one diplomat admitted, landing was not without risk, as: "All similar attempts since 1066 had failed, and William only succeeded on account of the weakness of the opposing army." But he insisted that "Although the English showed great aptitude for sport, and readiness in times of danger, it could not be said that they cared much for arms. Their only opportunity to make use of them was in war, and once that was over they forgot all about them." [3]

While the armada is generally regarded as the most serious invasion threat since 1066 and down to Napoleon's time, it was not the last contemplated, even in Philip II's day. There was a small-scale Spanish descent on Cornwall in 1595 and still another was feared for 1596, when an armada being assembled at Cadiz was partly destroyed by the English in a preventive naval action off that port. Among the Catholics in England and Ireland there were always abundant and active fifth column elements useful to invaders, even if Papist plots were never quite as serious as they were painted. English military unreadiness provided a constant temptation to such projects. While the defense of England was generally maintained through her ships, cruising and raiding against Spain, there nearly always lingered apprehension and, amongst the enemy, hope that when the stars in their courses might not fight for England, this mobile wall formed by the superior number, quality and ability of the fleet might not suffice, or be available for effective combat. In this was rooted both temptation and hope for the invader and apprehension lest the English people fall prey to invasion panic.

Preparing for the Future

In moments of national alarm it would so clearly appear that sea power alone was not enough to prevent a landing that measures would be set in motion to stiffen English land power, organize and reorganize the militia and strengthen the standing army, when one existed, "in hope we should be better provided hereafter not to be thus taken *tarde* on the sodain," or suddenly out on a limb, as an Elizabethan letter writer, hearing news of yet another Spanish armada, expressed it.

> A week after this had been written "came news (yet false) that the Spaniardes were landed in the yle of Wight, which bred such a feare & consternation in this towne (London) as I wold little have looked for, with such a crie of women, chaining of streets and shutting off the gates as though the ennemie had ben at Blackewall (on the Thames below Greenwich). I am sorry and ashamed that this weakness and nakedness of ours on all sides shold shew yt self so apparently as to be carried far & neere to our disgrace both with friends and foes. Great provision is made for horse as being the best advantage we are like to have yf the ennemie come." [4]

In a speech from the throne subsequent to 1588, Elizabeth scolded and threatened her subjects: "Some upon the sea coast forsook their towns, fled up higher into the country, and left all naked and exposed to his (Philip's) entrance. . . . If I knew these persons, or may know of any that shall do so hereafter, I will make them feel what it is to be fearful in so urgent a cause." However justified such reproof of the panicky elements might have been, the fact should not be overlooked that at nearly the same time the handling, or even instigating of invasion panics became, and remained, one of the standard tricks of governments. Governments early in the modern era found out that an invasion panic may be a powerful instrument to win over a people to the raising of military forces which they might otherwise oppose. Invasion scares were, in modern language, strong shots in the arm of the body politic, and were so employed by the Elizabethans.

Sir Francis Bacon tells how England was again alarmed in the autumn of 1599 by rumors of an approaching Spanish fleet, so that an army was hastily raised in defense of the kingdom. "But there was no such thing" and the discovery of the falsity of the moves provoked only cynicism. In fact the preparations were made with regard to treasonable moves, later brought out into the open, of Essex in Ireland. As Bacon, who was well acquainted with the workings of Elizabethan government, wrote:

> And it is probable that the Queen had some secret intimation of this design. For just at that time there grew up rumors (such as are commonly spread when the sovereign is willing they should circulate) and went abroad all over the land, that a mighty and well appointed Spanish fleet was at hand. . . . And yet these devices of the Queen were even by the common people suspected and taken in bad part; in so much that they forbore not from scoffs, saying that in the year '88 Spain had sent an Invincible Armada against us and now she had sent an Invisible Armada; and muttering that if the council had celebrated this kind of Maygame in the beginning of May, it might have been more suitable, but to call the people away from the harvest for it (for it was now full autumn) was too serious a jest.[5]

Fear of "Dark Invaders"

More often, however, panics spontaneously originated among the English people, even if the various governments did not share them. One of the sundry suspected "dark invaders" who preoccupied the English mind was Count Gondomar, the Spanish ambassador in the 1610's and 1620's. While he was always on the best terms with James I, the people were persuaded that he was making surveys and other preparations for a Spanish invasion. Pamphlets were sold in which Gondomar was represented "in the likenesse of Matchiauell" and "his treacherous and subtile practices to the ruine as well of England as the Netherlandes" (1620) uncovered.[6] He was the Dark Knight in a political play by Thomas Middleton, "A Game at Chesse," in which Loyola himself, the founder of the Jesuit Order, speaks the prologue, proclaiming his wicked designs concerning England. The anti-Spanish and Puritan element may have been particularly incensed at Gondomar, not only because he had been instrumental in bringing Walter Raleigh to the block, but also because he had been active and successful in his native land in warding off several English and Dutch raids at an earlier time. Another "dark invader" coming more directly from Rome, was Archbishop Rinucci, sent as nuncio to Ireland in the later stages of the Civil War, landing there with arms and money in October, 1645. For three and a half years he made the confusion of that war worse confounded.[7] This, like any intervention in Irish affairs from Rome, or elsewhere, only increased the ferociousness of the English towards the Irish.

Less Protestant, more Renaissance in style, was the "invocation of the Druids to the gods of Britain, on the invasion of Caesar" in a play of 1633 by an unknown author, "The True Trojans," who were of course the Britons. It called on the ancient deities to

Help us, oppressed with sorrow,
And fight for us tomorrow.
Let fire consume the foeman,
Let air infect the Roman;
Let sea entomb their fury,
Let gaping earth them bury;
Let fire and air and water
And earth combine their slaughter!

The Moon Goddess, "commandress of the deep," was implored:

Drive back these proud usurpers from this isle . . . .
Protract both night and winter in a storm,
That Romans lose their way, and sooner land

At sad Avernus' than at Albion's strand . . . .
Shed light on us, but lightning on our foe.[8]

The martial strength of Elizabethan England lay in its navy rather than its military forces, which at times hardly even existed, and in its aggressive, rather than home defense, as it is safe to say, even though the latter was never really tested. Whereas the Spanish, even with the armada, sought close combat, trusting to the known superiority of their infantry on board these ships, English naval tactics, as if in continuation of insular tendency towards keeping the enemy at arms' length, adhered to a policy of distant combat, *Fernkampf,* relying on superior gunnery and sailing. With such various tactical tendencies it is understandable that in England army and navy grew unevenly and apart, little inclined towards and as little fit for cooperation in war.

DRAKE REPULSED AT LISBON

Lack of military and naval cooperation was amply demonstrated by the failure of the expedition of Drake and Norris to Lisbon in 1589. While the Cadiz expedition of 1596 under Lord Howard of Effingham, Raleigh and Essex succeeded in defeating a Spanish fleet and taking and looting Cadiz, Elizabeth was still dissatisfied that the enterprise had not been pushed to a more conclusive end instead of being allowed to terminate abruptly. Effingham as lord high admiral had from the outset wanted to restrict the expedition to naval operations and it was Essex who insisted on landing as part of the venture, which would deprive the fleet of the opportunity of profiting materially by taking further prizes. Realizing this sympathetically, the council of war decided against further land operations and voted to return home.

The often excessive individualism of Elizabethan land and sea commanders was little conducive to the success of combined operations in which every one of the participants must know and fill his place. Such operations, in particular landings, were more apt to be successful when undertaken by sailors with aid from armed men, more or less forerunners of marines. One is almost tempted to say that the most successful of the Elizabethan wars were unofficial actions like those of Drake in 1572-3 against Nombre de Dios and on the Isthmus of Panama, or against various Spanish places during his circumnavigation of the globe (1577-80) although Drake took and burned several

Spanish strong points in the more official war of 1585, including St. Augustine in Florida. In appraising the importance of these victories it must not be forgotten that most of them were gained against small and poorly fortified places.

One of the best detailed landing operations undertaken in private *régie* during these English raids on the Spanish Main was that which occurred in "the well governed and prosperous voyage of Mr. James Lancaster, began with three ships and a galley-frigat from London in October, 1594, and intended for Fernambuck in Brazil." The leader, later Sir James Lancaster, commander of the first fleet sent out by the English East India Company in 1601 and one of the chief directors of that Company, had in earlier years combined fighting and trading in Portugal and now resumed this mixture of "commerce, privateering and war, triune, inseparable," as Goethe calls it.

As admiral of the enterprise of 1594, which was financed by several aldermen and "others of worship in the city of London," Lancaster led an original force consisting of three ships of altogether 470 tons and 275 men and boys. The frame and other timbers for a galley was carried on board "of purpose to land men in the country of Brazil" and this landing craft, with 14 banks, mast and sail, in her prow "a good sacar and two murdering pieces," were put together in a friendly harbor not far from Pernambuco. After this place had been reached, Lancaster with some 80 men embarked in the galley which at daylight was to go in ahead of the rest and obtain control of the harbor. But when the sun arose the ships had drifted away from the entrance and were now forced to work back, thereby losing the advantage of surprise. After an exchange of shots with the fort at the entrance, with neither party doing any harm, Lancaster got ready to row in with his galley, three great Hollanders anchored in the harbor mouth obligingly moving out of his way and ridding the Admiral of what he considered the chief obstacle to the progress of his invading craft.

Operations at Pernambuco

While the English were still waiting outside the harbor for the flood tide, the governor of Pernambuco sent a messenger to inquire their intentions. Lancaster answered that "he wanted the caracks' goods, and for them he came, and them he would have, and that he should shortly see." The inhabitants, organized in three, or four troops of about 600 men, manned the fort at the harbor entrance to repel the English whom they now "perceived to be enemies." So soon as the flood set in the latter attempted to land with their galley and several boats which had orders to be run "with such violence against the shore, that they should all be cast away without recovery, and not one man to stay in them, whereby our men might have no manner of retreat to trust unto, but only to God and their weapons." Although the fort began to play on galley and boats and guns carried away a piece of the galley's ensign, the English kept on moving in and ran galley and boats ashore right under the fort's walls with such speed and shock that they broke up and sank at once.

> At our arrivall, those in the fort had laden all their ordinance, being seven pieces of brasse, to discharge them upon us at our landing; which indeed they did: for our admirall leaping into the water, all the rest following him, off came these pieces of ordinance: but almighty God be praised, they in the fort, with feare to see us land in their faces, had piked their ordinance so steepe downewards with their mouthes, that they shot all their shot in the sand, although it was not above a coits [quoit's] cast at the most betweene the place wee landed & the face of the fort: so that they only shot off one of our mens arms, without doing any more harm; which was to us a great blessing of God: for if those ordinances had bene well levelled, a great number of us had lost our lives at that instant. Our admirall seeing this, cried out, incouraging his men, Upon them, upon them; all (by Gods helpe) is ours: and they therewith ran to the fort with all violence.

The garrison of the fort then retired inland and the admiral signalled to the ships in the roadstead to come into the harbor. He had the fort's ordinance directed against the high town from whence the most danger of a counter-attack threatened and, leaving a small garrison in the fort, marched the rest of the men towards the low-lying part of the town where the warehouses containing all the desirable goods were situated. These taken and the beachhead sufficiently well secured for the purpose of the expedition, to collect and transport away the riches in the storehouses, the lading of these goods was begun. Some prisoners were employed to haul the carts to the harbor, "which was to us a very great ease. For the countrey is very hote and ill for our nation to take any great travell in."

His foresight, which seems to have been as good in the military as in the commercial field, caused the admiral to take measures against fire-ships, lest the Portuguese might send them down the river flowing into the harbor to his damage. Such alarming contraptions need not be dreaded by the wary, as our relation puts it: "When it cometh upon the sudden and unlooked for, and unprovided for, it bringeth men into a great amazement & at their wits ende. And therefore let all men riding in rivers in their enemies countrey be sure to looke to be provided before hand, for against fire there is no resistance without preparation."

Lancaster seems in all respects to have been a very rational person, by his indication of what perhaps might have been done towards rationalizing the economy of war if business acumen and practice had been more generally applied to it, for after a stay of a month he decided that he had accomplished all that he came to do, had all the more valuable loot aboard his ships, and that the resistance of the Portuguese might get him into hot water. So telling his band, as became a later East India Company director, "that it was but folly to seeke warres since we had no neede to doe it," he prepared to depart. Some of the more militant among his young bloods wanted to go inland after enemy troops that had shown an inclination, although ineffectual, to prevent their departure. Sallying far beyond reach of support from the beachhead, many "by their forwardness came all to perish." After this final sacrifice to Mars by an expedition that rather was dedicated to the god of trade, Lancaster called it a day, turned his craft toward England and reached home safely with his plunder.[9]

NOTES, CHAPTER 19

[1] Cit. Sir William Temple: *Works*. Ed. Edinburgh 1754. II, 303.

[2] Champigny ca. 1590. Cit. A. W. Tilby: *The American Colonies*. London, 1911, 141.

[3] Lewis Einstein: *Italian Renaissance in England*, 217-8.

[4] *The Letters of John Chamberlain*, ed. McClure. I, 78 and 81.

[5] *The Works of Francis Bacon*, ed. Spedding-Ellis-Heath. Boston, 1860. XII, 57 ff.

[6] Catalogue of Prints and Drawings in the British Museum. I, no. 88.

[7] For R. see G. Aiazzi: *The Embassy in Ireland*. Translated by A. Hutton. Dublin, 1873.

[8] Charles Lamb: *Specimens of English Dramatic Poets Who Lived About the Time of Shakespeare*. London, 1887. 515-7.

[9] Richard Hakluyt (1552-1616): *Voyages*. Everyman ed. VIII, 26-44.

Chapter 20

LANDING OPERATIONS UNDER THE ANCIEN REGIME

IN THE PROGRESS OF SPECIALIZATION in war since the end of the Middle Ages, the development of navies preceded that of the other service. A more or less complete separation of war vessels from civilian commercial craft took place, together with their grouping into navies—*"l'armée de la mer"* or *"l'armée navale"* in the older French usage—but it was several centuries before a corresponding specialization of leadership on land and on the sea followed. Despite some early tendencies towards a split, unity of command was originally the rule in what the 18th century called "conjunct operations."

During about three centuries, then, we see, on the whole amphibious leaders like Drake, who during the expedition to Portugal in 1587 commanded a regiment as well as a ship, while his military partner, Norris, had a ship in addition to his regiment. Essex, the dazzling and disturbing favorite of Elizabeth, in the unsuccessful attempt against the Spanish fleet in Ferrol, the so-called Island Voyage of 1598, was "generall both at sea and land," having fought in several continental wars. Another outstanding land-and-sea general of the 16th century, happy in his activities, was the Spanish Duke of Osuna. Even in the 17th century the union of generalship and admiralship was exemplified by such figures as Wallenstein, generalissimo on land and sea, Wrangel, the Swede, who in 1658 was at the same time field-marshal and grand-admiral [1]—we do not know whether he opposed the crossing of the ice into Denmark in 1658 in the first or the latter capacity—and Prince Rupert who, beaten as a cavalry general in the field, went to sea as a royalist admiral.

In later centuries Apraxin (1671-1728) in Russia, where many institutions connected with warfare appeared belatedly, might be mentioned as "admiral-general," this being Apraxin's title. He built ships, ports and fortresses while he fought Charles XII in the field, contributing from the sea-side to the conquest of Finland in 1713 and by his landing and raids in Sweden in 1719 and 1720 brought the war into that kingdom, which had become accustomed to fighting its wars on enemy soil, thereby forcing it into the peace of Nystad (1721), by which Russia obtained the better, if not the larger, part of Sweden's trans-Baltic possessions.

The majority of the English Commonwealth's naval leaders were both land and sea commanders, with the usual title of admiral and general-at-sea, like George Monk, Richard Deane, Edward Montagu and above all, of course, Robert Blake, who had never served at sea before he was fifty, with only Sir William Penn and Popham as the more outstanding commanders who were sailors by profession. The presence of these landsmen in the navies of their times was necessary for several reasons. They had to discipline the navies after the armies had been brought to order first. They had to restore the unity of war conduct, in combined operations among other things, hold the services together and force them to act in cooperation. Thus they achieved the unity of command which had been lacking in such enterprises as that of Drake and Norris in 1589.

Land Officers Did Best at Sea

Cromwell's commanders were above the petty attachment to professional specialization that threatened to prevail in the English navy and which would have rendered it unfit for the ambitious and more complicated naval enterprises of the Commonwealth. To use the words of a naval historian, "The men who best served her at sea were, by training, land officers and consequently men of wider attainments and more general education and experience than belonged to the regular sea officers of the time." [2] The sociological circumstances, often of wholesome influence in the armed services, that these men were outsiders and therefore anything but slaves to routine, had probably even more to do with their energetic command and successful conduct of land-and-sea operations including landings, than the factor of education. This is not to say that the Commonwealth proved uniformly victorious in such difficult undertakings. The failure of the West India expedition of 1654 and after was disastrous. It sailed under a sea general, Penn, and a land general, Venables. They disagreed and after the land troops had ignominously failed in the attempt to take Hispaniola, they returned home contrary to orders and were put in the Tower.

The enterprise was left in charge of a vice-admiral Goodson, and a major-general Fortescue, whom Cromwell admonished "to consult together how to prosecute your affairs with that brotherly kindness that upon no color whatsoever any divisions or distractions should be amongst you, but that you may have one shoulder to the work; which will be very pleasing to the Lord; and not unnecessary considering what an enemy you are like to have to deal withal." The major-general in charge of the forces in Jamaica which the English had taken after the failure of Santo Domingo, was instructed by Cromwell, in one of the few orders wherein the Protector concerned himself with landings, to secure that island first by fortifying it and then to form a body of horse "as may, if the Spaniard should attempt upon you at his next coming into the Indies with his galeons, be in a readiness to hinder his landing; who will hardly land upon a body of horse; and if he shall land, you will be in a posture to keep the provisions of the country from him, or him from the provisions, if he shall endeavor to march towards you. [8]

The West India enterprise was about the worst of those planned and undertaken by the Commonwealth. Sea and land command were divided and were in addition saddled with critical and carping civil commissioners. Venables, as Carlyle puts it, "lay six weeks in bed, very ill of sad West-India maladies; for the rest a covetous lazy dog, who cared nothing for the business, but wanted to be home at his Irish Government again" (III, 114). The land forces were very uneven in quality, veteran regiments being combined with unreliable Royalist and Leveller (the Levellers were a revolutionary politico-religious sect, so called in mockery of their supposed purpose to make all men level in rank, according to Isa. 40:4) elements whom Cromwell preferred to have die beyond the seas instead of live in England, later to be replenished by rogues and vagabonds from Scotland —"We can help you to two or three hundred of these" (III, 124).

This armada of 60 ships and 9,000 soldiers "saw Hispaniola, and Hispaniola with fear and wonder saw it, on the 14th of April 1655: but the armament, a sad miscellany of distempered unruly persons, durst not land 'where Drake had landed,' and at once take the town and island: the armament hovered hither and thither; and at last agreed to land some sixty miles off; marched therefrom through thick-tangled woods, under tropical heats, till it was nearly dead with mere marching; was then set upon by ambuscadoes; fought miserably ill, the unruly persons of it, or would not fight at all; fled back to its ships a mass of miserably disorganic ruin; and 'dying there at the rate of two-hundred a day,' made for Jamaica."[4]

BLAKE BEST AMPHIBIOUS COMMANDER

Cromwell's best land-and-sea commander was, of course, Blake. Following up his earlier good management in land sieges, where he appeared to advantage as artilleryman and engineer, and in naval battles, he took the Scilly Islands from the Royalists in May, 1651. Not sharing the ever increasing awe—of a kind one hesitates to judge to be wholesome, or unwholesome—in which sailors held the effect of land-based ordnance against ships, Blake, sent on a Mediterranean mission in the winter of 1654-5 to restore there respect for the English flag, impressed the Dey of Tunis by destroying two of his fortresses by gunfire from the sea, "and though the shore was planted with great guns, he set upon the Turkish ships and fired nine of them."[5]

One of the greatest actions of ships against land guns was undertaken by Blake in 1657. Learning that the Spanish plate fleet was at anchor in the bay of Santa Cruz, the Canary Islands, well protected by a castle and several forts equipped with guns, Blake sailed straight into the bay, silencing the gun fire of the ships and the fortifications, setting the ships on fire and then, favored by a fortunate change in the wind, sailed out again, suffering only very negligible losses. One almost thinks of his record as incompleted, because of his failure to follow up Cromwell's hint of 1656 to take Gibraltar, "which if possessed and made tenable by us, would it not be both an advantage to our trade and an annoyance to Spain?" But here Cromwell himself realized that "nothing therein was feasible without a good body of landsmen" who were not then and there available.[6]

The strategic and tactical concepts of sea power and its use, as they were entertained in Spain, Holland and England at this era were at first very similar to, if not in imitation of, land tactics and land strategy. The Spanish in their zest to board enemy ships and thus seek decision in hand-to-hand sea battle followed this inclination most strongly. But gradually the two North Sea land and marine powers, England and Holland, evolved separate concepts of sea fighting, and then the day came when they were tempted to apply these to land warfare or make them predominant in combined operations. Sea power from time to time became overweening in its self-conceit and long before Mahan the conceptionists of sea power were liable to neglect vital considerations that fundamentally ruled military enterprise. They would insist on combining land and sea operations entirely on their own terms, thereby endangering the frictionless cooperation of the two arms which must be based on each partner working in, and to a certain extent from, his specific element.

Such forcing of naval strategic conceptions into army operations, and vice versa—with Napoleon the most flagrant in foisting military notions on his navies—is apt to occur when one arm is in political ascendancy over the other. This is always a dangerous situation and one which the Dutch, for example, tried to meet by making William the Silent and after him his son Maurice stadtholder, captain-general and admiral of Holland, or of the Union, as a means of constitutionally providing unicommand for combined operations. Even so, the interests behind sea power in the Netherlands in the year 1600 were able to impose their views on the conduct of the campaign in that year upon the unwilling military men.

Spanish fortunes in the Low Countries were then running low. Their financial distress and inability to pay their soldiers promptly made them mutinous. Although they still held Flanders, they had nearly lost the coast where only Nieuport and Dunkirk were still theirs, through which communication with the Peninsula ran. The ports were also dens of privateers whose activities were costly to Dutch commerce, that had grown despite the constant war. In view of this situation the maritime interest, largely identical with the urban burgher-manufacturing-commercial element represented by Oldenbarnevelt (Dutch statesman, 1547-1619), the latter becoming increasingly hostile to the military power for which the Orange family stood, clamored for an enterprise against the coastal places remaining to the Spaniards. The mirage, old by now, of sea power arose from the waves, the illusion that all lands can be conquered and kept under control from the sea. It was no mere accident that the same year 1600 saw the founding of the Dutch East India Company. Did not this company make its lucrative conquests by way of the sea?

Dutch Against the Spaniards

In vain did the land soldiers point out the unsoundness of

Sixteenth Century Projected Underwater Tank for Attacking Fortifications. (Bettmann Archive.)

the scheme that not only bared the lower part of the Meuse front and Holland, but risked practically all Dutch land forces which in case of a defeat would be without a line of retreat except the precarious one into Dutch-held Ostend. But fearing to lose the army command, Maurice (of Nassau, 1567-1625) and his officers yielded and accepted the plan, though it envisaged throwing the bulk of Dutch military forces from the north across the Scheldt estuary. Thence they were to march across Flanders to Ostend in order to operate against Nieuport and Dunkirk.

The maritime-civilian interest who knew that one reason for the good discipline in the Dutch army was regular pay, also believed the financial situation of the Spanish army was so desperate that their mutinous soldiery would not march if imperiled in Flanders. The military men pointed out that at the very threat against the survival of Spanish power in the Netherlands, Spanish corps d'esprit and religious zeal would rally the disaffected enemy soldiery to the banners of Archduke Albert. And in fact the Spanish troops, in a revulsion of feeling from mutiny to emulation for the privilege of volunteering for the vanguard, were restored to discipline so soon as the news of the Dutch advance came.

On June 21 and 22 the assembled Dutch forces were ferried across the Scheldt estuary from Zeeland, landed at Sas Van Ghent and marched to Ostend past Bruges, which was held by the Spanish. Maurice had just begun to blockade Nieuport when the disturbing report arrived that the Spanish were on the march and quite near. In strength they proved about equal to the Dutch, and threatened and finally severed Maurice's land communication with Ostend. Turning to receive them, the Dutch were now facing towards Ostend, their base, and the Spaniards towards Nieuport which was held by their own contingents.

Between Ostend and Nieuport, for nine miles runs a chain of dunes and a beach exceptionally broad at ebb time. The battle front was thrown rectangularly across this coast line, with the seaward wings close to the low water mark in the morning of July 2. In fact, the larger part of the opposing forces was placed on the smooth beach below high water mark, a feature which made the battle of Nieuport one of the most littoral in history. As the day proceeded the rising flood forced both parties to shift inland. But not only did the rising flood flank the Spanish right wing; it was also outwinged by the Dutch men-of-war that had been escorting Maurice's transports and were now brought closer to the shore by the flood tide. This movement of the Spanish forces into the dunes, which would seem to have ended the interference of the Dutch fleet due to the flat trajectory of their guns, did much to tire out the Spanish infantry which bore the brunt of the initial attack. The day was won by Maurice.

This was the only field battle of Maurice, who was essentially an organizer, and who would also seem to have lost all his zeal for battle through the experiences of the campaign of 1600. Infuriated by having barely escaped a danger into which other men's obsessions forced him, Maurice would not go on with the plans of his political foes; instead he hated them more and more until at last in revenge he sent Oldenbarnevelt to the block.

Victory Costly to Winners

Strengthening the garrison of Ostend he retired into Holland before July was over and could not be brought to enter coastal Flanders again even after the Archduke had begun the long drawn out siege of Ostend which the Dutch successfully resisted by sending small reinforcements by sea. After more than three years Ostend fell to the methodical siegecraft of Ambrosio de Spinola, a Genoese banker who had become a Spanish general and commanded the siege for the third year.

This siege proved ruinous to all concerned, and to Spinola personally. This banker had turned to the speculation involved in a *condotta* at the same time that Dutch capital was diverted to the infinitely more profitable investment of speculating in the limited liability shares of the East India Company; he risked his all, pledging his private fortune in order to obtain loans for the Spanish crown to carry on the war for which he was never reimbursed except by a title. The exhaustion on both sides brought on a twelve years' armistice concluded in 1607 and made more final in 1609.

While the British Commonwealth had no special marine regiments, its fleets still carried soldiers, their number depending on the tasks assigned to them. Occasionally infantry regiments were assigned to fleets as in 1652 and 1653, when two served with Blake. In fact, ever since the Middle Ages ships likely to encounter enemies were armed and provided with fighting men in addition to the seafaring personnel, and eventually gunners were added. On the Hanse ships the sailors, like those of ancient Athens, were provided by the burghers themselves and their sons, whereas the heavy armed men were mercenaries, knights, or men able to do knightly service, called as a group *"rutere tor see,"* riders at sea; in addition light armed men were taken from the common people, *de populo vulgari,* like the bakers and cooks, and, for entertainment during the long sea voyages, pipers and jesters (*joculatores*).[7] All these sea soldiers participated in between-ship battles as well as in landing enterprises. By the time of Elizabeth the largest ships would carry a complement of 450 seamen, 50 gunners and 200 soldiers, with a similar ratio in other ships, according to size. In the Cadiz expedition of 1590 the Queen's own ships carried no soldiers, but in their stead gentlemen volunteers who had been attracted by tales of treasure to be gained and who served without pay, whereas the hired ships, far more numerous on this occasion, had from 50 to 150 soldiers each on board. Drake's Lisbon expedition of 1587 counted with 3,200 English and 900 Dutch sailors, 17,000 soldiers and pioneers and 1,500 officers and gentlemen volunteers.

The need for soldiers to serve on board ships led to the forming of distinct, though not immediately permanent, corps of marines such as the regiments Vieille Marine (1627) and Royal Vaisseau (1635) in France, to guard ports and provide forces for ships. Even earlier the Maltese Knights had employed marines. Marines, although not called by that name, might be said to have won the battle of Lepanto. In Holland the States-General resolved in December, 1664, to add 25 men to each company of infantry and have 4,000 of these additional men serve on board the orlog ships; not enough were raised at once and the fleet that went to sea in the spring of 1665 had land soldiers on board in addition to these new *zee soldaten.* In that year, after some unpleasant experiences with such soldiers at Lowestoft which made Admiral Jan Evertsen write about both kinds that "they are so bad that it is unspeakable, creeping away from the enemy by twenty-five and more at a time," it was further resolved that the sea soldiers were no longer to return to their original infantry regiments but would be united in special regiments, one of them being headed by Col. Dolman, a Commonwealth man, and a special corps.[8]

MARINES ORGANIZED IN 1664

So close was Dutch-English naval competition at the time that both nations date their marines from the same year, 1664. An order in council of October 28 authorized the Duke of York and Albany's Maritime Regiment of Foot "1,200 land souldgers to be forthwith rayzed to be in readinesse to be distributed in His Majesty's fleete prepared for sea service." Recruiting proceeded very slowly at first; the service proved unpopular and the authorities believed as late as 1673 that a malicious whispering campaign was going on to discourage men from enlisting as sea soldiers.

On the average, the marines formed in England and later in the United States (1798) comprised for long periods of time about one-fourth of the total personnel strength of their navies. An English squadron commissioned in 1512 included 1,750 men who might be termed marines, in a total of 3,000; one of 1600 had 2,008 (plus 804 gunners) in a total of 8,346. For the later periods the following forces voted by Parliament which, of course, represent paper, not actual, strength, are indicative of the purposes for which marine corps were maintained, including landings. In the period from 1800 to 1934 the U. S. Marine Corps counted 180 landings, although not all were of a warlike character.[9]

	Total personnel	*Marines*
1762	70,000	19,061
1775	18,000	4,354
1783	110,000	25,291
1784	26,000	4,495
1786	18,000	3,860
1812	113,000	31,400
1816	24,000	9,000 [10]

The changing history of the marines in the various nations can concern us here only in so far as they played a rôle in landing enterprises, one of their three main functions, which have been defined as "fighting in ships; seizing and holding land positions necessary or advantageous to the naval and not also to the military operations of war; maintaining discipline of the ships and by expertness in handling arms to incite our seamen to the imitation of them." To these might have been added their special function in Anglo-Saxon countries of acting as if they were not soldiers, but part of the politically less offensive naval forces. While officers could change from the guards and the line to the marines in England and back again—as did John Churchill, the later Duke of Marlborough, and George Rooke, the later admiral, both of whom served together for a time as ensigns in the Marine Regiment—it can hardly be said that their existence formed the matrix of forces and of thought specially suitable for, or inclined towards combined operations. For on the whole marines everywhere were appendages of the various navies, to do the navies' landing work and spare them, except on larger occasions, the necessity of cooperating with armies. It was this customary friction that produced the Royal Marine Artillery, which was founded in 1804, strongly recommended by Nelson who, himself, was never very good as a cooperationist, after friction between artillery and naval officers had grown unbearable.

Most officers preferred this arrangement of marine forces, whether of light infantry (R.M.L.I.), or of artillery, and the proposal of Admiral Lord St. Vincent in the first years of the 19th century to have every regiment of the line serve afloat for a time, was in its spirit quite exceptional. The services remained estranged and the marines were less of a connecting link than they might have been, and even though they were naturally thrown together with the army forces in most of the subsequent landing operations, as in the taking of Gibraltar in 1704 where the actual assault, however, was made by the seamen of the Navy, after a landing had been effected on a sandy beach some six miles distant from the Rock.

British Taken Unawares

The first glorious landing in which marines participated occurred during the Dutch descent on the Medway, June 10 and after, 1667, where Dutch marines and land soldiers under the command of Col. Dolman were disembarked to wreak destruction among the English ships and shore establishments and to take Fort Sheerness from the land side while the fleet bombarded it from the water so effectively that the garrison evacuated it before the real assault was made. This enterprise had been forced upon the unwilling admirals, De Ruyter included, by the De Wits, one of whom, Cornelis, was civil commissar with the fleet and nominally in chief command. It was Cornelis who forced the captains up the river,[11] proving once more the usefulness of the outsider in war, even should he belong to that much decried modern category, the political commissars.

Jan de Wit had insisted in the deliberations of the Lords' States that an entry into the Thames "would touch the enemy more sensibly than any other enterprise. They should commit every hostility there. If the English desired peace, they would grant a more advantageous one to the Provinces if they had suffered from the arms of Holland." An English secret agent, the famous Aphra Behn, (1640-89, poet, dramatist, novelist) employed on espionage service in the Netherlands by the English government, had learned late in 1666 of the Dutch design, but her report had been disbelieved in London.

The descent which could have been carried much further, militarily speaking, if more troops had accompanied the Dutch fleet, was of far reaching political impact abroad. It led to the Peace of Breda and in England, where the "roar of foreign guns" was "heard for the first and last time by the citizens of London," [12] to the fall of the all-powerful Clarendon. A gibbet was set up before his house by the incensed people and, "as usually happens under similar circumstances, the Government is hated and the people desire a change." He was accused of having sold England when he sold Dunkirk to France in 1662. "The shame and loss inflicted on England by the foray of the Dutch fleet into the Thames are attributed to him, since he delayed preparing the armaments which had been voted by Parliament." [13]

A few weeks after Medway the first encounter between marines on either side took place in connection with the Dutch attempt on Landguard Fort near Harwich (July 2). Some of De Ruyter's ships being placed exactly to windward, such clouds of smoke from them rolled upon the Fort that the Dutch landing troops were hidden from the sight of the English marines. Thus protected, the Dutch brought scaling ladders and hand grenades and came close to the Fort which repulsed them after half an hour. A repeated, but half-hearted attempt an hour later was no more successful and then the whole force reembarked, not pursued by the unready English, or as Pepys remarks, "otherwise we might have galled their foot." As an

Admiralty man, Pepys incidentally ascribed the Dutch's lack of success to "our great guns."

NOTES, CHAPTER 20

[1] For Wrangel see Lefevre Portalis: Jean de Wit. Paris, 1884. I, 253.

[2] Clowes II, 97.

[3] Thomas Carlyle: *Cromwell's Letters and Speeches.* Everyman ed. III, 118, 122.

[4] Ibid III, 114. Carlyle cites *Journal of an English Army in the West Indies;* by An Eyewitness. Harl. Miscell. VI, 372-90.

[5] Carlyle III, 98.

[6] Carlyle II, 132, 145.

[7] Jähns: *Handbuch,* 1264.

[8] For the history of the Dutch marines see P. A. Leupe and F. A. Van Braam Houckgeest: *De Geschiedenis der Mariners van het jaar 1665 tot op heden.* Nieuwediep, 1867.

[9] Captain Harry Alanson Ellsworth, U. S. Marine Corps: *One hundred eighty landings of U. S. Marines 1800-1934.* First ed. 1934, n.p.

[10] *Encyclopedia Britannica,* XIth ed., vol. XIX, 305

[11] Sirtema de Grovestines: *Guillaume III et Louis XIV.* Nouv. ed. Paris 1866. I, 424.

[12] Macaulay: *History of England.* I, 193.

[13] Calendar of State Papers and MSS relating to English affairs. *Venice.* London, 1935. Vol. 35. no. 206, 228, 230.

Chapter 21

VAUBAN REPELS A LANDING ATTEMPT

ONE OF THE LEAST KNOWN of the more than 300 actions in which Vauban (1633-1707. French military engineer, famous for the system by which he fortified the cities of France for Louis XIV.) participated, was the repelling of an Anglo-Dutch landing attempt at Camaret, near Brest, in the summer of 1694. It provided the great engineer with an opportunity for a practical and personal test of his views on landings. He had already advised the introduction, or rather re-introduction, of the essentially Mediterranean galleys to the Atlantic, to serve as patrol vessels, mobile screens for heavier ships when inshore, or for descents upon the Orkneys, or even the English coast.[1] In a memoir of 1683 on a landing made then at Belle-Ile-en-Mer, he wrote:

> If I am not the most mistaken person in the world these are of all actions in war the ones most exposed to bad accidents, where the chances (advantages) are least equally divided and where the odds are always two to one in favor of the defenders, provided that good care is taken to make use of the advantages of the locality and not to submit to the risk of an instantaneous fight which might lead into a general affair. . . . I call risking a general affair and to expose oneself foolishly if a party undertakes to engage itself headlong with all the troops available and without any further precautions, to expose itself to the landing corps and under the cannon of his ships. . . . But rather should one wait for the enemy in good retrenchments and well beyond the reach of his advantages, in that I see little risk, and I am so little filled with the small measure of risk which must be run on such occasions that if I should command in such a region where this sort of thing might happen and should have but 200 men at my disposal, I should be found with one hundred men at the landing place.

These observations of Vauban's must be judged within the framework of the strategical thought of his time, and in particular with reference to the available battle and the danger of taking risks in war. "We must conduct our business in such a way that it will not be subjected to the hazard of a battle and that we do not proceed to battle except on the extreme necessity," was the advice tendered to Count Maurice of Nassau by his cousin, William Ludwig.[2] The risks run by the great French engineer at Camaret in 1694 were considerably reduced by the circumstance that the French government, thanks to the treasonable information given it by a person high in the English counsels, possibly Marlborough, had known since early May that an attempt on Brest was being prepared.

So soon as this was learned, Vauban was despatched to Brest, provided with full powers to summon the local ban. Regular troops were also set to march to that region where there was already a garrison of 1,500 men, including two battalions of marines. Vauban arrived on May 23, setting to work at once. He could not hope, he said, "to prevent the landing altogether, but he was decided not to allow the enemy to put his foot on land without making him draw his sword." Soon concluding that the beach of Camaret offered the greatest temptation for a landing, and hence the greatest risk to the French, he placed there one battalion and made the militia and the soldiers dig positions for infantry and artillery which, comparatively speaking, was amply provided and well served by marine artillery.

On June 16 a large fleet was signalled from Quessant (Ushant), the westernmost of the islands off the coast of France, approaching slowly against an eastern wind and anchoring eventually in a bay between Bertheaume and Cameret. Despite their late arrival (June 17, at 5 p.m.), the English at once undertook a reconnaissance of the bay around which Vauban had now assembled all his forces, but they were repelled by the lively fire of the batteries, the existence of which surprised them, and further activity was postponed until the following day.

French Repulse English

Fortunately for the French rather than for the English the morning of the 18th was foggy until 10 o'clock. With the lifting of the fog the English attack began with ship's guns opening up against the beach batteries in order to cover the landing parties. A first detachment attained the beach safely, but fell into the enfilade of a battery of two guns which brought some disorder into their ranks. At this moment of uncertainty, the ideal moment for the counter-attack, the French marines, stationed nearest the disembarkation point, fell upon them, supported by the militia and small bodies of horse. The English were thrown back to their landing boats, the greater number of which had been run aground on the sands.

The tide was now subsiding. The few boats still afloat were pulled down by the weight of the fleeing soldiers who crowded into them. And practically all of the 1200 or 1300 men who had set foot on land were either killed, or made prisoner. The troops in boats who had not yet landed, and who had to witness the misfortune of their comrades, hesitated to carry on the attack and turned back, not without suffering losses from the fire of the land batteries, mounting 31 cannon and 2 mortars, with a range that reached the ships, and which caused one Dutch vessel to run aground.

After a council of war the enemy, having suffered nearly 1700 casualties, as against 45 in the French force, resolved to abandon the enterprise and make no further use of his vast superiority. For while Vauban had only between 11,000 and 12,000, of whom hardly 3,000 were well trained, the English fleet of 156 sail, or 217 if small craft are included, carried from 6,500 to 7,029 guns and 39,947 men, or 41,455 if the small craft crews are included. The experience, which corresponded to the failure of a speculation in which considerable funds are wrongly invested—stock speculation had just got under way, hence the comparison—was disheartening to the British strategists whose enterprising spirit was dampened by it in the wars immediately following.

While the British as partners in the various coalitions against Louis XIV carried troops regularly to the continent, by way of Holland and into Spain, there were no serious attempts to land on French coasts. And Daniel Defoe did little more than provide some war-journalistic camouflage when he wrote in June, 1706, that the French "are fortifying as well where you will not come, as where you will come. The whole country is harassed; vast expense, intolerable obstructions to the business of the season, infinite loss both to landlord and tenant is the effect. . . . And all this while you are in the Isle of Wight. When you are embarked and sailed, as you approach one place, another is cleared; they know not what to do—but while you

BRIELE.

hover thus about 'em, and they know not where to expect you, they are in the utmost confusion—and thus you are influentially upon action; and really, speaking of consequences, I know not whether they lying thus at an uncertainty, and expecting you they know not where, is not one way as fatal to them as your landing will be, though you should have all the success you can expect. Then you'll possess a part of France, now you perplex the whole." [3]

Loss of Surprise Costly

To Vauban it appeared that "the enemies have gone the right way to work, for they have not lost a moment's time. As soon as they arrived they attacked in the place where I had always feared them; in short, they have thought very well, but not executed so well." On the whole, the English proved well enough informed of the weakness of the garrison of Brest as well as the topography of the landing beach, but with the secret of their ultimate destination betrayed, this availed them little as compared with the loss of the element of surprise. The chosen landing spot had the further disadvantage, a handicap weighing especially against the invaders, that the beach was of such small extent that the attacker could not land on an extended front, but had to disembark in successive waves against a narrow front.

Still, there remained chances for success. Vauban's weak forces were distributed over a distance of less than three miles and a resolute landing body might have taken the entrenchments, which were of no particular strength, and have held them if supported by the second wave of the landing corps. But the disorder rarely separated from an enterprise of this kind, heightened by the flanking fire from the beach battery, the fact that the leader of the landing corps who "contrary to true principles" had gone with the first wave, had been mortally wounded, and the general discouragement caused by the discovery that the intended surprise had failed, all these factors took the starch out of the landing corps.

While the affair of Camaret was a small one for a great soldier like Vauban to engage in, it still showed the versatility of the man who had proved his remarkable talent both in the defense and in the offensive, and his grasp of first principles involved in the defense against landings. He was aware of all that entrenchments—even if ever so weak, they are still the first symbol of the strength of the earth, the essence of all geotactics—can mean either to attacker or defender. To the attacker they must signify further strengthening of the already formidable appearing land which he aims to approach and conquer from the water.[4]

NOTES, CHAPTER 21

[1] Henry Guerlac: Vauban; in: Earle (ed.): *Makers of Modern Strategy,* 39.

[2] Hans Delbrück: *Geschichte der Kriegskunst.* IV, 353.

[3] Quoted *New York Times,* 5 May 1944.

[4] Largely based on A. Fouché, col. du génie: *Au sujet de diverses tentatives de débarquement faites sur les côtes de Bretagne. Revue du génie militaire,* vol. 19 (1900), 441-472. And an article in *Revue Militaire, Archives Historiques,* no. 13, April, 1900. Vauban's notions about landing enterprises and coastal defense seem not to have been treated systematically as yet. See, however, his *Projet d'ordre et de précautions que M. de Vauban juge qu'on peut prendre contre l'effet des bombes au Havre et qui peut servir pour les autres villes et ports exposés au bombardement,* in his *Mémoires inédits.* Ed. Augoyat. Paris, 1841.

Chapter 22

CHARLES XII'S BAPTISM OF FIRE IN A LANDING

NOT OFTEN HAS A GREAT MILITARY LEADER received his baptism of fire in an amphibious enterprise. It was reserved for, and characteristic of, the often wrong-headed genius of Charles XII of Sweden to find it on such an unusual occasion. Charles was yet undergoing his politico-military appreticeship at only eighteen years of age when a coalition was formed against Sweden by her enemies, Denmark, Saxony-Poland and Muscovia, the coalition of the Great Northern War, which forced the young king to war for the rest of his days. Turning first against Sweden's nearest foe, Denmark, he equipped a fleet which, unopposed by the timorous Danish admiral, and headed by an admiral almost equally so, was compelled by Charles to attempt the passage of the Flinterend, the eastern channel of the Sound hitherto considered unnavigable.

A landing was achieved at Humleback, three miles north of Copenhagen, on August 4, 1700. There a Danish army was assembled in opposition. Before his ships had touched shore, Charles, sword in hand, threw himself from his sloop into the sea which reached above his girdle. The Swedes, as a French 18th century Dictionary of Sieges and Battles has it, "eager to follow him, flew to the shore, in spite of the discharges of the enemy troops." [1] The king who, according to Voltaire, had never before heard musketry fire, asked one of the bystanders what the little whistling sounds meant. "That is the noise of the bullets fired at you." At the same moment, the speaker and another man in the king's entourage were hit. "Good!" said the king, "That will henceforth be my music."

After they had killed some thirty enemies, the Swedes made themselves master of the Danish entrenchments and their artillery and "thanked God for the victory." Receiving 12,000 reinforcements, they threw up at once redoubts about Copenhagen. Deeply impressed by the success of the Swedes, who seemed on the verge of succeeding where they had failed 50 years earlier under Charles' father, in the attempt to win Copenhagen and with the capital the whole double kingdom of Denmark and Norway, the Danes sued for peace. This sector of the Great Northern War was quiet after less than six weeks.

Although the career of the Swedish king brought him so far away from the sea, in his fight against Russian land power, he nearly wound up with another amphibious enterprise. "Continental entanglements," the acquisition of the former Swedish-owned bishoprics of Bremen and Verden by George I as Elector of Hanover, earned England the threat of a Swedish invasion following on the heels of an attempted Stuart restoration. Charles allied himself with the Stuart Pretender and his diplomatic representatives at London, Paris and the Hague entered upon and maintained a correspondence with the disaffected element in England. They wrote and published pamphlets of a seditious character in which the English people were urged to insist on a reduction of their army and sending home George's hated German auxiliaries. George caused the papers of the Swedish ministers in London and the Hague to be seized and laid a selection of them before Parliament which was designed to show that a conspiracy had been formed to dethrone him and put a Stuart in his place, and that it had been averted by his own vigilance.

The plot included the purchasing of ships of war and transports. These were to assemble in Gothenborg in March, 1717, when the eastern winds would be steadiest and proceed to England with 8,000 foot and 4,000 horse, besides arms and ammunition for 15,000. The discovery of the plot and the death of Charles in 1718 ended this danger, even if it had never been as serious as George and his ministers pretended.

NOTES, CHAPTER 22

[1] *Dictionnaire historique des sièges et batailles mémorables.* Paris, 1771. I. 477.

Chapter 23

18TH-CENTURY FAILURES IN CONJUNCT OPERATIONS

BRITAIN'S GLORIOUS REVOLUTION that put William of Orange upon the throne began, militarily speaking, with the landing at Torbay (Nov. 5, 1688), undertaken in "the fittest place in England for landing of cavalry" (Molyneux). During the century from then on to the French Revolution there were few landing operations or plans for them to which England, or France were not exposed or of which they were not objects. As the foremost sea power England almost monopolized the starting of littoral war, a war that, as an English military writer of the time put it, "when wisely prepared and discreetly conducted, is a terrible sort of war. Happy for that people who are sovereigns enough of the sea to put it in execution! For it comes like thunder and lightning to some unprepared part of the world." (Molyneux). By count from Sir Walter Raleigh's time to 1759 England made more than seventy such enterprises. During the 18th century, in England these, then usually called "conjunct operations," were what we today term combined operations.

The foremost impression derived from a survey of these enterprises, is that the majority seem to have sprung from the imaginations of civilian governors, fascinated by the thought of Molyneux's "thunder and lighting" which a successful landing might visit by surprise upon some unready part of the world, far oftener than from the considered proposals of either generals, or admirals or both, who apparently thought them neither feasible, nor decisive. Consequently and for that reason alone, one might be tempted to say, failures abounded.

The first special treatise on *Conjunct Operations; or Expeditions That Have Been Carried on Jointly by the Fleet and the Army, with a Commentary on Littoral War.* By Thomas More Molyneux (London, 1759) was itself a sign that such enterprises had come to be visualized as specific military problems, but that they were still managed with much confusion. "The many gross miscarriages," said Molyneux, sprang "from bad management," and although the author, about whom little seems to be known, described himself as an English soldier, it was his opinion that "in general the Army has been more deficient on their part than the Navy."

The book itself was conceived in "the humble hope of lessening in our future conjunct expeditions the effusion of British blood, by sending our armaments more prepared, with terror for the enemy, safety to those who compose them," with the intention "to reduce if possible this kind of warfare to some safe regular system, to leave as little as we can to Fortune and her caprices, we may say here: to wind and water." The author endeavored to apply principles of operation which he seems to have derived from the general rationalism of the time rather than from other better managed branches of warfare, to "the conducting of a military, naval, littoral enterprize" which was "never rightly pursued."

Breaking down the available statistics of English landing enterprises after 1600 to 1758 into large armaments employing 4,000 and more soldiers, or marines, and small ones below that number of participants, Molyneux strikes the following balance:

Theater of War	*Total Number*	*Miscarried* in ( ) large-size ones	*Successful* ditto in ( )
Africa	14	5 (4)	9 of which 7 were undertaken by sailors
N. American and W. India	23	13 (5)	10 (2)
Europe	30	19 (17)	11 (5)

Failures were most common in Europe and along the French coast, with 19 miscarriages out of 30.

James' Failure in Ireland

The superiority of British sea power between the British and the French Revolutions was not always sufficient to prevent French forces from landing, or making attempts, on the shores of English possessions and even on the homeland. Disorganization, or reorganization of the English forces after the Glorious Revolution enabled Louis XIV to land an army, accompanied by the dethroned James II, in Ireland in March, 1689, without immediate opposition, while an indecisive action later fought in Bantry Bay allowed his army to maintain its lines of communications. The Battle of the Boyne (July 1, 1690), where William III in person led the Protestant British against the

Popish invaders, sent the invaders home. They, including James, voyaged back safely enough since on June 30 the Anglo-Dutch fleet had been defeated by Tourville off Beachy Head.

At that time the French, in command of the Channel, had no army ready to be carried across to England and when they had one prepared, on the eve of the battle of La Hogue (May 19, 1692), the naval conflict went against them. After this sea fight, the English in pursuit of some of the fleeing French ships sailed into the anchorage of La Hogue where the troops destined for England, with James II, were encamped. Within their sight thirteen ships were burned by the English who brought to an end the project of landing in England through a battle which in its final stages included horse-to-ship fighting.

The two services steadily continued to distance as the 18th century continued; in formulation of ideas there was complete specialization and separation and very little, if any, interchange of conceptions between the armies and the navies. "Service interest," in England usually represented by officers of either arm in Parliament, was barefaced, and more so in England than in France where a strong administration would from time to time pull the two together, but to the almost entire neglect of the navy.

In England the naval officer body, once divided into the "tarpaulins," or officers trained from youth as sailors, and the untrained gentlemen who had obtained their rank through influence at court, or in Parliament, joined forces and every naval officer became both a gentleman and a sailor, a consolidation which did not contribute to open-mindedness towards the military and their argument, or receptive to grand strategical designs, or even better tactical cooperation. The armed forces could not often be regarded as one communicating system. Where and when they were thrown together in "conjunct operations," cooperation oftener than not was poor and its absence or weakness a frequent cause, main or contributory, of the many failures in this field. Molyneux, the 18th-century specialist in such operations, found the English enterprise against Port Royal, now Annapolis in Nova Scotia, in 1710 practically the only case of successful cooperation between English army and English navy in the first fifty years of the century (II, 39), a judgment which might be exaggerated.

If politics of one kind or another, not to mention the private interest of sailors eager for prizes, or soldiers for opportunities to loot on land, in the days when discipline was definitely worse in the navies than in the armies, interfered with successful co-operation between the two branches of a national force, how could coalition warfare at sea, or overseas be expected to go smoothly? While Marlborough and Prince Eugene stood out as shining examples of how to lead a war of coalition on land, their example, or desires were not strong enough to extend their genius to the theater of war at sea. English, Dutch and German sea and land captains quarreled off Cadiz where the requirements of coalition in the naval part of the war, as conceived by William III and inherited by Marlborough, had brought an armament that included an English and a Dutch fleet, 10,000 English infantry, of whom 2,400 were marines, and 4,000 Dutch soldiers. This armament arrived before Cadiz on August 12, 1702, and next day some twenty admirals and captains and generals met in war council in Admiral Rooke's cabin and wrangled until they had settled on the most innocuous course, which was the worst. Rooke was content to have it so because the whole enterprise ran counter to his narrow notion that the Grand Fleet belonged in the Channel and not, in or near the Mediterranean where the strategy of William and Marlborough would have placed and kept it.

The more daring naval officers proposed, or offered to sail into the inner harbor of Cadiz with a squadron and destroy the French and Spanish fleets there before a boom could be thrown across the entrance. The more courageous soldiers made a suggestion, vetoed by Rooke, that they should land on the narrow neck between Cadiz and the mainland, under cover of the fire from the fleet and from there storm the as yet poorly prepared city. Again, the whole armada might have sailed away to another Spanish port as their instructions allowed, if they found Cadiz too well defended. Instead, the cautious sailors prevailed and insisted that the military forces be landed across the bay from the city.

British Fiasco at Cadiz

The landing was effected against very little resistance and with the loss of but twenty men, drowned in the surf. After that, the army did nothing for 26 days beyond robbing and plundering the surroundings of Cadiz, raping, and by their misdeeds estranging all substantial Spanish men from the Allied cause; and stealing from the English and Dutch merchants who had long been doing business at Cadiz under Spanish firm names. In the meantime a treasure fleet had safely, but unknown to them, arrived at Vigo.

After an ineffective attempt at taking a fort closer to Cadiz where the garrison had been reinforced and reinvigorated during the delay, it was resolved in the middle of September not to land elsewhere in Spain, as some of the military leaders proposed, including Prince George of Hesse, who later took Gibraltar. Instead the armada was to return home before the equinox, insisted Rooke, who never exerted his authority except for the safety of his ships. With him voted the military officers who by then had looted to their satisfaction. While this majority settled the decision, admiral and generals had practically broken off relations, the military accusing the sailors of having deprived them of the chance of taking Cadiz and Rooke charging the soldiers with wasting their time ashore on plundering. The army was re-embarked minus some Irish troops who after taking their share of the plunder, elected to enter the Spanish service.

If redemption can ever be partial, the inglorious armament achieved this on its way home. The treasure fleet, escorted by a French squadron, had been brought into Vigo instead of a safer French port, for the Spanish had insisted that even in a coalition war the gold must be deposited in one of Philip's ports and not in one belonging to his grandfather. By the time Rooke's fleet was on its way back the royal share of the precious metals had been brought ashore and carried inland, the ships and the private freight of metals and merchandize still remaining at Vigo when Rooke passed by the harbor mouth without looking in. But before he entered the bay of Biscay he learned of the fleet inside the deep Vigo bay. A council of war was convoked and the majority, the Dutch admiral voting with it, decided that an attempt should be made to force Vigo despite protective boom and shore batteries.

In reversal from apathy to energy, the fleet sailed into the deep and narrow bay under and against the fire of batteries and the Franco-Spanish fleet behind the boom. An infantry force of 2,000 was landed, marched against the strongest of the forts, scattered guerrilla defenders, taking the fort with hand grenades and then assisting in breaking the boom. The Franco-

Spanish fleet was totally undone, destroyed, or captured. But Rooke could not be induced to stay at Vigo for the winter and all sailed home, in the smaller, yet not inconsequential, victory trying to forget the greater shame of the Cadiz failure, to face parliamentary inquests in England combined with thanksgivings.[1]

Finding such difficulty in working with soldiers in uncongenial company, sailors often preferred to undertake overseas enterprises with their own forces. The sailor-politician Admiral Vernon, who had boasted in 1739 that he would take Porto Bello with but six ships, was sent out later in that year with just that many ships by the Ministry, who hoped to call his bluff. But he made good his boast, took Porto Bello, castle and town, and thereby inspired a wave of jubilation and hero-worship in England. But as Molyneux sums up his story of conjunct operations: it is "rarely in our power to entertain the readers with two successful expeditions one after the other" (I, 160), and Vernon's next projects were failures. The general public at home put all the blame on the military partner as did Vernon himself, although his imperious temperament was at least half responsible for the debacle. A satirical print on the Cartagena expedition of 1741, in the coarse style of the time, carried these lines:

> . . . . one Godlike Admiral *confondu*
> With Rage and Shame, dat all's *perdu;*
> Of conjunct Expedition sick,
> And dis damn'd poltron, Army-Trick:
> In *Grand Mepris* he turn de A-se
> On those dat play dis sh-tt-n Farce . . . [2]

When a similar enterprise was launched against Santiago de Cuba in July, 1741, the result was equally disastrous. Vernon publicly ascribed this to divided command, as a good Tory laying the responsibility at the door of the Whig Ministry, and not in the least to his own "brutal insolence."

Novelist Describes Cartagena

On the Cartagena enterprise we have the observations of a participant, in Smollett's *Roderick Random*. Smollett had been present as a surgeon's mate and saw and judged things from the sick bay of a man-of-war in a way accepted by historians as essentially true. He describes the arrival of the fleet which carried some 12,000 soldiers on board; this was a greater strength than was usual on such expeditions which, as participants complained, were often too small, possibly on the basis of the principle of limited liability transferred from business to military enterprise.[3] It was in March, 1741, that, says Smollett:

> We came to an anchor off Cartagena and lay at our ease ten days longer. Here, again, certain malicious people took occasion to blame the conduct of their superiors, by saying, that in so doing they not only unprofitably wasted time, which was very precious, considering the approach of the rainy season, but also allowed the Spaniards to recollect themselves from a terror occasioned by the approach of an English fleet, at least three times as numerous as ever appeared in that part of the world before. But if I might be allowed to give my opinion of the matter, I would ascribe this delay to the generosity of our chiefs, who scorned to take any advantage that fortune might give them even over an enemy. At last, however, we weighed, and anchored again somewhat nearer the harbor's mouth, where we made shift to land our marines, who encamped on the beach, in despite of the enemy's shot, which knocked a good many of them on the head. This piece of conduct, in choosing a camp under the walls of an enemy's fortifications, which I believe never happened before, was practised, I presume, with a view of accustoming the soldiers to stand fire, who were not as yet much used to discipline, most of them having been taken from the plough-tail a few months before.

According to Smollett these raw troops had been sent out and seasoned troops left at home, either because the Ministry were loth to risk their better men on such desperate service, or because the officers, "enjoying their commissions as sinecures, or pensions, for some domestic services rendered to the court, refused to embark in such a dangerous and precarious undertaking for which refusal, no doubt, they are to be much commended."

After this waste of valuable time and the utter surrender of surprise when "the better half at least should be over before the surprise subsides of its having begun" (Molyneux II, 47), the English determined upon a "systematic siege," another three weeks going by before their land batteries could open fire on the forts. When the Spanish had been driven from the outer forts, on whose strength they had mainly relied for the defense of Cartagena, by the combined fire of ship and land batteries, the rest of the fortifications seemed ready for the taking, or so it seemed to the inmates of the sick ward for, as Smollett narrates:

> Indeed, if a few great ships had sailed up immediately, before they had recovered from the confusion and despair that our unexpected success had produced among them, it is not impossible that we might have finished that affair to our satisfaction, without any more bloodshed; but this step our heroes disdained as a barbarous insult over the enemy's distress, and gave them all the respite they could desire, in order to recollect themselves.

Army Nearly Destroyed

The sick bays were overcrowded. Bodies of the dead drifted about the harbor. Men wasted under the tropical sun, badly fed and lacking water, "with a view to mortify them into a contempt of life, that they might thereby become more resolute and regardless of danger," according to the satirist. After having put garrisons into the outer forts and re-embarked the rest of the soldiers and artillery, which took another week, the armament proceeded to force the inner harbor and a castle commanding the town.

> Whether our renowned general had nobody in his army who knew how to approach it in form, or that he trusted entirely to the fame of his arms, I shall not determine; but, certain it is, a resolution was taken in a council of war, to attack the place with musquetry only. This was put in execution, and succeeded accordingly; the enemy giving them such a hearty reception, that the greatest part of the detachment took up their everlasting residence on the spot.

By this time the army was reduced from 8,000 who had originally landed, to 1,500 fit for service and "the Daemon of Discord, with her sooty wings, had breathed her influence upon our councils; and it might be said of these great men as of Caesar and Pompey, the one could not brook a superior, and the other was impatient of an equal; so that, between the pride of one and the insolence of another the enterprise miscarried, according to the proverb, 'Between two stools the backside falls to the ground.'" After some ineffectual cannonading from the ships, the rainy season set in and drove the English from their incomplete conquest to Jamaica.

The successful attempts on Porto Bello in modern Panama, and Louisburg, Nova Scotia, April-June, 1745, where the military force of the expedition was provided by the colonial militia under Col. Pepperell, and also the failures at Cartagena and Santiago de Cuba fell into and formed part of the British contribution to the American sector in that complicated set of wars called the Austrian War of Succession (1740-48). Another landing attempt was made on the French coasts, in 1746, opening a long series of either unlucky, or inconsequential ventures which came to be called "breaking windows with guineas" on account of their costly futility. They were usually directed at some limited objective on the French coast of the Atlantic, undertaken in the hope of drawing, or keeping away considerable French land forces from the main theater of war by the appearance of ships and comparatively weak landing bodies. This aim was usually not achieved.

The first of these undertakings was directed against the French competitor of the British East India Company, the Compagnie des Indes, which had its dockyards at Lorient, two kilometers from the mouth of the Blavet. When an English squadron under Admiral Richard Lestock approached it on September 30, 1746, the place was in a good state of defence from the waterside although almost open from the land. Finding the landing impractical at Port Royal, across the water from Lorient, the fleet of 38 ships sailed along the coast as far as the Baie du Pouldu, disembarking here a force of 7,000 men who were not opposed, the weak coast guards retiring to Lorient, 12 kilometers away over a much intersected terrain and poor roads. The English were unable to transport more than four guns and one mortar under the circumvallation wall of Lorient.

Some ineffective shots were exchanged over two days. About the time capitulation was being considered by the defenders, they saw the English in great haste evacuating their camp, leaving their artillery behind and returning towards their ships. It appeared that the Admiral had signalled that he expected bad weather, while at the same time news was received of concentrations of French militia to raise the siege. After burning several villages in the neighborhood, the English re-embarked and on October 8 set sail for home ports. Col. Fouché, of the French *Génie,* who has passed in review the various descents on the coast of Brittany, finds the operation very poorly managed so far as intelligence work is concerned. The English could, and should, have known of the poor state of the land defences of Lorient, that the season was too far advanced for the ships to anchor long in the Baie du Pouldu and that the distance from the ships to the siege camp, 12 kilometers, was too great to allow a full supply with artillery and ammunition.

The Seven Years' War

On the seas the Seven Years' War brought a real season of "window-breaking-with-guineas." The great Pitt's small projects, beginning late in 1757 and becoming most numerous in the summer of 1758, mostly came to grief. The best to be said for them was that they relieved the British nation of the fear of invasion. The objective was usually a relatively unimportant place like St. Malo, or Rochefort, with troops landed at a safe distance from the guns of those strongholds, which was on the whole easier on the navy than on the troops engaged. The typical enterprise was based on speculations on the greater mobility of the troop-carrying fleet over enemy troops moving over bad country roads, where such existed. But once the British troops were landed, they were apt to move even more slowly than the French, or Spanish, waiving the possibility of surprise as if times were still feudal, and relapsing into sluggishness, indecision, carelessness, licentiousness, or occasionally toying with "method" and pedantry. Notions already becoming obsolete, such as precedence and post of honor, were allowed to interfere in these highly technical enterprises. "The absurdity of picking out particular corps to make always the first debarkation under the notion of the point of honor" is mentioned by Molyneux, a participant with James Wolfe in the Rochefort expedition of 1757, who also thought it was:

> Highly necessary to disregard this phantom notion of honor which has been so strictly adhered to during the last

> descents on the coast of France, though we believe at the affair in St. Cass' Bay it was pretty well beat out even of those who respected it the most. By quitting this shadow we gain these solid advantages: we preserve for the generality our regiments whole and entire: no particular companies nor particular regiments are harassed and fatigued more than their share. Nor shall it be the fate of the tallest, stoutest and perhaps the best troops to be exposed to every ordinary and extraordinary danger as well as labor, because it is their fortune to be the tallest and stoutest men. (I, 105).

So far as we see, all these landings, disdaining surprise as if by intention, took place in daylight. Indeed, disciplinary and tactical cohesion of the armed bodies of the *ancien régime* was generally too precarious to allow nightly movements of any kind, and less so nocturnal, or even early morning landings, so suitable for creating that state dreaded by commanders, disorderliness, or even panic which frequently seized soldiers during this kind of enterprise.

Operations at St. Malo

The first of the landings of 1758 in Brittany was directed at St. Malo, the base of daring and troublesome French privateers. The armament, under the command of Charles Spencer, third duke of Marlborough, sailed on June 1, transports in the center and men-of-war on either side, although the slower transports soon fell back. Since the place could not be assaulted directly, the fleet first anchored 10 kilometers from St. Malo on June 4, to weigh anchor again on the 5th, arriving towards noon off Cancale where the governor of St. Malo had followed them with 300 out of his 500 infantry and a company of dragoons.

The coast at Cancale is steep, up to 60 meters high, inaccessible except at three or four points where narrow roads lead from the beach to the plateau where the village is situated. The French troops, to whom had been added coast guard militia, did not resist landing and retired from the fire of the English ships which silenced the weak coastal batteries in the vicinity. The English occupied Cancale and the French, in order not to risk all their forces in that part of the province, drew back along the shore unmolested by the English.

The latter brought 14,000 men to the shore, including 1,000

cavalry and some light artillery. With, and for, these troops they prepared a beachhead to fall back on should they meet with reverses on the march to St. Malo. Not before another two days did they reconnoitre in that direction and towards Saint-Servan, today a sister town of St. Malo. Off the latter place, poorly defended, they burnt a merchant fleet estimated to be worth 30 million francs by some, but at only 3 million by others.

St. Malo itself, on a granite island connected with the mainland by a causeway 650 feet long, was hard to take by storm, but was comparatively helpless against artillery fire from the surrounding banks of the Rance, particularly since many of the houses were timber-built. But the English never brought artillery all the way across the 12 kilometers from Cancale, transport difficulties proving too great, and remained inactive until the 9th when they retired slowly to Cancale.

The French forces, not greatly reinforced in the meantime, followed with caution and refrained from hindering the re-embarkation which took from the 11th to the 13th and which was systematic, the invaders leaving nothing behind excepting a few scaling ladders. Held up by bad weather in the well-protected bay of Cancale until the 17th, the delay forcing them to give up other landing plans, for by that time fodder for the horses on board was exhausted and other provisions were running low, the fleet sailed away for Jersey, or the Isle of Wight.

The absence of French resistance to these operations seemed to invite a repetition with practically the same forces. They appeared on the 5th of August off Cherbourg which was defended by batteries and a garrison of 6,500 men, and where intermittent work on the fortifications had been halted for some time. After an insignificant cannonade on the 6th the fleet sailed along the coast and disembarked some 5,000 men on the beach of Urville, 10 kilometers from Cherbourg under the eyes of the French forces pinned down by the ship artillery. The French commander in the Cotentin who happened to be at Cherbourg, despite numerical superiority in his favor did not combat the British and retired inland. Since Urville had been thought inaccessible and was left without entrenchments, he fell back as far as Valogne to wait for reinforcements. The English marched on Cherbourg, levied a contribution from the town, destroyed the jetties, burned, or sank a number of ships and reembarked on the 16th, their movements being retarded by sloth and drunkenness among most of the landing force who had guzzled over-generously on wine of the country. Only after one soldier had been executed and a number of marauders were killed by the enraged peasantry, was something approaching camp discipline established. The French reentered Cherbourg on the 17th, their incomprehensible inaction being explained by the general dejection in the entire French army after the defeats at Rossbach and elsewhere.

On September 3, a Sunday, the English sailed once more within view of St. Malo. A French frigate stationed near Cape Frehel at 5 A.M. fired alarm shots which were echoed by the coastal batteries and the church bells of the villages along the shore. By 6 P.M. the fleet was anchored in sight and hearing of the people on the beach. This time the enterprise was directed at the left bank of the Rance in which general direction the fleet proceeded on the 4th, silencing the fire of the coastal batteries within the range of which the ships were. At 1 P.M. the disembarkation began on the beach of La Fosse, southwest of St. Lunaire, by use of specially constructed "flat boats with a platform which could be let down for the landing," as a description says, of men, cannon and animals. Between 10,000 and 13,000 men including 200 dragoons with their horses were put ashore.

Topographical Data Lacking

On the following day the troops committed all manner of excesses and then reconnoitered in the direction of Dinan, two kilometers from St. Malo and the Rance, a deep stream and 1,000 meters wide at its mouth, which flowed between them and the stronghold. They soon realized that the original operational plan of their commander, General Bligh, was based on utter ignorance of the topography involved. To march further inland and up the river, as Admiral Howe suggested, and cross the Rance at Dinan was a scheme declined by Bligh, who wanted to undertake nothing without the support of the fleet or his own artillery. During this exchange of views Howe was forced to shift his anchorage to the vicinity of St. Cast. "From this moment on it seems that Howe and Bligh ceased completely to understand one another." Once protected in his new anchorage Howe informed the general that he was ready to receive him with his troops and that reembarkation at St. Briac seemed quite feasible. To this point Bligh set his troops marching while the fleet used the day of the 7th in a leisurely manner to store loot aboard. Between this anchorage and the army ran the Arguenon river, fordable at low tide.

By this time French defensive forces had been concentrated and the long term defense measures inaugurated by the military governor of Brittany, the Duc d'Aiguillon, were tested. The Duke, a scion of the Richelieu family, representative in Brittany of French centralism and thoroughly hated by the proud and independent estates as the personification of "the invasion of Brittany by France, of the province by the Court, of Rennes by Versailles," had done much to make the extended coast line of that province defensible in the expectation of British attacks. He had reorganized the militia, of a nominal strength of 90,000, and had installed an alarm and warning service.

In order to mobilize a maximum of troops within a minimum of time, roads were improved and some constructed, for aside from the Rennes-Brest route hardly any then existed in Brittany. Postal stations were increased from 47 in 1748 to 84 in 1759 and a system set up for requisitioning relays of horses and for *étape* stations. All this was done at the expense, and mostly against the will, of the provincial estates. But military centralism insisted that this was necessary to overcome, or match on land—and incidentally on the inner line—the advantages which ship-borne troops had over troops dispersed and marched overland. When the British landed this time, they were first met by militia and Breton volunteers, a force of noblemen and royal troops of 15 battalions, with a paper strength of 800 men each; four troops of horse and some artillery were also stationed in the province. All these forces were ordered to converge on Matignon, to the left of the route of the retreating English.

At noontime on the 8th the English began their march towards St. Cast, first spending the morning in drill and other exercises. The garrison of St. Malo, having crossed the Rance, followed them closely, taking prisoners from among the marauders who by their outrages did their best to keep the Breton wrath at fever heat. Very insignificant local forces for some 26 hours prevented the Brgitish from crossing the Arguenon at Le Guildo where the river is fordable at low tide. Only after a bribed inhabitant had given away the secret of

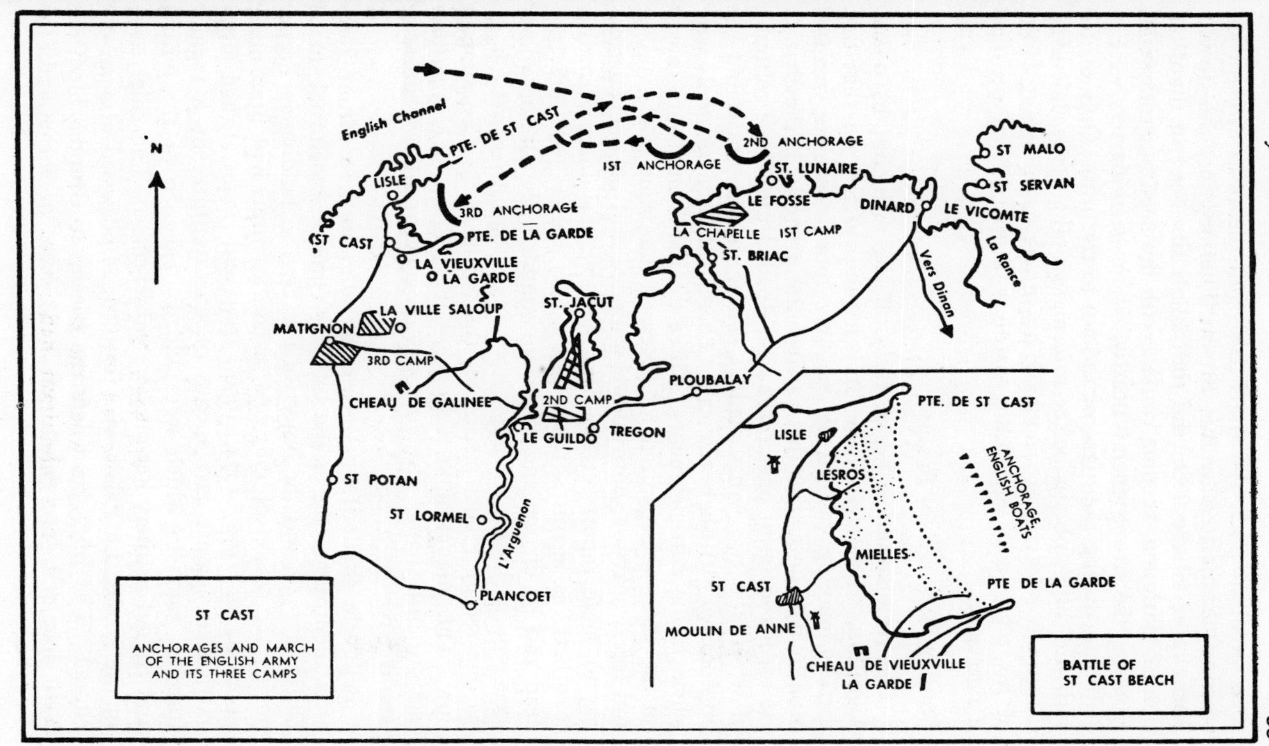
English Channel
N
PTE. DE ST CAST
LISLE
3RD ANCHORAGE
ST CAST
PTE. DE LA GARDE
LA VIEUXVILLE
LA GARDE
1ST ANCHORAGE
2ND ANCHORAGE
ST. LUNAIRE
LE FOSSE
LA CHAPELLE
1ST CAMP
ST. BRIAC
DINARD
LE VICOMTE
ST MALO
ST SERVAN
La Rance
Vers Dinan
LA VILLE SALOUP
MATIGNON
3RD CAMP
ST. JACUT
CHEAU DE GALINEE
2ND CAMP
PLOUBALAY
LE GUILDO
TREGON
ST POTAN
ST LORMEL
L'Arguenon
PLANCOET
ST CAST
ANCHORAGES AND MARCH OF THE ENGLISH ARMY AND ITS THREE CAMPS
PTE. DE ST CAST
LISLE
LESROS
ANCHORAGE, ENGLISH BOATS
MIELLES
ST CAST
PTE DE LA GARDE
MOULIN DE ANNE
CHEAU DE VIEUXVILLE
LA GARDE
BATTLE OF ST CAST BEACH

their small number, did the British gain heart to attempt the crossing. Enfuriated by this Breton Thermopylae, the invader burnt more villages. On the 10th Bligh advanced in the direction of Matignon, keeping contact with the fleet, but completely ignoring the strong enemy nearby, whose headquarters were at St. Potan during the night of the 9th to the 10th. Only on the 10th did Bligh obtain better information about the defenders, thereupon resolving to reembark on the 11th, informing Howe of his plans and demanding artillery support from the fleet if necessary.

Invaders Routed on Beach

The reembarkation began early on the 11th, but by 9 A.M. Aiguillon, who might have broken camp earlier on that day, thereby making his success possibly more complete, reached Moulin d'Anne whence he watched the progress of the British. He set his forces at once on march, divided into three columns, of which the two wings made a pincer movement around the unprotected British beachhead. There had been no tools with which to dig entrenchments. Three thousand men were caught on the shore unorganized for defense, their rout was complete and losses heavy—2,500 in dead, wounded and prisoners according to some reports; 1,200 dead and 732 prisoners according to others, as against French losses of 7 officers and 148 men dead and 57 officers and 283 men wounded, the Breton volunteers not included, among whom losses had been very high. As one recent French review of the battle of St. Cast sums it up: "The victors of St. Cast were the roads which Aiguillon had put into shape; without roads no concentration whatsoever would have been possible." [4]

Belle-Isle (Belle-Ile-en-Mer), eight miles southwest from Quiberon Peninsula, the scene of that unique nocturnal naval battle of 1759, was the object of the English landing in France during the season of 1761. As several attempts had been made upon it, in 1573, 1674, 1696, 1703 and 1746, it had been garrisoned against easy success of such ventures by a regular force of 3,000 to which were added 1040 men of the militia and coastal artillery (*cannoniers gardes-côtes*). The chief town in the island, Le Palais, was fortified; it possessed also an old citadel. The only good landing beach, at Grand-Sables, was well defended; everywhere else, the coast of the island, the area of which was 33 square miles, was steeply escarped and considered inaccessible.

On July 7 an English fleet was signalized in the roadstead of Le Palais, where it anchored beyond the range of French cannon. Finding the coast immediately opposite unfit for a landing, they moved to the beach of St. Andro where an attempt was vainly made with some 1,000 men "through boatloads," as an English report has it, of whom nearly all were lost.

While the English were forming along the beach three volleys were fired in quick succession by the French who were strongly entrenched on the side of a hill that had been made even steeper by scarping the foot of it. After the confusion and failure of the British on the beach, there followed a storm lasting three days and three nights, which threw the landing craft against the sides of the ships where they were moored, and shattered many of them. The fleet remained within sight of the island people until the 22nd, repairing damages to the landing craft, the officers informing themselves at leisure about the topography of the island and surrounding waters.

A bombardment was then opened against Kerdonis Point where the coast, although rocky, allowed a foothold at several stretches of beach. Due to the poor watch kept by the defenders, who could have repelled, or contained an enemy at this point with a very weak force, the English were enabled to bring 3,000 men ashore before their intention was understood. The choice of this landing point showed considerable perspicacity and boldness and may explain, or excuse the long delay on the part of the English. They at once took cover behind the stone walls which are a feature of the island and repelled a counter-attack of the French who suffered heavy losses and retired into Le Palais, their counter-attack having come either too late, or too early. It was early insofar as too many English had gained the shore to be thrown back by infantry alone and without the artillery which would have given the French superiority, since no English cannon had as yet been landed. The French were besieged for some time in Le Palais and at last forced to surrender. Thereafter the English held the island until 1763 when they handed it back in exchange for Nova Scotia.[5]

NOTES, CHAPTER 23

[1] The above is based on George Macaulay Trevelyan, *Blenheim.* London and New York, 1930. Ch. XII.

[2] *Catalogue of Prints and Drawings in the British Museum,* Vol. III, pt. 1, n. 2493.

[3] Admiral Sir Cloudesley Shovell complained in 1702: "The misfortune and vice of our country is to believe ourselves better than other men, which I take to be the reason that generally we send too small a force to execute our designs." Trevelyan, *Blenheim,* 254.

[4] Chef d'escadron P. Loyer, *La Défense des côtes de Bretagne pendant la guerre des Sept Ans; La bataille de Saint-Cast.* La Revue Maritime, Dec. 1932 and Jan., 1933.

[5] The descriptions of this landing from the opposite sides, by Col. Cyril Fields *(Britain's Sea-Soldiers. A History of the Royal Marines.* Liverpool, 1924. I, 101 ff.); and A. Fouché, Col. du génie *(Au sujet des diverses tentatives de débarquement faites sur les côtes de Bretagne.* Revue du génie militaire, Vol. 19 (1900), 441-472) differ vastly.

Chapter 24

18TH-CENTURY INVASION SCARES

POLITICS OFTEN ENOUGH have provided motives for landings, the military interest of which might be slight compared to the psychological impact produced by them. Many amongst peoples threatened with such an invasion would seriously dread it, others hope for it, some regarding it perhaps with mingled fear and hope, while fifth columnists would actively prepare and conspire for it. Perhaps of all the various historical fifth columns the one that lasted longest was formed by the Jacobites in the British isles.

The expeditions of the Old Pretender, (James Francis Stuart, son of James II) in the spring of 1708 and December, 1715, and of his son, the Young Pretender, July 1745, to Scotland are generally devoid of novelty in their tactical amphibian aspect. Both arrived and found friends to welcome them who were numerous enough to form at least a fifth column making for a safe landing. The first of these Jacobite expeditions, for which Louis XIV provided 6,000 men, was under the command of Claude de Forbin, one of the boldest French naval commanders. He was unwilling to go, protesting to the Minister of Marine that "this descent is not feasible; that everything is quiet in Scotland, no one having taken up arms in it, nor any city declared in favor of the Chevalier; that there is no harbor to shelter the fleet; that there is no place for the king to land his forces securely; and that to set 6,000 men upon the sand, without asylum or retreat, is to send them to certain destruction, and to be cut to pieces, not to say worse."

The Minister replied that the sailor spoke "too much in the tone of a philosopher; his Majesty will have it so, and that's enough." Forbin thought the 6,000 would be much better employed if the government should let him use them for a coup against Amsterdam, letting that rich city go up in flames. But the government was undeterred, their answer to his protests about the chimerical nature of the enterprise being invariably that "the loss of those 6,000 gave them no trouble provided it gave satisfaction to the King of England."

All that Forbin obtained was fast ships for his squadron, the best privateers in Dunkirk from where he had to keep away the soldiers before preparations were completed lest their assembly plus that of most English émigrés in France who were to accompany the Old Pretender would betray the attempt to the British government. The latter were in fact well informed about it. When the expedition was at last ready, the Chevalier fell ill of measles, the English meanwhile assembling ships at Gravelines. Forbin won a further postponement and permission to unrig some of his ships, but members of the expeditionary force, "sure to be gainers by the departure of the fleet," induced the Pretender's mother to gain a new order from Louis to send Forbin off. The Ministry of War now pressed for the embarkation of all in order to be quit of its own share in the project and leave to the Marine the responsibility for all further delays and possible failure. The generals, one of whom expected to receive the baton for his deeds, went on board before they were wanted and Forbin had to depart, giving them as far as was in his power a miserable time on the water.

French Retreat to Base

The coast of Scotland was reached without incident. Anchoring before Edinburgh in the Firth of Forth the expedition sent signals to landward, but received no response of any kind. Instead came word of the approach of English ships that had followed it from Dunkirk. The expedition might have been in danger of being bottled up in the Firth if Forbin's ships had not been swift sailers. The entourage of the Pretender was still in favor of landing him and the troops, but Forbin would not hear of it and a return to Dunkirk was voted in a council of war. After the loss of some ships, one to the English, others to the elements, with many men dead from privation, the squadron reached its base after three weeks' absence. That allowed Marlborough to recall the troops he had sent over from the continent to march to Scotland; they had suffered not a little while on the way to England and back, with no fit accommodation provided for them on the water. In France the return was followed by a long strife between the Ministries of War and Marine over the responsibility for the failure. Among the victims of that strife was Forbin himself whose zeal had been distinctly more for saving his ships than bringing his wards to land.[1]

The expedition attempted by France early in 1744 to carry the Young Pretender across did not get anywhere. A French squadron sailed, but was dispersed by gales with heavy losses in men and ships. Consequently transports with 7,000 troops on board who had been mobilized at Dunkirk and were to be led by Marechal de Saxe were detained. So Bonnie Prince Charlie had to go to Scotland in the following year with only a handful of friends. While Saxe had thought the enterprise feasible, he had written (March 3, 1744) to one of the ministers of Louis XV on the eve of the projected departure, that the greatest obstacle which he could forsee to its success, an obstacle meriting all of the minister's attention, was the opposition that might be encountered from British forces at sea.[2] Beyond that commonplace consideration this military thinker's ideas on combined operations do not seem to have gone; he had never before been employed on them.

The chief significance of both landing attempts lay in their political aspect. When John Wesley learned about the intended descent of 1744, when the French "were expected to land every hour," he exhorted his congregation "in the words of our Lord (Luke XXI, 36) 'Watch ye and pray always, that ye may be accounted worthy to escape all these things that shall come to pass, and to stand before the Son of man.' "[3] Others found in the rumors and the subsequent reports of the Pretender's landing less consolation. There were runs on the banks which could only be stopped at the Bank of England by paying out the withdrawals in sixpences.

The evolution of the banking system had created a new source for panics, or a new form of them, in English life. Almost so long as the Bank of England had existed, since 1694, scares had made trouble for it, and the menace of invasion, if sufficiently serious, would cause runs on all the banks. When Louis XIV threatened to land the Old Pretender in 1708, partly to make use of the dissatisfaction which the enforced Union with England had produced in Scotland, a run started after the Queen had informed Parliament of the scheme. A number of credit measures were taken to stop it, including a resolution passed by the Commons: "That whoever designedly endeavored to destroy or lessen the public credit, especially at a time when the Kingdom was threatened with an invasion was guilty of a high crime and misdemeanor, and was an enemy to her Majesty and her Kingdom."

Public Ready to Take Alarm

So dangerous grew this feature of public scares through the 18th century that the military took notice and General Lloyd, nearly the only English military writer of consequence in his time, in his plans to meet a French invasion of 1779 would "recommend to the stockholders not to be alarmed and not draw on the bankers and attempt to realize their securities—they will infallibly bring ruin and destruction on themselves and families." The result of such a panic would be that "the monied men are ruined and the whole nation is thrown into a convulsion, which may prove incurable, particularly if the enemy is in the country. Whereas if they remain quiet and repose a proper confidence in government, we shall no doubt be able to repulse the danger which threatens us." [4]

The propensity of the English people to fall into panics in face of invasion threats was great, and perhaps only commensurate with the usual ignorance and neglect of military affairs in England. A loyal Scottish lord would calmly write, in March 1744, to the Earl of Stair, commander-in-chief of the English forces on the continent in 1742 and at Dettingen: "A great many people here are highly feared about the invasion. . . . I cannot persuade myself there is great hazard in the French landing such a body of troops as to do more than put us in a little confusion for some time." [5]

But more persons seem to have been seriously afraid rather than possessed of such equanimity The letters of Horace Walpole mirror these panics and cool judgments and cynical attitudes in British society in the most illuminating and entertaining way through the years of the second part of the eighteenth century, even providing us opportunity to judge who may have surrendered most freely to fear, or profited by these mass fears. In 1775 when a futile French landing in the Isle of Man occurred, it set off a series of panics like successive rocket explosions which the government utilized to carry through the Militia Bill of 1757,[6] one of the several reorganizations of that force in the face of real, or imagined danger rather than after a realization of its unsatisfactory state. On October 19 of that year Walpole wrote a correspondent:

> I proceed to tell you in plain English that we are going to be invaded. I have within this day or two seen grandees of ten, twenty and thirty thousand pound a year, who are in mortal fright; consequently, it would be impertinent in much less folk to tremble, and accordingly they don't. At court there is no doubt but an attempt will be made before Christmas. I find valor is like virtue: impregnable as they boast themselves, it is discovered that on the first attack both lie strangely open! They are raising more men . . . The invasion tho not much in fashion yet, begins, like Moses' rod, to swallow other news, both political and suicidal.

And a week later:

> You will ask me, if we are alarmed? the people not all so: a minister or two, who are subject to alarms, are—and that is no bad circumstance. We are as much an island as ever, and I think a much less exposed one than we have been for many years. Our fleet is vast; our army at home, and ready, and two thirds stronger than when we were threatened in 1744. . . . Roads bad . . .

In the middle of November, 1755, he reported to his friend Horace Mann at Florence that there was "some suspicion that the invasion was only called as an ally to the subsidiary treaty; many that come from France say that on their coast they are dreading an invasion from us. . . . If you imagine that the invasion is attended to, any more than as it is played off by both those parties, you know little of England."

The Cynics and the Scares

A month later Walpole was really convinced that the scare "was dressed up for a vehicle (as the apothecaries call it) to make us swallow the treaties," that is to say, the agreements for a subsidy which had been concluded by the King and his ministers to avert the threat of an invasion by hiring continental bravos, behind the back of the most likely invader, France, from whom the danger seems not to have been serious in 1755. Walpole did not believe in it early in 1756 after the war, the Seven Years' War, had actually opened, for he wrote to Horace Mann (February 23): "The reigning fashion is expectation of an invasion; I can't say I am fashionable; nor do I expect the earthquake [which had recently destroyed Lisbon] though they say it is *landed* at Dover." According to the malicious Walpole there were Medeas in the fashionable world of his day: "Last night, at my Lady Hervey's, Mrs. Dives was expressing great panic about the French: my Lady Rochford, looking down on her fan, said with great softness, 'I don't know: I don't think the French are a sort of people that women need be afraid of.'" One wonders if this justifies us in calling her a rococo Medea.

Actually, the combined doings of the two governments kept the English people in a state of panic. The French, assembling 50,000 in various Channel ports, threatened, or seemed intent upon, a swoop across the waters, while the Newcastle government concentrated warships on the coasts and issued an order (February 3, 1756) that in the southern counties cattle were to be driven inland in case of a French landing. Instead of making a drive across the Channel, for sufficient transport was never collected for the 50,000, a descent took place against British-held Minorca in the Mediterranean. There the military had proved as under-anxious for security as their countrymen at home had been over-anxious. Minorca fell and when the half-hearted attempt of retaking it failed, Admiral Byng was sacrificed to the wrath of the English people who like other panic-stricken masses are apt to visit their scare upon such a scapegoat. Byng, while not wholly blameless, was made a political scapegoat for a criminally inept government. He was courtmartialed and shot on the quarterdeck of his flagship.

In the third year of the War Walpole found the invasion "not half as much in fashion as Loo," a card game, although Pitt was very positive about it in order to make the militia and the duties it entailed more acceptable. Walpole's common sense even detected the danger of inspired panics, inquiring as he did about the secret of how "to frighten your neighbors and give them courage at the same time." Considering the British successes in America, he found it "very comfortable, that if we lose our own island, we shall at least have all America to settle in." While it seemed to him throughout 1759 that nothing was talked of but the invasion, he declined "to grow more credulous. . . . We are still advised to believe in the invasion, though it seems as slow in coming as the millenium" (Nov. 16). By the end of the year he thought that "our sixteen years of fears of invasion are over—after sixteen victories." The next panic that swept England was, according to Walpole, one induced by roving mad dogs.

Yet to the end of the war certain worried patriots, like Lord Chesterfield, remained anxious about England's security. He, as Walpole reported, "with the despondence of an old man and the wit of a young one, thinks the French and Spaniards must make some attempt upon these islands and is frightened lest we should not be so well prepared to repel invasions as to make them; he says: 'What will it avail us if we gain the whole world, and lose our own soul?' " [7]

Gordon Riots of 1780

The European implications of the American struggle for independence later made the military threat to England, if measured by the French intentions and plans for an invasion, far more real than during the Seven Years' War. Yet so far as such popular sentiments are measurable and comparable, the English people do not seem to have feared it then so much as earlier. Not that fear and panic were absent during that period, but the incubus was a different one. It was dread of an invader already in the midst of England—"Popery," the Catholics who were to be admitted to Parliament as the price for their more active support of the foreign war (Savile Act of 1778). The popular terrors were organized in the Protestant Association and culminated in the Lord George Gordon riots of 1780,

> *When the rude rabble's watchword was—destroy*
> *And blazing London seemed a second Troy.*
>
> Cowper

Only after parts of London had been burnt and ransacked by mobs did the sense of a foreign peril return and then hand bills were issued by different political factions, asserting that "the horrible riots had been prompted by French money." American agents were also believed to have been at work, but there is nothing to prove the activity of either beyond a certain satisfaction on Benjamin Franklin's part that Lord Mansfield's house had been burnt and that "thus he who approved the burning of American houses has had fire brought home to him." [8] Gordon led the riots, but escaped punishment by successfully pleading insanity. Dickens vividly described the riots in his *Barnaby Rudge.*

The more sophisticated Englishmen, although frightened enough at the riots at home, with Black Wednesday (June 7, 1780) "the most horrible night" Walpole ever beheld, expecting half of London to be reduced to ashes by morning, remained sceptical about the foreign danger. They saw the bogey being manipulated by the government or court who in March 1781 took pains to spread reports of invasion in order to lower the premiums of a recently placed government loan which had given "enormous jobs" to some and had caused resentment. Though a French squadron had left a card in Plymouth in 1779, another French-Spanish squadron being on the loose in September 1781 did not greatly disturb the people. There were plenty of disciplined soldiers in the land at the time and Walpole trusted the foreign ships would soon be driven home "thanks to our only ally, the Equinox." [9]

NOTES, CHAPTER 24

[1] *Memoirs of the Count de Forbin.* Translated from the French. London, 1731. II, 249 ff.; George Macaulay Trevelyan, *Ramillies.* London and New York, 1932. 341-7.

[2] Capitaine J. Colin, *Louis XIV et les Jacobites.* Paris, 1901.

[3] John Wesley's *Journal.* Everyman ed. I, 453.

[4] Henry Lloyd, *A Rhapsody on the Present System of French Politics, on the Projected Invasion and the Means to Defeat it.* London, 1779, reprint of 1793, 135. For the background of this work see Jähns, *Kriegswissenschaften,* 2102.

[5] John Murray Graham, *Annals of the Earls of Stair.* Edinburgh & London, 1875. II, 310.

[6] Traill & Mann, *Social England.* V. 488.

[7] The passages quoted are from *The Letters of Horace Walpole.* Ed. Toynbee. III, 355-6, 370, 377, 399, 409; IV, 268, 270, 274-5, 276, 284, 305, 327, 359-60; V, 184-5.

[8] For details see Percy Colson, *The Strange History of Lord George Gordon.* London, 1937.

[9] *Letters of Horace Walpole* XI, 417-8; XII, 46, 59-60.

Chapter 25

CONJUNCT OPERATIONS, FROM FAILURES IN PRACTICE TO SOUND THEORY

WHILE THE LAND BATTLES of the *ancien régime* were most carefully planned as a rule, its descents on hostile coasts, which required even more preparation, were generally poorly devised and thought through. This sort of war "has never been studied enough to put us on an equality with the enemy," as Molyneux summed up the faults of Pitt's landings. More often than not, these expeditions might be started merely on strength of that statesman's hunches, like the futile and humiliating voyage to Rochefort in 1757.

The general encouragement in that case seems to have been furnished by knowledge of the weakness of the French forces along the western coast and specifically by the report of a young Scottish officer who had recently traveled in France that the works of Rochefort, only weakly garrisoned, could easily be taken, that the town's dockyards, arsenal, foundry, rope-walks made the enterprise particularly worthwhile from the standpoint of naval warfare and that the large Protestant element in the vicinity would prove helpful. Three former unsuccessful enterprises against the place, one by Tromp in 1674 and two by the English in 1690 and 1703, failed to act as deterrents.

One can readily see the temptations thus offered to a statesman of mercantilistic outlook of great gains promised from limited liability and investment and profits of a measurably useful character. Besides, it would draw away French troops from the main theater of war in Germany where lately England's ally Frederick, who was counseling Pitt to keep the French occupied by such a descent, had suffered the defeat of Kolin and the Duke of Cumberland the more specifically English defeat of Hastenbeck, to be followed by the Convention of Kloster Zeven, where England's continental army was forced to capitulate.

Alarmed by these continental setbacks, Pitt rushed the Admiralty into preparations for the enterprise. They were reluctant, but he was not to be refused, remembering perhaps that British sea and land officers had foiled a landing plan against France after La Hogue in which both the English and Dutch at the time had put great hopes and large funds. "The plain source of all this confusion was that the Ministers of State were unwilling to take upon themselves the direction of an affair which they were apprehensive would miscarry, but were willing to put it upon the land and sea officers, that they alone might remain accountable for what happened." The officers near the end of July came to the conclusion that the season of the year was too far spent to admit of the fleet's sailing to the French coast (Molyneux I, 73-5). But Pitt would admit no such shunning of responsibilities.

For military commander Pitt favored the more youthful Conway, but the king, imbued with patriarchal notions about military command, insisted on Sir John Mordaunt, a nerveless general, more "deserving" than qualified for the task at hand. The commanding admiral was Hawke who had no zeal for the enterprize, but managed to escape the public wrath visited upon the heads of the military participants. Conway was second in command, with Cornwallis next to him in seniority, and following him Wolfe as quartermaster general, in modern language chief of staff. The last two emerged from the affair with their honor untarnished. Wolfe was perhaps prepared for the amphibious enterprise, insofar as he had in very early years accompanied his father, a military man, on the Cartagena expedition, and had taken an interest, as he wrote in 1754, in "fleets and fortifications . . . . as strongly as [in] architecture, painting and the gentler arts," a versatility which, if it did not go very far, may have kept his mind open sufficiently to learn from the negative results of the enterprise.

Wolfe Learns in France

The armament, the secret of its destination well sealed, left Spithead on September 8, 1757, consisting of sixteen men-of-war six frigates, six bomb ketches, two fire ships and two hospital ships, six cutters and 42 transports with some 9,000 men on board, the equivalent of ten infantry regiments, some light horse and numerous artillery. Wolfe, as if to console the heroes of many later amphibious enterprises, was miserably seasick on the voyage, writing home: "If I make the same figure ashore I shall acquire no great reputation on this voyage." On the evening of the 19th the fleet anchored off the mouth of the Charente, which was guarded by the Ile d'Aix, with Rochefort nine miles up the winding river and 20 miles from La Rochelle. On the 23d Aix island was taken. Drunken soldiers and sailors pillaged and had their little *sacco di Roma,* running around in church vestments, disgusting the austere Wolfe who wrote of them: "Nothing, I think, can hurt their discipline—it is at its worst. They shall drink, swear, plunder and massacre with any troops in Europe, the Cossacks and Calmucks not excepted."

After that neither military, nor naval commander seemed to know how to follow up this easy small initial success. Instead, they fell to quarreling and holding war councils while the French on land strengthened their positions and forces. Wolfe was impatient at the delay which wasted what little opportunity was left when, according to his view, Rochefort could have been taken easily if a landing was made in time. He reconnoitered ashore and found a suitable beach halfway between La Rochelle and Rochefort. But there remained a fort on the way up the Charente which the military declared could not be attacked successfully on land and which the sailors said could not be approached closely enough for proper shelling.

This obstacle made the council of war forsake Wolfe's pre-Quebec landing plan and substitute another, after the arrival of fresh French forces had been observed from shipboard. But this went no further than ordering on the 28th that 1,200 men were to be in the ship boats by midnight forming a first wave to go ashore and hold the beachhead for six or seven hours, when the boats were to bring the second wave. After four hours in the boats on a cold and windy night the 1,200 were ordered back into the transports, "to the astonishment and disgust of all," as Wolfe says, although the scheme involving the two waves with their long interval had inspired but little confidence.

Thereafter the collective intelligence of the council could hit on nothing more promising than to blast the fortifications on Aix, where some men were wounded. This disgusted Hawke who, according to Horace Walpole, befriended Wolfe, the two contracting "a friendship like the union of a cannon and gun powder," which might be taken as a sign that, given mutual confidence, land and sea power could on occasion get together in the 18th century. Declining to attend any more councils,

Hawke informed the military that "if they had no further military operation to propose, considerable enough to authorize him detaining his squadron, he would immediately return to England." The army men having exhausted their capacity for planning, the fleet carried them away on October 1, to face a commission of inquiry and a court-martial for Mordaunt. The commission came to the conclusion that Wolfe's proposal if adopted, "certainly must have been of the greatest utility towards carrying your Majesty's instruction into execution." The court-martial acquitted Mordaunt, rather condemning Pitt and his whole plan. Wolfe on the witness stand made an excellent impression and it would seem that Pitt took mental note about the young officer as suitable to direct one of his subsequent landings.

Lessons of Rochefort Failures

Professionally Wolfe here seems to have learned much, and posterity will see Louisbourg and Quebec prefigured in these and other passages from his letters following the trial:

> As to the expedition, it has been conducted so ill that I am ashamed to have been of the party. The public could do no better than dismiss six or eight of us from the service. No zeal, no ardour, no concern or care for the good and honor of the country . . . The true state of the case is, that our sea-officers do not care to be engaged in any business of this sort where little is to be had but blows and reputation; and the officers of the infantry are so profoundly ignorant, that an enterprise of any vigor astonishes them to that degree that they have not strength of mind nor confidence to carry it through.
>
> I have found out that an admiral should endeavor to run into an enemy's port immediately after he appears before it; that he should anchor the transports and frigates as near as he can to the land; that he should reconnoitre and observe it as quickly as possible and lose no time in getting the troops ashore; that previous directions should be given in respect to landing the troops and a proper disposition for the boats of all sorts, appointing leaders and fit persons for conducting the different divisions.
>
> Nothing is to be reckoned an obstacle to an undertaking of this nature which is not found to be so on trial; that

> in war something must be allowed to chance and fortune, seeing that it is in its nature hazardous and an option of difficulties; that the greatness of an object should come under consideration as opposed to the impediments that lie in the way; that the honor of one's country is to have some weight, and that in particular circumstances and times the loss of a thousand men is rather an advantage than otherwise, seeing that gallant attempts save its reputation and make it respectable, whereas the contrary sink the credit of a country, ruin the troops, and create infinite uneasiness and discontent at home.

Even if these ideas be incommensurables, one still wonders whether the technical lessons drawn from the Rochefort failure were not of greater consequence in connection with the subsequent events at Louisbourg and Quebec in a century slow to draw technical consequences, than the moral ones evolved by Wolfe.

Still another participant of the Rochefort expedition brought together and published, without making apparently anything like a stir, the technical lessons while Wolfe was applying them. This was Molyneux with his *Conjunct Operations* of 1759, the year of Quebec. To him the English defeats and disgraces, only partly expiated by Wolfe's success in America, suffered in littoral war, spoke plainly: "We have no right understanding there . . . . It is a palpable demonstration from the number of conjunct armaments these kingdoms have fitted out and the many fruitless attempts that have been the issue of them, that there has been no right industry, no skill or watchful observation. That is, we have never employed our minds in the study of this war, till we have been called upon to make use of our bodies also; thus when it is too late, by knowing nothing beforehand, we doubly fatigue our mental faculties, with the vain hope of retrieving lost opportunities."

Special Landing Craft Projected

From the ill success of the Rochefort expedition he dates the origin and introduction of special landing craft. "The sluggishness, awkwardness and different sizes of the transport boats used there were so apparent to the whole fleet that it plainly pointed out a great imperfection in that part of equipment." Consequently, the Navy built and employed in the landings of the following year a flat bottom boat of a type that was exclusively constructed from naval plans.

> Notwithstanding this boat was made wholly for the use of the soldier, to transport him at a very critical time from the ship to the shore, and vice versa, we take for granted, it owed not its origin to any gentleman of the army; for there was no military order of the platoon preserved among the soldiers in the boat; nor was there any of the regiment preserved among the boats on the water. Confusion was as ever among the men, and as much as ever among the boats. It differed only in these respects from the common boats of the Fleet, it was constructed to go in shallower waters, and being all of a size, they contained the like numbers. Each had two sails and was full of benches . . . ten rowers on each side.
>
> Between every rower and the edge of the boat sat a musketeer to defend him, by which method each was deprived of the liberty necessary in his occupation that a few soldiers on the side might be in a position to fire very bad, the rowers obliged only to paddle. The contrivance of this piece of mechanism seemed, as if one main aim had been, to render it as difficult as possible for the soldiers, when they reached the shore, to get out of it; during which performance, the oars being tied with cordages sloped down the outside of the boat like the fins of a fish, which was the ingenious part of the construction. Each boat when freighted to the utmost, contained twenty soldiers, besides the twenty rowers.

The faulty one-sided conception of such boats leads the author to some very sensible, and even today largely valid, remarks on inventions and innovations in the field of amphibious enterprises: inventions relevant to this field must be amphibious, bi-elementar in character, just as other great military or other inventions have been. "Gun powder was not found out by a soldier, nor was the art of printing by one of the literati." Martinet, the epitome of land soldier, and not a sailor, was the inventor of the copper-covered pontoon, used first in the wars of Louis XIV against the water-protected Dutch; and Bernard Renaud, the inventor of the bomb vessels which were first used at Algiers in 1681 was likewise not one of "those who had their occupations on the water." Charles XII in crossing the Düna in 1701, employed boats of a new construction, "the sides of which being higher than the common fashion could be pulled up and down like draw bridges; when they were up, they covered the troops which they carried; and when they were down, they served instead of a bridge for their disembarkation."

IDEAS OF A PIONEER

Though not a shipwright himself, Molyneux insisted on laying down certain specifications for landing craft. "Inventions relative to a particular profession must not owe their birth to people of that profession." Rather would cooperation be propitious for invention and execution. "When they can be brought together, so much happier is it for that conjunct armament." He evolves his own notions about such craft, "notwithstanding the thousand difficulties we are certain will be started by those who are lovers of plodding in the old way; who have geniuses like that of a mill horse."

He was sufficiently a "functionalist" to make it the first consideration for all boat construction always to remember who is to use it, men "much more accustomed to the land than to the water; therefore to have it resemble as much as possible the land, it must be made as steady as possible," a boat with a flat and broad bottom, with seats to make the soldier feel even more as if on land, "by which he obtains one of the things principally to be considered, steadiness." His illustrations recall to mind the dictum of a sound 18th century military author, General Lloyd, that the battle order of the 18th century was similar to a collection of porcelain on a mantle piece which one hesitates to take down for fear of breaking it. The soldiers taken from the mantle piece of *terra firma* are carefully seated for the transit lest they be disordered. But why, Molyneux goes on, should not a soldier fire from the boat while seated, just like a duck hunter? "There can be no reason to imagine why a soldier may not hit a line of men on a shore which, instead of a swift flying snipe, is like a standing wall."

Molyneux's standard landing craft or "subdivision boat" held 84 men, plus officers, plus rowers, a few men to serve a two-pounder gun at each end of the boat and one man at the rudder, altogether 113 men, including 24 "irregulars," all-purpose soldiers who were to do odd jobs during the transit. The irregular must be men of some ingenuity who could handle sails when the wind permitted their use, replace rowers or gunners, handle boathooks, let down the landing ramps, be the first to land, afterwards joining their subdivision and forming the rearguard in case of reembarkation.

The side walls were to be high, but not sufficiently to exclude firing over them. The ramps were to be broad enough for platoons to march out of the boat in full front. The length of this type of boat was to be 66 feet and it should not draw more than three feet of water when fully freighted. Other types of landing boats were a smaller platoon boat to proceed on the wings of a landing waver, drawing but one and a half foot; an artillery ranger boat, etc.

While some of these proposals appear to verge a little on the pedantic, it should be remembered that the 18th century could conceive of war only as a methodical enterprise. This character seemed to disappear only in landing enterprises when the need for tactical cohesion would be ignored and the men shipped like so many potatoes, regardless of their original membership in platoon, company, or other formation. Molyneux emphasized the preservation of this cohesion as long and as strongly as possible. It must not be endangered more than was unavoidable and one temptation to disorder should be shunned above all others—alcohol. "The men got so excessively drunk at Cadiz in 1625 . . . . it put their officers into such a fright they were actually obliged to re-imbark their forces."

> The knowledge and skill of his profession which a private soldier acquires by constant practice, and which often is not much, ought not by any body to be lessened or taken away . . . . still much less when a soldier is required to exert that knowledge and skill most . . . . At such a crisis as this to tie down the arms of a soldier when they ought to be most at liberty; to annihilate his whole regimental exercize and cloud and confound at once all his military ideas, of regimental order, distinction, and regularity, by sticking him promiscuously in some crowded boat, and wafting him from the ship to the shore (over as it were, the River Styx), should the fleet be not able to protect him, and should an enemy be waiting in good order to receive him; the commander who does this, makes a waterman or bargeman of the soldier, and no general of himself . . . .

> The present business is to shew in what manner this division may be debarked in the face of an enemy, without breaking their rule and order, without derogating the least from the word regiment; in short, how they may be landed like regulars instead of irregulars, like soldiers instead of sailors . . . . Therefore the whole mystery at once not to drown a poor soldier's understanding as soon as he is put in a boat, is to let that boat be a subdivision [of the regiment] boat, which is to carry a subdivision, as the platoon boat would be too small and trifling, therefore too numerous."

MILITARY PSYCHOLOGY FIGURES

A regiment of the time consisted of 18 platoons in subdivisions of two platoons and the regiments of ten subdivisions, including two for the grenadiers. This method was "to balance the confusion, or rather the unaccustomed motion the winds and waves will naturally put the soldier in, as well as to make the affair more equal between him and the enemy who stands firm on the shore . . . To a regiment then and regimental order a soldier has been accustomed; in short they may be said to be the essence of him, for he knows for the generality little of anything else. Wherefore though we are going to put him on the water, we will not take him out of his element. We will not rob him the least of his military knowledge, perhaps (till he has been on the enemy's shore) the only wealth about him." To preserve the soldier in his regimental context, the boats themselves must be regimented, that is "put under rule, form and order," with ten boats to each regiment and the boats of the diverse regiments in distinct colors, numbered and otherwise marked.

It was pointed out that this numeration and other regularization served a very important psychological purpose: in both embarkation and reembarkation the soldiers "pass from one extreme to another."

> In the first (at least according to our system) they are taken out of darkness and confusion and placed in their boats in the exactest order and regularity, to meet perhaps immediately an enemy ranged in no less order and regularity than themselves. In the latter they quit an enemy's country, where their own preservation obliged them to

> observe the above order, to rush immediately to their ships and be plunged again among their friends into their old darkness and confusion. For a debarkment is a quick transition from darkness to light, confusion to order, water to lands and from friends to enemies: the re-embarkment is vice versa.
>
> The difficulty and danger during this critical performance of each of these parts of a conjunct enterprise must appear to a very common discernment to lie between the boats and the land, not the boats and the fleet; the former containing the enemy, the latter the friend. The inference we would draw from this, is, when we make the descent, it is upon a strange land, prepared by the enemy to receive us; but when we re-embark, it is from a place we ourselves have or ought to have been assiduous in rendering capable to defend and cover such an operation; for though we come strangers to it, we do not return so from it.

Hence orderliness is not quite such an absolute requirement for the return to the ships; boats may be crowded to capacity in order to quicken the business, each subdivision boat receiving up to an extra 31 men, every platoon boat 24. Re-embarkation should preferably begin during the night, with cavalry being shipped first, then invalids, baggage train, artillery, etc. The numeration and other markings of the boats serve a very important purpose during re-embarkation as well; soldiers are to be told beforehand to which boat they are assigned and all soldiers who try to cling to, or enter, the wrong boat are to be ruthlessly turned away.

DUNKIRK CONDITIONS FORESEEN

One chapter deals with the equivalent of Dunkirks, or re-embarkations in the face of a formidable army. Never are forces thrown more upon the defensive, never are they more on the retreat than when taking to ships under fire. "To retreat," says Folard, "is to fly, but it is to fly with art, with the greatest art."

> He speaks here of a land war. But in the littoral war the difficulties and dangers double upon us, and to fly with art on the land is not sufficient; for after we have retreated to the shore, we are to retire to our ships; the art of flying

> is to be preserved on the water as well as on the shore . . . Should we sum up the disadvantage attending an army that are to make their re-imbarkation with the bayonet pressing at their backs, we shall find them as follows: the uncertainty of winds and tides; the almost certainty of finding through irregularity and variety the situation of the grounds against them; their leaders having a kind of cloud spread over their senses (the frequent consequence of a multiplicity of difficulties).

To this would be added the absence, or insufficiency of artillery, and/or cavalry, "with scarce a possibility of re-imbarking more than 2,000 soldiers at a time. Beachheads must be formed and strengthened, heights in their neighborhood be occupied and held, though on the whole a flat terrain seems preferable where there is a choice, for such beachheads can receive the better fire support from the fleet. "The only place of safety we have left us here is our fleet; the fleet is our fortified town, our friendly country, in short our everything at such a crisis."

As to the size of a landing force Molyneux remarks that the English have seldom known how to proportion their strength according to the nature of the object. The author's ideal expedition would consist of three regiments with a firing power of 2,340 muskets plus those of the rowers of his system of boats, making together 3,000; artillery to accompany them should be 32 twenty-four pounders, 8 nine pounders and 60 two pounders. "Every gun should have a little buoy fastened to it, by which means they would very seldom be irrecoverably lost," by the capsizing of boats. Small bodies are easier to conduct and make surprise more feasible, particularly where no system of regularity has been introduced; losses are usually lower for small units in proportion to the total strength as well as absolutely—all these are considerations that play their rôle in modern commandos. On the other hand, most of the English enterprises of this kind have suffered from being undersized. "We have never sent scarce half forces enough" on such missions and have hence suffered from "a continuous round of bad management" whereas the foes of England have more often used forces proper to the task; the size of the Great Armada was not what was wrong with that attempt.

A conjunct operation calls also for conjunction between the several land forces. "Without some few strong cavalry we may as well lay aside all thoughts of making conjunct expeditions against any powerful country; for by being without cavalry of this sort, we are in effect without artillery." It is the cavalry that is to provide the horse for the landed guns. Realizing that this proposal which "carries with it the air of degradation among the gentlemen of the horse," is hardly short of revolutionary, the author thinks that a *douceur* offered the latter might "soften the harsh sound of occasionally converting the war-horse into a draught-horse, and the rider into a driver."

Such a conversion cannot be improvized, but must be worked out in time when horse and men learn both uses. For "a conjunct armament thus equipped with cavalry would be free from the lumber of mere draught horses, their drivers"—who were at that time usually civilians and not always easy to manage—"and forage, and consequently the addition of ships and expense to the Government, of time and trouble in connection with the landing and re-embarking." The land forces to be employed were to know the other element to a certain extent before they set forth. Regular 18th-century drill was insufficient preparation. Troops must be exercised in crossing and fording rivers and other water courses, a variety of which they were sure to meet along the landing coasts.

Littoral Elements Considered

So little is Molyneux a Navy-phobe that in the many failures of English combined operations he finds the Army "more deficient on their part than the Navy." It takes both to make the perfect littoral war which he at one time defines as a war waged "whenever the army receives its whole sustenance and support from the fleet." He knows the obstacles to their cooperation. But "when they both can be brought to cooperate, so much happier it is for that conjunct armament." But the army must not be made slavishly dependent on the fleet and all its measures on land must be so arranged "to leave ourselves as free to make our movements as the enemy shall be to make theirs." The fire support from the fleet will not always be available in littoral war due to the conditions of waters, winds and weathers. To replace its fire the author wants to see his artillery rangers employed, flat boats with twelve guns mounted which can go wherever the other landing craft are brought. The manoeuvring of such boats can very considerably raise the element of surprise.

With such provisions, after exact studies of distance and tides, of the chosen localities for landings, with feints where the real descent is not intended, "to carry the littoral war to perfection" the combined operation "goes against the enemy, like an arrow from the bow. It gives no traces where it has passed. It gives no warning where it is to come. It must wound too where it hits, if rightly pointed at some vulnerable point."

The forces envisaged by Molyneux were, on the whole, army troops, although these would in many cases include marines. Marines were usually available only in numbers so small that they could not go through larger landing enterprises by themselves alone. By themselves they might, however, be employed for such purposes as "to dislodge a small body of the enemy, while a sea-officer is burning some vessels in a creek or harbor; to attack a small fort on the landside, when a ship is battering it from the sea; to secure an advantageous post, till some troops, landed in another place, can take possession of it; to burn a village, attack a battery, nail up the cannon, carry off some prisoners, to gain intelligence from the enemy, support a body of troops already landed." Thus were the uses of marine troops epitomized by John MacIntire, author of a treatise on the subject, who was an English marine lieutenant at the end of the Seven Years' War.

Judging from this treatise several of the proposals voiced by Molyneux were only too much in need of reiteration, such as emphasis on keeping the men that were to fight side by side on land together in transit, "by which means they will the sooner be able to form on their landing." Before they were ordered into the boats, they should be drawn up "on the quarter-deck, or poop in close ranks and files, three deep, and every precaution taken to regulate them in the best manner, that they may be able to form immediately on their landing. They must be told at the same time the nature of the service they are going upon," which seems a little more information than was customarily given to land soldiers before the attack at that time.

Conflicts between land soldiers, who in most countries would include the marines, and sailors were by no means restricted to the higher commands, but ran deep down through the ranks. To avoid at least the most common *Kompetenz* conflicts, MacIntire lays down the rule which seems to have existed in his time, although often violated, that from the moment the boats are ready to put off, "the whole command of them is given to a sea officer who conducts them to the place of landing; and from this time, the Marine officer has little to do till the men are out of the boats, for then is the time for him to show his judgment."

> When the sea and marine officers don't agree, a detachment can never be properly regulated. The want of harmony between two corps which in fact ought to be the same when together, and mutually assist each other for the good of service and society, will occasion such animosities, as are best known to those gentlemen whose actions are not guided by the laws of prudence; and whose inexperience in the service makes them think that whatever is said, or done, is meant to affront them: such boys are to be found in every service, who know nothing, nor ever will know, because their presumption makes them believe they know everything. For my own part, I think it an easy matter to support harmony, if men will but reflect they are acting in the station of officers and gentlemen.[2]

NOTES, CHAPTER 25

[1] Humphrey Evans Henry Lloyd (1729-83), one of the very few outstanding British military authors of his century, author of *History of the Late War in Germany* (1766) and *Military Memoirs* (1781).

[2] John MacIntire, Lieutenant of Marines, ***A Military Treatise on the Discipline of the Marine Forces, when at Sea: together with short Instructions for Detachments Sent to Attack on Shore.*** London, 1763.

Chapter 26

LANDING OPERATIONS IN NORTH AMERICA; COTTON MATHER TO WOLFE

ISOLATION, AS THE ATTEMPT, or desire to give one's own country an insular character, springs from many motives among which the military factor concerns us here in the main. On the whole, the hope of bringing it about seemed in times gone by most liable of realization outside Europe, in Japan, China and above all in the new world. In the Americas ties with Europe seemed to many weak and impermanent enough to allow of a severance, of the adoption of policies of exclusivism and isolation.

Spain tried to isolate her vast colonies from Protestantism and trade with all but the mother country, with the Asiento of 1713 as an attempt to regulate interlopers, and avert war. In North America the governors of New France might hope and try to drive out the English altogether, plans which Louis XIV was ready to approve and at times assist, even if with insufficient means. This totalitarian expansion tendency provoked a counter-tendency on the part of the English colonists to rid themselves of the French and make all North America an expansion of New England, which led later to the removal of the Acadians from Nova Scotia. This policy was a forerunner in method of such measures as the transfer of the Greeks from Turkish territory after World War I. Both New Englanders and New Francists wanted to see anti-Christ removed from the New World and the Kingdom of God more fully established there. Cotton Mather called Canada "the chief cause of New England's miseries" and a crusading spirit would readily discover Saracens on the other side of the uncertainly drawn frontier.

In 1654 a fleet sailed from Boston to Port Royal in Nova Scotia where the British colonists landed to destroy this center from which their shipping and fishing had been much harassed. It later had been taken several times from the French when in 1690 a New England expedition under William Phips, governor of Massachusetts, captured Port Royal for the sixth time. Phips, a ship carpenter and excellent sailing captain, treasure hunter, replete with instincts both piratical and Puritan, thought it his duty "to venture my life in doing good, before a useless old age comes upon me." Since the French garrison was too weak to offer serious resistance, the tactical part of the enterprise was uninteresting, but it whetted the Protestant appetite for more and by August 1690, Phips had collected an armada to attempt Quebec, regarded by the Puritans as something in the nature of a northern Babylon.

This consisted of about 40 ships, mostly small, of which only four were sufficiently big and well armed to undertake a bombardment of the city. By the middle of October they arrived before Quebec. Phips was sufficient of a master mariner to bring the fleet all the way up the St. Lawrence and frighten the French, but as a land warrior he was not equal to old Count Louis de Frontenac, the defender. His fire was ineffective against Quebec's masonry walls and the heavier fortress guns. His landing troops, advancing across a river and broad mud banks, were met by French ambushed in the bushes, of whose strength they remained ignorant. Cooperation between English ship and land forces was a task too difficult and untried for Phips' raw formations, as in fact it would have been for most troops of the time. A week after his arrival Phips raised the siege and sailed away, leaving some of his guns behind, and not all his ships reached home. But, said Cotton Mather, the way to Canada was now learned: it went by sea to Quebec. Jubilant Quebec declared itself saved by "Our Lady of Victory," the Virgin who thus became more venerated than before in Canada for having repelled the heretics from Boston.

European War Reaches America

The repercussions of the War of the Spanish Succession (1701-14), resulting from Louis XIV's attempt to gain the Spanish throne for his grandson, reached the New World's north, but in such a way that the inhabitants of New France and New England had to fight each other largely with their own men, and by their own cruel, or crude methods which often failed to use to the best purpose and with highest economy the weak forces at the command of both. The absence of military talent and authority among the English colonists ended in failure of three expeditions in 1704 and in June and August, 1707, against Port Royal which by this time had again reverted to France. The first armament of 1707 met the garrison unprepared, "but rivalry of soldier and sailor, jealousy and disputes did their work and on June 17 the baffled English sailed away to Casco." (Wrong, 567).

Their own weakness or military ineptitude led the colonists to petition to London for assistance by regular forces. Help was promised, but eventually was sent to Portugal instead (1709). The next year, however, brought English support of ships and marines. They together with colonial forces, a fleet of five men-of-war and 30 transports, 400 marines and 1,200 colonial troops on board, sailed to Port Royal once more. Left unsupplied by Quebec for years, the town capitulated at sight of these superior forces. As an earnest of serious occupation of the place it was renamed Annapolis. Its hinterland, Canada, remained to be conquered and for this forces became available at home in England through shift of politics and policy to Toryism and, with respect to the European continent, isolationism.

An alteration in the London court gynocracy in 1711, from Sarah Marlborough to Abigail Hill, as the favorite of Queen Anne, gave the military command of an expedition against Quebec to Abigail's brother, General Hill, little better than a carpet general and as such ill fitted for overseas enterprises. When the arrival of the armament at Boston was announced in London, Secretary of State Bolingbroke wrote confidently: "You may depend on our being masters at this time of all North America." But this rounding out, or enlargement of New England was not so soon to be.

New France vs. New England

With the incompetent military leader was yoked an incompetent and timid naval leader, Sir Hovenden Walker, who believed that the St. Lawrence would soon freeze to the bottom and crush his ships. He lost his bearings near the river mouth and piled ashore, or on reefs eight troop ships and two supply ships, drowning nearly 1,000 redcoats. That left the leaders still with some 10,000 soldiers, many of them seasoned fighters from Marlborough's armies, but without courage to carry on. The forbidding face of the unknown and uncharted frightened the chiefs and their ignorant pilots who declared, or believed the river to be unnavigable; and although the representative of New England urged them on, a council of war voted to give up the enterprise and return to England.

The Peace of Utrecht, while giving Hudson's Bay and Newfoundland to the British, left still a New France to face a New England. both nourishing claims and hopes for a North America exclusively their own. Both were preparing for yet another battle, France founding Louisbourg in 1713, commanding the trade from England to New England and beyond as well as the fisheries, and provided with a permanent garrison of at least 1,500 men, while England, much later, set up Halifax as its countering sea fortress in 1749.

Both New England and New France continued to nourish their Monroe Doctrines, to speak anachronistically. This in military terms meant rounding out their own holdings and expelling the natural, or unnatural enemy, thus ending war in the north of the New World where the system of economy logically made war undesirable, rather than the reverse. Meanwhile their home governments endeavored to minimize the colonists' complaints against their neighbors and preserve peace in Europe and across the seas. War was contemplated in the New World as a means to end war and the threat of war, including landings from the sea and the Great Lakes, and overland depradations, so much dreaded. But in 1744 war erupted again in Europe and promptly spread to North America.

Receiving the news first, the French at Louisbourg took Canso, or Canseau, a small English outpost, but of some importance to New England fishermen. They had that summer in Louisbourg harbor some twenty ships, including three men-of-war and four privateers. Boston, knowing and fearing old French threats and bearing in mind as well its own plans against the northern stronghold, apprehended a French descent upon the Massachusetts coast. The representative of New England in London urged the home government to take Louisbourg and end this peril, at least for the time. With the menace of a French and Stuart invasion directed at Britain, such request found no favorable hearing in London.

But so earnest was the desire of the colonists to rid themselves of the menace of Louisbourg, which incidentally was reputed to be as rich in prospective loot as some of the settlements of New Spain, that early in 1745 they decided to proceed on their own. They formed plans thought to be "rather wild" only by the more cautious elements in America, like Benjamin Franklin. The crusading spirit was by no means absent, although it may have suffered by the war, as it usually does, and

George Whitefield, leader of the Calvinistic Methodists, one of the great evangelical "awakeners," proposed for the army the motto *"Nil desperandum Christo duce,"* no despair under Christ's leadership.

Military leadership for the combined operation of the American colonies was available in the colonies themselves, and with the many failures of the professionals in this kind of enterprise the belief of the colonists that they could undertake it with at least equal ability and prospective success was not unjustified. Their success against Louisbourg might go far to explain as jealousy on the part of the professional soldier Wolfe's subsequent condemnation of the Americans as "in general the dirtiest, most contemptible, cowardly dogs that you can conceive."

Leadership of the expedition was provided by New England, whence came also the soldiers and most of the sailors, in the person of William Pepperell, an energetic, wealthy Maine merchant and colonel in the militia, 49 years old at the time, later made a baronet and the only New England colonist ever so honored. A squadron from the West Indies station was to convoy the transports for the four regiments, artillery and stores, all of which were assembled within the comparatively short period of seven weeks. On the eve of departure, however, word was received that the squadron would not come. Undaunted the colonists decided to proceed.

Expedition Against Louisbourg

The armada consisted of some 90 transports, carrying a land force of 4,070–3,170 from Massachusetts alone—and a dozen armed vessels. On March 24 it set sail for Canso, which was retaken on April 5, and arrived at the theater of war almost too early as the shores around Louisbourg were ice blocked until late in April. Three weeks were usefully spent around Canso, fifty miles from the main objective, in drilling the militia while the New England cruisers captured some French ships with supplies for Louisbourg.

It had originally been thought that the place might be taken by surprise at night by landing troops "while the enemy were asleep." But that was assuming over-much carelessness on the part of the foe and more skill than could be displayed by raw, or for that matter trained, troops operating on strange and difficult terrain. The alternate plan was to proceed to manage more conventional landing, siege and blockade operations. To strengthen the latter a squadron from West India had arrived after all and in time other ships straight from England.

While the naval units proceeded to blockade Louisbourg, the transports landed the first New Englanders on the morning of April 30, promptly after their arrival on the scene. The landing took place somewhat over a mile west of the place on the rocky beach of Gabarus Bay and was uncontested probably for the reason that it was made swiftly. The water through which the men waded and dragged their few cannon, some 34 in all, was frigid and ice was still piled on the strand; tents were few and the first nights cold and cheerless. On May 2 a first detached battery of the fortress, the Royal, was captured from the rear, after an overland march of some 400 New Hampshire men, who threw the garrison into a panic by setting fire to some naval stores. The enemy without firing a shot and after having spiked thirty heavy guns retired to the town. The mechanics of the New England force managed to redrill these guns which were then turned against the town, loaded with the cannon balls which they had brought along in the anticipation of this contingency—so closely calculated was war undertaken by the colonists.

Of the progress of the siege not much need be said beyond that it was far from "regular," but still effective, probably for the reason that the besiegers had this project very much at heart, whether their hearts were set on loot and plunder, in which disappointment was to await them, or on confounding the Roman Catholics like the parson who armed himself with an axe and after the surrender used it to hew down images in the parish church. Cooperation with the naval commander was better than the usual discordant relations between colonials and British officers, possibly because this commander happened to be married to a New York heiress and hoped one day through the power of money and influence to become governor of the province.

Capitulation of the French

A first effort to take the Island Battery in the center of the harbor entrance with a landing force and without the cooperation of the naval commander miscarried—an inevitable punishment for non-cooperation, as some writers insist—and caused the loss of nearly two hundred men, most of whom drowned. The blockading ships captured several French vessels with supplies, among them a full sized man-of-war, a loss which disheartened the defenders. The perilous fogs along the coast made the naval commander anxious to hasten a decision by a more closely combined operation. He urged Pepperell, who was hesitating: "For God's sake, let us do something and not waste our time in indolence," proposing to take a large part of his troops on board, force the entrance to the harbor and attack the water front of Louisbourg. A simultaneous attack by water and land sides was actually in contemplation while the bombardment of the town continued with unabated fury, when on June 26, after a siege of two months the French indicated their willingness to parley. They marched out with full honors two days later.

The taking of Louisbourg, which became known in London shortly after the British defeat at Fontenoy by Marshal de Saxe, proved the best success that English arms, wielded by New England farmers, mechanics and fishermen, had been able to gain in the War of the Austrian Succession. As a siege it was particularly remarkable for its slight cost in lives and materials. Due largely to favorable weather both sides together lost only a few hundred men, although a pestilence among the colonials who garrisoned the place afterwards claimed 900 victims.

Despite the cry of *"Delenda est Canada"* raised in the Massachusetts Assembly the Louisbourg enterprise was not complemented by a mightier and more vital one against Quebec. For this purpose military forces were actually gathered in the colonies and a British fleet in Portsmouth, but the latter was eventually employed on a descent against Brittany. The Peace Treaty of Aix-la-Chapelle (1748) returned Louisbourg once more to France in exchange for Madras, a Presidency of the East India Company which its forces had taken. The peril offered by New France in the north was revived thereby and the American view confirmed that there was not room on the continent for both English and French.

Always stronger than in London was the conviction on the part of the colonials that France was aiming at a "universal monarchy on land and sea" and that the colonies must unite against such a danger as it regarded America. Hence the Albany Conference of 1754 to debate a union among them with such features as a central legislature that would have charge of defense affairs, or the removal of the Acadians from Nova Scotia whose essential neutrality and primitive isolationism neither side would admit, or recognize. When war came again, it was more than ever a world war, with America and India and much of the territory in between involved.

Muss denn die Welt so im Verhältnis stehen,
Dass Deutschland büssen muss, was Indien versehen?

(Must the world be so interconnected
That Germany must suffer for what India has done wrong?)

This was the plaint of a German poet of the time, weary of *Weltpolitik*. His India would include America, where war had begun even earlier than in Europe and where at first it went none too well for the British. An armament against Louisbourg, in 1757, reached Halifax in July, instead of April as Pitt had calculated, and never went further as the French forces in Louisbourg were considered too formidable to tackle. The expedition then sailed back to New York, having wasted through disuse the striking power of that part of the British forces for practically a whole season, while Montcalm profited by their inertia to strike at the English once more overland to the southward from Quebec, by way of Lakes Champlain and George.

Taking the negative lessons of Rochefort much to heart, Pitt made his preparations for 1758 and for what eventually took shape as the strategic aim for the British in North America, the conquest of Canada. While he realized that his energetic policies and the character of war in those distant theaters called for strong commanders, he could not at once avoid the use of elderly generals imposed by King George II with his penchant for military patriarchalism.

Young Men in Command

An age-worn gentleman might still hold supreme command in North America, but the colonial war was in reality carried on by the young men whom Pitt sent out, like Amherst, who at forty-one was entrusted with the Louisbourg enterprise and under whom Wolfe, then thirty-one and about as old as was Clive at Plassey when he won India for the British, was to serve together with two other younger brigadiers. The armament assembled at Halifax, 157 sail and 12,000 troops—the war more nearly approached the proportions of a world war also in the size of participating armies in distant theaters—all British soldiers except 500 colonial rangers, the ones who were to provoke Wolfe's angry judgment on the quality of the colonials. Leaving on May 28, they arrived in Gabarus Bay on June 2, instead of the end of April, when Pitt had sought to have the siege begin.

For five days the bad weather forbade landing, eliminated opportunity for surprise and gave the French a last chance to tighten up their defenses, of which they made measurable good use. Wolfe directed the real attack against Freshwater Cove, while three landing feints were made. Wolfe's force, pushing off shortly after daybreak, was in the character of a modern task force. He had five companies of grenadiers, the light infantry of several regiments—the same light infantry, good only for war, which the 18th century had to form and use so much against its own will and style—some American Rangers and Highlanders, with eight companies of grenadiers in support. They were observed from the shore and came under sharp enemy infantry and artillery fire, despite the heavy fire curtain provided by their own fleet, as soon as they arrived within range.

Finding the surf strong and no landing beach feasible, Wolfe—not at his best in marine activities, as he admitted—signalled a return to the ships, but some of the boats with light infantry, misunderstanding, or ignoring his order, pushed on, followed by the others, part of the grenadiers effecting the first landing. A number of men were drowned, thrown from the boats, or crushed while scrambling to the shore by the craft surging back with the swell. Wolfe, carrying a cane—to point, or punish?—was among the first to jump into the water and climb the rocks which were considered Louisbourg's outmost defense, and the first French shore battery fell to the bayonets of the attackers. By noon, all the landing corps was ashore and the French driven to the woods and the town, from whence its guns brought the British to their first halt.

Their situation was at first sufficiently precarious, for the continued bad weather made it difficult to get provisions, tents and siege artillery and munitions on shore, and a determined sortie might have pushed back the invaders. But the French defense was purely passive. As the journal of an officer of the garrison states: "The country which we had to defend, was known but superficially by the majority of the officers that were in charge of it; one thought that it sufficed to be a grenadier in order to resist a descent."

Army and Navy Cooperation

With Wolfe, an enemy of all military vices, almost to the point of priggishness, in the camp the virulent vice of dissension between army and navy could not arise or spread. He reported "great harmony, industry and union" between admiral and general. During the whole siege of 52 days, favored latterly by good weather, relations between the British land and sea forces were harmonious, to the very end when sailors rowed into the harbor at night to take the last of the French ships-of-line. Wolfe was the soul of the offensive, and grieved to see it end at Louisbourg instead of being carried to Quebec, a proposal vetoed by the naval commander on account of the lateness of the season and also of the ancient dread of navigation in the St. Lawrence.

In retrospect, to Wolfe the siege seemed to have been conducted in an overcautious manner: "Our measures have been cautious and slow from the beginning to the end, except in landing, where there was an appearance of temerity," if not overboldness. The formula *"Toujours l'audace!"* which might easily clash with the veracity of military history or memoir, was not embraced by Wolfe without qualification. Perhaps he was sufficiently intellectually honest to remember the terror of the landing in the surf for he wrote home afterwards that it had been undertaken rashly and succeeded only "by the greatest of good fortune."

Wolfe, wrote Horace Walpole, was exactly the type of officer "to execute the designs of such a master as Pitt," who now threw out the principle of promotion and bestowal of command by senority which had failed almost universally in the colonial wars and other enterprises dear to Pitt. He appointed the 32-year old Wolfe to command the army against Quebec in the year after Louisbourg and Amherst commander in chief in North America. "I am to act a greater part in this business than I wished or desired," Wolfe wrote in a family letter, humbly or mockingly, we are not sure. "The backwardness of some of the older officers has in some measure forced the Government to come down so low."

He was young which is to say that he was not prematurely

aged by dissipation, an enthusiast, a term which 18th-century aristocrats like Shaftesbury, or Wellington would not use without disparaging implications, perhaps because they were aware it involved "faults of manner," or even fits of depression like the one in which Clive ended his own life, or those that weighed upon Wolfe before Quebec. His enthusiasms at times puzzled even Pitt, who realized, however, that superannuated regulars were unsuitable for certain enterprises he wanted carried through. Since strategical, and in part even tactical concepts, spring from a soldier's mixed socio-technical outlook and may consequently run counter to those of others, also in part non-technically conditioned—a conflict which probably tends to lessen in more technological military periods—it would seem that Wolfe as a professional, in addition to his abilities, was considered just about "regular" enough by England's governors to be entrusted with the enterprise against Quebec.

The frictions between him and his three brigadiers at Quebec, all sons of peers, were mild, and luckily professional army-navy jealousy was completely absent. How hurtful such dissensions can be was demonstrated on the French side where the governor, Vaudreuil, representing a sort of northern creole, or native-son sentiment and interest, clashed with the military commander, Montcalm, and where the best naval officer was prevented from giving his best because he was not of gentle blood and had come from the merchant marine.

Where Wolfe Was Strong

Wolfe was fully cognizant of the vital and difficult quality of the task set for him, completely aware of the risks and also of the chances involved, suprapersonal and personal. Landings oftener than not are gambles, wherein adverse chances may be reduced and favorable ones enhanced by method and application, but which call for the staking of everything. Soldiers must die purposefully rather than cavalierly, and that is what Wolfe meant when he wrote to a relative: "You may be assured that I shall take all proper care of my own person, unless in case of the last importance when it becomes a duty to do otherwise." To the English people his death represented the apotheosis of the genius who had landed on a distant enemy shore to vindicate for his King and Pitt an imperial desire, only dimly aware that a danger nearer to the homeland, a possible French landing in Kent or Sussex, might be thwarted by him in command on a distant coast, or river. He was like a Phaeton who knew how to drive the sun chariot and paid with his life for the successful performance.

Speaking technically, Wolfe was one of the very few in his age who had acquired sound and reflective knowledge of, and ability for, the directing of landing enterprises, and indeed combined operations. This must not make us forget, however, that due to the lessons derived from the serial misfortunes of the British landing attempts of 1757-58 against France, there was a greater administrative understanding of the problems involved in London also. There even the Hanoverian king now was determined that "we must keep Cape Breton, take Canada, and drive the French out of America."

An efficient admiral, Saunders, was selected as Wolfe's naval teammate. Plans were worked out by, and in greater detail for, Pitt, Wolfe himself remaining in London for that purpose during the winter of 1758-9. The necessary troop strengths were estimated and promised, although Wolfe in the end got only about 8,500 men instead of the 12,000 he wanted, which made it doubly fortunate that most of his men were veterans of the American war, instead of such recruits as had formed the majority in Braddock's unlucky expedition into the Pennsylvania wilderness in 1755. Time tables were worked out which easily went wrong over the long distances and periods involved, due to administrative inertia and the then more powerful deflective force of the elements. Although the part which Wolfe took in the preparation of the expedition, or his own orders is not known to us in detail, it is clear that his vision went beyond the barriers which had been erected between the army and the "senior service." The judgment of a naval historian like Sir Julian Corbett that Wolfe was the greatest master of amphibious warfare since Drake is true, at least according to the saying that among the blind the one-eyed man is king, and that before and after him there were mostly blind men, professionally employed in this business.

From Louisbourg to Halifax

The fleet sent out from England for the Quebec campaign with Wolfe on board was hurried off as early as mid-February and found Louisbourg, its rendezvous, still ice-bound, so the ships moved on to ice-free Halifax. There, to Wolfe's dismay, it was found that a small observation squadron which was to blockade the mouth of the St. Lawrence, had unfortunately allowed some French ships with supplies and a hundred recruits to slip by and take information to Montcalm that a large English expedition was under way against him. On June 1, instead of May 7 as planned, the ships began to sail from Louisbourg harbor. It might have given Napoleon at Boulogne food for much profitable thought had he known how protracted a process the sailing of an invasion fleet is apt to be. With no interfering winds and a far more favorable point of departure than Boulogne it took the fleet of some 250 ships, 49 of them men-of-war, six days to clear from Louisbourg. At sea the fleet stretched out for more than fifty miles, in three divisions, corresponding to the three brigades of Wolfe's force, the transports in each being under the command of one of his trio of brigadiers.

To most intents and purposes the landing from the St. Lawrence around Quebec was from oversea and from the sea. The three hundred miles of river were broad, unbridgeable, uncharted, to many navigators a veritable Sargasso Sea of mystery, fearsome legend and uncertainty. Their dread was "inconceivably great" as Wolfe put it. But thanks to the labors of such British navigators as Captain James Cook, the discoverer of the Sandwich Islands, who served under Saunders, it became almost at once better known to the bold and enterprising invaders than to the French who had held control over it for more than a century and had trusted to its inaccessibility for the safety of Quebec. Wolfe's armada was the greatest yet seen in America. As it sailed up the St. Lawrence its coming was announced to the defenders of Quebec by signal fires flaring along the banks.

The admiral, ignoring the almost superstitious awe of the great river prevalent among the men of his fleet, steering it without mishap through the Traverse, 30 miles below the town, considered by the French as a particularly dangerous navigational hazard, brought his ships much closer to Quebec than he had given Wolfe hope to expect, or indeed than his instructions obliged him. His sailors and marines, twice as numerous as Wolfe's soldiers, were to be of the greatest assistance in the work of the siege, more than living up to the orders given to Saunders at home that the fleet was "to act in conjunction and cooperate with our said land forces," and that sailors and marines were to assist in such tasks as the manning of batteries.

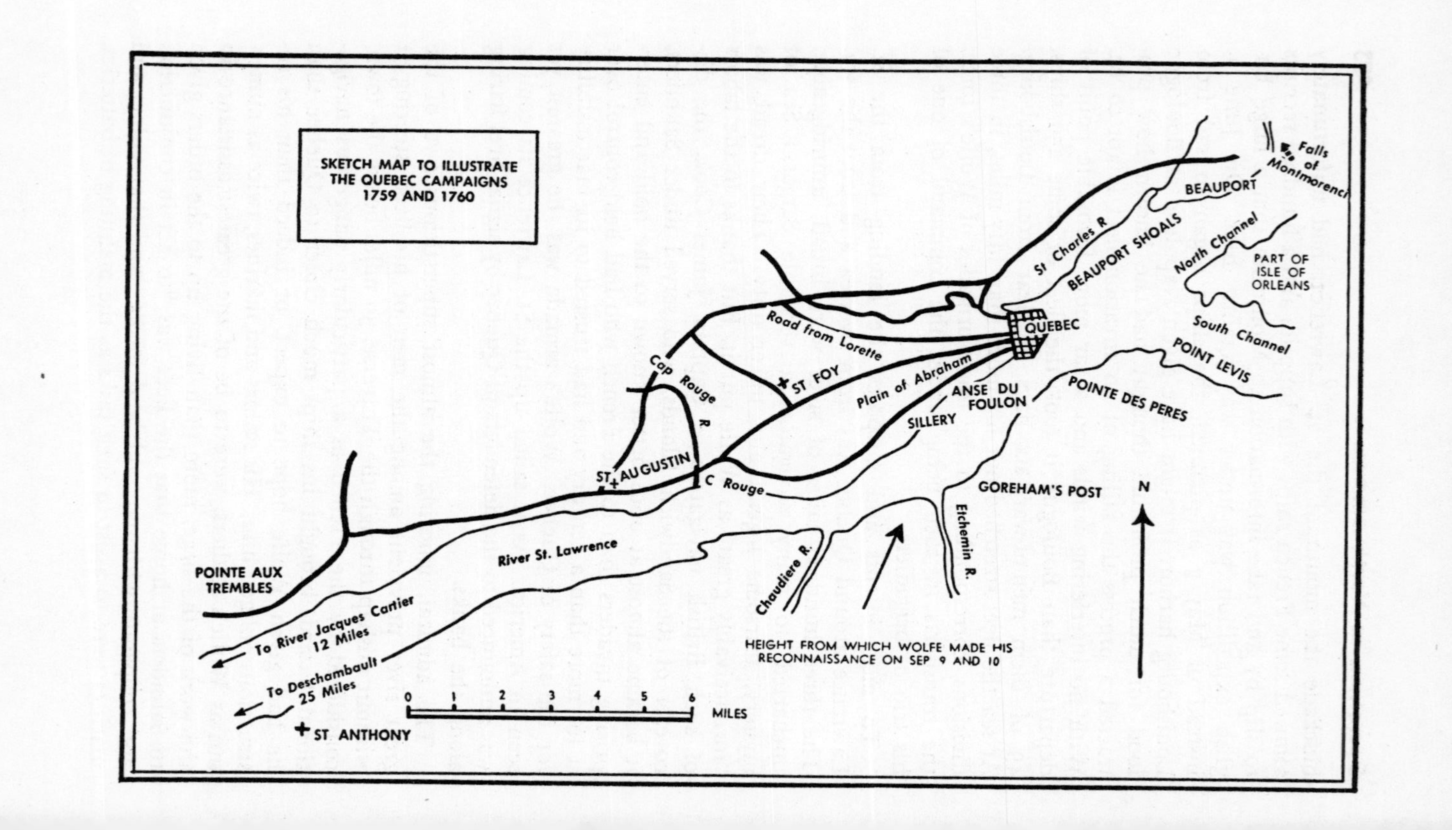

SKETCH MAP TO ILLUSTRATE
THE QUEBEC CAMPAIGNS
1759 AND 1760
Falls of Montmorenci
BEAUPORT
St Charles R
BEAUPORT SHOALS
North Channel
PART OF ISLE OF ORLEANS
South Channel
QUEBEC
POINT LEVIS
POINTE DES PERES
ANSE DU FOULON
Plain of Abraham
Road from Lorette
ST FOY
SILLERY
Cap Rouge R
ST AUGUSTIN
C Rouge
GOREHAM'S POST
N
Etchemin R.
Chaudiere R.
River St. Lawrence
POINTE AUX TREMBLES
To River Jacques Cartier 12 Miles
To Deschambault 25 Miles
ST ANTHONY
0 1 2 3 4 5 6 MILES
HEIGHT FROM WHICH WOLFE MADE HIS RECONNAISSANCE ON SEP 9 AND 10

What specific form their tactical cooperation would take, neither Saunders, nor Wolfe seems to have thought out in advance, although the latter wrote even before the departure from Louisbourg, clearly reckoning with operations above Quebec and the narrows which gave Quebec its name, that "it is the business of our naval force to be masters of the river, both above and below the town"; they should be capable of "stealing a detachment up the St. Lawrence and landing them three, four, five miles or more above the town." It was probably fortunate for them that Montcalm did not choose to defend the river below Quebec like another Dardanelles and Gallipoli. It is interesting to speculate upon the similarity of the chances of defense offered by these three natural sites—down to the parallel between floating mines at the Dardanelles and Gallipoli and the fire ships which the French sent down with the currents against the English fleet two nights after its arrival before the city.

Sailing Up the St. Lawrence

The voyage of 25 days to the Island of Orleans in front of Quebec was happily marked by the absence of events. Wolfe had been instructed to make his men available for service on board the ships if necessary and had actually told off 300 soldiers to complete the crews of ships which had wintered in America. Like Belisarius on the long voyage to Africa, he took care to keep them healthy in body and spirit, in the open air as much as possible, making them eat and exercise on deck, providing them with fishing utensils, ordering them to air their bedding, etc. He prepared them for the strict order he wanted preserved on land when, as a besieger as well as an occupant of foreign lands which were to be annexed, he kept severe discipline, punishing even thefts with death.

In the early morning of June 26 the first landing took place on Orleans Island, unobstructed by Montcalm from the other side of the river, the north bank which alone the French intended to defend as if the St. Lawrence were truly a seashore, a *litus,* and not a river bank, or *ripa.* The defence system of Quebec was essentially a one shore line, not a true fortress according to the standards even of its own day, but six miles of earthworks paralleling the river behind broad mud flats and high above the shore, comprising the so-called Beauport lines, with the right wing on Quebec, a poorly fortified place itself due to the corruption and inertness of French colonial officialdom, and the left wing on the Montmorenci river. The closest parallel from military history might perhaps be found in the Stollhofen lines of the Spanish War of Succession (taken by the French in May 1707). Again the situation reminds one of the Dardanelles, with only Gallipoli fortified by the Turks, but not also the Asiatic shore.

Montcalm's intention was to keep the enemy from landing on the north shore in proximity to Quebec on either side of that town which, as the capital of New France, policy dictated should be saved above everything else so as to leave France a remnant of her North American possessions, on which in the future peace negotiations she might base a claim for the return of the whole. The line of the steep bank beyond the town remained unfortified; any threat to it was to be met by a sideward deployment of Montcalm's forces through and around Quebec. It is obvious how easily they could be outmarched by the British moving on the water with the tides, once the latter had determined to use the river beyond Point Levis which the British occupied on June 29, and whence their cannon bombarded the town, and Cap Diamond where the river was less than a mile wide.

Unlike the Hellespont, where the flow of water is steadily towards the Mediterranean, the St. Lawrence is tidal considerably beyond Quebec where the tide is so strong that it reaches four knots. The regularity of the tidal flow enabled the power which held control of the river to determine the movements of its ships with a precision rarely attained in pre-steam days, and add to their speed by the use of oar, or sail. In the later stages of the siege when the British moved freely up and down the river, one of Montcalm's subordinates, being detailed to make a defensive demonstration wherever they might attempt a landing, was forced to march his men back and forth, in pace with the ships, no less than 42 miles between dawn and midnight of one day.

The British craft would drift along with the tide like freight cars in a switch yard rolling down an incline into their sidings, and land wherever they might find an unguarded place. Urged on by Wolfe who, the admiral thought, betrayed impatience and who long before leaving Louisbourg had thought it was "the business of our naval force to be masters of the river both above and below the town," parts of the fleet on July 18 during a dark night, aided by favorable tide and winds, went as far upstream as Cap Rouge, much to the consternation of the defenders.

British Rebuffed on Land

Remembering the later passages of Farragut and Porter past Vicksburg, Port Hudson and other points on the Mississippi, one might be tempted to call this a typical incidence of warfare in America, although the difference of proceeding under sail and under steam should be kept in mind, together with the progress in ordnance during the intervening century.

Preceding this successful control over river navigation, Wolfe tried to turn Montcalm's left by marching up the Montmorenci and fording it, but he failed with losses. This setback was balanced by a vain attempt on the part of the French against the harassing British artillery on the south bank. A force of twelve to fifteen hundred men, including a group of Catholic seminary students whose spiritual belief was clearly stronger than their nerve, took off from a point above Quebec, but so soon as they had landed showed "signs of a nervous agitation," opening fire with their guns when still several miles from the enemy. After this had occurred three times panic developed into a wild stampede in the direction of their boats which carried them back to Quebec by dawn, "overwhelmed with despair and shame." Evidently strong religious belief, although it had in the past often successfully animated irregulars in formal and guerilla warfare, in this case proved but poor preparation for such an enterprise calling above all for exactitude in timing and resolutely martial behavior.

Wolfe for a time seems to have contemplated a landing above Quebec where the fleet continued to keep ships anchored, but he thought, as he subsequently wrote to Pitt, that "the body first landed could not be reinforced before they were attacked by the enemy's whole army." Montcalm had indeed detached a force of 600 to guard the few paths leading up the cliffs along the river over the twelve miles from Quebec to Cap Rouge. But further he could not be budged in moving out of his Beauport position which suited the character of his forces, with the militia providing the greatest number, even after Wolfe had undertaken a raid on Point-aux-Trembles, still further upstream.

Irritated at his slow progress and in bad health—as bad as his generalship at the time, was the ill-tempered comment of one of his brigadiers—Wolfe moved a council of war to agree to a further attack on the Beauport lines to be made on July 31, consisting of a frontal assault across the river from Point Levis and Orleans Isle against Montcalm's left wing, combined with an attack against this wing across the Montmorenci which was fordable below the falls at ebb tide. The fleet was to support the attack from floating batteries and by fire from Saunders' flagship in the North river channel, all the men-of-war throughout the siege showing a healthy contempt for the land batteries.

For several hours the ships and the Point Levis batteries hammered Quebec and the batteries on the east bank of the Montmorenci at the left end of the Beauport line, while boats with soldiers hovered opposite the right wing of the line as if planning to attack there. The moment the ebb tide came, making the Montmorenci fordable, the boats set off for their assigned landing place opposite Montcalm's left, with a redoubt as their first specific objective. The conjunction of the British forces was too finely calculated, an undetected reef in the river caused a considerable delay and by the time Wolfe had decided on a fresh landing place, Montcalm had assembled nearly his whole force on the left wing.

Although a dozen grenadier companies and 200 Royal Americans landed first and rushed the redoubt they got no further for the redoubt, as Wolfe had not anticipated, was commanded by the French entrenchments higher up. From there the French fire raked the massed British troops. By now darkness and the returning tide approached, the latter threatening to cut off the retreat across the Montmorenci. Wolfe gave orders to retire everywhere, leaving behind his dead and many of his wounded, both being promptly scalped by Montcalm's Indians and some of the Canadians. The attempt cost him some 450 casualties, including 33 officers. The French were jubilant, the Governor writing at the time: "I have no more anxiety about Quebec," and Wolfe, his self-confidence deeply shaken, spoke of himself as a "man that must necessarily be ruined," and came close to quarreling even with Admiral Saunders.

The faith in him of the officers and soldiers was not restored for several weeks. During this time a combined army and navy force went up river to destroy French stores destined for Quebec and the crops in the fields, in order to draw off more of the regular forces of Montcalm and disconcert some of his militia men who had received, or taken, leave to harvest their grain.

WOLFE ACCEPTS SUBORDINATES' COUNSEL

Calculating on the disintegrating effects among the French, Wolfe still persevered in his plans for yet another storming of the Beauport lines. Debility caused by nervous strain and worry over the situation of the besiegers, for the summer was wearing away, sent him to a sick bed for weeks, where he brooded over present failure and future uncertainty. Still adhering to his own plans, he decided to take his three brigadiers into his confidence, although hardly all the way. He communicated to them for the first time his instructions from Pitt and ordered them to meet by themselves, debate his plans and possibly evolve proposals of their own. The procedure savored something of the decision of a sick man to request three specialists to discuss his own case and perhaps decide on a treatment, reserving the final decision for himself.

The generals, one of them recorded as believing that their leader had shown little fixed purpose, met, mulled over Wolfe's proposals and unanimously rejected them. In their counter-proposals of August 29 they advised leaving the Beauport lines alone, thinking it better to concentrate the attack above Quebec, establish themselves on the north shore there, cut off Montcalm's supplies and force him to give battle, or at least make him withdraw into the town a larger part of his force than before, thus rendering a later attack on the Beauport line more promising.

Wolfe accepted the advice of his subordinates. He harbored no resentment, frankly admitting to Saunders that he alone was responsible for the failure of the 31st of July. But he was not at once certain of his own capabilities, or future plans, writing to Pitt on September 2 that "in this situation there is such a choice of difficulties that I own myself at a loss how to determine. The affairs of Great Britain, I know, require the most vigorous measures; but then the courage of a handful of brave men should be exerted only when there is some hope of a favorable event."

On September 3, the British troops were skilfully moved from the Montmorenci camp to Point Levis whence on the 5th seven battalions marched upstream to embark on a detachment of the fleet. Wolfe was with them, for he had now regained his health and stamina, although combatting deep melancholia. His movements drew more French troops away from Quebec to guard against landing attempts, Montcalm being especially concerned about a steep road, or path which led from the Anse du Foulon, a small cove, to a point on the Plains of Abraham a mile and a half west of Quebec. Wolfe, studying intensively the opposing shore, considered this approach not fully closed. For did he not see women wash clothing by the river and later hang out garments to dry on top of the cliffs! In military intelligence, even more than in detective work and judicial processes, such indicia can never be too trivial to be overlooked and underestimated.

His own observation indicated to Wolfe "the path of glory" which often leads but to the grave, lines he cited while being rowed up and down the river at the foot of the cliffs on the 9th or 10th, to reconnoiter the route which his men and he were eventually to take to the Plains of Abraham. After that Wolfe' plans were quickly, although carefully, made, while not disclosed in detail to anyone until almost the last moment. In fact, his brigadiers formally protested that they were insufficiently informed about the exact landing spot, whereupon Wolfe at 8:30 of the evening before the battle gave them exact detail and his operational directions, assuming for himself, who wa to lead the enterprise, all responsibility involved to his King and country.

These plans called for perfect coordination between arm and navy, first for a feint down the river against the Beaupor lines to be undertaken by naval forces in order to pin down Montcalm so long as possible there; second for another feint upstream to draw the French detachments of some 3,500 place above the town still farther up river and leave undefended a much as possible the approach up the cliff from the Anse d Foulon. This was the logical landing-point which Montcalm had surmised as early as September 2 when he wrote that h believed "that Wolfe will act like a gambler who, having trie his luck on the left and on the right, will try the middle."

The troops for the attack, 400 light infantry as vanguard t be followed by 1,300 men, together with artillery and suppli under the command of two of the brigadiers, were to assembl at Cap Rouge and drift down with the ebbtide which bega about 2 A.M. on the 13th, to the Anse du Foulon where the were scheduled to arrive about 4 A.M. There the task force was to halt and wait for one officer and 24 volunteers to scale the heights in the dark and to follow them when the signal should come that they had succeeded. These 1,700 were all the available boats could hold, but 1,900 were to follow later in the men-of-war and transports which were to anchor near the landing point.

To the Plains of Abraham

Things proceeded according to plan, with a piece of luck thrown in by the fortunes of war. A French sentinel challenged the leading boats, in one of which Wolfe rode, while they were gliding along in the dark, and a British officer with presence of mind answered back in such good French that he convinced the sentinel that the boats were part of a French convoy expected from upstream that night. That the convoy had been cancelled, neither Wolfe, nor the sentinel had learned.

The strong tide bore the leading boats somewhat beyond the chosen debarkation point, but the 25 volunteer pioneers got ashore, scrambled up the cliff and rushed the post at the end of the Anse du Foulon path, their shouts and huzzas telling Wolfe of their success. The path was cleared of a barricade and the foremost troops gained the plateau while the boats were bringing over soldiers from the shore opposite the Anse as well as those coming from up the river in the second division formed by frigates and transports. Before the sun was far up Wolfe's whole force of 4,500 were ashore and ready for the decisive battle on the Plains of Abraham.

The situation which was forced upon Montcalm was typical of that which usually confronts the defenders. Should he attack at once with the men immediately available to him, taking advantage of the confusion that commonly attends a landing, but which in the case of Wolfe's disciplined troops was very slight and soon rectified, or should he wait to counter-attack until all his disposable forces could be assembled? Should he, in contemporaneous military phrase, give battle or withhold it? Montcalm was convinced that "we cannot avoid a battle, the enemy is entrenching; already he has two cannon; if we give him time to establish himself we can never attack him with our few troops."

By 9 A.M. he had some 4,000 men in line against Wolfe who outnumbered him slightly. Some of the usual advantages of the defender, inherent in unity of command, were denied him. Governor Vaudreuil interfered with his orders and the commandant of Quebec refused him more artillery. Besides, he thought it too risky to wait for the arrival of the considerable French force on detached duty above the town, although he might have expected them to arrive shortly after the opening of battle; or that of other troops from the distant left wing of the Beauport lines. Wolfe had finally forced Montcalm to join battle at once against the British "fire machine," that had been brought the long way across the ocean.

British and French faced each other. The French advanced briskly in three columns. The British held their fire until the French were within forty yards. Then from their musketeers "with one deafening crash the most perfect volley ever fired on battlefield burst forth as if from a single monstrous weapon, from end to end of the British line." The French broke, and after barely a quarter of an hour of actual battle the British had won. Both Wolfe and Montcalm fell. Quebec itself held out from September 13 to the 17th, and then surrendered. Montreal did not yield until September 8th, of the following year, 1760.

The cost to the British in casualties was astonishingly light, ten officers and 48 men being killed with 37 officers and 535 men wounded. Toll taken of the French was far more heavy, running to around 1,200 in killed and wounded.[1]

NOTES, CHAPTER 26

[1] Sources: John J. W. Fortescue: *A History of the British Army,* Vol. II; F. Parkman: *Montcalm and Wolfe.* Boston 1884; R. Wright: *The Life of Major General James Wolfe.* London, 1864; Beckles Willson: *The Life and Letters of James Wolfe.* London, 1909; George M. Wrong: *The Rise and Fall of New France.* N. Y. 1928; Lt. Col. F. E. Whitton: *Wolfe and North America.* Boston, 1929; Sir Edward S. Creasy: *Fifteen Decisive Battles of the World.* Ed. by Robert Hammond Murray. Harrisburg, 1943, 499-511.

Chapter 27

FRENCH PLANS OF REVENGE FOR QUEBEC BY CROSS-CHANNEL EXPEDITIONS

FOLLOWING THE CLOSE of the Seven Years' War, France consistently contemplated and schemed revenge against Britain, revenge for Quebec and Quiberon in particular. Her foreign minister for twelve years from 1758 on, the Duc de Choiseul, was, so to speak, the minister of revenge. It was with this purpose in view that he sent one of his secret agents, M. de Pontleroy, to the American colonies of the enemy to study and investigate the tendencies there towards emancipation from England as well as the hydrography of the coasts. For these purposes the agent took service aboard an American merchant ship (1764 and 1766). Another agent of Choiseul's in America in 1768-9 was Col. de Kalb who warned him, however, that if the Americans should secede they would try to do so without the help of a foreign power.

Partly on account of these and other reports from America there was a swing back in France towards the older projects of an invasion of England itself, the last of which had failed at Quiberon in 1759, towards *"une lutte directe,"* as Durand, one of the French ambassadors in London, termed it, saying: "Our wars with her would last less long if so many of the citizens of London who are enriched by the war and who desire it, would see from nearby the horrors which it entails." The theory was that an invasion would within a few hours create a financial panic that would go far toward putting the country at the mercy of the invader, something that had almost occurred after the young Pretender landed, causing George II to meditate a return to his German states. "This, monseigneur," said Durand, "is something that escapes foreigners who see nothing of England but immense fleets and prodigious warehouses. Those externals impose upon one, and few people consider that a nothing, a false rumor, the mere audacity of the enemy would embarrass credit, put disorder into a complicated machine and reveal a weakness which is not well known except to the interested parties." In his invasion proposals Durand, recalling earlier conversations with Bolingbroke, terms senseless all notions of attacking the colonies—they are mere members of the body of the country; one must aim at the heart.

Many, if not most, of the French agents returned from England convinced that "the English are never weaker than at home," that they were filled with a sense of this inward weakness, that "they feel very well that a descent would bring about a terrible financial perturbation and consequently in the government as well. Hence they are living in a continual state of alarms. This is for all enemies of England the best argument for succeeding with a descent." These were the conclusions at the end of 1770 of a Col. de Beville, whom Choiseul had entrusted with a military espionage task along the English coasts, one of several agents whose instructions generally told them to inspect "all the points of the coast of England where it might be possible to disembark," their accessibility for small and larger ships and squadrons. They were also told to investigate roads leading inland, resources, the spirit of the inhabitants, military forces, the usual budget of spies' objectives. At this time these agents were apt to be educated gentlemen, before the spying task was divided into a high-toned role for the military attache and other diplomatic personnel and one of lesser dignity for the venal spy.

To Strike Without Warning

One of the former category, Grant of Blairfindy, a Jacobite and lieutenant colonel in a French infantry regiment, beginning in April, 1707, stayed for several periods in England and made voluminous reports which were subsequently examined by Bonaparte. He proposed among other things not to precede invasion by a declaration of war and thus strengthen the element of surprise. The English themselves had begun war in 1754 and 1755 without a formal prior declaration. Ambleteuse, where a new naval port would have to be built, Boulogne, Wisant and Calais were suggested as embarkation ports for an army of 50,000 and Deal the central debarkation point. For transports he recommended fishing vessels, which were thoroughly good sailers and could carry 100 to 200 soldiers each.

Beville favored Lyme in Dorset, Dartmouth in Devon, Fowey and Lowe in Cornwall as vulnerable spots since the English would be less on their guard there than in Sussex and Kent. "A man under arms thinks rather of protecting and making sure of his heart without considering his thighs or his legs.

Here it is that England must be tripped up by the heels, if I may use this expression, which is the surest thing to bring her down. From all or any of these points the advance should be made on Bristol, which would be like threatening to put fire to the city of London itself." And soon all the trade corporations of London would distractedly importune King and Parliament to purchase peace at any price.

Realizing the still prevailing weakness of the French fleet, Beville thought a Channel crossing, forced by a naval battle, out of the question. Even the Homeric gods had made use of the clouds on occasion, if necessary, for their own protection, hence: "Let us pass over to England like smugglers; there is enough over there to make us conquerors." *Croc-en-jambe,* contrebandiers, *Helden* of France against *Händler* of London, the struggle was considered bitter enough, the technical difficulties sufficiently great to warrant discarding chivalric notions in war which are said to have prevailed as recently as Fontenoy. After thorough meteorological observations Beville proposed that use be made of the south and southeastern winds prevailing in the autumn and towards the end of winter, as well as of fogs and the darkness of long nights.

As the author of the Family Compact, between the French and Spanish branches of the Bourbons, Choiseul naturally planned for combined Spanish-French action at sea, for a French fleet of eighty ships of the line and forty frigates which he hoped France would have by 1770. However, he was brusquely dismissed after prolonged court intrigues against him who had enjoyed the potent protection of La Pompadour until her death, but then faced the hostility of La du Barry. "He would project and determine the ruin of a country," wrote Horace Walpole, apparently unaware that England might have been among the subjects of ruination, "but could not meditate a little mischief, or a narrow benefit."

Louis XV himself had for years a special agent to prepare for the *revanche* against England, the Count Charles François de Broglie, the head of that special diplomatic service, the Secret du Roi, maintained by the king apart from the regular establishment. Too well known as one of the grands seigneurs of his time to render it expedient for him to go spying abroad himself, de Broglie had his agents in England and elsewhere, including that strange figure, the Chevalier d'Éon; the later

General Dumouriez, or an officer of the engineers who procured data for de Broglie's grand invasion plan on which he worked for four years, from 1763 to 1766.

The engineer brought home good charts and maps and declared as the result of his invasion studies that a landing was quite feasible if only the king desired it; he thought also that the French navy and merchant marine were in far better shape than usually presumed. In June, 1765, de Broglie submitted to the king his memoire about the general disposition for the project, involving several feints, but with a descent on England as its principal objective *("C'est à Londres qu'il faut aller")*, with a landing corps of 60,000 to be disembarked at Rye, Winchelsea, Hastings and Pevensey. Everything was calculated down to the cost which he estimated at 33 million francs.

PUZZLING TACTICS URGED

As in several of the French projects of a descent during the later 18th century, there was a clear and sound functional distinction made between a battle fleet and an invasion flotilla and their respective uses, the first to attempt to battle, or scatter and decoy away the British fleet and make it inferior in the Channel, thereby rendering feasible the invasion, which should always be the principal object, according to de Broglie. "All other projected expeditions must have the essential object of puzzling the enemy, of occupying his attention and dividing his forces, so that he shall be unable to prevent the passage of the Army and its disembarkation on the English coast."

Tired of the fruitless work for the king, who treated all these vast labors as merely part of his *amusette,* de Broglie at last allied himself with Choiseul, despite the water tight compartment system, preferred by the king to keep them apart. But before the common work of the two foreign service heads had gone very far, Choiseul was dismissed. In 1777, after Saratoga and when the alliance with the American colonies was being discussed in Paris, de Broglie submitted his old project to the new king whose government, on the whole, preferred the indirect invasion of England, by way of the colonies or Gibraltar, without discarding all thought of a direct invasion.[2]

Dumouriez had been another, humbler member of the *Secret du Roi,* also disgraced for a time and even temporarily an inmate of the Bastille after the downfall of Choiseul. But he was fully reemployed after the accession of Louis XVI. He would seem to have owed his place as commandant of Cherbourg which he held for ten years, dating from the end of 1777, to plans he had submitted on the defense of Normandy-Brittany on which he had been working since 1776. From defensive plans his mind, early foreseeing that France would become involved in England's American War, was turned to such schemes as providing within a month enough shipping to convey 50,000 men to England, or for a French descent on the Isle of Wight, which was guarded at the time by only 150 pensioneers, preliminary to a larger project of assault that was to follow (winter 1778-9).

The enterprise seemed the more feasible to him since England at the time had 50,000 men in America and only 10,000 at home, while the militia was not embodied, and the fleet under Admiral Keppel in home waters counted but 20 sails, no more than its French opponent. In fact, an Anglo-French action fought on July 27, 1778, with Keppel commanding the English fleet had resulted in an ignominious draw.

To Dumouriez an advantageous embarkation point for an enterprise against Wight, was Cherbourg where the required expeditionary force, 24 battalions of infantry, one regiment of dragoons and eight companies of artillery could be housed, and all necessary preparations made quickly and unostentatiously to embark them within two hours, "after a previous rehearsal." He did not greatly concern himself about supplies as Wight could feed an expeditionary army, particularly if the inhabitants should be evacuated to the mainland. The dragoons were to take bridles, saddles, etc., but no horses, for they were to find mounts in England. For transports Dumouriez planned to take a sufficient number, perhaps 200, of the 1,500 oyster ketches, or *chasse-marées,* available in the ports from Brest to Dunkirk, decked craft, sometimes sternless and double-prowed, of shallow draft and excellent sailers, each to carry sixty men. Thirty of them were to be converted into gun sloops to be fitted with a pair of false keels so as to be beached without careening. In the history of landing craft this idea deserves notice.

Landing Plans Considered

The most propitious time of the year was thought to be November, leaving in the evening at ebb tide in order to reach the shores of Wight at full flood. Numerous pilots could be found among the smugglers. Landings should take place all along the southern coast of the island. Marches inland from the landing beaches would nowhere exceed four hours. From Wight the operations were to be carried to the English mainland in the following year with the support of a combined Spanish-French fleet.

Looking back on these plans of his from the vantage point of 1803-4, when the vicissitudes of revolution had brought him to England, Dumouriez thought that if they had succeeded, the political and commercial credit of England would have been shaken. Still, it would have been by comparison a nice war, a chivalrous *ancien régime* war. "As war was then waged humanely between completely civilized nations, neither carnage, nor devastation would have had to be feared. Things are now altered. The French armies consist of looters led by brigands. The fell sentence *delenda est Carthago* has passed Bonaparte's lips. With him, then, there can be no negotiation, no compromise. It is a life and death struggle, and nothing else." [3] However, Dumouriez was at one time close enough to the Revolution in its military sector to think landing in England, revolutionary in character and consequences even under the conditions of the *ancien régime,* a quite feasible project.

These French spying and scheming activities did not remain unperceived in London, where from time to time "the probability of an intended invasion, its practicableness, the dangers that must ensue, the means of guarding against them," were considered. In the autumn of 1771 a French *coup-de-main* on Portsmouth was discussed if not feared and Captain Guy Carleton invited to comment on its possibility and likely character.[4] He feared above all the panic that would be produced by the sudden appearance of a French fleet on the English coast. "The people will naturally fly," with only a few officers able to preserve "their sense amidst the first panic terrors." Portsmouth, therefore, must have a determined and calm commander who "by his own example and by clear and judicious orders quickly draws the mob from their distracted state into good order and confidence."

The place with little, or no permanent garrison must be at once filled with numbers of troops "for without hands no great work can be speedily performed." Dragoons stationed in the neighborhood should compel all the men in the region to go to Portsmouth, intimidating them by violence if other means failed. In addition to this anti-invasion dragonnade, he recommended that officers in and around the place be trained for such emergencies in time of peace, acquiring a knowledge of the country and discussing in writing cases and situations of invasion. Everything on which the enemy might draw for his subsistence should be destroyed, or removed.

BRITISH SEA STRENGTH FEARED

As a general proposition Carleton thought the French attack on England, the attempt to establish the seat of war there, rather more likely from the general system of French policies at the time than distant conquests such as rewinning Canada, "if they should not find out a new system of politicks more pernicious to Great Britain," which may be read as a British military forecast of the French support of the American colonies.

> The only difficulty France has to fear, in carrying a plan of invasion into execution, arises from our superior fleet; could her transport pass the seas unmolested, our land forces, were they much more numerous than at present, could not guard our extensive coasts, nor prevent an enemy's getting ashore: great as the obstruction of our navy may be, tis more than probable France will attempt to invade this country at the opening of the next war, as the least difficult plan for weakening that power, so formidable to her in the last; by experience she must now feel, what good sense ought to have dictated, that, our Austrian alliance dissolved, and the armies and arsenals of that, and of every other great power removed to a considerable distance, from her frontiers, the most judicious way of carrying on a war with Great Britain, is to establish the seat of war in England.

Despite all necessary detachments the English fleet was likely to remain master of the Channel. But the moment of danger would arrive in winter when south winds might confine the English ships in Plymouth, Portsmouth, or the Downs and at the same time carry the French transports within twelve hours from the ports between Dunkirk and Cherbourg to the coast of Kent and Sussex, the 80 mile stretch from Arundel to Hythe, where most of the strands were suitable for landing. Carleton assumed a French landing corps of 40,000 veteran troops, with the transports under orders to put men ashore as near to their respective rendezvous as the wind permitted without losing time by seeking too great precision.

Should they gain a foothold, there was nothing to oppose them, as arrangements stood in England at the time, but the very few regular troops and the militia. The latter in spite of all rhetoric in its praise would be unable to withstand the seasoned troops of France. To lessen the danger Carleton recommended the maintenance of more regular troops in Britain in time of peace, to use the militia mostly in entrenchments, "as it cannot be expected from those troops, however brave they may be, that they should move in line, in presence of an enemy, with so much order as those who have no other profession." And above all both Regulars and Militia should be made excellent marksmen.

> This is the most difficult, the most essential part of exercize, without which all infantry becomes nerveless, yet 'tis the most neglected by all the armies in Europe. The light infantry should be very active men, and practized to move with great celerity; they should be trained up in the partizan war, that of the savages I think the most formidable: they might also be taught to mix with the cavalry; this country is in general so much enclosed, the latter will seldom be able to approach the enemy, without assistance from the infantry. After all, when we have formed and disciplined an army, and selected our best troops to compose it, it may not be amiss to consider that the enemy may and ought to have the best troops of France to oppose us, without a recruit or awkward man, that when we come to a decisive battle, France hazards these troops, so does England, but England hazards besides her Empire of the Sea, and the desolation of her whole country. Common sense therefore requires we should not precipitate an event, generally very uncertain, where we have so much to lose, and the enemy so much to gain. The plan of an able officer therefore should be to take such strong encampment as will not allow the enemy to attack him without great disadvantages; but he must also know how to change his position, so as still to cover those important places, whatever movements the enemy may make to turn his camp, and endeavor to ruin him by degrees without risking the fate of this country

> on the cast of one die for the hopes of gathering the brilliant splendors of a great victory, this caution should have no timidity, but be the result of a strong judgment and firm mind, equally qualified to discern when circumstances change and of assuming a different conduct when it is required by good sense.

The limitations imposed on warfare in the 18th century weighed upon both sides and stopped, *pari passu,* the progress towards greater totalization of war which would have been required for either successful attack across the Channel, or proper defence against it. That is merely confirmed by the ideas of another military adventurer, older by a generation than Dumouriez, who studied the problem of an invasion of England at the same time as, and also earlier than, the latter—General Lloyd. He had served several of the European armies, that of his home country not included. "Like a red thread runs through the man's whole life, whether justified or not, the accusation of having traded with the diplomatic and military secrets which he learned among the various powers" whom he had served (Bülow).

Lloyd's Discussion of Invasion

These services included participation in the enterprise of the Young Pretender and two terms in the French army, during the second of which, 1754-57, he was employed to study the English coast for the purpose of a landing. He died in the Netherlands in 1783 and it is said that a commissar of the British Government took possession of his papers, among which was a detailed plan for the landing of a foreign force in England. Whether this is true, or not—a life of Lloyd is indeed one of the great desiderata of military biography [5]—this plan can hardly be identical with his rhapsody on *The Present System of French Politics, On the Projected Invasion and the Means to Defeat It.* Illustrated with a chart of the opposite coasts of England and France, published in London in 1779 and reprinted in 1793.

Lloyd's invasion discussion proceeds from the politico-military situation of the autumn of 1779, when England had to consider Weltpolitik and fight a world war. In this, however, "we act upon too narrow a scale, like traders, and seldom as a powerful nation. In forming treaties, a minister should have the whole globe before his eyes, and by no means confine himself to this, or that province, or branch of trade." He emphasizes the military danger for England of isolationism, of the negative answer to the much debated question "whether continental connections are useful, or otherwise to this nation. . . . Since we have abandoned the continental system France has acquired an unlimited influence in the different courts of Europe." In her world war England then had no allies; none to help her resist a landing on her shores.

The military implementation for world politics for Lloyd lay in marines, who had undergone numerous changes and ups and downs in England during the 18th century, at times even non-existence. "Land forces are nothing. Marines are the only species of troops proper for this nation; they alone can defend and protect it effectually. During the peace they garrison all your ports in each quarter of the globe; in time of war your fleet is instantly manned; and by employing many of them, fewer seamen are wanted in proportion." Besides, a fleet with from 12,000 to 15,000 marines on board "is equal to almost any enterprise against the enemy's settlements and keeps them in continual anxiety in every part of the world. . . . A powerful fleet and 30,000 marines will save us from destruction and nothing else."

The additional strength required depends on what we call war potential today, a concept which has been slowly evolved since that time and to the formation of which Lloyd was contributing when he wrote that although "no author that I know of has given any data which enable us to calculate the force of nations, ... I think the power and strength of a nation depend on the number of inhabitants and the quantity of their industry," etc.

Lloyd next discusses the possibilities, difficulties, requirements of, and for, landing enterprises in his time, on the whole with reference to a French descent on English shores and, in the military terminology of the time, to the shaping of which he himself had contributed such concepts, as "line of operations." The first conditions for success are superiority at sea and "a land force sufficient for any purpose." With such force at his disposal the enemy may have three objects in mind, a landing of 15,000 to 20,000 men in Ireland, probably in the bay of Galway, or against the western possessions of England and thereby ruin her trade, or the most to be feared, the setting of a powerful army in England itself.

Fleet Insurance Against Invasion

"If success should follow their operations, it is evident we must conclude a peace on any terms, and the war is soon brought to an end. This enterprise is decisive and preferable to any other." But it cannot succeed, he argued, "while our fleet is entire, though we may for a time be forced out of the sea, no invasion can take place. . . . It is not enough that they debark an army, it must be continually supplied and protected from France, otherwise, however numerous, it cannot make any progress, or penetrate into the country."

Despite the great length of England's coast line, it can only be attacked at a few points, "fixed and determined by the nature and position of the countries at war. An army, like a traveler, must necessarily depart from a given point, and proceed to a given point, in the enemy's country. The line which unites these points, I call the *line of operations*."

All deviations from this line, all delays on the march along it are so much time lost and in the end will force an enemy to turn about, either for want of subsistence or by bad weather, etc. War except in the form of Tartar forays is only possible along such a line. It must be protected as well as possible against enemy attacks which would seem hardly feasible if the line of operation runs across a sea. In fact, Lloyd's rhapsody is an application of the ideas of his main work, *Military Memoirs of 1781*, on war made across the sea, which if followed would prove too complex and impractical to succeed.

"When the frontiers of the contending powers are contiguous, the magazines formed in the country which attacks may, for some time, supply the invading army until victory enables him to take some capital fortress and secure a part of the enemy's country sufficient to form a new depot to support the whole or a great part of the troops during the winter." Anything like Napoleonic war or *blitzkrieg* is of course out of the question. And should the invader not achieve this minimum of success, he is forced to retire to his starting point. In case the belligerents are separated by the sea "so many difficulties will occur in such an undertaking that it is almost impossible it should succeed."

"An enemy who acts over a branch of the sea, must occupy

some convenient and safe harbor, gain a great and decisive battle, or by skilful manoeuvres force the enemy to abandon such a tract of country as will support the assailant; for if he depends in the smallest degree on shipping and precarious navigational aids for supplies, he cannot prosecute any solid operations, and successful campaigns will be defeated by fruitless and unmeaning excursions. Troops must, however, return to the shore to take up their winter quarters, and at last his men and money being exhausted, he perishes totally, or abandons the enterprise with loss and ignominy . . . .

"An offensive war must be prosecuted with the utmost vigor and activity and, vice versa, a defensive war with caution and prudence; and above all a general action must be avoided. You oppose the enemy in front by occupying strong posts and with the remainder of your forces you act on his flanks and rear; which in a short time will reduce him, though much stronger, to fall back and approach his depots. If King Harold had followed this doctrine, it is probable we should have known William the Conqueror by his defeat only . . ."

Defensive Concepts Considered

Applying his general principles to the situation of 1779, Lloyd makes Brest the starting point of the enemy—he in effect uses the idea of the *base,* though not the word for it—and Plymouth his most likely first object. That place, he thought, was not as well secured as it should be, partly due to "an unreasonable respect for private property," something with which the fortress planners henceforth often collided. A landing on the south side of the Isle of Wight he thought feasible, although convinced that "while England exists as a nation, an enemy cannot keep the isle a month though there were 30,000 men in it." Distinguishing between a raid and an invasion, he emphasizes that "when a *coup de main* is intended, you must debark as near the object you have in view as possible, because the success depends on secrecy and surprise, but when you propose to wage war in a country you are to land your troops at a distance, that you may have time to bring your stores on shore, fortify a camp, take some capital position and then proceed gradually towards the point you have in view."

The most likely line for English forces to defend against an invader would run from Plymouth to Dover with Ports Down as the central point; there, one third should be posted, another third on Haldon Hill, near Exeter, and the third along the boundary of Sussex and Kent. In the choice of such positions Lloyd reveals himself as one of the most pronounced representatives of 18th-century strategical thought, according to which the possession of certain points, or lines means *eo ipso* controlling vast regions around them, the conviction which survives in geopolitical strategy of today or yesterday ("Who rules East Europe commands the Heartland: Who rules the Heartland commands the World-Island: Who rules the World-Island commands the World" Mackinder).

It is the strategic doctrinarism about which Clausewitz wrote: "These tomfooleries have carried on down to our own times; the learned military men have not been able to resist the temptation to demonstrate for themselves a particular importance for certain points and lines and with secretive mien lay them down as those quanta on which things depended above all and the value of which only the genius could ascertain. . . . General Lloyd has probably been one of the first who started this foolishness." (Jähns, op. cit., 1872-3).

All in all, Lloyd would conclude that an invasion of England could not succeed simply because the order and orderliness required for all war that the ancien régime, practically shutting its eyes to light infantry and similar disturbing phenomena from, or at the frontiers of, civilization, the supply or magazine system, thought to be absolutely necessary for it, would not be available or maintainable. His conclusion about landing or invasion is practically a case of "something that cannot be because it is not allowed to be." The next set of planners for such an invasion would on the contrary build and rely upon the very disorderliness, the revolutionary effect and impetus of an overseas invasion.

NOTES, CHAPTER 27

[1] In *The Strange Career of the Chevalier d'Éon* it is told by J. B. Telfer that the first exploit of that notable was a secret mission to Russia in 1755, disguised as a woman, which enabled him to gain the confidence of the Empress Elizabeth. Later he served in the army and went to England as secretary to the special French Ambassador. After Louis XV's death in 1774 the new government paid him to surrender his official papers and return to France. Revival of the stories of his Russian disguise led to bets being made as to his sex, and just before he left England an English jury officially adjudged him a woman. For reasons not made public the French government insisted upon his wearing female garb after his return to France,

and he was known as the Chevalière d'Éon. He retained it when he went back to England for good in 1785. In his latter years he supported himself by teaching fencing. After his death it was definitely established that he was a man.

[2] Sources: G. Latour-Gayet, *Les projets de débarquement en Angleterre à la fin du règne de Louis XV. Le Correspondant,* t. 202 (1901) and the same author's *La marine de la France sous le règne de Louis XV.* Paris, 1901. These historical writings are not without a parallel between then and now, the now being Fashoda. See also R. Castex, *les idées militaires de la Marine du XVIIIe.siècle.*

[3] J. Holland Rose & A. M. Broadley, *Dumouriez and the Defence of England against Napoleon.* London and New York, 1909, 37-9, 50-9.

[4] Sir John Fortescue, the editor of *The Correspondence of King George III* (London, 1927. II, 287 ff.), identifies Carleton with "the future commander-in-chief in Canada, later Lord Dorchester," Wolfe's quartermaster-general in 1759. But C. was appointed to Canada as early as 1766 and held the rank of brigadier-general since Quebec.

[5] For Lloyd see still Max Jähns, *Geschichte der Kriegswissenschaften,* 2102 ff. The *Rhapsody* was translated into French (1801), German (1802) and Italian (1804). The text used is that of the English reprint of 1793.

PART 4

The Age of Steam

Chapter 28

REVOLUTIONARY AND COUNTER-REVOLUTIONARY LANDINGS, PLANS AND LANDING EQUIPMENT

EVEN WITHOUT OTHER INDICATIONS, the landing attempts and plans of the French Revolution and the Empire against Great Britain and those of Britain against France showed unmistakably that some military traditions and concepts of defense had survived the fall of royalty. The younger Pitt's expeditions against the continent showed improvements, but in other ways repeated the elder Pitt's wasteful and inefficient undertakings.

So early as December 30, 1792, before the war between France and England, a circular of the Ministry of Marine addressed to the Jacobins of all countries, reading much like a Comintern ukase, declared: "If King George wants war, we would make a descent on his kingdom, would throw there 50,000 liberty bonnets and establish on the debris of his throne the power of English republicans." The next day, the Convention, forming a Committee on Defense, proclaimed that it was not afraid of war with England, "for our fishing vessels are ready to transport 100,000 over to England who would finish the struggle on the ruins of the Tower of London."

Sharing these defiant sentiments, numerous projects were submitted to the men in power including one or more by Thomas Paine who had written in 1790, during the excitement over the Nootka Sound episode, from London: "I know the character of this country so well that nothing by carrying a high hand can manage them. Yet they are the greatest cowards on earth as all Bullies are, if you impress them rightly. Unaccustomed to wars at home, or on their own coast, they have no idea of war but at a distance, and that they are only to read the accounts of it in the newspaper. Of this sort of war they make a mere trade and ever will." [1] (Nootka Sound, a natural harbor of Vancouver island, British Columbia. A clash there in 1789 between British and Spaniards caused a serious controversy, which was settled by Spain paying an indemnity.)

Paine's revolutionizing proposal against England of 1792 did not envisage sending out a French fleet, which would only give England an occasion to use hers, but merely the granting of £200,000 to Ireland to rouse that country against its oppressors. In most of the earlier projects the revolutionary fervor blinds the originators to such circumstances as the want of stores, of disciplined troops, of commanders able to keep troops firmly in hand, or later on the activities of the Chouans (peasants who rebelled against the Revolutionary government in 1793) who had made large parts of the western coast a poor jumping board for attempts across the Channel.

While Jacobitism now was too weak and also too unpopular to rouse in England even the rudimenta of a Fifth Column, the over-sanguine revolutionaries in Paris depended upon and far overestimated the strength of their sympathizers in the British Isles. Even French military men were inclined to do that. So late as May 29, 1801, General Humbert, who had been in charge of an expedition to Ireland, wrote to the war minister: "Let us be persuaded of one truth: it will never be with a strong army that we can operate advantageously; it must be 100,000 Irishmen or 100,000 Scotchmen with whom we subject the pride of that power and avenge *la patrie.*"

The excesses of the Revolution estranged some of the English revolutionaries and those of Napoleonism most of the rest. Thus the French themselves destroyed the potential Fifth Column easily, as much as did the repressive measures of the British government. In all probability the latter stood in fear of the danger far longer than necessary. As late as December 1803 the Duke of York ordered the forming of militia companies from older men for the special purpose of protecting private property and to combat a Jacobin uprising.

British Opposed Revolutionists

For a number of years France was far too busy along and beyond her land frontiers actually to think militarily of great enterprises beyond her coasts. It was the British support of the opposition to the Revolution in the Vendée (along the Atlantic coast, south of the Loire) which started among the French military men new schemes for the invasion of Britain. Not that the British-aided landings of émigrés at Quiberon in 1795—the aid including "invasion money," false French bank notes printed in London—had been very impressive, or dangerous, but under revolutionary circumstances even ineffective measures are apt to leave a deep impression on the invaded country.

The French government had known of the preparations for this descent, although not its exact destination, at least after the sea fight off Belle-Isle on June 23, 1795. Their troops under Hoche were spread over all Normandy and Brittany to protect the coast and keep the Chouans down. On the 25th the English fleet with several thousand French émigrés and equipment for an army of 40,000 men, anchored in Quiberon Bay, but not before the 27th did their landing on the beach of Genès near Carnac take place. In the locality the Republicans had only some 400 men who could not be joined by the garrison of several hundred men at Aceray because the latter were attacked by the Chouans on the way and forced to retire to Aceray. The landing force was nominally under the command of two émigrés, one Pitt's man and the other distrustful of the English, and consisted of 4,500 émigrés and soldiers who were at once joined by 4,000 or 5,000 Chouans under the Bishop of Dol, papal vicar for Brittany. But the émigré regiments were of doubtful morale, for among them were 3,000 French ex-prisoners of war, the organization of the peasant Chouannerie poor and the chiefs of the expedition at loggerheads. Thus the first propitious moment for successfully combatting the Republicans remained unexploited.

At last the military leaders resolved to begin operations by making themselves masters of Quiberon peninsula. Liaison with the fleet was neglected and the possibility, or necessity of reembarking altogether lost sight of. They were unable to take Vannes. Not until the 3rd of July did the land force supported by the fire of the fleet attack Fort Penthièvre which controlled the neck of the peninsula. The 700 men of the garrison without stores and ammunition, surrendered without firing a shot and were promptly enrolled in the Royalist force, part of them even being left to guard their old post, a very dubious addition to the strength of the invaders.

Hoche who commanded in the Vendée, after having dispersed the Chouans in his way, marched on Carnac, the émigrés retiring into the seven kilometers long peninsula which Hoche blocked by a fortified line across the neck of Sainte-Barbe. On July 16 the émigrés, reinforced from England by an echelon some 2,000 strong, made an attack on this line, but were repulsed with a loss of three hundred and their best general, d'Hervilly. Their *déroute* was complete although a mass of the Republicans were pinned down by the fire of British gun sloops which directed broadsides at both sides of the isthmus, while the émigrés retired to the fort.

Dejection spread among the cooped-up aristocrats as Hoche learned from deserters. He resolved to attack and possibly cut off their retreat to the ships. In the dark and stormy night of the 20th, when the English gun-sloops had been drawn off, Hoche set in march three columns, one designed to assail the line of the émigrés across the peninsula and the others, guided by deserters, to move along the beach at low tide and thus outflank the line at both ends. These two columns reached the foot of the cliffs on which the fort is built, escaladed the steep scarp and surprised the poorly guarded place.

With the taking of Penthièvre the whole enterprise collapsed. the majority of the émigrés surrendered in sight of the British fleet and 750 of them were subsequently shot. The Republicans made some 8,000 prisoners, of whom 1,000 were émigrés, mostly noblemen and former marine officers, 3,660 Chouans and 3,000 ex-prisoners of war, and captured immense stores landed by the British. The attack by the Republicans had been audacious although the risk was amply justified, considering Hoche's information, and it made them the keener for an assault on England.

Generals like Hoche, Humbert and even Carnot by 1796 were flushed with the idea of turning the tables on England and planting a *chouannerie* on her shores, all of them vastly overestimating the inclination of Englishmen to carry their republicanism to such lengths. It was the scheme of the "organizer of victory" to throw on the enemy's coasts partisans, "veritable forlorn hopes [*enfants perdus*], to worry the government, keep employed part of its forces, and hand it back the evil which the *chouannerie* has done us."

Failure of Desperadoes

Humbert, who wanted to revolutionize Ireland by means of such a counter-revolutionary organization turned revolutionary, at least realized that such an expedition would require good troops, brave, robust, young and determined, led by competent and imposing officers and thoroughly loyal to the revolutionary government who were to explain to the troops that the English had started the civil wars in France and had thereby induced the French desire for revenge. Prisoners of war were to be made on such occasion only if the enemy did; but should this displease the troops, no quarter was to be given, except to civilian inhabitants.

To inspire particular exaltation he proposed that each leader should assume a *nom de guerre* and that leaders should confer such fighting names on the bravest of the rank and file: "This new baptism imposes upon the peasants and thereby inspires terror." Still another general wanted the Chouans in England to begin their work by attacking all public conveyances and pay offices, explaining to the people that they made war only against castles and not against huts. They were to open the jails and arm the prisoners, to burn all naval stores, "to speak much of liberty, but not to say anything positive; only the project of destroying and not constructing."

Every individual who was to go as a Chouan to England, woud do so with the project of pouching some 100,000 francs worth of loot, "in order later on to finish his career quietly and in ease." This French *rentier* ideal, the hope of retiring on an interest-bearing capital, provides quite a novel element in the history of soldier's pay. The plan was to have 1,000 to 1,500 men go, acting in small groups, including desperate characters like deserters and criminal prisoners who could be depended upon to make good on the job. A legion of free men for Ireland and Cornwall had actually been prepared when the Directory gave up the plan in July, 1796.

The desperados did not feel half as desperate, half as desirous of rehabilitating themselves, by such a burglarous invasion of England, as these generals expected them to be. The invasion by military convict labor did not come off until February, 1797, when a force of 1,400 or 1,500, principally convicts, under the command of Captain, or Colonel, Tate, an American adventurer, landed in Fishguard Bay in Wales. So poorly were they informed about the local topography that they chose for beachhead a specially difficult spot, with steep shores and four or five miles from a good road. Three miles to the south and two to the north they would have found very convenient landing places. Their plundering was on a small scale, and instead of terrorizing the countryside they meekly surrendered without having fired a shot to some 500 or 700 rural militia.

Societies cannot expect to win wars, as the French Revolutionists were inclined to think, by first placing in the field their worst elements; they need their best for that purpose. Nor was this the only occasion when the criminal outcasts of society declined to be used for martial vivisectional purposes. In the summer of 1796 a flotilla of a new type of flat boats for guns were built and tested by Muskeyn, a Flemish ship lieutenant, which were said to have proved their value in the recent war of the Swedes against Russia. They were to be manned in part by foreign deserters and Flemish sailors. The soldiers of the land army chosen to go with them displayed extreme repugnance towards the whole enterprise. Although they were taught to swim and in other ways trained for the enterprise, desertions increased and nearly everybody became discouraged long before the start, including the officers.

In two nights no less than 1,500 deserted and a battalion commander reported "that those who remain declare loudly that they will rather let themselves be hacked to pieces than put a foot on the flat boats." The employment of the deserters and Flemings enabled the French soldiers to raise a constitutional point. Such use of non-French violated the constitution, they said. And indeed success seemed doubtful enough since the English were well informed about the attempt. When the flotilla sailed at last it was at first becalmed, then thrown back by storms, one of the vessels foundering. The soldiers on another boat, badly fed, forced the captain and the foreigners to return to port. (November 18 and 19, 1796). The type of boat itself was much better than these recalcitrants allowed and they proved their value in a fight with some British men-of-war in June, 1798.

Irish Leaders Doubtful

The revolutionaries from the British Isles, mostly Irishmen, like Wolf Tone, were appalled at the idea of starting a *chouannerie* in their home lands. Instead of arousing sympathies, that would estrange substantial elements of the population from the cause of the Revolution, although Tone favored landing a thousand jailbirds in England for the burning of Bristol and doing what other damage they could. The first expedition actually reaching British waters could well dispense with Chouans, since it never set foot on land. After protracted preparations it sailed from Brest on December 14, instead of June 19, 1796 under the command of Hoche, comprising 43 sail and carrying 14,000 to 15,000 men in addition to large supplies of arms to be distributed among Irishmen. When Hoche was 40 nautical miles at sea, the Directory sent him word, that it was cancelling the whole Irish expedition, seemingly because they had used it merely as pressure diplomacy in dealings with the English which they thought were now taking a favorable turn, hence not requiring the enterprise.

The Revolution had brought the army and the navy services somewhat closer together in France, but not enough to exclude misunderstanding and quarrels. With regard to the Irish expedition the navy had managed that the debarkation was to take place in Bantry Bay. There conditions were most favorable from a mere nautical point of view, but it was rather remote from the regions where the United Irishmen, the revolutionary organization founded by Tone and others, had its strongholds. As against this Hoche had not strongly enough urged the military interest. The squadron was not greatly molested by British ships en route, although suffering losses by their attacks and more from the weather and bad seamanship. It lost sight of Hoche and the admiral, who turned back to France eventually.

Under the command of Grouchy, for 17 days, it lay inactive at anchor in Bantry Bay, where it arrived on Christmas, battling the winds and the waves. No troops were disembarked. When the food on board had been largely consumed the naval officers, pretending, or believing that they were shortly to be used for an expedition to India, brought about the return to France. A precious occasion was wasted, for (according to Leckie) the French were very close to a success, which is to say that a general uprising was impending and there were practically no English troops in Ireland at the time, although other auspices were less favorable, the expedition, for example, being without any land transport which was practically unobtainable in this part of Ireland.

When the Irish uprising finally came, in 1798, Tone could obtain little help for it from the Directory. Bonaparte, who had never shown and never would show a sincere interest in the Irish movement—another proof of its revolutionary character and the general's own increasingly anti-revolutionary outlook—had started for Egypt, and Hoche, true friend of the Irish cause had died. French policy allowed to pass the moment of

Britain's greatest weakness, during the movement in Ireland and the mutiny of the fleet at Spithead and the Nore (April-June 1797), without using it for a determined landing attempt. On one of the five small-scale and futile expeditions sent from France to Ireland during 1798 Tone himself was captured, to suffer the fate of Sir Roger Casement during World War I.

NAPOLEON'S PRE-EGYPT IDEAS

Before starting with his 35,000 or 40,000 men for Egypt, Bonaparte had given the Directory his opinion of the feasibility of a landing in England proper. They had appointed him general of the Army of England which remained, however, largely in Italy, and already on his way to Paris, before departing for Egypt, he had advised them to manufacture cannon of the same caliber as those used in the British Army so that captured ammunition could be used by the French invasion artillery, a consideration which tends to show that the general then had a more favorable attitude toward the project than later.

His opinion (of February 23) represents the great general's first considered judgment on such a descent. "Without being master of the sea"—and France would not be that for several years to come—"that is the boldest and most difficult operation that could be undertaken. If at all possible, it would be by making the crossing through surprise, either by escaping the squadrons blockading Brest, or Texel, or by arriving in small boats, during the night and after a crossing of seven to eight hours, at one point or another in the counties of Kent, or Sussex. For this operation long nights and hence winter are required. Once April is over, nothing can be undertaken any more. The expedition does not therefore seem possible before next year." Since the Corsican's views of the various factors involved in the enterprise varied, it might be emphasized that at this time he thought Brest, or North Holland suitable starting points and winter the only possible season.

The relations between army and navy in France were far from conducive to the success of their common enterprises. After the Revolution the navy was in worse condition than the army, almost naturally so as the arm, in which technical equipment, at that time even more than today, played a larger rôle than the manpower factor. It had suffered greater neglect through revolution, in which *man* gets the upperhand, or so he thinks, over the *machine* which is therefore neglected, together with all other equipment. Morale in the navy with no victory to support it, changed from revolutionary elation to a general depression among men and officers.

The admiral who was to be joined with Hoche in the expedition to Ireland in 1796, Morard de Galles, confessed to the Marine Minister: "I must tell you I possess none of the qualities necessary in a good general. My bad health and the pain and grief I have experienced have considerably affected my mental faculties; and the weakening of my eyesight which hardly enables me even to distinguish objects at four paces' distance, puts an invincible obstacle in the way of my directing the manoeuvres of a squadron."

Flushed with their victories, the soldiers were at best patronizing in their relations with the navy and often displayed exasperation at what they considered its shortcomings. "What is the Navy?" asked Hoche. "That problem is to be discovered. God forbid that I should have to do with it! What a miserable outfit! A large body, of which the parts are disjointed and disunited; contradictions of all kinds, indiscipline organized within a military body. Add to this proud ignorance and stupid vanity, and you have the thing complete."

The admiral first chosen for the Irish expedition had considered it foolish and impractical, and after violent quarrels with Hoche had been removed from command. The general was then offered the choice of his own admiral. His selection was Morard de Galles, who was most unwilling, because of impaired eyesight. After he was forced to accept, Hoche wrote: "He is the very man for the business. We will see for him if his eyes are weak!" And somewhat later, when the work of fitting out the ships had begun to tell on the Admiral: "Poor Morard de Galles! He is already twenty years older. How I pity him!"

Divergent Army and Navy Views

The generals from Bonaparte down showed little understanding of the difficulties experienced by the navy in helping them to materialize their bold projects. Roughly speaking, the military thought that the navy should take care of the marine part of a landing enterprise—to transport an efficient victorious army across the water—after which it would be dismissed. There is little, if anything, in the directives of Bonaparte about fire support in the moment of landing, or maintaining rearward connections. In a way it was an attempt on his part to infuse the navy with the army spirit when, in planning for the invasion of England, he at first turned his back on the traditional transports and accepted the new constructions which were to be navigated largely by the soldiers of the camp of Boulogne.

While these boats were not a soldier's invention, they were at first favored by them over the traditional navy craft. They appealed to the army for, as expressed by Dumouriez, who had first served in invasion schemes and later, as an exile was consulted in the defense against them, such vessels would avoid: "(1) the dilatoriness of embarkation and landing; (2) the difficulty of making the coast exactly at the premeditated spot; (3) the slowness and roughness of the passage; and (4) the need of an equally large fleet to guard and defend this ponderous convoy." [1]

They would make it possible to preserve on water the order for battle on land. It has always been the hope and wish of the continental military men either to overcome the British by better organization on their own part, when they were hostile, or to induce in the British, when friends, greater system and care. Gneisenau coming to England in 1808 with the plan for a landing in North Germany found the British army the worst commanded in the world and declined to take service in it, writing home: "In this country governmental business is carried on in the most miserable way. This people would perish if its geographical situation would not prevent it."

Bonaparte's own expedition to Egypt had the dazzling effect which good fortune rather than previous planning so often gives, which seems reason enough to dispense with any closer analysis of the enterprise itself as a military measure. It was an overseas enterprise involving the passage of a large-scale force over a long distance and, possibly, enemy-controlled waters. It was undertaken as a substitute for the landing in England itself which Bonaparte, appointed commander of the Army of England, that was still largely in Italy and unpaid except by promises of the spoliation of London, money capital of the world, considered unfeasible, given the state of preparations along the western coast of France.

When a fleet was assembled in the Mediterranean ports of France and Italy in the spring of 1798, the English, who had retired from that sea with their fleet since November 1796, believed first that it was intended for either Ireland, or Naples. Surprise and the comparative unreadiness of the British fleet, away from the Mediterranean and only recovering from the mutinies, were factors in enabling the French fleet to leave Toulon and the other ports. For not until after April 20 did the English learn what the French were up to and sent Nelson with a squadron to the southern coast of France.

On May 17 he was driven by storm from his patrol beat in the Gulf du Lyon for more than a week and on the 19th the French Toulon fleet, not aware of Nelson's proximity, set sail, being joined by contingents from other French and Italian ports, forming an armada of 13 sail of the line, 14 frigates, numerous small men-of-war and some 300 to 400 transports with 40,000 soldiers. The mere sight of the fleet and some earlier French intrigues among them sufficed to move the Knights of Malta to surrender what was considered the strongest fortress of Europe (June 13), after a few days of less than half-hearted resistance. Nelson's presence in the Mediterranean, had they but known about it, might have led them to resist a little longer.

Leaving a garrison of 4,000 men behind, Bonaparte sailed for Egypt on the 19th. When off Candia, he learned that the English were returned to the Mediterranean and that Nelson had been off Alexandria with 14 ships of the line but had left on hearing that the French had not arrived there. On July 1 the latter reached Alexandria and were moved by the reports about Nelson to go ashore at once. It must have seemed to Bonaparte that he had exposed himself to chance long enough. As Vivant Denon, a French officer, wrote after the discovery of Nelson's movements: "The presence of the English has hovered like shadows on our horizon. When I recalled that three days ago we had deplored the calms holding us back, and that without them we should have gotten amid the enemy fleet, . . . I vowed myself thenceforth to fatalism and commended myself to Bonaparte's star."

Bonaparte and His Luck

The day was nearly gone, the sea stormy and the swell high. But realizing that Nelson might soon return, Bonaparte insisted on landing. Although they suffered some losses from drowning, the troops were brought ashore; and the morning after Alexandria, but weakly defended, was taken by storm. "The urgency of the peril and the presence of a formidable enemy on the coast left us no choice in the measures to be taken," as a participant wrote. Following the landing, the fleet took refuge in Aboukir Bay, later to be disastrously beaten by Nelson.

In ascribing the responsibility for failures and errors in his wars to others than himself Bonaparte was as quick as in his marches: The loss of Aboukir he blamed on Admiral Brueys who, conveniently for the general, had been killed in the battle; or the navy generally, forsaken by the fortune of war, as in the end those who are inefficient usually are. Fortune had favored the navy in the sea passage, had allowed the expedition to land in the midst of a roaring storm, enabled 3,000 tired soldiers without cannon and almost without cartridges to take Alexandria, and five days in which to possess the coast. These five days should have sufficed to make the fleet secure. "Only when Fortune sees that all her favors are in vain, only then does she leave our fleet to its fate." [3]

In military history enterprises favored by, if not based on luck, as Bonaparte's Egyptian adventure so essentially was, and being part of the gamble involved when revolution turns to counter-revolution, may inspire the gambling element in later generations of military leaders. But they on the whole have endeavored to limit the factor of chance in war, which is probably still greatest in landing enterprises, by endeavoring to bring into their possession and into play all that is known and knowable about the objective and applying to the task scientific knowledge and methods. Only these permit mastery over the problem of transports and stores, impedimenta that proved rather difficult to handle during Bonaparte's crossing, and narrow the province of the unknown. Modern leaders are certain to have relevant maps and charts whereas Bonaparte had no trustworthy map of any roadstead in Egypt.[4] He himself, who had to sneak home from Egypt like a gambler after a night wasted in rooms with an Oriental *décor,* came to realize the need for more method in the preparation and execution of a landing enterprise and later would try to apply them in and from the camp of Boulogne, although probably only for the purpose of cloaking with science and method an attempt secretly abandoned long before marching out of that camp against Austria.

This opinion of their foremost general did not keep the

Directory from at least continuing the appearance of threatening the English shores. The Dutch yards went on producing invasion craft of various kinds. In order to escape the vigilance of the English blockaders these shallow craft were usually conveyed over the inland waterways system which extended from Flushing to Ostend. The vital point in this canal system was the Ostend locks which were guarded by only a small French detachment.

Pre-Zeebrugge Tactics

To destroy them the British sent out a task force whose doings and methods recall in some respects the enterprise against Zeebrugge in 1918. The force under the command of General Coote, having left May 16, 1798, at 4 A.M., anchored off Ostend on the 19th at 1 A.M. Even though the weather appeared threatening, the debarkation was undertaken after a pilot had assured the British that the garrison was weak. During an hour's bombardment of Ostend some 1,300 troops were landed, considerably less than were aboard. An infantry regiment, several companies of infantry including some Guards, a troop of dragoons and six cannon, reached shore without being discovered by 5 A.M.

The port was taken after the resistance of some tirailleurs had been overcome. By 10 A.M. mines were laid. Their explosion destroyed the locks and other works and a number of ships, although according to the French version the explosion produced more noise than damage, a disagreement much resembling that between the two accounts of Zeebrugge. The British objectives had been attained and the troops retired to the shore. But wind and the tide had interrupted communications with the ships. They were forced to dig in and defend themselves with their back against the sea. The French in the meantime had concentrated troops from Bruges and elsewhere, although hardly more than 500 in addition to the Ostend garrison. In the fight which began with great vivacity at daybreak on the 20th, the colonel of the British infantry was killed and Coote wounded. His successor in command capitulated, with some 1,500 men according to the French version designed to gloss over the original carelessness of the defenders, for scarcely more than 1,300 had been landed.

During Bonaparte's absence in Egypt yet another expedition was sent to Ireland under General Humbert. Including as it did four Irish "regiments" of very low strength, the force must be considered as having been cadres for an Irish revolutionary army. The troops were generally in good spirits, but the financial embarrassments of the Directory and delays in paying the men did much to dampen the morale and postpone the departure. On July 15 five demi-brigades of 102 officers and 516 men were embarked. A fortnight later non-payment of the overdue pay made it necessary to postpone the sailing date although just then the expedition would have had an excellent chance to escape the vigilance of the British cruisers. By August 5 arrearages of pay including a three month's advance that had been promised the grumblers kept the expedition from leaving, but finally it started after the moneys had come on the 6th, 80 officers, 939 non-com's, 1,019 privates strong. During the voyage Humbert aroused great dissatisfaction by declining to pay the advance. When a grenadier claimed it in the name of the protestors the general was unable to exercise authority to have him locked up. He was forced to pay and then retired behind the barred door of his cabin.

French Success in Ireland

The landing took place near Killala, on the north coast of county Mayo, being diverted there by unfavorable winds from Donegal. Humbert achieved an initial success, defeating with 800 men some 6,000 English at Castlebar under circumstances rather inglorious to the latter who fell into a panic, resulting in the ill-famed "Castlebar Races" which made Cornwallis, at the time Viceroy of Ireland, write that while one of the hardest blows he ever had to sustain was the news of Burgoyne's surrender at Saratoga, he never had experienced such an affliction and disagreeable surprise as when learning of the catastrophe of Castlebar. The success of Humbert's force strengthened the revolutionary movement for a while, but since the Directory did not promptly support his troops, completely tired out, had to surrender on September 8 to the 30,000 which, according to Humbert's estimate, Cornwallis had by that time assembled against him.

There was no dearth of French projects to support the Irish cause throughout 1798. From May to October the Directory itself formulated plans for no less than seven expeditions and actually prepared six, of which five did depart. Two failed, being intercepted en route by English naval forces, two proved useless even though they reached Ireland because they arrived too late to bring aid to Humbert's corps, and the fifth alone, the first to get under way, could be termed faintly successful from a military and nautical point of view.

The largest of the expeditions was that under General Kilmaine, larger than any force destined for Ireland since Hoche's; concentration for it began on September 26 and from about October 8 on the newspapers wrote about it and its destination, so little care had the Directory taken to preserve the secret. But the events and checks of the next year put a stop to further French landing projects—the landing of Russians and English in North Holland, beginning in August 1799; new outbreaks of *chouannerie* supported by nightly debarkments of émigrés from English ships, among whom was Georges Cadoudal; a descent at the mouth of the Vilaine River during the night from the 29th to the 30th of November by a force of 4,000 to 5,000 men, part émigrés and part English, who were joined by some 7,000 chouans, but which was soon beaten by General Hardy.

In fact, a civil war raged in France since October which carried the Vendéans in a regular style attack close to Le Mans and which was terminated by the 18th Brumaire. The setbacks overseas and the events at home kept all schemes of storming British shores in abeyance from December 1798 until the autumn of 1800, as far as they came under serious official consideration, although there were still enthusiasts with landing plans, including one that proposed capitalizing the English Christmas spirit. "The night which precedes Christmas and the eight days of feasts which follow are the best for this enterprise." There is hardly a proposal thinkable that could have shown more clearly that war was no longer fought with respect for common institutions such as the highest holiday in Christianity, or common conventions in warfare.

Every invention is directed against a political status quo no less than technological and the interests based on it. The invention of the steamship and of the submarine, first that of David Bushnell in the American colonies, afterwards that of Robert Fulton, both exponents of America's anti-status quo function in world affairs, threatened the British command of the sea, including as it did the practical inaccessibility of the British isles to invaders. It was under this aspect that Fulton's

Nautilus was considered and examined by the men in power in France. For might it not enable them to break with the past in naval warfare and strike an annihilating blow at British sea power and enable them to get across the waters to England at last, without outbuilding England in the traditional types of ships?

FULTON'S OFFER TO AID THE DIRECTORY

From the end of 1797 on Fulton had offered to construct the *Nautilus* for the Directory. For the destruction of ships of the line he was to receive 4,000 francs per cannon on board such vessels and 2000 francs per cannon in the case of smaller types; prizes to go to a company founded by Fulton with American capital. Fulton and others on board the *Nautilus* were to hold commissions to ensure their being treated as regular belligerents in case they should fall into enemy hands, a condition which strangely enough greatly exercized the same Directory who had been so much less scrupulous in their other schemes against England.

In July, 1800, the *Nautilus* was launched at Rouen and found manoeuvrable, a commission reporting: "The weapon contrived by the citizen Fulton is a terrible means of destruction because it acts in silence and in a manner nearly inescapable. It is particularly suitable for the French; for having a navy weaker than that of their adversary the latter's total annihilation is favorable to them." The commission which included scientists to whom total destruction in war seemed but part of the very logic of war, advised the Ministry of Marine to aid Fulton, although only a mild interest was taken by Bonaparte, whose concern with things technological was always slight and essentially non-progressive.

A credit of 10,000 francs was granted Fulton in March, 1801, for further experiments and he was able to demonstrate the potentialities of his invention by blowing up an old sloop in the harbor of Brest with a torpedo released from his submarine. But his suggestion of proceeding against one of the English frigates which were often anchored off Brest for periods of time, was vetoed by both the commanding admiral and general, the latter, Caffarelli, writing to Paris: "A still stronger reason has led the admiral and myself to this refusal, and that is that this manner of making war against one's enemy carries with it such a condemnation that the persons who undertook it and failed in it, would have been hanged. Certainly, that is not a death for military men."

French military commanders were no longer revolutionaries like Hoche, or that Scot by the name of Watson, "a great partisan of the French Revolution in England" and imprisoned for his activities by the British government for several years, who supported Fulton's projects and proposed to have the vehicle specially directed against the "enemy of liberty," Nelson. Since he could not impress Napoleon by either his submarine, or steamboats, Fulton moved to England in 1805. There the new inventions had at times caused great alarms and inspired fears that British naval supremacy was threatened by them, Lord Stanhope, Pitt's brother-in-law, writing to William Wilberforce (December 5, 1794):

> This country is vulnerable in so many ways that the picture is horrid. By letter I will say nothing upon the subject. One instance I will, however, state. The fact is this. I know (and in a few weeks shall prove), that ships of any size, and, for certain reasons, the larger the better, may be navigated in any narrow or other sea, without sails (though occasionally with), but so as to go without wind, and even directly against both wind and waves.
>
> The most important consequence which I draw from this stupendous fact is that it will shortly, and very shortly, render all the existing navies of the world (I mean military navies) no better than lumber. For what can ships do that are dependent upon wind and weather, against ships wholly independent of either!
>
> The boasted superiority of the English navy is no more! We must have a new one.

English Cool to the Submarine

Although Fulton was able to demonstrate practically the striking power of the submarine the British government, holding control over the surface of the waters, saw no need to add to this the sovereignty of the subsurface, a demoniac realm in which it might be challenged in the future. To accelerate this evil day by opening up submarine navigation was hardly in the British interest. Thus, non-progress in the final balance added to the security of the British Isles, access to which would still be denied by the superiority of her surface fleet. Those outer bulwarks and floating forts a submarine fleet might conceivably have swept away, as mine sweepers remove mines, to open the way for an invader. It remains exceedingly doubtful, however, whether then the manufacturing facilities of France for building either submarines, or steamboats were sufficient, or the technical perfection of both inventions great enough to make them of decisive general value in landing, or other martial enterprises of that age.

The peace of Luneville between France and Austria (February, 1801) and the armed neutrality of the northern Powers isolated England when she least wanted it. Once more, all the Continental naval power left, the sum of the "secondary navies," was hostile to, though as yet not closely combined against, her. The conquest of England, as it seemed then and at later times, would begin with her political and economic exclusion from continent combinations. The murder of Czar Paul I (March 24, 1801) and Nelson's trouncing of the Danish fleet at Copenhagen broke that exclusion.

To know whether the First Consul had relied on the effectiveness of his blockade against the prime blockader to bring England to her knees at last, would largely answer the question of how serious his landing plans were against England during the time preceding the Peace of Amiens with England (March 26, 1802). If he did thus depend on the effects of the exclusion of England from a continent forced into unity under French-Russian domination, he could well afford to take his invasion preparations, that had been continued in some form, or other, less seriously. This would also speak in favor of Captain Desbrière's conclusion from the study of the Paris archives that no serious intention to undertake an invasion existed on the part of the First Consul in 1801, and in fact that there is no trace of "a general plan adopted and followed by Bonaparte in his struggle against England and that there is all reason to doubt that he had ever drawn up one that was really sincere" (II, 273 ff.).

A fairly large flotilla, although comparatively few troops, were assembled in and around Boulogne in 1801, including some 90 gun boats, and public anxiety in England was again aroused. To quiet it, an attack was made on the French flotilla inside Boulogne port. This was hardly even a half-success. Nelson who was in command in the Channel, was not present, or else he might have seen a repetition of his failure of Santa Cruz de Teneriffa (July 24, 1797) when his exclusively naval forces, presuming that a Spanish treasure ship was inside the harbor, had attempted a most daring landing against a fortified place during the night. Nelson himself had led and found the Spanish on the alert. They shattered the troops that had reached the mole, while others whose boats had missed it were killed on the beach. As it was, Nelson wrote the Prime Minister on August 4 that he believed he could assure him that the French would not embark at Boulogne in order to invade England. But after observing the preparations made in England against an invasion he attempted another night attack on the 7th, which failed.

Weather Conditions Weighed

Professional opinion upon the most favorable time of year for a crossing was much divided at this time, both English and French admirals preferring the season before September 22, whereas Bonaparte still adhered to his opinion of 1798 that the long nights of winter were best and that nothing could be done after April. By the beginning of autumn he had actually foregathered only part of his navy; nothing had been done to assemble the necessary transports from the merchant marine; the larger warships were under orders that had little to do with an invasion. The distribution of the army then was not in the least indicative of the enterprise and the troops concentrated in the western provinces, where the rumblings of civil war could still be heard at times, were merely adequate for a strictly defensive stand, to make the coasts of France safe from "insults" by English ships and landing forays. Any further building up of fleet and army in the west of France was suspended with the peace of Amiens.

The breaking of that treaty was not prepared by Bonaparte in a military and naval way, whatever the British government maintained in defense of its own armaments and when calling up the militia (March 10, 1803) and sending out a cruiser squadron to observe the Dutch coast. There was a sincere note in the First Consul's famous tirade directed to the British ambassador, Lord Whitworth, on March 13: "Must I not wish for peace and is not that obvious? Show me, indeed, what I would have to gain by a war against England. A landing is the only offensive means there is in this respect, and I am determined to try it, if necessary, putting myself at the head of an expedition. But why suppose that after I have brought myself to the height on which I stand, I want to risk my life and my reputation in so hazardous an enterprise without being forced into it by necessity when it is likely that myself and the greatest part part of the expedition will be lost and sink to the bottom of the sea? . . . The odds are a thousand to one that I shall not succeed." A few weeks later Whitworth reported that "the chief reasons for delay are that they [the French] are totally unprepared for naval war."

NOTES, CHAPTER 28

[1] To William Short, U. S. chargé d'affairs at Paris, June 22, 1790. *Journal of Modern History* XIII, 369.

[2] Rose & Bradley, *Dumouriez*, 49.

[3] The best documentation for the Egyptian expedition is in Capitaine de la Jonquière, *L'expédition d'Egypte*. Paris, 1900.

[4] J. Holland Rose, *The Indecisiveness of Modern War*. London, 1927, 104.

Chapter 29

THE CAMP OF BOULOGNE

Like one that stands upon a promontory,
And spies a far-off shore where he would tread,
Wishing his foot was equal with his eye.
SHAKESPEARE, *Henry VI*

THIS IS NOT THE PLACE to decide whether war was desired after the Peace of Amiens more by England, or by France. The French were accused of sending consuls and commercial agents to England "with instructions to obtain the soundings of the harbors and to procure military surveys of the places where they resided." [1] The open rupture came in mid-May. Then, with no other war on her hands, France's bellicose energies were thrown into preparations for an expedition against England, the alleged peace-breaker. French economy was flourishing once more, money was ample and the enterprise popular, many districts offering invasion ships as gifts and regiments presenting, or being made to present, a day's pay for war purposes.

The Directory had maintained some 240 craft in preparation for an invasion. But many more now were ordered, 1,121 in the summer of 1803, a number raised subsequently to 2,000 warships and 500 to 600 transports after Bonaparte had paid an inspection trip to the coastal towns; that is to say, a ratio of four warships to one transport, the inverse ratio as compared with older maritime enterprises and projects. The whole expedition was to be more militarized than any before, and while the transports were to carry only the impedimenta of the soldier, all the fighting personnel, or at least the infantry, was to sail on board the warships which were equipped with cannon and prepared to fight en route. The various invasion craft types, all flat bottomed and of shallow draft, were *prames,* ship-rigged vessels, 35 meters long and 8 wide, fitted with 12 guns; gun-sloops, with 5 guns, brig-rigged, 24 meters by 5; and gun-boats (*bateaux cannoniers*), mere open boats, 19 meters by 1.56, with 2 guns, carrying 100 men and costing no more than 4000 or 5000 francs. The soldiers were to be far from passive passengers. "They train themselves to swim," wrote Bonaparte to Berthier, minister of war, from the coast in

August, 1803, "and become quite familiar with the use of oars. They are to serve the cannon and altogether render all the service necessary for the ships' manoeuvering," that is to say, all boats were to be rowed; and so in a way the galley was once more returned to the Atlantic.

After one year of war with England, whose shipyard output Bonaparte wanted to surpass by enjoining a working day of 14 hours, two-thirds of the planned 2,000 boats had been constructed, in large part thanks to the enforced exertions of Dutch craftsmen and laborers. In February, 1804, there were 535 transports available, including craft for 2,577 horses, and by July 19 craft for 2,225 more horses. The soldiers were assembled in camps along the coast from where, they were told, "they were to carry to the territory of England the war which that power had wanted to make against the Republic."

French Pre-Invasion Concentrations

By the summer of 1804 the concentration of the Grande Armée against England, ordered early in December, 1803, was ended and the total strength of the soldiers mobilized in the camps was 100,000, with a sick rate of six, or in Holland ten per cent. In the autumn of 1804 all troops had repeatedly been practiced in embarking and debarking. A few had even been in a small engagement on the sea with English units in November, 1803, under the eyes of the First Consul himself. The affair was much puffed up in his letters and the *Moniteur,* by which it was presented to the public as a thumping success. Stage-managed as a spectacle the camp of Boulogne was flawlessly impressive.

> Sincere or not, Bonaparte has acted so well that in the eyes of the French people he incarnates the ideal of the landing, an idea dear to the country, the realization of which, however, no government since 1793 has really been either able to undertake or wanted or desired. Though it does not know the truth of this whole lamentable story, the nation suspects that for some ten years it has been fooled or cheated and that never has one seriously attempted to rid it of its most redoubtable enemy. This time, conscious of power, proud of its glory, it counts on Bonaparte, on his genius, to bring all this to a finish and it is not going to haggle with him over its blood or its money or even its patience.[1]

The camp was, and so regarded by many, a supreme make-believe, creating a deceptive impression of gigantic preparations for a departure that never was to take place, in size so enormous that, if for no other reason, those who trusted in it would be slow to suspect a fraud, never dreaming that it might be designed not to come off, excepting, perhaps, instead of seaward, in the direction of creating the future Napoleonic Empire. Although he was fully aware that the ships could not be ready for a considerable time to come, Bonaparte had the troops concentrated along the Channel coast transformed from a thoroughly Republican into an Imperial army, with the Legion of Honor founded as one of the decorative embellishments of the transformation. This great army was assembled, so it seemed to the outward looking land-lubberly French, in face of an enemy who in reality remained as unattainable as a mirage, so far as subjugating it by invasion was concerned.

The populace might admire medals struck in Paris, inscribed: "Invasion of England. Struck in London, 1804," bearing a Hercules strangling a sea monster; but while the masses may not have detected that the whole demonstration partook of the character of a mirage, could Bonaparte fail to capitalize to his profit the belief that he meant business? He knew that it was a good thing to create and encourage images of war—and Fulton might have had better success with him had he offered his panorama as well as his boat inventions. Could the First Consul even afford to leave the French shores in 1803 when, according to his own brothers and others close to him, his régime still rested on shaky foundations? Or could he afford to stay behind and have another general, such as Bernadotte, perhaps, head the expedition and in case of victory become a second William the Conqueror, overshadowing even Bonaparte?

Ardent Bonapartist Propagandists

A spy hovering about his entourage reported to London in July, 1803, that those around Bonaparte had propagandized, built him up and enlarged him into the puissant public figure he was in the minds of most Frenchmen. "The English, by fearing him, have provoked him, exasperating his self-love. In short, so much has been done to make it impossible for him to recede or draw back. . . . Bonaparte is equally afraid to go or to stay behind; if he goes and should he be repulsed, he runs the greatest risks on his retreat."

The sceptical frequenters of Paris *salons* received the invasion plans with ridicule, in which traveling Englishmen joined. The story went around that Bonaparte had jailed a general because he had joked about playing with nut-shells in a wash basin and calling that "working on his little fleet." To Las Cases, one of the very few incidentally to escape from Quiberon, who reminded him of this story in St. Helena, Napoleon insisted that he had for himself reserved decision upon the possibility of a debarkation.

> I possessed the best army ever, the one of Austerlitz. Five days would have sufficed me to find myself in London. I would not have made my entry there as a conqueror but as a liberator. I would have renewed William III but with more generosity and disinterestedness. The discipline of my army would have been perfect, it would have behaved in London as if it were in Paris, as brothers who had restored to them their liberty and their rights. . . . Within a few months, these two nations, such violent enemies, up to a recent date, would have been nothing but peoples identified in their principles, their maxims, their interests; and I would have departed from there under the Republican colors, to operate, in the direction from south to north, the regeneration of Europe which I was later on the point of operating from the north to the south under the monarchic forms. . . . How many evils would have been spared that poor Europe! Never was a project so large in the interest of civilization conceived with more generous intentions and never closer to its execution. And, very remarkable thing, the obstacles which made me fail have not come from men, but from the elements—in the south the sea ruined me, and the conflagration of Moscow and the ice of winter wrecked me in the north.[3]

While this represents the legend born in St. Helena of the Emperor's landing intentions, an exploit undertaken, it is claimed, in the interest of that highly questionable Pan-Europe which he bequeathed to the "continental politicians" after him, it is also almost the last of the Corsican's *ex post facto* statements about his plans—to London in five days!—and the most incorrect one, historically speaking. But what were his true designs at that time? Under the conditions of 1803, did he really want to invade? Subsequently he declared both that he intended to and that he held no such notion. His disclaimer to Metternich in 1810 was that never had he been fool enough to think of a descent on England before revolution had erupted in that country; that the army assembled at Boulogne was always designed to act against Austria, "and you," he added ironically, "ought to know how close Boulogne is to Vienna."

Boulogne and its Realities

But to turn from the later Napoleonic glorification of the camp of Boulogne and what it imported, which have much exercized historians, to the step-by-step reconstruction of the reality of its military technicalities, we find even this somewhat beclouded by Bonaparte. The first orders that might be said to have included something suggestive of a plan were issued on July 21, 1803, addressed to the admiral who was to be the leader of the invasion flotilla, "forming part of the army which he, the First Consul, has the intention to command in person and which is to carry the war into the very lap of England. All the ports of the Republic and the banks of its navigable rivers, all those of the Batavian Republic (Holland. Organized in 1796 as a republic by revolutionary France) are at this moment the theater of the preparations directed against this power . . ."

Then follow the disposition of the military forces, whose concentration was to be finished by January 15, 1804; details of the embarkation of men, horses, stores, tonnage requirements which were to be larger by one-fifth than the space required by the mere personnel so that the articles immediately required for each company could be shipped with the latter and hence available the moment of landing such as complements of food, arms and tools. It also contained instructions about loading technique which not all landing expeditions before and after Boulogne have considered. Every general officer was to have under his orders a naval officer appointed by the admiral, an officer charged with all that concerned the sea service and in particular with the transmission and explication of the admiral's signals and maneuvers. While settling the vexatious problem of army-navy relations in combined operations, this arrangement established the superiority of the army in most theoretical situations connected with the landing expedition.

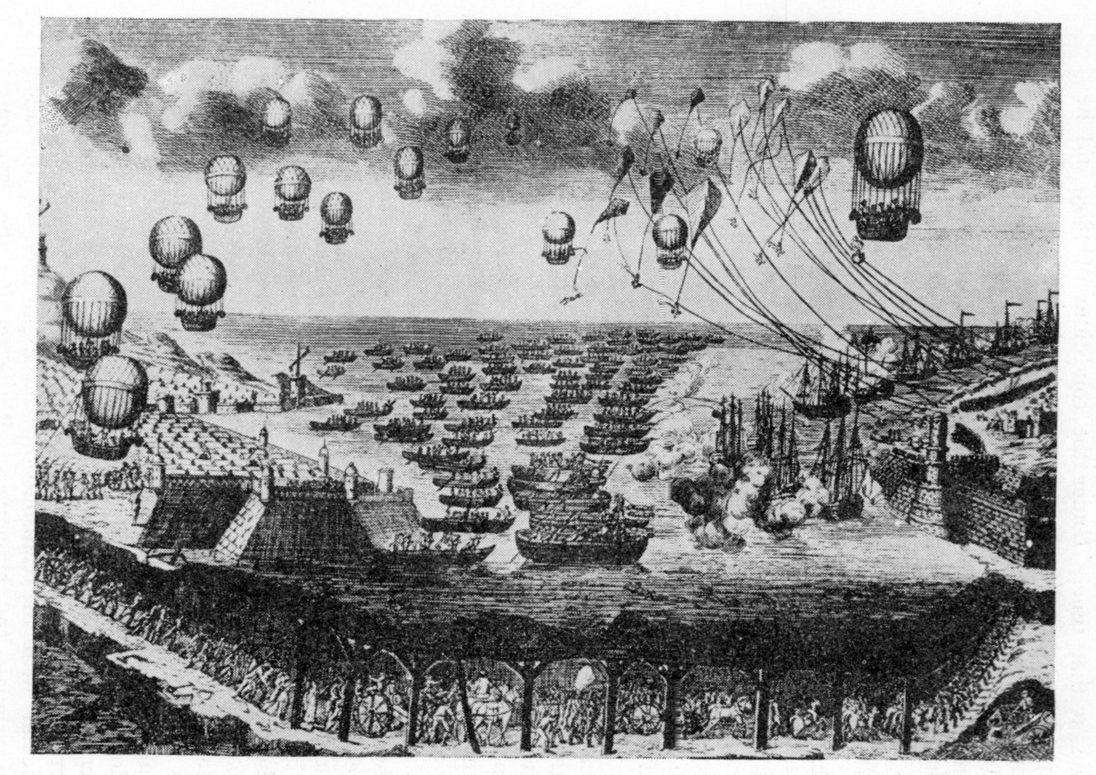

Napoleon's Proposed Tunnel to England. The supposition was that he would march troops into England under the Channel, while other French forces would descend upon the island from balloons and by parachute. (Engraving, 1804. Bettermann Archive.)

The plan centered on Boulogne as being better suited for navigatory reasons than Calais, despite the closer proximity of the latter port to England. It envisaged winter as the ideal season, with fog, or long nights assisting escape from the English cruisers. To surprise the enemy, or at any rate to avoid battle at sea, was the main feature of the earlier plans. Aside from a few technical arrangements concerning liaison and signalling, this plan did not reveal much military genius, for at times it verged on the impracticable, as in its direct interference with naval activities. It insisted, for example, and against all naval advice, in providing that a number of ships of considerable draft were to be brought to the shallow harbor of Etaples where they were aground most of the time. Whether detailed queries of the First Consul as to the ability of certain craft to take aboard an extra caisson, for instance, show the best, or worst employment of his genius may long be discussed by those who maintain that genius is "the transcendent capacity for taking trouble first of all."

Little difficulty was presented by the problems of the land army, its assembly, special training or composition—including a corps of 117 guide-interpreters, men who spoke and wrote English well. The naval end of the enterprise, however, lagged and was deficient in organization, assuming, that is, that a winter campaign for 1803-4 was ever really intended. The engineering work in the four assembly ports, including in particular the excavations for dry docks, etc., fell behind schedule, and the defects of Boulogne as a base soon became obvious to those who wanted to see them. All through the two years and two months in which the preparations went on the engineering work proceeded unsteadily, alternating from feverish speed almost to complete stoppage. The calculations of the engineers often proved wrong, adding their share to the general mess. The engineers seem to have fallen between the stools of military and naval demands and requirements, for the reports to Bonaparte on the various mishaps during the practice sorties from Boulogne and on their return, which depended much upon the engineering features provided there, varied a good deal, according to whether they proceeded from military or naval sources.

Collecting Invasion Craft

The invasion craft were built in many places, even as far inland from the Channel as Paris and Lyons. Their assembly in the four ports of departure was seriously interfered with by English cruisers, although these could not approach the coast closely enough to drive home crushing attacks. Eventually the assembly of the boats was achieved, but they proved to be clumsy craft, not very seaworthy and in their management providing great trouble when they left port and anchored for days in the roadsteads as Bonaparte, always invoking "the offensive spirit," insisted they should. An order in March, 1804, directed that fifty craft should stay outside the harbor at all times, when wind and weather permitted, which was not often, for during the 176 days from November 4, 1803, to May 1, 1804 parts of the flotilla had been able to leave harbor only thrice and in no case did their stay exceed three days.

The lesson thus derived was that from November to May anchorage outside the ports was practically impossible, and hence winter did not seem so favorable a season for a descent upon perfidious Albion as the First Consul had thought. There could have been no failure on his part to take notice of these adverse conditions; or for that matter by many of the Englishmen who went through the worst of their panic in the autumn and winter of 1803-04. The fact was also brought out that Boulogne could never send out more ships at one time than could be carried by a single tide through a narrow and tortuous channel, or one hundred of the 1200 craft assembled there. This meant that under the most favorable conditions all the boats could not assemble outside the harbor in less than six days. To put it differently, in order to get his flotilla to British beaches, Bonaparte needed six days of undisturbed control of the Channel. But not before March, 1804, did he make up his mind to give permission and provide funds for the improvement of the Boulogne port entrance by the digging of an inlet-sluice.

Difficulties in Concentrating Forces

The transport fleet formed during the winter of 1803-4 was put in command of a military officer as director-general. Besides for other freight it was to provide space for 6,800 horses of the Army and 2,000 of the Guards. By February 25, 1804, 535 merchant vessels sufficient to carry 6,004 horses were collected, but of these only 170 craft were reserved for the transportation of 2,677 of the animals. By July 19 tonnage for 4,800 animals was ready. To concentrate the whole transport force in the four ports proved impossible on account of the narrowness of their channels and the shallowness of their waters. "The insufficiency in the organization of the basis of concentration came to exercise once more its tyrannical influence over the projects of the First Consul," as Desbrières puts it. One might also say that his own style of warfare, largely dispensing with a system of bases and magazines, worked badly in face of the widely different requirements of sea and ultramarine warfare.

The plan of July 21, 1803, clearly was a complete failure, even though no attempt was made to carry it out fully. Neither the work on the ports, nor the shipbuilding, nor the assembly of the craft and the formation of squadrons had approximated what was theoretically foreseen by that program. It became apparent that winter was not the season for employing such a flotilla as had been constructed, and that Boulogne harbor, crammed with ships, was not a practicable point for departure. The few times when bad weather drove off the British observers, it had been too foul for the French to come out. "This country resembles much the country of Aeolus," Bonaparte had written from Boulogne, during one of the visits he paid in the winter of 1803-4. Perhaps he may have resented the competition offered by the god of the winds to the ventose quality that characterized many of his proclamations and orders of the day. Even if meteorology had not yet become scientific, local weatherwise seamen could have told him what to expect of the channel winds.

Bonaparte's insistence on a winter crossing possibly sprang from extreme reaction against ancien régime warfare which was not an all-weather enterprise, but largely restricted to the better seasons. Again it causes one to wonder whether such details were not purposely overlooked, consciously or semi-consciously, including such obvious handicaps as those presented by the harbor works of Boulogne, which remained neglected until the spring of 1804. The sufferance of impossibles, of factors that by themselves would have prevented a success, should go far to support the view that so far as concerns his policy of 1803-4, Bonaparte did not seriously contemplate an invasion. All that he did, or undid, or did not do, in and around Boulogne better served any purpose other than such an attempt. It did help him to disquiet the enemy beyond the

Channel, to keep employed and win over more completely to his personal ambitions and fortunes the army, which at the outset was more republican than the French population generally, and turn it to use for the founding of his empire. The first public cries for and acclamations of "Emperor Napoleon" came from its ranks at Boulogne. Caesar and Caesarism and their ancient glories were invoked when workmen preparing the foundations for his pavilion found Roman remains and even, it is said, a battered legionary eagle which was offered to the First Consul for his meditative inspection.

The mirages conjured up and manipulated by Bonaparte during 1803-4 gave him his empire in May, 1804—the mirage of honor, the mirage of fear such as that of the much inflated royalist conspiracies supported from abroad; the mirage of economy, produced by the tributes extorted from Spain, Portugal, Hanover and the 60 million francs paid him by the United States for Louisiana. Beyond these was the mirage of hope for permanent peace, once the last remaining enemy of France had been brought to her knees. No sense of disappointment had yet spread, although a whole year of war along the Channel had produced nothing of tangible, permanent value. In face of the expectations of his people, patient as they continued to be, something had to be done during the summer of 1804.

Difficult Transport Problems

As early as the autumn of 1803 memoranda had emphasized the advantages of the summer season for managing a cross-Channel adventure. "In the calm weather the men are not becoming sick and the whole energy of their souls is supported by the strength of their body"—that is to say, they would not get seasick "and all would be able to row. Calm weather would make it easier to carry and unload artillery and horses, to land them at the chosen points without haste, instead of their being forced to disembark quickly to the accompaniment of disorder and risk."

The first active measure ordered by the Emperor was completion of the concentration of invasion craft in the Channel ports. Those coming down from the Netherlands were to be protected by land batteries able to keep the English blockading units at arm's length. As far as concerned passage from Holland, the small bottoms displayed a certain degree of seaworthiness and combat value. Ships from other places were harder

Invasion Barges, Designed by Napoleon for Invading England, 1804. (Bettmann Archive.)

to handle. Le Havre was blockaded most of the time by the enemy, or closed to navigation by contrary winds, and the craft built, or armed in ports as far south as Bordeaux could only in part—35 out of 231—reach Boulogne. During May and June a few sallies to sea rather than against the enemy were made by a few ships, those permitted to go out on one tide. This produced nothing save a little experience and some losses for which Napoleon, who spent July and August in the camp, was himself responsible.

Against the opposition of Admiral Bruix, who pointed to signs of oncoming bad weather, he insisted that a sally be made on July 21. Two hundred men were drowned and the incident made an evil impression in the camp as Napoleon's valet, Constant, noted: "Agents loaded with gold run all over the town and camp in order to stop the murmurings which are on the verge of breaking out." The Emperor himself wrote to Josephine about the incident: "I thought all were lost, bodies and goods, but we have succeeded in saving everything. This spectacle was grand: cannon shot for the alarm, the shore covered with fires, the sea wrathful and roaring, all night long the anxiety of seeing the unfortunate men either drowned or saved. The soul was between eternity, the ocean and the night. By five o'clock in the morning all had cleared up, all had been saved, and I went to bed with the sensation of a romantic and epic dream" *(Correspondance de Napoleon,* 7861).

While preparations of this kind for invasion may have to approach the realistic more closely than other "images of war," one still feels that incidents of this kind, or meretricious descriptions of them, are far from justified. But Boulogne was a romantic stage set by Napoleon. Here historians like Thiers, following uncritically Matthieu Dumas, the greatest drawer of the long bow among the Napoleonic generals, represent the Emperor on the 16th of August as conferring the Legion of Honor and interrupting himself in the handing out of crosses in order to watch a unit of French craft come in from Le Havre battling the blockaders. Actually no such fight took place on the 16th. The few minor clashes that did occur outside the ports were indecisive and French and English versions usually disagree about them, the former ignoring the support given by the land batteries, the latter attributing to these alone their *échec.*

Although experiences around Boulogne during the six months from November, 1803, to April, 1804, had shown it was materially impossible to send out a large expedition at that time of the year, the better weather from May 1 to November 1, 1804, in all 185 days, still enabled the French to make only 13 sorties with parts of the flotilla, of which seven were accompanied by more or less serious misadventure. They could keep ships anchored outside the harbor for some 75 days in all, or merely 60 days if only those are counted when it was not necessary to recall part, or all of the ships. Not once had it proved practicable to send out more than 100 craft on one tide. While summer, therefore, offered slightly better chances for the employment and manoeuvres of the flotilla, these were far from sufficient. The craft had shown a certain defensive value when supported by fixed, or mobile coastal batteries, but no offensive force of any degree. We have no way of knowing the impressions made on the Emperor's mind by the experiences at Boulogne, but by now he must have seen the sheer technical impossibility of leaving the French shore, even in summer, and navigating the Channel with sufficient speed to escape the vigilance of the English cruisers.

The idea of combining the action of the invasion flotilla with that of the high seas fleet and thereby making himself master of the Channel for some time, days at least, seems to have occupied the First Consul since December 1803. After all, France still had such a fleet, even though Napoleon had no great confidence in it. It is well to remember that there is usually reason to be suspicious of actions in which a commander makes use of forces that he does not trust; they are generally not decisive, or pivotal.

This fleet was in 1803-4 widely dispersed, with one squadron at Toulon and another at Brest, and individual ships at Lorient, Rochefort, Ferrol and Cadiz, all bottled up by British blockaders. Could they ever be united before sufficient British forces could be assembled to stop them? And how could the British be induced to take large parts of their fleet away from the Channel by the time the French were to concentrate there?

All, or many of the orders which Napoleon gave for the movements of the French fleet would seem to fit into a plan to mislead and decoy the British so far away that they would return too late to prevent the French fleet from uniting

One-man Landing Party, With Waterproof Suit. Swan headgear designed as camouflage, held emergency rations. (Bettmann Archive.)

and briefly controlling the Channel. His plans seem to grow in scope, with the Spanish fleet becoming available to him after the declaration of war upon England by Spain in December, 1804, and in logical, or evolutional consistency, to the later historian, if perhaps considerably less so to him who might have attempted to look forward from the summer of 1804.

The first of the plans was vague and tentative. The Toulon admiral, Latouche-Treville, was ordered to sea on July 2 with ten sail of the line at a moment when he could escape Nelson, who might be driven off from patrol by a northerly gale. He was to take along one French ship from Cadiz, join Villeneuve in the Aix roads, at the mouth of the Charente, and after that rendezvous with Ganteaume's 21 ships in, or off Brest. These movements were to be timed so that he should arrive at Boulogne in September "when the nights are already reasonably long and when the weather is not bad for long periods of time." The Emperor's hope was that then the British ships in the North Sea would combine with the Channel squadron, thereby making the way free for the use of the Dutch vessels in the Texel. The plan was so hazy that the death of Latouche-Treville near the end of August provided enough pretext for dropping it.

The Corsican mind outmarched his ships almost too quickly, even if it appears to some historians that his intentions against the English coast were firm. On September 27 he wrote Berthier that he had resolved upon an expedition to Ireland. "There is at Brest embarkation space for 18,000 . . . The Grande Armée of Boulogne will be embarked at the same time and will do all that is possible to get over into Kent county . . . The Navy gives hope that it will be ready October 22; the land force will be prepared at the same time," including 25,000 men under Marmont commanding in the Netherlands.

Plans Known to English

A squadron was to be formed at Toulon, the soldiers being taken aboard. The ships going to Ireland were to return to Cherbourg after unloading and go thence to Boulogne, or sail around England to the Texel. The orders for the execution of the new plan of Sept. 29, 1804, were issued at Mainz and Trier and for a time became lost in a most mysterious way, whether intentionally or not—Desbrières thinks contrary to Napoleon's wishes—and became known to the well organized

British intelligence service. "While Napoleon could not always hinder treason, numerous indications show that he very often and on occasion very quickly learned that treason had taken place." He would seem to have known then that his latest plan was uncovered.

The next orders enlarging the Emperor's designs were for the fitting out, first, of the Toulon squadron, now commanded by Villeneuve, and the Rochefort squadron under Missiessy, the former shipping large numbers of soldiers. They were to fight in America, some maintain, to mislead the English into thinking that the war would be prosecuted more vigorously there. Eight or nine ships were to proceed from Toulon to the Antilles on November 1 and a second squadron of five was to follow, both returning to Ferrol by March 22 and thence to Rochefort, all this, it is commonly thought, to mislead Nelson and bring him away from European waters. But how could this shake the firm conviction of the British statesmen and admirals that the concentration of superior enemy power in the Channel, the first condition for a landing in Britain, must never be allowed? Napoleon could hardly have believed that they would have bared the Channel of English ships to chase his fleet in distant waters and at the end of the chase allow him to concentrate his own power near England.

According to those who believe in this spuriously final great plan, Napoleon had thought as early as September, 1804, of drawing large units of the English fleet away to the Antilles and had consistently adhered to this design, with constant improvements, by means of which he might have won, had his fleet, or merely his admiral, been better. Such a plan could not have existed. Not only did the presence of large numbers of soldiers on board the America-bound ships point to fighting there, the orders for Villeneuve and other naval commanders to return to Ferrol and thence to French ports contain not a word about their cooperation in a descent on England.

To this enterprise itself the general political situation in Europe, at the rear of the camp of Boulogne, gave an increasingly gloomy outlook. At this point, Captain Desbrières concludes, "the great enterprise of a descent on England by means of the flotilla had been condemned in his mind. The only objective which the political situation of Europe at the end of 1804, which the organization of the fleet permitted to aim at, is the taking or destruction of small English colonies. That is the paltry result of the efforts and sacrifices of eighteen months . . . At this moment (Sept. 29), all ideas of an invasion of England have been completely abandoned," whatever some historians may say about "the constant character of the imperial will with regard to the invasion of England."

Discouragement in the Armée

As if aware of this, the morale of the waiting army at Boulogne deteriorated considerably during the following winter. Little was now heard of the songs of summer 1804 about sailing for England—like the boastful German chant of 1940, or 1914: "Denn wir fahren gegen Engeland"—where there was more money than shells on the seashore. Impatience, or nervousness among officers and rank and file was expressed in numerous duels and acts of indiscipline, desertion, absence without leave, thefts, and the command was forced to adopt sharp measures of repression. In February, 1805, a soldier was executed for having held opinions likely to weaken the confidence of the army in the Government and its chiefs. "This Army so devoted, so ardent at the outbreak of the struggle against England, has come to feel the uselessness of its efforts."

The entrance of Spain into the war, giving a dubious addition to the naval strength available to the Emperor, hardly contributed to the consistency of his plans, including as they did for a moment the idea of sending all ships, French and Spanish, to East India, a destination so dreadfully remote that desertion in the fleet grew in direct relation to the distance from home that the men expected to be sent. But the general situation in Europe called for new and more defined schemes.

Already by December 15, 1803, the Prussian minister in Paris, Lucchesini, had written: "A continental war would offer to the general and the statesman chances far less dubious than maritime enterprises against England," and in May and again in July 1804 the same diplomat was more than ever convinced that such a war was "the secret desire of the First Consul. . . . It would release his pledged honor which is compromised in the descent announced with so much solemnity; it offers the generals employed along the coast a new prospective of glory and above all of wealth."

In October, 1804, he was told by Joseph Bonaparte that he thought it "useless to continue the war with a power which is nearly unassailable, without a war navy such as France does not possess. . . . He [Joseph] knows the firmness of the Emperor and his unsurmountable repugnance to go back from what he has perhaps imprudently pronounced in the face of Europe. Without admitting the impossibility of the landing, he [Joseph] has not even tried to hide from me the difficulty. The disgrace of an attempt that should fail would make upon the Army a far greater effect than a new emperor could risk." Officials of the civil government thought in August, 1805, that the number of malcontents amounted to three-quarters of the population, but hoped that a landing in England with a force of 60,000 would still be effected; and if it should hold the beachheads for only twenty days that would suffice to ensure its triumph.

These, then, were the salient elements in the situation of Napoleon by the turn of the year 1804-5: A powerless flotilla, a weak, or mediocre fleet; a new ally from whom little could be hoped, an army no longer believing in the landing and generals no longer keen about it, an entourage afraid of it. England was stronger than ever on the seas and much improved in her military organization, Russia had become hostile and Austria waited for an occasion to seek revenge for her defeats.

The difficulties of the enterprise against England would continue to grow. At one time Napoleon gave up completely the idea of invasion, while the army was ready to march against Austria to gain from her satisfaction for the failure of the plans against England. It was at such a moment that the Rochefort squadron, five sail of the line, slipped out undetected by the British, reached the Antilles, managing some successful *guerre de course* on the way out and back, and returned safely from the round trip on May 20. The voyage was achieved without loss, but was totally unconnected with any large plan of the Emperor. When the British discovered this escape, they sent in pursuit only a small force, being careful not to weaken their control of the Channel.

High-Seas Naval Activities

When Bonaparte returned to his invasion muttons at a time when peace on the continent seemed once more attainable, the Rochefort squadron was out of reach. As the new orders for these ships, dated February 27, 1805, drawing them into the great design, arrived at Martinique, they were already bound for Europe. The Toulon squadron under Villeneuve, after a false start in the middle of January, sailed on March 30, unseen by Nelson, adding to its 11 ships one French and 6 Spanish at Cadiz, and reached Martinique on May 14, where it was to wait for the Brest squadron under Ganteaume. The latter, which the British never allowed to sail, had orders of March 2 to proceed to Martinique via Ferrol and return from there to Europe along the most unusual routes in order to avoid discovery, arriving together with the Toulon squadron by the 10th of June or the 10th of July at Boulogne. The two in unison were to sweep the Channel clear and guarantee the undisturbed crossing of the transportation flotilla. Battle in the West Indies, where Nelson might eventually follow them, was to be avoided. Hence, Villeneuve, learning on June 8 of Nelson's arrival at Barbadoes, decided to sail for Ferrol at once. Assuming that he would return to Toulon, Nelson set course for Gibraltar, but one of his brigs with despatches for London detected Villeneuve's fleet of twenty ships headed for the Bay of Biscay and brought warning to London. The Admiralty acted promptly, keeping Villeneuve away from the Channel.

As this plan resembles a piece of grand strategy and combined operations on a world-wide scale, it has for that reason fascinated many writers. Others may suspect in it the final alibi for Napoleon's abandonment of the Channel crossing—that the navy failed the army. In any case, it involved the greatest hazards including two sorties from ports blockaded by a superior enemy, a junction in distant Martinique, a common entrée into the Channel with 45 ships of the line. Essentially it represents one of the Napoleonic *"Thèmes à deux fins,"* themes with two outcomes. In the worst event the result would be the defeat of some mediocre squadrons, that the public mind could be caused to forget by great victories on land, and in the most fortunate case, the conquest of England. Either way the military credit of Napoleon seemed sure to be preserved.

The tempo and amount of labor, shipbuilding, engineering etc., preparatory to the embarkment for the Channel crossing reflect to a certain extent the various decisions about the disposal of the fleet. While the winter of 1803-4 had been marked by considerable activity, the summer of 1804 saw a lag. By the beginning of the winter of 1804 nearly all labor was interrupted, partly due to shortness of funds. Only in February, 1805, had Napoleon caused the work to be resumed with a feverish impatience. He was infuriated when he learned in April that many of the ships of the flotilla, built largely from unseasoned timber, were rotting. Indeed, from the reports he received it would even appear that practically all the craft were unfit to go to sea. New organization merely hid, if not also produced, fresh uncertainties. When in May, 1805, a division of the flotilla into 8 French and 3 Dutch escadrilles was arranged, it did not provide for the embarkation of the whole army, and left undisposed of a relatively large number of vessels. Numerous apparently aimless embarkment exercises made the vagueness of the situation only more evident, and certain fundamental questions such as the final distribution of men in the different types of vessels remained unsettled to the end.

Strength of French Forces

On August 3, 1805, the strength of the Grande Armée along the Atlantic from Texel to Brest, with certain elements quartered as far inland as Paris, Cambrai and Compiègne, and the heavy cavalry ten days march from the Channel ports, was 1064 officers, 163,800 men, of whom 8,709 were in hospitals; 19,635 saddle horses and 3,742 artillery horses. In the immediate vicinity of Boulogne were some 90,000 men, 13,000 at Wimereux, 15,000 at Ambleteuse, 20,000 at Etaples and 45,000 at Boulogne. In addition there was a division each at Calais and Dunkirk, together 23,000 men who need not be taken into account in connection with the first crossing since Napoleon always admitted that an operation from the ports west of Gris Nez could not be combined with one from points east of it.

By this time, the transports available were in excess of those required for the troops and horses stationed at or close to the ports, the ratio for Boulogne being as follows: tonnage available for 73,000 men and 3,384 horses, of which there were actually ready for embarkation 45,000 men and 1,500 horses.

This, then, Captain Desbrières concludes, is "the sole result which the successive organizations, the complicated and precise orders and the whole correspondence exchanged with regard to organization and the assignment of the various troops to the craft of all kinds have produced."

If the ships under Villeneuve had arrived just then, Napoleon would have had 90,000 men at once available for the crossing. In the second or third week of August, historians say, Napoleon was pacing the seaside at Boulogne in hourly expectation of the coming of his fleet, which was to free the Channel for some days, or merely for some hours. But these 90,000 men, according to Desbrières were to be distributed by the greatest military organizer the world has ever seen in a manner not markedly conducive to the success of his enterprise, considering that order and rapidity of embarking were features of prime importance. They were to be assigned according to a system that was simply symmetric and in keeping neither with the state of the troops, nor with the numbers of the invasion craft in the various ports, all of which raises once more the question, "whether he ever seriously wanted to effect a sudden departure."

On August 3, Napoleon received the first news that Nelson had returned from the West Indies and that the French squadron had arrived in Vigo from where it later shifted to Ferrol, and had fought an indecisive action near the latter place on July 22, the very day Nelson had reached Gibraltar. It could not be hoped that it would arrive on time in the Channel. The orders sent Villeneuve at Ferrol from Boulogne were based in part on consciously wrong presuppositions and were above all calculated to bring on a sea battle as soon as possible, whereas before all effort had been directed to avoiding an engagement.

The Emperor endeavored to persuade Villeneuve that the English were numerically weaker: "If you should appear here for three days, or for 24 hours merely, your mission would be fulfilled. . . . Never has a squadron run risks for a higher aim and never have my soldiers on land and on the sea been able to shed their blood for a greater and nobler aim. For the great object of facilitating a descent near that power which since six centuries has been oppressing France, we all could die without regretting our lives. Those are the sentiments which should animate my soldiers. England has at the Downs no more than four ships of the line which we annoy every day with our prams and our flotillas."

What was the use to France of a navy if with 30 ships of the line she could not beat 24 of the English? This was hardly the true ratio of strength, but the Emperor continued with a final appeal: "The complicated combinations have failed; war with

Austria is imminent; there remains nothing to do but to play all for all and appeal to force. . . . A naval victory might be hoped for and might provide a remedy for all the mortification suffered." And if all were lost at sea, still not much damage would be done to Napoleon's military establishment as a whole.

Blame Placed Upon Admiral

Admiral Villeneuve, who was to be ordered displaced for his "excessive pusillanimity" just before Trafalgar, shrank from being made "the arbiter of the greatest interest." He also objected to the lack of confidence placed in him by the Emperor, a feeling general enough in the French navy, whose minister, Decrès, wrote to Napoleon on August 22: "It is my misfortune to know the naval profession because this knowledge does not gain me any confidence and does not produce any result in your Majesty's combinations." Too thoroughly aware of the poor shape of his ships and crews and of those of the Spanish in particular, Villeneuve could not bring himself to face a British fleet in the Channel which he believed would be superior.

Instead he left Ferrol on the 15th of August and sailed, not north, but south, to Cadiz, bearing in mind a letter of Napoleon's of July 16 that "a mass of imposing forces" would gather there. It is held by many that through this move, turning around when so close to the Channel after having sailed nearly halfway around the world, he wrecked the Emperor's whole elaborate scheme for the invasion of England and that the latter then at once marched from the camp of Boulogne against Austria. Is this thesis which the Emperor embraced, or rather promulgated himself, tenable?

On August 22d, or even on the 23d, Napoleon still expected to continue with his original designs, so his writings maintain, and as he wrote to the Minister of Marine: "If Villeneuve has been at Cadiz, my intention is that he proceed to the Channel after having joined to himself the six ships which are there and having taken provisions for two months." Which is to say: in the eyes of the Emperor the campaign was not as yet given over because Villeneuve had gone to Cadiz to strengthen his fleet by adding Spanish vessels. But this granted, it was impossible for the admiral to be at Boulogne before the middle of September, at the earliest.

Actually, Napoleon could no longer have been thinking of the landing in England, even if Villeneuve had appeared with fifty ships off Boulogne on August 25. For the Grand Army of 150,000 was needed for other purposes now. In a French note to Vienna of August 3 it had been said: "The attitude of Austria, says the Emperor, represents a veritable diversion in favor of England. . . . Unable to sustain the maritime war, he will march in order entirely to pacify Austria." (*Correspondance,* No. 9032 and 9038). And beginning on August 23 orders went out to the Grand Army for the movements in the direction of Austria. On September 1 all troops except those marked for local defense had left Boulogne, while all invasion craft were retired into the port by an order of August 30.

The troops from the camp of Boulogne and other places along the seaboard proved supremely ready for the war across the land frontier, a state of fitness that raises the final doubt about the true intentions of Napoleon with regard to England. It was in the camp of Boulogne that he formed the basic higher organization of the Grand Army, the army corps, as the normal formation composed of divisions, of as yet unstandardized numbers, for the original army corps consisted of 2 to 4 infantry divisions, one division of light cavalry and 4 to 8 batteries with 24 to 48 cannon.[4] It is dubious whether this army corps organization was created with the thought of landing, or post-landing operations, or whether it was better suited for these than the division, though it might be argued that the order to the dragoons who could not be mounted at the time, that they should march on foot and obtain their mounts in the conquered country, had been conceived with England and its abundant supply of horses in mind.

Why Was Invasion Abandoned?

Napoleon, the great leader of soldiers and misleader of civilians, would not truly say then, or afterwards whether the landing plan was given up as technically unfeasible, or politically inadvisable, or for both reasons. To Metternich who, as a sound conservative, had never believed such a hazardous and doubtful overseas operation possible, Napoleon declared that the descent on England was a project he never dreamed of undertaking and that it was a mere pretext for his real aim, to make war on Austria.

As for Villeneuve, the eventual scapegoat, he expressed at first only regret—"What a chance Villeneuve has missed!" (letter to Decrès, Aug. 29). He made no outright accusation. And

Decrès himself, writing to the Admiral in Cadiz on September 1, has no word putting the onus of failure upon the Admiral's shoulders. To do this was reserved for Napoleon himself, who always preferred to be the first historian and judge of his own actions and thus guide and form opinion for posterity. In two letters to Decrès of September 8 and 13 the case is made complete and the villain detected. The master of war on land and sea had built up, piece by piece, a nearly infallible design which the cowardice of a single man ruined. In the note of the 13th *(Correspondence,* No. 9209), the great design is drawn up, as a post factum rationalization of all the preceding orders and counter-orders, changes and shifts in plans with a few, although weighty admissions thrown in and made here for the first time.

"I wanted to unite 40 or 50 ships of the line in the port of Martinique, through the combined operations of Toulon, Cadiz, Ferrol, and Brest," wrote Napoleon, although Villeneuve actually had 33 sail of the line at Trafalgar, or even less if Napoleon's order of September 14 be taken literally that two such Spanish ships be counted equal to one French. To reach the stipulated number he would have had to include nearly all the ships in Brest, Rochefort and Cartagena. And, according to Napoleon, they were all to "come back with one stroke to Boulogne" and make him for a fortnight master of the sea. This requirement was not stated before, and it should be remarked that such a great soldier as Napoleon never gave a consistent answer to the technical question of how long he would have needed control of the Channel. He had spoken of six hours: "Let us be masters of the Strait for six hours and we shall be masters of the world"—or three days at earlier moments.[5] At Elba he told British officers he would have needed only three or four days of undisputed supremacy in the Channel.[6]

Had Villeneuve kept to schedule, Napoleon would have had his desire, as he wrote, "to have 150,000 men and 10,000 horses encamped along that coast, 3,000 or 4,000 flotilla craft and, as soon as the arrival of my fleet was signalled, to debark in England and make myself master of London and the Thames. This project has failed to succeed. If Admiral Villeneuve instead of entering Ferrol had contented himself with joining the Spanish squadron to his own and had set sail for Brest to unite there with Admiral Ganteaume, my army would have debarked and it would have been over with England."

The military weakness of the flotilla, its incapacity for acting by itself, of keeping at anchor in bad weather, of getting out on one tide or of achieving a crossing by surprise—all this, taught by bitter experience, is now frankly admitted. But there was still sense in the whole business, he insisted: the costly construction of the 2,000 craft equipped with cannon was merely to hide the long-time project of the combined operations of the ships of the line and the flatboats. One is almost surprised not to hear him say: Of the Atlantic sail of line and the Mediterranean galley, manned by the free French soldiers. Without the construction of these invasion craft, the English would have discovered the intention of using the high seas fleet in combination with transports. As it was, they were duped when he made it appear that the guns of the flotilla craft were intended against battle fleet guns. Thus, after the event, the invasion flotilla, composed of types of ships proved to be worthless, compared to traditional transports, is made out to have been mere camouflage. Even the vast expense, more than one usually pays for camouflage, was justified. The camp of Boulogne paid its way, if considered as a threat to England, pinning down English forces, and also was of value from the standpoint of French home politics, as well as forming an impressive force in continental affairs.

Perhaps this, although in reality it is not much, was all that Napoleon really sought or could wish for. And his grand design was largely an after-thought, produced either because the enterprise was discovered to be impossible, or the public needed a dramatic explanation.

Captain Desbrières studied the Napoleonic plans and its predecessors in days close to the Fashoda incident (in 1898, when a clash of colonial interests in the Sudan between England and France threatened war)—and general staff histories are usually written for purposes of application—with an eye on "the real means and the conditions necessary for a landing in the British Isles." From his studies and documentation, he concluded: "Just as for the battle of Marengo, but after a quite different outcome, Napoleon bequeathed to history not what had been, but what he wanted us to think and say. Systematically, he has created out of all pieces a legend which has contributed not a little to obscure the study of the real means and the conditions necessary for a landing in the British Isles." (IV, 832).

As psychological weapons, the demonstrations at Boulogne and other Napoleonic threats were successful enough for a time in the war between England and France, which had become a struggle for England's existence, as Pitt told the Commons in 1803, as no more continental alliances could be formed by her at that time. In all likelihood, Fox represented minority opinion, as he did in so many other respects, when he declared that he did not believe Bonaparte would venture an attack on Eng-

English Artist's Conception of French Rehearsal of an Invasion of Britain by Napoleon at an Island off the French Coast, May 7, 1798. British Troops shown Repulsing French. (Drawn by Rowlandson.)

land; that if he did, he would be destroyed; and that if he landed, he would frighten the English more than hurt them. A state of anxiety obsessed the nation, ready to develop into acute mass panic.

A series of these panics occurred in October and November, 1803. A fleet of 100 sail arriving unexpectedly from America, in Torbay, was mistaken for a French squadron. On October 27 there was an alert in Dublin; another on November 4 in the camp of Danbury, precipitated by bonfires and fireworks in Boulogne during a visit of Bonaparte. Many inhabitants of the coastal towns from Sandgate to Folkestone fled into the interior.

This explosive nervous and mental tension, in general a state of nerves and minds not warranted by the actual readiness of the French for invasion, was markedly different from previous official and civic indifference and inertia displayed under similar menace. With the growth of literacy and increase of literary industry, national apprehension became perhaps even more widespread than was beneficial, for the sensational in word and picture made more impressive appeal to the public than reassuring cartoons, like that of Gillray, "Buonaparte, 48 Hours after Landing," showing John Bull with Boney's head on a fork, saying: "Ha my little Boney!—what doest think of Johnny Bull now?—Plunder Old England! hay?—Make French slaves of us all! hay? ravish all our wives and daughters, hay?—O Lord, help that silly Head!—to think that Johnny Bull would ever suffer those Lanthorn Jaws to become King of Old England's Roast-Beef and Plum-pudding!"

Invasion terror was heightened by engravings on sale in London in 1798 showing fantastic Wellsian mechanical contraptions which theoretically might land on England's coast at any hour. A "Giant craft, drawn from the original at Brest" had on it four windmills and a battlemented wooden fort, with batteries of 40-pounders at the four corners. Other rafts, 300 square feet in size and fit to accommodate 4,000 men, were reported to be under construction at Calais. These were exact imaginary English counterparts of bizarre French projects, such as that submitted to the Directory by an artist of Lisieux, for a whole floating entrenched camp of 73,440 men, complete with cannon, horses and stores, which was to be towed to England by frigates.

Histories of past invasions of England and invasion attempts, putting the number of these at 49 down to 1798, were published. One of these was *A Sketch of all the Invasions or Descents, upon the British Islands from the Landing of William the Conqueror to the present Time, to which are prefixed Thoughts on the French Invasion of England by General Dumouriez* (Fifth ed. London 1803). Another called *A History of all the Real and Threatened Invasions of England from the Landing of Julius Caesar, to the Present Period; including the Descent on the Coast of Wales, in the year 1794* (2nd ed. Windsor 1798) [7] defended in particular the suspension of the Habeas

Corpus Act by a backward glance at the earlier invasion threats when "the invaders have generally been invited to such attempts by traitors in the bosom of this country" (p. 177), like the Jacobin sympathizers of the French Revolution.

Hints to Assist in the General Defence of London included the building of barricades in suitable places and the equipping of the inhabitants of street corner houses with hand grenades, the removal of all boats and barges on the Thames from the Surrey side. One doubts that anti-Bonaparte propaganda appeals by refugees in England to the French soldiers on the other side of the Channel could have proved successful, even if signed by "An Old French Soldier," saying:

> Your despot, that Corsican who sets no value on the lives of Frenchmen, calls on you to be prepared for an expedition against England of the most desperate nature. Reflect seriously, brave Frenchmen, what has been offered to your consideration. Tell the Corsican tyrant that you are soldiers, and not robbers: that you are warriors and not thieves and assassins, that you know how to engage with an enemy in the field of battle, but that you cannot murder him in his bed. Tell him that, if he will fit out a fleet to protect you in your passage, and cover your descent on the British coast, and if he will furnish an army to engage with British troops on their own ground, you are ready to embark, but that you are not willing to be sent to disgrace yourselves by plundering and wantonly murdering peaceful citizens and farmers, and laying waste their habitations, exposing yourselves to a vengeance which a conduct so mean and execrable would most justly deserve.

The first great relief which the news of the battle of the Nile gave to the English public should go far to explain the tremendous popularity of Nelson with the lower classes rather than with those in the higher ranks from whom he came. The Boulogne-fear complex had begun to diminish some time before the camp was evacuated, at least so far as certain elements of the British people were concerned, for a Bristol merchant wrote to an American correspondent in March, 1804; "The so long threatened invasion of the country has not yet been attempted, but we expect that in a few days it will be, and there seems to be no dread about it. The whole nation seems confident of being able to defeat the attack whenever it is made. It is a very fortunate circumstance that our good old king is so much recovered."

INVASION FEAR OVER BY 1810

This is tantamount to saying in modern psychoanalytical terms that in the anxiety state under the invasion threat the political father-complex was so strong that it trusted to the country's father even though he happened to be out of his wits from time to time. By 1810 any design for an invasion of England had come to be considered impractical by most Englishmen.[8]

Geographical insularity, despite its great natural strength, has one fundamental weakness in face of attack. It lies in "encirclement." That is to say, the extent of an island's coastline is equal to the length of its potential front, along which it may be attacked and where it must be ready to defend itself. In a way its situation is similar to that of a continental country assailed from all sides, when the diplomatic terms "isolation" and "encirclement" or *Einkreisung,* take on a strategic meaning.

Except in the case of two-front or multi-frontal war, however, considering a continental country in the schematical shape of a square, side *a,* things being equal, may be expected to offer for attack and defense a front of the same length as *a,* whereas an island country of the same shape and size would offer an invasion front of $4a$ length. The relationship, so far as concerns elements unfavorable to an island becomes even more disadvantageous if the usually indented character of the coastline, which lengthens it considerably, is taken into account. Practically speaking, this theoretical threat loses much of its force through the forbidding character of certain parts of the coastline, or because of the greater distance of certain parts of the coast from the enemy's bases.

Still, fundamentally coastal defense is based on the greater, that is to say more extensive, all-around exposure of the coastline and on the practical necessity of restricting defensive preparations to places where danger is most intensive, to coastal points most exposed to attack. As Lord St. Vincent wrote to Lord Keith (Oct. 21, 1803): "We have to defend such an extension of coasts that it is materially impossible to put everything into a state of defense. . . . Our great resource is the vigilance and activity of our cruisers. To reduce them in number would cause

English Coastal Defenses Against Napoleon. (Thomas Rowlandson. Bettmann Archive.)

disaster." [9] On this view the activity of the English front line fleet in the Channel, and especially against the camp of Boulogne, was based from October 1803 on.

It was this constantly active fleet, a fleet not merely "in being," not merely the "silent pressure" of sea power, that made Napoleon forsake his plan for an invasion of England. The outcome was disappointing to the military historian and to the specialist dealing in landing operations in particular, whose sentiments were voiced by Captain Desbrières of the French General Staff: "It is annoying in any case (in putting oneself merely in the place of pure military art) that the general system of British defense did not undergo the decisive test of battle against the method of Napoleonic tactics." [10] Untested remained the British military reforms, whether of the militia, or of the regular army, including the system of coastal fortifications, the 74 martello towers erected along the Channel shore at strategical points. These towers were masonry structures of uniform shape, usually circular 31 feet high, 9 feet thick on the seaward and 6 on the landward side, equipped with a swivel gun and howitzers. Equally untried was the effectiveness of "driving the country," as they called then what is today termed the "scorched earth policy"; and indeed that had been given up before the camp of Boulogne was evacuated because, as General Moore put it, England was "a country so well stocked that no effect will remove to any distance the means of subsistence."

The shock to Britain of the threat of French landings was sufficiently lasting to determine for something like a century British policy with regard to the Netherlands. While France could not be kept permanently disarmed—no such illusions prevailed after 1814-15 among the British peace makers—at least the Netherlands must be prevented from providing additional springboards for French descents on England, and must therefore be made and kept independent and neutral. To have Antwerp in the hands of France, wrote Castlereagh, who as Secretary of War had presided over the unfortunate expedition to Walcheren, in his *Memorandum on the Maritime Peace,* would fall little short of imposing upon Great Britain the charge of a perpetual war establishment.

NOTES, CHAPTER 29

[1] Declaration of King George III, May 18, 1803.

[2] Capitaine E. Desbrière. *Projects et tentatives de débarquements aux Iles Britanniques.* Paris, 1900 ff. III, 292.

[3] *Mémoire de Sainte-Hélène.* March 3, 1816.

[4] For details see W. Rüstow, *Die Feldherrnkunst des 19. Jahrhunderts.* Zurich, 1857, 202 ff.

[5] *Correspondence de Napoléon,* no. 7832, 8998.

[6] Rose & Broadley, *Dumouriez,* 377.

[7] The author would seem to have been one Charles Knight.

[8] For a good picture of the British state of mind see Carola Oman, *Britain against Napoleon.* London, 1942, from which most of the quotations above are taken; also Frank J. Klingberg and Sigurd B. Hustvedt, *The Warning Drum; The British Home Front Faces Napoleon.* Berkeley, Cal., 1944.

[9] Cit. Desbrière, III, 274 10) Ibd. III, 287.

Chapter 30

ABERCROMBY AND MOORE, BRITISH LANDING GENERALS

SO FAR AS THEY WERE CONCERNED, the British were in an ideal position, or so it seemed, to wage amphibious war against France, revolutionary or Napoleonic, or for that matter against the United States when the War of 1812 started. To Andrew Jackson this war was centered on Pensacola, Florida, which the British occupied in 1814, landing soldiers to begin their operations from there. Pensacola, he wrote from Mobile to the Secretary of War, August 30, 1814, "has become the strong point from which Great Britain can and will annoy our military operations. Commanding the Ocean, having transportation at will, she can assail any point on our coast either East or West in a few days. This will necessarily compel us as soon as practicable to strengthen all our weak points. Not having sufficient transportation for a disposable force (if we had one) to enable us to move rapidly to the relief of any point that may be invaded. Hence results the propriety of dispossessing the British of Pensacola. This in our possession would give the advantage to us and leave us a disposable force always ready to cover any point which might be assailed by our enemy."[1] So obsessed was Jackson by the rôle of Pensacola, that he took it in 1818, in the midst of peace, from the Spanish to whom it had reverted, precedent to the final cession of Florida to the United States in 1819.

The ideal position for the British in amphibious warfare was stated from another angle by Gneisenau, one of the military reformers of land-locked Prussia after Jena, in a memorandum to the British government in December, 1812, as follows: "If one possesses the sovereignty of the sea, one can undertake offensive war against all the enemy's coasts, and in multiplying these attacks one forces him to have his troops run from one end of his empire to the other. This, it seems to me, is the *true use* of the trident and constitutes the nature of its superiority."[2]

The reality of British amphibious warfare was far less inspiring. This was in part due to a dichotomy between the politico-strategic views of the various ministers, who would have subscribed to Gneisenau's views more readily than most of the later English generals, and the constitution of the British army which made what the Prussian postulated unsuitable for such enterprises. The conservative policy of the cabinet called for warfare against a revolution-born power and thus for an alteration, almost revolutionary, in the methods, drill, composition of its own army, a change which conservative civilian ministers could hardly have effected even if they had realized its necessity, as they failed to do. For a time, before 1809, it would appear as if a liberal army, as the resultant of conservativism and revolution, might emerge from within the army itself rather than from any civilian propulsion, with generals like Abercromby and Moore as its representative figures, proposing a new discipline and general arming of the civilian population and founding of a light infantry. But after the death of both, conservativism gained efficiency in Wellington and its spirit took full possession of England's war against Napoleon, fighting it out along its own regular lines to a victorious end in 1815.

Most of the landings preceding the war in the Iberian peninsula were failures, some distressing in their inefficiency, all due to the deconcentric nature of the younger Pitt's strategy which was as outdated as the mercantilism from which he derived his war concepts. In addition British army organization and leadership were poor, while cooperation was usually weak between the proud navy and the often humiliated army. Whenever British soldiers were landed on the continent of Europe, the initial superior speed given them by sea transport at a time when roads were still little improved, was soon overcome by the rapidity with which French troops moved on land; or those disembarked in such unhealthy places as the Antilles, or Walcheren (1809) were ravished by disease. Evacuation was in order almost as often as landing and may even be called more satisfactory in performance.

Moore and Abercromby to Fore

Adversity being the severest, even if not the most constant, military instructor, British disasters in amphibious enterprises produced at last such specialist generals as Sir Ralph Abercromby and Sir John Moore. The former had commanded the rear in the disastrous British retreat from Holland in the winter of 1794-5. As commander of the British forces in the

West Indies, where Moore served first under him, he had done some successful and some less successful island-hopping (1795-7). The amount of credit due to the army for the West India campaign is not to be found in the pages of Mahan, who is generally deficient in his treatment of combined actions, largely because his own time was not greatly concerned in such enterprises, but merely in general "navy interest." "He applauds the cloth he wears himself because *he* wears it," Major-General Sir J. F. Maurice felt constrained to write in the days when the fame of Mahan was at its zenith, intending this probably to be an understatement.

Army-navy relations as Moore experienced them were far more often jarring than harmonious. While serving in Corsica, he found Hood "so false and so unmanageable, that it is impossible for any general to carry on service with him." When as second in command Moore landed in St. Lucia, Abercromby wrote "that the Admiral was not ready to co-operate, and he therefore wished the landing to be deferred. It was already in part executed; to reembark was impossible. . . ." Moore was deeply shocked by his observations in the West Indies—"so little system, such neglect in the higher orders, and such relaxation in the lower, neither zeal, nor spirit anywhere, that I am convinced that the sooner we make peace the better. Against the spirit and the enterprise of the [French] Republic we have no chance." The losses due to sickness were frightful and due not only to the climate, as he found, but to

> A total want of discipline and interior economy in the regiments. The discipline of modern times, which consists of parades, firelock exercizes, etc., is easy to the officer, as it takes but an hour or two in the day. . . . The discipline of the ancient consisted of bodily exercizes, running, marching, etc., terminated by bathing. The military character of sobriety and patience would completely answer in this country; but the officers and men, in following them, would be completely occupied with their profession, and could pursue no other object. The military spirit is now, I think, gone. . . . I see this so strongly, that I fear if the war continues much longer in this country we shall be beaten by equal numbers of the black. (Aug. 31, 1796).

Moore's ideas concerning a freer discipline and a lighter infantry, deriving in part from his liberalism, were also deduced from his employment and experience in and with combined operations. They took him into Ireland where he served under Abercromby at the time of the French landing at Castlebar and on the "secret expedition" to the Helder (1799). This expedition, with its destination changed in the very last stages from Walcheren to Goeree and Voorne at the Meuse mouth, was again but poorly prepared in London. Information about the French forces in Holland and their defences was scanty. As Moore noted during the crossing:

> "The expedition has undoubtedly been hurried beyond reason, but the country having been put to the expense of assembling it, it is necessary that we should be sent to attempt something. We are now upon a voyage of adventure. The intention of the ministers is to get possession of Holland, for which 17,000 Russians and 17,000 British are assembling. Should this armament be able to establish itself, the whole will then be under the Duke of York. There is a chance that the Dutch may rise in our favor," which of course they did not, even though printed proclamations from the Stadtholder and the King of England were brought along for them to read.

Barren Success in Holland

After days at sea in stormy weather which lasted for two weeks and forbade approach to the Dutch coast, the difficulties foreseen for a landing at Goeree made the Admiral and Abercromby alter the plan once more and they determined to attack further north, at the Texel, and land south of the island at The Helder. They had been so long off the coast that little secrecy about the armada could be preserved. Still the enemy did not oppose the landing which was effected on August 27 "with great confusion and irregularity."

Moore was put on shore with only three hundred men of his brigade, "and these a mixture of very different regiments; the ground was such as to render the fire from the shipping of no avail. Had we been opposed we must have been beaten with little resistance."

The weak enemy forces retired southward, evacuating The Helder, and the Dutch fleet inside the Zuyder Zee surrendered to the British fleet as it came in without firing a shot. For a moment even Moore was jubilant.

> Thus the greatest stroke that has perhaps been struck in this war has been accomplished in a few hours, and with a trifling loss. The expedition though it has terminated so successfully, began with every appearance against it. We were a fortnight at sea, and the enemy perfectly apprised of our design. It showed great enterprise in Sir Ralph to persevere in the attempt, and he has met with the success he deserved. The chances of war are indefinite. The number which were against the success of this expedition were incalculable.

After this beachhead had been established more British and Russian troops poured in a fortnight later, with the Duke of York as supreme commander. From the time of his arrival things went wrong. The conduct of coalition warfare is a difficult diplomatic-military business in any case, and at least doubly so in combined operations. The first Russian general in command, Moore noted, "despised all assistance, held everybody cheap, and certainly had too much boast and pretension for a man of sense, and was at last taken prisoner, as some suspect purposely to cover his misconduct." By the 28th of September Moore was completely discouraged:

> The natural strength of this country is such that without a general rising of the people in our favor it is vain to hope to conquer it. Government would have done well to have withdrawn the army after the destruction of the Dutch fleet, making that the object of the expedition. The arrival of the Duke of York with strong reinforcements makes it necessary for the honor of our country and for our own as soldiers that we should make another attempt to force the enemy. If we are successful we shall probably be able to secure winter quarters in North Holland. If we are beaten we shall have no option but to re-embark.

Confusion and Fiasco At Cadiz

Re-embarkation proved impossible without the enemy's grace, so to speak. It was granted by a convention under which the Allies withdrew after promising to return the prisoners they had made.

Next in the series of the Abercromby-Moore team play was an attempt against Cadiz in October, 1800. Their force had come too late to be of assistance to the Austrians fighting in upper Italy. In the search for further employment for them someone picked Ferrol as the object of a landing enterprise. The attempt failed, in September, 1800. "The town was surrounded by a high wall with bastions. It is to be regretted that this was not known before the troops were landed," writes Moore who seems only half as much surprised as we moderns are at the failure of reconnaissance, or the absence of other information about an objective. The English landed, took a look at the fortifications and evacuated, without hindrance.

Following that performance Cadiz was chosen, the orders from London saying that the arsenal and shipping in the harbor were to be destroyed by a landing party, provided the latter's reembarkation was assured. Little was known to Abercromby about the opposing forces, or the defences of the Spanish who, as he learned, had been expecting a descent on Cadiz for some time.

Abercromby was willing to undertake the landing even though an epidemic was raging in Cadiz at the time, being certain that his troops could, and would, do their required part. He was ready to take upon himself responsibility for the land operations and even, as he said, to share with Admiral Keith "half the naval" responsibility. But the naval officers in the war council declined even such sharing of responsibilities; for they figured that if during the land operations a southwest wind should spring up, the transports would be lost and the men-of-war would have to put to sea. It is not clear by what ratio the responsibility was finally divided, but the landing was ordered, although poorly prepared for.

Under such conditions Moore was not in favor of the enterprise and in his Diaries he describes the confusion. About 10 A.M. on October 6, the signal was given to prepare for landing. He himself went on board the vessel of the ship captain who was in charge of the naval part of the enterprise. When he arrived, the second signal, for the troops to get into the flatboats, had been given. Moore found the landing director surrounded by numerous other captains who were to superintend the landing, asking for directions; he "was extremely busy, but confessed he was as ignorant as themselves." The fleet all this time was under way, seven or eight miles from the shore. When it signalled to the transports, asking whether they were ready, they had to answer no. For it proved that not enough boats had been provided; for Moore's brigade of 5,000 men who were to land first, boats holding only 2,550 were available.

At that stage, Moore proceeded to the admiral to explain the situation and "found him all confusion, blaming everybody and everything but attempting to remedy nothing. He said that he could not help the want of boats, and that his orders had not been obeyed." The Admiral then signalled the fleet boats to row to the frigate inshore where Abercromby was and receive their orders from the latter. By 11 A.M. it was clear to the general that before the troops already embarked, 3,000 in all, could land and the flatboats return for more it would be dark. He therefore sent word to the admiral to have the troops go back aboard the ships, which were to be anchored near the shore, and that after proper preparations the landing would be made next day at dawn.

The admiral declared that anchoring nearer the shore was impossible, that things would turn out no better if repeated, and uttered "much more incoherent nonsense." Moore assured him that Abercromby, so much nearer inshore, must be better able to judge the situation and that it was impossible to leave 3,000 men unsupported on shore until the next morning and that "this was the first time any person had attempted to land an army from a fleet under sail." At last Keith was persuaded to signal the troops to re-enter the ships. After a night of indecision and orders and counter-orders, promising no better management for the coming day, a southwestern storm blew up and decided the issue for the commanders. The attack on Cadiz was abandoned. "The figure we have cut is truly ridiculous, but the shortest follies are the best, and it is lucky we did not land."

Through these disasters much experience was gathered by Moore and Abercromby who felt strongly "the disgrace which had undeservedly fallen upon the army because of the ridiculous exhibitions at Ferrol and before Cadiz." They evolved methods which bore fruit in the first British conquest of Egypt in 1801. "Our business," Moore had written after the return from Holland, "like every other, is to be learned by constant practice and experience; and our experience is to be had in war, not at reviews."

Following Cadiz, the larger part of Abercromby's force was

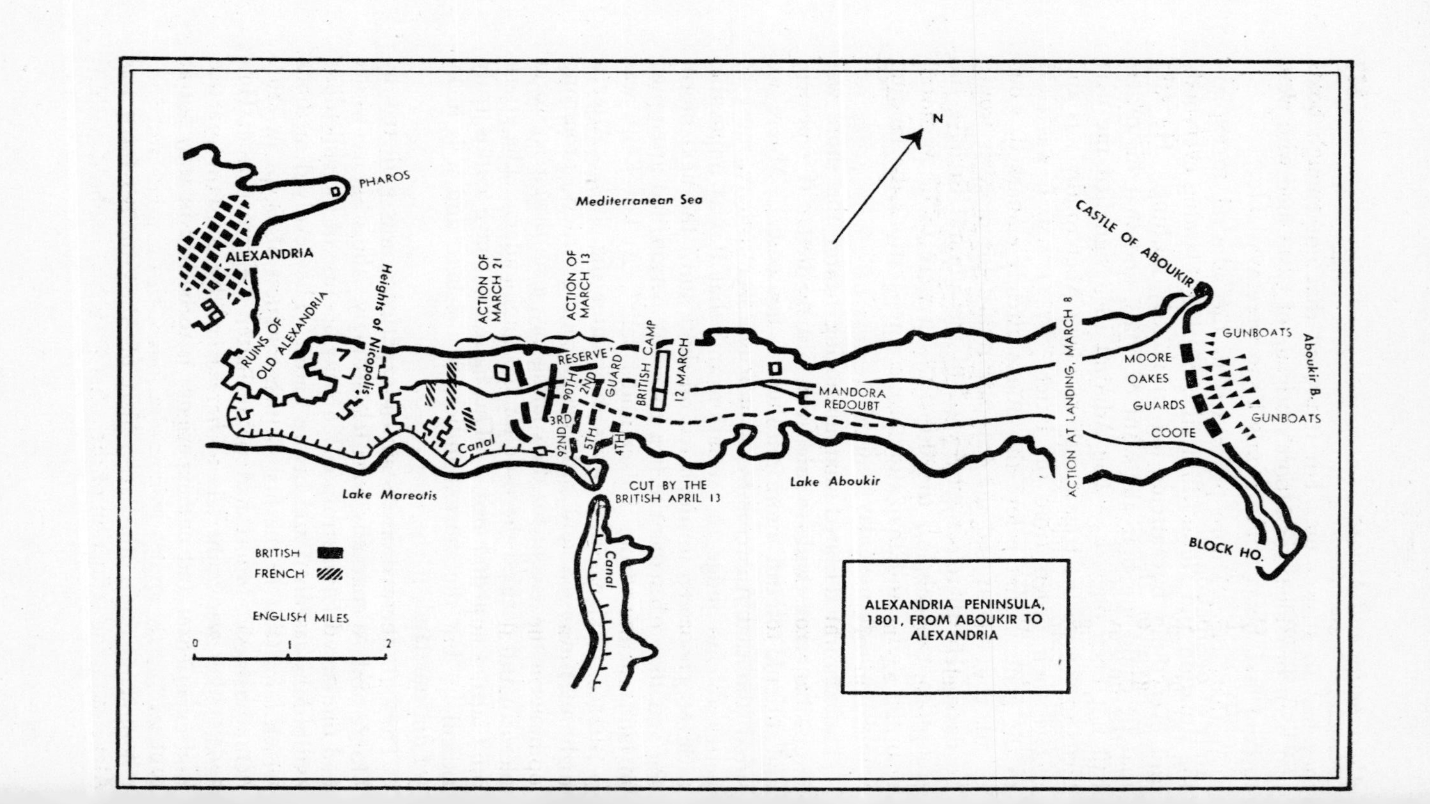
ALEXANDRIA PENINSULA, 1801, FROM ABOUKIR TO ALEXANDRIA
PHAROS
ALEXANDRIA
Mediterranean Sea
N
CASTLE OF ABOUKIR
RUINS OF OLD ALEXANDRIA
Heights of Nicopolis
ACTION OF MARCH 21
ACTION OF MARCH 13
RESERVE
90TH
2ND
GUARD
BRITISH CAMP
12 MARCH
3RD
92ND
5TH
4TH
MANDORA REDOUBT
ACTION AT LANDING, MARCH 8
GUNBOATS
Aboukir B.
MOORE
OAKES
GUARDS
COOTE
GUNBOATS
Canal
Lake Mareotis
CUT BY THE BRITISH APRIL 13
Lake Aboukir
BLOCK HO.
Canal
BRITISH
FRENCH
ENGLISH MILES
0
1
2

ordered to proceed to Egypt in order to recover that country from the French army left behind by Bonaparte which the Ministry in London, relying on intercepted letters from the army of occupation, thought was ready for the kill. They proceeded by way of Minorca and Malta, gathering information about Egypt on the way, trying various expedients to improve on the improvization of "Ministers ignorant of military affairs and too arrogant and self-sufficient to consult military men" (Moore). Moore reflected on the way as if he had known some of the unfortunate experiences of the Crusaders in Egypt:

> If we are able to force a landing at Alexandria, and make ourselves masters of the port, the business will soon be done; but if we are forced to land at Damietta, and to move first against Cairo, the operation will be more tedious. The season of the year is much against us. The north-west winds, which blow strong and directly upon the coast, make it dangerous with so large a fleet to approach it, and very difficult to land upon it. There is not a single port upon the coast but Alexandria. (Nov. 28, 1800).

In an exchange of views Abercromby and Moore discussed the problems offered by the proposed landing, as in most other attempts of the sort:

Weather. They agreed that the season from November to early April was not as adverse as many thought, relying in this opinion on the word of the younger naval officers and the most recent nautical observations that gales were neither violent, nor long lasting and that ships would be able to ride them out at anchor.

Locality. To a frightening degree *terra incognita.* "The greatest misfortune was the total want of information respecting Egypt. Not a map to be depended upon could be procured, and the best draught from which information could be formed and which was distributed to the generals, proved ridiculously incorrect" (Wilson). As a suitable beach in the vicinity of Alexandria, they selected Aboukir Bay, 18 miles away. Alexandria itself was to be taken afterwards; it was the only place where ships could remain in safety; its possession would interrupt completely French communications with the homeland and relieve the British of irksome blockade duties. Starting from Alexandria, the reduction of the other French strongholds would follow. Damietta would be a poor second choice and would necessitate changing the process of the reconquest of Egypt, beginning with the fight up the Nile and reducing Alexandria from upriver by cutting off her water supply. Marmarice Bay and the opposite island of Rhodes, where the armada was to join the Turkish fleet if the Turks had a fleet, were to be bases of operation and storage places. There the quartermaster-general was sent ahead of the army to organize a supply system and assemble such vital articles as horses for the army which included 2,500 dismounted dragoons, but not one horse for cavalry and artillery, nor one wagon.

British forces, some 14,000 strong. Both generals considered their quality high, the regiments among the best in the service, the general officers in vigorous health and experienced. Great care was taken to preserve and improve the men's health which was far from usual in those days of manpower wastage. While at Rhodes and Marmarice Bay the troops were "landed in the order that it is intended to be observed when we land in Egypt; this was done by way of practice. . . . The troops, particularly those intended for the disembarkation, were placed in the boats in which they were intended to be disembarked, and arranged with the guns in their proper order, and the landing was practised several times in the order in which it was afterwards executed." To make landings a part of the *imago belli* which manoeuvres are supposed to be, was unusual enough for the time and in this the British preceded Napoleon's preparations in the camp of Boulogne. Dismounted cavalry which had been sent from Portugal on an afterthought of the Ministry, could not be of much use. "It was not taken into consideration that this expedition could only last a few months, and that if cavalry could be at all of use to us it must be in the beginning, and, therefore, if the expense was what prevented the horses from being sent, it would have been better to have saved the whole and not even sent the men."

Aboukir Bay the Objective

Navy. Landing troops would have to leave transports at a distance of 4 to 5 miles from the shore. The boats of the fleet would carry 4,000 men and light Levant craft to be procured would take 3,000 more, enabling 7,000 troops to be landed simultaneously and within three hours after having pushed off from the ships. Two more trips would bring the rest of the men, cannon, stores, etc., to the shore, all this requiring some ten hours of favorable weather. In councils with naval officers Aboukir Bay was settled upon, the fleet declaring itself able to bring ashore sufficient quantities of water and provisions for a fortnight after the first landing. The order of debarkation was agreed upon and "everything respecting it is supposed to be so well understood that it is to be hoped that no time will be lost in instantly carrying it into execution the moment we arrive at the coast," as Moore's somewhat apprehensive diary entry puts it, at the departure from Marmarice Bay. The fleet itself consisted of 180 vessels of all kinds.

The Ally. Turkey provided not very much more than a basis of operations and a market for the pound sterling in which to buy horses, provisions and so on. Turkish troops and navy were in a wretched condition. Abercromby was at first ready to throw over his original notion of landing close to Alexandria and instead go ashore at Damietta and from there work in close cooperation with them, as the vizier doubted that he could get his army to act at all by itself and at such distance from the British. Moore was then sent to this army in the vicinity of Jaffa to inspect their fighting value and arrange for details of cooperation, including a Turkish liaison officer with the British command, and also to assure the Turks that the British intentions in Egypt went no further than expelling the common enemy. Finding the ally's army was merely "a wild ungovernable mob, incapable of being directed to any useful purpose" and destitute of everything, Abercromby could see no reason to throw over his first plan of landing close to Alexandria.

The Enemy. The French troops in Egypt were originally estimated by Abercromby at not much less than the 40,000 who had left France a year and a half before; that is to say, far too high to please him. In point of quality he considered his own force superior. Later he revised his estimate to 13,000-14,000, plus bodies of Greeks and natives whom they had armed and drilled. Of these they would at the most be able to spare 10,000 men, the garrison of Alexandria included, to throw against the British. Some reinforcements of artillery and ammunition for the French were brought in while the British were waiting for their horses. The latter could no longer count on surprise, as they might have done had they sailed straight for Egypt from Malta. On the whole, the generals on the spot in the Levant thought more highly of the morale and other qualities of the French troops than the Ministers in London, who relied strongly on their intercepted letters full of complaint and homesickness.

Post-landing. Planning thought extended beyond the moment of landing and took into consideration the tactics of the first movements following the gaining of the beachhead. Since cavalry would probably not be available—some little was, in fact—the order of the battle and on the march must be compact, with movable columns composed of steady troops and commanded by judicious officers on the flanks. That is to say, the cohesion of the troops, loosened during the wait and by passage over the water, could not be expected to remain fully intact and must therefore be maintained by elbow touch and by the light troops, of whom Moore was to make so much use subsequently and who could be expected to cohere in less close order.

After numerous delays, "with people growling at the tardiness of our operations, judging that the enemy may be strengthening their situation on the coast," as one naval officer wrote from Marmarice Bay,[3] the armada set sail on February 22 and showed itself before Alexandria on March 1. Moore could not see the reason why the Admiral did that. On the second of March it stood into the Bay of Aboukir, the men-of-war anchoring at a distance of six or seven miles from the shore in from 6-8 fathoms; the bottom sloped so gradually that even at three or four miles from land there were only 2 fathoms of water.

The coast itself was sandy and in its formations favorable to the enemy, enabling him to conceal himself and his strength well. It was easy to defend against a landing, particularly with the help of gunboats, of which the enemy had one or two anchored close to the shore. By 2 P.M. the English generals had reconnoitered and settled on a plan for landing the next day, provided the wind would abate, but this did not happen, giving the French more time to work on their entrenchments and slip in another gunboat, stationing it so that it would rake the beach. The British admiral declared he could not spare ships to go after it.

Landing Under Fire

On March 7 the weather had sufficiently improved to warrant the issuing of orders for a landing on the following day, to begin at 2 A.M., when the first wave of troops were to leave their ships, six or seven miles from shore, by flatboats and launches. Two smaller vessels were anchored closer inshore, one to mark the right wing of the landing, the other as an intermediate point in the same direction and both to serve as rendezvous for the landing craft so soon as they had been loaded, and as the command post of the naval director of landing, Captain, later Admiral, Cochrane.

Here, needless to say, out of reach by French fire, the boats were to be ranged in accordance with the battle order of the army. The troops of the second landing were transferred in the evening to smaller ships anchoring further in so as to give quicker support to the first disembarkation. Orders were that "the officers commanding the boats must take particular care that none of the troops stand up, as, on many occasions, it may endanger the safety of the boat."

Soon after daybreak most of the boats were assembled at the rendezvous where it took a considerable time to arrange them in battle order. Moore was there to fix the landmark for the right wing to go by, a high sandhill, the rest of the boats being told to dress by them.

> It was now eight o'clock; the enemy had for two hours been spectators of our movements, and we could see them drawn up with their cannon to oppose us. Some gunboats proceeded to engage their attention; the signal from Captain Cochrane's boat was made to advance; we were fired upon from fifteen pieces of artillery as soon as we were within reach, first with round shot, afterward with grape, and at last by the infantry. The boats continued to row in steadily, and the sailors and soldiers occasionally huzzaed. Numbers were killed and wounded, and some boats were sunk. The fire of grape-shot and musketry was really most severe. As soon as the boats touched the land the officers and men sprang out, formed on the beach, and landed.

Moore, himself with the right, which suffered two-thirds of the total British losses, took the marked sandhill and drove the enemy some distance inland to ground apparently more favorable to defense. The left had found the enemy ready to receive them; after it had formed expeditiously, the landing party was attacked by both cavalry and infantry, which it repelled and drove into the plain behind the dunes where the English took three cannon.

There was some confusion due to the landing of boats in unassigned places, but that was soon remedied as the first troops landed moved over the shore and to the flanks. The want of cavalry and horses for the artillery, which was only slowly dragged across the sands, kept the British from pursuing and annihilating the enemy. During the afternoon of the same day the rest of the army was brought ashore. The British losses, 600 killed and wounded, were not high considering the character of the enterprise, a landing in the face of the enemy and disputed by him. As Moore sums up: "The enemy had had eight days to assemble and prepare; the ground was extremely favorable for defense. Our attempt was daring, and executed by the troops with the greatest intrepidity and coolness."

Reward of Careful Preparation

With this judgment the senior service and some of the enemy were willing to agree. The commander of one of the ships to cover the landing and therefore an eye witness, wrote: "As the boats approached the shore, grape and canister shot were flying at no allowance. But nothing could intimidate them, many were drowned by boats being sunk and numbers were shot getting out of the boats and on the beach while forming."[4] In St. Helena, fourteen years later, the landing was still remembered in Napoleon's entourage as admirable. "In less than five to six minutes," said Montholon, "they had presented 5,500 men for battle, it was a movement as on the opera stage." Still later, careful military judgment is summed up by Sir F. Maurice: "Abercromby's great success had been due to the careful preparations which circumstances had enabled him to give to his army. Everything that could be neglected had been neglected at home." (II, 95).[6]

The military credit won by Moore on distant expeditions caused Pitt to tempt him in November-December, 1804, with the command of yet another landing force. The objective chosen for this once more was Ferrol, for information seemed to make this a feasible enterprise, to be undertaken in anticipation of a war with Spain. Moore studied the available scanty reports and somewhat more copious opinions about the place and found them far from sufficient—"very little real information, with respect to the defenses of Ferrol, very little even of that sort of information which might have been expected from an intelligent naval officer on points of which, professionally, he must have been a competent judge," and none from a landsman's point of view. The available data did not even include such as might easily have been obtained before Spanish suspicion had been aroused.

Appalled by this paucity of indispensible intelligence, Moore was able to persuade Pitt and his colleagues to desist from the attack until he himself could go to the squadron stationed off Ferrol, privately, and learn from its officers what they actually knew about the place. This work he tried to supplement by going on shore with the admiral in command, in disguise and under the pretext of hunting. So poorly organized was intelligence service in Britain at that time that the commander in person was forced to go to the intended theater of action and by personal spying and exposing himself to arrest, obtain data on which to base his decisions. Moore found the Spanish far from unready and a surprise enterprise, the only one feasible, quite out of the question. On the strength of his own observations he succeeded in persuading Pitt to give up the assault on Ferrol.

Moore also managed to prove to Pitt that the attempt proposed on what was left of the camp of Boulogne in the autumn of 1805, including largely the invasion flotilla, was "attended with too much risk to justify the experiment" (Pitt to Castlereagh, Oct. 6, 1805). It had been proposed to undertake this in modern commando fashion and with the help of the latest, still mainly untried, inventions such as submarine boats and Congreve rockets, and with naval personnel only, although later army cooperation had seemed advisable to the sponsors.

However, after the death of Pitt, Moore failed in dissuading the Government from yet other expeditions, which they had not sufficiently well meditated from the military angle. In the "eminently foolish expedition to Sweden" (Napier) in 1808, they had "no specific plan and had come to no determination beyond sending a force of 10,000 men to Gothenburg to be ready to act if occasion offered," to cooperate with the Swedes, of whose military strength they had no advance knowledge and which they overrated vastly. The next overseas expedition, directed to the Peninsula in 1808, was again conceived politically rather than militarily and was based to a large extent on the state of public feeling in England, calling for a second front, to use the language of the 1940's, rather than on reliable military knowledge and just estimation of the forces stirring in Spain. This ignorance brought on the retreat to Corunna and the death of Moore, whom England could ill spare, even if the Tory ministry thought it could dispense with him.

When Moore had taken over command the army to be employed in northern Spain was around Lisbon. Rather than embark it, as he was at liberty to do, and go to Corunna and thence advance to the eastward, he marched by land. To him "the passage by sea was precarious"—it was October—"and embarkation unhinges." Besides, he hoped to acquire "means of carriage" which London had not provided, along the land route. This might have been tactically advantageous, but it also meant entering into the perimeter of French overland marching.

Peninsular French Reverses

Actually, the undermining of French military strength in the Peninsula was to begin at the maximum distances from Paris, at Corunna, in the lines of Torres Vedras, before Cadiz, where the transport superiority of sea power supported England's now largely concentrated landing military strength, making it and keeping it superior to the French military power which was accustomed to live on the occupied country. The Portuguese peasantry was made to lay waste the land and retire with the soldiers into the lines of Torres Vedras (1810-11), which enabled Wellington to reverse the usual conditions of a siege by starving out the besiegers under Masséna. He was kept supplied by the English fleet which landed food and stores and replacements on friendly shores. There was indeed in the whole five years of the Peninsular War practically no combined operation of any size that involved a landing against a shore held by the enemy; nothing until the passage of the Adour in February, 1814, where there was again some tactical cooperation between the two armed forces. The rôle of the fleet consisted in supplying Wellington with the means for carrying on what he called "the war of magazines."

An island nation cannot always be best defended along its coastline, or on the surrounding seas, or at points where the juxtaposed continent is nearest. Traditional British policy, since the loss of Calais and Dunkirk and the end of the Anglo-Dutch wars, has endeavored to keep the Flemish-Dutch coasts in hands not hostile to Britain. Hence Castlereagh's insistence, before and at the Congress of Vienna, that the Low Countries must be made independent and that to have Antwerp in the hands of France would impose upon Great Britain the charge of a perpetual war establishment. While this represented sound enough defense policy, the attempt to carry war there was not always necessarily sound.

It was a fatal blunder when its cabinet detached the fighting force under the Duke of York in the anti-French coalition of 1793 and directed it against Dunkirk, thereby throwing away a fair chance of winning Paris. Practically all English enterprises against revolutionary France in the Lowlands were politically rather than militarily conceived and—in consequence? ——bungled in military and naval execution. While there seems to be some disagreement as to whether the Walcheren expedition of 1809 was wise in conception, there can be none that it was one of the most inept amphibious operations in its execution.

Only appreciation of the worst tendencies of conservative politics can explain the choices of military leaders by tory governments. The Earl of Chatham demonstrated that among the scions of the great families of England military talents rank low. He was bossing a military force of 40,000, England's best at the time, as against the mere 22,000 under Wellington in Portugal who at the same time was waiting impatiently for the reinforcements promised him. The force was carried and convoyed to Walcheren at the end of July by the most powerful fleet at England's disposal. The expedition numbered in excess of 100,000 men. It was the largest that had left the country up to that time. The objective was a large naval yard set up by Napoleon on the Scheldt where ships for naval rather than amphibious warfare were being constructed. Safely brought across the channel and landed at Walcheren, from whence Norsemen once had harassed the Lowlands in the 9th century, and with Antwerp at its mercy for a whole week, the army was allowed to sit on that miasmic island and be ruined by inertia and malaria, which killed 4,000 soldiers. The best result for England of the Walcheren expedition was that it hastened the concentration of further military effort on the Peninsula.

NOTES, CHAPTER 30

[1] Bassett (ed.), *Correspondence of Andrew Jackson.* II, 37.
[2] Pertz, *Gneisenau.* II, 493; Wilhelm Dilthey, *Schriften.* XII, 95 ff.

[3] *The Naval Miscellany.* London 1912. II, 399.

[4] Ibd. II, 343-4.

[5] Las Casas, *Mémorial de Sainte-Hélène.* Réimpreesion de 1823 and 1824. I, 303. Sept. 1815.

[6] Sources: *The Diary of Sir John Moore;* Fortescue, *History of the British Army, vol. IV, chapters* XXIII and XXVII; Bunbury, *Campaign in Egypt;* Sir R. Th. Wilson, *History of the British Expedition to Egypt;* G. R. Gleig, *Life of General Sir R. Abercromby.* (*Eminent British Military Commanders,* v. 3).

Chapter 31

STEAMPOWER AND LANDING PROBLEMS

UNDER THE NAPOLEONIC SPELL, many military minds after 1815 scornfully termed that post-war period "the halt in the mud." Few campaigns were fought and progress in martial art and science seemed to have stalled indefinitely. Retrogression rather than progress characterized combined operations. Although technical improvements went on within the services there was little, if any, in their actual cooperation. Rather to the pained surprise of the directors of both, what mutual action there was had to be freshly organized from time to time, from war to war, much as if armies and navies were merely fighting wars of coalition, without any vigorous, over-all conception involving the control, employing, apportioning of both army and navy. Euphemistically speaking, in the post-Napoleonic era, "the effect of sea power upon land campaigns was in the main strategical," as an English work carrying on the theme of Mahan beyond the days of the Corsican, concluded. This statement glosses over a vast number of sins of omission.

The two great influences in the 19th century making for military progress were the American Civil War and the building up of Moltke's Prussia. In the first case, after prolonged fumbling, a fairly efficient, but watertight division of war tasks between navy and army was achieved rather than a community of labor. Prussia-Germany was so overwhelmingly a natural land power that its navy was correspondingly weak. So far as it existed, however, it was used, if not exactly planned, for one all-embracing scheme of strategy. There was no special naval ministry there until the pressure and demand for specialism pried it loose from the War Ministry after 1871 and set it up in business on its own hook.

As another land power, Russia also kept her fleet relatively small. In the war of 1828-9 against Turkey she fought not so much across the Black Sea as around it, with one army advancing over the Danube and the Balkans and the other into Asia, to Kars and Erzerum. The fleet, however, cooperated in the investment of Varna in 1828 and in 1829 opened the campaign by capturing Sizopolis, near Burgas, which gave the Russians their first foothold south of the Balkan range. In the

Russo-Turkish War of 1877-8 Turkey rather than Russia had a fleet in the Black Sea, a circumstance that contributed to making this particular war more of a land than a sea affair.

The great military writers of the first half of the 19th century had almost nothing to say about landings, or combined operations generally. Both Jomini, who had been present in the camp of Boulogne, and Clausewitz, who thought that landings only had a chance of succeeding when surprise, that product of speed and secrecy, was complete and the landing party found support in the invaded region, (possibly the assistance of a province in rebellion against its government), wrote on war too much from the viewpoint of army men to give amphibious operations their due. Besides, their technical thought and theory were those prevailing in pre-steampower days. The revival of military and lay discussion of landings on enemy shores were in part a by-product of the recrudescence of Napoleonism which in turn might be called a psychic by-product of industrialism.

Industrial Revolution and War

There was much in the industrial revolution that threatened to revolutionize war and also much in a revolutionized war that menaced traditional concepts of security. The collective and diversified anxieties of the foremost industrial nation, England, were largely connected with the ups and downs of the business cycle. These preoccupations might be worked upon by statesmen honestly concerned with their country's safety, but as well by others, including spokesmen of service interests merely intent upon manipulating scares as a means of persuading the masses of the existence of military crises and the necessity of meeting them by war. The danger of a French, or German landing on English shores, "the oldest and most profitable bogey in the armory of our fighting services, is always brought out, dusted, and repainted in flaming red whenever the Generals and Admirals want to retain or increase their estimates of men, money or machinery." [2]

Discussion of the influence of the new industrialism on the more narrow problem of overseas operations and landings, including a descent on England, was started by a scion of the royal house that was most threatened in its survival by Napoleonism, an Orleans. This was Prince de Joinville, Louis Philippe's third son. The Prince, brought up as a naval officer, had served during the French expedition to Mexico in 1838. Following the bombardment of San Juan de Ulloa he headed a landing party and personally made prisoner of a Mexican general. In 1844 he commanded a French squadron operating along the Moroccan coast, bombarding several places and occupying Mogador. It was he who brought home the remains of Napoleon from St. Helena, which the British government had permitted as a good-will gesture. In 1845 he discussed in an article for the *Revue des deux Mondes* the shortcomings of the French navy and its chances for a renascence by the full application of steam power.[3] He also explored the possibility of thereby in a naval race outbuilding England, with which the Guizot government (1840-48) always sought to maintain cordial relations, and perhaps invade that island, or at least inflict considerable damage on its coastal towns.

In the history of diplomacy the incident figures as one of the first caused by a military, or naval expert who felt the urge publicly to air specific technical problems of war and preparations for war, although unfortunately de Joinville's observations had not only a technical aspect but proved highly embarrassing to the official diplomacy of Guizot, who aspired to cause France to live peaceably with the country overtly mentioned as a potential, or possible enemy. The Prince wrote:

> A fact of enormous import which has come about since some years, has given us the means of relieving our decayed naval power, to make it reappear in a new shape, admirably fitted to our resources and our national genius —the establishment and the progress of steam navigation. With the steam navy the most audacious war of aggression on the seas has become permissible. We are certain of our movements, free in our actions. We calculate on the basis of fixed day and hour. In case of continental war, the most unexpected diversive actions are possible. One can transport within a few hours (?) armies of France to Italy, Holland, Prussia," something which had already been done once, in the case of the Ancona troubles (in 1831).
>
> Next the Prince evolved a kind of *Risiko* fleet theorem resembling that of Tirpitz:
>
> I do not belong to those who in the illusion of national self-love believe that we are in a position to fight on equal terms against British power; but I will not hear it said

> either that in no case can we resist her. Who can doubt that with a steam navy strongly organized we have the means of inflicting on the enemy coast losses and sufferings heretofore unknown to a nation that has never felt all that war carries in its train? And in consequence of these sufferings she would also experience the evil, equally new to her, of the loss of confidence. The riches accumulated on its coasts and its ports would cease to be in safety.

In the thought of Palmerston and in the personality and characteristics of the man himself were embodied Britain's uncertainties about the consequences to her defense of the industrial revolution. Hearty, reckless, nothing in him of the neurotic, technically ignorant and essentially unwilling to learn, he was the modern type of leader who when questions of defense are on the tapis is always ready to put himself at the head of still more ignorant and often hysterically inclined masses and pander to their confused disorientations. He would tell the country in the summer of 1844, during a diplomatic tiff over distant Tahiti, that the Channel was now merely "a river passable by a steam bridge," or in 1846 that France could throw twenty or thirty thousand men over the Channel in steamboats overnight.

Wellington and Public Sentiment

Wellington, putting aside his own concepts of warfare, followed the spirit of the British public and in a letter published early in 1848 confessed "his inability of spurring the Government to adequate efforts; the development of steam navigation had exposed all parts of the British coasts to invasion and insult and from Dover to Portsmouth there was not a spot on the coast save immediately under the fire of Dover Castle on which infantry might not be thrown on the shore at any time of tide, with any wind and in any weather."

However, the economic repercussions of a crisis year, opening with the February Revolution in France, silenced the alarmists for a time. But when, as earnest of their intentions, the new French government (the Second Republic, 1848-52) in January, 1849, proposed a mutual and proportionate reduction in naval armaments, Palmerston answered that Great Britain could never fix the size of her navy with reference to the force maintained by any single power.

These controversies and pronouncements were the mental reflections of various developments in naval construction, rather than in naval warfare, of which latter there was as yet little, or none. This fact added in turn to the general uncertainty. The developments referred to were in the main:

1. The introduction of steam into the navies, beginning with Fulton's *Demologos,* built for the defense of New York city and launched in October, 1814, but which never saw action. The process was as much slowed up by post-Napoleonic reaction as by technical difficulties, many of which were obviated only by the development of the screw propeller whose location below the water line gave better protection to one of the most vulnerable parts of maritime mechanism. However, it was apparent that navies under steam would doubtless make landings on hostile shores more feasible by rendering such enterprises largely independent of weather and tide, allowing far more exact timing and permitting for the first time dependable logistic arrangements.

2. The introduction of shell-missile guns aboard ships. This was begun in the French navy in 1827, after considerable agitation, with the British soon following. While this innovation might not have a direct influence on combined operations—"expéditions mixtes," as Joinville called them—it seemed to its sponsors in France like Paixhans, inventor of the shell gun, that the shell would help to challenge England's naval superiority. The steam navy would require fewer skilled seamen, one of the elements of British manpower strength, and allow the use of more soldiers on the sea, behind the shell-missile guns. Of such men France, as the more populous country, had a greater abundance. Thus her naval power would become at last proportionate to her total population rather than to her seafaring population as heretofore. The shell gun unavoidably called for protection of ships by armor. This however, was not easily managed owing to serious technical difficulties. To British susceptibilities it seemed nearer its application during the deeply disturbing reign of Napoleon III.[1]

"Napoleon the Small" had once attempted to establish himself in France by landing on the shore of Boulogne, his Napoleonic idea being that the garrison there would join him and that the resurrected eagles, once more flying from steeple to steeple as on the Corsican's return from Elba, would precede him on the march to Paris. The magic of the parallel failed to come off and the enterprise ended in farce. Several of Louis Napoleon's 56 followers were shot on August 6, 1840, and the rest arrested by the very regiment they had hoped would desert the Orleans monarch and follow them. Perhaps the adventure was too un-Napoleonic to catch on with the French public, for a pretender to land at Boulogne from England and thus reverse imperial Bonaparte's idea. Rather would practical imagination seem to call for a reprise of the camp of Boulogne and with the help of steampower then at the disposal of the modern warmakers.

Napoleon III's Threat Real?

The English were repeatedly in fear of precisely this. A resolution in the House of Commons, February 29, 1844, declared that if Napoleon I had had steamships at Boulogne his undertaking might have succeeded and that recent developments in steam navigation had shaken that confidence "which our insular position inspires us with at present." That Napoleon III when he finally came to the throne in 1852 *had* to make war, seemed to many people on either side of the Channel "as clear as the sun"—so Engels wrote to his friend Marx early in 1852—and the question was merely whether Russia, or England would be the chosen enemy. In April Palmerston, not then in power, told the House of Commons that 50,000, or 60,000 French troops "might be collected at Cherbourg before you knew anything of the matter" and be transported to British shores during a single night. "All our naval preparations, be they what they might, could not be relied upon to prevent the arrival of such an expedition."

Napoleon III made the threat to Britain's naval supremacy far more real than had any Orleans, or so it seemed to the British. They never trusted him and his asserted friendly intentions, but thought rather that he was out to seek "revenge for Waterloo." Even after the British and French martial partnership in the Crimea against Russia the poet laureate, Tennyson, fleered at him: "True that we have a faithful ally, but only the Devil knows what he means."

Serious, although unconventional, thoughts as applied to the problem of a military invasion of England in the early 1850's were expressed by Friedrich Engels, the friend of Karl Marx and, as a former Prussian Guard bombardier, the military half in that politico-philosophical co-partnership; both of them were living in England at the time. While Engels judged the British army and naval forces to be in a very inefficient state, he still considered England's fear of invasion vastly exaggerated, although not wholly unjustified, since Napoleon must not be given any chance of conquering England.

Discussing in a letter of January, 1852, the prospects of a French landing, he made a careful distinction between a raid and a regular war. Considering the poor condition of the English coastal fortifications, he deemed the former quite feasible if directed against a coastal town, possibly even Woolwich. But that would be the worst that the French could do. In a real, full dress war everything depended on how quickly the attackers would act and how many troops they would be able to throw across the Channel. He fancied the English would be in a position to concentrate enough of their widely scattered fleet and gather an armed force before the French could land an army of sufficient size to conquer England.

In the beginning the French could count themselves lucky if their naval forces were able to ward off English attacks on their transport fleet. In the later stages of the enterprise they would be unable to keep the British from interrupting their line of sea communication. Time was leagued with the British, as the French would find to their chagrin, even if they should be able to land an army of any size. While the English would not at once have large numbers under arms, their people had plenty of martial spirit and potentially useful military elements. In particular the British mechanics and machinists seemed better prepared for work in armament factories and arsenals and for service with the artillery and engineers than any similar element of craftsmen in the continental nations. Nothing short of an initial army of 400,000 men would possibly enable the French to subjugate England from Dover to the Clyde, where a final front might be formed.[5]

British-French Naval Strength

Until the introduction of steampower into the navies, around 1840, the relation in strength between Great Britain and France, and in the former's favor, as measured by ships of the line and frigates had been as 2:1. In 1852 Britain had 17 steam line-of-battleships, France but two. From Napoleon's *coup d'état* on, the French navy rapidly forged ahead in construction and to the end of 1858 France added to her fleet 38 steam liners against 33 by Britain, making the relative strength in screw-propelled line-of-battle ships, 40 French to 50 English. In addition France at the same time had acquired 46 steam frigates against only 34 by England.

Maritime steampower's utility in long distance troop transportation and combined landing operations was early and impressively demonstrated in 1847 by the Americans at Vera Cruz, Mexico. There, under Scott, to quote from the legend borne by a contemporaneous drawing of the event: "The colors of the United States were triumphantly planted ashore in full view of the City and Castle, and under the distant fire of both. The whole Army reached the shore in fine style and without accident and loss."

In smaller vessels, like corvettes and gunboats, which were thought particularly suitable for combatting transports with invasion troops, England was vastly superior, somewhat in the relation of 82 plus 162 to 28 plus 28. Considering the heavy requirements for service on distant stations, that left the English home fleet about equal to that of France alone, at the outbreak of any cross-Channel war that might find France in an alliance with Russia. At that point of naval competition, by the late 1850's, the French threatened to gain a further lead by introducing ironclads, while England remained hesitant about following suit.

On the technical side the fundamental instigation to progress was provided by the application of steampower to naval navigation, both in peace and war. Steam power made planned maritime movements, including military, almost as feasible as movements on land, if not more so in some respects. For it emancipated such movements in large part from wind, weather and currents. These, it was now recalled, had played the chief determining part in the success or failure of the expeditions against England:

> When the Normans attacked England the winds aided the invader, first, by compulsorily delaying his voyage till the English fleet had left its post; and secondly, by blowing in his favor when the English King and his army were absent in the north of the island. On that occasion England was conquered. But when Charles VI designed to repeat the exploit of the Norman Conquerer over us, and

American Forces Under Scott Landing at Vera Cruz, March 9, 1847.
(N. Currier lithograph. Bettmann Archive.)

> when England lay almost defenceless before him, the northern gale blew steadily against our foes, until, in weariness and fatigue, they abandoned their armament against us.
>
> At the ever memorable epoch of the Spanish Armada the English nation gratefully acknowledged how much their preservation was due to the tempest that first delayed the enemy off Cape Finisterre, and gave this country time to complete her defences; to the state of the weather when the Spanish fleet was in the Channel (being eminently advantageous to the tactics of the English); and to the storms which completed the Armada's destruction. . . .
>
> Still later, the storm which drove Hoche from the Irish coast, when all our fleets had failed to bar his passage, saved us from the loss at least of the greater part of Ireland for a time, and from a disastrously costly struggle to regain it; for Hoche assuredly would have ventured the disembarkation in Bantry Bay, which Grouchy flinched from effecting.[6]

Mankind and war were in process of undergoing an added emancipation from Nature's control. In the last analysis, or psycho-analysis, this change was reopening once more the old trauma of invasion-fear in England. In the midst of a secure conviction about the country's immunity from invasion, "we are suddenly called upon to consider whether modern science, of which we ourselves have been to the world the practical expositors, has not done more against us than for us as regards this life-or-death matter," someone wrote anonymously in Frazer's Magazine in December, 1859. Discussion of the technical military problem as to whether the steamship was enhancing chances of invasion, or of defense against invasion was not purely technical enough, or of too vital general importance in its implications to remain restricted to the narrower circle of those who were professionally, or officially interested. They were carried to the masses by journalism, specific professional interests of armed forces, or economic groups and by politicians such as Palmerston, who as Prime Minister declared that "steam has bridged the Channel."

Palmerston and His "Follies"

On such occasions it was demonstrated that peoples are more obsessed by, or enamoured of, their fears than with reasonable diagnoses of, or promises for, the allaying of their apprehensions. While free trade was the rational product of industrialism, panics might be considered their irrational, spiritually toxic consequence. After Cobden had negotiated his famous treaty with Napoleon III in January, 1860, establishing free trade so far as Great Britain was concerned, suspicions of the motives of the Emperor on the part of the sovereign, the government and large masses led to the new volunteer movement, to bills for the fortification of Portsmouth, Plymouth, Chatham and Cork. What were dubbed "Palmerston's follies," appropriations amounting to the unheard-of sum of £11,000,000 were brought in by the prime minister in July, 1860. He frankly bound up his plans with "our immediate neighbors across the Channel, and there is no disguising it."

The panic-mongers and the panic-inclined would not accept remedies like proportional limitation of armaments to which Napoleon had long agreed; they refused "the only sedative" which, according to Gladstone, Cobden's treaty of commerce offered and which he defined as "the counter-irritant which aroused the sense of commercial interest to counteract the war passion." [7]

The inward and outward conditions of England's defence—and one is tempted to say that the inward motivations were easily as strong as the outward threats warranted—raised once more the question, periodically coming up in England's modern history, at a turn in the technological revolution of war: Whether, as the writer in Fraser's Magazine put it, England's resources and labor had not been too exclusively dedicated to works of peace, considering the vicinity of great military powers on the continent which had habitually in view a state of war, like France. That country, he thought, then in the later fifties, was England's "equal in naval steam machinery, and to a formidable degree our superior in the producing power of her dockyards."

Given time, England could make up for this lag, as she had done in all her past wars, for which she was almost always unprepared; winning in the end only because of her ability to apply her great economic resources to war effort. The Crimean War had shown "that such a change of direction in the energies of an industrial people—what we today call industrial mobilization—requires time." Now the great change for England, so far as concerned war, was due to its having become "more than ever continental in its character by the virtual subjugation of winds and waves which has been effected by steam."

The danger in this development was that an invasion might interrupt, or make impossible Britain's preparation for war after the outbreak of hostilities, thus giving continental enemies a headstart. This, in other words, was then the problem for Britain at the outset of the two World Wars. As in 1914 and 1939, a deep-reaching technological change in warfare was going on, more on the sea even than on land, where novelties were constantly being tested. Whereas on the sea, steam, heavy ordnance, shell guns, ironclads, enormous tonnage, mass transports, were rarely put to the test.

Command of Channel Vital

How far in Palmerston's time was the danger to England increased, or altered by these changes in sea warfare? If France, England's prospective enemy, should obtain command of the Channel, for some lengthened period, she could bring over as many troops as her available shipping allowed and with as much certainty and speed as if French armies were manoeuvring on land. Former French invasion plans had considered as obstacles not only the English fleet, but also contrary winds that frequently kept squadrons in port and might even interrupt communications once a foothold was won.

Should an enemy be ready to sacrifice his communications, he would probably find it easier even than in Napoleon's day to live on the country. He could make a dash across the Channel in steam frigates and transports, each carrying 2,000 men and some horses and guns, from the various embarkation ports whence they could be directed concentrically towards a given point and at a given time, provided the British fleet were but briefly withdrawn from those waters.

Against this increase in the mobility of the invaders stood an equal increase in that of the defending fleet; like the invader, it could leave the port at almost any moment, regardless of the weather. If the aggressor was able to hasten his troops to the coast by rail, so could England with her even better developed railway system. Something would also depend on advance planning for these troop movements. The introduction of the electric telegraph was probably not more favorable to one than the other; while the aggressor with its help could time and synchronize the movements of his forces. The defender could use it to warn and mobilize his military and naval forces which happened to be not near, or in the Channel. Whether steam men-of-war, with their constant need for refueling, were better blockaders than sailing ships seemed a dubious point, although not likely to handicap the defender greatly.

Another change in sea warfare was the shift in emphasis from seamanship to gunnery, with the increase of rifled cannon with larger caliber and shell guns, all of which would greatly tend to reduce the duration of naval actions, a circumstance apt to favor the aggressor who would probably have his forces more concentrated than the defender. In the absence, until the American Civil War, of battle experiences with iron-clads, it seemed as yet uncertain which of the two sides was more likely to be helped by their introduction, once they had been universally adopted. In 1859 England was behind France in this respect. Indecisive battles in the Channel, or elsewhere between two battle fleets, probably forcing both to withdraw to their harbors, seemingly would on the whole favor the invader who could depart without requiring many escorting warships.

Mass transports, although involving certain delays in landing, would assist the aggressor. At the end of the 1850's shipping space for troops to be conveyed over a short distance was calculated at one ton per man. That is to say, sixty frigates or transports of 2,000 tons each would carry 120,000 men.

While France, in case of a naval war, was not likely to make use of her steam frigates for transport, she was building up a special transport service, with a type of craft able to carry 2,000 men, 150 horses and some guns. A program envisaging 72 such ships by 1871 had been laid down. To this must be added France's mercantile steamships of which she was building more and more and which would be at her disposal. The growth of ships in size and number multiplied the power of transporting soldiers in large masses to any given spot.

The practical outcome of these considerations can be estimated by imagining England as more than ever since the days of Henry VIII a brick and mortar fortress, with the sea as a broad moat, equipped for passive defense, except in the country's food supply which, somewhat illogically, remained dependent on overseas provisioning, without allowing for accumulation of reserve stocks. The army would then be the dominant service, instead of the navy, a "blue water" navy which, in the words of John Fisher, would carry the war where "our maritime frontier must be—the territorial waters of the enemy."

Progressive Spirit in Navy

Not until the late 1880's did a change start, undissolvably connected with alarm over England's security, a new outward, imperialist urge, a driving service-interest in the navy—while the army remained floundering in a morass of traditions and technological backwardness—due to the importation of a suitable sea power ideology from America. From then on the "blue water" school insisted upon, and put over in an interdepartmental fight lasting from about 1888 to 1892, the view that the navy must be the true shield to guard England and the army its spear with which to strike. Or, in the blunt language of "Jacky" Fisher, the army was a "projectile" to be fired—and directed?—by the navy which would also carry it to hostile shores to round out wars gloriously begun by the navy sweeping away enemy fleets.

The War Office in 1888 had considered possible a French landing, given complete control of the Channel for three weeks and sufficient tonnage to carry 150,000 troops, divided into two waves, and supplies. The Admiralty, not quarreling much with the techniques of the transportation problem, still emphasized its flaws. The massing of so much shipping could never remain secret and would give the English at least ten days warning. If this concentration should take place before a declaration of war, England's naval strength in the Channel and other home waters would have to be reckoned with. The possibility of the French decoying them away was as unlikely as the chance of their being far from the British isles for three weeks, or more on peace time cruises, or maneuvers. The Admiralty would not even admit that certain contingencies which the War Office wanted to discuss were within the limits of reasonable expectation. For, as the navy men put it: "The landing of an enemy on these shores without interference on the part of our Navy—without the annihilation of the Channel fleet, our coast defense vessels, torpedo boats, and armed merchant cruisers, [is] a contingency so remote that it would hardly appear to come within the range of speculation."

Still, the Army was insistent on discussing with the Navy what it considered a possibility, even though remote: a French descent on the Thames-Portsmouth stretch of coast, with the Navy theoretically represented as absent in considerable force from the scene of operation. It posed four questions to the skeptical Admiralty: "(1) On what sections of the above coast is the attempt most likely to be made? (2) Supposing the enemy had six weeks time for preparation, what troops could be landed on any of these sections? (3) What would be the minimum time in which a division could be landed? (4) The whole maritime resources of France are assumed to be at the disposal of the enemy for performing these operations."

Considering one, or the other of these questions as too absurd to be discussed at all, the Admiralty declared (May 1890) that under no circumstances would an English naval force not be present, seriously to hamper and harass such an invading force.

> Composed as it must be of a number of merchant vessels varying in size, speed, and draft of water collected at different ports with no rendezvous of sufficient magnitude to enable them to be properly organized, drilled and all arrangements and contingencies provided for,—overloaded with troops, horses and the paraphernalia of an army of 100,000 men, and the ships commanded by Merchant Captains who had never acted in concert with other vessels but have been accustomed simply to look after their own individual ships. It would take a very large force of Men of War to guard such a fleet, and no force could guard it against the number of light craft of various sorts that could be sent amongst them at night. This large protecting Naval force could not be collected quickly and without knowledge on the part of the English Navy, and it can hardly be supposed that the British fleet would be decoyed away by an imaginary foe.
>
> My Lords have no hesitation in asserting that an invasion of England with 100,000 men would require weeks, if not months, of very careful preparation, and the very magnitude of the operation would increase the chance of failure. The accidents of weather must also be borne in mind, as however carefully plans were prepared a strong wind from S.E. to S.W. would seriously hamper a landing if it did not stop it for a time.

The outcome of the wrangles over the possibility of an invasion provided anything but a unified conception of British defense. Rather was it a bootless side-by-side survival of two conceptions having to do with the threat of invasion and the defense of Britain in general. The Army thought that there must be large home forces and fortifications able to withstand a formidable attack. The Navy considered that maintenance of sea supremacy was paramount, supplemented by a military force in the British Isles able to deal with a maximum probable landing force of from 5,000 to 10,000, assuming that it succeded in eluding the vigilance of the ever watchful navy. Each was intent on using the other arm in what *it* considered the right way, by use of the right strategy and in the right strength.

"Command of the Sea" Defined

Thinking of the Army in terms of its practical and traditional use, the Admiralty in 1901 defined the vague, ancient term, "command of the sea" as being that state of things when the Navy should be "able to transport across the waters commanded a large Military Expedition without risk of serious loss." And the Army thought in 1897, as Wolseley, the Commander-in-Chief wrote, that "we still have to convince the Navy that they can't win a war by themselves, and that we are not trying to nab the money they ought to have, but want to make our own power, what it must be to be effective, amphibious. If we can get the sailors to come in with us in this, we shall have some chance with the politicians." [8]

At about the same time Fisher believed that there was not a fair apportionment of funds between Army and Navy. Unfortunately, he thought, the military element was too powerful in Parliament and the people in mass had not yet come to realize the value of the Navy. But the Chinese-Japanese war then going on drove home the value of sea power.

> A Royal Commission of enquiry could point out as early as 1890 that "little or no attempt has ever been made to establish settled and regular inter-communication between them, or to secure that the establishment of one service shall be determined with any reference to the requirements of the other. . . . No combined plan of operations for the Defence of the Empire in any given contingency has ever been worked out or decided upon by the two Departments."

But the Committee of Imperial Defence which the Commission envisaged and recommended was not set up until 1903, and then under pressure which did not stem equally from the professional philosophy and ambitions of the two services.[9] It was quite different in Germany. There the watertight compartment system of the two services was allowed to persist. No over-all conception of defence, no combining permanent board, or bureau made a team out of the two. For with the Teutons the supposedly all-combining function of the Emperor as supreme warlord was nothing but a cloak for rampant, narrow-minded departmental bureaucracy.[10]

Every discussion of an invasion of England brought at once to the fore the problem of how to organize her defenses. More than those of any other great power must they be coordinated, or so it seemed to the majority of Englishmen who were qualified to express reasoned opinion, and in order to keep the home base intact—*Britannia intacta,* the island-fortress, "built by nature for herself against infection and the hand of war," the soil of which the foot of an invader had not pressed since 1066. Continental countries might be invaded and still win the war in which they were involved. Some, like Russia, must indeed be prepared to fight wars on their own territory, unless against Turkey. Only German strategy managed to avoid the violation of the frontier and, in a way, this created among the German people also an island mentality, since enemies never invaded Germany, except by air, from 1815 to 1944.

In treating of landings this passive side must be dealt with in summation up to the present war, when long stretches of accessible beaches on the south and east coast of England bristled with barbed wire defenses. Englishmen were mostly agreed that any attempt at defending the whole coastline, even on the east and south, was hopeless, wasteful and contrary to the principle of meeting the concentrated form of a landing corps with a counter-concentration. Of the latter an essential part must always be formed by the British fleet, able to concentrate in less than 24 hours, the time which, after the Crimean War, was usually considered to be the period required by a well prepared landing force to get ashore.

British Naval Superiority Imperative

To meet successfully an invasion force, the British fleet as a whole must be superior in strength, on the Two Power, or on any standard, to any other fleet, or combination of fleets. The threat of an invasion would usually bring up for discussion, or decision the question of whether conscription should be introduced for either the navy which had been forced to forego impressment of men, or the army, or both. Without impressment and with but moderate inclination of the seafaring population to serve in the fleet, could the latter be promptly manned in case of war as France and Russia were able to do, thanks to compulsory service?

Experiences with obtaining sailors provided an outlook none too hopeful. During the scare that swept Europe at the start of the war of 1859, (France and Italy against Austria), it took the British navy months, even with the offer of a bounty, to recruit 10,000 sailors. The situation was dangerous, considering that France had assembled that number within her ports in a fortnight, ready for embarkation. The strong and unsurmountable resistance of Victorian England against conscription would necessarily lead to dependence on various voluntary services by raw soldiers, accepted and organized with much misgiving by the country's military chiefs, doubtful whether they would be able to cope with the professional troops of the continental powers. Passing of the scare would again disorganize most of what had been set up.

The question of how strong the regular army should be and what part should be stationed at home to meet a likely invader with 150,000 or more troops came up, resulting in some decisions, but in even more indeterminate discussions. The necessity of improving the lot of the common soldier was appreciated to a degree that resulted in opening officer careers to the sons of middle class families who during the Victorian era were practically excluded from the services since they were reluctant to serve in the ranks and found no favor in the idea, voiced at times, that they should function as non-commissioned and warrant officers. Some class-conscious spokesmen, however, openly questioned whether the talents of soldiers coming from outside the aristocracy might not include those of a Garibaldi.

England's fortifications in the 19th century were specifically designed with reference to an invasion. They were few in number and served to protect on both the land and sea sides the great dockyards like Portsmouth. In case of an invasion these would serve as bases of naval operations against the enemy's communications, and in case of a defeat, or temporary inferiority of the English navy for assembling and refitting the old fleet and building and outfitting new vessels. In the late 1850's, the usefulness of Portsmouth, for example, was seriously limited by the circumstance that it was not also a naval arsenal, but was dependent for ammunition on Woolwich from which it might be cut off by the earliest operations of a French landing corps. This interruption would practically paralyze English warfare in the Channel and leave control of these waters to an invader even if an English fleet of appreciative combat value should still, or later, be in existence.

While the seaboard fortresses of England resembled in many ways the continental frontier fortresses, the question was raised *sub specie* of an invasion whether her defenses should not also include fortresses in the interior to protect the capital and other important centers of population, geographic lines, or fortified camps. They were intended by those who thought them necessary to offer resistance to an enemy who had succeeded in landing, and to gain time for English defense forces to form, compelling the enemy to waste precious time against them until a strong counter-attack could be prepared.[11]

The English invasion scare of the 1850's had its technical causation in the rise of ironclads with which France was experimenting. They would, so it seemed, sweep all wooden ships before them and prove even superior to land fortifications. Leading British naval officers testifying before a Commission on the Defences of the United Kingdom in October and November, 1859, were nearly unanimous that "forts could not prevent ironclad ships from forcing a passage through a clear channel unless the ships came under a concentrated fire at close range"; whereas the inspector general of fortifications had "no great confidence in iron-plated vessels . . . Improvements in artillery will go on faster than improvements in fortifying the sides of ships; and even with iron-plated ships the decks remain open." [12]

Steam Power vs. Inertia

War's "halt in the mud" could not forever withstand the pull, or push of steam power, however strongly inertia was implanted in the minds of the governing society, of which armies together with diplomacy and police were the mainstays. The application of steam power to maritime movement in war tended to enhance the mobility of land forces over that of sea power aided as well by large scale construction of metaled roads which had been begun by Napoleon. On the other hand, the movements of troops on land would be restricted to a degree by available railway, road and river transportation; whereas, theoretically at least, the introduction of steam power into sea transport would heighten very considerably the mobility of seaborne power. Actually, this mobility was lessened by the increasing draught of ships, which also increased their dependence on ports, quays and cranes for loading and unloading. Where the nature of the theater permitted, and where undisputed control of the sea was assured, sea transports could be used together with railroads for mobilizations at the start of a war. This was demonstrated by France in the war of 1859 when one-half of the army which was to operate in Italy was transported overseas to Genoa in order to advance from there into Lombardy against the Austrians. The mobility of sea power in war, its penetrative power was further reduced by the increasing shyness of ships to land-based artillery, although this relative strength was more taken for granted than tested, or by the non-evolution of special landing craft.

The Crimean War was the first of the various wars served by but also caused by steam power, by industrialism and its toxins, and by the opinion of the masses, so far as England was concerned. These were deeply disturbed by the progress towards an oppressive rationality and away from the feudalism of their world of feelings. As an overseas war it was the greatest of its kind in the 19th century, that is, if measured by the logistic product of the number of soldiers engaged, multiplied by the length of waterways over which they and their supplies were to be carried and the number of days that the war lasted. Everything connected with the conflict was on an unprecedentedly large scale and called for much experimenting. The steam age had only partly come into its own, and in order to keep fleet and convoys together the steam-propelled craft had to take the sailing ships in tow. The English navy as a rule declined to take army personnel or materiel on board as the French did. Generally speaking, transport proved the best feature of supply and rearward connections. Other combined operations were far less fortunate, or effective.

After a flotilla of some 150 war vessels and transports convoyed by battle fleets had brought the original invasion force safely across from Varna to the Crimean coast, which took a whole week, obsessed with the idea of providing the utmost safety for the naval vessels and an undisturbed debarkation the Allies disembarked the troops on a beach north of Sebastopol where, however, no harbor, or bay usable in subsequent operations was to be found. The original landing in Old Fort Bay was uncontested, the secret of the destination of the fleet having been well kept from the Russians. While it took less than one and one half hours to land the French First Division, and the fleet brought the 45,000 men, 83 cannon and several hundred horses to the shore in eleven hours, the more difficult task of getting the trains to the shore and poor weather prolonged the undertaking for five days until September 18. On the 19th the Allied army, more than 50,000 strong, marched south, paralleled by the movement of its fleet.

They were met on the next day at the Alma river by the Russian commander who had been unwilling to believe in the Allies' intention to land and whose strength was only half as great as theirs. Allied warships were within range of the battle, but since no detailed arrangements for their cooperation had previously been made their presence was an inconsiderable factor although the situation bore some resemblance to the battle of Pinkie, which was a better fought action. According to the Prince de Joinville, some artillery support on the part of the fleet was rendered, "a thankless service, however, for given without incurring danger it did not bring in any glory." [13] Instead of trying to throw back the Russians to the seashore by attacking their right wing with the main force, the Allies allowed them to retire to the interior of the Crimea which forced the invaders to get away from their own secure base, the Allied fleet. Not until the lapse of another three days did the Allies resume their march.

Sevastopol Siege Begun

If they had originally hoped to take Sevastopol by a *coup de main* from the north, strongly supported by the fleet which was expected to stand in for the harbor, this hope was now blocked by the defenders' sinking several of the ships of their Black Sea fleet in the harbor's mouth, as much as by the attackers' own timidity. They perforce came to realize that a siege could not be undertaken from the north, for there was no base for extensive operations there, and also that reembarkation in order to attempt a new landing elsewhere was impossible.

They therefore resolved on the 24th to march around Sevastopol and start a regular siege from the south, based on good harbors, Balaclava for the British and Kamiesh for the French. Instead of attacking at once from the south, where no works of a serious character offered opposition, the Allies decided to develop a systematic siege. This gave Todleben his opportunity of building up unorthodox defense fortifications, which had not really existed before, consisting of earthwork largely armed with cannon taken from the Black Sea fleet. All combat-effective Russian sea power was thus converted into land power.

Several attempts by the Allies to abandon the Sevastopol front by making use of their vastly superior fleets and to try their luck elsewhere were either given up, or proved futile. The Turkish forces along the Caucasus frontier proving unable to withstand the Russians. The British and the French considered supporting them there, but were forced to realize that the maintenance of supply lines and the problems of basing would be even more difficult there than in the Crimea. Another plan, conceived by Napoleon III himself, intended to attack the Russians posted behind Sevastopol, near Simferopol, where the line of communications with the interior of Russia was maintained, and envisaged a new landing north of Sevastopol and a concentric attack from at least three directions.

Since no common supreme command between the several Allies had been established, or seemed likely to emerge from an understanding among the generals themselves, the Emperor was ready to go to the theater of war in person and take over supreme command. The troops worn out before Sevastopol during the winter of 1854-5, were judged unequal to this grand design. Instead, as far as the cooperation of the navies was concerned, an expedition was carried to the Straits of Kerch and Kinburn. The entrance to the Sea of Azov was forced and the Allies dug in at Jenikaleh and Cape Paul while a squadron went in to interrupt the Russian supply lines as far as they ran through the Azov Sea, destroying supplies and bombarding coastal towns. But they did not succeed in interrupting traffic across the Bay of Arabat, the main line of supplies for the Russians next to the one across Perikop isthmus, although those lines were strewn as much with Russian dead as the works of Sevastopol.

The main, or at any rate new, vehicles in these successes, as far as they went, had been three French iron-clad floating batteries used against Kinburn, opposite Ochakov. Firing some 1,000 shot and shell each at distances ranging from 875 to 1150 metres during some four hours, they had been able to set a fort on fire, silencing most of its guns and breaching its walls in several places. Much impressed by the performance, the British admiral reported: "You may take it for granted that floating batteries have become elements in amphibious warfare, so the sooner you set about having as many good ones as the French the better it will be for you." [14] However, progress in naval construction favored the ironclad for high sea fleets and war strategy veered rather away from than towards amphibious enterprises.

The Russians gave in after they had ascertained that in the long run it was hopeless and too costly to expose themselves to the material superiority of western industrialism, although it only very slowly shifted to war production, as in England. "If Russia had had a railway to Sevastopol in 1856, the war would certainly have had a different outcome," as Moltke subsequently wrote.[15]

The events in the northern theater, in the Baltic, had but little influence on the result, although perhaps a good deal on one aspect of combined operations, the comparative strength of land-based guns and ship artillery. The lessons from the two campaigns of the Anglo-French squadrons in the Baltic, which had sailed out under high expectations on the part of the public, have usually been summed up by saying that the ships were unable to do anything against the fortresses along the seaboard. "It is a fallacy which this campaign was not the first to demonstrate, to suppose that ships alone can wrest great maritime fortresses from a determined foe." [16]

Actually, nothing of the kind was even attempted and hence still less demonstrated. The naval authorities, the Admiralty and the admirals in command, concluded before the fleet sailed that the granite fortresses of Cronstadt and other Russian sites were impregnable. "We feel—no one more than myself—that nothing can be done against such places as Helsingfors," wrote Admiral Berkeley, one of the Lord Commissioners, in June to Sir Charles Napier, commanding the fleet during the Baltic campaign in the summer of 1854, when he refused to attack Cronstadt.

After a few months of fruitless cruising around and the definite refusal of the naval officers to expose their fine big ships too close to the artillery of any fortress, a French land force division was brought to the Baltic and landed on the Aland Islands to attack Bomarsund. The Russians did not oppose their landing and the invaders proceeded to batter Bomarsund exclusively from the land side with siege guns. "On all sides the greatest disgust was expressed for the modern system of naval warfare, the principle of which seemed to be to keep out of gun-shot," as a participant expressed the general sentiment.[17] After five days Bomarsund capitulated. Nothing much more happened in the Baltic except a few exchanges of shots between the Allied fleet and the Russian land batteries, before which the 70 Allied warships engaged in the campaign of 1854 cautiously retired.

While the participants of the expedition were impressed chiefly by the tactical-technical difficulties, some observers of the Crimean War, such as the Prince de Joinville, dazzled by the strategical temptations which combined operations have at all times offered, believed that the experience was promising as regards the future of such operations. Even despite the improvements in communications, including better spying, Joinville was convinced that transport of troops overseas could easily be kept secret at least as concerns its destination, which had been the case during the undertaking against Kinburn. "There is hardly a region in Europe which does not have its Kinburn today," he wrote in 1859, after the transition from sail to steam had become ever more complete, "which is for the peoples without a navy a manifest cause of inferiority. . . . Mixed operations bring into play on the military chessboard new pieces whose sudden, secret, irregular move upsets all calculations and promises to him who disposes of such means of action undeniable advantages." France was such a power; it would have 50,000 troops and the necessary naval equipment available for such an undertaking in a continental war. Hence asked the Prince:

> Is it saying too much to presume that it would spread terror? Would one not be on the lookout every day and on all points for this redoubtable fleet and would not this by itself be a very real evil for the enemy who expects it? If the population on the littoral are devoted to their governments, they will be consumed by a painful anxiety; if

> they are discontented and unkindly disposed, they will expect the arrival of the French armament as a signal of their deliverance. In either case, all will be victims of the agitation and the threatened government can do nothing that will not tend to increase it. . . . The telegraph which says everything today, has nothing more to say about the movements of a fleet on the high seas.

The ex-Admiral concluded that France, provided she had England's tacit permission, a necessary premise, could hope to disembark her troops freely on such coasts as those of Prussia and Austria, who possessed no fleet to speak of, without fear of being disturbed and rely on surprising the enemy, as successfully as had been the case in the Crimea. Railroads were no check for a freely moving navy, this great and novel element of military force, by transporting troops in sufficient numbers to the landing shore. Neither could troops direct damaging fire against a fleet that controlled the shore with its own guns.

In his enthusiasm Joinville even thought operations feasible of such a fleet on the great rivers of Europe, although actually river wars at best were possible only on the mighty rivers of China and America. Thought and speculations as to the uses of new weapons like steam navies are above all fascinating, and apt to be misleading as well, when opposed to the actualities of real battles fought with these weapons; battles that were not to take place until after de Joinville wrote. There was indeed among the French studies for a war against Prussia-Germany before 1870 at least one dealing with a line of operations from Cuxhaven to Berlin up and along the Elbe River (dated Oct. 10, 1867).[18] But the French forces were not to be put to such uses.

19TH CENTURY EXPERT'S OPINION

One of the few military writers of the century to note and discuss the influence of steamships on landings in war was the rather unconventional W. Rüstow (not to be confounded with his brother Alexander, author of a work on coastal war, *Der Küstenkrieg*. Berlin 1848). In his book on 19th century generalship *(Die Feldherrnkunst des neunzehnten Jahrhunderts.* Zürich 1857, 491-2) he wrote:

> Large scale landings, the transportation of whole armies far across the seas into foreign countries, have always been

considered extremely difficult operations, and their difficult character has by no means been relieved altogether by the introduction of steam shipping and the daily growth of steam shipping, but as far as this difficulty was formerly rooted in the imperfection of the means of transportation, the slow movement of the latter and dependence on wind and weather, it has been removed for a very large part. Not only are the steamships themselves independent of the wind, they can make a whole fleet independent by taking sailing ships in tow. The quickness of their movements makes surprises on the part of the enemy more feasible than before, surprises from which the success of landings so often depends; the same quickness makes possible quick communication, safe rear connections—as far as this is possible under such conditions at all—of the army debarked on a foreign shore, with its mother country. Into the midst of the enemy country, however, the effects of a navy cannot aid the land army even today; the former remains bound to its element, the sea.

How great, therefore, the advantage is which arises for landing operations, and large scale operations from the fact that sailing fleets have been replaced by steam fleets, which in turn enlarge trade and are enlarged by trade, will depend on the question whether the objective of such an expedition is to be looked for immediately along the coast or farther inland, whether with other words landing and transportation across the sea is the main thing or rather the operations which must follow the landing. . . . To all this must be added the easy possibility for the holder of the sovereignty of the sea of making diversions, by which he constantly divides and misleads the enemy's attention.

This facility is clearly enhanced by steamshipping which not only makes surprise possible in a much higher degree but also allows landing larger corps of troops than before at one time and, in case of need, quickly retire them—in such a way that the disadvantage of diversions, weakening the main force at the really decisive point, can be obviated since under a number of circumstances steamships would allow to bring the troops used for the diversion back to the main force and for the decisive moment. (The use of steam shipping to effect diversion, which in the beginning was expected by several writers with some reason, never really came about.)

Much further, generalizations about steam power and landings should hardly proceed at this point. Specific problems obtrude as they arise from specific military situations and the defense policies of the various great powers, such as: Did the introduction of steam locomotion on land strengthen English defense relative to a steam-propelled invasion from France? The latter seemed a likely threat when conservativism in England hesitated to go ahead with the construction of a steam navy, or an iron-clad navy. A great deal would depend on the density of a railway net and its orientation with regard to the coastline. Since Russia did not possess a railway to Sevastopol at the outbreak of the Crimean War she would have done far better if, instead of exposing her weakness at the terminus of a poor highroad, through what was in part desert country, she had retired into the interior and drawn the Allies after her.

Naval Demonstrations and Landings

The vast majority of warlike landings in the century after 1814 were incidents of colonialism and imperialism. Often landings would be undertaken in order to back up diplomacy as a second step after the customary first, the *persuasion en rade,* after a threatening demonstration by warships off the coast of some recalcitrant overseas government, or even a bombardment had proved insufficiently strong as arguments. In most cases exotic peoples were unable to resist such a bombardment as that of Lord Exmouth against Algiers in 1816 which set afire the Dey's fleet and arsenal and smashed all seaward defences. This was one of the last examples of the era in fighting and reducing land defences with ship artillery. It had been preceded by Admiral Duckworth's forcing of the Dardanelles in 1807, an enterprise more fortunate at the start than on the return when the British suffered considerable losses. The resistance of exotic peoples against landing forces would mostly take place not along the shore, but in the interior, in bush fighting, where topography, climate and disease would prove the strongest allies of the defenders, such as the Ashantee, Burmese, Maori and various South African opponents of the British; the Tonkinese and Filipinos, Hereros and Abyssinians. The French conquest of Algiers beginning in 1830 met its chief resistance in the interior, after a very carefully planned landing had brought the invaders to the shore before the Algerian forces had arrived there.

On one of these overseas expeditions towards the outer rim of western civilization steampower first figured in a landing operation and with a tactical application of the additional mobility which the steamship gave to sea power. Diplomacy united England and other Turkophile powers of the moment in an expedition to support the Ottoman Empire against the rebellious Mehmet Ali, Viceroy of Egypt, in 1839-40. An allied fleet operated along the Syrian coast against Mehmet and prepared for a landing on September 9, 1840, in the D'Jonnie Bay near Beirut.

> At dawn of day the fleet weighed, and followed by the steamers with their cargoes of troops stood towards Beyrouth Point. The Egyptian forces at once hastened to the threatened point, and as their dark-clad columns concentrated in that direction, shot after shot from the ships fell among their ranks. Having succeeded in drawing the greater portion of the enemy in this direction, about noon, when the sea breeze set in, the steamers were despatched full speed to the other extremity of the Bay, some ten miles distant. . . . It was impossible for the Egyptian troops to change their position soon enough to oppose a landing so far distant and the whole of the troops were on shore by 4 P.M. without firing a shot.[19]

This outmanoeuvring of land power had become possible through the greater mobility of sea power in the absence of railway power, and its increasing emancipation from control of the elements. It enabled the allied powers to force Mehmet Ali to surrender and retire into his "original shell of Egypt."

The several wars fought for the preservation of the "original shells" of the European state system, whether of the Vienna settlements of 1814-15 or even previously, were largely undertaken by the conservative powers against their intention. They tried to stem the forces of liberal political progress who, in their own way, were more guilty of precipitating war than their opponents. This is as true of the Crimean War, which the British people wanted more than many of their leaders, as of the first phases of the American Civil War, of the wars growing out of the famous Schleswig-Holstein imbroglio, or the Italian question. Whether in these wars the more modern force was represented by the participating armies, or by navies raises a rather interesting problem in connection with the social history of war.

So far as there was any cooperation between the two it can hardly be called intimate, or under a clearly defined strategic concept, while tactical cooperation proved difficult enough. It was exceptionally effective in the Danish defense of Fredericia in 1849 against a force of Schleswig-Holsteiners whose cause and belligerent efforts were hamstrung by adverse diplomacy. The Danes, pressed back from Schleswig and Holstein to the island of Alsen and the northern tip of Jutland, concentrated their greatest strength first on the island of Fünen opposite Fredericia and then transferred it to the town itself. The German troops stationed opposite Alsen failed to observe the withdrawal of half the Danish forces and those posted farther north remained equally oblivious to the fact that they were faced by only one Danish battalion instead of a brigade as before.

Danish Ruse Against Prussians

Since Fredericia had no bomb proof casemates, the Danes unmolested by the besiegers had often relieved the troops within by boat and enabled them to rest in Fünen. The besiegers mistook this traffic in and out of Fredericia and did not suspect that the fortress was being heavily reinforced. Following a two months' siege and after 24,000 Danes had been brought into the place, the latter made a sortie against the 14,000 besiegers who were quickly forced to retire after heavy losses—3,000 killed or prisoners—leaving their siege train behind.[20] How far the diplomacy of Prussia, who wanted to retire from the war which it had fought much as did the intervening powers in Spain's civil conflict in the latter 1930's, and how much the stupidity of German leaders had to do with this stunning defeat, history seems yet not to have fully settled.

Various landings on European and American shores favored liberal principles and efforts, as they were carefully advertised by their entrepreneurs in both hemisphere; or as they more cautiously proclaimed, were directed against advocates of the *status quo,* such as the Holy Alliance, or other aggregations of conservative powers. These landings, in part supporting nationalistic movements, had various strange forerunners in the 18th century, when the British lent support to Paoli in Corsica, or the Russians to the independence movement of the Greeks in the Morea during the early advance of Russia into the Mediterranean in 1769 "for the sake of liberty." Orlov, at the time Catherine's favorite, brought a fleet to the Aegean and landed in the Morea to organize a revolt against the Turks. The attempt failed, Orlov retired to his ships and left the Greeks to the far from tender mercies of the Turk, although subsequently (1774) Russia acquired a vague sort of protectorate over their co-religionist Greeks in the Ottoman Empire and with it an entering wedge for future intervention and wars.

Louis Napoleon's abortive landing at Boulogne in 1840 was by many Frenchmen assumed to have had British support and evoked an inclination to retaliate with invasion plans against England. So did the Orsini plot of 1858 on the life of Napoleon, who was bombed on his way to the Paris opera, which had actually been prepared in England, "the lair of assassins," as it was called in an address of his colonels to the sovereign, that might one day have to be cleaned out.[21] The Carlist civil war in Spain (1833-39) drew many military adventurers from overseas, but also varied foreign governmental support, openly, or clandestinely, in much the same way as that extended by Hitler, Mussolini and Stalin in the 1930s to the Spanish parties, although a hundred years earlier liberalism as an aggressive force gave it aid openly, and England then knew where her interests lay better than she did later. According to Palmerston his government "by assisting the Spanish people to establish a constitutional form of government, were assisting to secure the independence of Spain and they had no doubt that the maintenance of that independence would be conducive to important British interests."[22]

America Vulnerable to Landings

In the western hemisphere small populations and still more inconsequential military establishments, unstable governments and poorly policed shores for long periods countenanced military landings either in favor of, or against political liberty, such as Miranda's, Bolivar's, San Martin's or William Walker's, the latter in military history being one of the last *condottieri* to attempt setting up a dominion of his own on foreign soil. Even the most stable government in the Americas, the United States, seemed highly vulnerable to amphibious enterprises on the part of the sovereign of the seas. So at any rate the blustering and imperialistic Palmerston was convinced. Hearing that Americans thought Britain to be weak in Canada, in 1855 he considered them "still more vulnerable by us in the Slave States, and a British force landed in the southern part of the Union, proclaiming freedom to the Blacks, would shake many of the stars from their banner. . . ."[23] It is hard to believe that this would have developed to be the profitable military, or political line to follow had Britain then launched an amphibious enterprise against the United States.

The 19th-century virtuoso of landings in liberty's cause, the maestro of combined operations for freedom, one of the few remaining warriors who would and could fight with determination and success equally on land and sea, as no conventionally trained soldier would, was Guiseppe Garibaldi. As a partizan of liberty and liberalism his main and broad strategic principle was simple and direct—vigorous and unremitting war on despotisms. This was based on the conviction, as his chief of staff in 1860, W. Rüstow, put it, that against despotically governed states one must always make aggressive war because their whole strength is in their professional armies, and that they cannot as a rule expect any considerable assistance from civilians among their fellow countrymen. If one allows them to attack first, they lose little, or nothing materially, but profit considerably in morale by the first successes they are permitted to gain. If one successfully attacks them, their strength is reduced with every step nearer to their capital accomplished by the liberals and once it is taken, resistance is broken and victory is attained with the downfall of the despot.[24]

Garibaldi's adult life had much to do with combined operations. Born in Nice, brought up in the coastwise trade and the Sardinian navy, he and a number of other members of the Young Italy movement attempted to seize a frigate in order to cooperate from the sea with an expedition Mazzini was to undertake from Savoy. The plot was discovered and Garibaldi forced to flee to South America where he fought for liberty, or at least in revolutions on land and sea during most of the years from 1836 to 1846. Returning to Nice in 1848, he served Piedmont which accepted his services and those of a volunteer corps, but reluctantly.

When Piedmont and the movement for Italian unity was overcome, he fled to New York where he made a small fortune.

With this money he bought the island of Caprera in 1854 as a base of his further military and political operations. He fought as the leader of Alpini in the war of 1859. When the diplomacy of Piedmont forbade him to invade the Romagna, he retired to Caprera to sulk and plot with Crispi and others the invasion of the Two Sicilies, the other remaining stronghold of despotism in Italy. An insurrection started in the island in April, 1860, and after some hesitation Garibaldi sailed with two ships to join the rebels in May with his Thousand, protected in his landing by two English ships and encouraged by English sympathies for this filibuster, Lord John Russell reminding the Commons that "we had once a great filibuster who landed in the month of November 1688 and to whom all the people of England flocked."

Having proclaimed himself dictator in Sicily, thereby arousing the suspicions of Cavour in Piedmont, since in those days filibusters were loose in the world trying to set up regimes of their own in Central America, or in China, Garibaldi at last routed the royalist troops and prepared an army for the invasion of Apulia and Naples. After all Sicily had been subdued and Messina taken, Garibaldi, instead of crossing the Straits, went secretly to Sardinia to lead an expedition prepared there against the Papal States. Too fearful of the international implications of the Roman question, Piedmont forced the expedition to go to Palermo instead.

Returning to Messina and asking Victor Emmanuel's "permission to disobey" without waiting for an answer from Turin, Garibaldi crossed the Straits and beat the Neapolitan troops at Reggio di Calabria on August 21, following the route which the British Eighth Army took in 1943 and fighting and by diplomatic persuasion making way much like the latter and at about the same time of year. He entered Naples on September 7 and on October 1 smashed the last of the Neapolitan armies on the Volturno, where the Italian government and troops kept him from entering the Papal States. Much to the relief of Cavour and his royal master, Garibaldi retired to Caprera again after having accompanied the King during his entry into Naples and disbanding his Red Shirts.

Garibaldi and "Rome or Death"

Soon out of sympathy with the time-serving methods of Piedmontese diplomacy, Garibaldi set out once more from his island

and landed in Palermo to gather an army under the slogan of "*Roma o Morte,*" Rome or death. As the Italian Government thought it still inappropriate to solve the Roman question with its religious complications they surrounded his three thousand with regulars and dispersed them, inviting Garibaldi to retire once more to Caprera. After having fought with his volunteers at Custozza in 1866, emerging from the Italian defeat with more credit than the regulars, he prepared once more a design on Rome, by now evacuated by the French troops, but the government arrested him on the continent and brought him back to his island whence, eluding Italian cruisers on the watch for him, he returned once more to Tuscany, this time with the connivance of the Italian Government, and marched into Roman territory on October, 1867. Papal and French troops defeated him and Garibaldi was once more returned to Caprera by his government, this time to write romances in order to replenish his funds. His last departure from thence was for the purpose of serving the French Republican Government in 1870-71.

The situation of the South in the American Civil War was in many respects similar to that of Russia in the Crimean War—spacious territory, thinly populated, a railway system underdeveloped, inferior in all that steampower, or capitalism confers for war purposes, largely cut off from what war materials might have been supplied by neutral countries. From these features stemmed the cause and effect of the landing enterprises of the North. When the Union forces took the principal ports of the South, they eliminated the mainsprings of its power. The landing operations of the North along the coast were therefore of great strategical consequence, although of small tactical interest since they took place against slight enemy resistance.

In the fighting across and on the seas, ingenuity not hitherto dedicated to war was applied to it by either side, producing ironclads and by the North the "double enders," reversible paddle-wheel steamers especially designed for service in narrow waters where turning around was difficult, or impossible, as in inlets and estuaries. The construction of such novel craft is always an indication that it has been designed specially for war, instead of making existing types serve the purposes of war, even if indifferently.

The first attack carried over the sea occurred as early as August, 1861, when by an expeditionary force of 1,000 Union men, proceeding from Fortress Monroe, the forts controlling the entrance to the Pamlico Sound were taken, thus establishing a powerful support for the blockade of the North Carolina coast. Early in November an army and navy force proceeding from Chesapeake Bay, including 35 transports carrying 12,000 men and convoyed by 17 men-of-war, bombarded and captured Forts Beauregard and Walker at Port Royal, S. C., driving the Confederate flotilla into the bay where the Confederates themselves set it on fire.

Early in 1862 a further number of sea coast forts in Georgia and Florida fell to Northern superiority on the sea, although Wilmington, Charlestown and Savannah and several ports on the Gulf coast remained open for blockade runners. Roanoke Island was taken, whence Federal troops advanced as far inland as Newbern (Feb. 8 and March 14), and at nearly the same time Fort Pulaski, controlling the river entrance to Savannah. Attempts against Charleston under Commodore Dupont and General Hunter with 7,000 troops that might have done better service at the time in Virginia, failed, first in May, 1862 and again in March-April, 1863, when a concentration of monitors was unable to knock out the city's strong forts. Subsequent siege also failed.

Farragut in the Mississippi

The greatest combined operation of the year under joint leadership of Commodore Farragut and General Butler, with a force of 46 ships carrying 300 guns and 14,000 soldiers, was directed against New Orleans. The flotilla forced its way up the Mississippi despite fire from permanent forts of masonry and brick construction mounting rifled and smooth bore cannon, and against a boom across the river. "The unexpected happened. At that time the eternal duel of ship versus fort seemed to have been settled in favor of the latter, and it was well for the Union government that it had placed its ablest and most resolute officer at the head of the squadron." [25]

By the North, McClellan was hailed as "the young Napoleon," preparing in the camp around Washington a stroke against Richmond similar to that which the Frenchman had never been able to deal England from Boulogne. Naval preponderance allowed the North to hope for a successful landing on the Virginia Peninsula to aid McClellan. Since the Confederate force north of Richmond seemed too strong and well entrenched to be taken by frontal attack, the main army of the North was to be carried by sea to Fortress Monroe, from thence to advance on Richmond.

The transport flotilla was assembled at Annapolis and the embarkation began on March 17. By April 6 more than 100,000 troops were embarked on upward of 400 steamers and sailing ships, a number indicating the quantitative rise in man and transportation power in warfare, for this was the largest single waterborne expedition to that time; to the 1944 launching of the Sicilian landing in fact. On the well known story of the Peninsula expedition only a few notes of emphasis are needed for our special purposes—the vicinity to the theater of war of Washington, which the governing powers did not want to see stripped of troops to the extent that McClellan had planned, or that he had been given reason to expect, the complete loss of surprise advantage, caused by the defenders of Yorktown who held up McClellan's progress for a whole month; this was partly due to the declared inability of the Navy to lend its support, as its main strength was absorbed by blockade duties.

Added to these was the improvised character of the whole enterprise, based on excessively imperfect knowledge of the terrain and of the enemy force, which most of the time was overestimated. The original landing was unopposed but the subsequent advance towards Richmond proved almost as difficult as Asiatic and Pacific jungle fighting in World War II, with the region as unmapped as few South Pacific islands can have been. The Northern army allowed itself to be drawn into this region of swamps and rivers and few, if any, roads and away from nearly everything that gave it superiority, suffering a dangerous increase in the vulnerability of its rearward communications which, being fleet-based, were already precarious. The most decisive stroke in the Civil War was not directed from the sea, but to the sea—Sherman's march to the sea. This was the great American anabasis, or rather katabasis.

The relations of land power and sea power were appreciably reversed during this war, a sign of its unconventional, or uninhibited character. The "Father of Waters" on his broad back carried sea power, or what was designed for sea power, which was then a synonym for Northern industrialism, deep into the traditional domain of land power to an extent that could only be equalled in China, in the Taiping War, or in the Japanese war of the 1930s. This created highly unconventional conditions in warfare, as when a Northern admiral during the siege of Vicksburg took ships, including iron-clads, through inundated woods where some stuck in the mud and had to be dug out by the army. The same admiral wrote to the Navy Department at the time: "Hurry up your small steamers that will run on the grass in the dew—I want a private one of my own." [26] This humoristic anticipation of amphibious vehicles was condemned to wait for some eighty years for fulfilment.

NOTES, CHAPTER 31

[1] Major C. E. Callwell, *The Effect of Maritime Command of Land Campaigns Since Waterloo.* London, 1897, 29.

[2] Lloyd George, *War Memoirs.* III, 222-3.

[3] The writings of Joinville are collected in *Essais sur la marine francaise, 1839-1852.* Paris 1853; *Etudes sur la marine.* Paris 1859 and 1870; *Vieux souvenirs.* 1894.

[4] The classic work about these developments is James Phinney Baxter, *The Introduction of the Ironclad Warship.* Cambridge, Mass., 1933.

[5] Gustav Mayer, *Friedrich Engels.* The Hague, 1934. II, 34-7.

[6] E. S. Creasy, *The Invasions of England from the Saxon Times.* London, 1852.

[7] R. W. Seton-Watson, *Britain in Europe,* 1789-1914, 415-9.

[8] Maurice & Arthur, *Wolseley,* 285-6.

[9] The above is largely based on Arthur J. Marder, *The Anatomy of British Sea Power.* New York, 1940, ch. V.

[10] For the details of this see Vagts, *Land and Sea Power in the Second German Reich. Journal of the American Military Institute,* vol. III.

[11] For the above discussions see J. E. A., *The National Defences. Frazer's Magazine,* Dec. 1859; General Shaw Kennedy, *Notes on the Defences of Great Britain.* 1859.

[12] Baxter, 160.

[13] *Etudes sur la marine.* Paris 1859, 324.

[14] Baxter, 83 ff.

[15] *Die deutschen Aufmarschpläne,* 1871-1890. 32.

[16] Callwell, *Effect of Maritime Command,* 180.

[17] Rev. R. E. Hughes, *Two Summer Cruises with the Baltic Fleet.* 1855.

[18] *Revue militaire,* July 1900, 513.

[19] Field, *Royal Marines.* II, 43.

[20] Rüstow, *Feldherrnkunst,* 714-6.

[21] Seton-Watson, 373.

[22] Philip Guedalla, *Palmerston.* London 1926, 181.

[23] *The Panmure Papers.* I, 403.

[24] Rüstow, *Feldherrnkunst,* 51.

[25] Encyclopedia Britannica, XIth, ed. XIX, 531.

[26] Bernard Brodie, *A Guide to Naval Strategy,* 2nd. ed., 171.

Chapter 32

OVERSEAS INVASION-FEAR AS A STATE OF MODERN MIND

"The cry is still they come"—*Shakespeare*

THERE ARE AMPLE REASONS to support the premise that mass fears and hopes, beliefs and superstitions are really the great constants that persist through all history, that although new technical circumstances and other fresh irritants may play on them, they still form the responsive instrument. Fears of sea invasions were as common after the last great military revolution in Europe brought about by the German victory over France in 1870-1 as before. In its way, the so-called international law of war preserved these emotions, for contrary to popular suppositions, the "next war" according to its rules need not begin with a declaration of war, England herself to then not having delivered one since 1762. War very possibly might begin with an abrupt invasion and still be legal.

Old though the fears were, the possible enemies of England, singly or combined, were new. And equally novel they also were in their ways of expressing, or provoking apprehensions. Napoleon III or Napoleon I, a Napoleon was *the* invader. Where conscious political consideration laid low this spook, spiritistic mediums would still insist that "Napoleon will invade us." [1] As soon as this invading emperor was gone and a new emperor was in the world, early in 1871 a Mr. Cowper-Temple, speaking before the Liberal Association of Portsmouth, declared: "We must contemplate the contingency of a combined fleet coming from the ports of Prussia, Russia and America, and making an attack on England." [2]

One of England's 19th-century minor poets, of the quite rightly so-called "spasmodic school," Sydney Dobell (1824-74), who during the Franco-German War wrote such violently Germanophobe letters that the London papers declined to print them, challenged:

Russian, Yankee and Prussian,
Wherever you be,
That stand by the shores of our sea
And shake your fist over,
This is the Castle of Dover,
You knaves! [3]

Not himself distinguished for emotional stability, Laurence Oliphant, reporting the Franco-German War for *The Times,* wrote to a member of the British aristocracy after a prolonged stay at German headquarters, that he had not found much sympathy there for England, nor expressed any. Without mentioning the widespread resentment in the German army against English ammunitions sales to France, without which Gambetta could not have carried on the war, he thought that:

> The official or Junker class detests England with a mortal hatred because they instinctively feel that the institutions of England strike at the root of their various class prejudices and bureaucratic system. The liberal party in Germany is only waiting for the war to be over to assert themselves and I think a German revolution will follow very closely upon the heels of the French one. . . . The feeling against England among the Germans is increasing every day, and it is amusing to hear them discuss plans for the invasion of England. They have worked the whole thing out: Blumenthal told me he had considered it from every point of view, and regarded it as quite feasible. On the other hand, the French are a danger to no one any longer. [4]

This is not convincing as far as concerns Blumenthal, the chief-of-staff of the Crown Prince's army; for both he and the Prince were the most Anglophile among the German officers. Perhaps it all amounted to no more than clumsy joking in German GHQ during the long winter months before Paris, on the part of officers who, without bearing any special love for England, did not consider her a likely enemy, since there were others, more pressing, on the continent.

The 19th century saw the rise of a new type of officer, the fluently writing military author, who, perhaps poorer in purse than most of his compeers, would eke out his pay, or income by literary industry. The elder Moltke wrote short stories and translated Gibbon into German; Roon published a best-selling geography; von der Goltz, aside from serious military history, wrote dime novels and articles on military affairs including parades, the latter, if necessary, days before the event, but correct in detail. Several times he was on the verge of abandoning his military career and entering the publishing business. Several of the military writers contributed to a new sort of what might be described as anxiety literature, writings about future wars, some of them technical, some in fiction form. Among the latter was Boulanger's son-in-law, Driant, who wrote under the thinly disguised pseudonym "Capitaine Danrit," and as such composed a five volume installment novel—5 centimes *la livraison*—under the title, *La Guerre de Demain. Grand récit patriotique et militaire* (Paris n.d.ca. 1900).

NOTABLE SCARE LITERATURE

One of these officer-authors seems to have been the originator of a new kind of thriller, the novel dealing with an imaginary invasion from the sea,—Lieutenant Colonel (later General Sir) George Tomkyns, with his *The Battle of Dorking,* first published in Blackwood's Magazine in July, 1871, and tactlessly at the time when Victoria was entertaining her son-in-law, the German Crown Prince, one of the most convinced German Anglophiles. The author, an officer of the Royal Engineers, impressed by the recent German victories, told how in 1875, when the British government with an insurrection in India and a conflict with the United States on its hands, would in addition quarrel with Germany over Denmark and declare war on the Reich. Germany would land an army in England, defeat what remained in the island of the British army together with the volunteers in the battle of Dorking, not far from London, and then impose a Punic peace on England.

This novel was altogether out of line with the official British policy of maintaining a good understanding with Bismarck. Other fictional productions might be considered more in keeping with the military policy prevailing at the Horse Guards, such as *The Siege of London* by someone who chose *Posteritas* (London, 1885) as his pen name. The story described the conquest of England by France, the siege and fall of London and the loss of the colonies. It is said to have been the last book that General U. S. Grant read before his death.

A subordinate of Tomkyns in the Royal Engineers, George Sydenham Clarke, later Lord Sydenham of Clarke, gave independent treatment of the theme of invasion, again a French one, in an anonymous novel, *The Last Great Naval War. An Historical Retrospect* by A. Nelson Seaforth (London, 1891). Here is an example of its general vein:

> After the great fleet action . . . . we began to strike hard all over the world. The number of troops employed

> was, from the continental point of view, obviously insignificant; but these British troops could be carried across the seas at will, while France could not transport a corporal's guard from Toulon to Algiers or Tunis without risk, and nearly a million of men lay perforce inactive though their comrades were being outnumbered over half the world . . . . France was unconquered. Yet every day the merciless pressure of the British Navy was more severely felt, and outside of France there was no place where the French flag was secure. Attempts on the part of labor on both sides of the Channel to fraternize failed. The whole incident appears exceedingly significant as a first attempt of the working classes of two countries to deal directly with each other during a period of war.[5]

William Le Queux's *Great War in England in 1897* (1894), first published in a penny magazine during the scare of 1893, and afterwards in book form with a preface by Lord Roberts, and reaching eleven editions by 1896, depicted the consequences of the *guerre de course* such as the French navalists contemplated in case of a war with England—tremendous price rises, distress and civil discontent, starvation, serious interference by domestic troubles with the conduct of the war. Its cover shows a red-coated soldier, presumably a volunteer, standing on the edge of a cliff and levelling a rifle against a French battle ship off Dover. In July, 1894, two such books in narrative form dealing with "the next naval war" were published in addition to Le Queux's thriller, Early-Wilmot's *Next Naval War* and the Earl of Mayo's *War Cruise of the Aries.*

Harmsworth Beats the Drum

Not always were such speculative endeavors to capitalize on public anxiety successful, either because they came too late, the scare being over, or they were too preposterous to fit the occasion. Such was young Alfred Harmsworth's attempt to win Parliamentary representation for Portsmouth in the general elections of 1895. The later Lord Northcliffe stood for the Conservatives who carried the elections by a large majority, but lost Portsmouth. It was a main part of the Harmsworth campaign to take over a local newspaper and fill it with installments of a melodramatic story of "The Siege of Portsmouth," describing the fate of the place in lack of a strong navy to protect it. To attract attention to the paper, posters showed a market square strewn with dead and wounded men, women and children, slain by enemy shells and bombs. "The most astounding story of the day" did not win over the voters who seemed to consider young Harmsworth's candidacy and compaign largely as a joke.[6]

Fashoda, which on the French side provoked some serious historical and technical studies of former landing attempts, and the animosity engendered against the French during the Boer War, produced *The Coming Waterloo.* By Captain Cairnes (London, 1901).[7] This novel of the future relates how England once more invades the continent, returning to the old alliances which had produced the glorious victories of Rossbach and Waterloo. An English army lands at Etaples and Berck-sur-mer in 1903 with 140,000 men. Nothing interferes with the crossing except two small submarines, "these little monsters, insidious and hypocritical enemies, products of the ingenious French spirit of invention," which evoke from an English naval officer: "A good sailor does not navigate below the water but only above." The German Emperor, having declined to receive President Kruger in audience, concludes an alliance with England and with the permission of the Belgian government violates that country's neutrality. With Kitchener at its head, as Wellington II, the English and Germans fight and win the battle of New Waterloo.

This roseate version of Anglo-German relations found fewer British to take stock in it as time went on. As early as the autumn of 1897, when Kaiser William's new German navy had hardly got off to a start, a Colonel Cornwallis Maude's imaginary story of *The Invasion of the British Isles,* published in a service journal, depicted Germany as no longer benevolently neutral, or allied with England, but joining an attack on her by France and Russia without a declaration of war and invading England.[8] There loomed the new potential invader, his ships painted an ominous and businesslike grey, and therefore difficult to sight, a camouflage device that was actually introduced by the German navy early in the new century.

In its way this was novelistic treatment of a new diplomatic theme previsioned by the cautious Salisbury in writing to Joseph Chamberlain in April, 1897, against a war with the Boers, which he thought would have "a reaction on European policies which may be pernicious. The Emperor William's dream is to have a strong Navy." He might give the young Queen of the Netherlands a German husband and "conclude a *Kriegsverein* [Salisbury's own German] with the Dutch Government. Now the one sentimental spot in the Dutch heart is for their kinsmen in the Transvaal. For reasons therefore unconnected with Africa I should look with something like dismay to a Transvaal war. It might mean the necessity of protecting the Northeast of England as well as the South." [9]

After German-English antagonism had become deep-seated and final, these fear-instigating and horror-filled stories, some of them fairly successful commercially and considered in line with patriotic endeavor and therefore in less need of apology, dealt only with a German invasion of England, like Erskine Childers' *Riddle of the Sands* of 1903 [10] or William Le Queux's *Invasion of England* of 1906, which was first published serially in *The Daily Mail,* always a sensational scare-monger, following advertisements with captions spread over a map of England, such as "London Invested, Bombarded and Sacked," "Parliament Finally Meets at Manchester," all this to take place in 1910.

The Liberal government in answer to a question in Parliament declared itself unable to stop mischievous publications of this sort, leaving the matter to be judged by the sense and good taste of the British people who were just then stirred by the campaign of Lord Roberts for compulsory service. He argued that this was necessary on account of the threat of a German invasion, which he believed entirely feasible.[11] The two most unimportant books by such a distinguished writer as Charles M. Doughty, of *Arabia Deserta* fame, were contributions to this literature of panic—*The Cliffs* (1909) describing the failure of the attack of a German naval and air force on England, and *The Clouds* (1912), depicting the conquest of England by the Eastlanders who before any declaration of war destroy the English navy with the help of U-boats, etc. England is in the end saved by her dominions, who come to the help of the motherland.

Swelling the Fear Chorus

Edgar Wallace, a profiteer from anxiety-complexes and arch-purveyor of thrills—a word which has as its bottom meaning that of pierced, bored, or hole. Psychoanalysts, please note!—employed the theme in his *Private Selby* (1912). He had been struck, says his biographer, during the Algeciras Conference with the thought of a war between England and Germany, the course of which he depicted, relying on his own army experiences, the whole thing beginning with a surprise invasion of England.[12] *When William Came. A Story of London Under the Hohenzollerns,* by Saki (H. H. Munro), published in 1913, has England defeated, after a German victory "won on the golf-links of Britain," Society listening to the National Anthem of the *fait accompli;* the work of quislings as we have come to call a grouping of humans who at that time only a satirical writer was able to imagine; and only the youth of the country stirring under a foreign yoke which seems a good deal lighter than that imposed by the Nazi on occupied Europe.

A German story of the victorious coming of the Kaiser to London had been written a few years earlier by a German, August Niemann's *Der Weltkrieg. Deutsche Träume.* These dreams were concerned with a new partition of the world by a continental coalition and at the expense of England. Russia invades India, Germany takes Scotland, unopposed by the British fleet, which is defeated by a German-French fleet off Flushing. French and Russian troops also land in England and the Kaiser makes his entry in London and dictates the peace. Some 25,000 copies were quickly sold in Germany and we do not know how many in England where a translation under the title *"The Coming Conquest of England"* was published in July, 1904, its "meaning and moral obvious and invaluable," as the translator asserted in introducing it into an atmosphere already overheated, or underchilled, by war talk and war scares.[13] On the stage Major Guy Du Maurier's military play *"An Englishman's Home. A Play of Invasion"* early in 1909 was "most excitedly acted," according to Lord Esher.[14]

From sampling of this literature it does not appear that military problems were their authors' prime preoccupation, but rather speculation in and profiteering from the anxieties of large numbers of the English and other peoples. In some of their light manners of making hearts heavier they may have exceeded in imagination the size and potency of contemporary navies and armies, none of which then ever got around to doing what the writers dreamed. Thus Childers, whose fancy in expression shows some of the military realism which his later life and death in the Irish Revolution was to confirm, describes some of his English spies "assisting at an experimental rehearsal of a great scene to be enacted in the near future, a scene when multitudes of sea-going lighters, carrying full loads of soldiers, should issue simultaneously in seven ordered fleets from seven shallow outlets, and under escort of the Imperial Navy, traverse the North Sea and throw themselves bodily on English shores."

To the best of our knowledge no such sally manoeuvres were ever undertaken, or even planned, if only for the reason that the sea-going lighter in tow was slower than the tempo of a successful sea-borne invasion would permit. For a flotilla of such craft to pass from the East Frisian island, where Childers had them start, to the Wash, or even to the flats on the Essex coast between Foulness and Brightlinsea where he thought they would land, would have taken two days, too long for them to remain undetected in those much traveled waters.

But that was a minor consideration for the literary war futurist, who could always rattle the skeleton of the "made in Germany" scare: "Germany is pre-eminently fitted to undertake the invasion. She has a great army in a state of high efficiency. She has a peculiar genius for organization. She knows the art of giving brain to a machine, of transmitting power to the uttermost cogwheel, and at the same time concentrating responsibility in a supreme center" (Childers).

Roots of Preparedness

Still, such writers performed a useful service within the total lop-sided framework of British defence policy. With their emphasis on the invasion danger they kept alive an anxiety state in the island which its governors, even if they exercised better tastes than these scare makers, thought indispensable. As Lord Esher, the civilian official longest concerned with British and Empire defence, who "for years thought daily of war—both by sea and land," wrote to Sir John Fisher in October, 1907:

> No state of mind in a nation is more dangerous—or as experience shows, more foolish—than overconfidence . . . . It is the discussions which keep alive popular fears and popular interest, upon which alone rest the Navy Estimates. A nation that believes itself secure, all history teaches is doomed. Anxiety, not a sense of security, lies at the root of readiness for war. Prussia lay fallow after the death and victorious career of Frederick the Great. Then came

> 1806 and Jena. Our people thought their army invincible after Waterloo. Then came 1854 and the Crimea.
>
> An invasion scare is the mill of God which grinds you out a Navy of Dreadnoughts, and keeps the British people warlike in spirit . . . . Your functions are not only to believe that you possess a Navy strong enough to defeat the Germans at all points, but to justify the belief that is in you, wherever and whenever required! . . . . Invasion may be a bogey. Granted. But it is a most useful one, and without it Sir John Fisher would never have got 'the truth about the Navy' into the heads of his countrymen.[15]

The threat of an invasion was thus admitted as one of the great AS IFs of British defence policy. The intellectual dishonesty involved in the pretence was apparent to at least some of the officials and politicians concerned with defence, who excused it by the apparent inability of the masses to understand any other threat of a military nature and their unreadiness to permit their representatives in Parliament to vote funds for purposes beyond those intended for the direct defense of the island, even if the latter then was better, or best protected by a continental first line with the help of an expeditionary force and a high seas fleet.

Nations in an insular position will above everything else react violently to an invasion threat. It is the most plausible military argument that appeals to them, even if such an invasion should be a contingency most unlikely to arise, as it was for the United States during most of its history. In fact, foreign military experts in the 1880's would consider it "the most fortunate of nations since it needed no army and navy," being "beyond the possibility of attack and invasion."[16] The danger which lurks in popular receptiveness towards the invasion argument is that it may interfere with measures actually required and intended in war, or in preparation for war, which might involve fighting far from the shores of the home country.

In order to win support for naval bills from American isolationism in the early 1890's suggestions were made as to what would happen to New York city if an enemy fleet should possess itself of the harbor. In his *Annual Report* for 1890 the Secretary of the Navy, Tracy, wrote: "Commerce would be annihilated. Communication would be absolutely cut off. The ferry boats would cease to run. The Brooklyn Bridge would be closed to traffic as the condition for its preservation. Finally, the railroad communications would be cut off, and the food supply of 2½ millions of people would come to an end."

Such flights of imagination could only help to create a strong fixation in the popular mind that war, if it should come, would result in an invasion. After such psychological preparation for war, could anyone be surprised that shortly before, and at the start, of the war of 1898 with Spain scares swept the cities on the east coast? That each insisted on getting at least one gunboat for defense against predatory Spanish torpedo boats and cruisers? That these demands interfered to a certain extent with the intended movements of the American forces against the Spanish in the Caribbean?

After the sinking of the *Maine,* Theodore Roosevelt, then Assistant Secretary of the Navy, wrote to his chief (February 17, 1898): "The Journal the other day and the *World* today have pictures of the *Vizcaya* [a Spanish armored cruiser at the time in New York harbor] shelling and destroying New York. . . . I wish it were possible in some way to point out the criminal character of such articles. Until the sinking of the *Maine* they amounted to mere hysterics, compounded in part of sensationalism, and in part of physical cowardice. Now they are more serious." Roosevelt appealed to his chief to make an authoritative statement stressing the improbability of these insinuations.[17] All such appeals were of no avail, and only to an objective, poised Bostonian like Henry Adams did it seem queer that in our day, good Bostonians should be still trembling, like their ancestors in the 16th century, for fear of a Spanish Armada![18]

Pleas for Channel Tunnel Vain

Unreasonable, or partly reasonable fears connected with questions of defense cannot be countered with arguments, no matter how sound. Against stubborn refusal to listen to them the English advocates of a Channel tunnel, regardless of the merits of their project in the economic and even the military field, have battled in vain, even to the present. Col. George Sydenham Clarke, later Lord Sydenham of Combe, in his time the foremost English specialist on coastal fortification, did not share, at least not after 1883, the general English apprehension of the dangers to which a tunnel might expose his country; but his expert advice weighed little against the fear-inciting and authoritatively conclusive verdict of the then commander-in-chief, Lord Wolseley, that "were a tunnel made, England as a nation would be destroyed without any warning whatever when Europe was in a condition of profound peace."

While Admiralty and War Office pronounced repeatedly against the tunnel and in favor of maintaining complete insularity, a number of officers did not share their views, particularly those who were aware of the obligations which Britain had assumed in connection with military arrangements with France. These officers would argue as Clarke did in 1914, a few days before the outbreak of war: "If ever we were compelled to send military forces to France, Belgium, and Holland, through railway communication would be of enormous importance. I need not remind you that, apart from any obligations to France which may exist now or in the future, we have definite treaty responsibilities as regards Belgium in certain contingencies." After the World War, the military opponents of the tunnel, according to Clarke, "shifted their ground, and as their present position has never been explained it is not possible to bring reason to bear upon the objections." [19]

The professional psychoanalyst may have individual cases of *insularitis* on record and a scientific language to describe them. Freud, at any rate, has written history to fit a theory of psychic insularity. He thought that:

> It may indeed be that with the present economic crisis which followed upon the Great War we are merely paying the price of one of our latest triumphs over Nature, the conquest of the air. This does not sound very convincing, but at least the first links in the chain of argument are clearly recognizable. The policy of England was based on the security guaranteed by the seas which encircle her coasts. . . . The moment Bleriot flew over the Channel in his aeroplane this protective isolation was broken through; and on the night on which, in a time of peace, a German Zeppelin made an experimental cruise over London, war against Germany became a certainty.[20]

The flaw in this theory of war causality is that no Zeppelin flew over London before January, 1915. A better case for a mass panic instigated by an invading airship would have been provided by the excitement of the French people due to a forced landing of a German Zeppelin at Luneville in 1912, near the Franco-German boundary and in the French fortress zone.

Whatever the sensational writers of invasion thrillers imagined and wrote, and whatever the great English and American public believed about the new German Navy, beliefs which after all provided the broadest basis for England's policy of forsaking splendid isolation in favor of alliances and ententes with Japan, France, Russia and, less formally, America, this navy was not built for any such specific purpose as the invasion of England. Its construction had far less to do with a future clear strategic purpose than with a past highly complex socio-economic causation.[21] It was not set up as an essential part of a total and centrally conceived strategy of the Second Reich, for there was no such strategy and war planning.

Germany's Complacent Military Directorate

The Army, including the General Staff, and even more so the Ministry of War, to the eve of the First World War remained as "saturated" as in the days of Bismarck, Moltke and Caprivi, not because it harbored friendly feelings towards England, or any other foreign power, but because it considered itself as having grown to a state of completion beyond which no further expansion was considered possible, or desirable. By about 1905 it judged that all elements in the Reich capable of providing army officers of the desirable type had been absorbed. Because it did not want to expand, because its governors did not care to admit new elements, likely to be imperialists and thereby apt to conceive new strategic concepts, the controlling Army cliques were little interested in Germany's expansion over regions such as the islands beyond the North Sea where it had never fought before. Tradition clung to the old enemies and the old fields on which to battle them. The German Navy's own notions about its concrete rôle in war were decidedly hazy. It drew heavily on Mahan's sea power ideas, about which certain things should be emphasized in connection with our theme:

(a) It was essentially a doctrine, supported by an arbitrary use of war history, of a naval, and not of a total, unified concept of war; it was in keeping with a naval interest which in certain countries like Great Britain and the U. S. may have more closely coincided with the national interest than elsewhere and the best distribution of the national strength for war, but did not do so in the case of Germany.

(b) It had but small appreciation of combined operations either in the past or in the wars to come, which are seen largely in a picture painted in or by the past and not by the technical possibilities. It centered on battle ships rather than any landing craft, on sea battle rather than any landing enterprises, on the "silent pressure" of sea power rather than the tactical or strategical cooperation with the country's army.

Should memorials to Mahan be reared in Britain and America, they should be inscribed, assuming that the truth ever is expressed in these circumstances: He misled Germany into spending a large part of her energies and immense sums on a useless navy and thus helped England and France to win the battle on the Marne, where German army corps were missing which went into the making of the navy. Unfortunately, paradoxes, however true, have no place on the pedestals of monuments.

Resortialism, or departmentalism was only strengthened by Mahan's teachings in Germany, perhaps more so than elsewhere. There was in the Reich no organ, or organization to hold the forces together and combine them in planning of war or for it. The Supreme Warlord on whom the Reich constitution conferred this rôle, to whom some 40 high military and naval offices stood in an "immediate reporting" relationship, was not the man to tackle this task, if indeed it was a one-man job in any country.

The large majority of the German army officers corps, and most clearly its directive men like Falkenhayn, thought "the Navy superfluous, have always done so," as Tirpitz realized early in World War I, if not before.[22] Bernhardi, as a cavalry general, paid the Kaiser the peculiar compliment of saying that "in a truly statesmanlike spirit he had powerfully pushed and steadily advanced the development of the German fleet without being forced to it by a political necessity."[23]

NOTES, CHAPTER 32

[1] 1865. *The Amberley Papers.* I, 348.

[2] Cit-in Ruskin's *Fors Clavigera,* 2nd letter. *Works.* New York, 1894. II, 28.

[3] *The Life and Letters of Sidney Dobell.* London 1878. II, 348 ff.

[4] Margaret Oliphant, *Memoir of the Life of Laurence Oliphant and his Wife, Alia Oliphant.* Edinburgh, London and New York, 1891. II, 77-80.

[5] Lord Sydenham of Clarke, *My Working Life.* London, 1927, 107-11.

[6] Hamilton Fyfe, *Northcliffe—An Intimate Biography.* London, 1930, 90-1.

[7] It was duly noted in France. See *Le Correspondant,* to. 202, 1901, 979-87. We use this article, not having been able to find a copy of the book in the United States.

[8] *Naval and Military Record,* Sept. 9 to Nov. 4, 1897. Marder, 291.

[9] Garvin, *Life of Joseph Chamberlain.* III, 141.

[10] This novel was published in New York in 1915 and again in two editions in 1940, one of them at the popular price of 25 cents. This was at a time when Childers' prophecy of a German invasion of England was considered as nearer fulfilment than ever.

[11] Woodward, *Great Britain and the German Navy,* 117.

[12] Margaret Lane, *Edgar Wallace. The Biography of a Phenomenon.* New York, 1939, 243.

[13] Marder, 476.

[14] Brett (ed.), *Journals and Letters of Esher.* II, 375-6.

[15] Ibd. II, 249, 251-2.

[16] Oswald Garrison Villard, *Fighting Years.* New York, 1939, 64.

[17] *Papers of John Davis Long.* 1939, 56, 135-7, 142, 225.

[18] *Letters of Henry Adams,* 1892-1918, 175. (May 5, 1898).

[19] *My Working Life,* 110, 205-6.

[20] The Coming of the Machine, in *New Introductory Lecture on Psychoanalysis.*

[21] The classical work on the rise of the German Navy is Eckart Kehr, *Schlachtflottenbau und Parteipolitik, 1894-1901.* Berlin, 1930. On it is based ch. II in Charles A. Beard, *The Navy—Defense or Portent?* (1932).

[22] Tirpitz, *Ohnmachtspolitik im Weltkrieg.* Hamburg and Berlin, 1926, 123.

[23] *Deutschland und der nächste Krieg.* 1911. 6th ed. 1913, 142.

Chapter 33

COMBINED OPERATIONS IN PLANNED WAR

PRACTICALLY EVERYTHING in the opening stages of the Crimean War and the American Civil War had been left to improvisation, a circumstance which tempts one to say that it was the armies and navies concerned which least wanted these conflicts, because they were least prepared for them, whereas the peoples were more eager for hostilities in their unawareness of the technical difficulties involved. It was otherwise in Prussia, however. There the organization of the general staff as directed by Moltke was calculated to leave as little as possible to fortuitous and unforeseen conditions. Incidentally the professional militaires allowed the people but little opportunity to say whether or when they wanted the wars that were cooked up for them by the diplomacy of Bismarck.

Among the tactical eventualities of war for Prussia, as they were studied from 1850 on down to 1882, the chief-of-staff included, for the year 1861, although the case was not considered likely, the problem of how to defend Brandenburg Province against an enemy corps landed in the Oder estuary while the Prussian main army was employed on the Rhine. It was one of the first imaginary visualizations of a situation in a two front war for the Germans.[1]

The necessity for Prussia to act as a landing power in one of her wars would seem to have been less well contemplated in advance. Considering a war with Denmark, Moltke had written in 1862: "As long as our navy does not make possible a landing in Zealand in order to dictate peace in Kopenhagen itself, there remains only the occupation of the Jutland peninsula; to be effective as an instrument of pressure, it must be somewhat prolonged, though that may provoke diplomatic intervention and possibly the actual interference of third powers." [2] Moltke somewhat feared, or rather considered as the most undesirable step to be taken by the Danes, their retreat to the islands without giving battle on the continent. Want of a war fleet by the Allies in 1864 protracted the war in which little Denmark at moments seemed superior to the two land powers of Prussia and Austria, until the Alsen Sound had been crossed and Austria had brought her fleet into the North Sea where it cooperated in the reduction of the North Frisian islands.

The crossing of the Alsen Sound, 520 to 1120 meters wide towards the northern end where the landing took place, and therefore more like the crossing of a river than of an arm of the sea, took place on June 29, after the end of a two months' armistice. It had been originally planned by Moltke for the night of April 2 to 3, before the armistice, when bad weather forced him to desist. In the meantime craft to be manned by engineering troops, enough to carry 3½ battalions across at one time, had been assembled. The Danish naval forces in the vicinity being far superior, the Prussians had to rely on beach batteries, speed and surprise. The latter was complete. The crossing began at 2 A.M. The troops who had been carried over by 3, rolled up the Danish entrenchments to the right and left of the beachhead. Danish counter-attacks, unsystematic, were repulsed. In the afternoon the Danish commander gave up the battle and retired into the easily defendable Kekenis peninsula, south of Sonderburg, and from there was evacuated to Fünen with the remainder of his two brigades. The Danes lost 2400 prisoners and 100 guns. Prussian casualties were 33 officers and 340 men—that is to say, in such uncommon enterprises officer casualties are apt to run high—and the Danish ones about twice as high.[3]

The taking of Alsen decided the war and drove the Danes, disappointed in all their hopes for more than diplomatic support from England or other great powers, to sue for peace. The invasion of their archipelago, which Moltke threatened with a landing at Fünen, stirred the Danish governors as much as did the invasion of Sicily in 1943 the governors of Italy.

Basis of Moltke's War Plans

In Moltke's first plan of operations against Austria the small Prussian navy did not figure, although it did so in the subsequent discussions about the cooperation with allied Italy in 1866, but not beyond generalities. "The two fleets would have to be set in motion for the purposes of war, that of Prussia so far as the protection of her coasts and of Schleswig-Holstein would allow this." But since relations with recently defeated Denmark remained bad, the Prussian fleet did not proceed to the Mediterranean; besides, the short duration of the war, a mere seven weeks, forbade closer Prussian-Italian cooperation, for which after the defeat of the Italian fleet at Lissa there would have been but small occasion.

Moltke's Sketch of Operations for a war against France was based on the numerical superiority of the German forces which by itself would make the detaching of French troops from the main theater of war for the purpose of a landing in northern Germany or Denmark highly improbable. The balance of military strength was so precarious that France in all likelihood would shun that operation, even though some temptation to her of a political rather than of a military character existed. And only in the beginning of war might a landing be feared, "once the French forces are attacked inside their own country, they will hardly enter upon such an undertaking."[4]

It is obvious, said the Sketch, "how very important it is to make use of the superiority which we enjoy from the outset, with even the North German Bund alone. The latter will be still more enhanced at the decisive point if the French should indulge in expeditions toward the North Sea, or into Southern Germany. For the defense of the former sufficient forces are to remain inside the country." Until the end of the war the Germans kept certain forces at home, originally three of the thirteen North German army corps, which were to guard the eastern frontier and the sea coasts and one of which might be considered as posted along the North Sea. By the time of the investment of Metz the three army corps had been added to the field armies and the duties of guarding the coast were left to second line troops.

French projects about combined operations in the North and possibly the Baltic Seas suffered greatly from hesitation and uncertainty, and those for a cooperation with Denmark seem not to have gone far. The war plans as worked out since the autumn of 1860 envisaged the building up of a transport fleet which, together with the part of the merchant fleet available at the outbreak of the war, would carry 40,000 men, 12,000 horses and the necessary artillery, engineers and supplies over a distance which ordinary packet boats would traverse in three times 24 hours.[5]

The French navy, much superior to the North German, but poorly organized and prepared, at the beginning of the war was under orders to prepare transports for 30,000 men, who were to be commanded by the enterprising Bourbaki. For this purpose, in addition to marines, certain land troops were earmarked. According to another version a landing with two brigades, two regiments of cavalry and eight batteries was planned, although "planning" would be a euphemism covering general uncertainty, much change and contradiction in orders from the imperial headquarters. Following the first land battles all earmarked troops including marines were distributed among the field armies or the forces defending Paris.[6]

Great General Staff's Views

The experiences of 1870 confirmed Moltke's earlier views about a French or any other landing attempt against Germany. In his various plans for assembly (Aufmarschpläne) until 1890 he mentioned combined operations only twice. Neither the Russian navy, which began to be enlarged after 1882, nor even the also growing German fleet were assigned any rôle in his plans for war. He reckoned damages to the expanding Reich merchant marine and ports as scarcely avoidable, but a landing in force as unlikely; a victorious enemy would more certainly pursue his purpose on land.

> France possesses in her fleet too powerful a weapon not to make use of it in a new war against us. When in the French harbors not only the armored fleet is being equipped but the transport fleet as well, something that cannot be concealed for very long, we have to conclude that an invasion is intended for the purpose of carrying the war into our country. Such an enterprise will probably occur in the very beginning of the campaign, for in the later stages the enemy can, if he should be victorious, pursue his purposes more safely on the continent—in the reverse case, he will have no forces to spare for far-reaching expeditions. Incidentally, one may, in a civilized, thickly populated country, crisscrossed with roads and railroads, foretell a catastrophe for the enemy with some certainty.

At that time, 1873, Moltke added: "The observation point given to Germany by nature, Helgoland, does not belong to us. Whether it might not be possible to obtain it through diplomatic negotiations against financial or other considerations may only be mentioned on this occasion." [7]

In his determination to keep the German land forces concentrated, Moltke went so far as to oppose for a number of years—until about 1885—the construction of the Kiel Canal, his main argument being that such a costly investment would require for its protection in wartime a considerable number of troops, 60,000 men, if Denmark should join an enemy landing on her shore with her connivance.[8]

That Germany could not and would not be invaded from the sea and should not herself attack overseas in Europe was axiomatic during the era of "saturation." On the whole Moltke agreed with Bismarck, who put the common conviction into the formula that a corps landing in the Reich would simply be put under arrest. This view was taken so much for granted that a German philosopher of Schopenhauer's school, Eduard von Hartmann, a former Prussian officer, declared that "the era of landing operative corps of land armies on the territory of great powers is as irrevocably over for today and all future as the era of Viking cruises." [9]

However, where even a philosopher of pessimism might be optimistic, Moltke could not avoid contemplating a bad contingency, and on his last great General Staff excursion in 1881 through Schleswig-Holstein he discussed the problem of how to meet the advance of Danish and French divisions against the North Sea-Baltic communications lines and Kiel, including the problem of whether Kiel should be built as a fortress on the land side or not.[10]

Moltke's second successor on the Great General Staff, Schlieffen, was no more alarmed about the threat of a landing on German soil. Of the fourteen large tactical-strategical problems which were to be dealt with by his subordinates from 1891 to 1905 on the annual "tactical rides," only one, that for 1896, embodied such a problem, and the German fleet played no rôle in it. The theme was based on the supposition that in a war with one of the great powers the larger part of the German fleet had been forced into Kiel and into Wilhelmshaven, both harbors being blockaded by the enemy. A northern hostile corps of 30 battalions of infantry, 12 troops of cavalry and 96 guns had crossed the German-Danish boundary, coming from Kolding, on May 1, and reports made an additional landing of enemy forces on the East Holstein and Mecklenburg coast appear as immediately impending.

The German field army was supposed to be employed in distant theaters of war. The commander of the German forces, also the commander of the forces remaining in the IXth Army Corps (Altona), had orders to defend the Kiel Canal and in any case to hold Kiel and Hamburg, the forces at his disposal being slightly superior to those of the invader, who should be promptly attacked lest other landing forces off the German coast, but not likely to land before the 9th, gave the enemy superiority. Battle was envisaged for the 7th. For this Schlieffen insisted that no forces be wasted on the defense of any Holstein towns, but all be made available for the battle and carried to rather far advanced positions by the railroads.

"According to the examples provided by the history of war, from Eupatoria, 1854, to Wei Hai Wei, 1895, there have always passed five days at least between the beginning of the debarkment and the advance of the troops. Even if one reduces this time considering the progresses made by the European navies in the meantime with regard to landing methods, it must still be kept in mind that the disembarked enemy corps— landing probably in the Hohwacht Bay near Lützenburg or in the Neustädter Bay—must first somewhat secure its beachhead, that it has to inform itself about the war situation," etc.[11]

As compared with the *tempi* of the German march through Belgium in 1914, under Schlieffen's plans, not to speak of those modern landings which cannot afford days of preparation for the advance lest they favor the defender overmuch, Schlieffen's time allowances for the latter would now seem ample indeed.

The common opinion of German officers as to the improbability, not to say impossibility, of large scale hostile landings on the shores of the Reich, or indeed those of any other of the great powers was expressed in public most forcefully by Colmar von der Goltz, later Goltz Pasha, in a widely read, much translated book on "army institutions and conduct of war in our own time"—*Das Volk in Waffen,* The Nation in Arms (first published Berlin 1883), a book in many ways serving to publicize Moltkean concepts.

Germans Did Not Fear Landings

He ascribes the comparative success of landings in the American Civil War to the fact that "together with the harbor towns the rebellion was robbed of the main sources of its strength and that it was impossible in the thinly populated country to quickly assemble new armies in order to reconquer what has been lost." The factor of density of population with respect to landing, which Moltke had also stressed, would seem well emphasized, forecasting one of the chief obstacles to the success of airborne landings which succeed more easily in regions thinly held, sparsely defended by military, or semimilitary bodies.

> The results of landing operations will only seldom equal the disadvantages caused to the field army through the weakening due to detachments. Before a landed body has not achieved considerable success and has not spread out considerably, before the fleet has not brought a number of coastal points under its control, its liberty of movement is very much restricted. Only enterprising boldness and very surprising procedures might balance these shortcomings, but for that purpose cavalry will be wanting. It is cavalry that an army landed on foreign shores would be much in need of in order to reconnoitre at once and in all directions, to destroy railways far inland, to hold up the approach of defenders assembling from all directions. But horses are of course even harder to transport by ship and harder to land than men or materiel, hence cavalry will be absent.
>
> The military strength of the great European nations is today so far prepared for use that even if all field and field reserve forces are already in battle along the frontier or in enemy's country, there will always be enough left to establish considerable superiority over landing forces. Replacement troops, fortress garrisons, improvizations, *Landsturm* levies find their use here and will do good work when they realize that threat to the home soil. Telegraph and railroads, unhampered in their full efficiency, will convey forces from the most distant provinces. Of course, the attacker may strengthen himself by bringing on a second landing corps, but before it arrives, considerable time goes by and the fate of the one first arrived will have been settled in the meantime. Landings and coastal enterprises, therefore, have not only to battle with great difficulties, but on the whole offer also but small hope for achieving considerable effect. They can be undertaken, therefore, only under exceptionally favorable conditions.
>
> Among the latter figures a surplus of strength. If Germany should be attacked by two great powers in the East

and in the West, the latter's navies and armies together might well possess the forces necessary to undertake a landing on our coasts in respect-inspiring strength. There would also be the possibility of concerting the movements of the landing army with those of the field army of one of the two allies. If Denmark in 1870 had been on France's side, the latter could have thought of landing troops on the easily accessible Danish coasts and, after combining with the Danish forces, of undertaking an expedition against the lower Elbe. The Danish cavalry would have made up for the scarcity of horse. The combined armies would have gained a numerical strength deserving of some respect. The whole Danish monarchy would have served as its base. Under such circumstances, however, the character of a mere landing expedition would no longer exist. What has come about is that a part of the total enemy strength has effected its separate assembly by means of sea transportation [that is to say, the landing expedition according to Goltz is characterized by being based on ships].

It will always be good caution, in order to win some freedom of movement on hostile soil, to obtain control of the coast to a certain extent. An island close to the coast may guarantee safe debarkation and assembly of the troops, but surprise would thereby be lost. The defender gains time to make his own arrangements.

It is obvious that large scale enterprises from the coast deep into the interior against objectives of importance or even with decisive intention against the enemy's capital become possible only after a long war has exhausted the strength of the attacked nation to the utmost and when even his last resources in men, horses and arms have been applied to keep hostile armies from invasion across the land frontier.

Possibly the attempt of landing might also be made early in the war when assembly has not yet been completed. Such landings carry, however, more the character of alarms, intended to disturb mobilization and to disquiet the people of the invaded country, rather than that of a serious attack. The great mass of the people will always be somewhat impressed if suddenly the enemy who was believed to be far beyond the frontiers, appears off the coast in ships. Bue suppose an army of 40,000 or 50,000 were thrown by surprise on that part of the Baltic coast nearest to Berlin, the Oder estuary, leaving quite aside the difficulties a landing would meet just there, the five or six days which it would require to get to the capital would quite suffice to oppose it with superior strength.

Landings, therefore, represent for a populous and militarily well organized state phantoms rather than serious dangers. (382-5)

This conviction was and remained dominant in the German army, including its General Staff, until 1914 and even beyond. One of its preceptors, Goltz, taught it at the Military Academy in Berlin (1878-83) and also Balck, major on the Great General Staff, teacher at the War Academy and author of *Taktik,* the pre-1914 standard work on that subject, which in a chapter on Sea Transports and Landings deals with these problems as questions of loading and unloading, since, says the author, in almost literal agreement with Goltz, "for the great continental states overseas enterprises take on the character of phantoms rather than presenting a serious danger."[12]

Value of Heligoland Underestimated

This view was also dominant in contemporary English army and navy circles until at least the end of the century,[13] and in the British government generally. For without it the Salisbury government would never have exchanged Heligoland for the German rights in Zanzibar. Besides, its reliance on the pro-English policy of the Caprivi government in the Reich, which had broken the misleading re-insurance treaty with Russia, was more justified than the complete minimizing of the military importance of Heligoland, the cession of that stronghold being justified by Admiral Mayne, a Salisbury man, by "ridiculing the idea that Heligoland had value, even for Germany." In case of an unlikely Anglo-German war an English man-of-war passing by "would probably throw half a dozen shells into the place for a morning's entertainment, and, if the Teuton showed his usual sense, he would as promptly get into his steam launch and make for the nearest port on the mainland."[14]

For Caprivi, the general and former head of the small German navy, and chancellor after Bismarck, the only war possible, or likely, for Germany was one on two fronts and on land against France and Russia, with England as a neutral. He could not think of England invading Germany or being invaded by her. But already in his time forces—among the navy people and the colonial enthusiasts—were shaping in the Reich, who were willing to take her on as an enemy, either in addition to continental ones, or in lieu of these. Caprivi realized the dangers threatening from these parties within Germany and tried hard to keep them under control.

An episode told in the *Memoirs* of one of these *Kolonialmenschen* may serve as well as any long exposition to illustrate the Chancellor's apprehensions of the consequences of their expansionism and their reckless policies, which they never thought through to their military implications. Liebert, one of this unsaturated group, an upstart, with his patent of nobility dating only from 1900, founder of the local group of the German Colonial Society in his Silesian garrison, employed in the supply organization for the German colony in East Africa, after a trip to that colony was received by Caprivi, who used to call him an enthusiast.

After a number of complaints by the Major of the opposition by some English against the German colonial administration, Caprivi asked him bluntly: "How do you intend to defend East Africa against England?"

The question was new to the Major. But since he had to give an answer of some kind, he said: "Then we will have to land at the mouth of the Thames and defend East Africa there." The Chancellor rose and said merely: "I thank you," dismissing him. "With that the table cloth between us was cut in two."[15]

Caprivi took care that the reckless army politician was transferred to the provinces where he might be supposed to do less damage with his oratory and writing powers than at Berlin. When he was pensioned in 1903, at the early age of 52, this Anglophobe in addition to directing a number of chauvinistic organizations like the Pan-Germans, the Colonial Society and the Naval League founded the Reich League for Combatting Social Democracy. He was the sort of irresponsible soldier who for his liking could not have too many enemies for his country and had no dread of leaving it almost friendless.

Liebert was by no means the only German officer to speak, or write about the invasion of England as a possibility, or necessity in the late 19th century. In the immediate entourage of the Emperor Major von Morgen, the "bloody Franz" of World War I, then one of Wilhelm's aide-de-camps, talked to a diplomat's wife: "If one should very quickly land an army corps in England, one could subject with its help that whole country."[16] In the *Militärwochenblatt,* the oldest service journal, Captain von Lüttwitz, then of the General Staff, much later the military leader of the Kapp putsch of 1920, published an article in February 1896, duly noted in England, in which he favored a powerful German fleet as an instrument of German world and colonial policy.

"Expansion overseas into a place in the sun would mean clashing with England, or striking a bargain with her; in either case Germany would be better prepared if she could use, or display sufficient force. A German fleet need be only strong enough to deal with the English squadron in home waters, a large part of the English fleet would always have to stay on foreign stations; the war would be decided quickly and without the interference of those far away ships.

"That England is unassailable is a legend. Due to the introduction of steam and electricity the situation has much changed since 1805, and to the disadvantage of England. The assembling and the fast and surprising transport of invading armies had therefore been essentially facilitated."

It was an example of the tactfulness of Wilhelminian days to appoint the writer, who was known to be a member of the English section in the General Staff, a little later military attaché in London, a post that had not been filled for years; also a sign of the German Army's *désinteressement,* in May 1898.

His various articles on the not impossible invasion of England, coupled with his appointment aroused much suspicion in English circles where it was thought, not quite illogically, by the offices and officers having to do with defense, that the captain would now proceed to round out his theoretical studies by some practical reconnoitring on the invasion spots. Some of the officers were thinking as early as 1898 of the day when Germany would have 17 battleships, abundant transport resources and a nearer basis for embarkation after the obviously planned conquest of Belgium.[17]

German Invasion Talk Political

As far as publications allow us to judge, this *fin de siècle* talk in Germany about an invasion of England was always more political than technical. It grew out of the hybris of power so prevalent in officer circles after Bismarck and Moltke had gone, out of the desire to give sense and reason to the new German Navy as an intrument for war, but which had been built for the Navy personnel, and for very concrete interests in the Reich, as well as to satisfy some hazy liberal feelings, but definitely not in order to fill a rôle in Germany's total equipment for a likely war such as it was the General Staff's business to contemplate.

At that time a representative of that world of demi-gods, as Bismarck called them, would have agreed in private with Winston Churchill's public pronouncement that the German fleet was a luxury fleet. As far as German military officers ventured opinions about an invasion of England, they were moved by their professional scorn for English army organization at the time, when it was backward and awkward. That a few General Staff officers were navyphile, that there was perhaps even a minority opinion in that august body that saw some usefulness in the new navy, must not lead us into believing that their office had called for this fleet for purposes germane to their plans for likely wars.

The army's technical approval of the navy was essentially *ex post facto,* dating from the time following the creation of the navy, or at least after the movement for this navy had been started by the Emperor, Tirpitz and others, but by no army officer of any weight. *Then* they might write in favor of the navy, something apt to please the Emperor who was able to declare on occasions that he wanted to make the navy as strong as the army. That there was a vast deal of irrationalism in this talking and writing, a strange disregard of the English fleet that might after all be more concentrated in the home waters and withdrawn from its stations in America, in spite of occasional incidents like that concerned with Venezuela in 1895, should not surprise anyone who reads the arguments of the Emperor himself preceding his Kruger despatch of 1896: "After the Jameson raid had started, H.M. evolved somewhat startling plans. Protectorate over Transvaal . . . Mobilization of the marines. Sending troops to Transvaal. And when the Chancellor protested: That would mean war with England, H.M. said: 'Yes, but only on land' . . . At last H.M. sends a telegram of congratulations to President Kruger."[18]

That should make clear why people in the Reich insisted on calling certain of their policies *Realpolitik,* even if that went no further than talking the Emperor out of his wild notions. The whole incident proved very popular as far as the German public knew it, and served well the agitation then getting under way for a big naval bill. With the actual rise of the new German navy the discussions of landings in England entered upon a new stage. It had been obviously political to the end of the century. Would the character of the new ships and their officers bring the discussions on the firmer level of technique and the technical means involved? That we shall endeavor to find out.

NOTES, CHAPTER 33

[1] *Moltke's Tactical Problems from 1852 to 1882.* Edited by the Prussian Grand General Staff. Authorized translation by Karl von Donat. London, 1894. Problem no. 14. The solution is missing. Problem 64 deals with the situation of an army operating beyond the Oder towards the east, receiving intelligence that a hostile corps was advancing in its rear from Stettin on Berlin. In his solution Moltke called the situation not very realistic.

[2] *Moltke's Militärische Korrespondenz.* Reconsiderations of Dec. 6, 1862, and letter to the Allgemeines Kriegsdepartment, Dec. 23, 1862.

[3] For details see *Der Deutsch-dänische Krieg 1864.* Published by the Prussian Great General Staff. Berlin, 1887.

[4] *Moltke's Militärische Korrespondenz 1870-1,* no 16.

[5] W. Rüstow, *Guerre des frontières du Rhin, 1870-1.* Paris 1871. I, 77.

[6] For further details ibd. I, 304-313.

[7] Graf Moltke, *Die deutschen Aufmarschpläne 1871-1890.* Berlin 1929. 30-3, 139.

[8] Bismarck, *Gedanken und Erinnerungen,* ch. 19, v.

[9] *Zwei Jahrzehnte deutscher Politik und die gegenwärtige Weltlage.* 1889, 389.

[10] E. von Liebert, *Aus einem bewegten Leben.* Munich, 1925, 88.

[11] *Dienstschriften des Chefs des Generalstabs der Armee,* Generalfeldmarschall Graf von Schlieffen. Die taktisch-strategischen Aufgaben aus den Jahren 1891-1905. Berlin, 1937, 31 ffff.

[12] Taktik. Berlin, 1898. II, 360 ff., 371.

[13] Callwell, op. cit., 15.

[14] Hansard, 3d. ser., vol. 347, 922-3.

[15] Liebert, op. cit.

[16] Elizabeth von Heyking, *Tagebücher aus vier Weltteilen.* Leipzig, 1926, 303; entry of Jan. 12, 1900.

[17] Marder, *Anatomy,* 291, 298-9.

[18] *Europäische Gespräche* II, 212-3.

Chapter 34

LANDINGS IN THE SPANISH-AMERICAN WAR

AMONG MODERN WARS the Spanish-American is entitled to peculiar distinction because the delay between the declaration of war and the opening of hostilities, whether on land or on the seas, was the longest on record. Nearly six weeks passed before the land armies came to grips. In an era wherein nearly all other wars were to a varying extent planned and prepared, this conflict revealed an unbelievable degree of American military unpreparedness, uncertainty, unfamiliarity with the problems of a war with Spain for which newspapers, politicians and masses of Americans had long been clamoring; or for a war with any other country, even a minor power.

Whether the absence of a planning staff, a Great General Staff, which was organized in the United States army only after the war, was an excuse, or an explanation for this failing need not be debated. In its absence the holder of the constitutional role of commander-in-chief of all forces, filled by a politician like McKinley, was even less able to provide, or guarantee a combined strategy for the army and navy. Without informing the War Department, the Navy, well ahead of the war, had sent Admiral Dewey to the Far East with a squadron to hold himself ready to wrest the Philippines from Spain. Having defeated the Spanish in Manila Bay on May 1, he telegraphed home: "I control bay completely and can take city any time, but I have not sufficient men to hold." This occupying and holding force did not arrive from the States, 12,000 miles away, for many weeks. Manila surrendered on August 13, in this sluggish theater of super-sluggish war.

As for the West Indies, it was originally intended at the end of April to send a reconnaissance force of regulars under General Shafter of 5,000 or 6,000 men to Cuba to obtain a foothold and support and furnish material to the Cuban insurgents. This was to be followed eventually by a larger army which then did not exist. Shafter's departure from Tampa was postponed from day to day since no convoying warships were available, with plenty of *ordre, contre-ordre, déordre.*

Finally reports of the disappearance into the blue of a Spanish squadron under Cervera stopped everything for awhile, except for two attempts, headed by a cavalry captain, to get supplies through to the insurgents. The first attempt in which two companies of infantry participated failed, largely on account of newspaper indiscretion which had announced the departure well ahead of time, as well as inability to locate and get in contact with the insurgents. A second expedition, starting from Tampa on May 17, was more successful and landed a Cuban general and 300 insurgents, arms and supplies at the eastern end of the island.

Spurred by word of what Dewey had done in Manila Bay, McKinley on May 2 ordered General Miles, commander of the Army, to send from 40,000 to 50,00 men to take Havana, with Shafter's troops to form the vanguard, and nearly a week later presidential orders went out to assemble an expeditionary force of 70,000. Shafter was to "move his command under Navy protection and seize and hold Mariel, or most important port on north coast of Cuba [some 20 miles west of Havana] and where territory is ample to land and deploy army." On the 10th he was told that the departure was to be before the 16th.

So long as Cervera's squadrons, which were supposed to be somewhere on the Atlantic west of Spain, remained unlocated, no expedition could depart for want of convoying ships, most of which were to be used either in endeavoring to find Cervera, or to quell the squawlings from American eastern coast cities, which were clamoring for protection and "rescue from the advancing Spanish fleet." After it had been ascertained that the Spaniard had dodged the American war ships and brought his craft safely into the harbor of Santiago de Cuba, the Navy told the Army that it would be "inexpedient to expose the Army, or any part of it on the waters of the vicinity of Cuba" and that therefore the orders to the Navy to convoy troops had been countermanded. This decision seems to have been arrived at by the Navy people without allowing the Army to have anything to say about it. There were at the time as well as later "the usual differences and bickerings among the officers of the Army and Navy, which in certain high quarters are altogether too apparent," as McKinley's private secretary noted (May 15).

Pre-Cuban Invasion Chaos

Postponement of the departure of the army contingents for Cuba served in a way to help perfect their outfitting and organization, although nothing was done during these weeks to train the raw volunteers how to embark and disembark. Assembling of the troops at the embarkation point and in camp had revealed an appalling incompetence and technical backwardness, showing that organization for warfare was the most backward sector in the most highly industrialized country of the time. Slight forethought had been given to the needs of an expeditionary force, such as preparation of embarkation camps, ports, including housing and water supply, sanitation and commissary; transports, equipment, tactical handling, strategical objects.

Tampa, chosen for its proximity to Cuba, proved a poor embarkation point, with only a single-track railroad running to the port and a water supply completely inadequate for large numbers of encamped men and the ships which were to carry them over to Cuba. No embarkation schedule, or program had been, or was, worked out while there was time in which to do it. No staff officer was set at logistic problems, no more elementary than needful. Troops and their equipment were distressingly unready when orders went out from Washington, where there was apparently complete ignorance, or disregard of the disorder and disorganization at Tampa, and reached General Shafter on May 30: "Debark at various points east or west of Santiago under protection of the fleet, as you think convenient; advance to the heights which dominate the port or to the interior, whatever permits you best to capture or destroy the Spanish garrison . . . or with the help of the Navy take or destroy the Spanish squadron . . . Cooperate closely with the Navy. When do you start?"

Confusion Worse Confounded

Thirty-seven transports had been assembled. It had been loosely estimated that they would carry some 27,000 men; actually, they could take only 18,000 to 20,000. Shafter on the 1st of June telegraphed that he would be ready for embarkation on the 4th; on the 4th that he would go on the 6th. Troops did begin to take ship on the 7th. There was the greatest possible confusion, especially after word had gone out for every unit to choose and obtain its own transport. There was no assignment of particular boats to regiments.

When the troops finally got aboard, the Navy, having previously notified all concerned that it was ready and its convoying ships waiting, decided that it could not sail in protection of the transports because some of Cervera's cruisers were reported to be on the loose. It took nearly a week to run these unfounded rumors down, during which General Miles, seemingly in sarcastic vein, generously offered to disembark the troops and equip the transports with artillery to help out the Navy.

Shafter finally got off on the morning of the 14th. His Fifth Army Corps included 819 officers, some 16,000 enlisted men and 89 newspapers correspondents—.0921 per officer—2,300 mules and horses and one captive balloon. The convoy included two water boats, a steam-lighter, two pontoons, one of which sank en route, and carried on board the ships 153 debarkation boats, sufficient for 3,034 men; in addition to the lighter, which was big enough to take 400 men ashore. In the choice of landing boats no special endeavors had been made to meet beach conditions, or much thought given as to who was to man the boats, apart from the ship crews, which would have meant exposing civilians to fire, something usually avoided in modern war. But the possibility of landing under fire seems hardly to have been even contemplated by the planners of the expedition.

The crossing to a point near Santiago under naval convoy, which had somehow been arranged, but not without acrimonious argument and friction, took six leisurely days, and without molestation by the Spanish. The troops had a thin time, with poor quarters and worse food. At the end of the six days the water supply was exhausted, which made it imperative to get the men ashore quickly. The question of who held supreme command over the expedition seems to have remained unsettled, including jurisdiction over the captains of the chartered steamers who, when nearing the island coast, showed considerable hesitation at exposing their owners' property to enemy fire, if offered.

Certain Disaster Predicted

There was no blackout at night, no silence imposed, and all danger from Spanish gunboats possibly darting out of Cuban bays blithely ignored. Foreign military attaches with Shafter huffed and puffed discouragingly about landing enterprises. They took no stock in them, for they then were not in style. "One, a French major, who was very friendly [perhaps the one who wrote a scathing report of what he saw in the *Revue Militaire* of the Paris General Staff] said it was certain to be a disaster," as Shafter remembered. He himself devoted considerable reflection to various earlier attempts at invading Cuba which had failed.

When the convoys arrived off the chosen landing beach, Admiral Sampson came aboard Shafter's ship, ready with a plan for the troops to land and storm the forts at the entrance of Santiago harbor from the land side, and thus allow the Navy to steam in and annihilate Cervera's squadron. This amounted to something on the order of a reversal of the opening stages of the Dardanelles enterprise. Shafter saw in it nothing but a scheme to let the Army in for bloody work and high casualties while the Navy collected all the glory at slight cost. His own plan was to make the taking of Santiago an army victory by assailing the city itself from the land side.

The landing, as he decided, would take place at Daiquiri, 18 miles east of Santiago, with the help of the Navy. After that, combined operations would be considered ended. Shafter's Quartermasters Department's landing preparations had originally been planned in the light of the Secretary of the Navy's warning to the Army that no naval personnel could be spared for the work, because it "ought not to be fatigued by the work incident to landing troops." But Shafter did manage to obtain from the Navy steam launches and other small boats. The landing itself was to take place under control of a naval officer as beachmaster. Fire support from the naval units was also arranged.

The landing was set for the morning of the 22nd, at an indefinite hour. No particular attention had been given to possible opposition from the Spanish forces, or to their positions and strength. But in the end this deadly military sin brought no punishment, for the Spanish conveniently and obligingly made themselves scarce, although they must have observed the sloppy preliminary work of the Americans and might have offered at least troublesome opposition. They marched off at 5 A.M., a fact which insurgents ashore tried unsuccessfully to signal promptly to Shafter. By this the Spanish, who later at San Juan Hill, really did display vigorous fighting spirit, "had carefully thrown away their best—indeed, their only—chance," as Walter Millis rightfully insists, to inflict serious losses on, or perhaps even stop the landing of, the American forces. These were the only ones the United States then had available, or would have for quite some time to come.

By 6 A.M. the naval launches set out for their assigned transports whose captains could not be induced to move their ships close to the shore. Some 600 men remained on board of one transport eight miles from the shore without anyone knowing where they were, with four launches "vainly seeking her far out at sea." By 9:40 A.M., after most lanches and boats had done their job of taking the soldiers aboard, the warships opened a "furious fire" of thirty minutes against the evacuated Spanish trenches. They killed two of the insurgents who had occupied them. There were no obstacles to the landing, aside from the poor debarkation facilities on shore. The troops were fearfully seasick, but were safely unloaded, except for two negro troopers who were drowned.

WIRE NIPPERS MISSING

By the evening of the 22nd altogether 6,000 mere were ashore. A number of artillery and pack animals had been made to swim to land, after they were pushed into the sea from the cargo ports of the steamers. But not until the 25th was everyone and everything landed. Or nearly everything, for on the 29th a cavalry regiment, dismounted, found that it was much in need of its wire nippers left behind on board its transport. Cuba was the first theater of war in which barbed wire was used on a large scale. It is interesting to conjecture what would have happened at Daiquiri if the Spaniards had strung wire along the shore, or under water and behind it had put up a fight.

The chief lesson which the Army drew from the landing in Cuba taught it to worry along without the Navy on its next similar operation, in connection with the expedition against Porto Rico. Instead of directly advancing upon San Juan, the capital, as at first intended, General Miles landed his forces at the other extremity of the island. Beginning on the 28th of July, it took nine days to get one brigade and its materiel on land. This inordinate time lag would have been more than sufficient for the Spaniards, had they possessed the required belligerent enterprise, to throw their entire considerable force between the Americans and the capital.

Their lethargy allowed the invaders to make the conquest of Porto Rico at their leisure, and the Army to do without any assistance from the Navy, worthy of mention, and also to avoid a cooperation in which the Navy, through its spokesman, Mahan, told the Secretary of War when it was discussed, that "he did not know anything about the use, or purpose of the Navy," and the Army generally how to "do the part which the proper conduct of war assigns to it." These expressions "rather amused the President, who always liked a little badinage," as the Secretary of the Navy wrote. But McKinley, the final constitutional, rather than technical arbiter of combined operations by American forces, could accomplish nothing, and was compelled to limit his activities to admonishing generals and admirals that "they should confer at once for cooperation."

Practically everything, politically, tactically, technically, or strategically speaking, in combined operations that could have been done wrong was done by the American authorities in the war of 1898. The stars in their courses fought for them, and luck was with them, as it so often is with the improvident and undeserving, rather than with the efficient who, incidentally, in this war were conspicuous by their absence. But luck in military operations teaches nothing and hinders rather than helps military program.[1]

NOTES, CHAPTER 34

[1] The above is based on the *Annual Report of the Secretary of War*, 1899; Walter Millis, *The Martial Spirit*. Boston and New York, 1931; an unsigned article on the Cuban War in the *Revue Militaire*, Rédigée à l'Etat-Major de l'Armée, Jan. 1900.

Chapter 35

AMPHIBIOUS OPERATIONS, RUSSO-JAPANESE WAR

WARS BETWEEN OTHER NATIONS and the manner in which they are fought are of limited value in their applicability elsewhere, and as guides to warring, or how not to wage war. Their usefulness depends largely upon the judgment of military observers, and even more upon that of military and other officials in home offices as to the degree to which information obtained from wars of outside belligerants may profitably be utilized. This applied especially to the Russo-Japanese War. Even though this was the single large-scale conflict between the Franco-Prussian War of 1870 and World War I, military and naval authorities of neutrals and of governments not-so-neutral, believed that campaigns from which they were far distanced could teach them little, and perhaps least so in their amphibian aspects, political, strategic and tactical.

The Japanese and the Russians fought a littoral war; more so in its early than in its later stages. But its amphibian features seemed too much bungled, as they often were, to afford valuable instruction, and negative features are not lightly to be taken as lessons. In spite of what proper cooperation of the services of the winning side and the errors of the losing side might have taught, the services in the western world remained as far apart as did the army and navy in Russia. This tendency the Japanese afterwards came to share.

The landward urge of the Japanese was old in its life span and in its nature, the land hunger behind it once more reversing the usual feudal *conquista,* that of proceeding from continent to island. Technical preparations for this war were started well in advance of the opening of hostilities, which in turn preceded the declaration of war. Diplomatic maneuvers, to veil the opening move of the war and to give the warmakers advantage of surprise, were as well-timed and diligent as before Pearl Harbor. The mobility required in 1904 and 1941 for Japan's violently abrupt strokes was furnished in both cases by the more modern arm, sea power at Port Arthur and air power at Pearl Harbor, and in either case with the intention of crippling at the outset the enemy's sea power, or sea *cum* air power;

of destroying, or reducing his mobility and thereafter proceeding with Japan's landpower.

The Russian forces were massed at Port Arthur, where the larger half of the Eastern Asiatic fleet was stationed, as well as some 22,000 soldiers, and at Vladivostock, where in those waters Russia had the lesser half of her naval forces and some 45,000 men. There and behind and between these two bastions were altogether 73,000 men, a small part of Russia's 4.5 million available fighting man-power. The Japanese army, around 800,000, had its forward units at Sasebo on the Tsushima Straits. Its problem was to transport to the continent more soldiers than Russia had immediately available and bring their weight to bear before the slow moving Russian reinforcements —it took a battalion of infantry at least a month to travel over the Trans-Siberian railway from Moscow to Port Arthur—could be assembled to press them back into the sea. The longest Japanese overwater approach, from Sasebo to Pitzuwo, was 750 miles as against 5,500 miles from Moscow to Port Arthur.

The Russian problem was whether to fight a littoral, or a continental war; or which happened to be nearly the same thing, mainly with their army, or with their army *cum* navy. For the latter they were even worse prepared than for the former. While on the Japanese side the concept of war was one and indivisible and the two services were held by the Emperor Meiji and his advisers to one and the same purpose, the Russian forces differed radically in their outlook and always were at cross-purposes. Their commander-in-chief, the Czar, was unfit to straighten out the muddle and unable to impose his authority on the side of order and organization.

The viceroy in his eastern possessions, Alexeiev, was an admiral and being in high favor with the Czar at the outbreak of the war was made commander-in-chief of all Russian forces in the east, including the army, commanded by General Kuropatkin. Kuropatkin had proposed an Antaeus-like strategy to the Czar—to retire from the Liaotung peninsula to the interior until sufficient forces had arrived from Russia to make it possible to assume the offensive. Meanwhile Port Arthur and the ships in Asiatic waters were to fend for themselves.

This proposal ran counter to the ideas of the Russian naval men who were convinced, and wanted operations to be based on their conviction, that a defeat of their own fleet, considering the ratio of strength, was "absolutely excluded and that therefore Japanese landings at Newchuang and in the Gulf of Korea were unthinkable." But as the ship losses mounted, doubts on Alexeiev's part of the invulnerability of Port Arthur increased and he insisted, with the Czar's approval, that more be done to aid the fortress to sustain itself. The navy heads finally got around to realizing that Port Arthur must be saved to preserve what remained of the fleet in the east. This fatally watered down Kuropatkin's earth-bound strategy.

Miscalculated Port Arthur's Strength

Weaker than the Russians had calculated, Port Arthur still proved stronger than the Japanese had estimated. To obtain naval superiority in the Yellow Sea, they had attacked the Port Arthur squadron in the night of February 8-9, did much damage and blockaded the surviving Russian ships in that fortress port. At the same time they destroyed two Russian cruisers at Chemulpo, the port of Seoul, and disembarked there the first echelon of their land forces, seaborne by three transports under torpedo boat protection. Landings on Korean soil nearer the Japanese home shores were inadvisable since the southern part of the Peninsula was practically roadless; highway conditions in the north during February-March, were not much better. In its slow advance from there towards the Yalu River the First Japanese Army could be based successively on the harbors of Tsinampo, Yonambo, Wiju.

On the first of May, the first land battle of the war was fought on the Yalu, giving the Japanese complete control over Korea. Their success set the signal for the landing of the Second Japanese Army at Pitzuwo, some forty miles from Port Arthur. This maneuver showed daring. Scarcity of transports allowed the Japanese to carry only two to three divisions at a time, and of these an energetic and enterprising commander at Port Arthur could easily have disposed. Japanese success here was largely due to Russian inertia during the three weeks it took to complete this landing.

At the beginning the first Japanese division on shore was isolated by storms from the rest of the landing force for a whole week, but still the Russians did not stir to molest the enemy. On the whole, the Pitzuwo landing, nearly four months after the opening of the war, was no logistic masterpiece. Actual performances fell far behind schedule. A divisional embarkation and debarkation figured to require three and five days respectively, used up as many as 15 days.

Even though the Japanese had carried two armies to the continent, their control of the sea still did not seem safely established. To make it more certain, they tried to block the entrance of Port Arthur by sinking in the channel no less than eight steamers together with their crews. But success was far from complete. While subsequent attempts of the Russian ships to break out and get to Vladivostock were foiled, the Russians were still able to sink three Japanese transports, while two Japanese battleships ran on Russian mines and went down.

Later combined operations of the Japanese land and sea forces in the tactical sphere were none too successful. Naval units which were to support a land attack on Kinchou, the first land fight for Port Arthur, arrived later than scheduled and the attack consequently failed and had to be repeated. This time, with Russian ammunition running low, the troops of the Japanese right wing pressed forward breast-deep through the water, thus turning the enemy left. The taking of Kinchou was followed by the storming of Nanshan, still closer to Port Arthur, on May 26, when on both sides gunboats of shallow draft delivered enfilade fire.

These successes gave the Japanese the port of Dalny with all its installations intact and ready to provide the much needed base for the further siege of Port Arthur. The fire of Japanese ship artillery against land fortifications was less effective than had been anticipated, but after having won a dominating height the Japanese were able to pot the remaining Russian ships in the harbor and they prepared with greater equanimity against the coming of the Russian Baltic fleet which had left Libau for its long voyage to Tsushima.

This fleet turned out to be a bad investment of steel and money in sea power, when instead it should have been put into rails and other components of land power transportation. It was fundamentally the same enormous error in the use of steel and manpower which Germany committed in building her 20th-century navy. But Russia in 1905 made the mistake of considering her defeats in Manchuria as final, a thing she had not done when opposed to Charles XII and Napoleon. An error which she was not to repeat in World War II.

Chapter 36

NAVAL COMPETITION BEFORE 1914 AND THE LANDING PROBLEM

WHATEVER THE MOTIVATIONS for their construction, none of the navies that came into being during the 1890s, with the exception of the British, were built for an imperative strategical purpose. The various legislative bodies had wanted and voted them. Naturally, the navy men wanted them also, for their great professional aggrandizement. The armies, wherever they had been in a position to do so and regardless of their jealousy of the rival service, had not used their influence to veto them. On the whole, the senior service had not cared much one way, or another. In none of the great powers existed machinery for expert discussion of, and decision on, purposeful division of the national land power and sea power strength. There were ample provisions, constitutional and otherwise, governing the division of men and money between armies and navies and for apportionment of their respective tasks. But this practically ignored the common product of the supreme purpose for which these tasks might be undertaken—the best development of the war strength of the nations.

Nowhere was this more apparent than in the Second German Reich. The German navy was not an instrument of war which the German army had conceived or desired, but it had not greatly opposed it, although some survivors of the Moltkean school became disturbed by the burgeoning of the fleet. Moltke's first successor, Waldersee, saw with pained surprise at the end of the 1890s that "inside the navy the idea is more and more cultivated that the wars of the future would be decided on the seas. What does the navy plan to do if the army should be beaten, either in the east or in the west? That far the good gentlemen prefer not to think."[1]

The politically more powerful army tolerated the navy, since to its directors the new arm appeared not to take away from the army anything that it needed in the way of funds and men, including desirable officer personnel. The navy prospered by this sufferance on the part of the army. The leaders of the latter thought but rarely of the consequences which the existence and constant growth and policies of the navy might and did entail, believing that it would be the task of the Reich's diplomacy to avoid the very worst that might be expected, a war not on two fronts—against that the land forces were prepared—but on three, or four. From about 1900 on the army would seem to have abandoned all thought of the dangerous political consequences of Germany's naval construction.

The price the navy paid for this sufferance was not cooperation, which was not called for by the army. What it was interested in getting was merely what one might describe as the Prussianizing of the heretofore somewhat more liberally-spirited navy—more drill, more parochialism instead of the wider outlook on the affairs and knowledge of the world which the German navy men had once shared with those of other navies. Only a minority of army men thought the navy useful, actually or potentially, and these were naturally inclined to reckon on its utility in combined operations, the combining principally to be initiated, if not also arranged, by the army.

Just before the end of the century it came to the attention of some specialists among the officers in Berlin that a more useful co-existence of the services might be worth discussing, at least as far as the strategical aspect was concerned, while the political side of the question was easily dismissed as not falling in the province of the officer. Most of the older officers who took part in the *fin de siècle* discussions in their public and literary sector—and some participated in official discussions as well—thought a war with England to be simply impossible, politically and technically. They would insist that the German Navy existed for the purpose of making a combined effort of the French and Russian fleets on the seas impossible.[2]

Army-Navy Cooperation Essential

The foremost specialist on army-navy cooperation, General von Janson (retired) demanded concordant action by the two. The navy must not act in complete independence, seek battle with the enemy on its own, driven by its zest for fighting; it must respect the general war plan and stay within its framework; a common strategy must prevail in every moment of the war. In the more technical field of tactical cooperation the General foresaw nothing beyond the existing extreme division of labors except perhaps in operations along the coasts. His book, *Strategical and Tactical Cooperation of Army and Navy* of 1900[3] was, according to reports of the U. S. Naval

Attaché in Berlin of, April, 1901, "read and studied by the German Army and Navy officers and since the war with Spain has been the chief theme of study in the General Staff in Berlin." [4]

While this is vastly overstating the actual German interest in these problems, they were doubtless somewhat engrossing in Germany and elsewhere until about 1903. H. von Gizycki, a writer on military tactics, dedicated three numbers of his *Strategic-Tactical Problems with Solutions* [5] to the up-to-date problem of landings. They may be said to have anticipated Sir John Fisher's project of a landing in Pomerania.

No. 14 dealt with the problem of the initial measures to be undertaken by the defense against a landing. Red has declared war on Blue, Germany, at a time when Blue was fighting in the east and west, has defeated the German Navy, blockaded the German harbors and has landed strong forces on the Island of Rügen on July 1. Blue, anticipating a landing, holds three infantry divisions in readiness at Hamburg, Berlin and Stettin. The reports reaching Berlin, the seat of the coastal corps command which united the three divisions, on July 1 make clear the intentions of Red, enabling Blue troops to be set in motion at once and concentrated, thanks to prearranged transport schedules. Data follow on time tables, assembly of the defender's troops and the situation up to July 4, the eve of the offensive.

No. 15 deals with the landing of Red, two army corps strong, and its movements to July 4, although not with the landing itself which is presumed to have taken place without hindrance. No. 16 brings the end of the drama. The landing corps is being encircled in its fortified camp and the landing attempt must be considered as a complete failure.

Elsewhere similar studies were undertaken. The American Naval War College went into the problem of the "transport of troops and armies generally by sea." [6] Russia in 1902 and 1903 exercised regiments along the Black Sea in firing on targets representing a hostile landing detachment, at distances of 1,000 and 2,000 paces. The firing was by company salvos and the results were considered unsatisfactory, partly because targets on the water proved unfamiliar to the troops.[7]

The proximity to England of France, where near the end of the 19th century Foch was directing studies on landings, made the French Britain's most dangerous enemy. An invasion from France appeared more feasible than from any more distant coast. French officers, such as Vice-Admiral Jurien de la Gravière, had long since proposed to make "the navy re-enter the game of the armies"; his proposals for this happy reunion included such technical innovations as the setting up of a "school of debarking." [8] However, on the whole, during the thirty years of Anglo-French rivalry after 1871, landings on British shores, or combined operations were not the methods by which the French intended and planned to fight a war against Britain and which oscillated between the methods of the *jeune école* and the *guerre de course* schools, including raids on the English coast towns and English trade and food supply.

Germany being the constant and main enemy, this would not allow the setting apart in any mobilizatilon scheme involving 30, 50 or 100 thousand men as an expeditionary force against Britain. Nor was an English attack on France, except perhaps the island of Ushant, much to be feared. Actually, British combined operations in case of war with France were to be directed against colonies such as Dakar, Madagascar, New Caledonia (1895). Still, French coastal fortifications along the Atlantic were kept up, even if somewhat spasmodically, and seemed to inspire respect from the British during the Fashoda crisis of 1898.

This was one occasion when a French invasion of England was seriously considered, and then more seriously, it must be said, by laymen than by experts. Warlike preparations both military and naval, offensively and defensively, were made in France, and went beyond the diplomatic denouement. These preparations were caused by English bluster that the opportunity might be propitious for bringing on the "inevitable war" now rather than later. French bitterness over her crushing diplomatic defeat caused talk in service circles that France should unite with Germany against Britain.

FRENCH CONSIDER DOVER BOMBARDMENT

The possibility of bombarding Dover from Calais was considered. Service journals and other magazines discussed invasion projects, even if merely historically at times, but that was also one of the methods by which the French General Staff studied the problem. Its own organ, the *Revue Militaire de l'Etranger*, published in March, 1899, an article on Moltke's plans for landing in the Danish isles in 1864, with the parallel scarcely disguised, quoting with approval Moltke: "If the enemy takes rational measures, success is uncertain, but one must not hesitate taking the offensive if one has no other means of reaching the end in view. . . . What can it matter if our ships are sunk so long as the game is won."

Writing for a more general public, the *Revue des Deux Mondes*, March 15, 1899, published a proposal, altogether too reminiscent of Napoleon and other would-be invaders, of constructing some 1,500 pinnaces of small draught and sending them forth from the mouths of rivers and canals of Northern France, each armed with a quick firing gun and a machine gun, and designed to carry an infantry company, or half a company and 24 horses; and enough of them to transport 170,000 men. The danger of their being sunk by gunfire was to be eliminated by building the boats with water tight compartments and that of being rammed by the enemy's warships by equipping them with torpedo tubes.

A little earlier, and perhaps not under the immediate impulse of the Fashoda crisis, the French Army and Navy had undertaken their first manoeuvre in combined operations in the summer of 1898, under the direction of Colonel, later Marshal, Foch. The subject was introduced in the École Supérieure in 1898-9. If the invasion preparations were not actually developed very far, they were still thought to serve as a deterrent to the British by immobilizing large parts of their army at home, keeping the latter from attacking French colonies, and holding a part of the fleet in the Channel. Even the school favoring the *guerre de course* thought that "the menace of a landing has such an unreasonable influence on the English" that it was advisable to make elaborate and not too secret preparations. Permanent preparations included doubling the railway lines in the northern departments and measures to ensure the presence of sufficient rolling stock on the roads leading to the northern ports.

Military Heads Unworried

While some of these French activities might have disturbed the British public, the naval and military offices were not greatly alarmed. The military attaché in Paris reporting upon the French preoccupation with combined operations in April, 1899, after the Fashoda episode had come to a formal close, called it "a very seductive subject, and it is my opinion that the French never have understood it and never will understand it, a fact that makes an eventual attempt all the more probable." Someone in the Admiralty agreed that in case of war with England "the French Army having nothing to do and being ignorant of the difficulties attending overseas invasion and driven by 'the popular outcry,' will probably revive the idea of invading this country," but some still higher officials considered it "too good to be probable."

The next invasion scare in England, that of 1900, came as a sudden realization that the recklessly undertaken Boer War had left England no friends and had practically deprived her of military and naval forces at home. It received part of its sustenance from French articles once more projecting a landing in England. These included writings by naval officers in *La Marine Française*, the organ of the *jeune école*, one of whom explained how 90,000 men could be carried across in fishing boats to be collected in a few hours in the ports from Le Havre to Dunkirk.

Other fanciful inimical schemes were imagined by the English press, inspired in part by the fact that the French fleet was holding manoeuvres in 1900 for which both the Mediterranean and Atlantic squadrons were concentrated in the Channel and the Bay of Biscay. This was unusual, since the two customarily held their manoeuvres separately, and coincided with another rare event. For the first time in thirty years, the great army manoeuvres, mobilizing four or five corps, about 150,000 men, took place in the northern departments instead of along the German frontier, and at the same time as the naval manoeuvres.

Being in the field, the corps might easily be entrained, embarked and then be despatched together with the fleet for an invasion of England and a new "Battle of Dorking." Some of the more apprehensive of the alarmists expected the invasion in November, even though more sober minds pointed out that this month was an unlikely time since then the old conscripts in France would be leaving their regiments for home and the recruits coming in, thus making the troops least fit for such dangerous service. The Ministry and the Admiralty in London declined to take the scare and the reasons advanced for it at all seriously, even when a former war minister and member of the *Conseil Supérieur de la Guerre,* General Mercier, in a speech in the Senate (Dec. 4) insisted on the practicable character of an invasion of England.

A NEW FEAR

The English newspapers, having regained their equanimity, now called such schemes on the part of the French "acts of national suicide." The latter from then on seem to have abandoned such belligerent projects and little was heard about them after 1902, except in connection with the fearsome new invention, the submarine, which at the outset was considered by the English and the French, who had promoted construction more than their neighbors, as somehow lessening the danger of a mutual invasion and strengthening the defense of both.[9]

At that same time naval officers would direct combined operations wherever the marines usually at their disposal might not suffice and army forces would be required. Where one would actually employ forces of the other, the chief who loaned them would consider himself but poorly rewarded. "I wish the Army appreciated the excellent work done for it by the Navy," an American naval captain wrote to his chief during the Spanish-American War, "but our sister branch of the service is a spoiled child and takes every exertion on our part as a matter of course. From its point of view the navy is but a handmaid of the Army. Some of the things done lately have not been calculated to soothe the nautical temper. Especially it is hard for us to put up with the irritating assumption of superiority." [10]

One would think that officers' thoughts on combined operations might have derived from technical considerations. A historical survey, however, of the problems involved as they were viewed by German land and sea officers, leads largely to the conclusion that the motivation was as often political-imperialistic as military-technical. The most highly placed advocate of combined operations in Germany, including an invasion of England was, strangely enough, a man who earlier had declared large scale landing enterprises altogether impossible—von der Goltz Pasha as he was now, having returned from twelve years' service in Turkey (1883-96), where he had reorganized the army which defeated the Greeks in 1897. Like a tropical disease, he brought home strong Anglophobe imperialism. In 1898 he was made chief of engineers and inspector general of fortresses. He brought about a much closer touch between army and navy in the Reich, notably in coast defense, the army garrisoning the four fortresses of Borkum, Swinemünde, Danzig and Pillau with two regiments of unmounted heavy artillery, and also in the study and practical demonstration of overseas transport and landings on hostile coasts, for which purpose engineer officers and 40 war academicians were put on board ships and took part in the exercises of the Admiralty staff and its studies of sea transports and landings.

In September, 1899, when he was still engrossed with these questions, Goltz wrote to a colonel of engineers that "a continuation of our experiments is urgently required and may perhaps unexpectedly soon take on a practical importance. . . . If we had positive aims and not merely the negative one of *paix à outrance,* we should have brought the Turks so far by now that they, led by German officers, could occupy Egypt and march to the Indian frontier along the Persian Gulf, and we ought to be ready, in an alliance with Holland, to proceed from her coast to cross over at the favorable moment to the coasts of Albion, like William the Conqueror and the third of the Oranians."

Invasion of England Tempting

The Boer War which was to make German Anglophobes still more virulent had not yet begun, and Goltz added cautiously: "If one does not want to be considered insane, one must not even talk of such things." He did, however, allow himself to be persuaded by Tirpitz to state publicly his views on *"Sea Power and Land War,"* a magazine article in March, 1900, in which he declared that "it would be a mistake to consider a landing in England as a chimera. The distance is short enough if an admiral of daring succeeds in securing supremacy on the sea for a short time." [11]

In October, 1900, and later there were discussions of landing exercises among the Berlin high authorities, in a committee including Goltz, Schlieffen, the inspector general of heavy artillery, Tirpitz, and Diederichs, as chief of the Admiralty staff, later replaced by Admiral Büchsel. Goltz thought the riding would be slow. Büchsel's views proved unusual; Goltz was satisfied that a "section for landing preparations" would be set up in the Admiralty staff; to have it there seemed better than having none, since no one wanted to let him create such a section in his own office. "In the General Staff one would allow it to be starved to death, if only for the reason that it has not sprung from the great tent show's own brain."

On one occasion in this committee a war with England was discussed in the presence of the Emperor. The majority viewed it with great concern. "All our sea lions present were of the opinion that in such a case nothing was possible except giving a grateful receipt for the beating received," and when Goltz pointed out that if Germany should make use of her influence in Turkey, England might very well be attacked via the land route in India, Grand Admiral Koester looked at him "as at a person about whom one suddenly discovers that he has become insane." But perhaps the Admiral had merely thought of Bonaparte. While Goltz was one of the candidates to succeed Schlieffen, he realized as early as 1902 that "a chief of staff who for example would urge the absolutely necessary first measures for the struggle of existence against England, would not long remain in his post," and this highest office in the German Army passed him by. And long before Schlieffen had retired in 1906, the discussion of landing enterprises between department chiefs in Berlin had been dropped, while his successor, the younger Moltke, was not inclined to take them up again.[12]

Only *after* harmony between Army and Navy had been brought about did attempts follow to make their co-existence in the Second Reich sensate, useful, purposive. Once created, Goltz wanted a big Navy, and appreciated its value because he had become imperialistic and anti-English. Whether as a military thinker and technician of some consequence he succeeded in vitalizing his ideas into an over-all strategy, with relative strengths and functions assigned to both Army and Navy, we do not know. Most of the published discussions of officers on the co-existence of the services, which hardly reached the level of Hans Delbrück's reminder of the interdependence of land war and sea war,[13] took two directions: (A) When army officers wrote about the navy, they thought it could, or might and should, help the army; (B) when naval men thought about cooperation with the army, they wanted the latter to help out with land fighting personnel in enterprises conceived of, and carried through, by the navy. Generally speaking, this was the state of affairs in all the great powers, although perhaps nowhere were people as outspoken about it as in Germany. This frankness, whatever its political folly, provides the historian with more material than he finds in the case of the other powers.

Only rarely did a general of Moltke's generation see the new Navy in a favorable light. An exception was General J. von Verdy du Vernois who wrote in its favor early in 1900,[14] saying that he remembered only too well from his fifty years of service how at the outset attempts at building up a navy had met with no particular sympathy in army circles, which were rather apt to insist that Prussia-Germany was a continental power. But the day had arrived when all powers of the first rank were competing with one another all over the world and felt forced to pursue a world policy. Whosoever participated in world policy, must have the necessary instruments, including a navy. While Germany still remained menaced from her continental neighbors, the possibility of a war on the world's seas and on distant continents had come considerably closer. Leaving aside pure naval war:

> In how far is the increase of the navy of importance for the tasks assigned to the army? The army may come into a situation to send not only detachments but also corps and greater masses to the theater of war by way of the sea. This, however, can be done only if the fleet is strong enough to keep the sea open for us . . . Whosoever admits the possibility of a conflict with great powers which can only be reached by way of the sea, must recognize the need for a fleet which, aside from the ships necessary for local purposes, is strong enough to take up the fight with the maritime forces of the respective powers. If we exclude for the time being England, this demand arises for us definitely with regard to America, Japan, and China.

Naval Strength Postulated

Her alliances, however strong, could not give Germany the necessary naval strength, thought General Verdy; her security required naval power of her own which as time passed "enables us to meet other great powers on a basis of equality." Germany's naval superiority over small neighboring states would never suffice, for they, like Denmark, might ally themselves with one or the other of the great powers, a suspicion that lasted through World War I and led to the occupation of Denmark in the second conflict. "Unless the strength of our maritime forces is sufficient to obtain through them complete security of ports and coasts, forces of the army will be called for to a considerable extent."

The power undertaking landings has great superiority, thought Verdy, due to freedom of movement and surprise which this superiority would supply. In any case, "the nation threatened with the danger of a descent needs as a rule considerably more forces to secure its own territory than the enemy plans to use for his own expedition. These forces, however, will to a large extent be taken away from the army of operation since it does not suffice to rely on garrison and replacement troops to an extent necessary for this purpose." (This, incidentally, concedes the navy far more usefulness than official German doctrine and the actual practice of the German General Staff did, for the troops left in Germany to defend the coasts in 1914 were second line troops, *Landwehr,* at best.)

In a war against Russia, Germany's operations would be favorably influenced should she dominate the Baltic Sea. A strong fleet would help to ensure the food supply of a modern mass army, in the feeding of which, and not in its direction, would lie the greatest difficulty of the future war, in addition to the problem of providing food for the civilian population.

> In how far is the strengthening of the navy also of value to the army? The present-day extension of spheres of interest on the part of all great powers includes as well the possibility of conflicts in distant seas and on foreign continents. Many of the tasks resulting from this a fleet must be able to solve with its own forces; others, however, can only be settled by the appearance of larger units of the army. All these tasks can only be settled through the existence of a strong fleet. Without it the strength of our army cannot be applied in this field of war-making. Conditions may force us into making war over seas, provided we want to maintain the position we hold today. This position we must maintain or else we lose our political significance and face economic decline.
>
> The strength of our fleet must in this respect be measured first of all by the maritime strength which the several extra-European powers are in a position to oppose to ours. . . . The same points of view remain valid when a war with one of the European powers is considered, whose ter-

ritory can be reached only across the sea. An offensive into England does not as yet come under consideration. In a clash with continental neighbors a strong fleet could offer a support for the operations of the army, which under certain conditions can even be of weighty influence, even though annihilation could take place only through battle on land.

Strategy and Irrelistic Policies

The views of this general, invested with authority from the fact that he had served as minister of war, bring out once more to what extent prime considerations of strategy were formed by extra-technical political moves on the part of the army officer, policies highly irrealistic regardless of how often they might be called *Realpolitik*. It is obviously necessary to have possible strategic objectives politically established by all relevant powers in a nation before specific military, or naval forces are built up and set in motion. The German navy came into existence without such deliberations. Largely its existence and growth represented nothing but a favor rather patronizingly conferred upon it by the senior service; sometimes merely a good turn done to the industrious Tirpitz who went around soliciting articles from generals, providing a general would give it his political-military, but above all political, sanction *after* fleet construction had got under way, instead of previously asking, and possibly himself answering, questions such as: When and where and over what distances are landing operations feasible? What type of craft are required? What is the maximum size of a landing corps, considering available shipping space, length of line of operations, etc.?

The political motivation obtruded even when these problems were limited in discussion to an ostentatious technical study of overseas operations undertaken by a temporary member of the Berlin General Staff, First Lieutenant Baron von Edelsheim. He published the results of his studies, obviously made while on duty with the General Staff, in an 80 page brochure, *Operationen über See,* (Overseas Operations) Berlin, 1901. It had slipped by the informal military censorship, to the fury of Schlieffen, and caused Bülow to write to the Kaiser protesting against such dangerous writings, besides embarrassing the Navy propagandists. They realized its bad timing, for it proceeded porcelain shop of diplomacy. It received a good deal of attention abroad at once and more in the later years of war propaganda.[15]

The little work was a reflection from four previous wars, the Chinese-Japanese, Spanish-American and Boer Wars and the Boxer Rebellion, all of which had involved large scale transport of troops overseas. These experiences presented to future army commands the problem of preparing and directing wars across the seas as one of their most important tasks. "There is no state in the world which for such enterprises of a landing war unites better forces, more powerful weapons, than does Germany. For one thing the quality and readiness and the speed with which we are able to mobilize considerable bodies of troops are not equalled by any great power; and secondly Germany disposes of the second largest commercial fleet in the world and possesses in the fast, large boats of her steamship companies an excellent transport flotilla which in its quality cannot be surpassed even by England."

(Actually as far as the transport of German troops by German steamships was concerned, the China expedition proved that they were not prepared for it—and why should they?—and that their charges were high with the connivance of the Reich authorities who had hopes to be fully reimbursed by the reparations payments of the Chinese. England, with an older experience of shipping troops, maintained a number of government-owned transports: they were cheaper, if not always the most seaworthy, craft.)

The then current enlarging of the German fleet would make it possible to undertake long distance maritime troop transportation with greater safety. "But this Navy alone could not solve all the problems with which an energetic world policy might present Germany. The latter makes it desirable to have such states across the seas see and feel the forces of our army, states which in the belief of being out of our reach have up to now regarded Germany as a state not much to be respected. Consequently, we have to envisage, aside from landing operations in connection with a continental war, operations against states which can be reached only across the water."

Basic Principles Considered in Theory

Having stated the political reasons as he saw them, as the more irate Agrarians and Pan-Germans saw them, for contemplating and planning overseas expeditions, ranging from war involving the most vital interests of the nation to small colonial expeditions, Edelsheim entered upon theoretical considerations of the basic principles of such undertakings since the beginning of steam shipping, which, due to the greater independence of its movements, allowed basing of more definite plans on it and the evolution of partly new principles for perfecting overseas operations. The young author was sufficiently open-eyed and unprejudiced to see that the main difficulty of such operations lies in the fact that sea and land forces have to be combined in them. More history than he himself had read bears him out. These frictions can only be overcome with ease if army and navy already have been made familiar with their mutual requirements in time of peace, and if all details of their cooperation have been ascertained, tested, exercised in advance of war. Operations overseas cannot be improvised; they offer chances for success only if the functioning of their complicated mechanism has been provided beforehand.

In fact, one might say in favor of Edelsheim that his demand for the treatment of overseas and combined operations in the methodical ways of general staff work, this to then being far from general enough to include combined operations, was at the time foremost among public discussions of this complicated military problem. If organizational progress for war and military purposes had in some respects overtaken industrial and commercial organization, it had not done so in dealing with combined operations.

Reason for this lag was provided, of course, by the prolonged departmentalization of war in the very era when war was becoming total again. The independent survival of army and navy departments without an overarching organization was like the co-existence and competition of two industrial companies whose capital was identical, whose personnel was necessarily specialized, but could not unite as to how they should handle certain tasks confronting them both. Rational capitalism would not have suffered that condition very long, but would have done one of two things—brought about amalgamation, or set up a holding company, or *Dachgesellschaft,* a roof-company as the Germans call it. In military language that would have meant either one ministry of defense, or of war while absorbing in it the earlier separate ministries for the army and the navy, or by creating a super-ministry above the army, navy and airforce ministries, as Germany finally did in the 1930s.

Surprise a Fundamental Factor

Edelsheim emphasized *surprise* as one of the first conditions for the success of overseas operations; for against surprise even a strong defender can hardly concentrate sufficient forces in time to oppose the invader. Hence the necessity of preparing for mobilization in peace time, thus insuring quickness of mobilization and transportation in case of war. The long delays in the sending of American troops to Cuba in 1898, and of British troops to South Africa demonstrated how difficult and comprehensive this work of organization can and must be. *Secrecy* concerning the objective of the operation is the next requirement and the enemy must be misled at least about the purpose of the first preparations. Bonaparte's expedition to Egypt offers a good example on this point.

Post-Light on Edelsheim's Views

In the light of events subsequent to 1901, palpable interest attaches to Edelsheim's views, expositions and deductions as they are given in digest below.

A landing operation against a hostile coast, he postulates, is feasible only if one has superiority over the naval forces which the defender can bring on the scene at the decisive moment and oppose to the landing. Once the landing has succeeded, a subsequent victory at sea is of no use to the defender unless he disposes of sufficient land forces to fight the invader successfully. From this it follows that the German navy must command sufficient strength at least to insure the transit of troops and defeat, or keep in check, that part of the hostile fleet which the enemy at the time of the intended landing has ready in the neighborhood chosen by him for his invasion. Hence naval operations have to precede the overseas transportation of troops and the possibility has to be envisaged that if these earlier operations do not succeed, the landing operations cannot be undertaken.

In principle all available units of the fleet must be used for such an overseas operation, either in the battle fleet which may have to fight the way open for the transport flotilla, or as escorts for the convoy. Again in principle, troops should not be transported aboard warships since their presence interferes appreciably with the fighting functions of the ships. Historical example: Admiral Ganteaume, in spite of his superiority, had to retire into Toulon in 1801 before a British squadron because the presence on his ships of landing troops had considerably reduced his fighting ability.

For transports only steam craft, and these as large as possible, should be used since large boats speaking absolutely and relatively, can ship more troops, usually possess a more extensive radius of action and, being less in number, need fewer escorting warships, thus providing the battle fleet with greater strength.

This is a consideration of importance for smaller navies, such as the German. (The question of the maximum, or best size and speed for troop transports would be one of the greatest intricacy, were it not for the fact that this problem is largely solved by the type and number of the ships actually available to the transporting belligerent. The danger of being torpedoed, or of striking mines militates to a decided extent against the maximum size, which limits the chances for a safe escape of troops in case of such a shipwreck. A governing factor also lies in ship building considerations which favor medium sized standardized ships.)

The capacity of transports diminishes with the increase of the distance over which troops are to be moved. Only in the case of very short routes can transports be quickly turned about and used twice for the same operation. But in principle the main strength of an expedition should go by the first transports so as to make the main strength available for the first landing stroke. In order to deal a strong initial shock the landed troops must be not only sufficiently numerous, but also have the highest possible combat value. There is no place for second line troops in such enterprises. Combat requirements following the landing demand a combination of all service arms. Some, however, like cavalry and artillery, are not transported and loaded and unloaded with equal facility, particularly if the landing is to take place on open coasts.

Results of Artillery-Cavalry Lack

In practice this has usually led to restricting cavalry and artillery to a minimum. As an historical example, in the battle of the Alma, St. Arnaud had only 100 cavalry and four horses per cannon, too few by far to undertake vigorous pursuit operations. Lack of cavalry must inevitably curtail reconnoitering. Without artillery attack quick reduction of fortification is impossible. Besides, artillery reduces infantry losses which, in landing operations, are particularly difficult to replace. With respect to command the views of Edelsheim were:

As a rule it will be better to entrust the direction of operations overseas to an army officer rather than a naval one, for the essence of the enterprise lies in the operation of the landed troops in enemy country, and all measures, those of the fleet included, must serve this purpose. As a matter of principle the supreme command of the transport fleet and the escorting squadron will have to go to this commander, for the critical moment comes about when the troops are disembarked; and to have a change in command occur at this very moment appears more than questionable.

As things are arranged with us in Germany the commander is subordinated to the chief of the escorting fleet during the transit, a rule which the history of war has shown us to be disadvantageous. For several times overseas enterprises have failed because the commanding admiral has been moved by unfavorable maritime conditions to refrain from instantaneous landing while in the meantime the previously favorable military situation changed for the worse. On the other hand Bonaparte, against the advice of his admiral, had his troops disembarked at once and thus preserved them from the fate that struck the fleet at Aboukir.

Even after debarkation it might be necessary to put the transport fleet together with the escort or even possibly the battle fleet under the supreme commander of the landed corps in case a change in the basis of operations or in the theater of war should become necessary or if an immediate (tactical) cooperation of land and sea forces along the coast shall take place. Regarding questions of naval technique, naval staff officers will have to serve with the supreme commander. This is of evident necessity in case of a combined attack on a coastal town, but is also required in operations along the coast or in actions on the coast. (Historical example where cooperation in such a situation failed, battle of the Alma.)

Planning for Overseas Operations

Peace time preparations must be made for overseas expeditions in the same way as for all other likely operations, including the laying of plans of operation which must allow for political and military conditions—the political considerations, of course, being those of the German General Staff itself whose directors would not consult the Foreign Office on such a point, and at best might condescend to inform the civilian authorities as they did belatedly in the case of the violation of Belgian neutrality. There are no indications that the Wilhelmstrasse was ever asked about the politico-diplomatic necessity of overseas operations, even though an occasional diplomat might voice doubts about the wars contemplated by the Army or the Navy. During peace time the necessary data for landing and subsequent land operations in the various foreign countries will have to be assembled through intensive scouting and spying, the former to be undertaken mainly by naval officers, the latter only by army officers.

The navy will have to ascertain: (1) what naval forces are necessary for the protection of the transport fleet, how rear communications with one's own ports can be maintained and how they are to be protected; (2) which parts of a hostile coast are most suitable for a landing and what are the influences of tides and weather; (3) what are the conditions of the ports along the chosen landing coast, including disembarkation facilities and how can the latter be improvised if necessary; what is the capacity of such harbors and how can they be defended on the land and the sea side; (4) what is the character of the enemy coastal defences and what forces are necessary to overcome them.

One does not quite see why some of this information might not also be obtained by the army as well as the following: (1) the local objective of operations and the theater of operations including communications, obstacles, etc.; (2) estimates of the most likely strength of the forces which the enemy will be in a position to oppose to the landing; (3) the resources which the invader can draw from the zone of operations, and the local conditions, including climate and water supply, particularly so far as they may require special equipment for the landing force. Short descriptions of the invasion regions and necessary maps can then be prepared on the basis of this information. All such data are conditional upon the seasons which will exercise a strong influence on the choice of the time of landing.

Landing troops will have to be chosen from among the army corps in the vicinity of embarkation ports where they can be quickly brought. The forming of special corps such as the German expeditionary force for China, is inadvisable, consuming too much time; new formations do not possess the necessary internal cohesion and are not sufficiently in firm control of their officers. To put whole regular divisions, or army corps on a war footing will not always prove feasible, and the most satisfactory procedure would be to put one division per army corps on such a footing, drawing only on the youngest classes of reservists. The division, equipped with its share of trains and engineers including also pontoons, of heavy artillery, signalling and balloon units, will prove the most suitable unit even for large scale operations overseas.

Special Pre-Invasion Training

Troops intended for such operations must comprise personnel for whose benefit every year special courses in embarkation, transportation and debarkation, in ports as well as along the open coasts, should be arranged, and not only for infantry and cavalry but also for forces handling heavy artillery and other bulky material. Officers and non-com's of the army should be detailed in large numbers to such exercises, thus forming within a comparatively short time a body of men familiar with the techniques of transportation.

During these exercises and instruction courses war materials obtained from abroad might be put to a test, general rules studied of the simplest way of transforming a merchant ship into a transport, as well as plans for the most practical organization of the lading, for the embarkation and debarkation of troops and material, for the construction of the required machinery, for the special equipment of ships to serve as horse, telegraph, or balloon boats; for maintaining order on board and regulating the day's routine while en route.

Mobilization plans must be prepared for overseas expeditions at least in as much detail as those for land war, including choice of troops, their transportation to the embarkation port, preparations for shipment; in short, everything calculated to guarantee the quickest possible execution.

Wherever necessary railway yards and quay facilities will have to be enlarged. Depots must be installed in these ports containing materials for refitting merchant ships as transports, as well as ventilating apparatus, bridges, large flat boats and prams, steam pinnaces, all highly necessary equipment for troop transports which is usually lacking on merchantmen, apparatus for distilling drinking water, oxygen inhalators to combat seasickness, etc. In addition depots must be prepared for food and fodder, coal and special light vehicles. Since Edelsheim considered war to the west of Germany as being most likely to call for overseas troop transport—he obviously belonged to the school of those able to contemplate war without Russia's participation —he discussed in detail the embarkation problems in the North Sea ports of Emden, Wilhelmshafen, Bremen-Bremerhafen, Hamburg-Cuxhafen and Glückstadt, in most of which after slight preparatory alterations a division each could be embarked within two days.

Preparations for Embarkation

One of the most important measures of preparation is military control of ships' movements through a permanent board of officers which would also take care of embarkation. This board would take measurements, or make estimates of the loading capacities of German merchant ships; their fitness for military transportation, keep listed their whereabouts and those of other ships which might be found in German harbors from day to day, including a certain measure of alertness in watching neutral ports, Amsterdam, Rotterdam, or Copenhagen, where neutral ships might be chartered immediately before the outbreak of war and from which German ships might opportunely be recalled.

Preparations at the outbreak of war must be directed toward preserving the possibility of surprise, not an easy thing to accomplish when movements of ships are so easily followed, and therefore to be undertaken not gradually and over a period of time, but suddenly and on full scale. Only then is there a chance to keep an enemy in the dark and surprise him in the midst of his counter measures. Surprise lies not in the operation per se, but in the speed of its execution. Nothing could ruin the chances of success more hopelessly than using the threat of operations overseas as means of diplomatic pressure.

Operations begin with mobilization of the units of the fleet and the army chosen for the enterprise, requisitioning of merchant steamers for transports and their outfitting. The navy, and particularly a small one like the German, will do best by concentration on the landing operations and their protection, leaving aside for the time being the *guerre de course* and even coast protection. A certain hiatus between the mobilization plan for the expeditionary forces and the execution might be difficult to avoid. The outfitting of steamers in England for the transport of infantry to Capetown took four days and seven for cavalry and artillery, which Edelsheim thought was longer than necessary, for German steamers going to China had taken only two days for unloading, outfitting and reloading, and that for a voyage of long duration. The preparations for a shorter distance—say to England—would require even shorter time, less than one day for infantry and 2-2½ days for cavalry and artillery. This was the time envisaged by English and Russian regulations for short distance transportation.

The units chosen for the expeditionary force must be on a full war footing. Ample equipment with machine guns adds much to the fighting strength, while not absorbing too much shipping space. Whether bicyclists are to be included must depend on the condition of roads in the enemy country. Cavalry must be made much stronger than the usual equipment of the division with horse; the addition of a special cavalry force, a brigade, or division, might recommend itself for promptly obtaining knowledge of the measures of the enemy.

As for artillery, the equipment of the German army corps with field artillery was happily very ample; whether heavy artillery was to be included must depend from the nature of the intended theater of war. Trains are awkward impedimenta for sea transports. While requisitioning in enemy country might be to a certain extent relied upon, certain kinds of trains and particularly artillery trains have to be taken along complete.

Loading and despatching is the business of the land army officers trained for this work. All materials must be loaded in the reverse order of their unloading after the landing and must be so stowed in the ships that as far as possible each unit has its own material with it, in the same ship. Amount of shiploads must vary according to the length of the voyage; on a short passage, up to 48 hours, according to German regulations the regulation load can be doubled. The Italian General Staff considered five days a short voyage. If soldiers can spend three or four successive days in bivouacs without suffering depreciation of their fighting value, there is no reason to stop at 48 hours as a maximum for short voyages. The usual tonnage rates per man and horse—2 tons per man and 5-7 tons per horse, the latter rates varying somewhat in the European armies—can still be reduced as shown by recent English experiences made on trips to Capetown, as well as American transports to Cuba, when fully a third more troops than had been anticipated were taken on board. For a German war against any European great power tonnage can safely be reduced to 0.6 per man and 2.5 per horse.

Loading and Crossing Problems

Loading time: If alongside quays, 10 minutes per 100 men, 1 minute per horse, 10 minutes per gun, or vehicle, including fastening down. According to Russian experiences, an 8,000-man unit, including infantry, cavalry and artillery, can be embarked within eight hours. During the German expedition to China, which meant a long sea voyage, a battalion with all its vehicles was put on board in 1½ hours. This was considered a record by experts, although there was no reason why, with experience in more preparation and practice, these periods might not be further reduced. For was it not reported that Bonaparte, after much trial exercises in embarkation in 1795 had put 132,000 men with their impedimenta on board within two hours?

Under normal circumstances there will be available enough shipping in Hamburg and Bremen alone to carry four infantry divisions on a short voyage, or six such divisions, or five infantry and one cavalry divisions, if the ships available in other ports as far east as Mecklenburg are included, not to mention the large modern liners, which may, or may not happen to be in their home ports, or recalled in time for use in an expedition.

Subsidiary Transport Problems

Next follow technical discussions of sleeping quarters and washing facilities for men, storing luggage, arms and ammunition. Artillery and machine guns, for instance, are to be put on deck as a matter of principle, and also field artillery to be used possibly against enemy light forces by firing from the deck. Quarters for horses and means of exercising the animals are entrusted to the careful attention of the cavalry officers. Other transport problems are presented by the drinking water supply, particularly in the tropics; and special transports for coal, munitions, etc.

During the *voyage* of a transport it is the supreme task of the navy to clear the way for it and establish superiority on the sea at least for the duration of the crossing. Since the battle fleet will always be more quickly mobilized than the transport fleet, it can enter on its offensive and defensive functions sufficiently in advance of the latter's start. It is to carry on these operations while the transport fleet is on its way, but without remaining in close contact with it later. The escort ships serve the same purposes as the battle fleet, only on a narrower radius. They must not be too numerous, in order not to weaken the battle fleet; their task is merely to protect the transports against the attack of single, or a few enemy ships.

As a matter of principle, the transport fleet should sail so soon as it is ready, and even before the battle fleet has reported success in its mission. While this involves a certain risk, the danger is not excessively great since the transport fleet will always learn of a possible defeat of the battle fleet in time to allow it to return to port. Only an early start will enable it to progress at the speed necessary for it to accomplish a successful landing. Given the speed of a modern transport fleet, it might proceed on its landing enterprise under daring and energetic leadership, even if the battle fleet has been beaten. At this point the cavalry man indulges in some *va banque* speculations about the self-protective power of transports through the use of transported artillery which must have made his superiors wince.

The *weather* is still an important factor, even though its influence has been so much reduced since the days of sail as to permit a certain degree of reliable calculation to be made as to the time between home port and destination. Scientific meteorology will play its role when short-distance expeditions are in view, which might be postponed for a few days if delay is made absolutely necessary by foul weather.

A *transport once at sea* must proceed by the quickest and shortest route to the landing point. Its speed is determined by the highest speed of the slowest ship in the convoy, which might possibly be increased by the use of tow boats. Sailing masters of the transports receive sealed orders, which must not be opened until after leaving port, appointing their places in the convoy and informing them of their destination. Vanguard troops, including engineers and balloon troops, are placed at the head to enable them to speed in advance of the rest of the fleet immediately before the landing.

At the end of the column a cable ship will lay a cable to establish and keep open communication with the home land. This proposal is questionable, considering the very slow speed of cable-laying craft, although it indicates a lively appreciation of the need of providing landing troops with communication facilities, later to be provided by wireless. The headquarters of the supreme command of the expeditionary force is best established aboard a transport in order to enable it to remain with the troops in case the escorts should be required to fight off enemy attacks.

Regulations for voyage regime have to be made known for the information of the ships' companies so soon as the ships are out of port. They will largely be determined by the length of the voyage. On shorter trips the troops, except those in care of the horses, will mainly devote themselves to gymnastics and receiving instructions of one kind, or another. "Generally speaking, it will be necessary to influence the troops through rigid discipline in order to overcome seasickness and boredom." Troops on ships are under the command of the ranking officer as is also the ship's captain who, however, retains nautical authority. If there are enough naval officers available one should be aboard every transport.

Debarkation, Early 20th Century

Edelsheim says of the *landing* itself: "It is an experience confirmed by the history of war that as yet it has never been possible to stop a landing actually undertaken." While this is not altogether true, it still brings out the most reassuring, or tempting features in such enterprises. He goes on:

> The defender either has to distribute his forces all along the coast to be protected and thus scatter them, or he can with his assembled forces cover immediately only one point, whereas the attacker, thanks to the mobility of his transport fleet which leaves the spot of the landing uncertain until the last moment, can move more troops to the land by surprise and under the protection of his far-reaching ship-guns than the defender can gather during the same period at the landing point.
>
> A simultaneous landing in several spots, is therefore, if the enemy disposes of considerable force, questionable. For even if the operation of separate units against one objec-

tive, strategically speaking, appears favorable, the size of an expeditionary force is seldom large enough to cause disadvantages from landing in one spot. On the other hand, the protection of several beachheads from the sea requires numerous men-of-war; the necessary reconnoitering inland is made more difficult and the enemy will be in a better position to attack the separate parts of the expeditionary corps with superior forces. Besides, widely separated beachheads offer great difficulties for the unified command at the beginning of the operations and time and means will be wanting to overcome this.

For debarkation purposes a harbor, of course, is most suitable; and next to it a closed and protected bay, and the least advantageous is an open coast. The latter, however, offers the greatest opportunity for surprises and hence will provide the least resistance to a landing attempt. Should the original beachhead selected be in the vicinity of a harbor, or bay, the vanguard troops will have to obtain possession of these better debarkation points for the unloading of the main forces, horses and materiel.

The possession of such facilities would speed debarkation considerably, heighten the mobility of the expeditionary force and make easier the protection of the debarkation against attack by land and sea. If such a coup does not succeed, the debarkation of the whole force has to take place without loss of time by means of boats which must be brought along in sufficient numbers and models to carry ashore even the heaviest materiel.

General requirements for a good beachhead are: Close vicinity to the objective of operations in order to shorten the line of advance and force the enemy precipitately to decide upon defensive measures; good roads in the direction of the objective; in the neighborhood of the beach itself a terrain open to observation as far as the range of covering ships' guns which are to guarantee the landing against interruptions; easy navigatory approach, deep quiet water close to the shore; and existing possibilities of sealing the beachhead off on its land and sea sides.

Landing within reach of considerable enemy forces, or fortifications is impossible. Open coastal fortifications, not protected in the rear, are of value only if enough troops are available for the defense of the coast and if they cover the only convenient landing beach. Should the attacker succeed in landing even a few troops on the flank of such fortifications, out of range of coastal artillery, these strongholds become worthless at once.

PRE-INVASION RECONNAISSANCE

Surprise always offers the best security for the first landing. Reconnaissance of the hostile coast by ships sent ahead of the expeditionary force is impossible; this would inform the enemy of one's possible choice of beachhead. Reconnaissance must be undertaken long before the start of the operations, or else the landing shore must be selected from maps and charts, at least as far as the vanguard is concerned. To guarantee a quick, orderly landing, unified direction is highly important and can be ensured by signaling from ship to ship and by telephone connections between ship and shore.

Each division requires two kilometers of coast line for landing, partly for the reason that the transports have to anchor at certain distances from one another so as not to interfere with their mutual freedom of movement and safety. As a rule, the landing is to be executed as follows: The vanguard together with some escorts steams ahead of the main force during the last stage of the voyage and lands in a way to ensure surprise, possibly at night.

The beachhead is at once to be strengthened by trenches, that is, if the main force is to land as well at the original landing beach. The vanguard is to be followed by the cavalry units who have to do the first reconnoitering, although captive balloons on ships, or ashore might prove helpful; next by the main force, trains, supplies, all proceeding at maximum speed in order to secure themselves against enemy interference. The best time for approaching the coast is the early dawn; it is the prime hour for surprise and leaves the whole day for debarkation, which might be prolonged into the following night by the use of searchlights.

To effect a landing of the whole expeditionary force along an open coast, ordinary ship boats are insufficient; special flat bottom boats suitable for the landing of horses and heavy materiel must form an essential part of the equipment of the transports. Their maximum size is determined by the ability of the transports to handle them and their minimum size by the weight of the heaviest equipment.

While Germany by the opening of the 20th century did not possess any special landing craft, other powers like England and

Russia had constructed them. The English had a folding boat type for 50 men plus boat crews, the Russians steel prams 13 meters long, 3½ meters wide, with a draught of a meter, carrying 200 men, or one cannon with crew and horses, a type which seemed worth imitating, or experimenting with. To row the boats to the shore and back would be a loss of strength and time; instead, steam pinnaces have to be used and the majority of the transports equipped with them. Horses should be carried ashore in boats and not forced to swim which is apt to cause losses—7% during the American landing in Cuba—as well as disorder and loss of time.

The time required for landing is considerably shorter than that for lading. For one thing, the natural urge of the troops to get on firm soil is a contributing factor. Lord Cochrane landed 18,000 men in America within five hours on an open beach. During Russian manoeuvres in the Black Sea it took 11½ hours to land a weak division from steamers which were forced to anchor at a distance of from 5 to 6 kilometers from the shore. A French naval writer has estimated that under average conditions the landing of 25,000 infantry, 1,000 cavalry and 60 cannon would take six hours; that could be decidedly reduced if the landing was in a harbor.

So soon as the landing force is disembarked, the further uses of the transport fleet will depend on the general maritime power relations of the belligerents. If the attacker controls the seas the transport fleet might remain along the enemy coast as the base for the landing corps, or else used for bringing up supplies and replacements from the home country. If the attacker cannot permanently maintain control of the sea, he will have to try saving his war and transport fleets from the enemy by a prompt retreat to home waters.

Post-landing operations will generally conform to the principles of land warfare. But some of the specific conditions confronting landed troops should be emphasized. As Mahan has pointed out, the main characteristic of a landing operation is the offensive. Even bold landings such as that of the English at Aboukir lost their initial advantage through over-caution and disregard of the fact that energy and speed of execution more than balance all the strategic disadvantages of the operation. Quick and vigorous action with closely concentrated forces along the line of weakest, that is shortest, resistance, are first considerations for the success of landing operations. A defensive attitude following landing is justified only if coming reinforcements will provide superiority over the enemy; in all other cases mere defence will only rarely result in permanent success, as did Wellington's position in the Torres Vedras lines.

First objectives are always the enemy forces first encountered. To fight them successfully, it is of supreme importance that men and horses are landed in good condition so that top service can be expected of them without excessive losses on the march among the limited, per se, number of combatants. During the advance towards the objective of operations, the resources of the land traversed can be fully drawn upon to round out equipment and provisions for men and animals, but care must be taken that the speed of the operation and the energies of the troops are not thereby reduced, for old experience proves that extensive requisitioning costs too much time and strength.

Living on the Country

In any case, the comparatively small size of a landing corps will often allow drawing on the resources of the invaded country. This, in combination with food and feed in modern concentrated form carried by the invader, will make permanent basing on the coast and the home country not absolutely necessary. Napoleon's army lived for years on a comparatively poor country like Egypt. In a civilized, well populated and rich land all this will be still easier, since feeding and acquirement of horses and materials, and even the production of ammunition, may be managed from local resources. The maintenance of a very long line of communications to a base on the coast would easily become dangerous to the invader by absorbing too much of his limited force. A landing corps must be economical with its strength. Expensive victories may in the long way develop to be tantamount to defects. Enterprises consuming time and strength, such as attacks on fortified positions, may better be avoided. For battle that turns on the objective of operations, however, all forces must be thrown in to the fullest extent.

When and if the battle fleet has established command of the sea, it will be in a position to consolidate the base on the sea coast as well as shift the latter by the transport fleet to another point on which the invasion corps may be based in a later stage of the operations, thus increasing the mobility of the latter considerably, as was demonstrated during Sherman's famous march. Still greater mobility would be gained for the landing corps if, after having obtained one objective, it should be carried by the transport fleet to another point along the coast in order to operate from there against another objective.

If the operations of the landed troops should take place along the coast, or if their objective is a harbor, or coastal fortification, operation and combat of both army and navy units must be under one command. During the march parallel to the coast, the fleet not only provides a certain protection and additional strength; it also forms a mobile base to assure freedom of movement to the land force. During combats near the shore, the men-of-war represent in effect heavy mobile batteries, so much superior to field artillery on shore, through their weight and armor, that action against them on the part of the latter seems out of the question.

The more flank support warships can lend, the better and more intensive can be their cooperation. In such cases communication between land and sea forces is of supreme importance, so as to enable the commander to influence action of the ships. The closest cooperation between the two will have to take place during combined attacks on towns and cities and coastal fortifications. Most of the latter no doubt will be unable to withstand an energetic attack from the land side. Even for a large coastal fortress the situation is considerably more unfavorable than for an inland fortress, since attack against it takes place from two elements at the same time. If, however, the battle fleet cannot maintain control of the sea and has to retire before the enemy sea forces, the operations of the landed forces change to land warfare pure and simple.

Observations on the *re-embarkation* of landed troops will be of some interest in post-Dunkirk days. It is possible only when the battle fleet controls the sea sufficiently to guard the transport of troops across the water. This will have to be done quickly if the enemy is in the field in great strength, and it is particularly difficult after a defeat, although by no means impossible, or hopeless, as is often maintained. It will be very difficult even for a superior enemy to make reembarkation altogether impossible, although he may perhaps force the leader of the landed troops to leave behind part of his materiel and vehicles in order to get his fighting men back on board. The main requirement for the success of a reembarkation is absolute maintenance of order among the troops.

Weight of Landing Threats

The applicability of these theoretical discussions on landings is then tested by Edelsheim with regard to *landing operations in connection with land wars.* These necessarily assume the character of side operations, or of final actions after the main battles on land have been fought. Edelsheim does not share the view common in German army circles that overseas operations in connection with the war on land are either worthless, or harmful, assuming that the troops needed for them are better used in land war. He thinks the Franco-German War of 1870-1 alone teaches something quite different, showing what considerable forces even the threat of a French landing was able to tie down in Prussian territory.

In a future German war on two fronts, a landing of strong enemy forces of two to three army corps in Jutland and their advance in combination with weak Danish forces would indeed create a very difficult situation for Germany. Both the French and Russian armies had recently undertaken landing manoeuvres. Such a threat is of course possible only after the German battle fleet has been swept from the sea. The latter, if undefeated, in turn would make German landing enterprises possible in either the North Sea or the Baltic, against France, or Russia.

A large landing operation against France would not seem effective during the decisive phases of the war in the field because there was no suitable objective, Paris due to its fortifications and geographic situation offering no hope for success even to a strong landing corps. Only in the last stage of a victorious campaign could a large scale operation against th French coast be of value to Germany, when it might help t combat French national resistance in the more remote regions Surprise landings by small detachments, however, may prov useful in the opening stages since they hinder the importatio of war materials from abroad which alone enabled France i 1870-1 to prolong resistance. They may force the enemy t leave along the coast considerable forces which might be wan ing in the main theater.

But always and ever a landing of German troops in Franc will be an enterprise of secondary importance. (If we reca that ostentatiously at least the new German navy was built against the traditional, old enemy France, it is rather disconcerting to see a navyphile officer reckoning it to be of such minor importance in the war against this power). The adequate enemy for this navy, the best landing ground for a German combined operation is Russia rather than her ally France. If for example after the conquest of Poland by German forces, Russia should be unwilling to make peace, but should rather try to prolong the war and tempt the Germans into a dangerous offensive in the direction of Moscow and Petersburg, the landing of strong forces in the Gulf of Finland and their advance on Petersburg would form a far more effective operation and would exert stronger pressure towards making peace.

Conditions for such an operation seemed favorable. The narrow waters of the two Belts and the Sound are easily blocked and support from the North Sea thereby excluded. Before a superior fleet the Russian ships might retire into the fortified harbors without giving battle. There they would find themselves unable to withstand a combined attack of the German land and naval forces very long and thus leave the command of the sea to the latter, enabling it constantly to bring up reinforcements and supplies.

The distances from the German to the Russian Baltic ports made close loading possible, while Russia's poor railroad system precluded moving of enough troops to guard the whole of her Baltic coast against landings. In connection with the fleet and based on the coast a German invasion corps might even operate for a considerable period against superior Russian forces, and the Baltic coast might serve as a base for offensive operations of an army against Moscow, which is only half as far away from the Gulf of Finland as from the German-Polish frontier, and offers far more favorable terrain than the direct land route. The strength of modern ice breakers will keep winter from interrupting such operations.

Conditions on the Finnish Gulf coasts favor landings, as neither high tides, nor surf hinder debarkation. There are a number of places where deep water is found only one to two kilometers from the shore, enabling battle and transport fleets to anchor close in and dominate the shore with their guns. Debarkation can take place on a broad front, and larger masses of troops can quickly be put ashore. It does not appear that when Edelsheim wrote either the Berlin General Staff, or Russia reckoned with a German landing in these regions. By 1910, however, Russia had come to consider them more likely, consequently removing two army corps in garrison in Poland "in order to escape a Sedan" and neglecting the fortresses in that region, and forming instead two army corps in Finland against anticipated German or Swedish-German landings.[16]

U. S.-German War Forecasted

Edelsheim's ideas about "landing operations against powers which can be reached only across the seas," meaning England and the United States, give his studies real piquancy and invest them with the character of a shocker to his own superiors and to German diplomats who were working hard to maintain "amicable relations" abroad, and later on to English and American susceptibilities. The Americans largely ignored the lieutenant's little book when it was first published, although, as Cambon, the French ambassador, reported from Washington in the generally strained status of American-German relations, "it suffices that some guard-lieutenant in Berlin occupies his leisure in organizing on paper an invasion of American soil by German troops to give a new growth of actuality to Admiral Dewey's saying that the first war of the U. S. would be with Germany." [17]

The political motivation for the discussion of these two wars seems a good deal stronger than that provided by insight into and prevision of the technical and even diplomatic difficulties of such enterprises, not to speak of the "causes" of possible conflicts with the Anglo-Saxon powers. A conflict with England, Edelsheim says, must be anticipated by Germany whose tremendously increasing trade offers as much of a threat to England as Russia's advance towards India. A naval war with England would be hopeless for Germany. Even if she should ally herself with Russia and France, England would still be able to outbuild them all, bottom for bottom. "But England's weakness lies in the same factor that makes for our strength, the land army," which neither in quantity, nor quality is in keeping with her position as a great power, or even the size of her home country, due to the conviction that every attempt against her soil can be warded off by the navy.

This conviction, however, is by no means justified, for whatever naval strength England will eventually assemble around her islands, the forces that she would have ready during the very first days of a war will not be so overwhelmingly superior to a weaker naval power that keeps its forces concentrated and ready, as to discount Germany's chances for gaining a temporary success.

Germany's measures for a war against England must be directed so as to create conditions which allow her to throw parts of her land army on the English coast and thereby transfer the decision from the sea to the land; or into the enemy's insular territory wherein German troops will prove superior to the English and where England's powerful navy can no longer have the slightest influence on victory. The English land forces, aside from the regulars, would be slow to mobilize and of inferior combat value. The larger part of the forces stationed in Ireland would have to remain there to prevent the potentially rebellious Irish from supporting the German invasion which would bring them freedom. Shades of Casement and Erskine Childers!

England has immediately ready for combat only six infantry divisions of some 10,000 men each and three cavalry brigades to which a German force of four divisions of 16,000 each and one cavalry division would be superior, not to mention the additional six divisions which Germany could throw into England within a short period. "How such an operation overseas against England should be conducted, can of course not be discussed in this place."

The crossing from German North Sea ports could be performed in a little more than 30 hours, given tolerable weather. England offers extensive beaches suitable for the landing of troops and the country contains such resources that the invasion army could live on it forever. On the other hand, the extent of the island is not so great as to wear down an invasion army once it is victorious by compelling it to exert itself unduly, according to its numbers, in defending, or maintaining its positons. It is unlikely that such a war would be long drawn out and call for replacements of man-power, whereas war materials may largely be available in the country. The problem of communications with the home country might therefore be dismissed. First operational objective will be the English field army; the second London, the two probably coinciding since the safety of the capital will call for its defence, and the pressure of public opinion will insist upon it. Once London is taken, England has fallen.

A war against the United States would have to be fought in still another way. Again, such a war is not unlikely. Recent years have brought a number of conflicts due to commercial reasons which up to the time Edelsheim wrote were mostly solved by German concessions. Since there is a limit to these, the question arises: Which means of power can Germany apply in order to fight such encroachments on the part of the United States or to impose our will by force?

Effect of Taking U. S. Ports

The first power factor is the German Navy which has chances to meet victoriously the sea forces of the United States, spread out over two oceans. But the defeat of the navy alone would not force the United States, so large and possessed of such vast resources, to make peace. A blockade of American ports would not prove decisively effective. This makes it obvious that a purely naval war against America would not suffice, but must be undertaken from the outset by combined action of land and sea forces.

Operations of the expeditionary force far into the interior in order to break American resistance seemed out of the question, considering the enormous size of the country. "There is, however, a reasonable hope that victorious enterprises along the Atlantic coast, cutting off of the most important veins of export and import, will create in the whole country such an oppressive situation that the government will readily agree to acceptable conditions in order to obtain peace."

If the preparations for a German transport fleet and landing force should begin at the same moment the battle fleet steams out, it might be assumed that four weeks later this force can begin operations on American soil. By that time the U. S. could assemble no adequate force in the field to oppose the Germans, since only 20,000 of the 65,000 men in the regular army would be available for a war on the eastern seaboard. Ten thousand alone were considered necessary by the Prussian officer, whose knowledge of the redskins seems to have been derived from Cooper and the Sitting Bull series of thrillers, to guard the Indian territories.

While a surprise against America is unfeasible on account of the long distance and the time required for the crossing, a blitz invasion would make victory still easy enough, considering the absence of well ordered American mobilization plans, the inexperience of the personnel and the weakness of the regular army. Surprise would figure only with respect to the choice of the landing beaches. Occupation of large areas of American soil would not force that nation into making peace; this was more likely to come about through the successive conquest of several of the populous Atlantic harbor cities and the enormous material damage done thereby.

The object of German operations would therefore be for the army and navy, working in close cooperation, to take a number of these cities and thus upset the whole economic life of America. It would be difficult for American defending forces to contest these operations successfully. The attacker will have finished his operations before the defender has brought sufficient forces to the front lines over the railway system, which is in itself so splendidly developed. "Germany is the only great power able to attack the United States all by herself. England might be able to fight through to a victorious end the attack on the sea. None of the other great powers, however, by itself alone has at its disposal the transportation fleet necessary to undertake such an operation." This statement appealed more strongly to the rabid Pan-Germans than to German diplomacy.

Here we take leave of Edelsheim and proceed to examine the views of other Germans.

In the American navy the German navy thought it had found its eventual enemy, one of its most likely enemies, when clashes developed between Admiral Dewey and the German admiral in command of a squadron of five ships after the battle of Manila Bay. These irritations of 1898 did not remain local and temporary, for there was no strong hand in the Reich government to sooth, or cauterize them. In the spirit of revenge for the premptory manner in which Dewey had called his Teutonic opposite number to order at Manila the German navy men happily contemplated a war with the United States. Such a war might serve the existence and continued growth of the fleet, besides justly showing the army, the supercilious and patronizing senior service, what the navy could do on its own, in uncombined operations—at least on paper. Thus, from 1899 on, the Reichsmarineamt discussed a question like this: "Which points on the Atlantic and the Gulf coasts of the United States, in the case of warlike enterprise against the latter, would be best for points of support?"

Cape Cod a Target

Some officers suggested Cape Cod as such a *point d'appui;* and after personally reconnoitering on the spot and along the New England coast, carried on at least until 1901, Kapitän-Leutnant von Rebeur-Paschwitz, the first naval attache that Germany detailed to the United States (since the early weeks of the Spanish-American War), agreed with them. Their opinions were divided, however, as to whether "an attack should proceed better from the South or from the North," although apparently no military help seems to have been contemplated on this occasion. The navy was to do it all.

As to what should be done "after the main purpose, the destruction of the enemy's dominion of the sea," had been achieved, the attaché believed that there was only one suitable course:

> To force the enemy to conclude peace—nothing but an energetic attack on the rich harbor cities of the East. If from any one point pressure is to be exerted on the directing authority, which will force it and therewith also the people to yield, this can only be done at the commercial centers of the East, where, at least today, the weight of the commercial and industrial interest still rests. Expeditions far into the interior of the country would meet with great difficulties; occupying the nominal capital of Washington would make no impression, since neither trade nor commerce is of any importance there.
>
> A blockade, such as could still be undertaken with success during the Civil War against the Southern states, would today, considering the much more developed conditions of traffic, be impracticable for us, if the whole coast of the United States is in question. An attack on the West Indian possessions, Cuba and Puerto Rico, would at the utmost have an importance only for those [Americans] who are interested in commercial enterprises in those regions and who represent but a small number. Therefore, I can actually see a possibility of success, particularly if we refrain from the use of large masses of troops, in nothing but a merciless and ruthless [*schonungs-und rücksichtlos*] action

against the commercial and industrial centers situated in the Northeast.

Boston and New York he considered as the two places that could most successfully be attacked, much more easily than Philadelphia and Baltimore, since the latter cities were situated on a river and bay, far from the sea.

Clearly, the representative of the *Reichsmarine* at Washington had convinced himself, and perhaps other officers in his department, that the powers which in the McKinley era largely dominated American peace time politics, commerce and industry, would be determinative in war. Equally obvious is the reliance of the German navy upon its own strength; in urging abstention "from the use of large masses of troops," von Rebeur-Paschwitz voiced the hope of the navy to make war upon the United States and force a peace by its action alone, without the help of the army. This conviction or hope was part of the more discreetly than overtly debated problem of whether the navy could fight through to the end any autonomous battle problem—a question which deeply involved the whole curious existence of a "Risiko fleet."

Views of German Diplomats

It is certainly no reflection upon the military judgment of German diplomats that the head of the embassy in Washington at the time, von Holleben, the temporary and official superior of von Rebeur-Paschwitz, viewed the prospects of Germany, *vis-à-vis* an American-German war, far less lightly and confidently than his subordinate. Discoursing at great length on the anti-German sentiment in the United States after the war of 1898, von Holleben, it is true, thought war between the two countries not at all unlikely:

> Only once before, on the occasion of the first great Samoan conflict, has there been serious talk of a war between Germany and the United States. Today the possibility of such a war lies much nearer, and it will be a good thing to keep this possibility in mind in all our intercourse with the United States. It is obvious that a real gain would not accrue to the United States out of such a war . . . . A successful attack on our coast is unthinkable. But America knows very well that Germany is actually the

> preponderating power [Vormacht] of European civilization. Besides, Germany alone in Europe stands on the footing of an intact martial glory. These ethical momenta are exactly what is tempting for America in order to inflict a wound on Germany and with it on deeply hated Europe. Spain was merely a *corpus vile,* though on the other hand in the war with Spain great material interests were involved. A war with Germany would be something more like a sport; it would cost money, perhaps plenty of money, but that would not matter as compared with the moral advantage of playing a trick on the ideologues and the world's schoolmaster and tarnishing the glory of the victor of 1870.

In such a war, however, and especially a war at sea, both navies, nearly equal, would suffer about equal losses, perhaps Germany a little less, but, he thought, "America, with its colossal capital power, would be superior with respect to the speed of supplementing its materiel." German trade would suffer considerably more, since the American trade was carried almost entirely under foreign flags. Germany's navy would have few points of attack and would soon have even fewer of them since the Atlantic coast of the United States was being fortified.

> The whole thing would be a war without a valuable bone of contention and, since our land army could not take any decisive action in it, would be without any conceivable end. . . . And besides all this presupposes that none of our European neighbors makes use of the occasion in order to settle old bills with us. That we should find an ally against the United States is very unlikely.

Ambassador and attaché in the German embassy at Washington, it thus appears, then stood in much the same relation as in 1914-1917, and they severally judged the chances of an American-German war in much the same manner, von Holleben and von Rebeur-Paschwitz paralleling Bernstorff and Boy-Ed.[18]

GERMAN NAVY CREATED ENEMIES

A survey of military landing operations through history is not the place to discuss at length, and possibly ascertain, the influence of Edelsheim's writings and those of other military and naval officers on Anglo-German and American-German relations, which in those very years of the new century were taking their turn to the diplomatic-military constellation of World War I. The German Navy was cultivating new enemies for the Reich while the Army could not possibly give over its confirmed animosity to its old ones. After the turn of the century England began to feel that she and Germany were possible, or likely enemies and that an invasion was a real contingency to reckon with, even regarded as possibly more of an irrational than a rational enterprise.

After all, top authoritative German generals like Goltz, or admirals like Livonius said so, the latter, long retired, maintaining that Germany had nothing to fear from England, but that the latter might have reason for concern, since invasion was quite possible.[19] German diplomats and civilian governors could not dispel the suspicions of the English, who viewed Germany as their most probable future antagonist, thus making necessary a higher permanent concentration of the English navy in the North Sea and the building of new bases in the north of the island from whence a German invasion fleet might be flanked.

So-called public opinion, a factor still very much in need of analysis to ascertain its soundness and volume with reference to defense questions, was more quick to react psychically than sensately, when it was told, for example, that large docks in Emden constructed in connection with the Dortmund-Ems canal were capable of embarking 300,000 (!) men. The old and periodically renewed English public fear of an invasion became lively first, the Admiralty next feeling the danger and the diplomats last. Diplomacy was repudiated by English public opinion in its endeavors to keep on good terms with Germany, as in the highly unpopular financial participation in the Bagdad Railway, or the German-British-Italian debt-collecting expedition against Venezuela in 1902-3. In this large masses of the American people suspected danger in the prospect of a German landing on the *verboten* psychic ground marked out by the Monroe Doctrine. England then turned to the entente with France, which incidentally, diminished the danger, probably not very real even during the Fashoda crisis, of a landing from a hostile shore nearer than the German.

NOTES, CHAPTER 36

[1] Mohs (ed.), *Waldersee in seinem militärischen Wirken.* Berlin 1929. II, 388. Diary entry of Jan. 25, 1898.

[2] General von Boguslawski in *Die Woche,* March 31, 1900.

[3] August von Janson, *Das strategische und taktische Zusammenwirken von Heer und Flotte.* 2 vols. Berlin, 1900. He published in 1905 *Das strategische und taktische Zusammenwirken von Heer und Flotte im russisch-japanischen Kriege.* For a French treatment of these German discussions see *Revue militaire,* no. 872 (1900), 457 ff.

[4] *Political Science Quarterly,* vol. LV, 57.

[5] *Strategisch-taktische Aufgaben nebst Lösungen,* hefte 14-16. Leipzig. 1902-3.

[6] *Political Science Quarterly,* vol. LV, 56.

[7] *Streffleurs Oesterreichische Militärische Zeitschrift.* 1903. I, 797-8.

[8] See his *Corsaires barbaresques.* Paris, 1887, ii.

[9] The above is largely based on Marder, *Anatomy,* chs. XV, XVII and XVIII.

[10] Captain Goodrich, USS St. Louis, to Secretary Long, July 10, 1898. *Papers of John Davis Long.* 1939, 154-5.

[11] *Seemacht und Landkrieg.* Deutsche Rundschau, March, 1900. The *National Review* noted the article in its May, 1905, issue.

[12] Generalfeldmarschall Colmar von der Goltz, *Denkwürdigkeiten.* Berlin 1929, 181, 220, 222, 226, 260, 329.

[13] *Preussische Jahrbücher,* Jan. 1896, 404.

[14] *Heer und Flotte.* Ibd., March, 1900.

[15] An English translation was made for the services and a commercial one under the title of *Germany's Naval Plan of Campaign Against Great Britain and the U. S.* London, New York and Toronto, 1915.

[16] Gustav Graf von Lambsdorff, *Die Militärbevollmächtigten Kaiser Wilhelms II. am Zarenhofe, 1904-1914.* Berlin 1937, 400.

[17] Nov. 19, 1901. *Documents diplomatiques françaises* 2e série, t. I. no. 517.

[18] For the documentation of these plannings see Vagts, *Hopes and Fears of an American-German War.* Political Science Quartery, LV, 58-61.

[19] *Die deutsche Nordseeflotte und die englische Seemacht.* Deutsche Revue Feb. 1902.

Chapter 37

ENGLISH AND GERMAN LANDING PREPARATIONS, PRE-1914

WITH A FEARFUL NAIVETÉ and disregard of consequences, political and also military, German officers pursued, or suffered the building of the Reich navy and the all too open discussions of its eventual uses. When confronted with the possible consequences, such as a war on two, or more fronts, they would merely answer that "into such situations our diplomacy simply must not be allowed to let us drift." [1] Corresponding groups in Britain could not share this naïveté, false, or genuine.

Until the early 1890s a military entente had existed between London and Berlin. It went so far that the two military intelligence divisions were constantly exchanging certain information about possible enemies, such as Russia. This was largely arranged by the later General Grierson, who was then on friendly terms with Schlieffen and Goltz, "a wonderfully clear-headed fellow," as Grierson called him. France was another enemy which the two Germanic powers held in common contemplation until well after 1895, the one power that might attempt a landing in England. To prepare against it Wolseley, as commander-in-chief in 1897, wanted the British Army strengthened, so that it "should, with the help of auxiliary forces, be able to protect this country against the largest invading force that France can be expected—under favoring conditions—to put across the Channel." [2]

But the unwritten Anglo-German military entente, which Moltke (himself English-married) had in mind when he wrote in his *Military Testament* that "among the great powers England necessarily requires a strong ally on the continent; she would not find one which corresponds better to all her interests than a United Germany that can never make claim to the command of the sea"; [3] and on which Caprivi relied when he dropped the "re-insurance treaty" with Russia, vanished before it was ever turned into a written instrument. By the late 90s the break in the good understanding between the armies on either side of the North Sea had come about, at least as far as Grierson saw and expressed it. At the end of 1897 he wrote from Berlin, where he was then military attaché, that "we must go for the Germans, and that right soon, or they will go for us later. A pretext for war would not be difficult to find, and I don't believe that even Russia would stand by them." [4] It was the same Grierson who in 1905-6 arranged the technical details of Franco-British military cooperation in case of war against Germany.

But those overt indications of hostilities which generally precede a war were far from general as yet between England and Germany. The first reaction on the English side was rather a swing back of the pendulum to stronger insularity in defense. On the eve of the Boer War it was officially stated that the British Army existed for these purposes: support of the civil power in all parts of the United Kingdom, providing men for India, garrisoning of forts and coaling stations, rapid mobilization of all forces in the country for home defence and, subject to the foregoing considerations, sending abroad in case of necessity two complete army corps. "But it will be distinctly understood that the probability of the employment of an army corps in the field in any European war is sufficiently improbable to make it the primary duty of the military authorities to organize our forces efficiently for the defense of this country."

After the animosities bred by the Fashoda crisis and the Boer War had blown over, the German peril had become so far fixed in English military minds that from 1902 on the War Office regarded war with Germany as within the bounds of possibility "earlier rather than later," [5] without, however, envisaging until the conclusion of the Entente with France in April, 1904, the employment of British troops on the continent against Germany, alone or in collaboration with a foreign power.[6] By the summer of 1904 the British Admiralty considered a war with Germany as generally possible, discussing its technical implications, including an attack on Heligoland and the technically impracticable blocking of the Elbe by sinking vessels in the channel.[7]

Reaction of British Opinion

Public opinion in questions of defense is inclined to react in one of two ways—either in accustomed grooves of thinking and feeling, or in the violent pendulum swings produced by panic, rather than in a third way, that of rethinking through a defense situation in practical terms of new technical and new political conditions. It was an exceedingly well managed movement by which the British public, although ignorant of the actual diplomatic-military obligations involved, was moved to discountenance a policy of splendid isolation and to think in terms of alliances and ententes. It began by invoking the spectre of invasion danger to England. This was completed by the time the danger ceased to exist. England gave up her own isolation in order to isolate Germany, as the conservative *Spectator* early in 1903 had urged that she do, joining France and Russia "in making a ring around Germany and in isolating her," [8] a move that when completed provoked the German wail about *Einkreisung*, or encirclement.

The most recent developments in naval warfare, including the experiences of the Russo-Japanese War, would seem to justify the abandonment of old-style British insularity along with heavy coastal defense; for it had demonstrated that the superior naval belligerent need not greatly fear either landings, or cruiser raids. Consequently, Prime Minister Balfour stated in May, 1905, that "serious invasion of these islands is not an eventuality which we need seriously to consider."

The smallest force with which a full-sized invasion might be attempted from France, that power being tactfully chosen for purposes of exemplification, would be 70,000. This remained the standard size of an invasion force which British defense officials might expect. Either the effort to by-pass the British Navy must be sudden, or be preceded by long and open preparations, in which case the Navy would be the best prepared to front the enemy. Should the attempt be sudden, there still was not shipping space immediately available in the French Channel and Atlantic ports to transport the 70,000. Were the necessary bottoms obtained, no convoy could hope to escape torpedo attack. Long before the enemy could reach English shores, alarm would be sounded from the Faroe Islands to Gibraltar, bringing every available British war vessel to assigned, or menaced points. Should the huge convoy, a helpless mass of transports exposed to the many and various craft always near English ports, still escape unscathed, it would need at least 48 hours of calm weather to disembark along the coast between Plymouth and Dover.

This marks about the end of the more alarming theories

promulgated of an invasion of England, freeing the British Army for use as a continental expeditionary force and ending vast expenditure on coastal fortifications in Britain. Incidentally or logically, it also put an end to most of the German invasion talk, whether expert or inexpert. Fear of invasion, however, was too deeply rooted in the British psyche not to suffer relapses occasionally, as during Lord Roberts' agitation of 1909 for conscription and a large army for home defense. At that time, the Committee of Imperial Defense once again examined the prospects of invasion and found that, given the country's naval supremacy, an invasion attempt on a large scale, undertaken with 120,000 to 150,000 men, was "an absolutely impracticable operation." [9]

Once the *Entente* had been formed, continental landings and invasion by the English became definite future military tasks continually confronting the London military and naval authorities, while civilian governmental offices and politicians were mostly inclined to ignore these possible consequences of diplomatic arrangements. Whatever the British public was told to the contrary, German landings on English soil, leastways at the beginning of a war, became ever less likely as the years passed, to 1914. The Berlin General Staff expected and planned to seek the decision on the continent and had as little as before any plans for carrying the war into England, being convinced that a war with the latter could not be fought with any chance of success on the seas.[10]

Tirpitz's ideas never really included a landing in England, and least so early in the war; it was only after possession of the eastern Channel coast had been gained, towards the end of August, 1914, that he thought "something might be undertaken with the fleet and guns placed there to fire over into England." [11] To a man of the General Staff like Seeckt even late in 1916 it was axiomatic that "America cannot be attacked by us, and until technology provides us with totally new weapons, England itself neither. It is in its limbs more vulnerable than at its heart. Hence must the way toward Asia be open." [12]

When Haldane visited the younger Moltke in his Berlin office in February, 1912, the latter told him that the Army was to Germany what the Fleet was to Britain, adding with a good deal of truth that there were in the General Staff archives no plans for an invasion of England. When the visitor asked whether this was also true of the Admiralty, Moltke said no; that the Admiralty naturally had worked out a plan for the invasion of England, but it was an uncertain business and would in the end only hurt trade between the two countries and redound to the advantage of the United States.

Aviation and Landings

Seeckt even in 1916 echoed the pre-1914 conviction of the German General Staff that the then relatively new dirigibles and planes were, as yet at least, unable to serve as transports for invaders either over land or sea frontiers. Count Zeppelin, indeed, was motivated strongly in his aeronautical experiments by Anglophobia. It provided the politico-military drive that worked behind his invention. When the War Minister at Berlin, in spite of the pessimism of the experts, in 1911 obtained government funds to enable the Count to carry on his experiments, he did so with the remark: "I believe, dear Count, that I know where things are drawing you. You want to go across the sea to the English coast and show yourself above the English harbors and the Grand Fleet."

The face of the air warrior lighted up: "Yes, Excellency, that is what I should want." Minister and constructor met for the last time during World War I. "All England must burn!" the old Count exclaimed and his eyes sparkled.[13]

The Franco-British *entente* served as a shock-absorber when "Bleriot's hawk in July, 1909, gained the cliffs of Dover for the world," as a German poet rhapsodized over the first crossing of the Channel by a heavier-than-air machine, the flight from Calais taking 31 minutes. Had it not been for that alliance, the shock to the susceptibilities of the British, who obstinately opposed the building of a Channel tunnel, of this bridging by air of the strait would have been far worse, all their super-big ships defiantly, or reassuringly dubbed "dreadnoughts" to the contrary notwithstanding. However, considering the limited radius of flight of the early airplanes, they aided the defensive, through their reconnoitering services, rather than the offensive.[14]

Germany's military planners reckoned far more with an invasion of their own soil than with a German invasion of England. Once the English army reorganization under Haldane was started, the Great General Staff, from 1906 on, considered an English landing operation against the Schleswig-Holstein coast, or Jutland a possibility calling for the earmarking in the mobilization plans of certain forces to guard against such a contingency. More islands and parts of the North Sea shores were fortified and garrisoned, although none of this was on an extensive scale, inasmuch as the region possessed great natural defensive strength.

"Against enterprises overseas the difficult navigatory conditions along the North Sea coast provided natural protection," the German official military history of the First World War puts it, "Along the Baltic coast, however, four times as long, landings of larger units were possible in many places." In his last *Denkschrift* Schlieffen discussed the problem of an English landing, and although he did not consider the danger of it to be serious, believing that a quick German success along the main fighting front would be the most effective answer to such a thrust in the German flank, he insisted that in order to be ready against surprises "all the forces remaining behind in the country—and they are still very considerable—be concentrated in order to smother the English invader troops."

Schlieffen's successor thought a landing of troops in Schleswig-Holstein possible, from the 15th day of mobilization on. But he said: "If we are forced to mobilize on our western and our eastern frontier, we cannot leave an army behind in Schleswig-Holstein which is equal to the English"; only a first protection could be left there (July 1910). He took such a landing somewhat more for granted than did Schlieffen, who in the meantime had made light of the idea in a public statement, saying that "after it has set the world on fire, Great Britain has something better to do than have its troops, as Bismarck had once said, put under arrest in Schleswig-Holstein."

A *Kriegsspiel* of the Great General Staff in 1910, followed by a staff ride, dealt once more with the same problem, the somewhat lame conclusion being reached that "it could not be foretold with any certainty whether an English landing was planned there, or in Jutland." In November, 1912, when Ludendorff was chief of the *Aufmarsch* division—the section which drew up plans for transport, disposition of troops and advance—he made Moltke demand a strengthening of the fortresses along the frontier, including Borkum. "Without question we must also be prepared in any direction for a sudden English attack. The protection of Sylt calls for energetic measures" as part of the general defense system repeatedly insisted upon by the Great General Staff.

> We cannot make the security of our frontier fortresses, or of the threatened islands which have high military value, as well as the execution of our mobilization and *Aufmarsch* dependent on the timely receipt of information about the enemy or the timely execution of extraordinary arming measures. Both can fail. We must look for security only to the strength which we give to our Wehrmacht in time of peace and find in it the guarantee that the enemy does not think of enterprises which bring out our weaknesses, so that we need not be worried when our enemies stir and can annihilate them or at least repulse them if they should come after all.

It was thought most likely that the troops which England would send to the continent, would go either to France, or Belgium. Against possible landings in Schleswig-Holstein, or Denmark some reserve divisions which would not be ready during the first days of mobilization, or which would not be sent off at once to the front, 3½ divisions in all, largely forming part of the IXth Reserve Corps (Altona) were to remain north of the Elbe. Once the situation had sufficiently cleared, they would be sent off either to the east beginning on the 9th day of mobilization, or to the west beginning on the 11th day.[15]

INDISCRETIONS OF FISHER

In a vague way, the German army heads were informed of the English plans for participation in prospective continental war and also of the two schools of thought concerned with, and disagreeing on, these plans. Admiral Fisher was too indiscreet a person to leave them long in the dark about his violent, but hazy notions of 1905-6, during the First Morocco crisis, and later, of "Copenhagening" the German fleet, or landing English forces 100,000 strong along the German shores of the North Sea, or the Baltic, or in Denmark.[16]

At the same time that Fisher offered "his" 100,000 to Delcassé, the Reich military men considered a preventive war against the *entente*. The military situation favored them since Russia had been defeated by Japan in the Far East and was hampered by the Revolution at home. "And since the intervention by England at that time was hardly to be feared as yet, we would have had to deal with France only," as Einem, Prussian war minister at the time, wrote in his memoirs. But the nerves of the emperor and the chancellor failed them, and the most favorable opportunity for starting a war with the odds in Germany's favor was allowed to pass.[17]

The British Admiralty, in spite of the existence of the Committee of Imperial Defense as an overarching arrangement for Britain's war plans, remained a water tight compartment so far as the rest of its partners were concerned. It kept on brooding over ideas about an army *and* navy war at least until the time of the Second Moroccan crisis of 1911. It shared Fisher's fears that the Germans would defeat the French on land with, or without British help. Whereas Fisher had a scheme for "scratching in the Baltic," as Repington called it, by having troops landed there by the British fleet, which he thought should be able to deter and tie down more German forces than a BEF could hope to do by fighting with the French and Belgians in the west. As far as known, all this lacked solid substance, no plans being worked out and no special vessels constructed, or blue-printed. One reason for this was that Admiralty lacked planning arrangements required for such undertakings, a medium something like a general staff, which the British army under Haldane was building up.

To do something more effective for France than merely blockading the Reich and taking some worthless German colonies, the Admiralty through Fisher in Franco-British naval conversations made promises of help. About this nothing quite definite is yet known, although they may have included a landing in Schleswig. Concerning the nature and scope of this assistance the commander-in-chief of the Channel fleet, the one to be used against Germany, proposed to his colleagues:

> . . . to devote the whole military force of the country to endeavor to create a diversion on the coast of Germany in France's favor; in view of the rapidity with which events moved in the War of 1870, any diversion to be effective must be made at once. . . . In order to make an effective diversion we should be obliged to expose our ships in the Baltic or on the German coast in a way that would not be necessary if we were at war with Germany alone. . . . The operations that it is possible to undertake against Germany are very largely dependent on the attitude to be taken by the neighboring nations, Belgium, Holland, Denmark, Sweden, Norway, and Russia. The attitude of Holland

> and Denmark are [!] especially important. With Holland neutral, a blockade of the German North Sea ports would be almost useless, and without the consent of Denmark to use her waters it would be almost impossible to conduct any operations in the Baltic.
>
> Supposing all the above Powers to be neutral, the course that seems to me most worthy of consideration would be an attempt to capture the works at the mouth of the Elbe and Weser by a combined military and naval expedition . . . As the main object would be to draw off troops from the French frontier, simultaneous attacks would have to be made at as many different points as possible. . . . If Denmark were on our side, a very effective diversion might be made by assisting her to recover Schleswig and Holstein and in that case the Fleet might operate very effectively in conjunction with a land force on the coast of the Little Belt or Kiel Bay in addition to the attacks proposed on the mouths of the Elbe and Weser. . . . (June 27, 1905)

In the case of such a war the admiral proposed arrangements between himself and the Army commander to prepare for combined operations on the largest possible scale, the assembly of all available small naval craft, the use of obsolete battleships for bombarding forts, the assembly, or building—it is not clear which—of numerous flat-bottomed, light-draught steamers capable of grounding without injury for operating over the sands inside the Frisian islands.[18]

It can hardly have been expected that the latter type of boat could have been made available at once, or soon enough to give the French the prompt support promised and needed by them. In fact, there was not a little contradiction between Fisher's dreadnought obsession and the neglect of smaller vessels, and his ideas on landing which would have demanded considerably of the latter.

In general, ideas were current for outrunning plans and the data and material required for execution and even planning, and threatened to end in nothing beyond improvisations. Much of the information on which such combined operations would have to be based was not procurable in London, where for some time "the invasion of the main territory of any Continental Power" and hence combined operations in this direction had not been considered "a practical scheme . . . . and no countenance has ever been given to operations which would endanger H. M. ships."

Superior Anti-German Naval Strength

This risk the Admiralty was now more ready to incur since the entente with France gave the anti-German side overwhelming naval preponderance, which could easily afford some losses. It thought that "to carry 60,000 British troops through the labyrinth of sandbanks and shoals shielding the German seaboard to a landing in Schleswig-Holstein would be an operation which might not be impracticable, but which would certainly be arduous and would need careful study and high organization. . . . It might be advisable to send officers in time of peace to study the country and the coast line; and a vast amount of secret work would have to be done." [19] This was subsequently done to a certain extent, as the Germans became aware when they caught an occasional English officer-spy.

It is not quite clear how strongly impressed the War Office was by the only too well justified opposition to a French Navy suggestion of early 1906. This involved arranging a joint landing of 100,000 British and 100,000 French on the German coast. But at any rate the Office, and following it the Committee on Imperial Defense, decided to throw the weight of British military intervention across the Channel and into France and Belgium. A breech of Belgian neutrality by the Germans was taken for granted, and so soon as they could possibly do it after the outbreak of war, and England would then fight side by side with the French and Belgians.

When the military came to the Admiralty to arrange for the protection of the transit of troops to Boulogne, Calais, etc., which was to begin on the 3rd day of mobilization, they ran into troubles with Fisher who wanted "to fight his own war in his own way" (Repington). And it was said in the Navy at the time that the soldiers merely disliked "getting their feet wet."

Since the diplomatic crisis of 1905-6 was settled without resort to arms, the conflict in British war plans was not brought to agreement at the time, both Army and Navy adhering to their views and separate preparations, as far as the latter went. In March, 1909, Fisher still clung to expeditions "involving hell to the enemy because backed by an invincible navy (the citadel of the military force), ... of inestimable value, only involving 5,000 men and some guns and horses, about 500—a mere fleabite—but a collection of these fleabites would make William scratch himself with fury."[20]

When Fisher with his picturesque and blustering flow of language about landings in Germany was invited in a meeting of the Committee of Imperial Defence in November, 1909, to state his concrete proposals—the most concrete proposals of the Admiralty in this respect have never been published and seem to have been destroyed—English officialdom, to whom defense questions were entrusted, became deeply shocked, and with good reason. For:

> Fisher told the Committee that if 120,000 English were sent to France, the Germans would put everything else aside and make any sacrifice to surround and destroy the British, and that they would succeed. Continental armies being what they are, Fisher expressed the view that the British Army should be restricted to operations consisting of sudden descents on the coast, the recovery of Heligoland, and the garrisoning of Antwerp. He pointed out that there was a stretch of ten miles of hard sand on the Pomeranian coast which is only ninety miles from Berlin.
>
> Were the British army to seize and entrench that strip, a million Germans would find occupation; but to dispatch British troops to the front in a Continental war would be an act of suicidal idiocy arising from the distorted view of war produced by Mr. Haldane's speeches, and childish arrangements for training Terriers after war broke out. Fisher followed this up with an impassioned diatribe against the War Office and all its ways, including conceit, waste of money and ignorance of war. He claimed that the British Army should be administered as an annex to the Navy . . . .[21]

At this point the Prime Minister adjourned the meeting, thus saving Fisher's ideas from closer examination. It could hardly be expected that they would convince any military, or lay mind. Considering the long and tortuous approach, even a sensate naval mind would doubt that such an expedition from the Channel to Pomerania had a realistic character. Which beach at a 90 miles distance from Berlin, or at 82 as he insisted on a later occasion, Fisher had in mind has never been cleared up. Fisher used to tell that he had received the idea from a former member of the Berlin General Staff who had viewed a landing of British troops with great trepidation. The distances from Berlin given by Fisher would point to the waters of the Haff, north of Stettin and behind a chain of islands guarding the mouth of the Oder, all in all not a very propitious region for landing purposes.[22]

Cross-Channel Transport Muddled

With or without Fisher, who left office temporarily in 1911, the Admiralty adhered to its views about the use of the Army, "a projectile to be fired by the Navy," according to Fisher.[23] During the Second Morocco crisis it protested against sending Britain's small army to France-Belgium to be chewed up in the conflict of millions of continental combatants. Rather should it be kept ready aboard ships for counter-strokes against the German coast, to be landed seriatim at points on the Baltic shore of Prussia, presumably Pomerania, and thus divert more than its own numbers from Germany's fighting force.

The Admiralty found general opposition to its ideas and plans—if plans, in a concise sense of the word, they were, which seems doubtful—and gave the War Office a staggering shock when the latter realized that the arrangements for the transports across the Channel, which it believed to be settled definitely and completely, were altogether one-sided and that the two departments responsible for war were not of one mind about their supposedly joint war plans.

The Army representatives in the Committee of Imperial Defense protested that the plans of the Navy for the Army were "from a military point of view hopeless, because the railway system which the Great General Staff of Germany had evolved was such that any division we landed, even if the Admiralty could have got it to a point suitable for debarkation, would be promptly surrounded by five or ten times the number of enemy troops."

Haldane thought that Fisher, from whom his successor had inherited the idea, had derived it from the analogy of the Seven Years War, when Russian forces had been landed in Prussia, much to the chagrin of Frederick at the time, and as Secretary for War insisted that "the mode of employing troops and their numbers and places of operation were questions for the War Office General Staff" and that they had been worked out together with the French.

The Chief of Staff, Sir William Nicholson, known for his sharp tongue, asked the Admiralty representative "whether they had at the Admiralty a map of the German strategical railways. He replied that it was not their business to have such maps. 'I beg your pardon,' said Nicholson, 'if you meddle with military problems you are bound not only to have them, but to have studied them.'"

Based on a common negative conviction, a somewhat greater consensus of opinion prevailed between Army and Admiralty on another amphibious project, an expedition to the Dardanelles. Whether it was posed to them "as an academic question" by the Committee of Imperial Defense, or whatever other occasion there was for its discussion in 1906-7, both General Staff and Admiralty agreed that no action by the Fleet alone was possible; the former even doubted that a joint enterprise was feasible. Only the Director of Naval Intelligence thought that this showed an under-estimation of the potentialities of a joint enterprise and that there was "no reason to despair of success though at the expense, in all likelihood, of heavy losses."

The CID decided that "the operation of landing an expeditionary force at, or near the Gallipoli Peninsula would involve great risks and should not be undertaken if other means of bringing pressure to bear on Turkey were available." [24] The practical consequence, as far as staff work was concerned, was failure to prepare for such an eventuality and for further study of the project. Hence the later lack of much topographical and other information about the Dardanelles in London and a fairly general belief in the impossibility of an enterprise against the Straits.

Subsequently, the Admiralty was forced into closer harmony with the War Office, in part through a reorganization which included a general staff that the Navy had never wanted, least of all Fisher, who "thought that if there were such a staff, the war plans of the Admiralty could leak out." According to Haldane, who personally would have liked to give the Navy such a scientifically conceived institution as he had already introduced in the Army, Fisher "did not realize that in the 20th century it was impossible to conduct military operations successfully either on sea, or on land without close preliminary study on an extended scale." But, according to the disappointed Haldane, Winston Churchill got in ahead of him and came to direct the Admiralty.[25]

The organization of the Naval Staff was slow enough, and only gradually was cooperation attained between the Army and the Navy who had so long maintained "naval claims to provide wholly, or in part, for Home Defense." [26] Only after two years, shortly before the outbreak of war in 1914, were exact plans for ferrying the BEF to the continent ready, after preparations which included embarkation and disembarkation trials and tests.

German Invasion Studied

To the outbreak of the war and even beyond there still remained one fundamental difference between Admiralty and War Office, between the Blue Water School and Lord Roberts' "Bolt out of the Blue"—the question of the probability that a German force might land in England and how much of a military force should be retained at home for such a contingency. According to the Navy, no large German army, "large" meaning something above 70,000, could be brought to England once the fleet was on war station. It was maintained that the latter could be depended upon to intercept, or break it up while landing and a lesser force could be dealt with by the Regular Army, if a sufficiently strong section remained at home in case of war instead of being sent in full force, six divisions, across the Channel at once after war began.

The crossing of smaller bodies, of 20,000 to 30,000 Germans, while considered unlikely by the Admiralty, could not be absolutely hindered; hence it proposed that against such a contingency two divisions of the six should be left at home until the situation had been clarified. This force, plus the Territorials, would compel the Germans to make an invasion attempt with a considerable body of men, which in turn would be more apt to be detected and caught in transit by the Fleet. "Only an army of a certain size could give the Navy a sufficiently big target on salt water." [27]

When war did come in 1914, Admiralty and War Office reversed their positions, the latter retaining at home two divisions, once promised to the French, and the former waiving its demand for the two divisions since the Navy would be able to guard the island against practically all landing attempts. The decision on the part of the War Department kept a force of around 100,000 men, including Territorials, at home.[28]

In view of the stalemate soon developed along the fronts, it might well be doubted whether the continued Allied decision not to land forces on, or near the coasts of Germany can be defended on all counts. The strategic temptation of doing so was certainly great. But not less weighty were the arguments existing against it before 1914 and to a large extent afterwards. They were technical, tactical, organizational and also political, the latter being inherent in the nature of coalition warfare such as both sides fought in the World War I.

The China expedition of 1900 had shown, for example that it was nearly impossible to bring the various national detachments on the scene under unified command, or unite in a common view the conflicting naval and military ideas on operations. The representatives of the two greatest land powers at the time, Germany and Russia, were unable to understand the importance of having the command of the water-way to Peking as far as it went, while the land communications were poor and hard to maintain.[29] Waldersee, sent out to China after considerable diplomatic exertion, as *Weltmarschall,* arrived far too late for any vital operations to be engaged in.

Some reasons working against combined operations in the First World War might also be called historical. Particularly the navies, although in their nature the more technical services had become too much history-minded. "Steam navies have as yet made no history which can be quoted as decisive in its teachings," Mahan had told them in the 1890's and much of this conviction remained even after the Russo-Japanese War. The American navalogue consequently proceeded to deduce the nature, if not also the working, of sea power from warfare in the pre-steam age, slighting the importance of practically all combined operations, when he wrote, for example, that all of Wolfe's operations "were based upon the fleet . . . The landing which led to the decisive action was made directly from the ships . . . In a word, the possession of Canada depended upon sea power," *and not also* upon soldiers.

Only in an age of extreme division of labor, when the common purpose of labor hardly even existed any more in the minds of the experts, could such a statement pass without the most incisive challenge. Naturally, this historical justification of the purpose and actions—and inaction as well—of the pre-1914 navies, nearly all of them embracing the Mahan doctrine, led to great hesitation on their part; they did not know how to steam and fire into the uncertain future, how to try and apply the surprises which evidently were embodied in the existing navies.

DEAD HAND OF MAHAN

This Mahan-induced hesitation many times during World War I worked like palsy: it helped to keep the German fleet inactive when it might have been most useful against the British cross-Channel troop transports in August 1914; it made Viscount Esher, no army man, but rather the long-time member of the Committee of Imperial Defense, write to Haig (May 6, 1917): "It is time that something was done at the Admiralty. There has been no critical, or creative movement within its antique walls since the war began. That celebrated phrase of Mahan's about Nelson's ghostly, storm-tossed fleet has been the navy's undoing." [31]

Departmentalization had worked out most disastrously in Germany, with consequences largely in the political-diplomatic field, but also to the detriment of the senior service, the Army; only the leaders of the latter did not see, or say this until Germany's defeat in 1918 had become obvious. There was no official or public protest on the part of any army chief against the "luxury fleet;" if there was protest, it went no farther than the pages of a diary, such as Waldersee's, the latter confiding to those secret pages early in 1903 that the Emperor "pursues unceasingly his aim of enlarging the Navy without, however, betraying it"—as if that were possible!—"which is perhaps just as well since the size of his plans would provoke alarm. It is dangerous, though, that in the long run the Army must suffer under it." [32]

Even though in England there functioned the Committee of Imperial Defense as a constantly working arrangement for the cooperation and integration of the services, a scheme preferable to the spasmodic treatment accorded defense questions by cabinets and parliaments, even so the services had drifted dangerously apart. And it would seem doubtful whether, without the perturbations provided for the English world by the moves of German diplomacy and the construction of the Reich

Navy, even that small amount of integration that came into play by 1914 would have been achieved. But this integration was never sufficiently close to insure discussion and planning in detail the problems involved in landing upon hostile shores. This in turn would have led to study and adoption of types of landing craft, specific landing shores, etc., such as the Dardanelles, which London had decided not to study and about which information was therefore practically entirely lacking in 1915.

While total war approached, there was no thought given to such combination of effort. All constitutional arrangements for the division of men and moneys were deceptive and largely contrived to establish the necessary compromise of interests, and not unity of doctrines. There was scarcely any interpenetration of thought between naval and military personnel. While specialization in the sciences had gone far, war-making, or preparations for war had gone even further in that direction. Among the great powers of the West and in Japan, each partner thought of its own rôle as "decisive," with intention to seek the decision when and where it alone intended.

This "no-partnership" instead of co-partnership, kept the littoral war out of the minds of pre-1914 war planners. The navy, to be sure, was to obtain and preserve for its own country "command of the sea," either at all times and in all places, or only temporarily, and in specific waters. This command according to British official and traditional naval definition was "the test determining whether it shall be possible to transport across the waters commanded a large military expedition without risk of serious loss."

War's Future Misviewed

But this was a definition which envisaged nothing but internaval war, leaving littoral war out of consideration, with its objective of winning the command of the *litus*, of the shore. This was to be gained, if not also to be contested, by the closely combined actions of army and navy. For such a purpose, however, no effective organizational arrangements existed before or in 1914 in any of the war-preparing great powers.

The firm and widespread convictions of the nature of the war to come, with armies fighting armies and navies only navies, expressed themselves in a most ludicrous manner in a field where the two services clearly should have viewed opposition in conflict with one another as something likely to occur, even if they did not want it. That field lay in the exchange of naval gunfire and coastal artillery.

The basic view of "mutually exclusive" war extended to this field, influencing such arrangements as the choice of forces, army, navy or something in-between, to man coastal fortifications, and it did more to settle the views about the relative firing strength, or hitting chance of the two sides than any of the few actual experiences that had recently been made, such as the bombardment of Alexandria in 1881 or of Port Arthur in 1904.

It was the general conviction of the incompatibility of the two kinds of war and the two arms, rather than experience and experiment or legitimate speculation, which confirmed ancient saws about *un cannon sur terre vaut dix sur mer,* that "a 6-gun battery could fight a 100-gun ship," or Nelson's dictum so often quoted by Fisher that "any sailor who attacked a fort was a fool." As far as that meant that no ship should attempt any fort, this was simply a *fable convenue* about warfare, " a superstition of old forts being stronger than new ships."[33]

As Winston Churchill rightly demurred, at least in 1915: "No general or absolute rule can be laid down about fighting between ships and forts. It depends on the ship: it depends on the fort."[34] Although ultimately in such a duel the land gun, with the earth as the stronger and more stable artillery platform, may always be able to outfire the naval gun and protect itself better, at certain times and in certain places the naval gun may well find itself in a position to outrange fort guns, carry superior protective armor and possess the advantage of mobility.

NOTES, CHAPTER 37

[1] v. d. Goltz, *Denkwürdigkeiten,* 87.

[2] J. E. Tyler, *The British Army and the Continent, 1904-1914.* London. 1938, 12.

[3] Cit. Winston Churchill, *The World Crisis 1911-1914.* New York, 1924, 12.

[4] D. S. Macdiarmid, *The Life of Lt. Gen. Sir James Moncrieff Grierson.* London, 1923, 86-7, 101 ff., 132-3.

[5] W. H. H. Waters, *Private and Personal.* London, 1928, 240-1.

[6] Tyler, 10-11, 16-7.

[7] Marder, *Anatomy,* 479.

[8] Ibd., 466.

[9] Ibd., 80-2.

[10] The younger Moltke's statement in 1909. Bülow, *Denkwürdigkeiten,* II, 436.

[11] v. d. Goltz, *Denkwürdigkeiten*, 352.

[12] Rabenau, *Seeckt.* I, 259.

[13] von Einem, *Erinnerungen eines Soldaten, 1853-1933.* Leipzig, 1933, 163-4.

[14] Churchill, *The World Crisis, 1911-1914*, 167.

[15] Das Reichsarchiv, *Der Weltkrieg 1914 bis 1918.* Vol. I, 4, 70-71; *Kriegsrüstung und Kriegswirtschaft.* Anlagen Zum I. Bano, 88, 110, 112, 146-7.

[16] Bülow, *Denkwürdigkeiten* II, 194-6.

[17] Einem, 110-7.

[18] Marder, 504-5.

[19] Ibd., 506-7.

[20] Brett (ed.), *Journals and Letters of Esher.* I, 375-6.

[21] Bacon, *Lord Fisher.* II, 182-3.

[22] For further criticism of Fisher's ideas see Sydenham of Combe, *My Working Life*, 309.

[23] Marder, 70.

[24] Winston Churchill, *The World Crisis, 1915*, 377-8.

[25] Richard Burton Haldane, *An Autobiography.* N. Y. 1929, 241 ff.

[26] Brett, *Esher.* II, 252.

[27] Churchill, *The World Crisis, 1911-1914*, 158-9.

[28] Tyler, *The British Army and the Continent*, provides the most convenient treatment of the English discussions about invasion and participation in a continental war.

[29] For the China expedition see Keyes, *Amphibious Warfare*, ch. II.

[30] *The Influence of Sea Power upon History, 1660-1783.* 13th ed., Boston, 1897, 2, 293-4.

[31] Brett, *Esher* IV, 115.

[32] Waldersee, *Denkwürdigkeiten*, III, 227.

[33] Sir Ian Hamilton, *Gallipoli Diary*, I, 87.

[34] *The World Crisis, 1915*, 96-102.

Chapter 38

1914-1918

IN THE FINAL BALANCE, in World War I there were a few more combined operations than the participants, the two, historically and technically divided services, had expected to undertake before 1914. In all likelihood, however, far more might have been carried out, given the strategical possibilities and the technical means at the disposal of some of the combatants. Conflict on such a scale involved vast transportation tasks which brought armies and navies into manifold contacts. From these nothing resulted but logistics, transportation under highly risky conditions, or speculations in logistics, such as the prospect of using Allied tonnage superiority by bringing Russian army corps to the western front from as far away as Archangel, or Vladivostock. Such speculations, like others about the combined use of the two services, sprang largely from civilian brains. While that war produced many specialists, on artillery, trench mortars, tanks, only one true specialist on combined operations emerged, the later Admiral Keyes.

Certain landings were unavoidable here and there over the globe. Most of them proved as uninteresting on the military side as evictions by sheriffs. Thus the Japanese took to the Asiatic continent troops to oust the Germans from Tsingtao and establish themselves there instead. To lessen risks they landed their siege troops on neutral Chinese territory, 150 miles from Tsingtao, and thence proceeded to the objective.

Combined operations included fairly constant fire against the land fortifications from the Japanese blockading fleet, which lost one cruiser, one destroyer and one torpedo boat. The fortifications were far too extensive for the weak defending force to man, and were surrendered after a week's siege. Australians and New Zealanders dispossessed the Germans from their Pacific holdings and the *Emden* retaliated by landing a force on Keeling Islands to destroy a wireless station. Transcending the work of this amphibious guerilla were the amphibious parts of the operations against German East Africa from 1914-18. There, Lettow-Vorbeck, commander of the *Schutztruppe,* held out in the interior against sea power which controlled the shore line, until after November 11, 1918.

Anglo-Indian troops had landed as early as November 2 and 4, 1914, at Tanga with a force at least eight times as strong as Lettow's who, favored by the tropical terrain and the maladroitness of his adversaries, knew how to choose the right moment after the landing for bringing on the peripeteia, the sudden change which lurks in a landing enterprise, not so much in the moment of gaining land, but during the early ensuing hours or days. The impact of this failure on the Anglo-Indian side was long and severe and not until 1916 did Imperial troops re-appear in considerable force.

Largely cut off from the sea, although an occasional blockade runner got through to Lettow, the Germans attempted to break the water blockade by air. A Zeppelin was sent out in 1917 with medical and other supplies for Lettow. It covered a distance of 6,000 kilometers, but did not land, being misled by a wireless message that East Africa had capitulated.[1] But it was another reminder that war and possibly landing operations as well had become tri-elemental.

With all the profound separation of work and thought prevailing and observable in one's own camp, it was still supposed that for the enemy no such separation existed; that on the contrary he must be suspected of scheming and preparing combined operations against one's own shoreline. The rival services did not know before and during the war how fundamentally similar they were, in their organization, in their outlook, in their intentions and non-intentions, in their strategy. In their technical opinion, for example, that "ships will not proceed against coastal fortifications," [2] Tirpitz and Fisher fundamentally agreed. In many ways they were identical, though hostile, twins.

Still, when the uncertainties about the other party's intentions of invading across the North Sea, or the Channel are compared, those of the British were always considerably greater than the German, if measured by the number of troops retained at home to guard against the contingency of a hostile landing. This was the case at the outbreak of the war and remained so during nearly all of the war years.

Germany at all times kept only inferior forces at home, particularly in Schleswig-Holstein, for the purpose of meeting an

English, or Allied landing force. The troops stationed there in August, 1914, were released for employment elsewhere before the end of that month, the Navy being willing to guarantee that no landing in that province or in Denmark would take place.[3] By the 20th of August "the danger of Russian landings which at first had not been thought absolutely unfeasible, seemed to have disappeared." Denmark's attitude was neutral. Nothing pointed to English intentions against North Germany, or Denmark, and by the 22nd English troops were known to be in France; hence the IXth Reserve Corps could be spared for operations in the east.[4] Troops in Schleswig were temporarily strengthened again when an Allied landing in Denmark, with or without Danish official approval, was later suspected. As to landing places there, Ludendorff had thought before 1914 that "more quickly and comfortably the English could not land anywhere" than at Ebsjerg.[5]

Invasion Fear in England

In the final reckoning the fear of an invasion worked upon the English like a diversion, even though the Germans never undertook it, it being the chief characteristic of a diversion that it diverts from the principal theater of war more of the enemy's troops than are employed on one's own side in the manoeuvre. The British defence authorities—at times the Army was more alarmed, at others the Admiralty—thought German landing attempts possible, if not also probable, for the greater part of the war, their estimates of a likely invasion force running all the way from 10,000 to 250,000.

At the outbreak of the war the working hypothesis in Britain was that the Navy would detect and annihilate any landing squadron of over 70,000 and the Territorial forces overcome any bodies up to that number that might have been able to elude the vigilance of the Navy. When the Germans reached the Flanders coast in October, 1914, invasion of a different kind seemed possible, one proceeding from the Channel coast and using self-propelled light draught barges, such as were in use on continental inland waterways, each carrying from 500 to 1200 men, proceeding alone, or in conjunction with deep water transport from the Reich ports. Employment of the German High Sea Fleet in connection with such a force was not held probable by the British. Its more likely purpose in connection with such an enterprise seemed to be to draw the British Grand Fleet away during an attempted crossing of transports, the latter receiving their protection from Germany's second rank sea forces. To provide against this danger English patrol flotillas of torpedo boats were stationed along the east coast in such a way that none was farther than eight miles away from any likely invasion point.

By the end of 1915, after a number of the larger ships had left the home station to be used in the east and some had been lost, with the stalemate in the west making an enemy landing attempt again conceivable, numbers like 135,000 or even 160,000 for such a project were thought credible in England. With Flanders as their debarkation shore, they might escape detection, or interruption on the sea for from 24 to 48 hours. This led to demanding the retention of ever larger numbers of British divisions in the country. Even after the German naval raids against English coastal towns had revealed their relatively innocuous character, it was thought that 450,000 men, a number never actually reached, were required for home defence.

After Jutland the Admiralty would give no guarantee against an invasion of 160,000, and only from the end of 1917 was a force of 70,000 considered the maximum to be provided against. It was better realized than before what difficulties the enemy would encounter in the enterprise; the task of bringing some 160 transports out of harbor, the time consumed in anchoring, navigatory difficulties, mine fields, carrying men to the shore from ships anchored some two miles to the sea, the interruptive strength of the airforce. Hence the return to 70,000 as the yardstick for an invasion force.

This uncertainty about a German landing, which temporarily included concern over a possible "German-Irish landing" in 1916, led in the end to an acrimonious debate between the military men and Lloyd George. The latter was accused of being responsible for keeping great masses of men at home—the number of soldiers of all sorts in the British Isles being put at nearly 1,750,000 by the end of 1917—because of his obsession about the German danger. The Premier countered by saying that he had always regarded it "as a bogy invented by those who wanted to re-establish permanent conscription. I agreed with the decision of the Asquith Government that the Germans could not possibly accomplish more than a rush and a raid without artillery support."[6]

Invasion was not contemplated on the German side, so far as is known. "It must have led to a first class disaster, as Tirpitz would know," rightly declared Lord Sydenham of Combe, a non-alarmist; and if Tirpitz had not known it, the Army, with Falkenhayn, a pronounced anti-Navy man, at its head, would still not have entrusted even the smallest number of men for the Navy to transport across the North Sea.

Still less military reality was behind the fear, or threat of a German invasion of America. It was then a phantom, although far-sighted minds occupied themselves with speculations of German attacks in a possible second World War, Ambassador Gerard writing from Berlin on November 1, 1915, that he was "afraid that after this war the Navy party will be all for attacking the United States in order to show the Navy is worth something—get revenge for the loans and exports of arms, a slice of Mexico or South America and money. And if Germany is successful in the war, the country and Army will agree to this raid."

So far as considering American participation in World War I as a preventive war to avert a second, it was a failure. But while such consideration might have played a rôle in some statesmanlike resolves for future preparatory action it was not an argument apt to carry in public, and impose upon a people with whom greater sense of the immediacy of danger is required to arouse them to the reality of danger. This the so-called preparedness movement was furnishing. Its fear-inspiring arguments were in large part pseudo-technical, and neglected true technical circumstances such as the limited coaling capacity and steaming range of the German battleships, which then were insufficient for crossing and recrossing the Atlantic safely. This might have been known in 1914, and was ascertained fully while the German fleet was interned at Scapa Flow before it was scuttled, much to the relief of American naval men since a distribution of these ships among the various victorious powers could only have favored the British.

Alarmist Literature in the U. S.

While from a technical point of view a German landing in the United States was highly improbable,[7] the authors of various kinds of alarmist fiction as part of the preparedness movement did their best to make it credible. To encourage the feeling of insecurity by all means was one of the main features of this movement. It took such literary forms as the letters of Walter H. Page from London [8] or of J. Bernard Walker's novel *America Fallen! The Sequel to the European War.*[9]

This dealt with a war by Germany against America, following the close of the European War, in which Germany was to have been beaten by the end of 1915 and forced to pay high reparations, although permitted to retain her navy. In the story, a *Kronrat* at Potsdam sets April 1, 1916, as *"Der Tag"* on which Germany, with the permission of Great Britain in return for acquiring Mesopotamia, is to start her test of the Monroe Doctrine, "the most magnificent bluff in all history and, so far, the most successful."

The German attack, arising from an American protest against the Reich's purchase of the Virgin Islands from Denmark, is directed against New York, Boston and Washington, thus combining the features of the earlier hypothetical plans of attack. A German submarine flotilla, attacks the Brooklyn Navy Yard. In the approved style of William Le Queux—"London invested, bombarded, and sacked"—a German fleet forces its way into New York harbor and bombards the city (a map shows how this is done).

The city capitulates and has to give as ransom 1000 tons of gold, thus enabling Germany to pay her reparations bill. The American fleet, weak and neglected on account of the fear of militarism on the part of Congress, is beaten in the Caribbean. More and more German troops land, occupying the regions where armaments are produced, Connecticut, New Jersey, Delaware, home of the du Pont works. After the fall of Pittsburgh, peace.

This novel of the putative dangerously near future was followed by another predicated by a slightly more distant, although equally menacing, if less dark prospect, in C. Moffett's *The Conquest of America; a Romance of Disaster and Victory: USA 1921; based on Extracts from the Diary of Jas. E. Langston, War Correspondent of the London Times.* Illustrated. (New York 1916). The then most novel modern vehicle of propaganda, the film, was made to serve the same fear-exciting purpose. A French financier saw in 1916 in New York what he strangely considered a German film, which was more probably based on Walker's novel, or on Hiram Maxim's *Defenceless*

America, which had been given a movie-version under the title "A Battle Cry for Peace." In this were depicted German battleships anchored off New York, their shells tearing down skyscrapers like castles of cards, all for the purpose of retaking the gold sent to the United States by the Allies.[10]

In the literature printed for the purpose of spreading insecurity, and thus paving the way for preparedness, was no less a production than Edelsheim's brochure of 1901, in two translations.[11] Probably it will be forever impossible to estimate its contribution toward making Americans war-minded. Perhaps, though, it helped to turn one American finally against Germany. To George Sylvester Viereck, Theodore Roosevelt explained the shift in his stand on neutrality on the ground that he had "positive information of a contemplated German attack on the United States." In that connection he showed "among his papers an alleged plan of the German General Staff, signed by some obscure general, a plan published shortly afterwards in the press" [12]—a plan that might possibly have been Edelsheim's, for he had been "In the service of the German General Staff in 1901," as the title-page of the British translation of his brochure emphasizes. It is doubtful, however, if anything literary proved quite so demoralizing to American isolation as the visits during World War I of German submarines, commercial and naval, to American ports.[13]

Obviously the two adversaries—Germany and the United States —could not meet at all, except by steaming across the seas, and the post-mortem on German plans against America, undertaken before the German fleet was scuttled at Scapa Flow, revealed that its bunker capacity would have allowed it to fight in European waters only.[14]

French Suspicious of British

But to return from the nightmare world of fiction and fancy to the business that came to hand in 1914: When after the Marne the race for the Channel began, the Allies' fleets were called in by Joffre (Oct. 16) to support the Allied left flank in Flanders from the sea by acting with long range guns on the enemy's right wing. This request had come after the French command had declined British proposals that British troops be sent into northern Belgium. The French reasoning against such a landing was of a largely political nature, based on the fear that in case of a success the British would never again be dislodged from the Channel coast—the ancient memory of the British in Dunkirk and Calais rising once more—or that after a failure they might altogether disengage themselves from a continental war. In any case, coalition politics kept the British away from the utmost left wing of the Allies and put Belgium and French troops between them and the Channel,[15] restraining them also from greater activity on and along the Flanders coast. Their Navy accepted the task, more readily at the outset than later, employing monitors and other craft, although supplying munitions for a prolonged bombardment of the Flanders coast constituted a problem from the beginning. While at first the British marine artillery was able to enfilade the German attack, with naval balloons on shore being used for observation, after some ten days German long range guns and also the submarine threat came into action.

The Germans placed their own Marine Corps, formed early in November, 1914, from surplus naval reservists, representing another army corps that had been missing on the Marne, on the Flanders coast, "in order to start from there as well the war against England," as the Chief of the Admiralty Staff wrote early in the war.[16] Churchill's hope to have the enemy "off the Belgian coast, even if we cannot recover Antwerp" had to be surrendered for a long time. There was to be spasmodic repetition of the bombarding of the Flanders coast, usually at the request of the military commanders,[17] until the submarine base at Zeebrugge became too much of a damaging nuisance for the British Navy to endure forever, thus leading to one of the few determined landing attempts of this war.

Neither Antwerp, nor the 70 miles of entrance to the Channel could be withheld, or wrested from the enemy by amphibious power; Flanders was to be surrendered only to the advance of land power. "I am shy of landings under fire—unless there is no other way," Churchill less confidently wrote to Fisher in December, 1914,[18] while in August he had informed the Grand Duke Nicholas that "the operation of sending a British fleet through the Belts to enter the Baltic is feasible." [19] But did a failure on a near shore such as Flanders bode well for attempts against more distant beaches?

In any case, his hint to Russia about naval support reads almost as if it was inspired by Fisher, whom Churchill brought back to the Admiralty in November, 1914, where he promptly revived his old dream of a landing on the Pomeranian coast only 82 miles from Berlin, this becoming in time Fisher's alternative to the Dardanelles expedition. The new feature of his old plan provided for landing in that vicinity a million of Russian soldiers from British bottoms, "612 vessels constructed to carry out this decisive act in the decisive theater of war." While this "unparalleled Armada of 612," the implementation of Fisher's amphibious plan, was ordered, it does not appear that discussions were ever opened with the Russians whose reservoir of cannon fodder seemed inexhaustible to the western Allies in the beginning. Nor were his ideas about Pomerania ever seriously discussed by the War Council or the combined staffs.

Baltic and Dardanelles

In a "war of missed opportunities" the Baltic scheme was definitely the largest "miss." But it was also the most risky to undertake, the one that called for the most detailed and meticulous preparations among three, or four staffs and governments, even for preliminary, or diversive enterprises like the taking of at least one of the German North Sea islands like Borkum or Sylt, or making the Kiel Canal unusable. That the long approach of British forces would leave any occasion for exploiting surprise either as to time, or space would seem highly dubious at a time when German airplanes regularly patrolled the narrow waters of the Baltic.

Both the Baltic and Mediterranean alternatives to "sending out armies to chew barbed wire in Flanders" were being studied at the London Admiralty rather than at the War Office, since the end of 1914. Departmentalization proved as strong as ever, except that the Navy clandestinely poached upon the other arm's preserve by studying the tank problem, while showing greater initiative in all planning for helping the Russian ally either near his left, or right wing. A Russian appeal for relief from the pressure exerted by the Turks on their left, in the Caucasus, passing as this need proved to be, shaped a decision in favor of some sort of amphibious enterprise against Turkey. For this, other political factors worked, such as the hope of winning over the Christian Balkan states to the Entente, while Russian desires for winning control of the Dardanelles, which ran counter to the aspirations of some of the Balkan nations, were believed to be in abeyance. Those secular desires, it was thought, would recede before the necessity of opening the way to Russia for the products of western war industries and also in assuring an outlet for Russian grain to help feed the western peoples.

Even Fisher in the first days of 1915 was temporarily in favor of the enterprise against the Dardanelles, presuming that considerable military forces would be available at once, including some 75,000 seasoned troops of the BEF in France, to be replaced in the trenches by Territorials. They were to be sped from Marseilles and landed in Besica Bay, while at the same time the Dardanelles would be forced by the Fleet, the Balkan allies would move on Constantinople and the Russions and Serbians and Rumanians on Austria. As for the direction, "You want one man," he wrote to Churchill, and no junta. (On the same day he wrote to Balfour that Balfour was the one man).[20]

Fisher was a "school" largely by himself. While a few French generals adhered to the "eastern school" of seeking a decision away from the western front, predilections and dislikes were on the whole divided between the military—Kitchener excepted—who preferred continuation of the main effort in the western theater and the naval men in favor of the eastern enterprise. The civilians, including Lloyd George and Winston Churchill, were tipping the balance temporarily in favor of the latter, with Lloyd George, as untutored in military history as a man could be, sounding an early warning that "expeditions decided upon and organized with insufficient care and preparation generally end distastrously." In general he favored Salonika.

Strategical ideas like others have their origin in definite sociological settings and that of the eastern school was, negatively speaking, original with the traditional thought of neither of the two British services. The two first originators would seem to have been the purely civilian politician, Lloyd George, who thought it hopeless longer to seek the decision by sacrificing still more Englishmen and Frenchmen in the west; and the secretary of the Committee of Imperial Defense and the War Council, Colonel Hankey, who appears to have derived from his services with these two overarching bodies a sense of the need for, and the possibilities of amphibious operations. The

British Empire possessed both arms, therefore, why not employ them together?

NOTES, CHAPTER 38

[1] For the operations in East Africa see *History of the Great War.* Military Operations, East Africa. Compiled by Lt. Col. Charles Hordern. Vol. I, Aug. 1914-Sept. 1916. London, 1941; and Lettow-Vorbeck's *Meine Erinnerungen aus Ostafrika.* 1921, and *Heia Safari.* 1922.

[2] As to Tirpitz see Admiral Hugo von Pohl, *Aus Aufzeichnungen und Briefen während der Kriegszeit.* Berlin, 1920, 22.

[3] Ibid., 26 (Aug. 22, 1914).

[4] Reichsarchiv, *Der Weltkrieg 1914 bis 1918.* I, 262, 432.

[5] Margarete Ludendorff, *Als ich Ludendorff's Frau war,* Munich, 1929, 231.

[6] *War Memoirs* II, 170, and V, 169-70; Churchill, *World Crisis 1911-1914,* 276-8, 409, 419, 490; Sydenham of Combe, *My Working Life,* 184-5; Robertson, *From Private to Field-Marshal,* 190-2; Admiral Sir Herbert W. Richmond, *The Invasion of Britain,* 73 ff.

[7] See for a discussion of 1916 Rear Admiral Bradley A. Fiske, *The Navy as a Fighting Machine.* N. Y. 1916, 295-301.

[8] Burton L. Hendrick, *The Life and Letters of Walter Hines Page.* N. Y. 1922, particularly I, 349.

[9] N. Y. 1915.

[10] Octave Homberg, *Les coulisses de l'histoire.* Paris, 1938, 176.

[11] *Operations Upon the Sea.* N. Y. 1914. See a review in the Harvard *Military Historian and Economist,* vol. I, 383-4.

[12] George Sylvester Viereck, *The Kaiser on Trial.* N. Y. 1937, 471. Some fifteen years later the Kaiser himself, with more aplomb than good memory, assured Viereck "that the 'plan' was a forgery. Never was any plan for military or naval action against the United States prepared or even contemplated. The so-called German plan for an attack upon the United States was evidently a fabrication intended to precipitate the United States into war."

[13] Walter Millis, *Road to War,* 345-6.

[14] Statement of Secretary of the Navy Daniels after his return from Europe. *The Nation,* May 24, 1919. Compare with this the ambiguous statements about the steaming range of German battleships by Churchill before the Dominions representatives in 1912. H. H. Asquith, *The Genesis of the War.* N. Y., 1923, 77 ff.

[15] Bacon, *Fisher* II, 184.

[16] Pohl, 45.

[17] *The World Crisis, 1911-1914,* 401-9, 491-2.

[18] Bacon, Fisher II, 178.

[19] *The World Crisis, 1915,* 23.

[20] Blanche E. C. Dugdale, *Arthur James Balfour.* N. Y. 1937. II, 90.

Chapter 39

THE DARDANELLES

BRITAIN'S GOVERNORS, offices, and services finally took a definite stand on the Dardanelles project. Since Kitchener, as War Minister, had no considerable number of troops available for the enterprise, he generously suggested naval measures against the Straits, his contribution being thus at first "purely theoretical" (Churchill). He and Asquith "were hopeful that the Navy would do the trick, in which case very few troops would be needed," [1] while Churchill and Fisher thought that old battleships might be used to open the Hellespont by bombardment. When they inquired of Admiral Carden, commanding in the Mediterranean, whether he considered such an operation "by ships alone practicable," he answered that this might be done by "extended operations with large number of ships." No land soldiers were, or seemed to be wanted.

The British commander on the western front, Sir John French, admitted that the success of an Allied breakthrough on that front was unlikely, and that therefore an attack on Austria appeared a desirable alternative, an admission which still left him in full control of all troops he had. Hence the War Council on January 13 declared for "a naval expedition in February, to bombard and take the Gallipoli peninsula with Constantinople as its objective." "So Lord Kitchener swung round to the Dardanelles plan, and that settled it," as Lloyd George described the contribution of the general to the naval enterprise.[2]

So far, the whole plan, as Churchill emphasizes, in its genesis and elaboration was "purely naval and professional. . . . Right or wrong, it was a service plan," with all the higher naval officers adhering to it, openly or implicitly, including Fisher, until about the end of January. It seemed a legitimate gamble: the risks were definitely limited; it was perhaps the best opportunity to invest "our surplus ships and ammunition" (Chuchill); and the results could be staggering. By all estimation of profit, it appeared a worthwhile venture, entailing but limited liabilities. In that it closely resembled similarly reasoned enterprises of Pitt, things done at a throw, like a throw, carrying the speed of the conception of the idea over into its execution, with even the necessary secrecy hardly preserved, for the forts at Dardanelles had already been bombarded once, early in November, 1914.

The decision of the 13th was based on the plan submitted by Carden which had been worked out in detail by a Marine officer on the spot and gunnery experts. And although it involved a considerable amount of "ships fighting forts"—in Fisher's eyes a crime—it was found technically sound by the naval officials in London at the time. It proposed to demolish the forts one by one, "as the Germans did at Antwerp" (Churchill)—a hope that betrayed quite clearly how little thought had been given to or learned in the Navy about, the strength of field fortifications—through the superiority of weight and range which the guns on board possessed over the guns of the Turkish forts. Their systematic reduction would require a few weeks; after that the minefields would be cleared and "the Fleet would proceed up to Constantinople and destroy the *Goeben*," as Churchill told the War Council on January 13.

In order to have a face-saving object in case of failure at the Dardanelles, Churchill as an after-thought proposed (Jan. 20) that Alexandretta, where the Bagdad Railway came closest to the sea, be taken and held and proclaimed to be the true main objective of the attempt, with the Dardanelles as a mere feint. However, Kitchener declared that for the time being no troops could be spared, none of the four divisions then ready in England to be sent to the field. His help went no further than gratuitous advice and praise for the "vitally important naval attack which if successful would in effect be equivalent to that of a successful campaign fought with the new armies," and which if unsuccessful would not involve great loss.

All divisions were pledged to the western front from which the French ally would not release any men, considering even the last available British division "as a symbol of Britain's future attitude with regard to the Western front," as the British *Official History* puts this point of coalition warfare. The Shylock in the combination intent upon holding on to every pound of his man-power flesh, could be and was, highly exacting. Besides, he had most of the British generals on his side.

The military fighting cocks were completely hypnotized by the mud trench-line drawn in front of their beaks from the North Sea to the Alps.

Opposition From the Navy

While preparations went on for this limited and purely naval enterprise, opposition to it got under way. Jellicoe, as Admiral of the Fleet, represented that the most modern dreadnought chosen to go out, could not be spared from the North Sea. Fisher, usually the protagonist of landings, evolved and promulgated some Mahanlike notions about the command of seas being endangered by "subsidiary operations such as coastal bombardments, or the attack of fortified places without military cooperation;" instead, England should continue "quietly to enjoy the advantage" of her naval superiority. These notions alternated with, and measurably contradicted his recurrent ideas on landing operations in the Baltic. But in the end he gave in once more to the persuasions of Churchill and Kitchener and to the mirage of easy success, rather than resign his office in open protest.

At this juncture considerations of diplomacy—the hope of impressing and bringing over to the Allies the wavering Balkan states and Italy—gave birth to the desire of employing military forces at the Dardanelles. Although all British, or French troops seemed pledged to the sacrificial altar of the western front, it was decided that a number of divisions could after all be released. They were brought into the plans not by military men of their own volition, but by the civilians and naval men like Fisher and others, who now (Feb. 15) thought military forces necessary safely and surely to garner the fruits of this heavy naval undertaking. "The naval bombardment is not recommended as a sound military operation, unless a strong military force is ready to assist in the operation, or, at least, follow it up immediately the forts are silenced."

On February 16 the momentous decision was made at London to use troops in the attack on the Dardanelles, not on a large scale, to be sure, but to the strength of several divisions, which were made available here and there, including one, the 29th, that was to go from France. At once however, supporters of the western school, French and English, began to work on Kitchener to induce him not to release this division from what they insisted to be the sole decisive theater. For the time being they succeeded.

While the question of making the Dardanelles expedition a combined operation was hanging fire, no thought was being given to the question of whether the opening of operations should not wait until military forces were assembled in the vicinity. In any case, the naval forces began with the bombardment of the outer forts on February 19, which they damaged so severely that the Turks evacuated them. English Marines were landed against slight resistance on the 26th and after, as demolition parties. Having done their work they were re-embarked. As the commanding admiral told General Ian Hamilton on the latter's arrival: "At the time of the first bombardment 5,000 men could have marched from Cape Helles right up to the Bulair Lines."

The Allies seemed now engaged irretrievably, and on the 24th Kitchener, always very much the Sirdar and in the part "felt that if the Fleet could not get through the Straits unaided the Army ought to see the business through. The effect of a defeat in the Orient would be very serious. There would be no going back." Two days later he authorized the use of the Australian army corps "up to the total limit of its strength" in aid of the naval forces, but declined to release even one division of British regulars from France, or England until March 19, when the transports assembled for the 29th Division and subsequently scattered had once more to be collected.

The change from a purely naval attack to a possibly extensive military enterprise had been made, so far as Kitchener could make it, not for military reasons originally but for those that were political-imperialistic. Political reasons and reasoning carried on to these decisions, not any profound preliminary examination of the problem by something like joint naval and military staffs. Kitchener's resolutions were never formed in that deliberate, or deliberative way. He was, and was allowed to be, a sport in military evolution, a general with out-of-date absolutistic inclinations and small use for staffs and their work. With such ways of reaching resolutions there was apt to be improvization on the organizational side, no one center of will, "divided councils, half-hearted measures, grudged resources, makeshift plans, no real control or guidance." (Churchill: *The World Crisis. Aftermath*).

The immediate moral repercussion of the first naval attacks in the various capitals was marked. The Turks and Germans feared for some days that Constantinople would be lost after the expected break-through of the English fleet, while Bulgaria hesitated on her way into the camp of the Central Powers and Italy and Greece edged closer to the Allied side. Russia, however, vetoed any substantial military cooperation on the part of Greece and possibly Bulgaria against Constantinople, as likely to interfere with her own ambitions. Diplomacy thus complicated the already complicated military enterprise. The various effects achieved by the fleet, not all of them encouraging, were poorly followed up. There was uncertainty everywhere; the new air arm was not at once able to furnish good aerial observation to guide the fire of the ships against the forts and the enemy's mobile artillery. When Hamilton before his departure from London asked Kitchener for more aircraft, he received the answer: "Not one!"

What the naval forces on the spot did after the 19th made a military observer sent out by Kitchener seriously doubt that they could force the passage unassisted by military forces. This warning reached London the 5th and 6th of March, but did not greatly accelerate Kitchener's moves in assembling military forces, or in having staff work applied to the various problems likely to arise in connection with them. "Such as, how much artillery and ammunition can be provided for the force? What is the extent of enemy preparation in the theater? What bases can be utilized for the forces? Are landing provisions adequate? Can the troops be provisioned and watered on the Peninsula? What is to be done in case of reverse? How will the contemplated operations affect the military program already agreed upon in France?" [3]

While transports were loaded and sent off, no care was taken to embark the troops in any order that would present them ready for battle on their arrival. It was assumed that they were going to friendly ports. Only on the 12th of March did Kitchener break a silence which sometimes bordered on idiocy, and ordered General Ian Hamilton, who was commanding a force at home, to head the military force that was to support the fleet.

Hamilton found that practically nothing was prepared for him. Officers for his staff were still being gathered; Kitchener wanted to allow him only a few for this complicated and unprepared enterprise. Adequate information about the terrain, including maps, and the enemy proved even more unprocurable. "Ten long years of General Staff . . . ; where are your well-thought-out schemes for an amphibious attack on Constantinople? Not a sign. The Dardanelles and Bosphorus might be in the moon for all the military information I have got to go upon. When I asked the crucial question:—the enemy's strength? Kitchener thought I had better be prepared for 40,000. More actually there were only some 20,000 in the peninsula. How many guns? No one knows."[4]

Questions like that involving supreme command on the spot were not regulated. Everything was left vague, including whether a military landing would be at all necessary. Kitchener sent Hamilton off with the hope that "you will not have to land at all; if you do have to land, why then the powerful Fleet at your back will be the prime factor in your choice of time and place. . . . We soldiers are to understand we are string no. 2. The sailors are sure they can force the Dardanelles on their own and the whole enterprise has been framed on that basis; we are to lie low and to bear in mind the Cabinet does not want to hear anything of the Army till it sails through the Straits. But if the Admiralty fails, then we will have to go in. . . ."[5]

Lost Time and Opportunity

Hamilton arrived off the Dardanelles, where the island base of Lemnos had been borrowed from Greece, some 30 miles from the entrance, in time to witness the second naval attack on the Straits. It had been undertaken on the spur of a telegram from the Admiralty, where all parties now suppressed their doubts about the feasible character of the adventure, temporarily. "Everyone's blood was up. There was a virile readiness to do and dare. All the will-power and cohesion necessary to mount and launch a great operation by sea and land were now forthcoming. But alas a month too late!" (*World Crisis* 1915, 219) The admiral at the Dardanelles, de Robeck, successor to one who had fallen ill, had been invited to run risks, considering the "decisive" consequences which a deeper entry into the Dardanelles might have on the whole operation, if not on the fortunes of the entire war. On March 18 the last purely naval attack was launched.

While the battleships made considerable headway against the ancient forts, silencing most of them, mines which had been placed and replaced by the Turks during the interval of fighting, struck three of the Allied battleships. None was of the latest types. However, things to be risked and gambled with were submarines, not battleships, which were capital stock reserved for one transaction, the battle with the German High Seas Fleet.

Psychological impact was greater than material loss. Naval mentality was not prepared for such seemingly big forfeiture of nearly obsolete materiel. The commanding admiral had based his action on the expectation of finding the passage cleared of mines. Were these dreaded mines drifting down upon his ships on the strong current through the Narrows? Above all, it was such uncertainties that moved him to break off action. The home authorities admonished him to carry on, so as not to encourage by an apparent suspension of operations the enemy, who, little as the Allies knew it at the time, was indeed once more near collapsing, with forces and materiel, including mines, largely spent.

On the Allied side, the casualties had been fairly slight and the losses in ships no higher than had been allowed beforehand. Churchill was still full of confidence and drive when he received word from de Robeck that he was now quite clear he could not get through without the help of all the troops which Hamilton was expecting by April 14. The general himself, without telling the admiral, "saying nothing for, or against amphibious operations," had been driven to the conclusion by the experiences of the 18th that the Straits were "not likely to be forced by battleships as at one time seemed probable and that, if any troops are to take part, it will not take the subsidiary form anticipated. The Army's part will be more than mere landings of parties to destroy forts; it must be a deliberative and progressive military operation carried out at full strength so as to open a passage for the Navy."

It is impossible to discover in Ian Hamilton the all too common service stigma of expansionism, or *cacoethes,* the tendency to reach over into adjacent fields of jurisdiction and labor. Rather would it seem that the witnessing of the scenes of the 18th from the bridge of one of the warships, had filled his generous nature with a sort of compassion for the sailors who seemed somehow out of their element (without reasoning too much or at once whether the army itself was in its right element), something of *la grande pitié des grands vaisseaux,* which should have been foreign to a soldier. The sailors who had made up their minds, in face of pronouncements of Nelson and Fisher that all forts are superior to all ships, even the most modern and strongest, still to fight from the water against land based artillery, had been upset by being blasted by mines from an unsuspected direction, from below the surface, the lurking place of danger that they thought was to increase with the soon to be expected arrival of German submarines.

This experience and expectation fairly unnerved the naval commanders and the admirals in London, while more unconventional and enterprising younger men like Commodore Keyes, de Robeck's chief-of-staff, proposed to go on minesweeping with the help of better personnel and equipment, now available. There was thus in addition to the other dissensions a rather clear-cut battle of generations raging, or latent within the British Navy, of Churchill-Keyes vs. Fisher et al. Of all possible conflicts open and latent, one was distinctly absent, although almost traditional in combined operations—that between soldiers and sailors. The latter, as Hamilton testifies, were indeed "eager to meet us in every possible way" (I,100). The combination was as sincere as at Quebec even if not as successful.

While Churchill, fearing the consequences of a three weeks delay before a full-size military, or amphibious enterprise could be staged, was still in favor of forceing the Straits by purely naval operation, the London admirals grasped at the proposal to take the Army in as a means of withdrawing the Navy from the affair or, bluntly, pass the buck to the Army. And Kitchener, instead of giving up the enterprise altogether, took it over, without any previous discussions on his part, or that of the Cabinet which silently approved the fundamental change involved, that of making the Army the main partner in the venture. The latter was not in a position to do anything at once to capitalize profitably the surprise and the effects upon the enemy which had resulted from the attack of the 18th. It had to make a new beginning.

As his first measure Hamilton shifted the base of his operations from Lemnos, where the facilities for unloading and reloading the chaotically laded transports and freighters were too limited, to Alexandria. He had found that the transports sent to him were laden on commercial rather than on military, if any, principles, much as if such processes had never been considered before. This meant another three weeks' delay.

GERMAN COMMANDS TURKS

On the 24th Hamilton left with his staff for Egypt. The same day the Turks offered supreme command over the defending force to General Liman von Sanders, head of the German military mission to Turkey. He took over in Gallipoli on the 26th. "The English allowed me four whole weeks before their great landings. . . . This space of time just about sufficed to carry out the most necessary measures," as he acknowledged in his memoirs. He changed the primitive Turkish tendency to defend the coastline with thinly strung out lines of men and concentrated his forces, leaving but weak detachments close to the shore.

His forces, six divisions with some 60,000 men, were made highly mobile by exercise and marching, new-built roads passable for artillery which had not heretofore existed in the Peninsula, and assembling barges for the transport of troops along and across the waterways. The field defenses were strengthened, including submerged wire entanglements along likely landing beaches. He had no aircraft until April 11 and the English airforce, which was exclusively the Navy's since Kitchener had denied military planes to Hamilton was ranging freely over Gallipoli, observing the more visible results of his preparations.

While strategical surprise was washed out, the possibility of tactical surprise was still with the attacker, who might try to win a footing in one or more of these three places: On the Asiatic side of the Dardanelles, which Kitchener, however, had declared to be out of bounds, since the limited number of Hamilton's troops and his insufficient land transport would be inadequate for sustained overland operations. On the Bulair Isthmus where "the prospect of a strategic decision" (Liman) offered itself and where Hamilton might have cut the communications of all Turkish forces in Gallipoli by land, or water. But these lines were fortified and besides Bulair was twice as far from Lemnos as the third of the likely landing shores, which was near the southern tip of Gallipoli.

These three possibilities at the outset dictated to Liman the division of his forces into three equal parts, and their positions.

He could not know conditions on the enemy side where the exigencies of the western front excluded the most promising landing, that on the Trojan shore. Therefore, each third was designed to hold up one land force, which in that specific theater was apt to have a triple numerical superiority, for two or three days until the other two-thirds could be shifted from the places where feints might be made and bring relief.

Hamilton had thoroughly been infected with Churchill's obsession that the greatest prize of the war was dangling before the eyes of England and her Allies. But he had become almost as much aware that "half of that unique chance has already been muddled away by the lack of secrecy and swiftness in our methods" (I, 94-5). So far as that involved reproaches, those against the indiscreet press are clearly less justified than those against the dispatchers in London. While the general "saw" Constantinople, the same cannot be said of the subordinates, whose vision was glued and limited to the beachheads.

When he left for Egypt Hamilton realized that "the operation of landing in the face of an enemy is the most complicated and difficult in war." To the many necessarily unknown quantities in the problem were added what might have been learned in peace time, such as the topography of the beaches and their hinterland, or the unloading and reloading of transports, about which the British officers found that "it takes three times as long to repack a ship loaded at haphazard as it would have taken to have loaded her on a system in the first instance."

Landing craft were of the most varied kind, much of it bought hurriedly in Greece. Special landing barges, possibly bullet proof, motor-driven lighters carrying 400 or 500 men or 30 to 40 horses, which the sailors told Hamilton were being constructed for Fisher's Baltic landing project, were denied him when he called for them. He thought to the last that a large number of boats and steam launches would have bettered his chances (I, 44-5, 58, etc.).

The troops forming Hamilton's motley army—five divisions, two from the United Kingdom, the 29th and the Naval Divisions, two from the Anzac army corps, one of French colonial troops—while waiting practiced landing manoeuvres, Hamilton finding the Senegalese as "awkward as gollywogs in the boats." However, "every hour's practice will save some lives by teaching them how to make short work of the ugliest bit of their job."

Weather was apt to interfere with preparations and schedules, storms in the end postponing the date of the landing by two days. "What a ticklish affair the great landing is going to be! How much at the mercy of the winds and waves! Aeolus and Neptune have hardly lost power since Greeks and Trojans made history out yonder!" In choice of the hour for the landing Hamilton was faced by another typical dilemma of the invader. He himself originally favored getting the first boatloads ashore before daybreak, but the nautical reasons for daylight landings and the military fear of too much disorder proved stronger.

Obstacles to Landings

With a very strong current running around the southern tip of Gallipoli and the exact lie of the beaches unknown, the unavoidable confusion would have been so serious as to make it preferable to run the risk of exposing the men in the boats to the aimed fire of the defender. Actual experience pronounced strongly against daytime landing. The shore configuration made the landing beaches few, most of the coastline being formed by steep cliffs with only a few narrow gullies running somewhat less sharply to the sea.

Changing his contemplation from the enemy shore to the enemy leader, Hamilton proposed

> . . . To try to move so that he, Liman, should be unable to concentrate either his mind or his men against us . . . . Sea power and the mobility it confers is a great help, and we ought to be able to rattle the enemy however imperturbable may be his nature and whatever he knows about us if we throw every man we can early in our small craft in one simultaneous rush against selected points, while using all the balance in feints against other likely places . . . There will and can be no reconnaissance, no half measures, no tentatives . . . . We've got to take a good run at the Peninsula and jump plump on—both feet together. At a given moment we must plunge and stake everything on the hazard.

The margin of hazard in war can be narrowed to a certain extent by the counter-hazard of surprise plus the most careful military planning and execution of plans. "On maps and charts the scheme may look neat and simple," Hamilton noted a fortnight before the venture. "On land and water the trouble will begin and only by the closest thought and prevision will we find ourselves in a position to cope with it . . . . So much dovetailing and welding together of naval and military methods, signals . . . . and the worst punishment should any link in the composite chain give way . . . ." [6]

After the reloading of the transports in the spacious port of Alexandria the harbor of Mudros on Lemnos Island was made the base of the Gallipoli enterprise. Thence at nightfall of the 24th the landing force started out over a "glassy smooth sea" in some 200 large and hundreds of small craft under escort of the warships, largely manned by sailors from the fleet. As the convoying and conveying sailors reported, "not one word was spoken or movement made by any of the thousands of untried troops, either during the transit over the water in the darkness or nearing the land when the bullets took their toll."

What was considered the best of the force, the 29th Division, was to land in daylight and after artillery preparation on five different beaches near Cape Helles, "S" and "V" beaches to the right and "W" and "Y" to the left of it. "W" and "V" beaches seemed most important as they promised the greatest usefulness for the subsequent operations and the enemy, aware of this, defended these beachheads with heavy wire entanglements, entrenchments and machine guns. "Y" seemed rather inaccessible, reminding the Navy of Wolfe's path up to the plains of Abraham, and was also undefended by the enemy.

The Anzacs were to land fifteen miles further up the Aegean coast, near Gapa Tepe, before daylight. These two forces were to converge while the feints undertaken by parts of the Naval Division into the Gulf of Xeros, where no landing took place, and of the French at Kumkale, near the remains of Old Troy, from where they were withdrawn after some fighting, were to keep the enemy guessing so long as possible. It actually took Liman about 48 hours to see through these feints.

Day Lost by Main Force Delay

At "Y" a force of more than two battalions of marines came unmolested to the shore before dawn, and so did the forces sent to "S" shore on Morto Bay and "X," all three actually to the rear of, and numerically superior to, the weak enemy forces defending "V" and "W." They remained, however, inactive all through the day waiting for the main force to come to the shore and advance. Keyes' proposal to divert to these more fortunate landings part of the main force was turned down by Hamilton, with misgivings. This had not been envisaged in the original plans. His chief of staff "was rather dubious from the orthodox General Staff point of view as to whether it was sound for GHQ to barge into Hunter-Weston's plans . . . . But to me the idea seemed simple common-sense"; yet "it was not for me to force his hands: there was no question of that" (I, 132-3). It was Moltke-ism gone wrong, in a battle where the supreme commander was actually on the spot and a witness, and when the original plan had clearly run into a hideous snare.

At "V" 2,000 men on board of the *River Clyde,* an old collier provided with a number of contrivances for disembarking troops, were practically pinned to the vessel which had been run aground close to the shore, by Turkish machine gun fire. Lighters were pushed in between ship and shore to bridge the distance, with gangways connecting them. They were mere bridges to death. Boats trying to make the shore, ran up against underwater wire entanglements and the same machine gun fire. Only a few reached the shore unhurt where they were immobilized by the Turks' fire and had to wait to move until sundown when at last troops could be landed freely. Ship artillery was unable to help since attackers and defenders were too close.

The experiences of the second main landing promptly confirmed Hamilton's premonitions about "getting ashore before the enemy could see to shoot out to sea." The Anzacs, although missing each of the designed landing by a mile, got ashore at what was later called Anzac Cove, practically without casualties, landing in several waves of some 1500 each. This brought to shore 20,000 men and a little artillery within 24 hours. They made but poor use of their opportunity against an enemy whom they originally outnumbered vastly, perhaps sixteen-fold, but whose counter attacks in the afternoon, led by Mustapha Kemal, made their commander despair of the undertaking and ready to re-embark until Hamilton told him "to dig, dig, dig, until you are safe."

The troops ashore were quickly worn out by the duties imposed upon them by the nature of their beachheads, without quays, cranes, or sufficient water supply, and lacking almost all local reserves. Energies of determined, but inexperienced troops were as quickly spent as were the Allied opportunities for surprise. The Turks everywhere held commanding positions. Where they had stripped their positions of men, as at Bulair, opportunity went begging because there were no Allied forces left to be thrown there. But Turkish counterattacks designed to throw the Allies back into the sea, failed with heavy losses. Hamilton had no reserves left and the few whom Churchill and Fisher obtained for him from Kitchener on the 27th, some 12,000 or 13,000, stationed in Egypt and heretofore denied to Hamilton, could not arrive before May 1 and 5. Meanwhile, the enemy brought the most modern arms into play, with an airplane dropping bombs on the *River Clyde*, which did more to foreshadow difficulties in future landing enterprises than immediate damage.

The Turkish munitions supply had long been low, but now (May 5) Hamilton was told by the War Office that supplies for him were not calculated on the basis of a prolonged occupation of Gallipoli, a warning which resulted in a new attack by the Anglo-French forces on and after May 6. The attack of the 50,000 with 72 guns failed almost completely against the defending force of 30,000 with 56 guns. Only a few hundred yards were won at a high price in casualties, 19,000 from April 25th to the evening of May 8th. These attacks and the daily trench warfare were in truly western front style, fought in an eastern theater where the eastern school had expected something very different. Tactical help from the Navy remained slight.

Whatever series of errors and wrongs the Gallipoli enterprise produced, there was on the spot practically no inter-forces conflict, either between the military and the naval men, or between the English and the French. Realizing the complete check which the military had suffered in those May days, the forward school among the naval officers, the "fire eaters" like Keyes, now proposed to try again for the Straits in ships, for which task they felt better prepared than before, thanks to the presence of two fast destroyers and perfected arrangements for the sweeping of mines, the item of material hardest for the Turks to replace.

Army Expected to Save Ships

Admiral de Robeck had at first resisted. "He thought, as we thought, that the Army would save his ships," Hamilton wrote on the 10th. "But our last battle has shown him that the Army would only open the Straits at a cost greater than the loss of some ships and that the time has come to strike home with the tremendous mechanism of the Fleet" (I, 214-5). This showed more magnanimity, or true sense of proportion as to losses than the Admiralty in London could produce when de Robeck submitted the problem for their decision.

As no one knew better than Churchill, the Admiralty was fundamentally averse to risking ships except in maritime battle against the enemy who was equally eager to spare his own vessels. This save-the-ships mentality, whether to be ascribed altogether to Mahan's doctrines or not, certainly helped the Turks to save the Dardanelles, through which the Allied squadrons were not to pass until a day after November 11, 1918. Never again, after March 21, had the Admiralty been in one mind with the Admiral on the spot as to the supreme naval issue—whether or not to force the Straits with ships alone. This time it was the Admiralty who took pity on the big ships. While Churchill was still willing to risk them, at Fisher's insistence and against the violent protest of Kitchener, who felt that the Army was being left in the lurch by the Navy, the strongest of all the battleships was recalled from a field thought to be increasingly dangerous due to the U-boats, which had not yet operated.

But when the mere word "submarine" struck imaginations before any torpedo had inflicted a physical blow, Hamilton "in less than no time saw a regatta of skedaddling ships." Transcending the local scene it signified for him the death of

> . . . the invasion of England bogey which, from first to last, has wrought us an infinity of harm. Born and bred of mistrust of our own magnificent Navy, it has led soldiers into heresy after fallacy . . . until now it is the cause of my Divisions here being hardly larger than Brigades, whilst the men who might have filled them are 'busy' guarding London! If one rumored submarine can put the fear of the Lord into British transports, how are German or any other transports going to face up to a hundred British submarines? The theory of the War Office has struggled with the theory of the Admiralty for the past five years; no more is left of a soap bubble when you strike it with a battle axe. Some other stimulus to our ter-

ritorial recruiting than the fear of invasions will have to be invented in future . . .[7]

The setback at the Dardanelles worked in recoil upon British politics. Both Fisher and Churchill left the Admiralty and the Government was regrouped as a coalition after an interregnum of three weeks which left the political issue involving the Dardanelles postponed for that period. To watch over it Churchill remained in the Cabinet and on the War Council. His energetic promptings obtained more divisions for Hamilton from a coalition which by its floundering was only adding to the diversity of opinions about the strategy of Britain and her Allies. These new divisions, however, were given only piecemeal instead of all at once as Churchill warned. This enabled the Turks and Germans to build their forces *pari passu,* strengthen their positions, and finally allow the submarine to strike and preserve the margin of defensive strength over the attacker.

"Too Little and Too Late"

The coalition government, far more under the influence of the western school among the military than its predecessor, could never believe that they possessed the means to take Constantinople, except when under the immediate, but not permanent, influence of Churchill's forceful persuasion. All their decisions, slowly maturing, consumed the time favorable to the enterprise: the forces and the opportunity for a new stroke were available four or more weeks before the decision to use them was politically arrived at. "We have always sent two-thirds of what was necessary a month too late" (Churchill).

After some two months the government decided to raise Hamilton's army to 12 divisions. By then that of Liman had risen to 15. They also granted him an ample supply of ammunition and a number of Fisher's specially built landing barges, or motor lighters. These "Beetles" as they were called, were motor-driven, bullet-proof, had landing bridges at their bow, moved at five knots and carried 500 infantry.

But Kitchener, who had survived the shake-up, denied him younger and more experienced generals and he thus had to fight with one of the worst combinations of World War I; raw and young and perhaps high-spirited troops and superannuated and inefficient generals, to tackle the most complicated tasks.

Unfortunately, Hamilton was no severe taskmaster and altogether unlike Liman, who knew whom to sack and whom to pick among the Turkish officers, with Mustapha Kemal Bey as his prime favorite, a man "who rejoiced in responsibility." Also regrettably, Hamilton had never served on the western front and thereby learned the futility of local frontal attacks on entrenched forces. He frittered away a considerable portion of his own men in the Peninsula by three times vainly attacking in the Cape Helles region, in June and July. This cost him 12,000 casualties, or nearly the effective strength of a division.

The new forces put at his disposal, which gave him none of that vast numerical superiority which the westerners demanded and obtained for their murderously futile offensives, decided him upon a double stroke aimed at getting a position astride the Peninsula, slightly above and dominating the Narrows. The attacks were to strike from Anzac Cove, where one and one half divisions were brought ashore at night, and from a new beachhead at Suvla Bay, five miles to the north from Anzac Cove. This region was known to be but lightly held by the Turks.

The attack, started on August 6, on a moonless night, fell far short of its objectives. The plans were much too complicated and exhausting for the troops and officers at Hamilton's disposal. Besides, they were given practically no opportunity to familiarize themselves with their tasks. Bitten by the bug of secrecy, Hamilton ordered maps to be handed out only on the eve of the 6th, whereas we now know that landing troops to be successful should be enabled to go over their terrain beforehand *in absentia,* helped by maps of all sorts and other information.

The columns proceeding from Anzac Cove while overrunning the Turkish outposts became bogged down in their own inefficiency, one pausing for breakfast a few hundred yards from a dominating crest which was then but weakly held, but was "bristling with rifles" by the end of the meal; another becoming lost in the wild terrain although all opposition in front had temporarily ceased. In both cases the troops moving up steep inclines were weighed down with equipment thus losing mobility where it was most needed. These troops were unable to hold their most forward positions, from where they had been able to see the Narrows below. British naval shells hit and drove back the very men who had reached the point where they might have shouted "*Thalatta!*" had any breath been left in them.

The rawest troops—20,000, of whom 12,000 were to be casualties within a few days—and the most incompetent general were employed at Suvla Bay where the greatest surprise, thanks to the use of sea power, and the highest accomplishments were possible. The Turks had expected no landing at this point, from where their forces fighting at Anzac would have been outflanked. The troops were brought to shore without any mentionable loss, whereupon the British commanders congratulated themselves and their troops on this feat. After that, they ate and either took a bath, or a nap, to which the commanding general, not in the best health, was accustomed from his long pig-sticking service in India. He did not even go ashore to encourage his troops, to push ahead to the high ridges, which were practically deserted for two days. To Liman this inactivity remained "incomprehensible," and so it did at last, on the 8th, to Hamilton himself, alarmed by the absence of reports.

Command by Remote Control

With all the mobility and communications afforded him by sea power, Hamilton had immured himself at Lemnos, if not quite as far at least as completely remote from the scene of action as the younger Moltke during the first Marne; and like the latter sending General Staff emissaries to the front to find out for him what was going on, although not providing them with any full powers. When he finally decided to go to the scene, he found that the naval vessel put at his disposal was not under steam. Another delay of four hours during which the supreme command was held incommunicado by the other partner to the amphibious enterprise. When he made his inert corps commander attack, after finding him "happy" in the opinion that "everything was quite all right and going well," it was too late.

Turkish reserves had arrived. Nothing was gained for the British lion, after the doors to the temple of victory had stood wide open for two days, but narrow strips of land, a few hundred acres, to hold to, but not to jump from. Everywhere in front the Turks held dominating positions while autumn storms were soon to threaten the invaders in the rear. "We have landed again and dug another graveyard," as one of the participants noted in his diary on the 12th.[8] Still not enough, for Hamilton ordered an attack on one of the spurs above Suvla Bay on the 21st and 22nd. This was another bloody and wasteful failure, resulting in nothing but 5,300 casualties out of 14,300 attacking.

The Allied troops on Gallipoli, at last supplied with better generals by Kitchener, who realized now that "this is a young man's war," were awaiting further plans of what was to be done by them, or to them, the while they wrote home and bypassed the censor with accounts of the waste and inefficiency that was consuming them, in addition to the winter weather, beginning in November and freezing to death men who had survived torrid summer heat. Losses through disease mounted, until the totals of all casualties rose to 252,000 out of 480,000 men landed, of whom 70,000 were Frenchmen, instead of the 5,000 whom the War Office had once considered as top price to be paid for a successful and decisive operation. Luck and dictative power were with the Central Powers for the time being. Moved from her wavering attitude by the obvious defeat of the Allies in her immediate neighborhood, Bulgaria decided to join the enemy.

Undeterred by the Entente attacks at Loos and in the Champagne in September-October, behind which cavalry was waiting to make warfare mobile once more, the Germans and Bulgarians launched their offensive against Serbia. The first troops of the western powers to be moved against them, or at least to Saloniki, were taken from among Hamilton's divisions, an indication that the end of the Gallipoli adventure had begun. Once more the forward school of naval officers on the spot, backed at home only by Churchill's waning influence, proposed to renew the naval assault on the Narrows. They were turned down by the Admiralty where Balfour was far less willing, or able to oppose the ranking admirals than Churchill.

Evacuation of Gallipoli was now considered and the General Staff was nearly frightened into staying and prolonging the misery when facing the estimated cost of evacuation, 50,000 casualties. Kitchener's attitude, according to Sir Henry Wilson, foremost adherent of the western school, was "that we can't get out of the Dardanelles without appalling disasters there and all over the East, without some success . . . He realizes that it is bleeding him white and dreads the whole thing, but favors operations ending in success and then withdrawal."

No chance for such a prestige success came Kitchener's stumbling way to cover up his retreat, which was more urgently demanded by political pressure at home where reports were circulated on Hamilton's poor staff work—under a man whom Kitchener had selected for him. The General still favored holding on to Gallipoli, believing that the cost of evacuation would amount to half of the remaining troops, plus most of the materiel.

"Damned Side Show"

Munro, a general of the western school, was sent out to replace him and "to put a stop to this damned side show," as the new chief of staff told his opposite number in the Navy, Commodore Keyes. Without much local inspection and interrogation—half of the local commanders were against evacuation—but with his mind firmly made up before his arrival, Munro advised London strongly to evacuate, even though it might cost as much as 30 to 40 per cent of the military forces still there, or some 40,000 men. The western school was clearly accustomed to high casualties.

Once more Keyes urged the naval advance through the Straits. But Churchill, now excluded from the newly constituted War Committee, could no longer put his persuasive powers behind him. Kitchener was half won over, but during an inspection trip to the Dardenelles, where he was sent by Asquith to have him out of the way, was dissuaded by Keyes' superior. What he still nourished as an alternative scheme was swept away by the now overpowering opposition to the enterprise and he agreed to the evacuation, hoping at first to preserve the toe hold near Cape Helles. The last to adhere to the enterprise were Admiral Wemyss, de Robeck's successor, and Keyes, and they had to be silenced by the Admiralty and cut off from free and open discussions with the military commanders by Munro.

A thorny crown of irony was thrust upon the Gallipoli business when the evacuation of Suvla and Anzac, on the night of December 19, and that of Helles on January 8, after most careful and detailed arrangements, were carried out without losses and with the luck that had so often previously failed. Perhaps this was Fate's inspiration for Dunkirk.

The general, strategical lessons to be drawn from the Gallipoli affair were that "de-landing" is as a rule far less difficult and catastrophic than expected; and that landing, obtaining a foothold on the hostile shore, is neither the most difficult, nor even the decisive thing. In the fever to which a landing enterprise might be compared, the landing itself is merely infection manifested through abnormal exhilaration. "We have made good the landing," Hamilton noted after the initial debarkations, "The thing seems unreal; as though I were in a dream, instead of on a battleship. To see words working themselves out upon the ground; to watch thoughts move over the ground, as fighting men . . . ." (I, 129).

The true crisis follows later, when the enemy is to be encountered and to be beaten and annihilated on his own ground. Therefore, planning and initiative must never halt even for an hour on the shore line, a line which many a general has mistaken for a goal, when in actuality an amphibious enterprise is a race in two elements. The first goal line, if adhered to, merely tells the attacked party where the invader is. The hydraulic pressure of invader command, working its will through the hierarchy of ranks and communications, must propel this landing force at once to a maximum of inland distance.

In a majority of professional minds the Dardanelles "proved" that such landing enterprises were practically "impossible." It was largely on this conviction that Douglas MacArthur based his ideas of 1936 for the defence of the Philippines, and not even on a strictly correct history of such an enterprise. It was simply not a fact, as MacArthur wrote at that time, that Turkish infantry "in many cases decimated whole divisions in their attempts to land." [9] However, the whole conviction was stronger than the historical facts, bad as they were.[10]

NOTES, CHAPTER 39

[1] Lloyd George, *War Memoirs,* I, 365.

[2] Ibd. I, 395.

[3] H. A. DeWeerd, *Great Soldiers of the Two World Wars.* N. Y. 1941, 118-9.

[4] Hamilton, *Gallipoli Diary* I, 13-5.

[5] Ibd. I, 8-9.

[6] Ibd. I, 100-1.

[7] Ibd. I, 243 ff.

[8] Ashmead-Bartlett, *The Uncensored Dardanelles,* 197.

[9] Frank C. Waldrop (ed.), *MacArthur on War.* N. Y. 1942, 345.

[10] The above is largely based on the following literature: Winston Churchill, *The World Crisis, 1915.* N. Y., 1923; Sir Ian Hamilton, *Gallipoli Diary.* London, 1920; The Dardanelles Commission, *Final Report.* London, 1919; General Liman von Sanders, *Fünf Jahre Türkei.* 2nd ed., Berlin, 1922; Admiral Fisher, *Memories* and *Records,* both London, 1919; Admiral Sir R. H. Bacon, *The Life of Lord Fisher of Kilverstone.* N. Y., 1929; Lord Keyes, *Naval Memoirs.* London, 1938, and *Combined Operations.* N. Y., 1944, ch. III; John North, *Gallipoli; the Fading Vision.* London, 1936; Compton Mackenzie, *Gallipoli Memories.* N. Y., 1930; Maj.-Gen. J. F. C. Fuller, *Decisive Battles.* N. Y., 1940, ch. XXVI.

Chapter 40

GERMAN ENTERPRISE AGAINST THE BALTIC ISLANDS, OCTOBER, 1917

THE NORTHERN END of the eastern front formed one of the flanks resting on the sea which, from 1914 on, suggested essays into amphibious warfare. At times before 1914 the Russians had feared German landings in Finland, but as far as is known, nothing like that was officially premeditated and, despite the constant temptation, little of the sort was undertaken during the first three years of the war. In London Admiral Fisher, deeply disappointed that his own Baltic schemes were not considered feasible at home (and apparently just as little in Petersburg) in March, 1917, on the very eve of the Russian Revolution, which was to change German views about amphibious enterprises in the East, endeavored to draw Lloyd George's attention "to the imminent danger of the German High Sea Fleet convoying a large number of transports and taking a German Army into the islands of Riga by sea, and into the vicinity of Petrograd, thereby endangering the Russian capital—a deadly blow to Russia; and our Grand Fleet, with its unchallenged sea supremacy, condemned to be a passive spectator of such an appalling catastrophe."

After America's entrance into the war had given the Allies a still more overwhelming sea preponderance, Fisher wanted to see this combination and "an immense armada of special rapidly-built craft" to carry a Russian army, not the English or French, to proceed via the Pomeranian coast to Berlin (July 11). It grieved Fisher no end that in October of that year the Germans, as he had warned in March—this providing for his friends further proof of Fisher's prophetic instincts which went to waste during his retirement—undertook the very same operation which he had foretold.[1]

That their success in even the slightest degree proved the feasible character of his own ideas cannot be fairly maintained, for they would have lacked the prime condition on which the success of the German enterprise was based, low morale on the enemy's side. As the Chief-of-Staff of the enterprise against the Baltic Islands admits in commenting about its "practically program-like procedure": "The Russian was no full weight

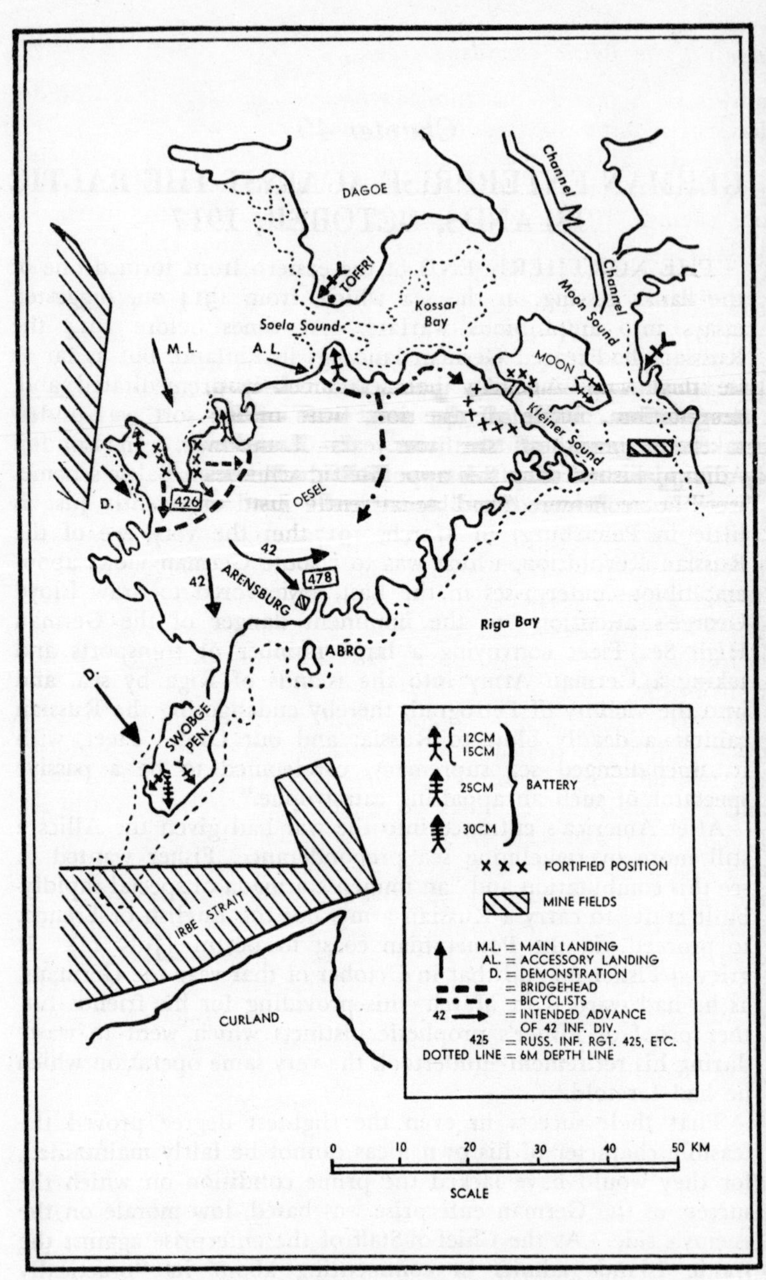

German Operations In the Baltic, 1917.

adversary so that against him much could be dared that would have been impossible against an equal foe."

For this reason the enterprise ranked somewhere between peacetime amphibious manoeuvres and an action against a serious enemy. It was a show staged by military experts, its perfection marred only by constant awareness that the risks involved, while considerably higher than in any peacetime exercises, were still limited. However, it was based on the fully, although quickly, matured planning, careful liaison arrangements between army and naval units and command, and observations and experiences made since the beginning of the war. It lacked specially constructed landing craft for which neither time nor materials were available.

The German Eighth Army (von Huitier) early in September had taken Riga and a bridgehead beyond. It felt threatened there so long as the Gulf of Riga remained controlled by the Russians, based on the islands at the two entrances, Oesel, Dagö, Moon, Runö, etc. And almost before anything had been proposed by any German command, London newspapers carried headlines about "The German fleet to assist the land operations in the Baltic."

Russian guns on the islands, some outranging practically everything that German warships carried, and minefields blocked the two entrances to the Gulf, the Irbe Sound between Oesel and the continent, 18 miles wide, and the four-mile width of Moon Sound between Moon and continental Esthonia; Oesel and Moon were connected by a causeway. Whether or not the idea of taking the island group was due to Hoffmann, the "idea man" in the Supreme Command East (Oberost), orders to proceed towards it were given on September 19. Ironically named, if Fisher had but known it, the enterprise was to be referred to under the code name of "Albion," and it was to be headed by an army man and army staff, General von Huitier and his AOK 8.

Many Problems Offered

Officers at landing corps headquarters had to think fast in their new element, somewhat along these lines: For all overseas enterprises complete control of the seas involved had heretofore been considered the first condition; however, such a supremacy was not as easily maintained as in times gone by, due to the submarines, of which eight British ones were known to be based on Hangö in Finland, bombing from the air, minefields, with mines and submarines considered the two most dangerous obstacles on this occasion, when superiority on the sea was definitely on the German side in the Baltic.

The effect of the high angle fire from shore to ship was feared, while not available in the opposite direction. Disembarkation had become more difficult than in the past, due to the infinitely strengthened defensive power conferred by the machine weapons. Defensive strength of a sea shore position was great, greater by comparison than that of a river shore where the attacker could use artillery on equal terms with the defender and machine guns about as well. Nautical conditions called for the use of relatively large-sized boats, offering larger targets, as compared with the small boats and pontoons used in river crossings.

Remembering the utility of a tank placed in the bow of the *River Clyde* at the Dardanelles, whose machine gun fire had held the Turkish fire somewhat in check, the Germans decided to equip even fairly small boats with machine guns. The use of smoke and artificial fog during the critical moment of debarkation was contemplated, although in the end neither was applied. Dependence on the weather was fully realized. Surprise was a main factor. How much would even an inattentive enemy know of and learn from, the preparations? There always remained to attempt obtaining tactical surprise by landing at various points, particularly helpful if convergence of operations of the various landing parties could be ensured. "The weak moment for the attacker when landed lies in the very first hours when he is still without guns and possibly without machine guns as well."

Waters and the islands in them were well charted; the topography of the islands was not unfavorable to the attacker, even where the shores were steep. Russian strength, of about one infantry division; gun positions, presence of warships in Moon Sound and Russian weakness due to the low morale of the troops were well enough known in German headquarters, and therefore they kept their numerical superiority within moderate limits. Libau was the chosen embarkation port, so as not to make the approach too long, and also to keep out of the perimeter of Russian air reconnoitering, which could have included the still nearer port of Windau. The High Sea Fleet units were based on Danzig.

The combined deliberations of army and naval officers, most of the time in the presence of von Huitier, began at Libau on September 21. They covered questions of equipment and time-tabling, no less than arrangements for the command during the various stages of the enterprise, about which an order of the AOK laid down that "the execution of sea transportation and its protection is directed by the leader of the naval forces. Under his orders are for the duration of the voyage and until the moment of the effected landing all the participating forces of the land army."

There was only one exception to this. Should the vanguard of the disembarked troops meet resistance on the way to the shore, the local infantry leader, and not the boat commander, was to decide as to the choice of the final landing spot and also as to possible slight delay, always bearing in mind the order that the landing must be effected in any case. After the landing the leader of the naval forces "has to comply to the extent of all available strength with the orders of AOK 8, or the leader of the landing corps," the XXIIId Reserve Corps, who would have to arrange for support of the landing force by ship artillery.

The task force consisted of one infantry division plus one infantry regiment, one infantry bicyclist brigade, two storm troop companies, two troops of cavalry, five heavy batteries, or in all 24,600 officers and men, each provided with a swimming vest; 8,500 horses, 2,500 vehicles, 40 cannon, ranging up to 21 cm. caliber; 220 machine guns, 80 trench mortars, with rations for 30 days. Seventy-five airplanes, of which 68 were hydroplanes, were earmarked for the operation; eight of the latter occupied Runö Island on the 13th and six others Abro Island on the 15th, arranging for the relighting of the lighthouse fires.

The naval forces numbered 300 vessels of all kinds, including 19 transports varying in size from 1,750 to 11,500 tons, enough to carry half of the force and equipment at one time. The steamers carried regulation horse boats, pontoon-like craft of light draft with let-down bows. Moored, they were to serve as the first landing piers.

The originally selected embarkation date was September 27. But storms and difficulties encountered by the mine-sweepers, not available in sufficient numbers, postponed this date to the 9th and 10th of October, and almost, it might be said, *ad calendas graecas,* for the Supreme Command, in need of troops for the battle in Flanders then going on, and the impending offensive in Italy, threatened to call off the expedition if the delay continued. At last the coal reserves at Libau were running so low that postponement for another 24 hours was impossible. Members of the AOK 8 thought none too highly of Ludendorff's interference on the occasion, for he was equally impatient with local difficulties and the meteorological hindrances.

Mines Offer Few Obstacles

The transports, under protection of the warships including two squadrons of the High Sea Fleet, left Libau on the 11th. "The officers and men of the Army embarked on the transports had probably no idea what difficulties had to be overcome; they were filled with unsuspecting confidence, and that was good," wrote the Chief of Staff of the landing corps. The voyage taking less than 24 hours, was not devoid of incidents. Two battle ships and one transport struck mines, but the former remained manoeuvrable and the latter was beached after the troops on board, without suffering casualties, had been shifted to other transports. The Russians had their waters amply mined, but it proved that the old military rule that mere obstacles without active defences behind them are of small value was almost as true in naval warfare. The enemy left the German minesweepers practically undisturbed.

Surprise was nearly complete; although the Russians had suspected a German coup in the second half of September, most of them believed it improbable, if not impossible. In consequence, the main landing in Tagga Bay on Oesel was effected on the 12th against very little opposition. The transports arrived outside at sunrise, 5:30 A.M., when the vanguard began to disembark in the steam, or motor pinnaces of the larger warships. At 6:45 the transports entered the Bay where at 8:15 the main debarkation, undisturbed by the Russians, started, the troops being carried by the life boats to the beach, and from 10 A.M. on to emergency piers swiftly constructed by the engineers.

Russian opposition in the interior of Oesel proved not much stronger than at the beach; some troops fled across the causeway to Moon which German torpedo boats took under fire, and by the evening of the 16th all Oesel was German-held. The Rus-

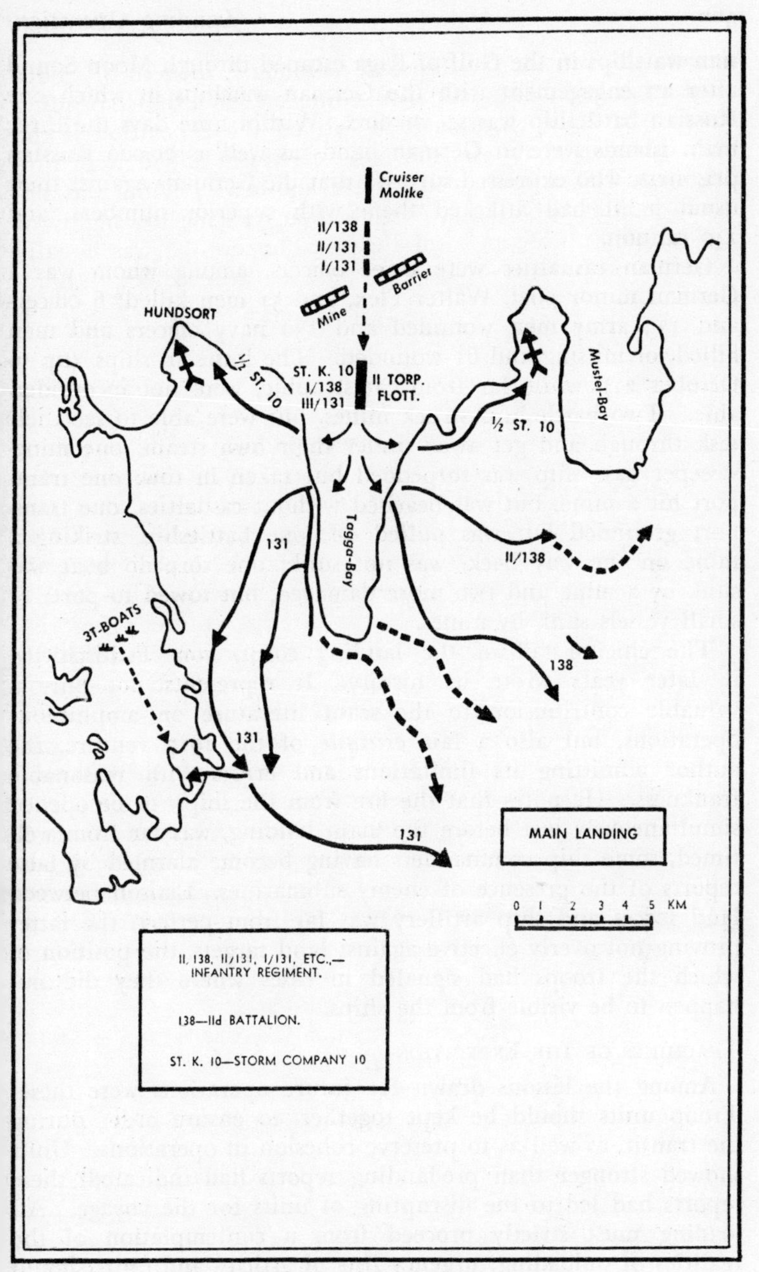

German Operations In the Baltic, 1917.

sian warships in the Gulf of Riga escaped through Moon Sound after an engagement with the German warships in which one Russian battleship was set on fire. Within nine days the three main islands were in German hands as well as 20,000 Russian prisoners, who expressed surprise that the Germans against their usual habit had attacked them with superior numbers, and 140 cannon.

German casualties were three officers, among whom was a German minor poet, Walter Flex, and 51 men killed; 6 officers and 135 army men wounded and 130 navy officers and men killed, or missing and 61 wounded. The losses in ships (up to October 21), while far from catastrophic, were not inconsiderable. Two battleships struck mines, but were able to see their task through and get away under their own steam; one minesweeper base ship was torpedoed but taken in tow; one transport hit a mine, but was beached without casualties; one transport grounded but was pulled off; one battleship, striking a mine on the way back, was not sunk; one torpedo boat was sunk by a mine and two more damaged, but towed to port; 21 small vessels sunk by mines.

The chief-of-staff of the landing corps, von Tschischwitz, in later years wrote its history. It represents not only a valuable contribution to the scant literature on amphibious operations, but also a fair *critique* of the 1917 venture, the author admitting its limitations and errors with reasonable frankness. He notes that the fire from the ships, to be opened simultaneously just before the main landing, was far from well timed, some ship commanders having become alarmed by false reports of the presence of enemy submarines. Liaison between land forces and ship artillery was far from perfect, the latter proving not overly effective against land targets, the position of which the troops had signaled in cases where they did not happen to be visible from the ships.

Teachings of the Expedition

Among the lessons drawn for future operations were these: Troop units should be kept together, to ensure order during the transit, as well as to preserve cohesion in operations. Units showed stronger than pre-landing reports had indicated; these reports had led to the disrupting of units for the voyage. All loading must strictly proceed from a contemplation of the realities of unloading; urgency lists of articles and units should be worked out in time. Medical units must by no means be placed at the bottom of such lists; they belong rather among the units to disembark early. Cavalry proved useless, the unloading taking an inordinately long time. Bicyclists proved to be infinitely superior.

Ships in tow should not be included in a convoy since they slow up movements. So long as the command is still on board the transports communications between it and the troops on land must be maintained by all means. Some reports stress that the difference of army and naval signalling methods contributed to faulty ship-to-shore communications. Troops landed must at once be moved away from the beachhead and drive forward, establishing a beachhead large enough to ensure subsequent landings against setbacks. Risks from minefields, or airplanes during the transit of the convoys should be more evenly guarded against than in the Oesel enterprise, when the largest transport carried no less than 74 officers and 2100 men —over one fifth of one wave—350 horses and 130 vehicles. To unload her took considerable time, since she could not be brought as close to the shore as smaller transports.

Partly for the reason that only a fraction of the Eighth Army strength was engaged in the Oesel enterprise, AOK 8, having assured itself that its influence was paramount, had preferred to stay behind on terra firma in Riga and boss things by remote control. Due to this and the breakdown of wireless connections, its influence on actual operations, which seems to have been called for, or on disputes between the services, suffered "a highly undesirable limitation." Its absence would have been felt even more severely if, as seemed feasible to the task force on the spot, the enterprise should have been carried on beyond the islands to the continent in the direction of Reval and possibly St. Petersburg.

The cardinal problem to Tschischwitz was whether the direction, or leadership of a conjoint enterprise should be military, or naval; OHL in the last instance, or Naval War Direction. While he attempted no general answer, since conditions might be too diverse, he thought it "only a natural, because healthy, phenomenon that in a field in dispute between army and navy, a rivalry spontaneously evolves of venturing its all in the service of people and country, of doing something extraordinary, of carrying off the crown. It would be a pity were it different.

But the psychological moment which comes to the foreground in this competition must not be allowed to decide one way or other. To find complete mastery of both land and sea warfare combined in one person who is at the same time a natural leader will always be a great rarity."

While according to English writers like Corbett, no supreme command should be established in combined operations, with adequate coordination between army and navy producing less friction than some form of super-ordination, the Dardanelles experience seemed to the German general not to favor such ideas. All military command in war, whether on land or on the sea, must in the last instance be one. With the German general the preference for the monolithic arrangement was unavoidable, and since the constitutional arrangement making the Kaiser (or the Führer later) commander-in-chief did not suffice in technical respects, considerations growing out of combined operations, among other things, led in the late 1930's to the device of an *Oberkommando der Wehrmacht.*

In a less extreme way the General considered "unitary command even in a combined operation *per se* practically an indispensable condition for success. The leader of it, however, must be in position to survey the potentialities and limitations of the possible uses of the sister arm." It would be of great assistance to such a commander if a definite number of army and navy officers should receive special training in combined operations.[2]

The conquest of the Baltic Islands was the last action on the eastern front between Germans and Russians before the armistice of December 15. "One more black page in the book of war had been turned" (Trotsky). The diplomatic report from London that the British Admiralty in spite of all urgings considered it impossible to help out in the Baltic, the Bolsheviki interpreted as a sign of the Allied willingness to let Germany strike a blow at Petrograd and thus send the turbulent seat of the Revolution "to school to Ludendorff and Hoffmann." The taking of the Islands was followed in time, and perhaps causally connected with it, by the Bolshevik coup of November 7. There would thus be a parallel between the landing of the Allies in Italy and the ensuing fall of Fascism in 1943, and the loss of Oesel, Moon, etc., and the fall of the Kerenski régime in 1917.

The loss of Riga had already endangered Petrograd and the

taking of the Baltic Islands had considerably increased this threat to the capital, which was then living "in a two-fold state of alarm" (Trotsky), under the impact of the political and military events. The military defeats added greatly to the arguments in favor of transferring the seat of government to Moscow, the Bolsheviki arguing at the time that the bourgeoisie wanted to leave Petrograd to the mercies of the Germans, who would take efficient care of the Revolution, whereas their own followers in the Army and the Baltic fleet were the only true defenders of the approaches to that nerve center of the Revolution.[3]

Confirming the Bolsheviks' worst suspicions, the western powers took a leaf from the German book when they landed forces against still more thoroughly revolutionized Russian troops, whom they considered no longer their Allies, in the following year. Originally landed for the purpose of saving vast stores piled up in the Arctic ports of Murmansk and Archangel from the Red Russians and the Germans who were now operating in Finland and later contemplated driving the Allies from these ports, until the defeat in the west forbade such moves, some ten thousand Allied soldiers were dispatched to both ports in May, 1918. Their subsequent operations included river boat actions on the Dvina during the summers of 1918 and 1919, the expedition not coming to an end until September, 1919. The invasion proved of more political than military significance, adding greatly to the distrust of Soviet Russia of the intentions of the western powers against her.[4]

NOTES, CHAPTER 40

[1] Bacon, *Life of Fisher.* II, 297, 193-4.

[2] The main source for the Oesel enterprise is General von Tschischwitz, *Armee und Marine bei der Eroberung der Baltischen Inseln im Oktober 1917.* Berlin, 1931.

[3] Leon Trotzki, *The History of the Russian Revolution.* N. Y., 1936. II, 186, 211; III, 62-4, 68, etc.

[4] Monograph on the Allied landings in the far north, by Leonid Strakhovsky, *Intervention in Archangel.* Princeton, 1944.

Chapter 41

ZEEBRUGGE, 1918

"It was left to Admiral Keyes to show at Zeebrugge that there were other ways of making war from the sea."

Churchill.

WORLD WAR I SAW NUMEROUS RAIDS—raids on land, raids by cruisers and other fast vessels and, most numerous of all, taking the place of cavalry raids of yore, air raids. But there was only one amphibious raid, that undertaken by British naval forces against Zeebrugge on April 22-23, 1918. It represented in its ambitious strategic aim an attempt on the part of the British Navy to do something it had shied from three years before, when an enterprise against Zeebrugge was considered in London as an alternative to both the Dardanelles and the Baltic schemes. The Navy aspired to succeed where the Army had failed, that is, in preventing the German submarines from longer operating from their base, Bruges, with Ostend and Zeebrugge as the sea exits of the Bruges ship canals.

Aside from the prevailing notions about attrition, the third Battle of Ypres in the summer and autumn of 1917 had for its more specific motive the desire of the Admiralty, and its clamors to the War Cabinet, to have the Germans driven out of Flanders. If the submarine bases were not taken and "if the army can't get the Belgian coast ports, the navy can't hold the Channel and the war is lost," Jellicoe had declared early in July. Shipping losses threatened to make it impossible to carry on the war in 1918, he argued. That clinched the decision of the War Cabinet to carry on the Flanders battle, while it made Haig, much to the amazement of his staff, stick to his plans which originally were predicated upon quite different reasons.

According to Churchill, this U-boat argument, which "seemed to throw the army into the struggle against the submarines," was "wholly fallacious. A grave responsibility rests upon the Admiralty for misleading Haig and his staff about the value of Ostend and Zeebrugge to the submarine campaign." For, however inconvenient were the submarines working from those bases, there were many others available to them in German home waters.[1]

The sparse territorial gains of the attackers and their heavy casualties had been disproportionate and the Germans' hold on the coast remained unshaken. They had in fact, by a small-scale surprise attack, taken a British bridgehead near the sea and thus "upset the coastal move which was intended to help Haig's frontal assault," a naval demonstration and a landing of the Fourth Army near Ostend, in which tanks were to participate. "This move was postponed indefinitely for strategic reasons."[2] The Germans regained so much confidence that they went ahead with their attack on Riga and the other operations on the Baltic flank of the eastern front, thus emphasizing that a communicating system of amphibious strategy was indeed in existence.

To gain their purpose in their own way, British naval men, some of them as early as November, 1916, and since both bombardment from monitors and airplanes had proved futile and use of the biggest naval guns against the submarine nests, as proposed by Lloyd George, was out of the question,[3] offered to block Zeebrugge and Ostend by sinking ships in the narrow entrances, which were only about 300 feet wide, and to do a maximum of damage to the two ports. Transferred from the post of Director of Plans, Admiral Keyes was made commander of the Dover Patrol and as such was to head the enterprise.

With a magnanimity on the whole rare in the conjoint history of the services, he was "determined that the Army should not be asked again to make sacrifices, in order to relieve the Navy from carrying out an operation which, cost what it might, would greatly handicap the enemy's submarine campaign." Whether this choice was conceived in the same expiatory sentiment or not, Keyes and the Admiralty resolved to employ only naval and marine personnel, largely volunteers, in all stages of the operations, including the landing parties. The latter, consisting of a battalion of 750 marines and 1030 sailors were given what is today called commando training.

Both Zeebrugge and Ostend were small, but vital parts in the vast fortress into which the Germans had transformed the Flanders coast. While this was engineers' work, the occupying forces were largely of the German Marine Corps placed there because of Tirpitz's conviction that "the Navy posted itself with a strong force of its own along the Flanders coast, as a symbol so to speak that Germany was resolved to come to close grips with the most bitter enemy and to maintain itself opposite the latter in a position of equal worth." [4] This rather empty rhetorical official statement about the symbolic purpose of the Marine Corps in Flanders could hardly shake the prevailing convictions in the German Army that the Marines fulfilled their duties carelessly, which seemed once more proved by the partial success of the British coup of April, 1918.

LAND AND WATER DEFENSES

The heart of this German defence system was a triangle formed by the coast from Zeebrugge to Ostend and the two canals, six and one-half and 11 miles long, running from these points to Bruges. The Zeebrugge canal, the more important of the two, was deep enough to carry the largest types of destroyers and submarines to their base at Bruges, and even light cruisers. The coastline, nearly 40 miles from the Dutch border to the mouth of the Yser, was studded with batteries mounting 225 guns, more than half of which were of 6 in. caliber and above, and of an extreme range of 23 miles, including Battery Tirpitz, west of Ostend with four 28 cm. guns, which was ready by autumn of 1915; battery Kaiser Wilhelm II near Knocke with four 30.5 cm. guns, ready by spring, 1916, and Battery Deutschland with three 38 cm. guns, ready in spring, 1917. All were installed without hindrance from the Allies.

In the later years of the War 28 cm. railway guns and five 17 cm. gun batteries were added. Nothing like a duel to the bitter end was ever fought between these guns and the British monitor guns, to settle the ever open question of the relative strength of land-based guns opposed to ship-based guns. According to a German engineer the "various British bombardments, in point of ordnance technique and nautically speaking, were nearly always carried out with extraordinary care and skilfulness; but from an obvious fear of risking losses of ships the will towards an energetic attempt, without which no real success in war can be achieved, was wanting."

It was only the use of artificial fog, for the first time attempted by the British on a large scale on May 12, 1917, in a bombardment of Zeebrugge with heaviest calibers, intended to smash the locks, which brought a change in the former firing methods. For a whole hour the German batteries were in the dark as to the position of the British gunboats until hydroplanes ascertained their approximate location. After this experience the German coastal defence introduced new methods, such as

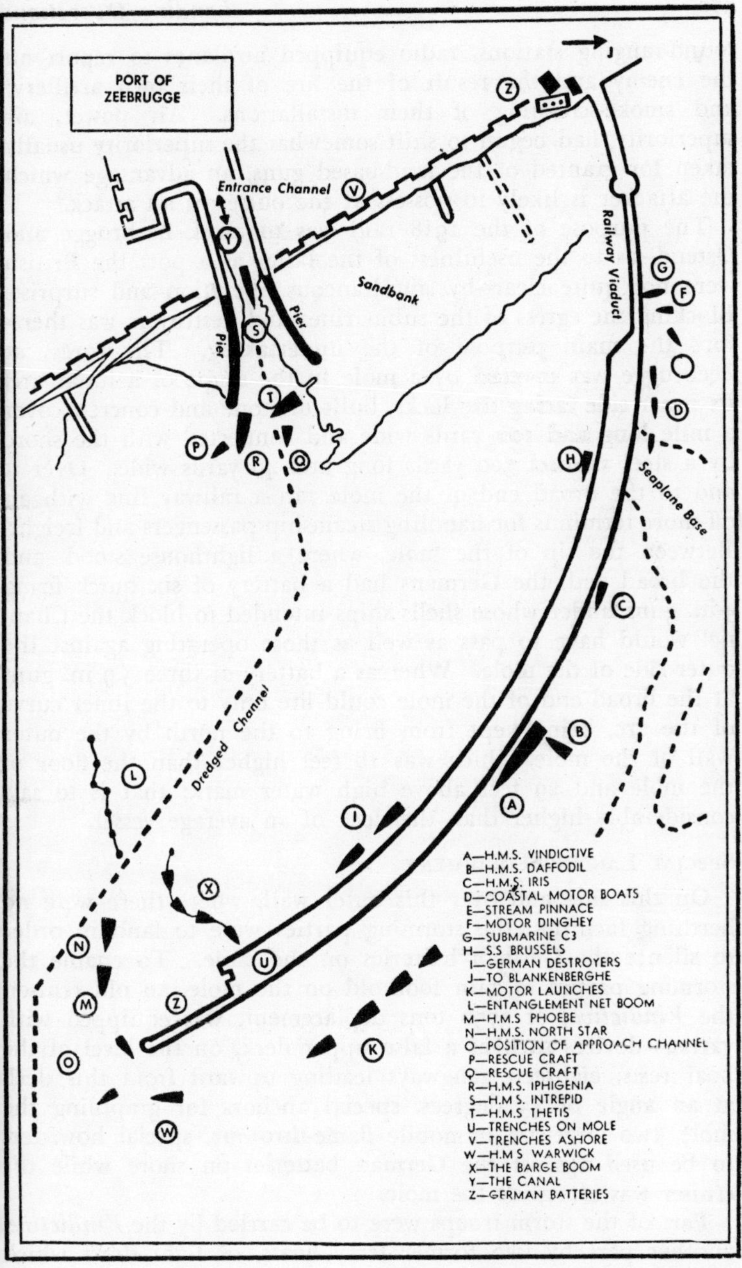

Courtesy, Houghton Mifflin Company.

sound-ranging stations, radio equipped airplanes to report on the enemy and the result of the fire of their own artillery; and smoke-screening of their installations. Air power, air superiority, had begun to shift somewhat the superiority usually taken for granted of the land-based guns, an advantage which the attacker is likely to possess at the outset of an attack.[5]

The purpose of the 1918 raid was to block Zeebrugge and Ostend—as to the usefulness of the latter as a port the British were not quite clear—by simultaneous operation and surprise. Blocking the egress of the submarines and destroyers was therefore the main purpose of the undertaking. The egress at Zeebrugge was covered by a mole in the shape of a lunar arc, its inner side facing the locks, built of stone and concrete, over a mile long and 100 yards wide and connected with the shore by a steel viaduct 300 yards long and 40 yards wide. Over it and to the broad end of the mole ran a railway line with an off-shore terminus for handling steamship passengers and freight. Between the tip of the mole, where a lighthouse stood, and the broad end, the Germans had a battery of six quick firing 4-in. guns under whose shells ships intended to block the Channel would have to pass as well as those operating against the outer side of the mole. Whereas a battery of three 5.9 in. guns at the broad end of the mole could fire only to the inner curve of the arc, being kept from firing to the north by the outer wall of the mole, which was 16 feet higher than the floor of the mole and 29 feet above high water mark; that is to say, considerably higher than the deck of an average vessel.

Special Landing Equipment

On this side and over this outer wall, where there were no berthing facilities, the storming parties were to land in order to silence the German batteries on the mole. To enable the storming parties to gain foothold on the mole, an old cruiser, the *Vindictive,* of 5750 tons deplacement, was equipped with various devices such as a false upper deck, on the level of the boat rests, eighteen gangways leading upward from this deck at an angle of 45 degrees, special anchors for grappling the mole, two large-size immobile flame-throwers, special howitzers to be used against the German batteries on shore while the cruiser was alongside the mole.

Part of the storm troops were to be carried by the *Vindictive,* another part by two former ferry boats, of light draft which enabled them safely to pass over minefields at high tide, and accustomed to taking the shock when they bumped against the mole. They were also to help the *Vindictive* to get close to and keep alongside the mole. Two submarines loaded with explosives were assigned to blow up the viaduct, one to replace the other in case the first should be lost, or thrown off its course. Aside from the demolition effect this was to keep the Germans from bringing support from the shore to their fighters at the sea end of the mole.

The sinking of ships to block narrow port entrances and landlock enemy naval units had been tried several times before, as at Santiago in 1898 and at Port Arthur in 1904, but never with conspicuous success. Such sinkings to be effective involve the reversal of the maritime salvage problem. They require destruction engineering. The same salvage company which advised the Admiralty on the planning to make salvage of the sunk vessels by the Germans nearly impossible, was subsequently entrusted with the removal of the two blockships, and needed over a year to do the job.

Five old cruisers, filled with explosives and cement, were selected as blockships, three for Zeebrugge and two for Ostend; or two and one more than the single one which would have sufficed to seal the channel in question if luck favored. Making it possible for the blockships to enter the channel, to proceed to their assigned places and go down as planned, required a number of diversionary and veiling measures, including temporary landing on the mole for the purpose of employing and misleading the German batteries, whether on the mole, or ashore, which could be expected to lay down an automatically released barrage.

Against this latter, in the last, or next to the final stages of the enterprise, a counter-automatism of fire was laid out, involving minute preparations envisaging all possible forseen contingencies, including encountering German vessels en route, and carefully worked out time tables, in which moon phase and tides were the supreme determinants. The time tables covered in distance a voyage of 100 miles for the units coming from the rendezvous in the Thames estuary, and somewhat less for those proceeding directly from the Dover Patrol.

This meant that command-influence prevailed before rather than during the attack, when Keyes could merely observe the running of his elaborate preparatory clockwork.

Command in this, as in most other commando raids, was severely restricted for still other reasons. The number of participating vessels and units was too great to ensure constant signal contact with all of them. Besides, the use of signals in the enemy's vicinity, whether optical, or wireless, exposed the attack force to premature detection. It was surmised, incidentally, that the Germans had sound detectors on the Zeebrugge mole. The arrival of the storming ships was timed to bring them to the mole twenty minutes ahead of the blockships, and give them only that period in which to fasten to the mole and land the storming troops whose task was to prevent the mole batteries from interfering with the blockships.

The other vessels employed, bringing the total to about 150, served various purposes. Some were cruisers to cover the transit, mine sweepers, monitors to bombard landward, fast rescue boats, etc. The blockships were given an extra number of men to serve them and their engines. Those intended for the final actions were resting; those above the minimum required for the fight were to be taken off at the edge of the danger zone, partly to relieve rescuing parties of work. Other measures were taken to have the men at their physical and mental best during the decisive hours. To make the enterprise tri-elemental two wings of the RAF, one for reconnoitering, to supply the latest aerial photographs of Zeebrugge, and one for bombing on the night of the attack, were at Keyes' disposal.

Numerous shallow draught craft, able to approach the coast closely, were to produce a smoke screen to cover the final prelanding movements. Smoke, already employed in naval battles such as Jutland, was here to be introduced in amphibious operations. At the Dardanelles Churchill had offered smoke producers to the admiral who had not seen fit to use them. Outsiders could not persuade the professionals to make use of artificial fogs and smoke until the Germans had done so. These serve to cover and camouflage movements on sea and land. They are modern chemistry's fulfillment of longings of ancient warriors, expressed in Germanic mythology by the magic "fog cap" *(Nebelkappe* or *Tarnkappe)* to make the hero invisible. To a large extent their use might shift the balance between attacker and defender in amphibious enterprises in favor of the former. It would cover from the enemy's observation and his aimed fire ships and certain areas in, or towards which, they might be operating. Under exceptionally lucky circumstances it might black out enemy means of observation, including searchlights. Smoke and fog screens could cover a premeditated retreat, as in case of a coastal raid, serving as a mechanical substitute for a rearguard, as they did with good results at Zeebrugge in 1918, the same year that the land war combination of artificial fog and tanks was introduced by the Allies.[6]

Exclusively a Navy Task

The personnel was all-navy, and as nearly as possible of volunteers to boot, who were willing to undertake extra-hazardous service. Beyond information that an especially perilous expedition was afoot, to ensure secrecy none was given. But shrewd guesses were possible, as seamen storming parties, marine storming parties and demolition parties, composed of both seamen and marines, were trained by day and night on a model lay-out of the Zeebrugge mole, something closely resembling the training of infantry storm troops by use of models of specific enemy positions.

From April 3 on, later than had been intended, but the vast supplies of smoke-producing materials could not be obtained earlier, the main force was assembled at an anchorage in the Thames estuary, sufficiently remote from the shore to remain unobserved. Postal and other communications were completely cut off, as the rank and file were now informed of their destination; this had previously been disclosed to the officers. Several more days were employed in further training the men on board their ships. Everything was done to feed and entertain them while waiting for the first of the five moonless nights in April.

After two days of unsuitable weather, the meteorological reports promised better and the flotilla departed on the 11th, timed to arrive at Zeebrugge at 2:05 A.M. the next day, at high tide. By time schedule that would make it possible to see the action through, for which one and a half hours were required for the *Vindictive* group, and to retire from the scene long enough before sunrise (5 A.M.) and before detection of the homebound craft by the German shore batteries, whose guns ranged from 15 to 20 miles to sea. When the armada, without trouble, reached a point only 16 miles from its objective, the wind dropped and soon came to blow offshore. Since success depended entirely on the help of the smoke screen rolling eastward, Keyes although sorely tempted to keep on after having proceeded so far and so well, signaled to his flotilla to turn west and, in the nick of time, to the monitors not to open fire that night.

Once more the flotilla set out, on the last day of the moonless period, the 13th, but the sea became too rough for the numerous small craft, and once more it had to return with its mission unaccomplished.

Rather than wait three weeks until the next moonless period, by which time both the secret of the enterprise might have leaked out and the patience of the Admiralty become exhausted, Keyes decided to have another try after ten more days when there would again be highwater at Zeebrugge near midnight. Fresh timetables for another seven days' period were worked out. On the 22nd the weather and forecasts seemed fairly promising and orders were given to set out once more. At about an hour's distance from the Flanders coast a slight rain began to fall, considerably reducing visibility. It affected the success of the enterprise in two ways, by delaying the opening of the fire of the monitors against the coast and by preventing the aircraft, which suffered heavy losses on the flight of that night, from spotting and bombing the mole. No German patrol boat was encountered and only the fire from the monitors gave the alarm to the Germans.

German star shells were bursting overhead, lighting up the scene, while the smoke-screening craft released a pea soup fog. Not long after, the wind changed to off-shore, driving the smoke screen over the approaching vessels and reducing visibility to nearly zero. Then this smoke lifted suddenly, and the *Vindictive* saw the mole 300 yards away and was seen by the Germans who promptly opened fire. Men and material of the *Vindictive* suffered severely. Most of the leaders of the storming parties who had exposed themselves prematurely, were killed, or wounded, more than half of the gangways blown away and the large flame throwers put out of business.

Returning the fire, the *Vindictive* proceeded to her position outside the mole. The proximity of the ship probably prevented the German gunners, trained to fire at longer range and with greater deliberation and better aiming, from inflicting more severe damage on the *Vindictive*. Swinging out of the cone of fire, the *Vindictive* put alongside the mole at one minute past midnight. Its officers were confident of having familiarized themselves with all its features by study of aerial photographs and other pictorial material.

In the confusion caused by darkness and noise of the artillery, they took her too close to the shore line and too far away from the German gun positions near the outer end of the mole, in fact 340 yards beyond her intended berth. Pushed into place by one of the ferryboats and held there, in spite of the heavy swell constantly bumping her against the concrete wall, the *Vindictive* sent off over the remaining gangways the first landing parties who were to fasten the tossing vessel to the mole by grappling anchors.

This effort failed and she had to be kept in position throughout the engagement by the ferryboat. Stopped, or delayed by wire entanglements and German machine gun fire, the assault force from the cruiser was unable to capture the guns, or obtain any commanding hold on the mole. They added to the resultless disorder, but did not come to close grips with the enemy. During this foray, which however was necessary to the success of the blocking enterprise, the blockships made their entrance as scheduled. The first fouled by barrier netting, could not reach her destination and had to be abandoned.

Results in General Successful

By then and through the movements of this vessel it became clearer to the Germans what the true intentions of the attackers were, that it was a blocking operation rather than a landing in force along the coast, but it was too late for them to concentrate measures against the blocking attempt. The other two blockships were sunk by their crews in and across the Zeebrugge channel at their assigned positions. One of the two submarines successfully placed itself with its charge of explosives under the steel viaduct. Set off by a time fuse, a hundred foot gap was blown in that part of the mole.

At 12:50 A.M. orders were given to sound the retirement signal and after 20 minutes waiting the *Vindictive* pulled away from the mole. Smoke screens helped her and other vessels to get safely away, while one destroyer was sunk by the German batteries. A surprisingly large number of survivors, wounded and unhurt, were carried to safety. The English casualties at Zeebrugge amounted to 170 killed, 400 wounded and 49 missing.

The attempt against Ostend during the same night had failed completely. The change of wind, having driven the smoke screen to seaward and the harbor entrance in consequence being obscured, the invading force found it necessary to rely on a certain buoy which, unknown to the English, had been moved by the Germans a mile away from its old station. Hence, the two blockships ran ashore three quarters of a mile off the channel entrance. Determined later to try again, Keyes met with but partial success, and only a third of the width of the channel was blocked (May 9-10), the two attempts costing 637 casualties (197 killed, 413 wounded and 27 missing). Intelligence reports that actually the Bruges-Ostend canal was being silted up and was only of small service to the Germans, which might have been known earlier, led to calling off a third expedition against Ostend, for which everything had been prepared.

Material and moral results of a specific enterprise in war cannot easily be ascertained and compared. In the case of the Zeebrugge expedition the latter seems far more certain. It was like a ray of light breaking through the dark discouragement hanging over England during the spring battles on land. It also confirmed an honest belief that amphibious enterprises of so pronouncedly a raiding nature were feasible, a conviction that luckily persisted into World War II. On the material side the English claim was that the Zeebrugge blockships had been sunk at their selected positions—Keyes later regretted that he had not selected points still closer to the locks—and that thus the first successful blocking operation in naval history in the face of an active defence had been achieved; that the hindrances created to German submarines and destroyers were considerable and long lasting, to be specific, that the harbor of Zeebrugge was completely blocked for about three weeks and passage was dangerous for U-boats for a period of two months.[7]

The Germans claimed, aside from the obvious lie in the first communique that *they* had sunk the blockships, that they were never seriously inconvenienced in entering and leaving Zeebrugge and that interference with their submarine warfare could not have been considerable since U-boats' sinkings during May, the month after the Zeebrugge raid, were higher than the month before. According to Admiral Scheer *(Die deutsche Hochseeflotte im Weltkriege),* the attack made with great pluck, found "our guards at their posts. Two old light cruisers . . . were sunk before they reached their goal—the lock gates which were uninjured. It was found possible for the U-boats to get around the obstruction, so that connection between the harbor of Zeebrugge and the shipyard at Bruges was never interrupted even for a day." Unfortunately for the endeavor to establish a more definite record, the German official naval history—*Krieg zur See 1914-1918*—does not yet include the Zeebrugge affair.

Zeebrugge was not so much a combined operation in the traditional sense that it brought together the historically separated services—for the British Navy had the job and did it all its own way, taking over infantry assaults, though perhaps omitting such necessary equipment as wire-cutters—as it was a combination of the three elements, sea, land and, to an extent less than had been hoped, air. So far as Britain was concerned, the inspiring experience of Zeebrugge threw operations of this kind into naval hands, such as those of Keyes, who proved a sounder instrument for transmitting World War I experience than, let us say, Gamelin.[8]

NOTES, CHAPTER 41

[1] *World Crisis, 1916-1918.* II, 43-4.

[2] Liddell Hart, *War in Outline,* 192; Duff Cooper, *Haig.* N. Y., 1936, 163.

[3] *War Memoirs of Lloyd George.* III, 83, 117.

[4] *Der Krieg zur See, 1914-1918.* Herausgegeben vom Marinearchiv. Vol. II (Berlin, 1922), 116-7.

[5] Major Klingbeil, *Küstenverteidigung und Küstenbefestigung im Lichte der Weltkriegserfahrungen.* Berlin, 1924.

[6] Churchill, *World Crisis 1915,* 65-83, 392-3.

[7] Churchill, *World Crisis 1916-1918.* II, 87, 44.

[8] The story of Zeebrugge is based on the following books and articles: *Zeebrugge and Ostend Despatches.* London, 1919; Captain Alfred F. B. Carpenter, *The Blocking of Zeebrugge.* Boston, 1922; Lord Keyes, *Amphibious Warfare,* ch. IV; Lt. Cdr. H. H. Frost, *The Attacks on Zeebrugge. U. S. Naval Institute Proceedings,* March, 1929; Captain W. A. Murley, *Zeebrugge and Its Lessons. Journal of the Royal Artillery,* April, 1936; further discussion Ibd. April, 1937 and April, 1938.

Chapter 42

1919-1939

IN THE INTERIM OF PEACE combined operations were almost everywhere out of bounds, remote from the military and naval minds. What war had wedded, in a sort of shot-gun wedding, peace divorced promptly; that is to say, more in the victorious nations than among the losers. The Anglo-Saxon powers preferred to avoid monolithic arrangements for command and staffs. In England, the RAF and the Air Ministry, both set up early in 1918, survived the peace, against onslaughts of economy as well as those proceeding from the older fighting services. Both the latter thought unjustifiable the largely untried, and at times extravagant, ideas and theories of strategical and tactical air force independence. No setup proposed for triangular coordination seems to have satisfied the three services.

In the United States, where later at certain times tendencies towards unification were inspired by the hope of averting air-force independence and unavoidable three-sided quarrels, compartmentalization and its reverse was at least as complete as in England. Coordination did not go beyond naval and military planning through a Joint Board, in addition to a Joint Munitions Board, without even a chairmanship. This Joint Board consisted of the Army chief-of-staff, the deputy chief-of-staff and the assistant chief-of-staff, War Plans Division, on the one hand and the chief of naval operations, with two similar assistants (as per 1938) on the other. Its duties consisted in "conferring upon, discussing and reaching common conclusions concerning all matters calling for the cooperation of the two services in war; conclusions which are then translated into the war plans of both," [1] that is to say, not into one war plan for both. By comparison with the German arrangements it was a mere conference body, without even a head, not one embodying, or even envisaging unity of command, which was only constitutionally, but not also technically provided for; a body delimininating mutual exclusive fields rather than circumscribing in common the vast totality of a war to come.

Only some twenty years after the World War I did the French, in 1937, come around to the creation of a general staff for national defence, with General Gamelin as chief, to ensure constant coordination of the army, navy and air force. Where the army was as a matter of course and traditionally in such a predominating position as in France, the choice of a military head was obvious.

While most of the Allies were thus reluctant to accept the lessons of the previous war, the Germans were taking them to heart. Laying down in their service regulations for the Conduct and Combat of Combined Arms *(Führung und Gefecht der verbundenen Waffen)* that "only if we keep alive the memory of the arms which have been taken from us, shall we find ways and means to stand battle against an enemy modernly equipped," they proceeded at once (1919) to set up a single Ministry of Defence *(Reichswehrministerium)*, which was replaced in the later 1930's by a High Command of the Armed Forces *(Oberkommando der Wehrmacht OKW)*, placed above the three high commands, army, navy, and air force, to which was added during the war that of the SS-in-Arms. This arrangement was to ensure unity of command, strategy and policy, hence a seat was given in the Reich cabinet for the representative of the OKW. (For diagram of set-up see page 580.)

Vary as they might in intensity, these unifying tendencies towards setting up supreme commands that are not merely an expression of constitutional law, but of technical war, became world-wide. For the first time since the war of 1894-5, the Japanese established Imperial Headquarters in November, 1937, a few months after informal, undeclared war with China had once more been started by the Nipponese. It had as members the Emperor as commander-in-chief, the chiefs of the Army and the Admiralty general staffs, the ministers of war and of the navy, and other high officials, military and civilian chosen by the Emperor, such as the inspector-general of military training. "Imperial Headquarters places the Supreme Command on a war-time basis. It assists the Emperor in the exercise of his command of the armed forces and coordinates military, naval, and civilian activities to attain the nation's strategic objectives." [2]

Planning After 1918

The German Army that survived and reburgeoned after 1918 was by no means willing to let the Navy "just grow" like Topsy once again, as in the Wilhelminic era, in the last months

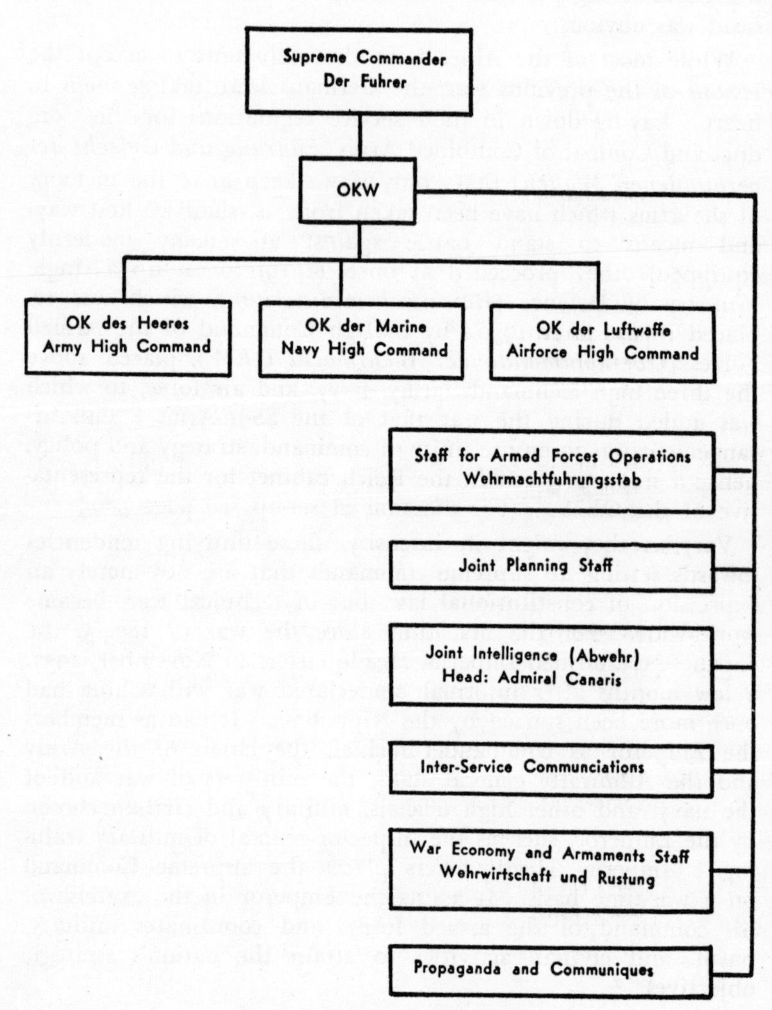

German Military Organization.

of which officers of the General Staff had actually discovered that it was the German Navy that had made England into an enemy of Germany.[3] Never must the Navy be allowed again to absorb strength more needed for other components of the unitarily conceived *Wehrmacht*. In addition to this strategic-economic-political cooperation, the effort was made to assure always ready operative-tactical cooperation of all three, or if need be of only two of these forces, through an overarching staff.

The two foremost national exponents of military reform—or revolutionaries, if military groups can ever be of that persuasion—Germany and Russia, had no great reason to study and train for amphibious operations. There are, however, some material indications on Germany's side that she was preparing against hostile landings. Such measures included the refortification of Helgoland, Sylt and others of the Frisian Islands, the improvement of overland connections in the direction of the North Sea, of highways rather than railroads such as that from Hamburg to Cuxhafen, and above all the change in the regional organization of the Army in northwestern Germany.

Whereas before 1914 Hanover and Schleswig-Holstein, including all the likely landing shores in the northwest, fell into two different army corps districts, the *Wehrkreis* organization of the later 1930's took care that one and the same *Wehrkreis*, the Xth, should include the whole littoral from Emden to Lübeck. At the same time all coastal artillery was manned by the Navy. In German military literature, after all something of an index of what was on military minds not given to much publicity, themes like "combined operations of army and navy under the influence of airpower," and the great tri-elemental war to come, were discussed in a manner superior to anything that could be pointed to in the military literature of Germany's former and future enemies. Her navy men had come to realize that combined operations presented the most effective form of naval warfare.[4]

The Italian War against Abyssinia was on the naval side merely a gigantic transport enterprise covering a distance of 2,500 nautical miles over which 16 divisions (360,000 men), 30,000 draft animals, 6,500 motor vehicles and 3 million tons of materiel were to be brought in six installments by a transport fleet of 154 ships. They sailed without convoy, except for light protection while crossing the Red Sea. Unity of command, it was emphasized, could not be better guaranteed than by the person and the functions of the Duce who, for the purpose at hand annexed a seventh ministry, that for colonies, to those he already held.

The whole enterprise was started and put through as quickly as possible, with the sufferance, or without the inteference, of the so-called democratic powers. But it had the adumbration of tri-elemental war. Nothing much was done to shake the illusion of the Italians—and perhaps of the Germans—that it was Italy's airpower which had enabled her to risk this campaign in spite of England's frowns. According to the Italians, there was a strong difference of opinion between the commander of the British Mediterranean fleet and the London Admiralty. While the former proposed to open the war, if it should come to that pass, with a naval bombardment of the major Italian ports in spite of the strong Italian air force, and the constant threat it exercised over Malta, the Admiralty removed the more important units from Malta to Alexandria, Port Said and Haifa. Against the threat of closing the Suez Canal, Italy demonstrated by assembling four divisions, including one motorized in Lybia, as a menace from there, as well as with air power from the Dodecanese, England's position in Egypt. The Italian Navy in such a case was to protect the rearward lines of the forces in North Africa.

British Practice Ineffective

Only after the Axis powers had begun to land on foreign shores as conquerors, from the sea and from the air, landings in which the Japanese employed special landing craft, armored vessels of shallow draft, did Britain resume her often interrupted, or lapsed tradition by reviving amphibious manoeuvres, beginning in 1936. But then in such a perfunctory manner that even a service journal, "considering their importance and the frequency of amphibious operations in our island story," was ready to call these exercises, held at Studland Bay on the Channel coast in 1936 and at Lee-on-Solent in 1937, "half-hearted and ineffective." [5] While in these manoeuvres the Navy predominated, the Army a year before World War II held large-scale manoeuvres in Yorkshire, with the problem that of an enemy landing on the coast, "when available troops were to be rushed northward by every possible means of transport to meet the invasion." [6]

The frontier zones of western civilization, or the regions beyond have at various times opened the path for the progress of war technique when it was blocked in the west itself by convention, or other military "restraints of trade." This has been done in ways which the military in the foci of the west have not always promptly observed or taken to heart. It happened as regards various experiences made during the Russo-Japanese War of 1904-5; it happened again in the 1930's, when certain topographical conditions characteristic of the Far East and a Japanese imperialism bent on upsetting the status quo in the East combined to bring about a new progress in amphibious warfare. In spite of the all too common western views on the wide and deep differences between the Army and Navy in Japan, it was their cooperation that opened a new phase in amphibious warfare. This was first brought about in river warfare.

Topographically speaking, river warfare is resorted to because of lack of adequate land communications, absence of roads and heavy, or exclusive dependence on river transportation. The common purpose of potamic warfare, involving amphibious enterprises along and across rivers, is, on the attacking side, to obtain control of the stream as the one, or virtually the only, means of transportation in a region; on the defender's part it is to block the river and the movements of the enemy on and adjacent to it. This leads to a predominance of action and movement either up, or down the river rather than crosswise. The push will be deep rather than broad. After the ending of the Dutch wars of independence, with the siege of Antwerp by Duke Alexander of Parma in 1584-5 as its most interesting fluvial operation, Europe, with its ever improving highway communications, witnessed but little river warfare. Consequently, Clausewitz's chapters on the defense of rivers deal only with crossing operations. The absence of land communications caused resort to river warfare in America, both before and during the Revolutionary War. In the latter the location of West Point on the Hudson as one of Washington's defensive strong points was due to considerations imposed by fluvial warfare. River warfare was waged on a still larger scale during the Civil War, when operations of prime importance took place on the Mississippi. In roadless China, foreign makers of war and diplomacies had used gunboats and other craft far up the great rivers, about 1,000 miles in the case of the Yangtse, in order thus to break the impenetrability of that vast Asiatic geographical massif.

Tri-Elemental War in China

The Japanese operations in China reopened river warfare in 1937, introducing into it at the same time the new element of air power. The tri-elemental war, the first in which amalgamation of all three components had really been achieved in formal combat, began with the operations against Shanghai in August, 1937. On August 11, promptly following an "incident" which they had been seeking as a pretext for attacking the Chinese, the Japanese brought from 15 to 30 warships into the Hwang-pu River and landed 4,000 marines. The latter had the situation none too well in hand and would have been thrown back into the river, by a swiftly arriving, well equipped Chinese division, had it not been for the strong fire support given them by ship artillery and airplanes. Twelve days later 10,000 more Japanese were landed to save the situation at Shanghai, where fighting took place in the densely settled oriental city and under the eyes of Shanghai's foreign colonies of commercially minded westerners.

The Chinese had some of their best troops in the vicinity, who were able to repel a further Japanese landing attempt in the Yangtse estuary on October 12, with heavy losses to the agressor. Only after the latter had occupied several islands off the China coast and had built airfields and prepared other bases could a more systematic landing attempt be made with success (November 5). It took place under strong artillery protection from the fleet and with the help of airplanes, starting from these island-bases. From forty transports 180 fast armored barges swarmed to the shore. Since the Chinese were caught by surprise and were without adequate forces to oppose them, the Japanese were able on the first day to occupy a beachhead five miles wide and twenty miles deep and to extend it the following day with the assistance of several more newly landed divisions. It was this campaign which ultimately led to the taking of Nanking (December 13).

This tri-elemental war was resumed in the following year, having for its objective control of the Yangtse at least as far up as Hankow, then the seat of the Chiang Kai-shek government, within the shortest possible time. After careful preparations it started early in June from the Nanking-Wuhu region where troops had been carried by boat from Shanghai, which to then was the center of Japanese operations. The forces originally employed were some 20,000 troops and 50 naval vessels, specially organized as a river squadron, to which reinforcements were constantly added until they counted approximately 100,000 troops and 100 vessels at the close of the campaign. Not a little care and time were devoted to the aspects of joint actions.

The operations were opened with the Japanese attempt to capture the first river fortifications at Tatung, on the south bank, 230 miles from Hankow, early in June, After 30 naval craft with medium caliber artillery had endeavored for eight days, to batter down the Chinese field fortifications around Tatung, on the ninth day a landing was attempted by forces concentrated on the left bank, preceded by artillery and aerial bombardment, the naval guns firing at a distance of a mile from the south bank, the transports moving up behind a screen of naval craft. Three or four hundred yards from the south bank, motor barges with troops were lowered from the transports and the naval artillery fire was ranged to the rear of the Chinese lines. The landing, although repeatedly attempted, failed.

After that emphasis shifted to naval action by which the barrier on the river was forced and a movement started for Anking, on the northern bank, 45 miles nearer Hankow. The place was protected by three old forts on the east side, relics of Chinese river wars of yore, such as the Taiping Rebellion, and by field fortifications on both sides of the Yangtse, rather weakly held by two Chinese divisions of 18,000 men. From Anking on, land movements seemed more promising and free of obstacles, hence the importance of taking the city.

Strengthening their troops in the Luchow area, nearly 100 miles to the north, for a larger enveloping movement, the Japanese sent a small force southward from there against Anking to occupy the Chinese troops in the eastern sector of the Anking lines, where the main landing was to take place. Against the latter was also directed a force moving on the left bank against Anking. A landing force of two infantry brigades, a marine landing battalion, artillery, engineers and a chemical corps, altogether 12,000 men and 80 to 90 guns, was mobilized to be taken near their destination by 20 transports and provided with air support by 100 planes, mostly light bombers, part of them operating from carriers.

Early on June 12, the river squadron forced its way past the barriers near Anking and proceeded to clear the channel of obstructions and mines, aided by planes in detecting obstacles from the air and in bombing the minefield. Disembarkation points were selected and reconnoitered. At 2 A.M. on the 13th, the landing operation started. The squadron, divided, took positions from whence it could cover the landings with its fire. A special force of 1,200 men in fast motor boats pushed ahead up the river past Anking in the dark in order to attack the city from above at daybreak. The main force went ashore just below Anking in barges and boats protected by artillery, bombers and smoke screens.

Execution Far From Perfect

Repelled by Chinese artillery and machine gun fire, the first landing attempt failed with heavy losses for the Japanese who retired their transports behind the curtain of the naval units. After several hours of renewed bombardment it was repeated and then a foothold below the city was obtained, albeit at a high cost. All through the action the main landing objective was screened by numerous feints at other points on the river bank. By nightfall Anking was surrounded; after heavy fighting on the following day it was taken by the Japanese. The landings were far from perfect in execution of plans; schedules were not kept and consequently heavy casualties were incurred.

The river armada made Anking its base for further operations. From there it was intended for gunboats and destroyers to progress on the water *pari passu* with the partly motorized land units moving on, or close to the river banks. But while the water borne artillery silenced the Chinese shore batteries, the advancing troops were halted 20 miles beyond Anking, due to bad roads, or their absence and the resistance of the Chinese, who made the best of that condition.

Pushing ahead of the slower land forces, the naval units on June 24, near Matang, ran up against a Chinese river block, formed of a chain of sunken junks loaded with stones. Attempts of the Japanese airforces to bomb away this obstacle and of the naval units themselves to break it failed. To by-pass it and the Chinese forces defending it, Japanese troops were landed 15 miles below. Supported by planes and the naval units, these detachments, in an air-river-land battle, slowly worked their way around the Chinese defenders, and a rise in the Yangtse allowed the Japanese craft to steam over the obstacles in the channel. Once beyond the river block, the Japanese could land troops higher up the river and demolish the Chinese land defences around the block.

The next obstacle upstream was encountered just below Kiukiang. There the Chinese had a number of batteries, none with larger than medium caliber guns, on an island commanding the Yangtse and the outlet to it from Lake Poyang. By-passing of the island by troops marching overland was particularly difficult in this region of broad swampy areas. After a wait of three weeks, the Japanese decided to risk landing troops on the island. This threat and that of a Japanese column which had got as far as Hwangmei, ten miles to the north, but separated from the river by lowlands that were presently to be intentionally flooded, caused the Chinese to evacuate the island. This enabled the enemy to move as far up as Kiukiang by the end of July.

Shaken in their confidence of waging a rapid fluvial war, the Japanese—after "the admirals and the generals had argued it out," as has been plausibly suggested—resolved to attempt the land route to Hankow via Nanchang. Progress in that direction became stalled almost as soon as the troops got beyond range of the guns of the Navy, the most mobile artillery for operations in this kind of terrain. Once again the Japanese tried outflanking by water when, during August, they attempted a landing at Singtse on the eastern shore of Lake Poyang which failed. At the end of the month they swung back to the river route in an effort to break the Chinese block between Matuchen and Wusueh.

Army and Navy reinforcements both supported by air, were set to work, the naval units to shell and contain the river block and adjoining positions, while the former, landed down the river behind the front lines of the gun boats, threatened the flanks of the Chinese positions. Gas shells also seem to have been used by the Japanese. Infantry, "working in exemplary cooperation with gunboats and the air force," in the official language, took Matuchen on the right bank.

Since Japanese forces further north threatened the Chinese left wing, the Matuchen-Wusueh block was evacuated on September 14. After breaking by the same methods another block 12 miles further upstream, which held up the Japanese for two more weeks, they came at last within sight of Hankow. Threatened at the same time from the north, General Chiang Kai-Shek evacuated his capital and the Japanese made their entry on October 25. During the campaign their progress averaged slightly more than a mile a day. They suffered heavy losses which reached 50 percent in case of some divisions.

In the light of what followed, the Yangtse campaign partook of the character of military experiment and field laboratory—China serving the Axis as one guinea pig and Spain for another—in which much was learned, although apparently only by the Japanese so far as amphibious operations were concerned. As the best available analysis of these inland amphibious operations summed it up:

"They give us examples, one after another, of the water-envelopment tactics which have characterized Japanese operations in the Philippines and, especially, in Malaya. In principle, of course, such envelopment is entirely conventional: there is the pressure on the front, and the thrusts against one or both flanks, or against the rear. The mechanics of the water-envelopment, however, are unique. The troops must be skilled in the technique of embarking and debarking under all sorts of conditions. They must be, in effect, marine-soldiers." The campaign illustrates "the Jap predilection for coordinated water-land operations" (Col. Thompson).

The land-bound characteristic of the Yangtse operations was at all times predominating, determining choice of objectives and much of method. Beyond Hankow the fluvial element became too restricted further to provide much aid for inland mobility. While the technical cooperation of the Japanese was not always perfect, there seems little doubt of the earnest intention of maintaining it on the part of army, naval and air forces, the latter meeting practically no Chinese opposition in the air. This was particularly important in all reconnaissance work on which the choice of landing sites depended, as well as the screening of movements from the Chinese. The Yangtse armada altogether was like a combination of armored train and transport train, at times carrying the soldiers behind its armor, at others facilitating their jumping off, usually in two or three echelons, the first formed of those amphibious soldiers, the marines.[7]

NOTES, CHAPTER 42

[1] George Fielding Eliot, *The Ramparts We Watch.* N. Y., 1938, 331.

[2] Hillis Lory, *Japan's Military Masters.* Washington, 1943, 83.

[3] For details see Arthur Rosenberg, *Die Entstehung der deutschen Republik.* Berlin, 1928, 220.

[4] See such articles by Admiral Gladisch in *Militärwissenschaftliche Rundschau,* 1936, No. 1; by Admiral Dr. Gross, ibd. 1937, No. 4; by Captain Sorge, ibd. 1938, No. 3.

[5] *The Fighting Forces,* Oct., 1936 and Aug., 1943.

[6] *N. Y. Times,* July 31, 1939.

[7] The above is largely based on *How the Jap Army Fights.* N. Y. and Washington, 1942, ch. IV by Lt. Col. Paul W. Thompson, and an article on *Japanese Landing Operations,* from Krasny Flot, April 14, 1939, translated in *The Axis Grand Strategy.* Compiled and ed. by Ladislas Farago. N. Y., 1942, 136-42.

Chapter 43

THE SECOND WORLD WAR, 1939-1945

THE FIRST PHASE OF WORLD WAR II was distinctly continental, with the Polish campaign, the fourth partition of Poland, the "phony" war in the west. Even before the war, on January 30, 1939 and again three months later, Hitler had assured the Americans that "the assertion that National Socialism will soon attack North or South America is on the same plane with the statement that we intend to follow it up with an immediate occupation of the moon." (This might well have been an answer to the suspicions sown shortly before by Major General H. H. Arnold, who, when asked by a Congressional committee what specific danger of attack the General Staff had in mind in building up an air force, had replied, "I don't think I ought to answer."[1]) But neither then, nor later were German promises that the war would be confined to the European continent entirely convincing to Americans.

Public opinion polls of February, 1939, and September, 1939, indicated that 62 or 63 per cent of those questioned—a rather small affirmative rise considering the intervening shock produced by war's start—thought that Hitler's ambitions did extend across the Atlantic. This belief was not shared by many military observers, who held that even if Germany should defeat England and France she would be too much exhausted to undertake a large scale trans-Atlantic offensive, or that the modern bombing plane had made the North American coast practically "impregnable to invasion."[2] But so far as this judgment really prevailed in the United States it was probably based less on military grounds than on a revival, or survival of the age-old dread of an Alexander, or a Napoleon, eager to rule the whole world. The 37 per cent who could not accept the idea that Hitler would war against the United States, thought that "America was too far away for invasion" or that "Germany would be too much exhausted to attack us" or that Hitler wanted "only Europe and would have his hands too full there to attack us." Sound or not, minority opinion was based only on military argument.

Perhaps one should not too hastily jump to conclusions, but so far as polls do establish such things, they show a connection between social security and military security. Persons in the lowest income brackets were in a greater fear of invasion than those in the upper brackets, for Gallup's 68 per cent were in the lowest income group (below $20 a week), 63 per cent in the middle ($20-40), and 55 per cent in the top income group ($40 and over). At the time when only a small minority of voters in the United States were ready to send American troops abroad, those in the most modest circumstances had at least the logic of their convictions when they pronounced their greater readiness to support the implementation of the offensive defensive; that is, to send American soldiers abroad.[3]

If mass feelings as to warlike threats are judged by the writings, largely journalistic, which cater to them, or exploit them, they are still much the same, in style, intent or disregard of military realities as in times gone by. From 1939 on, if not a little before, the American people were repeatedly told by journalists, radio broadcasters, politicians of the more irresponsible kind that their country was threatened by an invasion, that "the United States is riper for an invasion than was Czechoslovakia, as fundamentally defenseless as was Poland," with the St. Lawrence River valley offering numerous landing places and "a happy highway for mechanized invasion," and the New Jersey coast a likely beachhead in the vicinity of New York.[4]

After the fall of France, a profound and alarming shock to the Americans, it seemed "the most popular editorial pastime to be imagining how the United States could [will] be invaded by foreign expeditionary forces," said *PM* (October 13, 1940), which had indulged in this forecasting, but was now taking issue with the alarmists and showed in pictures, *biblia pauperum,* how an invading German force would actually be more likely to fare, suffering defeat southwest of Harrisburg, Pa. Upsetting as are war's developments to expert opinion, it is still more so to prognostications offered by inexperts.

While not differing essentially in type from the invasion literature of the past, at present there would seem to be less in quantity of such military romancing. This is definitely true of that in book form which was almost solely represented by the late Hendrik Willem van Loon's *Invasion* (N. Y., 1940), a piece of it-can-happen-here fiction, modelled on what the invading Nazis had already done to the neutral countries of

Europe. The pseudo-military details of the ocean crossing, the landing in New York, etc., are among the most unrealistic features of this short-lived thriller.

The films, as an art developed in the main since World War I, made less use of the smashing island-invasion theme than might have been expected. The movie based on Alice Duer Miller's poem, *The White Cliffs of Dover,* did not make the survival of the society depicted behind those cliffs a matter of sufficient excitement. *Forty-eight Hours,* a British production, shown in New York in June, 1940, told of a small-sized German invasion of the isle in a vein that British propaganda has repeatedly tapped. It went worse and worse for the British side until, bar a complete collapse for them, it could no longer fail to become better for them and fatal for the intruder.

In an altogether different category fell the documentary film dealing with landings, *Attack!* composed from Signal Corps photographs of the New Britain campaign (Summer 1944). In its depiction of landings and subsequent battle it was a great improvement over *At the Front in North Africa* (released March, 1943), which included scenes of tankmen unloading their giant machines in Bone Harbor; since it was done in blurring technicolor, however, its documentary character was much diminished. The German film *Battle of Crete,* like *Feuertaufe* and *Blitzkrieg im Westen,* seemed designed to impress the audience with fear of an irresistible machine approaching and invading an island objective across the water and through the air.

Woodring a Good Prophet

Experts and defense officials were more reassuring as to the invasion peril. While some, like Secretary of War Woodring, warned Americans not to rely on "weary soldiers of Allied armies to hold the lines while Americans prepared" and that the next war might come to them "with the awful rapidity of a Midwestern tornado," [5] preparedness in whatever form could not seriously be based on invasion fear. Isolationists were to a degree justified in minimizing the danger of an invasion of the continental United States, maintaining with Herbert Hoover that, even considering the unprepared state of the country, it would take an Axis army of a million men and an armada of 22,000,000 tons of shipping, only about half of which was Axis-controlled, to invade the United States.[6]

Military men whose memory served realized that an invasion scare, like the one of 1898, could have serious repercussions; that in case of another war there might conceivably be a demand to protect urban points of fancied danger along the coast or elsewhere by forces that in the grand strategy of the United States would be required elsewhere. Colonel Frederick Palmer, retired, pointed to "signs that we no longer shall have to depend upon the false scare of a land invasion to arouse our people to the need of preparedness," that to indulge in scares meant nothing but waste motion in defense measures, and with such a better poised state of the public mind prevailing the United States Army would "be freed of a lot of motion for public effect, which will be waste when war comes" and would be available for its broader task, the "offensive defensive." [7]

Between the fall of France and Pearl Harbor public opinion was hardly ready for more than the defensive defensive. The first rearmament measures of the American government in their more obvious aspects were governed by it, with the President in the election year of 1940 assuring the nation that no American soldiers would be sent abroad, unless we were attacked; with the setting up of the Canadian-United States Defense Board under the Ogdensburg Agreement of August, 1940; with the leasing of naval-air bases in the western hemisphere and beyond its confines (with assumptions underlying these acquisitions which were not always correct).

It is a subject for wide speculation whether without the actual invasion of American territory by the Japanese the departure of American forces from the western hemisphere would ever have been approved of in the mind and psyche of an American majority. As it was, the Japanese attack on Hawaii, an almost needle-point invasion, apparently too little expected by the heads of the Navy and the Army, provided the stimulus for America's now inescapable entry into the war. It could now be declared and had to be declared and undertaken as a counter-invasion.

It is already known that this aggression in the Pacific was decided upon, timed and undertaken without Hitler's approval. In any case, it was the most serious error in the Axis' psychological warfare, comparable in its consequences to that merely diplomatic invasion which Germany attempted during World War I through the Zimmermann despatch which promised

Mexico American territory in return for an alliance and which went far toward preparing the public minds for the American counter-invasion of Europe in 1917; and rendering that invasion certain.

NOTES, CHAPTER 43

[1] *N. Y. Times,* Jan. 19, 1939.

[2] Lt. Col. Thomas R. Phillips, of the Army General Staff, in *Army Ordnance,* Sept., 1941.

[3] *N. Y. Times,* Sept. 29, 1939.

[4] *PM,* Aug. 25, 1940.

[5] *N. Y. Times,* Feb. 17, 1940.

[6] Ibd., Nov. 1, 1940.

[7] Ibd., Oct. 6, 1940.

Chapter 44

TRIANGULARITY OF SERVICES, PERFECT AND IMPERFECT; NORWAY, 1940

AMONG THE PARTICIPANTS in World War II the Nordic belligerents, Finland, Norway, and Denmark, were highly unwilling partners. They wanted to have preserved to them in total war what the older diplomacy had called "the tranquillity of the North," their aloofness from the active and passive play of the balance of power. But all the three great powers and power groups were bent on drawing them in, Russia first, by attacking Finland; England-France next and Germany last, she for the time having been satisfied with the *status quo* which was favoring her.

The Allies supported Finland, in a half-hearted fashion, with war material, but were willing to go much further. They assembled on expeditionary force of 60,000 in their harbors early in March, 1940 and promised more help, all that seemed indispensable to the Finns, in fact. This force was redistributed, when Finland gave up the struggle and made peace with Russia (March 12), including the *Chasseurs Alpins* who went back to their mountain posts.

Anxious to emerge from the inertia of the "phony war," the Allies next determined (March 28) to make the use of Norway's territorial waters impossible for the German ore carriers sailing to and from Narvik. Plans to cut off Germany's oil supplies, possibly by an expedition to the Caucasus, were also entertained. While Churchill was doubtful, Chamberlain accepted French proposals for prolongation of the "phony war," with the blockade theoretically depended upon to be its most potent weapon.

The military consequences of this second Nordic venture seem inadequately to have been thought through. When on March 28 Daladier asked Gamelin: "Are you ready for the *riposte?*" the latter answered yes, but actually no military force had been organized after the Finnish expeditionary was redistributed, like type of a book that had not been printed. Everything had to be reassembled after the *coup* of the Germans, who had more than an inkling of the state of affairs described above. Their troops were under way (from the 6th, if not the 5th), when the Allies laid mines in the Norwegian waters on the morning of April 8. Their troops were also prepared for such an expedition, and far beyond mere improvization.

Close triangular relationship of land, sea and air forces was not at once in great tactical evidence when World War II, fully prepared for on one side and lamentably unprepared for on the other, exploded, to the dismay of the Chamberlains in the British government. There were glimpses of their cooperation in the German taking of Gdynia (Gdingen), among naval stations the one with the distinction of having been estimated to be "impossible" of holding out, even before the start of the war. The first great demonstration of its working details came with the German conquest of Norway, when, justifiably enough, German communiqués and other propaganda played up their cooperation.[1]

The strategic control of Norway was important enough to either side to contemplate ways of securing it, though it might be said that the threat to Germany from a British-held Norway would have been far greater than that to Britain from a German-controlled Norge. Hence, presumably, the tremendous dispatch shown by the Germans. Allied control would have menaced Germany in the Baltic, and also her ore imports from Sweden and her airplane manufacturing which was in large part located near that sea.

Neutrality Did Not Save Norway

Maintaining that she was merely forestalling Britain and her ally, who had been in diplomatic negotiations for the use of Norwegian territory and whose staffs were preparing plans for landing troops in Norway, Germany, on April 9, invaded Norway. She took all the important harbors in the kingdom from Oslo to Narvik, from the sea, only little less hampered than the Allies by an international morality which might still render more than lip service to neutrality. Many of the German troops were carried in the holds of merchantmen, like the men of Maurice of Nassau who, hidden in peat barges, landed at the town's quay and captured Breda in 1590. They took by surprise the Norwegians, who thought they were at peace, their ports, their airfields and part of their shipping, and hence against but little initial resistance.

The Reich navy, gladly or not, assumed the risk of sailing

through waters which, in a term grown almost Byzantine, were British-controlled. They "had defied our naval superiority and taken great liberties in Norwegian waters" (Keyes). The secret of the long advanced and immediate preparations had been well preserved, although Blomberg, shortly before his own downfall, had paid the regions in question a rather spectacular visit. This proved once more that British naval and other intelligence was far less reliable than between 1914 and 1918.

Careful preparation, secrecy and flagrant breach of neutrality worked together to stun the Norwegians and also the Allies, among whom Chamberlain insisted at first that the Germans had not seized Narvik, far north, but merely Larvik, in southern Norway. The surprise took other forms as well, re-awakening the ancient *post eventum* reaction of "Treason!" or in modern nomenclature, "Fifth Column," or "Nazi espionage," not only in Norway, but in British and French high quarters.[2]

Through the vacuum created by surprise the attacker, although numerically weak at the outset and unprovided with those heavy weapons which a modern force requires from the outset for practically all operations, artillery and tanks, passed swiftly, thanks to the mutual cooperation of the various arms. They demonstrated in a new combination Moltke's somewhat ambiguous definition that strategy is a matter of improvizations *(Aushilfen,* the word can also signify, mutual help), in the tactical field.

Ship artillery furnished the necessary *Ersatz* for land artillery and the navy in supplying it was quite ready to have its vessels sacrificed in the performance of that duty, as happened at Narvik on April 10 and 13 where seven German destroyers were shot to pieces by superior British forces. Aircraft also substituted for artillery in its bombing activity, and in addition to reconnoitering supplied a large part of the transportation from Germany to Norway and between the German detachments in that kingdom which were separated by wide distances, difficult terrain and poor overland communications.

The three components of the Wehrmacht were as three factors in a multiplication problem, visualized and solved from day to day by one competent war-arithmetician. The three corresponding components on the Allied side were by comparison quantities forming a sum, in the adding up of which several persons took a fumbling hand. "Anybody with half an eye

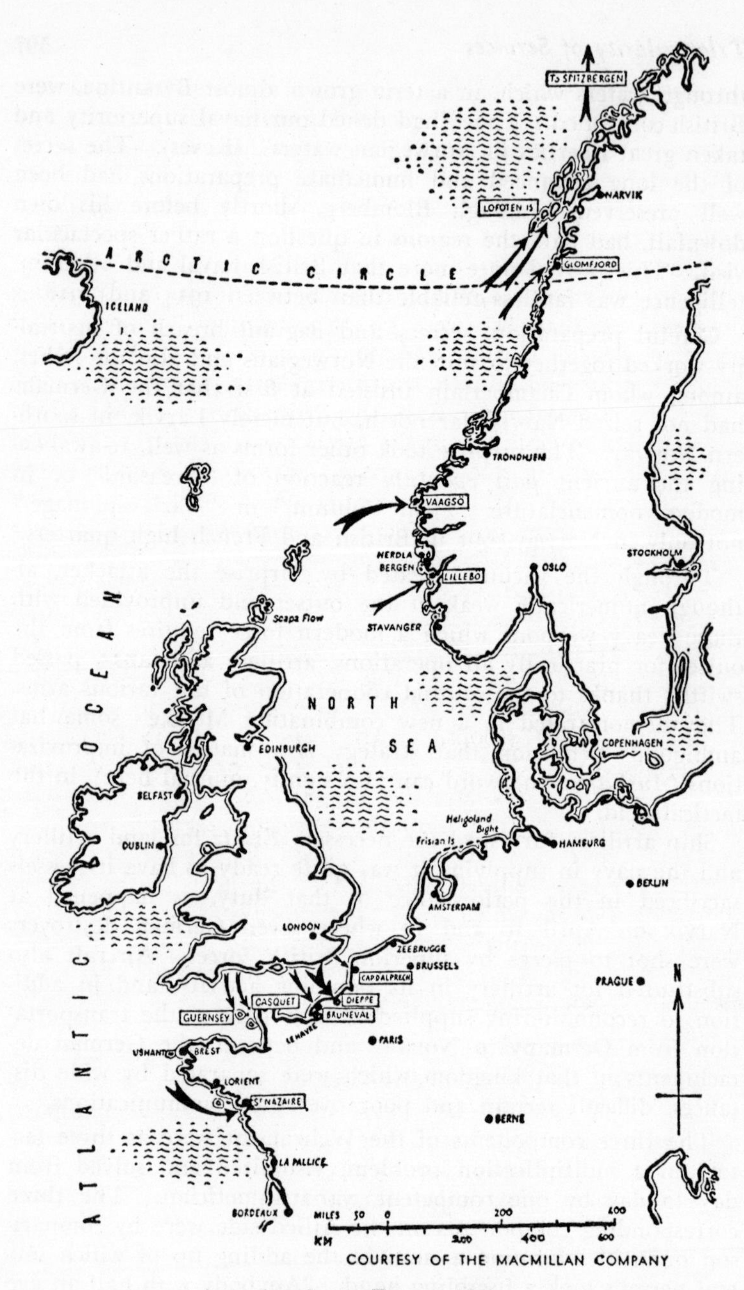

Attacks on Europe, 1941.

on the British Supreme War Command can see every now and then the evidence of this constant attitude of contempt and quarreling among the services," wrote an American correspondent from Scandinavia at the time.[8]

Again First Lord of the Admiralty, and given additional power after a Cabinet reshuffle, although once more running up against the professionals, their hesitations and disunity, Churchill declared overhopefully in the House of Commons on April 11, that "we were greatly advantaged by what had occurred, provided we acted with the necessary vigor to profit from the strategical blunder which our mortal enemy has made."

But the essential disunity of the three British services which had bungled at Gallipoli in the attack, was to bungle this time in the defense, in the attempt to dislodge the invader by counter-landings. Although the details of this part of the coalition war seem not to have been disclosed, it is evident that as before, at the Dardanelles, the French left all initiative, as far as it went, and command in this enterprise to the British.

With reasonable alacrity British troops from the United Kingdom were carried to Norway, their first debarkation taking place on the 14th. By the 16th they had entered upon cooperation with the Norwegian forces, and on the 22nd the arrival of French troops, which included Alpine Chasseurs, was announced. On the 23d and 24th operations of these troops against Narvik and Trondhjem were under way. Obtaining control of the Trondhjem region seemed most important. Here was not only a good harbor, where heavy materiel could be landed, although at the head of the narrow waters of a fjord which under the downward pressure of air power were increasingly narrowed, but also an airfield which the Germans as yet were not in a position to use since it was dominated by a Norwegian-held fort.

German Gamble Won

Trondhjem would have given Allied air power, weak as it was, a much needed *pied-à-terre* in Norway. The Germans were fully aware that, as Shirer was told in Berlin by some one from the Oberkommando, that "the whole issue in Norway hangs now on the battle for Trondhjem. If the Allies take it they save Norway, or at least the northern half of it. What the Germans fear most, I gather, is that the British Navy will get into Trondhjem Fjord and wipe out the garrison in the city, before the Nazi forces from Oslo can possibly get there. If it does, the German gamble is lost." *(Berlin Diary,* entry of April 21, 1940).

Allied forces successfully landed at Namsos and Andalsnes, to the north and the south of the mouth of the Trondhjem Fjord, were to receive support, as their commander was assured, from British naval units—he was unaware that German naval units were in the Fjord. British ships were to force the Fjord in improved Dardanelles style on April 25. The entrance of the Fjord was more weakly defended than the Dardanelles outer forts twenty-five years before, and the defending guns might well have been battered by one battleship of superior range and weight. In vain, the two survivors of the belief in the essential soundness of the idea of opening the Straits, Churchill and Keyes, endeavored to shake disbelief in this and similar enterprises. "Inter-Service Committees were in full swing in Whitehall, irresolution reigned, time passed and golden opportunities were missed" (Keyes).

The naval part of the combined operation, the water-borne ram, was in the end cancelled, for one reason because it was thought in Whitehall that the attack overland was making sufficient headway and thus did away with the necessity of risking possible losses by a direct naval action, although as Churchill insisted on May 8th the Admiralty never retracted its offer. The most serious danger seems to have been expected from the Luftwaffe, for the German naval forces in the Fjord amounted to only two torpedo craft, and in the words of the official explanation, "it would not have been justifiable to undertake to force Trondhjem Fjord for clearing up that very small item." [4]

The latter, however, proved sufficient to block the approach from Namsos, at the place where the Allied troops had to use the shore road along the Fjord; and here the British were taken under flanking fire from "that very small item" and their vanguard cut to pieces by German infantry landing behind them (at Steinkjer). Assistance to the ground forces from the RAF also proved absent, or haphazard. There were no planes to support the first landings at Namsos or, worse, at Steinkjer, where the tri-elemental cooperation was most urgently required and was working only on the German side, for the Luftwaffe and its bombers held the aegis over the infantry.

On April 26th the Allies faced about on the way to Trondhjem, and reembarked at Namsos on May 2. Chamberlain, shortly to lose office for his failure to save Norway, pointed to a rising tension in the Mediterranean as the reason for the evacuation; he tried to dismiss Norway as a mere "sideshow," operations then being restricted to the far north, the Narvik-Tromsö region. Narvik was taken by the Allies on May 29 and evacuated on June 10, a week after Dunkirk and the day of Italy's entering the war. "England was, more than we could imagine, in need of her ships," wrote a member of the Foreign Legion regiment that had been sent to Narvik. "Besides, the campaign in Norway cost her dearly in ships, and the Germans, something we also did not know, were climbing gradually, fjord by fjord, up to us and could not be stopped." [5]

NOTES, CHAPTER 44

[1] For details see Ernst Kries and Hans Speier, *German Radio Propaganda.* N. Y. 1944, 343-4.

[2] Despatch of James Aldridge. *N. Y. Times,* May 7, 1940.

[3] Ibd.

[4] Keyes, *Amphibious Warfare,* 79.

[5] Much of the above is based on Keyes, *Amphibious Warfare,* 78-81; Col. George Soldan, *Militärischer Rückblick auf den Feldzug in Norwegen.* Deutsche Allgemeine Zeitung, May 12, 1940; Capitaine P. O. Lapie, *La Légion étrangère à Narvik.* London, 1941; Alfred Vagts, *Germany's Northwall. The New Republic,* May 6, 1940; Pertinax, *Les Fossoyeurs,* Vol. I; the debates in the two houses of Parliament on May 8, 1940.

Chapter 45

THE INVASION THAT MIGHT HAVE BEEN—BRITAIN, 1940

"We'll fight on the beaches"

Churchill in 1940.

WHEN THE BATTLES in the west drew to a close in 1940, the British were aware, far more than the victors, that the latter might, in fact might be forced, to attempt, an invasion of Britain. They were consequently unwilling to fall in with any of Weygand's last line resistance schemes and sacrifice to them their remaining ground and air forces. They were not even convinced that the Gallic generalissimo did not think that England was as hopelessly lost as was France. It seemed to them by the end of May that "London might be struck before Paris. It would not be reasonable to gamble the British divisions on Weygand's shaky strategical combinations. First came the safety of the British troops. Later, they would come back to aid their Allies if there was still time, if the German furor spared British territory . . . . For the British the battle of Flanders was over and the battle of France concerned them only in as far as there would not simultaneously be a battle of Britain." [1]

Over this supreme issue the unity of inter-Allied command went to pieces. Britain evacuated through Dunkirk 224,000 of her own troops and 112,000 Allies (June 1-4). She tried by all means to keep the RAF intact for whose assistance Weygand called again and again. England would never surrender, declared Churchill in Parliament (June 4), and if the island, or a large part of it should be subjugated, as he would not believe for a moment, the battle would be carried on from the outlying Empire, to the day when the new world would come and liberate the old. At the same time Weygand and Pétain, seeking an armistice, saw no hope in transferring France's government and center of resistance to North Africa, or elsewhere. England could neither save France, nor herself, they thought. England's remaining power lay in the navy and the air. "Neither the one nor the other can reconquer a country but only contribute to finally destroy it" declared Weygand on June 14.[2]

At the end of the blitz war in the west, on and after June 16, when the last French line along the Loire cracked and the French had to seek an armistice, the Germans were thoroughly amazed and stunned by their overwhelming victory. It was so great and unexpected that the logical follow-up step, an invasion on the heels of the British evacuation from Dunkirk, remained untaken. It had not been planned in any detail, if in fact much thought had been given it; it does not seem to have occurred to the Germans in the interval of June 1 and 16. The traditions of Seeckt's setting of objectives for the *Wehrmacht,* that Britain and the United States could not be reached with the means at Germany's disposal, that Germany should rather make herself blockade-proof and invasion-proof in a German-controlled Europe, or Greater Europe, still prevailed. Since Seeckt's days great technical changes had intervened. But even though there had been, just before the start of the war, some military and semi-military writing upon the possibility of invading England,[3] including parachute raids,[4] nothing that has subsequently become known points to more than speculative discussion of long-term German landing preparations against Britain. The proof of present-day landing intentions on a larger scale necessarily lies in the construction of special landing craft, and up to 1939, or even the summer of 1940, these the Germans had not built.

Rather than a landing after Dunkirk, when there was not one fully equipped division left in Britain, Hitler, as his Reichstag speech of July 19, 1940, indicated, was inclined to seek peace with the British conservatives, hoping that they would no longer fight when the fight was "really over." Should his overtures to "common sense" be turned down by their leaders, he warned the British people, it would be because the latter "probably have no real conception of what it will mean once the German offensive begins in earnest against the British Isles."

"Best People" Wanted Peace

Only after the expected receptive echo failed to follow his speech, if not also other offers to the "best people in England," who, according to United States Ambassador to England, Joseph Kennedy, wanted peace,[5] did Germany gird herself for the Battle of Britain. She chose to fight in the air above Britain rather than against her shores, in spite of many subsequent threats such as Goebbels' (of December 11, 1940) that the war would soon end with a great lightning offensive against the British Isles and that the Channel was no better guarantee of their immunity from invasion than the Maginot Line had been for France. In the High Command's communiqué of August 8 it was announced that "in connection with the German air attacks, large air battles developed," with a Berlin broadcaster commenting that "the great German victories announced today and yesterday have made it quite clear to the English people that this war is essentially to be decided by the Air Force."[6]

The British, necessarily, prepared equally against invasions by water and air. Their propaganda perhaps played up most vociferously the more historic of the two threats. Many of their defense measures, such as setting up and training, so far as it went, of the Home Guard; removal of road signs, internment of aliens and possible quislings, blocking of certain passages, concentrated on any routes an invader might use. It was like an attempt at outdoing the Nazi in Wagnerian images, specifically apostrophizing the zone of flames surrounding the place where Brunhild lies banished before Siegfried frees her—

> *"Appear, wavering spirit, and spread me thy*
> *Fire around this fell . . .*

when the British in June-July, 1940, arranged for a barrier of oil defenses from Scotland to Cornwall. Following Dunkirk gasoline was stored along the coast, particularly in fuel tanks on hilltops, which in case of invasion attempts were to be released and ignited, sending cascades of flame down the slopes, or shooting tongues of flames 200 feet into the air. In operatic terms the story of this feature of the defense of Britain as told by the then Minister of Petroleum, Geoffrey Lloyd, is very impressive.[7] How far such installations would have survived an intensive bombing is quite another question; and it does not appear that it was one of the main deterrents to the German invasion plans.

Other features of the defense of Britain, like the acquisition of fifty destroyers from the United States, which the British Purchasing Mission had proposed before the fall of France and again after Dunkirk, pointed to the need of vessels to control the ever narrowing protection afforded by the waters of the Channel. Britain found once more, as in her earlier wars, that the numbers of small naval craft were insufficient. The repercussive effect of Germany's across Channel threat, as felt in America even during an election year, gave sanction to the deal for the destroyers, informally linked with the lease to the United States of naval base sites in adjacent British territories.

While the Luftwaffe was to prepare for the invasion by softening up British morale, talk in Berlin in the late summer of 1940 was allowed to reach foreign reporters that Germany had built a fleet of suicide speedboats capable of crossing the Channel in 42 minutes and carrying 200 men each as spearheads of an invasion, to be followed by larger transports proceeding from French and Belgian ports, both covered by barrages from long range guns along the Channel coast which would keep the British from bringing their heavy fleet units into these narrow waters. The Luftwaffe was to hold its aegis above the beachheads, and the then still new parachutist terror-arm would act in and around these landing places as well as in destructive raids in the interior.[8]

While no special invasion craft were built on a large scale, the Germans, according to British publications, did assemble a "formidable collection of barges"—it is not stated from where—in Rotterdam, Antwerp, Flushing, Ostend, Dunkirk, Calais, Boulogne and Le Havre. Their number has been given by the British as 2,500 which would indicate that a very large part were not sea lighters, but inland craft from the Rhine and other rivers and the canals connecting them. Against these ports the RAF made 536 attacks from June, 1940, to May, 1941, by which time they had completely dispelled the threat of a seaborne invasion.[9]

Invasion Stories Hazy

An imposing saga has been built up concerning a Nazi invasion attempt of 1940—and 1942, as some have declared—one version maintaining that a full-scale expedition actually reached British soil and had been promptly smashed; another that the Nazis had been damagingly strafed by the British Navy, or the RAF, during exercises preparatory to a landing, or that exercises undertaken by the Nazis along the Channel coast had cost them hundreds of lives. Some of these stories smacked rather of parallels with what Napoleon did in Boulogne. On September 17, 1940, the London Air Ministry announced that equinoctial storms had intervened in favor of Britain and had done the work for the RAF, "for concentrations of German barges and supply ships, presumably poised for an invasion of Great Britain, were scattered by the winds." In addition, the announcement inspired far-fetched historical reminiscing about storms having wrecked Spain's Armada in these same waters.[10]

Left altogether without any clues from the Germans as to their real intentions, a British cabinet member declared in the House of Commons (July 29, 1943) that "it is well known throughout the world that the enemy's preparations for invasion were frustrated by the RAF." The RAF has supplied its own version of this frustration. According to its statements the Germans planned to work up a five weeks' program, geared closely to tides and the weather, to culminate around mid-September in an actual invasion in force. They intended to knock out the RAF, which the Germans outnumbered by 4 to 1, then bomb London and thereby try above all to break British morale, and perhaps provoke irresistible demands for reprisals against Berlin. This the Germans would have welcomed since it would take protective planes from England, when the Germans would bomb coastal towns in the vicinity of the beachheads and then airfields.

The Germans adhered throughout to this time-table, without doing more than initial damage in any case.[11] A better balanced version, which might be called that of the Fighter Command, was provided by Churchill's statements in the House of Commons (June 20, 1944) when he answered a member's query whether in 1940 "the enemy set in motion the apparatus of a seaborne invasion," by saying:

> I do not quite know what is meant by setting in motion. Setting in motion in the sense of crossing the Channel—no. But setting in motion in the sense of making very heavy concentrations of troops and ships to cross the Channel—yes.
>
> Did any of this shipping ever emerge from ports across the Channel?
>
> Not to my belief. A great deal of it was sunk in ports and then they changed their minds.

This declaration, confirmed by Foreign Secretary Eden on Sept. 28, 1944, who added that he knew of no German invasion attempt in 1942, may indeed be considered the "partial unveiling of one of the war's greatest secrets." But it does not settle questions like these: When were these invasion craft bombed to pieces? Where were ships and troops concentrated? Were the great number of barges not assembled, perhaps, (and soon again dispersed) to carry off the vast amount of materiel captured in and near the ports mentioned and thus to ease the load for the German controlled railways of western Europe? Would the mere two hundred bombers at the disposal of the RAF have been sufficient to smash the assembled craft, which in case of an actually intended invasion would have been much better protected by flak? Was the concentration, perhaps, merely part of the terror tactics which had been so amply promised by Hitler on July 19 and by German propaganda after him?

Air Defeats Ended Invasion Plans

On the whole, the tactics of the *Luftwaffe* in the Battle of Britain ran more in a pattern of misfired frightfulness, of which the threat of a waterborne invasion formed part, this merely fitting into the entire invasion scheme; with the carrying over of troops as its central feature. While allowing the *Luftwaffe* its fling at England, letting it fight "the first great absolute air war over Europe", as the foremost *Luftwaffe* radio commentator put it on August 24, in the German army mind such an invasion had in all likelihood nearly always remained in the category of "an impossible war", as compared with those possible wars for which it subsequently prepared against the Balkan nations, the British in North Africa and eventually the Russians. *"Denn wir fahren gegen Engeland"* was never as much the Army's slogan as it was, for a considerable time, that of Nazi propaganda or indeed of the then most Nazified component of the *Wehrmacht,* the *Luftwaffe.* The latter had fallen for the Douhet doctrines and had temporarily obtained decisive influence over the resolutions of the High Command. Across their maps and plans Goering's broad shadow fell during an eclipse of the regulars.

After the utmost limit in *Luftwaffe* losses—2375 planes in three months, by British count—had been reached, the enterprise was called off by the High Command, whose "unheard-of strength of nerve in carrying out preconceived plans" the *Luftwaffe's* own radio commentator had praised earlier in the battle. October 22, 1940, has been given as the date of this resolve. It would be too much of an admission of plutocratic thought to explain this by an admission of the failure of a speculation, in which a considerable, although definitely limited part of the *Luftwaffe* had been invested. Hitler in the image prepared for the German people was never a gambler in war, or peace; he was an infallible prophet, plethoric with military foresight. He admitted it himself.

A year after the Battle of Britain, which the Germans were at pains to declare had never taken place, while the British proclaimed to the contrary, although not always with complete consistency, Hitler explained that the breaking off the battle was a credit to his own perspicacity in detecting a new enemy: "In August and September last year, one thing could be realized—a settling of accounts with England now, which in the first case would have tied down the whole of the German Air Force, was no longer possible, because in my rear there was standing a state already getting ready to go against us at such a moment" (Speech of October 3, 1941).

Only by German propaganda was the Battle of Britain still carried on for a while. It was revived from time to time, as after the capture of Crete, which proved to Goebbels' satisfaction that the successful parachute technique there could be used against Britain as well. "If today the events of Crete are hotly debated in England, one need merely substitute the word England for Crete to understand what is meant," the Propaganda Minister wrote in an article. "If the Churchill clique in England doesn't discuss the theme publicly, it is not because invasion isn't feared, but precisely because it is feared."[12] But somehow the theme did not sound good enough for German ears, for something occurred that had never happened to the Fra Diavolo of German propaganda—the paper with Goebbels' article was suppressed.

What became known of German invasion intentions after the fall of the Third Reich, was not fully satisfying to military historians. It was reported that in liberated Brussels elaborate German invasion plans begun in August 1940 and kept up to date as late as 1943, including "hundreds of thousands of dossiers," had been found.[13] Among them were, it was said, designs to reach England by way of Ireland.[14] What we have been allowed to learn about the explanations of captured German generals to their captors about the non-invasion of Britain would rather tend to confirm our surmises—that Hitler at the outset, for political reasons, had not thought it necessary to invade Britain; that she was bound to fall without it; that Brauchitsch and Keitel had considered the enterprise unfeasible and the Navy "very dubious" so long as the British fleet remained intact.

While "the army was ready," weather conditions hampered the Luftwaffe. In the division of minds in the High Command, which for the first ten days after the Armistice, issued no orders whatever relevant to an invasion, only Kesselring, always somewhat of an energetically unconventional commoner among Hitler's field marshals, advised that the attempt be made, pointing to the critical state in which Britain's defences were at the time. Clearly, neither Führer nor Staff were prepared for the invasion at the moment of the unexpectedly swift victory in France.[15] In the late summer of 1940 ten divisions were thought sufficient to bring about a decision in Britain, provided they could be landed; but that with the insufficient shipping actually available and the failure of the *Luftwaffe* to obtain air superiority over the Channel the attempt should not be made. Thus one of the most decisive battles in history remained unfought, as the Germans insist, or fought, as the Britons would have it, and won by the RAF; but, fought or unfought, it was lost.

NOTES, CHAPTER 45

[1] Pertinax, *Les Fossoyeurs.* I, 259, 261.

[2] Ibd. I, 324.

[3] See the rather amateurish discussion of invasion routes, etc., in Ewald Banse, *Germany Prepares for War.* N. Y., 1934.

[4] *Rear Admiral Gadow Holds Parachute Raids On Britain Still Possible. N. Y. Times,* July 16, 1939; *Reich Army's Plans in Case of War With Britain Believed Changed Now.* By Augur. Ibd., July 31, 1939.

[5] *1941 Britannia Book of the Year,* 264.

[6] Kries-Speier, *German Radio Propaganda,* 389, with an excellent analysis of the propaganda woven around the Battle of Britain.

[7] *N. Y. Times,* June 4, 1945.

[8] U. P. despatch from Berlin, Sept. 10, 1940.

[9] *Bomber Command.* Official Publication of the RAF.

[10] *N. Y. Times,* Sept. 18, 1940.

[11] A. P. despatch from London, Sept. 23, 1943. A somewhat different account of the Battle of Britain is to be found in *The First Great Air Battle in History, the Battle of Britain.* An Air Ministry Record of the Great Days from August 8th to October 31, 1940.

[12] A. P. despatch from Berlin, June 13, 1941.

[13] *Time,* Nov. 6, 1944.

[14] *N. Y. Times,* Oct. 18, 1944.

[15] General Marshall's *Biennial Report,* 1943-45.

Chapter 46

NARROWING OF ITALY'S MARE NOSTRO

THE STRATEGY OF THE ITALIAN WAR against Greece has so much in common with that of colonial wars oversea that it might well be called a colonial war inside Europe. In keeping with it, the expansionist power in possession of naval superiority first acquires a foothold in the close vicinity of the object of attack and after that proceeds to the inland conquest. This was the Italian formula for subduing Greece as it had been for Ethiopia.

The starting point for this latest venture across the Duce's *mare nostro*—whose since 1940?—was Albania, which Italy had taken over in April, 1939, ahead of World War II, much as the taking of Tripoli had been the storm signal of World War I. Usually the more important and critical second stage of colonial conquest is that of inland penetration where the defender, even if materially or numerically inferior, is often aided by the terrain and the logistic difficulties of the aggressor. Italy during the autumn and winter of 1940-1 bogged down before the Greeks' valiant resistance, while the powerful backers on either side, Great Britain and Germany, were as yet unready to play an open hand in the war during which the two peoples of classic origin reversed the ancient roles of conqueror and conquered.

In order not to provoke Germany, with which she was not at war, Greece endeavored to get along with as little as possible overt help from Britain, her ally since the conquest of Albania. What she did receive consisted largely of aircraft and air, rather than naval, support against the Italian supply lines across the Adriatic. Crete was occupied by Britain in November, 1940, although not put in a state of defense, while Italian possessions in the eastern Mediterranean were on the whole left undisturbed. There was a definite limit on what Britain could offer to assist the Greeks, and her fighter planes could be spared least from the coasts of Britain where the invasion threat had by no means passed. No bases for such planes were built, or prepared by Greece, according to the British official version, because the Hellenes shrank from provoking the Germans who had provided themselves with a number of air bases in non-belligerent Bulgaria. Land power, or Germany, moved towards the sea by means of its air power, but sea power, England, at the time could not summon up sufficient air power to stop it. Land power, through the use of air power, was narrowing the space in which sea power alone which, historically speaking, had been the mobile, evasive power, could act, and this at the same time increased land power's own mobility and evasiceness.

British Makeshifts in Greece

What force Britain thought she could release to the Greeks, once the threat of the German intervention began to loom in February, 1941, when the Reich moved into Bulgaria, was part of her system of compulsory Mediterranean makeshifts. The troops for Greece, three divisions and an armored brigade, were borrowed from Wavell's command in North Africa, where at the moment things were going well for Britain. This makeshift troop movement began early in March, with which the Italians tried in vain to interfere (sea battle at Cape Matapan, March 28), and was unfortunately paralleled by a shift of German forces into Africa. This double shift might have been compared to the working of a revolving door: while parts of the British North African forces left for Europe, more Germans went into Africa. Due to the withdrawal of naval auxiliary craft and aircraft to escort the transports for Greece, the maritime watch in the central Mediterranean had been relaxed, allowing the Germans to slip through and thus build up the force that was to sweep the British back to Egypt.

The supremacy of the British Navy in the eastern Mediterranean was sufficient to carry 58,000 troops and mechanized and other equipment safely to Greece. Most were disembarked at the Piraeus and a smaller contingent at Volos, some 40 miles south of Larisa, which was the advanced base for the British, rather than at Salonika, which was obviously too exposed to German air attack based on south Bulgaria to be used as a naval and air base. As was to be expected, it fell early (April 8) to the Germans from whose grip a sizeable portion of Greeks were evacuated from Salonica by sea and carried to Greece and islands that could not be held for long.

The Imperial forces had to move inland from the sea in order to stop the Germans, who had broken through the Yugoslav army before the latter could cooperate with the Greeks. They were thrown back, by superior armor first and then increasingly by air power based close by. Fighting constant rearguard actions, one of them near Thermopylae, where they opened the sea dykes to render the shore line impassable, after the main Greek army had capitulated in the Epirus (April 22), evacuation became unavoidable. The question was how far the Germans could be kept from interfering with it. The threat to embarkation from the *Luftwaffe* was acute and so far as air protection went, conditions were worse for the British than at Dunkirk.

The German land army was not quite close to their heels, but its approach was rapid. Setting out to the south, from the Epirus, the SS. Panzer-Grenadier Division Leibstandarte "Adolph Hitler," to give it the full title later accorded to it, a motorized division, reached the Gulf of Patras at Missolonghi in six days and thence crossed over into Morea. From there they threatened several of the British re-embarkation ports. Evacuation had begun during the night of April 24-25 at various points near Marathon and further to the southward.

In order to interfere more directly with these evacuations, the Germans dropped parachute troops near Corinth to block the isthmus and thus keep the British from entering Morea and make use of the ports south of Corinth. These paratroops were only slightly too late. Another attempt at interfering with British departure was made by a German mountain division landed at the northwestern tip of Euboea, from whence Khalkis was reached on April 25; from there the division recrossed to the continent, menacing by this double crossing the route of the British retreat from Thermopylae at Thebes, where the Germans were slowed up on the 26th.

While the British sought and fought their way back to the seaboard, their movements were persistently interfered with by the Luftwaffe. For lack of air protection of their own, they could not use a large harbor like the Piraeus, which had been under attack since April 7; or the roads to the ports during the day, and could only embark at night being in hiding during daytime, and then only in the small harbors, from above Athens such as Marathon, Raftis, down to the island of Kithira. Three nights, from April 26 to 29, saw the majority of the remaining British being taken off, around 25,000 in all, part to Egypt and the remainder to Crete. Some of the evacuating craft were bombed and strafed by the *Luftwaffe,* and not all who reached the sea could be taken off because the enemy finally entered ports where the British ships had intended to fetch them.

Ancient Bone of Contention

After Sicily, Sardinia and Cyprus, the next battleground, Crete, is the largest of the Mediterranean isles, three of which have re-entered the history of war since 1940, after various intervals of peace. None proved as strong as out-moded conceptions of strength presumed them to be. Thalassocracy, dominion of the sea, had been exerted from this Cretan island seat in the past, if never to the extent of that preponderant influence over Greek affairs which, according to Aristotle, its natural position should have given it. But people are apt to spoil their best geopolitical chances, as the Cretans did by fighting more wars among themselves than with foreigners (Polybius).

They successively became the object of Roman, Saracen, Byzantine, Venetian, Turkish conquest and dominion, with the Turkish siege of Candia from 1648 to '69 as perhaps the longest siege of military history. This consumed more years than the Germans did days in taking the island in *Blitzkrieg* style, and thereby demonstrated the end of thalassocracy pure and simple. It represented also the full swing from the myth to the machine, when the air machine came back to Crete, where it had first been built by Daedalus, the artful engineer, and builder as well of the Labyrinth for King Minos.

The essential weakness of Crete in the present war consisted in its proximity to the European mainland—180 miles from the Piraeus to Khania (Candia), 200 miles to Iraklion and only 50 miles from the island's western end to Kithara, which meant flights of only minutes from the Peloponnesus as well as the enemy-held Dodecanese—and its much further distance from British-held North Africa and Egypt, 400 miles. This put it beyond range of British fighter planes.

While three airfields on Crete, at Maleme, near Khania, Rethimnon and Iraklion, all on the northern coast, were at their disposal, the British were forced to abstain from using them, once the German attack had started seriously, since planes based there would have been at the mercy of the *Luftwaffe*.

Crete was completely within the latter's striking sphere, but could only be reached by British bombers from Africa and these, except at nights, were incapable of action without fighter support. While Crete in all its zones would therefore be open to the attack of, and by, air power, it had to be defended mainly on the sea and on, and from, land. New and surprising as it may seem, the island while never blockaded was invested and besieged from a distance of 50 to 200 miles by the *Luftwaffe,* which made it exceedingly difficult for the British to bring in reinforcements and supplies through the curtain of bombing which Goering's flyers let down over the island fortress. To create worse conditions for the defenders, there were no good harbors on the southern, or nearer, side and what little men and materiel were landed there could not be taken across the island because of almost total lack of land transport. Indeed, the British, temporarily the land power, were nearly without roads, while the unobstructed air gave the Germans all the roads and channels of approach they required. There was only one thoroughfare running lengthwise through Crete, following the north coast, and three crossroads suitable for motor traffic.

The garrison of the island amounted to 27,000, or 28,000 at the end of April, plus eleven Greek battalions, poorly equipped and officered. Before May 20, 18 antiaircraft guns, four 3.7 howitzers, six medium and 16 light tanks with their crews were landed together with another battalion of infantry. Men and machines were divided into three detachments, one each around Khania, Rethimnon and Iraklion, with hardly any reserves, although in any case these could hardly have served effectively due to the want of roads and transport. The British were dangerously weak in tanks and anti-aircraft, the decisive weapon, if we may use this much abused term on the occasion, for the breach into the fortress had to be made from the sky. The troops who had fought in Greece, were still being reorganized when the onslaught came and others who lost their equipment were reformed into infantry units during the battle. The German siege bombardment preceding the 20th of May was primarily directed at British shipping and not aimed so much at vanquishing the ground forces.

Crete Campaign Rushed

As if realizing that the best chance for an invasion of the British isles in 1940 would have been in the days immediately following Dunkirk, the Germans did not allow the British to put themselves into a full state of defense in Crete, in which endeavor they were handicapped by the reverses in North

Africa. Within three weeks after the conquest ot Greece, the Germans prepared for the Cretan invasion. They assembled what was probably the majority of their parachute troops and transport planes in Greece; as a ground force to operate in mountainous Crete they retained the 5th and parts of the 6th Mountain Divisions who had fought well in the Greek campaign.

The assault on the fortress of Crete began on the morning of May 20 with a bombardment of Maleme airfield. This blitz breeching from the air was followed by the appearance, 45 minutes later, of low flying planes, each with from two to five gliders, in tow. Each glider carried a dozen men provided with considerable fire power in the form of light machine guns, etc., most of them coming down on the Maleme field. Some 3,500 parachutists were dropped over an area ten miles by three, to the east of Maleme, from carriers and at heights hardly more than 300 feet during the morning hours and in the late afternoon of the 20th.

They took their first cover and jump-off positions in the craters produced by the German bombs and had the airfield's AA guns silenced within two hours. By the end of the day they had driven off the battalion defending the Maleme field. The Germans thus had breached the Cretan fortress and it did not greatly matter to them that the parachutists had suffered heavy casualties elsewhere, up to four-fifths in some places while they were coming down, and were contained by the British.

On the 20th and particularly the 21st, Maleme served as what corresponded to a beachhead for more and more air-transported troops, most of them brought in by carriers, of which the Germans employed about 650, according to British estimates, each bearing from 20 to 30 men. They came in and went out with shuttle-like regularity. While many enemies were killed, or wounded in crash landings, far more survived until thousands swarmed where hundreds of paratroops had crouched originally. The time for counter-attacking had definitely arrived; the 22nd might be already too late, and it was consequently fixed for the night of the 21st. It was undertaken by a New Zealand brigade unfamiliar with the terrain, not informed about the position of the enemy and his lines, if any. Two hours late in starting, the thrust made some headway during the night, but daylight saw it fail. The enemy remained in possession of Maleme airdrome.

Air transport except over well functioning termini is notoriously wasteful, and the Germans were anxious to get troops and materiel in by the orthodox sea routes. Using Italian naval units and captured Greek craft they sent off a transport flotilla to Crete. But in this they were thwarted by the British Navy. During the night of May 21-22 and on the 22nd, by destroying two invasion flotillas and causing the drowning of most of the Germans aboard, it kept the invasion of Crete almost exclusively an air-borne performance so long as the battle for possession of Crete lasted. The price for keeping the Germans in the air and off the water was not cheap. It cost the British two or three cruisers and four destroyers during the invasion and an additional cruiser and two more destroyers during the evacuation. The loss of merchant shipping during the supplying of the Crete garrison, the carrying of reinforcements, some of whom arrived during the battle, and the evacuation, in which flying boats also participated, seems never to have been stated by the British.

By the 26th of May all hope of holding Crete was ended. Even if the garrisons of Rethimnon and Iraklion maintained their hold in their close perimeter, the enemy elsewhere was bringing in fresh troops from the 27th on. The breach at Maleme could never be sealed. After nine days of fighting the battle of Crete was over, and 15,000 officers and men were carried in retreat to Egypt, as on a receding wave of civilization which had once been carried forward via Crete. Some were taken off, as in Greece, by the specially equipped vessels and landing craft of a British commando force, who were waiting in Egypt for something to do and whose craft were used in the evacuation of Crete before they found themselves employed on their original purpose in Syria, early in June, 1941.

British and German Losses

Landings and, for that matter, withdrawal embarkations have long been dreaded on account of the usual high casualties, particularly drownings, entailed. These casualties in actuality have probably never been as high as feared or anticipated. Data published on the battle of Crete from both sides allow a comparison in costs even if the figures of the Germans' losses might be more correct in their comparative than in their absolute numbers. The British have given the number of defenders as 27,550. Of these 14,580, including wounded, were safely removed, which brings the non-evacuated casualties to 12,970, or 40 per cent.[1] The German estimates, putting British and Greek casualties, dead plus wounded, but not including drowned, at 5,000 plus 10,700 British and Greeks taken prisoner,[2] can be made to square with the numbers given out in London, although they leave some 8,000 British and Greeks unaccounted for. The Germans have reckoned their own casualties in connection with the Crete show as follows:

	Killed	Missing	Wounded
Army units			
Officers	57	18	13
Other ranks	301	506	274
Luftwaffe incl. parachutists			
Officers	105	88	104
Other ranks	927	2009	1528
	1390	2621	1919

Conclusions to be drawn from these numbers are: (1) officer fatalities were high, which is only in part explained by the higher percentage of commissioned men in the *Luftwaffe,* in part by the necessity of higher exposure and therefore greater danger to leader personnel in unfamiliar situations and terrain; (2) the missing on the German side must be considered as having drowned, since the Germans remained in possession of the battlefield and had thus occasion to count the dead, and since the British under the given circumstances could hardly be expected to make, or take away many prisoners.

The German strategic dictum on Crete, on the first airborne invasion of an island, as delivered by Göring, reich marshal and commander-in-chief of the *Luftwaffe,* was that it had proved the truth of the Führer's word (which we had not heard of earlier) that "There is not an unconquerable island," a word so confidently resounding that it was said to have disturbed even the distant Japanese.[3] He told the "Crete Fighters" in a special order of the day on June 2 that "this first and audacious operation over the sea has crushed the enemy like a thunderstorm within a few days . . . . In an old comradeship of arms from the great days of Narvik, aviators and mountain troops conquered the island and thereby threw England out of an important position in the Eastern Mediterranean. The entire German people feels the deepest wonder and unending thanks . . . ."[4]

Since statements of this kind in German propaganda were often heralds of similar actions to come, Göring's declaration was in many places understood as indicating that Hitler might now turn to Britain and use against her the Cretan methods. But while the "lessons of Crete were still being studied in Britain" and the idea spread that "Crete was a dress rehearsal for a more gigantic spectacle yet to be staged" against Britain, which was so much better provided and organized for her defense than a year before,[5] the Führer, if Crete had any large looming rôle in his strategy, decided that Russia must be attacked first, although Reich propaganda might go on harping of the pregnable character of the British island. He affected to believe that Russia would strike Germany in the back should he prefer to attack England first, and consequently turned from contemplation of Crete and the temptation of trying the methods of Crete against England, much as Napoleon in 1805 had turned his back upon the cliffs of Dover, and prepared to strike in the East (June 22).

Possession of Crete enabled the Germans to hold control over the northeastern Mediterranean, including to a large degree Turkey, and keep the latter out of the war. Why they did not try to capture Malta in the Cretan style, particularly after they had submitted it to the most concentrated aerial bombardment or aerial siege, beginning early in 1942, the intensity of which recalled to the Maltese the siege of 1565, is anybody's guess. It proved a fateful error, similar to that of neglecting to invade England in 1940, or after. Malta relieved and militarily rehabilitated, became one of the bases from which the North African and Sicilian attacks of the Allies across the sea and through the air were launched. The Axis had failed to narrow the Mediterranean sufficiently to dominate it.[6]

NOTES, CHAPTER 46

[1] A. P. despatch from London, Aug. 2, 1941.

[2] *N. Y. Times,* June 13, 1941.

[3] Ibd., June 6, 1941.

[4] U. P. despatch from Berlin, June 2, 1941.

[5] *N. Y. Times,* June 15, 1941, despatch from London.

[6] The above is largely based on the communiqués of the parties involved and on *The Campaign in Greece and Crete.* Issued for the War Office by the Ministry of Information. London, 1942.

Chapter 47

AMPHIBIOUS GUERILLA-COMMANDOS AND RANGERS

BLITZ WAR WAS CLAIMED BY GERMAN propaganda as "a German monopoly that cannot be imitated in Moscow, London, or Washington. The designer of Blitz wars was Adolf Hitler and the spirit with which he had filled the German people and the Armed Forces is the secret weapon which will never be available to the enemy."[1] In the past there might have been monopolies in warmaking, but no more. Patents protecting and covering methods no longer run in war. Given time and leadership, men and materials, the essential identity, that is to say, above all the organizability of the great modern industrial nations, any invention, strategical, tactical, or technical can be taken over, improved upon, multiplied and sooner, or later turned against the original inventor. The Marne of 1914 marks the breakdown of the first attempt of Blitz monopoly, Stalingrad that of the second, while the failure to invade Britain in 1940 constituted the fatal omission to extend the Blitz to the sphere of amphibious warfare.

The first, then largely experimental and hence small-scale, wielding of Blitz methods against their original conceiver, in the field of amphibious warfare, were the British commandos. As their name indicated, the organization of these special troops, with their also special weapons, craft and tactics was regarded by the regulars in Britain as something rather on the novel, foreign side, for the term was borrowed from those formations of the Boers who had carried on the war after their main army had been destroyed, in guerrilla style. Deneys Reitz' *Commando. A Journal of the Boer War,* first published in 1929, had put the term back into circulation. The American term "Rangers" sprang from like experiences in the past.

Guerrilla means two things not easily differentiated in the public mind, or for that matter in most dictionary definitions: (1) war carried on by irregular bands in an irregular way; or (2) war undertaken on a small scale within the framework of a big war, *la petite guerre,* or *Kleinkrieg,* by regulars, if not by special troops, often in outlying theaters. After Britain had lost the great war temporarily, or appeared to be doing so, after Dunkirk and Narvik, she gave it a fresh lease on life, while preparing for resuming it on a larger scale, by guerrilla tactics in the amphibious field. The actions of the Commandos were largely in the tactical sphere and of tactical importance until, with greater forces employed, the strategic level was reached, in the North African landing where the Commandos formed part of a battle force.

The amphibious guerrilla idea was broached to, and favorably received by Churchill and the chief of the Imperial General Staff, who only a week after Dunkirk ordered Lt. Col. D. W. Clarke, R.A., a member of the I.G.S., to work out a scheme of organization for troops to undertake this kind of war. Having had personal experience in guerrilla war as fought by the Arabs, Clarke proposed that an amphibious small war be undertaken by units called Commandos, raiding over the water instead of deserts, against an enemy holding the rim of the water and some oasis-like islands. Unless the search for the paternity of military ideas be prohibited, that would discover to us the original idea as an offspring of the thinking of land soldiery.

Keyes Placed in Command

Slightly later, the Prime Minister, mindful of Admiral Keyes' services at the Dardanelles and Zeebrugge, and who during the Belgian campaign had observed the workings of the Wehrmacht as a unified force, made him Director of Combined Operations (July 17, 1940). He was to be in charge of raiding operations undertaken by bodies up to 5,000, plus the necessary landing craft to carry them across the water and if necessary through the air, as well as of the training and development centers of these special troops.

To the British way of thinking "Combined Naval and Military Operations must necessarily at the outset be the Admiralty's responsibility," as Keyes himself put it. Whether or not this view augured well for success, after all this might be said for it: The Navy in peace time being more nearly on a war footing than the Army, had comparatively more officers available for new enterprises than the other services whose expansion coefficient under the turn from peace to war had been infinitely higher.

Besides, part of the personnel and craft of the small vessel flotilla, originally designed for defense in home waters, might well be turned to the offensive, once the hour of the greatest danger was past. But it would appear by Keyes' own story that one of the consequences of this Admiralty dominance over combined operations was the resistance they ran up against in the various inter-service committees and sub-committees. In these Keyes found the unhealthy traditions of the Dardanelles period much alive; and to them he ascribes the fact that, although Churchill himself approved and declared of one of his proposals that "it would electrify the world and alter the whole strategic situation" at least in a certain area, he was allowed to engage in no combined operations, none of which "of any moment . . . were projected during the first three years of the war." The old spirit of hesitation found new reasons against it, as it did in face of the menace of air attacks (Keyes, 83-5).

Keyes' impatience, more vocal than the hesitations of the Committees, led to his dismissal and the reduction of his Directorate of Combined Operations to a mere Advisory Committee, in October, 1941. His successor was a much younger man, Captain Lord Louis Mountbatten. Under his leadership and following Pearl Harbor, which gained a new and powerful naval, as well as land, ally for Britain, combined operations emerged from eclipse and on March 18, 1942, Mountbatten was made Acting Vice-Admiral, Acting Lieutenant General and Acting Air-Marshal, in order thus to emphasize unity of command over all component parts of landing forces; and chief of Combined Operations and head of the Combined Operations Command, with a personnel composed of officers of all three services.

Things usually do not really begin with their formal start, and the first British troops raised for raiding purposes, the so-called Independent Companies, were formed for the purpose of fighting the Germans in Norway. They were composed of volunteers from nearly all British army regiments. Their training could barely have started when half of them were sent to battle in Norway, side by side with regular English and French regiments, as stop-gaps and not for the special purposes to which they had been dedicated. After Narvik they resumed their training together with the rest of their fellows "as a force to supplement the Royal Marines, in whom reposes the tradition of amphibious warfare" (Combined Operations, 5). These Independent Companies, who were placed in ships as their barracks and as a fighting base to insure maximum mobility, were subsequently transformed into Special Service Battalions and these at last into the Commandos, an amphibious body largely assembled from infantry, which "meant, first and foremost, that they must learn to cooperate with the Royal Navy" (ibid. 5).

The training of the British Commandos, as of the American Rangers whose birthday is given as June 19, 1942 and who were originally sent to learn from, and with, their British counterparts, is basically that of shock troops become amphibious. They are units of volunteers, in the best physical condition, hardened and inured to the special hardships produced by an additional element, or two, water and air; accustomed to the hazards and uncertainties of night actions in which much, if not most, of their fighting would have to be done. Their exercises and maneuvers, as the simulacrum of war, approximated war far more closely than those of regular troops. Shocking as it seemed to minds willing to accept without concern hecatombs during battle rather than exposure to risks before, "real bullets enforced lesson at army amphibious training center."[2]

Building Up Confidence

With such an uncertain equivalent of a jump-off trench as the sea, it is doubly necessary to familiarize a landing party more thoroughly with its objective, limited and subdivided as it is apt to be, through help of maps, air photographs, intelligence reports. Individual initiative had to be restored to, reanimated and affirmed in men who, by nature of the combats before them would perhaps at most decisive moments, be unable to see, or hear their officers and non-com's. Sound psychological observation—not to allow men to feel "all alone" in sudden new and hostile surroundings—as well as the intention of providing a mutual replacement system which was extended to many commands, led to the employment of the men of the commandos by twos. Self-reliance thus was strengthened by mutual reliance of teams of two, one member of which might still carry on and succeed where two might fail. The way of the commandos nearly always leads into the dark. In order not to make it appear darker than it is, not to discourage commandos, or encourage the enemy, Britain, the United States and Russia as a rule did not mention failures, or comment on enemy announcements that hostile commandos have landed and been wiped out.[3] Whatever past ages may have preferred, our own would never call "forlorn hopes" by that ominous name.

Familiarizing land troops with the other element meant also introducing them to the conditions under which a sister service works, the craft and engines it uses, and if necessary to the extent of learning how to operate them. This included frequent sea trips in special craft in order to make men immune if possible against demoralizing *mal de mer*. Officers learned by living together with personnel of other arms the potentialities and limitations of the various services.

A most logical step from this closer association towards coordination on the staff work level was the establishment in the United States of a joint Army-Navy Staff College, beginning June 1, 1943. This, called an "agency of the joint Chiefs of Staff," was to train senior officers of the Army, Navy and Marine Corps, with one third in each case coming from the air arm, "in all phases of joint or coordinated operations involving land, sea and air," with stress on "air operations and the logistics involved in combined operations." [4] It forms part of the Command and General Staff School at Fort Leavenworth, the Army's highest educational (or thinking?) institution after the war had resulted in most illogically shutting down the Army War College at Washington, "and really corresponds to a college for the study of combined operations—a college for the higher study of war—and represents the latest and most advanced of the services' professional educational systems." [5]

In such institutions of military learning the question might be settled whether the tasks set for the naval partner are not more unprecedented than those of their soldier partners. Consider, for example, that whereas a regular sailor's endeavor will be to keep his craft from being stranded, or beached, those who have to navigate landing craft, which while highly expendable must yet be preserved as long as possible, have to invent and learn the whole art of beaching such craft without stranding, of following and allowing for the tides as they rise, fall, or run, of staying along an ever changing beaching line, of keeping propellers from being stripped—all essential to the job of standing by, ready to re-embark troops. It is obvious that in landing training this is by comparison the more difficult and hence longer part of an amphibious training program.[6]

Commandos and Rangers are elite troops, and hence limited in their numbers as well as the magnitude of the objectives they are designed to tackle. Their original purpose was not to manage landings, but to undertake raids, of a duration not usually beyond a day. To obtain the troops necessary for larger amphibious enterprises, in which the Commandos would yet play their own rôle, bigger formations were required which underwent considerably less intensive and specialized training.

SPECIAL TRAINING IN THE U. S.

The United States Marine Corps, whose tasks lay traditionally in this double field, received a new training, and a special amphibious force within the much enlarged Corps, known as the Atlantic Amphibious Force, with headquarters at New River, S. C., was set up in September, 1941, in order to be "trained and equipped with every modern device for lightning attacks on enemy shores." [7] Once the United States was fully at war, this establishment was enlarged into a Unified Invasion Force under the jurisdiction of the Navy, with bases on both coasts in which "thousands" of men were soon undergoing training, while craft at the same time were increased in a highly experimental way,[8] until it counted its men by divisions and its boats by thousands.

This synthesis of warfare could only be achieved by a new diversification and specialization of schooling. The American "invasion mill" on the Atlantic coast in the autumn of 1943, precedent to the grand landing in France, included a camp for basic amphibious training which the majority of the personnel of an army division chosen for such duty would undergo. A second sea shore camp provided schools for commanders and staffs, for transport quartermasters, shore fire control officers and communications and air liaison personnel. In a third camp landing craft crews were trained, furnishing craft for the sea-going training taking place in the other camps; and in a fourth the officers and crews for handling the LST's, LCI's and LCT's. A fifth gave intensive schooling to scouts and raiders, intended for reconnaissance work, and demolition squads. Special training for that hitherto rare maneuver, outflanking an enemy whose one wing is resting on the sea, and training in what is called shore-to-shore landings, was given in still another camp.[9]

Carrying the war back to the continent from Britain, aside from the air war, began on the smallest possible scale and in the commando style of re-landing and de-landing, with the first enterprise that took place, during the night of June 23-24, 1940, less than three weeks after Dunkirk. Some 120 men went across the Channel in four parties with orders to ascertain the nature of the German defenses along the nearest French shore, between Boulogne and Berck. The information gathered was hardly as valuable and certain as the conviction thus established that the project itself was feasible.

The first Commando raids were for the most part short and violent nightly visits to the German-held coasts, beginning at nightfall and usually ending before daybreak, and dependent for timing largely on the tides. Their objectives were limited, except in the effects they had on British morale. German morale was probably less affected by them, and their propaganda advertised these attacks as mere martial petty larceny forays into the Festung Europa, even though local garrisons presumably did not especially enjoy being surprised and mauled by black-faced nocturnal intruders, armed with unpleasantly lethal weapons. A few times these landings were opposed; more often they were not.

Along this periphery vital installations were located which offered valuable targets for raiders, such as light houses (raid on Casquet Light House, Channel Islands, Sept. 2, 1942), stations for radio location (Bruneval, February 27-28, 1942), and war-industrial plants including power stations, coal and pyrite mines, fish oil factories and aluminum producing plants. German merchant ships used in transporting the product of these plants were occasionally encountered and sent to the bottom.

In these undertakings the Commandos penetrated, as no earlier shock troops could have done, the sector of industrial warfare. Not always were the objectives of these raids apparent at once, as for instance in the case of a craft with raiders trying to land near Cherbourg (Sept. 12-13, 1942), which was sunk by the Germans who made prisoners four officers, including one de Gaullist naval officer whose presence might point to hydrographic reconnaissance as the raid's actual purpose.

These raids mostly were restricted to regions where approach to the shore objectives was not too complicated, as it would have been in the tortuous waters of Holland and Denmark, or of Germany proper, all of which the raiders seem to have left unvisited. Irruption into the Fortress Europe from the air was added to, or combined with, that from the sea, when parachutists were employed against important objectives situated at somewhat greater distance from the shore to which they would have to find their way and from which they were forced to fight their way back, if they could, for reembarkation (Bruneval and, considerably farther inland, the Tragino aqueduct in the Italian Campagna province, February 9, 1942).

Sharp Work at Lofoten

The one enterprise above small-scale raids permitted in home waters conceded to Keyes by the powers that rationed shipping in London, during all his fifteen months as Director of Combined Operations was the raid against the Lofoten on March 4, 1941. In these islands the Germans were making use of the catch of the Norwegian fisheries, producing fish oil and fish meal, important articles in enemy economy. These installations were to be destroyed as well as the shipping employed in carrying the product to Germany. Prisoners from among the Germans and Norwegian quislings were to be made while loyal Norwegians were to be given an opportunity to go to Britain and join their own country's forces.

A brigade of ten commando troops was formed, with a Norwegian force of fifty men attached to them. Two passenger ships served as troop carriers with landing craft on board, two of the latter of a type capable of landing a 20-ton tank, or 100 men. They were given an escort of five destroyers. Surprise was complete. All navigational lights were burning inside the fjord when the flotilla steamed in, serving as guides; besides, each craft had a Norwegian pilot on board. The landings were unopposed, although there was subsequent street and other fighting. The demolition work could be done without haste; 225 prisoners including some quislings and 315 Norwegian volunteers were carried away. Another valuable item of destruction such as fill the books of war, could be entered upon the ledgers to the credit of the British.

However, these various small-scale enterprises, including the raid on Rommel's North African headquarters in November, 1941, which struck at one of the enemy's operational brains, a thing that had not often been done in past warfare, were episodic. They were not evolved into something resembling in scale, or method German, or Japanese amphibious warfare. The Commando seemed overtrained and under-employed. In his disappointment over this comparative inactivity Admiral Keyes, early in 1943, was even willing to concede Russians precedence in this field over the British. Both the Black Sea, and in particular the Crimean and the Kuban bridgehead, and the Gulf of Finland as well as sea-like lakes like Ladoga offered ideal theaters for amphibious operations.

Absence of detailed information on this kind of warfare in the east allows us merely to state that on the whole there was more initiative in such enterprises on the Russian than on the German side. While the former constantly stressed close cooperation between the Red Fleet and the Red Army, the Germans and Finns were usually satisfied to announce in their communiqués that Russian landing attempts had completely, or largely failed due to the watchfulness and activity of their own coastal defenses, acting in cooperation with their own light naval forces.

A typical communiqué of the Germans would say, as on Dec. 8, 1943, that in the Crimea their land and air forces, with those of Balkan allies, had "smashed the enemy landing bridgehead south of Kerch. In three days of heavy fighting the Soviet formations which had landed were wiped out, some 2,000 prisoners being captured. Light German naval forces participated in this success. Under difficult conditions they prevented the regular supply of the landed Soviet troops. All attempts on the part of the Soviets to evacuate the landing on the night following December 6 were frustrated, while seven enemy vessels were burned up." The Russians in turn stressed their successful landings across the Sivash or Putrid Sea into the Crimea, or the effective cooperation of the Baltic Fleet in the resumption of the offensive against Finland in June, 1944.

British African Offensive

The British offensive in North Africa, beginning in November, 1941, did largely without amphibious help although there was then available in home waters a considerable amphibious force including at the time a regular brigade that had been trained together with Commandos and Marines, a force for which General Montgomery had at one time been suggested as leader. Only a small part of it was employed on such occasions as the raid on St. Nazaire (March 28, 1942), which in its purpose and nature suggested a close parallel with the blocking of Zeebrugge in 1918. Success of this enterprise was as strongly claimed and denied as on the earlier occasion.

But what was done Keyes thought, and according to him Churchill as well, seemed to be much less than might have been achieved through use of the Special Service troops at Britain's disposal at the time as a striking force. Both had suggested that it should be employed so soon as possible after the entry of Italy into the war, or by the end of November, 1940, at the latest, in order to obtain island bases along the line of British communications through the Mediterranean, to hold the sea lanes open which the Axis would soon endeavor to cut and narrow, and thus save shipping and reduce freight distances. But according to Keyes, the Admiralty which should have demanded the armor, etc., to turn Crete into an airbase when Greece put it at Britain's disposal in the autumn of 1940, believed that shipping was not available for such more ambitious enterprises, even if they were to save shipping eventually, and denied it to both Keyes and Churchill.

It was like a return to mobility of the Symplegades, those cliffs moving on their bases that, in antique legend, crushed all who attempted to pass between, and brought the Argonauts to a standstill, when late in 1940 a British convoy for Malta was badly mauled by Stukas in the Sicilian Narrows through which it tried to pass in daylight. This threw such a scare into the Admiralty that it cancelled an already much delayed sailing of Commando forces which were split up in the fall of 1940 between a northern and a southern theater of operations and with a considerable Commando enterprise for an unstated objective. The three Commandos destined for Egypt to fight with Wavell, were eventually sent off in three ILS's by way of the Cape of Good Hope and did not reach the Mediterranean until March, or April, 1941, when they were joined with two more locally raised Commandos, of a strength of 200 men each.

They found there but little to do in the field for which they had been schooled. One Commando of some 500 men was added to the garrison of Cyprus from where it was eventually sent to participate in the Syrian campaign against the Vichy French. In the Litani river action (June 8-9, 1941), the one successful amphibious operation fought in the middle east, they landed with their own craft from the sea on the further side of the blocked river above the estuary, in a double crossing, so to speak, of sea and river, and by attacking the defenders from an unsuspected direction demonstrated their decisive usefulness within the framework of the general advance. The price paid in casualties was not too high, although amounting to 25 percent in dead and missing alone, considering the higher casualty rate to be expected in Commando enterprises. Instead of landing, the other Commandos in the near east had to assist in the evacuation of Crete. There was not even much of their special landing craft at hand, some of it having been smashed when pressed into the service of evacuating British forces from Greece.

The four Commandos invested in Crete lost 75 percent of their strength. The rest was used up in North Africa where they had occasion to demonstrate to the Italo-German forces that the thin strip of coast on which the fighting took place, could be outwinged and flanked and raided from the sea no less than from "the other desert."

The time of perhaps the greatest usefulness of Commandos and rangers arrived when they came to fight within the larger framework of the big invasions, such as Sicily and Normandy, when their place was in the vanguard of the vanguards and on the outer wings of the beachheads.

NOTES, CHAPTER 47

[1] Broadcast of July 12, 1941. Kris and Speier, 361.

[2] *Newsweek,* March 22, 1943.

[3] See for such typical announcements and the "no comment" attitude U. P. despatches from London, Oct. 2, 1943, and Apr. 11, 1944, A. P. despatch from London, March 27, 1944, a Tokyo broadcast about a small Allied landing in Southern Burma, recorded by A. P., Oct. 11, 1943.

[4] A. P. despatch from Washington, May 13, 1943.

[5] Hanson Baldwin in *N. Y. Times,* Nov. 3, 1943.

[6] Hanson Baldwin, ibd., Oct. 20, 1943.

[7] Ibd., Sept. 11, 1941.

[8] First public announcement about this program on the part of the Navy Department. Ibd., June 20, 1943.

[9] Ibd., Nov. 16, 1943.

Chapter 48

LANDING CRAFT

"The landing craft, a wholly new type of ship, one we didn't dream of two years and a half ago . . ."
President Roosevelt, speech of August 13, 1944.

THE RELATIONS BETWEEN WEAPONS AND WARFARE and vice versa may vary between two theoretical extremes: Either weapons, offered and sold to, or inherited by, the armed forces, coming largely from the outside, determine forms of warfare; or war, recognizing its latest demands, calls upon its own, or other countries for adequate weapons, and sets planners and constructors in and out of the services, to work, and following them the domestic, or foreign industries. For long periods in the history of war the makers of amphibious warfare did not call for, or receive vessels specifically suited and built for this kind of war. Only during World War II was a situation reached, sufficiently serious in its imperative requirements, which forced the leaders to demand special landing vessels and have them constructed in conformity with the requirements of modern amphibious operations.

These requirements call, in general terms, for structures that make landings possible independent of harbors, on practically any type of coast except the most precipitous. Hence the feature of flat bottoms, usual in such craft, that allow the carrying of landing troops to the shore at great, or medium speed, and which therefore are generally motor propelled. This enables early, or simultaneous landing of heavy weapons, artillery and tanks, in order to meet on equal terms an enemy who possesses these, or to overwhelm him with their help if he does not. Such craft in addition must supply a modicum of fire power during the assault and provide for the maintenance of an initial supply system.

World War I saw little use of special landing craft with new features, in spite of the then considerable advance in specialization of war tools. For those, however, who had eyes to read its lessons and used them it was clear by the close of the war, that "by the introduction of landing craft the navies have become far more independent in the choice of their debarkation points and that one must reckon with the landing of smaller enemy units along the open coasts." [1] The development of such craft remained slow.

The landing barges of the Japanese, the first in the field and setting the ball of military imitation rolling around the world, were too primitive to offer much inspiration to Allied constructors. They likewise preferred not to copy the so-called Siebel ferries of the Germans, used between Italy, Africa and Sicily, and around the Crimea. These stem from European river ferries, consist basically of two large, much subdivided pontoon hulls with a deck between. They are of extremely shallow draft and torpedoes are likely to pass under them. They are driven by Diesel engines, can develop a speed of 9 knots, are about 80 feet long with a 50 foot beam, carry 300 tons of cargo, or 250 troops and are armed with AA equipment up to three 88 mm and two 22 mm guns.[2]

The historic situation, in fact, the collective catastrophe great enough to impress Allied war makers with their necessities in the amphibious field was Dunkirk and what came after. Then, for the first time in centuries, Britain had no harbor on the continent from which to resume war. Conditions for her were more serious in this respect than in Napoleonic times. In this hour of the slump of sea power, it was most clearly recognized that means of transportation in addition to the traditional ones would be required if Britain were to return to the continent in puissance.

What had been evolved by the British in their Combined Development Center at Portsmouth, established in 1936, seemed still inadequate for the purpose. This was an armored landing craft, self-propelled, usually powered by American engines, giving protection against rifle and machine gun fire, carrying 36 men plus crew, drawing only 9 inches forward, that could be lowered by davits from decks of a merchantman, or other ship. New types were needed. They were not to be designed, blueprinted and built over night.

All of the invasions, actual and attempted, by the Allies to the end of 1942, including the North African landing, still made use of the customary landing craft, among them the American Higgins boat, a 36 foot speed type with a dropping bow for disembarking, also built in 50 and 100 foot models as cargo carriers. Whatever the preceding study and experiment, whatever the building programs for landing craft in the

United States, sometimes traced back to 1935, or at least a month before Pearl Harbor, when the program was said to have reached "definite form," the actual large-scale building did not get under way before mid-1942, when the joint chiefs of staff in Washington gave it a top place in the U. S. Navy's construction time tables.[3]

GENESIS OF THE LST

As to the genesis of any of these boats we seem to be best informed about the LST, Landing Ship Tank, among the landing craft the *Megalobrachus maximus* (the largest surviving amphibium). It dates, generally speaking, from soon after Dunkirk and derived from the necessity of landing heavy tanks in face of the defender. Early models proving unsatisfactory, Churchill himself, the old hand at amphibious warfare and eager still to rehabilitate this kind of warfare after the failure, or misreading of Gallipoli, is said to have taken a personal and active interest in the kind of ship that might have to exert a decisive influence on the war. He proposed to construct craft of 7 to 8,000 tons, with forward ramp, of shallow draft and thus capable of beaching and retreating, but also seaworthy, able to cross a sea, or even an ocean on its own power.

Such a craft, adopted by the British Admiralty, was among the Lend-Lease articles of which the British in November, 1941, asked the American Government to produce 200. It was at this time, late in 1941, that most of the several landing craft were put on the drawing boards. Pooling their own ideas with those of the British, American naval constructors worked out final types of landing craft and escorts, more than sixty variants by the time of the Normandy invasion, to be used in Anglo-American coalition warfare. On the original LST, Roosevelt and Churchill, the two statesmen with admiralty service in their past, are reported to have agreed in principle during one of their early meetings.

The LST is described as "a low-lying, shallow draft, seagoing craft, of compact design." Its central feature is a set of water-tanks, like the ballast tanks of the submarine: when filled, they make the LST sit low enough to cross any sea. The tanks were to be blown out, as the craft neared its destination on a hostile beach, lifting itself out of the water until it was near enough to most shores to send off its heavily equipped men and machines like clanking medieval knights and horses under their own power.[4] This "landing ship"—in official language all other types are "landing craft," carried by ships on davits and in other places and used in ship-to-shore operations; or "landing vehicles," motor vehicles able to move in, or on water and land—is 327 feet long, of 4,000 tons displacement, or 5,500 tons when loaded; of medium speed, four decks, heavy AA equipment, and can carry tanks, trucks, bulldozers and other heavy machines. These are driven off over a heavy cleated ramp, raised and lowered by motor power, situated in the bow which opens by two giant doors. While simplicity is the basic feature of its construction, its operation calls for no less than 142 motors and 850 valves. Such a ship could if need be proceed from an American port directly to any trans-oceanic landing beach. It would seem to have been first used on the occasion of the trans-Mediterranean invasions in 1943.

A subsequent development of the landing ship type is the LSM (Landing Ship Medium), apparently used first in the Normandy landings. Faster, perhaps less seaworthy than the LST, slightly over 200 feet long, they were designed to serve as complements for LCI (L)'s with which they can keep pace so as to put tanks ashore immediately following the infantry. Their cost has been stated at about $750,000 apiece.

Simultaneously evolved with the LST, the first actual models of the three being completed in October 1942, were the LCI (L), Landing Craft, Infantry (Large), for which the British are reported to have applied under Lend-Lease in May, 1942, when they were originally designed and ordered; and the LCT, Landing Craft Tank, now called LCT (5) and (6). The LCI (L), of about corvette size and provided with accommodations for a crew of 30, is large enough to navigate by itself, across seas and oceans, is of shallow draft, but highly manoeuvrable. It is 157 feet and above long, has a displacement of 246 to 300 tons, can carry more than 200 fully equipped troops. These are sent off over twin landing ramps, not opening at the bow, but as companion ways on each side of the bow which are lowered from the deck into the water. It is primarily designed to meet the comparatively short-haul requirements in European landing operations and was first used in large numbers in the Sicily, Salerno and Anzio landings, and later in Normandy and southern France.

The LCT (5) and (6), of about 100 feet length and 163

tons displacement, carry a smaller number of tanks, or trucks and also soldiers; they have bow ramps and accommodations for a crew of 18. They are transported to or near the zone of operations, from the United States to Britain, for example, complete, or in three sections, aboard an LST, or a cargo boat. Beginning with an initial assembly schedule of 53 days to put sections together, the same operation was brought down to nine days by the summer of 1944. Like some other craft, these types are provided with means to pull themselves off a beach when stranded, with the help of kedge anchors and anchor cables.

The LCVP, Landing Craft, Vehicle, Personnel, and the LCV, Landing Craft, Vehicle, are used for bringing infantry and lighter vehicles such as jeeps, trucks, light tanks from ship to shore. They are carried on LSTs, or transports into the landing zone. Their length is 36 feet, their speed fairly high, somewhat less in the case of the LCV. They have bow ramps, mount machine guns, their sides are fairly high and lightly armored.

The LMC (3), Landing Craft Mechanized (Mark III), some 50 feet long, is primarily employed for landing bulldozers, medium tanks, guns and trucks in ship-to-shore traffic and is equipped with bow ramps. The LCC, Landing Craft Control, is designed to bring directive personnel, including presumably the Beachmaster (Navy), who are to see that time tables are kept in the movements of the ship-to-shore boats and their proper place in transit and on the shore.

Landing Craft, Support, LCS, are lightly armored craft, miniature destroyers, or watergoing tanks with closed decks, equipped with machine guns in twin turrets, or 37mm, or 40mm guns, or rocket launchers, or also smoke producers. They are to advance in the van of a landing flotilla to break down, if necessary, light enemy fire resistance. Later types evolved about which only scant details have been published, are the Tank Landing Ship, Medium, LSM, and the Landing Craft, Industry, LCI.[5]

The LCR (L) and (S), Landing Craft Rubber (Large and Small), are used for landing infantry groups of 7 to 10, or upward; deflated and easily stowed away in small space, they can be inflated in eight seconds and are propelled by an outboard motor, or paddles. They are essentially of the type employed by the Germans in their Blitzkrieg for river crossing and lend themselves to use by reconnaissance and commando units.

AMPHIBIOUS PIONEERING IN CHINA

The true amphibia in this amphibiology could not well have evolved in the days of the screw-propelled ship. The latter had actually put a stop to the slight beginnings of amphibious vehicles while the paddle-wheel steamer was in vogue. Such "a species of amphibious boat, which possessed the power of moving upon land as well as upon water, for she could drive over the bed of a creek upon her wheels when there was not sufficient depth of water to keep her afloat," was the steamer Hyson which did good service in that great river war under General "Chinese" Gordon against the Taipings in 1863.[6]

The real amphibia in World War II are the LVT, Landing Vehicle Tracked (Unarmored), the so-called Alligators, the LVT (A), Landing Vehicle, Tracked (Armored), the so-called Water Buffalo, and the Army Amphibious Trucks, or Ducks. The Alligators grew out of a Roebling-designed machine used in Florida swamp lands. They are about 20 feet long by 8 feet wide, weighing 16,000 pounds when fully loaded; as amphibious tractors they can run from an off-shore transport through the water, up the beaches, across swamps, fields, rivers and other inland waters, climb at angles to 55 degrees, are equipped with machine guns and more lately light armor, can carry cargo or troops in a space of 8½ by 7½ feet. They have been much used by the United States Marines in the Pacific on such occasions as Tarawa and Hollandia.

The Water Buffalo, LVT (2), is a super-Alligàtor, carrying a 37mm gun and two machine guns in a turret. Both use their flukelike tracks, or curved treads for movement in the water and on firm or wet ground, whereas the Duck is not tracked, but wheeled and propelled by screws while in water. It is used for landing troops, carrying twenty men, or four tons of cargo, weapons, ammunition or supplies; it averages six knots in a moderate sea and is not armored.

These truly amphibious craft have their limitations. Primarily designed to bring troops through the seas as dry-shod as the Israelites and unhurt to the shore, the Water Buffalo has been used as in New Guinea to support from the water the flank of an advance on land, to lay telephone wires along the ocean shore, or overland, to reconnoitre enemy territory from the sea, that is to say, from the flank.[7] However, their maintenance presents definite problems and one terrain in particular they do not seem able to cope with fully, the Pacific coral reefs. These are of highly irregular surface, partly emerging from the sea and partly submerged with perhaps only a few inches, or more than twenty feet of water above their sharp crusts. The reefs, which often form rings around the island objective, have extremely serrated and sharp surfaces, apt to tear the tires of the Ducks to shreds and also to interfere with the tracks of the amphibious tractors.

Of all landing craft the largest is the LSD, Landing Ship, Dock, of 450 feet length. It is like the AVI, a Diesel-motored small cargoship, a mothership for smaller craft. Instead of hoisting them over the side, they can be flooded, sunk deeper and then through a pair of doors at the stern can send out the Water Buffaloes, Ducks, Tank Carriers like so many viviparous offspring. After a beachhead has been secured, the LSD may also be used as a dock for the repair of landing craft.[8]

Among other craft aside from the traditional types used, or usable in amphibious operations is a portable tugboat or "marine tractor," evolved, or employed by the water division of the Transportation Corps, United States Army Service Force. This is an all-steel craft constructed in four sections which can be carried by plane and put together within three hours. The two forward sections are pontoons and two aft sections contain the engines. The boat is 40 feet long, 15½ wide, weighs 25 tons and will tow, or push up to 400 tons at a maximum speed of 9 miles an hour as against a maximum speed when travelling alone of 14 miles.[9]

UTILITY OF PONTOONS

Of the most manifold usefulness, like the building block to the child and its appeal to his imagination, the pontoon, particularly in its all shut-in, unsinkable form, has proved of late one of the most versatile pieces of equipment in amphibious warfare.[10] Most, or many of the experiences in use, or experiment with landing craft, up to the summer of 1944 are to be embodied in a new fleet of several hundred "attack-transports, each capable of landing 1,000 men with a complement of tanks and heavy guns and of operating on its own if convoy protection is weakened" and hence heavily armed against air, or undersea attack. In spite of their **large size** these ships will be enabled to operate close to the shore and around coral reefs. They were reported as under construction in June, 1944.[11] By then, the time of the invasion of France, more than sixty "variants of these landing craft and escorting vessels" were in use on the Allied side, providing "not only for the landing of the Army, but for everything that Army needs." [12]

As a very modern part of war, landings shared fully in the speed-up that characterizes war in our day no less than industry. While combat, hydrographic and weather conditions will continue to influence the speeds of such craft, it will provide an idea of the tempi involved if we quote a classic witness, Winston Churchill, who shortly after the invasion of France watched a group of six medium landing craft "charge up in line until they were stopped by a sloping sandy beach. Down went their drawbridges, out poured their vehicles and in under five minutes the whole heavy battery was drawn up in column of route along the road for almost immediate action. In less than fifteen minutes these craft had pushed themselves off the shore and were returning to England for another assignment." [13]

While in the designing of these craft the experiences and ideas of the two partners in the war against the Axis were pooled, it is not known how the work of building the craft was divided between Great Britain and the United States. In order not to complicate the work of the established shipyards both maritime powers to a large extent employed, or set up special shipyards along inland waterways. The American building program for amphibious craft as it stood at the beginning of 1944 provided for no less than 65,000 vessels plus around 15,000 miscellaneous small craft such as rubber boats, rafts etc., for which Congress had allocated $5,000,000,000, not including the funds for their ordnance equipment.

Of the 65,000 craft, 20,000 had been actually built by the beginning of 1944, 4,000 prime contractors and 30,000 subcontractors being employed on this program which received much priority during 1943,[14] when for example during October contracts for 427 anti-submarine vessels were cancelled in order to make possible an increase in landing craft production.[15] This production shift was closely paralleled in Britain where by the middle of October it was announced that greater emphasis would be placed on the country's output of invasion craft, bridging materials, transport vehicles and other equipment "needed for the rapid movement of armies by sea, land and air, and particularly for large-scale amphibious operations." [16] Congress was liberal in its grants for this type of vessel, construction of another million tons being authorized in May, 1944, when it was told by the experts that the vessels would be used "to open new fronts on foreign soil wherever necessary." [17]

BASIS OF WASTAGE COEFFICIENT

We do not know on what a wastage coefficient, such as had been worked out, for instance, for materiel in the Sicilian campaign of 39 days, where it varied from 13 to 54 per cent according to types of guns and carriages landed there,[18] the construction program for landing craft at various times was based. It must have been rather high originally since it was based on the losses of this type during the North African campaign which were, as the Chief of the Bureau of Ships of the Navy Department declared before a House Committee in May 1943, "extremely heavy." [19]

But happily it turned out that in the American Navy's landing craft program during the six months to May 31, 1944, production due to the "combined operations" of government, industry and labor and owing also to the "gadget-free," simple construction, surpassed schedules, and that besides in the Normandy and other landings the losses in landing craft such as the LCI (L) were lighter than expected. For the time being this allowed a lessening of the pressure for building much more of this type. Hence emphasis could be shifted to articles in more urgent demand at the time, such as tanks,[20] while it was possible even to cancel certain contracts beginning in August, 1944. These are examples of the working of fluctuating demand and supply in amphibious warfare.

By this time, at last, the "critical shortage" as it is termed in General Marshall's *Biennial Report*, 1943-45, was over. While it lasted, it had dictated two major decisions of the Allied staffs—the postponement of the Normandy landings from early May to early June, even though that threw the operation into a period of more unfavorable weather, and the separation of the intended simultaneous landing in southern France from that in Normandy. Instead, the Mediterranean landing was delayed for months so that the Allies could first use all available craft in

Normandy and then rush to the south whatever was needed there.

The invasion craft have nearly revolutionized the navies in their personnel set-up if only by the sheer mass of them which have been built and used. It was stated that by the middle of 1943 already more naval personnel served in landing craft than in any other category of the British Navy.[21] This percentage can only have increased since then. In the United States Navy, which in the summer of 1944 counted 48,000 landing boats, part of the landing craft are operated by the Coast Guard who managed completely those used in the invasions of Guadalcanal, North Africa and Sicily.[22] The new types of "water bugs" are mostly assigned for handling to the "90 days'" or "180 days' wonder" type of officer, or officers of the Royal Naval Volunteer Reserve, the so-called H. O. (Hostilities Only). None of these groups from their "standing" are apt to influence operative thought greatly, and whether and how the operation of landing craft will affect that naval thought that once centered in Dreadnoughts and super-Dreadnoughts, and then swung to aircraft carriers, in part at least, remains to be seen. These officers, says a British report, "feel that they are opening up a new branch of the Navy, and that at the end of the War they will be able to hand over to the Regular officers a new branch of the Service, well established with a technique of its own. 'For don't forget that twenty or thirty years ago no one would have thought that these ships would have any use at all,' said one officer. 'Beaching is a skilled art, and has to be learned. If it is not done properly, the ships will just become stranded wrecks, and then they are as good as lost.'"[23]

NOTES, CHAPTER 48

[1] Major Klingbeil, *Küstenverteidigung.* Berlin, 1924.

[2] See art. *Amphibienfahrzeuge* in *Deutsche Wehr,* Nov. 27, 1942.

[3] U. P. despatch from Washington, Sept. 29, 1943.

[4] For more details see Lt. Earl Burton, *By Sea and By Land.* N. Y., 1944.

[5] *N. Y. Times,* June 1, 1944.

[6] Andrew Wilson, *The Ever-Victorious Army.* Edinburgh & London. 1868, 150.

[7] A. P. despatch of June 16, 1944; *N. Y. Times,* June 26, 1944.

[8] The above is largely based on well illustrated articles in *Life,* Dec. 20, 1943, and *National Geographic Magazine,* July, 1944, as well as articles by Hanson Baldwin in *N. Y. Times,* May 15, 1944 and Dec. 3, 1943.

[9] A. P. despatch from Washington, Sept. 4, 1943.

[10] For details see *Innovations of Amphibious Warfare.* By Rear Admiral Lewis B. Combs. *The Military Engineer,* Feb., 1944.

[11] U. P. despatch from Washington, June 25, 1944.

[12] Churchill in House of Commons, Aug. 2, 1944.

[13] Ibd.

[14] Statement of Undersecretary of the Navy, Forrestal. *N. Y. Times,* Jan. 10, 1944.

[15] *N. Y. Times,* Oct. 30, 1943.

[16] Statement of the Ministry of Supply, House of Commons, Oct. 15, 1943.

[17] A. P. despatch from Washington, May 14, 1944; *N. Y. Times,* May 20, 1944.

[18] *N. Y. Times,* March 18, 1944.

[19] A. P. despatch, May 14, 1943.

[20] *N. Y. Times,* June 25, 1944, and Aug. 23, 1944.

[21] The Observer, July 4, 1943.

[22] *N. Y. Times,* Aug. 8, 1943.

[23] *A Day With the Naval Commandos. The Fighting Forces,* Aug., 1943.

Chapter 49

JAPANESE PACIFIC LANDINGS, 1941 AND AFTER

IN DISCUSSING the comparative defensive strength of various types of coasts in littoral war, General Haushofer, the German geopolitician, in 1932 came to the conclusion that "in the repulsive strength of mountainous and steep coasts, easily observed from the sea, and of flat coasts, partly or temporarily flooded, swamp-protected, not easy to observe, a great change has occurred. The former have been devaluated, relatively speaking, the latter have essentially gained in their supporting strength for defense, and the earlier danger of small landings from shallow draft craft has been reduced. . . . Only try a nocturnal approach to a mangrove coast and see what happens to troops not thoroughly tested." [1] A reading of this statement causes one to wonder if the general-professor's friends, the Japanese, since 1941, had taken it as challenge and subsequently proceeded to show him what could be done on mangrove coasts by conquering British Malaya.

Mangrove beaches, the *non plus ultra* of military inaccessibility, several miles deep, are to be found along most of the western coast of Malaya, much cut up by rivers, creeks and swamps. Whereas the eastern coast, sandy and gradually sloping, swept clean by the monsoons, is more approachable from the sea. The numerous small harbors along the west coast are usually crowded with the craft of the Malay fishing and shipping fleet. The two coasts of the peninsula are separated by a mountain barrier not high, but rugged; the tropical forests on the coastal plains have been cleared for the cultivation of rubber and rice. Roads are few, but paths for those who know them are passable for infantry and animal-drawn transport, for forces sure of their objectives in and through the wilderness.

So far as they were not guided by local Japanese in Malaya, the invaders were presumably helped in this particular milieu by what may be called a re-barbarization for military purposes which the Japanese governors had systematically encouraged in their people. This was what a British general commanding in Northern Malaya meant when he called the Japanese "very formidable opponents. They combine the cunning and resourcefulness of the tribesmen of the northwest frontier of India with the discipline and direction of a modern army." Once the mountain divide between Thailand and Malaya had been crossed, the few railroads and the Perak and other rivers, forming "the most lavish water-system in the world," were used for southward transportation by the Japanese, who requisitioned the native craft and food.

Japanese-controlled Thailand, with the two east coast harbors of Songkla (Singora) and Patani, both at the head of rail lines running over passes and through tunnels into Western Malaya, furnished the base for the west coast drive, which was to be the main one, undertaken with some two-thirds of the available strength. The base for the secondary drive along the east coast was Kola Bharu on British territory near the frontier. It was on the way to these ports that the British, provided with slight air protection, lost the *Prince of Wales* (35,000 tons) and the *Repulse* (32,000 tons) on December 9, which were intended to interfere with the Japanese landings. When the Japanese sent them to the bottom, the British abandoned their already dubious naval supremacy in the South China Sea. On the western coast, along the Malacca Straits, there were as yet of course no Japanese warships to protect landings as they did at Kota Bharu.

The British ground forces were numerically equal to those of the invaders, but being spread all over Malaya from Singapore to Kota Bharu, they proved inferior in strength in all local clashes, and unequal also in jungle and amphibious warfare. When jungle warfare might halt, the Japanese would shift to an amphibious manoeuvre. Requisitioning Malay shipping, they would use small units to envelop the British left wing, landing repeatedly behind their front. Here they put in play something they had learned in China, repeating it here on a smaller scale, i.e. with smaller units, the landing forces usually running from a company to a battalion.

In addition to the native craft, some special landing craft of their own, brought overland from Songkla, was utilized. As eyewitness accounts tell "the landings are made at night from barges, towed down from farther north by tugs or launches in convoys of twelve or fourteen. The landing craft use their own motors only when fairly near the shore." When they met with strong local resistance they would grope for better landing sites farther down, or up the coast, much as did the Norsemen more than a thousand years ago on the coasts of Western Europe. On the whole, they seem to have eluded the British Navy, which "controlled" the Malacca Straits and therefore according to conventional formulas for amphibious enterprises should have made them impossible, largely by moving at night, sometimes hiding in bays and creeks during the day.

Infiltration Tactics Efficient

The frontal push of the Japanese, while persistent like tropical ants, was never of the banzai-bayonet style. It was accomplished by infiltration through lines that were kept from being continuous by the tropical forests and British numerical weakness. Psychologically more impressive attacks came through threats to the British rear where landings would carry the foe. "In most wars," wrote a *London Times* correspondent, "the soldier has been reasonably certain that his rear was secure, even if there was danger to his flank. In this campaign, the soldier has always felt, even at night, that danger lay all around him, and that he was liable to be cut off at any moment of the day or night."

Not frontal attack but constant, round-the-clock menace to flanks and rear threw the Empire forces back to Singapore. To the defenders the sea no longer provided, as in most military history, a line upon which to rest a wing, and rivers no longer presented perpendicular barriers to an attacker. Both had become elements of infiltration and disruption, running the invader's way.

The defensive strength of naval stations has never been so great as the belief in them has gladly attributed to them. In not many cases has it actually been tested, such as Malta, Gibralter, Sebastopol, Port Arthur. Since 1905 no serious testing of the resistance of a naval station had been attempted. Still, trust in their inherent strength was weakening after the close of the World War I when the French Admiral Castex put the case against them in this formula: "A naval station which cannot be secured against *all* modern weapons, must be considered as non-existent," a conclusion in which the Germans were ready to share soon after the start of World War II.[2] For long, such stations seemed to be most secure when their hinterland was occupied by weak, although perhaps hostile nations, or comprised an impassible tropical wilderness as in Panama or at Singapore.

Serious strategic errors were committed by the Tory governments in Britain when they allowed the potentially aggressive and inimical Franco government to take over the hinterland of Gibraltar, Spain; when they suffered Italy to undertake the Ethiopian campaign, which incidentally threatened Malta with landbased air power, and when they declined to interfere with Japanese aggression in northern Asia, whence it spread south and eventually to Hongkong and Singapore. Neither of these strongholds was specifically prepared against the threat from the air and still less against menace by land. "The Imperial plan for the defense of the island of Singapore appears to have been a makeshift. The situation in which the island now found itself had never been foreseen, or at least had never been provided for" (Col. Thompson).

Neither it, nor Hongkong had even a clear glacis. Singapore was closed in by the edge of a primeval forest. At Hong Kong encroaching population density impeded the range of firing and detection. Their harbors, destined for the largest ships, had not been defended against storm by small "special landing craft." Although both naval bases, to which might be added Penang in the Malacca Straits, which the Japanese neutralized in their sweep along the western coast of Malaya about December 18, were situated on islands, the intervening waters were but narrow, 500 yards at Hongkong and from 1,000 to 3,000 yards in the Straits of Johore, 30 miles of the waters which separate Singapore Island from the peninsula.

The captures of these places upset one conventional belief after the other. As if to mark the end of Christianity grown lazy in the East, the Japanese started the assault on Hongkong on Christmas Eve, at the unconventional hour of 9:30 instead of shortly before dawn, as was usually expected in case of a landing. By 9:50 the landing on Hong Kong island was effected and by 7:30 A.M. enough Japanese had been brought over to take this white enclave in the Asiatic yellow co-prosperity sphere.

By blasting the causeway to the mainland on the night of January 30, Singapore was once more made an island, for a few days. Promptly the Japanese deployed their divisions along the opposing shore where the Straits are narrowest, throwing across a heavy and constant artillery and aerial bombardment on the weary defenders who had to guard a 45 miles long coastline, much indented, and although in part of mangrove beach character, offering numerous good landing shores for large forces. Confident that the troops who had fought successfully in the gloom of the jungle were prepared for full nocturnal combat, the enemy during the night of February 9-10, before the moon's rising at 1 A.M., started crossing in force, using once more that "special landing craft" which the Japanese had employed before without having moved the Western powers to imitate it—for weren't the Japs, according to legend, original and constant imitators?—but of which they now began to take serious notice.

The landing, undertaken on a 5,000 yard front, was effected and the British forces were now so spent that no daylight counterattack on the 10th was possible. On that, or the next day, the causeway was repaired and tanks were brought over by the Japanese, who pushed the defenders steadily back to and beyond the naval base, the RAF airport, the Radio Station and into the city, leaving them little chance to evacuate. There was no alternative save unconditional surrender on February 16.[3]

"Insecure naval stations," "insecure islands," the Axis powers rejoiced as the Japanese continued to land on numerous islands of the Pacific, a process they had begun on December 9, two days after Pearl Harbor, and for which they had meticulously prepared still earlier. "We will always strike first, we will always deal the first blow," Hitler declared jubilantly in the Reichstag (Dec. 11, 1941).

The "we" was a distinct euphemism for Axis disunity in their so-called world conquest. For the Japanese had not struck in unison with Hitler, or in keeping with any long range combined planning. But it was not then the time to state, as some Germans have done subsequently, that they erred in not following up the first blow with the logical second, particularly as regards Hawaii, where the air attack should have been complemented by a landing. Or to put it in General Marshall's words: "The Japanese strategic plan initially failed when she missed the opportunity of landing troops on Hawaii, capturing Oahu and the important bases there, and denying us a necessary focal point from which to launch operations in the Western Pacific."

The Germans and the Japanese, as the makers and, in fact, the theoretical and practical conceivers of this war demonstrated almost all these first strokes and, allowing their enemies more time than they had originally assigned them, taught the latter how to strike back, how to specialize in equipment and training, even surpassing the aggressors. And the Allies had more material for such warfare. Both the Axis powers assumed a *j'y suis j'y reste* attitude, in a German-controlled Europe, in a Japanese Asiatic co-prosperity sphere, which their antagonists, unlearned in the arts of landing, were supposed to be incapable of penetrating.

Loss of the Philippines

As the emancipator of the Philippines, the intention of the United States was to give that commonwealth security in addition to independence. This security was to be built over a period of ten years, to 1946, during which time a United States force was retained there to enable the Filipinos to effect an orderly transition to independence in all its aspects. The former Chief-of-Staff of the United States Army, General Douglas MacArthur was lent to the Commonwealth to develop "a reasonably adequate defensive system in the Islands, important to the United States as well as to the Filipinos themselves," comprising an army, rather than a navy able to repel attack; a navy being out of the question because the young republic was too poor to afford either that, or a considerable air force. The army was, however, to be complemented eventually, by 1944, with small torpedo craft for the purpose of denying a hostile fleet the use of territorial waters and maintaining communications between the islands. If his general defense plan were followed, the Philippines, according to MacArthur, could "rest in perfect security. It would take 500,000 men, $10,000,000,000, tremendous casualties and three years time successfully to invade the Philippines." [4]

Looking forward to the results of his ten years' plan of defense, MacArthur found at once available "the first element of strength" in "the geographic isolation of the Philippines" (isolation from what?); exposed, with the possible exception of air raids, only to an overseas expedition with all the difficulties characteristic of such enterprises.

"All these difficulties of transportation, supply, protection, and maintenance are intensified the moment that there is to be anticipated even slight resistance at the shore line. Tactically, the most difficult of all operations is that of attacking, from small boats, a force defending a coast." While a prospective enemy would try to land in undefended areas, making use of his higher manoeuvrability on the water, the Philippine Plan was intended "to assure an active and carefully planned defense of every foot of shore line in the inhabited islands of the archipelago." (We do not know whether the total shore line of these islands has ever been measured and hardly dare to raise the question how many soldiers there would be, in 1946 or earlier, per miles of coast).

Each such island was to be supplied with a garrison of sufficient strength, providing concentration of strength that could be thrown against an invader over a system of roads and other communications. In how far MacArthur's plan for the security of the Philippines, which had run half of its allotted time, and perhaps had been accelerated by the outbreak of the war in 1939, in turn gave the Japanese aggressor the idea of starting a preventive action against the commonwealth at its halfway station to independence and security, we shall not know for some time.[5] The approximate strength of the United States Army that had been mustered by December, 1941, was 19,000 troops, including 8,000 airforce men with 250 planes, 12,000 Philippine Scouts and 100,000 active soldiers, only partially trained and equipped.

United States naval forces in the Far East were one heavy and two light cruisers, 13 destroyers, 20 submarines, etc., and some thirty bombers.

A state paper for public information, like MacArthur's, is not the place for high authority to admit even tentatively the possibility of a hostile invasion of the archipelago, or the mere delaying function of the island military group in the total strategy of the United States. This would have vitiated its purpose. Other American officers were openly doubtful about the issue of an attack on the Philippines. While some army men, particularly among those in the United States, thought successful attack not impossible, others, on the ground in the Philippines, as well as officers of the United States Asiatic squadron, believed in the possibility of successful defense, the latter basing their belief on the insufficiency of Japanese shipping required for the transport of an army of 150,000 men over a distance of 1,000 to 1,500 miles—which would presuppose a basing of the invasion on the Japanese homeland, a rather unwarranted supposition by the autumn of 1940—taking at least 6 tons of equipment and ordnance per man for two months.[6]

The surprise, based on "this breach of international faith and honor" in the language of the Roberts Report on Pearl Harbor, which the Japanese had in store for the American nation, and especially for the defenders of the Philippines, was quite as stunning as to the commanders at Pearl Harbor. Although the former were closer to the main mass and bases of the one possible enemy and had about four hours warning about what the Japanese had done in Hawaii, Japanese planes were able to knock out MacArthur's largest airbases, such as Clark and Nichols fields near Manila, where they found all planes grounded and in parade formation. With most of the American airpower eliminated during the first 24 hours, although some planes survived to sink a Japanese battleship—perhaps—etc., the invaders could begin to land troops from the sea without much immediate interference. They began with two landings on the northern-most tip of Luzon, the main island, on December 10, but were driven off with heavy losses in an attempt on the western coast of the same island.

The landings in the north were preceded by air bombardment, the bombers, based on Formosa, paying particular attention to the small airfields in the vicinity. There was practically no resistance on the ground, since very few troops were stationed in northern Luzon, separated from the Manila region by high mountains. These first beachheads gave Japanese airpower its initial *pied-à-terre* in the archipelago, allowing pursuit planes to shift from the carriers to land fields and enabling their land forces to spread out along the western coast in the direction of the Lingayen Gulf.

Encirclement of Luzon

By landings in the latter region as well as south of Vigan and north of San Fernando (both on the west coast), by parachutists operating in the extreme north and north east of Luzon, by small landings in the extreme south, by naval concentrations off Zambales Province on the west coast, all reported on the 12th, the Japanese indicated an encirclement of this most important island, to be followed by a concentric movement on the Manila region. The forces landed were met in their inland advance by American and Filipino troops who repulsed a larger landing attempt in the Lingayen area on the 14th. These original landings on the outmost perimeter of the Luzon defense were followed a week later by landings much closer to Manila, its center, which was intermittently subjected to heavy bombing. Little seems to have been published of what happened during the week after the 14th, as to whether the Japanese paralleled the amphibious tactics which they pursued in Malaya, though they lived off the invaded country from the outset. At Lingayen, 150 miles north of Manila, they staged a landing in force which was to be the main one.

Escorted on the water and in the air, a transport flotilla of 80 vessels estimated to carry from 80,000 to 100,000 men, approached the coast and promptly sent a large number of 150-man barges into Lingayen Gulf. "Some of them succeeded in getting ashore . . . . ., met with fierce resistance" by greatly outnumered American and Filipino troops (Dec. 22). In this fighting, on the 23d, the invaders began to use light tanks and came to within 100 miles northwest of Manila, at the head of the Pampanga valley. During the night to the 24th large forces were landed 75 miles southeast of Manila from some 40 transports while the presence of other troopships presaged landing attempts south of the capital that were to take place later in the day near Nasugbu, on the west coast, 50 miles from Manila. Altogether by Christmas eve some 100 enemy transports divided into several convoys, each protected by naval and air escorts, were in the waters around the islands, among which Mindanao was also attacked.

Pressure was constantly increased, supported by new reinforcements, including tank regiments and horse cavalry, from the Japanese transport fleet in the Lingayen area and on the coast southeast of Manila which, after undergoing stepped-up bombing, was declared an open city by the end of the year. MacArthur's strength was divided between the Pampanga province to the north and against the landing force to the south of the capital (Dec. 29).

On the first day of the new year all remaining American and Filipino forces were united north and northwest of Manila, the city falling on January 2 as "no close defense within the environs of the city was possible." The island fortress of Corregidor and other places in and on Manila Bay remained in American-Filipino hands, thus denying the enemy ships the use of the bay. When the Japanese land forces from the north and southwest met back of the bay, the defenders had evaded them, moving into "previously prepared positions" in the Bataan Peninsula. The remaining bombers were ordered to base themselves on Mindanao and from there support the defenders of Bataan as best they could.

Final Stages of Defense

This ended the first stage of the defense of the Philippines. The second, fought in the Bataan Peninsula in the vicinity of coastal waters including Subic Bay, where the Japanese landed reinforcements on January 22 and 24, was hardly less amphibious in character, and included jungle infiltration, trench and siege warfare and a very effective blockade. There were on many occasions heavy infiltrations along the beaches as well as in the mountain passes of Bataan, the former undertaken with naval and air support which cost the defenders, although they successfully counter-attacked, heavily. Two night landing essays on the west coast of Bataan on February 2, the first made by special shock troops early in the evening and the other at midnight in the customary "specially built barges" proceeding under naval escort, were completely frustrated, the second being detected by American night flyers.

Not until the last day of March did the Japanese, meanwhile grasping at Pacific islands as far west as the Andamans and east as Midway, although that was held against them, resume their attacks, in the meantime forcing the besieged to consume the half rations on which they had been subsisting since January 11, with a further cut at the end of March. Conditions under which the siege of Bataan was waged, involving a fairly dense military occupation of the peninsula, made the success of further Japanese landing attempts ever less likely. Those undertaken during the two nights from April 4 to 6 with a considerable number of barges, carrying troops and mounting 75mm guns which were used to shell the beach defenses, failed. "They shall not land." The crushing of the defenders' strength in Bataan was due rather to hunger and weariness, artillery and air bombardment which forced them to surrender on

April 9th. Corregidor fell on May 6, 1942, after small parties of foes had landed, and Mindanao a few days later.[7]

Humiliating as were these defeats in Malaya, the Philippines and elsewhere in the Pacific, with their high losses in men, territory and materials, if they are regarded as delaying actions, fought while mobilization by the United States slowly got under way, they served a great purpose. They enabled the Allies to establish a front on the perimeter of Japanese aggression from Burma to the Aleutians, a nearly perfect semi-circle around Tokyo, as if to please the geopoliticians, or Japanese symbolism. Further, beyond this arc, the rising sun was not to rear itself.

NOTES, CHAPTER 49

[1] Karl Haushofer, *Wehr-Geopolitik.* Berlin, 1932, 79-81.

[2] See *Unsichere Stützpunkte,* by Konteradmiral Gadow. *Deutsche Allgemeine Zeitung,* Dec. 23, 1939.

[3] Lt. Col. Paul W. Thompson, in: *How the Jap Army Fights,* ch. 5.

[4] A. P. despatch from Manila, May 29, 1936.

[5] For the text of the Philippines Plan see Waldrop, *MacArthur on War,* 308-48.

[6] *N. Y. Times,* Oct. 8, 1940, despatch from Manila.

[7] The above is largely based on the communiqués of the U. S. War Department, General Marshall's *Report on the Army, July 1, 1939, to June 30, 1943,* and Lt. Col. Allison Ind, *Bataan; the Judgment Seat.* N. Y., 1944.

Chapter 50

ATTU AND KISKA

AMONG THE VARIOUS REGIONS of the earth that seemed too forbidding before 1939 ever to become theaters of war, was Alaska. Such islands as Attu were not even mapped, beyond the shore line. Still it received a little attention by exponents of bold and unconventional strategic thought, to an extent which it perhaps never quite deserved. "Billy" Mitchell was no less fascinated by its potentialities than a Nazi general, like von Reichenau, who told an infantry officers' school of the Reichswehr in 1938 that he considered that the full weight of the Japanese attack on America would be carried by a "hop-hop" attack on Alaska and farther south, advancing from one port to another, from Dutch Harbor to the mainland. There they would divide their forces, into a northern one to seize Nome and the Bering Straits and thus interrupt all supplies to Russia, while the main body would move on the continent in the direction of Vancouver Island. Such a plan, he thought, would be in keeping with the general Japanese strategic idea of infiltration, and avoid any serious operations on the American mainland.[1]

The Japanese clearly had no such intentions and their subsequent landing in the Aleutians proved nothing of the sort that the German had mapped out. Among the reasons to the contrary may well have figured their own conviction that they were far better tropical than Arctic fighters. While many of them considered the tropics as the ancient home of their race, where they were able to exist on a lower subsistence level than their enemies, at the same time they knew fighting in the far north would favor enemies who could procure better protection from their industries against the cold in the way of clothing, shelter, food, etc., than against tropical conditions.

The Japanese occupation of the Rat Islands in the Aleutians, Kiska and, somewhat intermittently, Attu, (June 7, 1942) at the tip of the Aleutians, and Agattu, can never have had any other purpose than to pin down considerable American forces in and near this region and keep them from other, more immediately important zones in the Pacific and elsewhere.

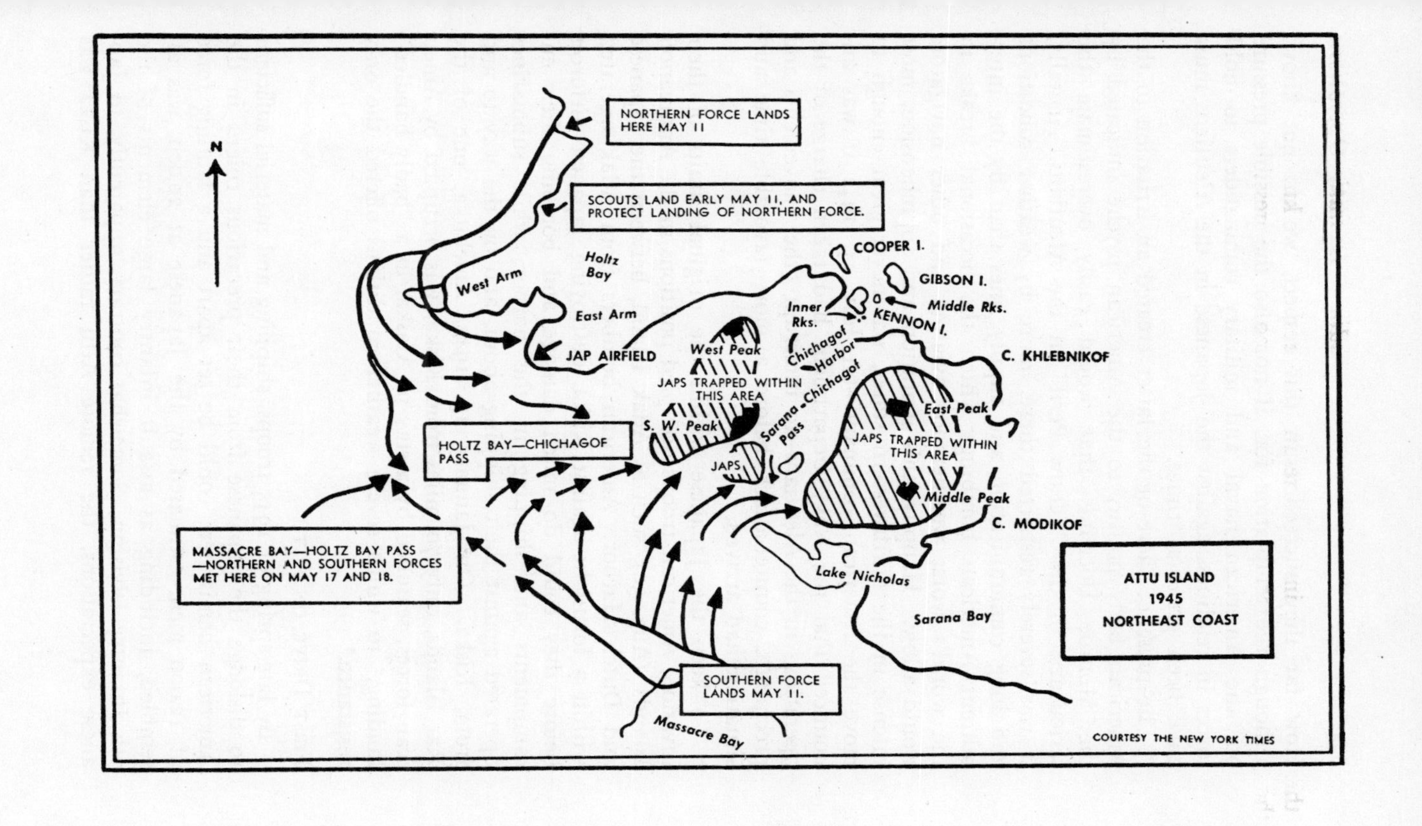

N
NORTHERN FORCE LANDS HERE MAY 11
SCOUTS LAND EARLY MAY 11, AND PROTECT LANDING OF NORTHERN FORCE.
Holtz Bay
West Arm
East Arm
JAP AIRFIELD
COOPER I.
GIBSON I.
Middle Rks.
Inner Rks.
KENNON I.
Chichagof Harbor
C. KHLEBNIKOF
West Peak
JAPS TRAPPED WITHIN THIS AREA
S. W. Peak
Sarana-Chichagof Pass
JAPS
East Peak
JAPS TRAPPED WITHIN THIS AREA
Middle Peak
C. MODIKOF
HOLTZ BAY—CHICHAGOF PASS
MASSACRE BAY—HOLTZ BAY PASS —NORTHERN AND SOUTHERN FORCES MET HERE ON MAY 17 AND 18.
Lake Nicholas
Sarana Bay
ATTU ISLAND 1945 NORTHEAST COAST
SOUTHERN FORCE LANDS MAY 11.
Massacre Bay
COURTESY THE NEW YORK TIMES

How far they succeeded in this respect, we do not know, although the temptation for, if not also the possible pressure on, the American naval and military commanders to hold forces in readiness against the Japanese in the Aleutians must have been great at times.

The presence there of the latter created an irritation to the American body politic, to the American psyche indicated by the Monroe Doctrine, that would easily overestimate the seriousness of the "Yellow Peril" in the Aleutians. Actually, it was severely restricted there, more by weather conditions and long communication and supply lines than by the intermittent American bombings which the occasional breaks in the world's worst weather for aviators and other navigators would allow. Happily, this irritation and apprehension, most intense in the northwestern states, was never great enough to move the American command—on July 24, 1942 it was announced that the Navy Department had taken charge of the operations in the Aleutians—to interrupt other movements and precipitate counter-attack before a more favorable time and situation had arrived.

To keep the Japanese within the original limits of their invasion, American troops occupied positions in the Andreanoff Islands (Aug. 30), such as Adak Island, between the Japanese and Dutch Harbor. American bombers from Adak operated within a fortnight against Kiska, destroying Japanese airforces before they could do much damage and bombing their encampments and shipping in the harbor, while submarines operated against their shipping about, and on the way to and from, Kiska. On January 13, 1943, Amchitka, one of the Rat Islands, only 70 miles from Kiska, was occupied by American forces, seasoned by a stay in Alaska, in a "badly handled" landing,[2] to which severe weather conditions offered the only resistance.[3]

First Drive on Attu

By late spring of 1943 troops, shipping and materiel sufficient to dislodge the Japanese from their precarious coigns in the American hemisphere could be set apart and a landing force of 12,000 men, estimated by the Japanese at 20,000, was assembled, including as assault infantry large elements of the 7th Infantry Division. Somewhat contrary, apparently, to Japanese expectations, the remote Attu rather than Kiska was chosen as the objective, probably for topographical reasons since the shores of the former, while formidable in their difficulties, appeared not quite as forbidding as those at Kiska which besides was far more strongly held, by an estimated 10,000 to 12,000 as against some 2,500 on Attu. In either case, the coasts are rocky, surf battered, the land rising immediately from the beach to more or less steep mountains which seemed to call for the employment of those mountain troops which the United States Army was so reluctant and slow to build up and even slower in using.

There was deep snow in spots and practically no roads existed in the island. The operation, like all those in the Pacific with the exception of MacArthur's later *Siegesallee* to Manila, fell under the jurisdiction of the Navy, with a major general of the Army taking over command once the troops had been brought ashore, and in the purely military sector of operations. The naval hierarchy of command ran from the commander-in-chief of the Pacific Fleet, to an area commander with "over-all responsibility for coordinating and implementing the various forces involved," and a commander of the Amphibious Forces of the Pacific Fleet, with the immediate responsibility of transporting troops over water to the points of attack and landing them on the beachheads. In reviewing their work the late Secretary Knox declared that it "emphasized the close integration of branches of armed services which must be effected in present-day warfare." [4]

After a long voyage, prolonged by three days through fogs, during which the men were exercised in callisthenics and by the study of relief maps made still more familiar with the terrain to be taken, the force was landed on May 11, instead of the 7th as expected, in weather bad for almost any other region except the Aleutians, and after a bombardment from ships and air by which, according to an order of the commander of the Japanese garrison of May 29, the battalions on the first lines of the land front were almost annihilated. As the number of missing Americans, stated as 58 in all, would indicate, comparatively few of the landing troops were either drowned, or hit by the Japanese while making their way to the shore.

The long Arctic day in spite of poor visibility gave the troops some light until nearly 2330, while fogs often limited vision to 100 yards and less. The terrain, with no growth of trees, a tundra which would only from July to September provide cover for man, favored a defender relying largely on small arms fired by troops in prepared and well camouflaged ditches and foxholes. While the Japanese in the island fought tenaciously, it was evident from their defense arrangements that it was their presence in the Aleutians that counted rather than the strength of any positions occupied by them. For nothing had been done by them during the 11 months of their stay that pointed to further aggressive intentions, such as the building of airfields might indicate.

Two pincer-like landings from the north (Holtz Bay) and the south (Massacre Bay), from whence the main force advanced, cut off the eastern fifth of the Island where the Japanese preferred to make their stand. Scouts landed early in the morning of May 11, to protect the landing of the main northern force, taking place later in the day more to the north along Holtz Bay. No opposition was met on the beaches. The two forces joining compressed the Japanese very slowly—taking seven days instead of the expected one and a half—but relentlessly in the eastern part. After a last counterattack in which nearly all remaining Japanese forces were rounded up and killed during the night of May 29, organized resistance ended.

If they had lived to see the investment against them of comparatively vast American forces and materiel, thanks to which a fighter strip was completed within twelve days after the landing, they would probably have considered their presence well justified.[5] As it was, very few lived to see and judge, only 28 prisoners surviving out of a garrison of some 2,400 men. Some few fought on until June 9, although the fighting in the main had come to a close on May 31, after three weeks.[6] American casualties were given as 342 killed, 58 missing and 1,135 wounded at the time, although General Marshall in his subsequent Report speaks of a loss of 512 soldiers. No warships were sunk or damaged during the operation, one transport was beached, but got off safely. Not many illuminating details have been published about the taking of Attu. But those in a position to learn some of them have declared that "the planning was confused and too little coordinated; the troops' clothing was not the proper arctic type."[7]

After the drama succeeded farce, for the Japanese in their

final communiqué on Attu insisted that Kiska was "definitely maintained by Japanese forces." The Allies believing this, sizeable assault force of fifteen thousand Americans and five thousand Canadians, and the necessary shipping, nearly 200 vessels, and airforces were gathered to dislodge them from Kiska. After a naval bombardment of some strength, thousands were landed on August 15 without finding one Japanese, none of the 10,000 who were expected to be encountered.

All Sunday from early dawn the search went on, and on Monday further thousands followed, proceeding from yet another bay. No enemy, no resistance, except from the *topos* and its nature. The Japanese, leaving behind installations which were declared by the conquerors to have been designed as "permanent," but which had not been damaged severely by our preceding bombardments from the air and the sea, had gradually and systematically evacuated the island after they had recognized that it had become untenable and that their invasion had served its purpose. They had taken away even their landing barges. Our aerial reconnaissances had not detected them, although they had taken place ten to fourteen days before the Allies' arrival, the Japanese radio having, as was later remembered, shut down on August 1. Still, Acting Secretary of War McCloy, upon returning from Kiska, declared it had been a question "of taking no chances."

The American tactical principle of safety first, often sound enough in the way of limiting risks, was in this case carried to an amusing extreme. Instead of sending ten thousand men, as it turned out, ten could have done the job. Instead of first sending a small waterborne patrol into the island to substitute for, or complement, aerial reconnaissance, and gather the information which the airmen had not delivered, the whole big enterprise was staged and performed as if desperate and bloody resistance was certain. "It may have been," declared Mr. McCloy, "that the operation could not have been changed even though it was wanted to do so." This reminds the historian of the younger Moltke's reaction to the Kaiser's request in 1914 that mobilization in the west be stopped: "Once planned, it could not possibly be changed." Thus, "the island was occupied, precisely as if the Japanese had still been there." [8]

After a slight hesitation on the part of those higher up, a bloodless victory, "beautifully carried out," a battle which the enemy "could have fought," was declared to have been won. It was said, judging from the strength of Japanese defense installations, to have spared the Army up to 7,000 casualties, a potential price for Kiska which incidentally came close to what the Japanese had claimed as the price paid by America for Attu.

But no saving of red faces seems admissible in spite of all ingenious statements, including one about the commanding admiral's "third Pacific victory." Kiska was not merely "a waste of munition"[9] but also of time in which the highly trained men and the ships tied up by the operation could have been more profitably employed elsewhere against the Japanese, who deserve credit for having so cleverly misguided a considerable force of their adversaries.

NOTES, CHAPTER 50

[1] Cit. *The Fighting Forces,* Aug., 1943, 126.

[2] Hanson Baldwin in *N. Y. Times,* Sept. 1, 1943.

[3] See despatch, dated Jan. 28, in *N. Y. Times,* May 9, 1943.

[4] Ibd., June 5, 1943.

[5] There were actually 50,000 soldiers, sailors and marines on Adak Island in August, 1943, shortly before the attack on Kiska. Foster Haley, *Pacific Battle Line,* 340.

[6] Some tactical details about the fighting on Attu are in *Infantry Journal,* Aug. and Oct., 1944, more in *The Capture of Attu.* Prepared by the War Department. Washington, 1944.

[7] Hanson Baldwin in *N. Y. Times,* Sept. 1, 1943; more details in Hailey, 367-8. Brodie, 183.

[8] *N. Y. Times,* Sept. 3, 1943.

Chapter 51

PACIFIC ISLAND-HOPPING, MOPPING-UP AND LEAP-FROGGING

AFTER THE LOSS OF THE PHILIPPINES, Guam, Wake, the Andaman Islands, the Netherlands Indies and the largest part of New Guinea and the islands around it, the initial obligation of the Allies was to establish a wide-flung Pacific front, far from Tokyo. While this front was vital, still it was, as the Churchill-Roosevelt conversations of January, 1942 decided, of secondary importance as compared with the front, or fronts against Germany, the prime enemy.

In January, 1942, United States air and ground forces began to arrive in Australia, threatened for a time by inclusion in what the Japanese called their "Co-prosperity Zone." In mid-March MacArthur was evacuated from the Philippines to Australia to become "the supreme commander in that region, including the Philippine Islands, in accordance with the requisition of the Australian Government," a statement of the United States War Department which the President at once elucidated by declaring that the General would command everything, including sea and air forces, east of Singapore in the Southwestern Pacific. This Southwest Pacific Command of the United Nations was sharing the ocean with a Southern Pacific Command, with a combat sector which included New Zealand, somewhat against that commonwealth's wishes as she would have preferred to remain closely linked with Australia in one defense area under MacArthur. However, "strategic considerations as determined in Washington," the New Zealand prime minister stated made this impossible. Not a few persons at the time were inclined to read "service considerations" instead.

For several months the Americans continued building up their bases, and while the air forces participated in the defensive struggle off Australia at an early date, not until July 11 were their ground forces reported as joining in the New Guinea jungle fighting. American troops in April had arrived in New Caledonia under agreement with the Free French local administration, and before that time in India. Since Madagascar presented "the definite danger to the United States of occupation or use by the Axis power, especially Japan," British forces early in May began to land in that island which was not fully wrested from its Vichy garrison until November.

There was but little interference on Japan's part with this laborious drawing up of a Pacific front until the naval battles in the Coral Sea (May 7-8) and at Midway, four weeks later, when the Japanese were en route to Midway and perhaps to Oahu. Having kept the Allies guessing for some time as to whether they would continue against India, or Australia, the latter direction seemed indicated by their occupation of the Kai, Taninbar and Aroe Islands; the advance across the Owen Stanley Range in New Guinea in the direction of the United Nations base at Port Moresby; the landing in Milne Bay at the southeastern tip of New Guinea (about August 25); and by the countering of first American landings in the Solomons by additional Japanese landings in the same archipelago (Sept. 3).

For the time, the two Allied main bases across the Pacific, one in India, practically outside that ocean, and the other in the South Pacific, were not unlike two millstones placed so far apart that then they could not soon begin grinding the enemy between them. The active opposition—too early to deserve the term offensive—on the part of the United Nations from the south got under way with the American landings in the Solomons (beginning August 7) and the confused battle for Guadalcanal. This investment of the then strictly limited American forces indicated that naval operative thought directed at an island-bound campaign had been given preference over the far more land-bound campaign for New Guinea; that Tokyo rather than Manila was the preferred ultimate objective. The raid undertaken by 200 United States Marines against Makin in the Gilberts presented an enterprise in the Commando style which on the whole has not been considered practical in the Pacific area, where raids were naval and above all naval-aviatory, dispensing with landings.

Following the period of the last centrifugal push of the Japanese along some of the radii centered at Tokyo, a very slowly increasing centripetal pressure was exerted along some other radii of the same circle. This led to campaigns, in China, along the Burma-India frontier, in New Guinea, which halted the Japanese and indicated that they had lost the race for India and Australia. After that, as subsequent to the dash for the Channel coast in 1914, war of *usure,* or attrition, began, wearing out men and machines—the Japanese often enough wearing out men instead of machinery. That in the long run could only favor the United Nations, advantaged by the infinitely greater industrial output which the United States could throw into the Pacific, as compared to the enemy. It was *usure* on land and sea, and in the latter element going so definitely against the Japanese that by the spring of 1943, they retired their battle fleet units towards the home ports, *usure* at the very far ends of the equidistant communications lines from Yokohama and Pearl Harbor to the Solomons, long distance being in itself a very potent element in *usure.*

FAST START; SLOW ENDING

Eventful highlights in this period were marked by the inclusive dates of August 7, 1942 and February 9, 1943, when Guadalcanal was finally taken. The conquest or reconquest began with speed and ended slowly. The first landing was a complete surprise which still held good on the 8th when the Americans captured for use of their planes Henderson Field, which the enemy had thoughtfully hewed out of the jungle. The expected Japanese counter-attack came during the night of August 8-9, when they caught off-guard the Allied naval forces that had retired their aircraft carriers behind the island for reasons of "security."

In the Battle of Savo Island, they sank four heavy cruisers and did other damage, without, however, attacking the American transports, and landed 1,000 of their best troops on Guadalcanal. With the help of these and other forces they came dangerously near to recapturing Henderson Field, as they tried to do, after a heavy bombardment from sea and air September 12-14. On its possession depended the air domination over the southern Solomons and the surrounding waters, as well as the umbrella of naval forces and aerial reconnaissance spread over and around them in these waters.

Here the United States Navy was to learn that, as Admiral King puts it in his report of April 23, 1944, in this war victory depends, first, on air power which "more often than not" predominates, and second on the closest coordination of land, air and sea power. About this much had to be learned since the first days of Guadalcanal, and not a little from the Japanese who owed, as the Admiral stated, their early successes to such cooperation, and who were far more ready to use ship artillery in support of land operations, as they did during the attack on Henderson Field on September 12-14, than were the Americans at the outset.

The Marines on Guadalcanal, who were left "without air cover even before disembarkation had been completed," [1] and who found little, if any, naval gunfire support, felt as if deserted by the Navy for days on end; a feeling of being expendable and doomed that appears in so many accounts and memories of Guadalcanal as well as in some psychoses acquired there.[2] As King's report frankly states, the Japanese counter-surprise of August 8-9 and the ensuing naval defeat were "the result of a combination of circumstances. Because of the urgency of seizing and occupying Guadalcanal, planning was not up to the usual thorough standards. Certain communications failures made a bad situation worse. Fatigue was a contributing factor in the degree of alertness maintained. Generally speaking, however, we were surprised because we lacked experience. Needless to say, the lessons learned were taken fully into account."

Shock, educationally absorbed, is generally not as expensive a schooling in war as *usure* is, of which the parties on, and around, Guadalcanal came to endure so much. For ample reasons the surrounding waters were called the "iron-bottomed sea," where 28 American ships of all types rested as against 57 Japanese, together with some of the 200 planes which were lost, as against nearly 800 of the Japanese. Casualties were high, even if they were only one-tenth of those on the Japanese side.

Beneath the tropical foliage of the scene and on and above the waters, one might detect many a feature of the attrition war waged on the western front of World War I. Like the occupants of the trenches of the western front, who were, however, largely under *overhead* fire from land artillery only and not also ship fire and air bombings, the Japanese lines across the islands received reinforcements, reliefs, materiel under cover of night. "The enemy has reinforced and supplied his units by means of small craft which approach the shore under cover of darkness. Despite opposition to these landings, it has not been possible to prevent them entirely," ran a typical Washington communiqué about Guadalcanal (Sept. 12, similarly Oct. 5 and 6).

Large scale attempts on the part of the Japanese to bring in reinforcements were frustrated, their losses on one occasion alone being estimated by the general commanding the United States Marines at 30,000 in twelve transports in November. After that encounter "the Japanese kind of lost interest in sending more men down there,"[3] and let the campaign be fought to the end by their men on the spot who were used up—as far as they were not evacuated—by February. Guadalcanal was an island arena in which the two antagonists slugged it out as if they had met there by silent agreement.

One still finds it impossible to see how far this *usure* at sea, on land and in the air close to the seaboard was avoidable, due to the insufficiency of forces on either side to reach a decision; while not quickly decisive, this was by comparison less costly to the United Nations than a larger investment in the fighting in New Guinea would have been, for that might have extended too far away from the sea to permit the Allies to bring their increasing water-borne strength into play against the jungle and the jungle fighting Japanese. It seems likely that inexperience and unwillingness to learn from earlier British experience, or shortcomings in cooperation between American naval and military forces following the landing surprise, which had been so perfect that not one marine was lost, may have prolonged the ordeal. Indeed, the Japanese felt so emboldened at times, as after the naval battle of November 13-15, that they declared that "the fate of the United States landing force on Guadalcanal is in the hands of the Japanese forces."

Returning from a tour of duty in the Pacific, Representative Maas, a colonel in the MC Aviation Reserve, stated that "things have not gone well in the Pacific, and there just is no unity of command in the Pacific" (November 12, 1942), an admission which did much to bring to an end the so to speak amphibious service done by American parliamentarians, in their double quality as civilians and soldiers, in itself not a bad institution for a democracy.

There were definitely anxious moments on Guadalcanal, as for instance during the night of October 13, when Japanese aircraft and heavy naval units bombarded Henderson Airfield and left only one bomber and a few fighters intact to oppose the Japanese landing operations on the following day.[4] It was a somewhat belated, or unwillingly recognized, discovery made on this and later occasions by American naval commanders of what ship-based artillery could do for, or against, one's own land-fighting forces. This hesitancy to use naval guns against shore-based guns still looms large in such a statement as Admiral Nimitz's after Tarawa and Kwajalein that: "You can't fight ships against shore-based naval guns, because you can't sink them, and they can sink you. We may expect such installations at some of Japan's major bases." [5]

This reluctance to face naval guns ashore, with no qualification as regards caliber, is more easily understood from the traditional shore gun-shyness of old-school mariners who either invoke Napoleon, or Nelson, rather than from the actual experiences in the Pacific where to the summer of 1944 American forces seem not to have encountered land based guns of much more than 8 in., and few enough of them.[6]

NOTES, CHAPTER 51

[1] Bernard Brodie in *Virginia Quarterly,* Autumn, 1944, 499.

[2] See Lt. Commander E. Rogers Smith's account, *N. Y. Times,* May 12, 1943.

[3] Statement of Maj. Gen. Vandegrift. *N. Y. Times,* Feb. 12, 1943.

[4] Brodie, *Guide to Naval Strategy,* 2nd ed., 224.

[5] *N. Y. Times,* Feb. 9, 1944.

[6] Ira Wolfert, *Battle for the Solomons* (Boston, 1943) and Foster Hailey, *Pacific Battle Line* (N. Y., 1944) are of limited helpfulness in illuminating the problems we are here concerned with. The same might be said of Frazier Hunt, *MacArthur and the War Against Japan* (N. Y., 1944), and Joseph Driscoll, *Pacific Victory, 1945* (N. Y., 1944), the former a MacArthur partisan, the latter a pro-Navy work.

Chapter 52

U. S. NAVY'S PACIFIC AMPHIBIOUS OPERATIONS

"It has always been the concept of the United States Chiefs of Staff that Japan could best be defeated by a series of amphibious attacks across the far reaches of the Pacific."
Gen. Marshall's *Biennial Report, 1943-45.*

FEW OF THE TRANS-PACIFIC ROUTES of war trended westward. Perhaps there were only two or three. It was probably necessary to divide the Pacific as a theater of war into two zones. But this quite unavoidably led to a difference of views as to "whether the main Allied drive should be made via the Central or South Pacific"; this in turn encouraging proposals that "without altering the operational control," MacArthur as the senior officer in the Pacific "should receive strategic control of these operations" and would thus be able to decide the timing of the hits in the various parts of the ocean.[1]

The greater land mass of New Guinea and what lay outside the huge island and measurably close to it, the shorter inter-island distances, would seem to have called for a military commander in the southwest Pacific. Whereas the broader water mass in the central Pacific, the larger inter-island distances, the original presence of Japanese fleet units in these parts, would all require the command of a naval man. But once the Japanese retired their battle fleet units from the southern Pacific, after Guadalcanal, fighting for, and on, islands came to be much the same, calling for nearly the same kind of troops with similar training. The Marines had grown from an authorized strength of 80,000 in 1942 to 450,000 on April 30, 1944, and to 590,000 by the end of the war, and were employed equally at Bougainville and Tarawa. The same mixture of forces was required for either form of combat, until at last MacArthur was even using battleships.

Army units fought under naval officers and naval units under command of military men. This marked the extent of "unity of command in the Pacific," something unavoidable and occasionally invoked to settle remaining differences of opinion about the strategy and tactics of the trans-Pacific progress. For there survived an occasional clashing of military and naval marine minds. In one arena, during the battle of Saipan, it grew so violent that the local commander, Lt. Gen. Holland M. Smith, USMC, later commander of the Fleet Marine Force in the Pacific, was "forced" to relieve Maj. Gen. Ralph Smith, U. S. Army, commanding Army forces under him, of his command, making use, as he later stated, of "one of the many prerogatives and responsibilities of a commanding officer operating under the principles of unity of command, the assignment and transfer of officers commanding subordinate elements in any operation."

The Marine general did not state his reasons, but military circles in Washington thought that the two Smith's had disagreed on methods of conducting the campaign, the Marine favoring quick, hard drives, even at the price of high initial losses which would mount up in the case of a prolonged campaign, whereas the army man desired a slow, methodical and loss-cutting campaign.[2]

Other differences, more objective in character, remained as far as the island targets in the two sectors of the Pacific were concerned; these objectives were farther apart in the lane of the Navy and beyond Guadalcanal and were more often of the small island type than of the larger island overgrown with tropical foliage that MacArthur was apt to encounter, and which allowed him to exult again and again that he had been able to surprise the enemy with his landings. Jungle in the vicinity of the beach handicapped the foe and favored the invader, at least initially, although it might cause him heavy casualties if he insisted on conquering the whole island in question, something which the Japanese appear to have expected us to do on Bougainville and New Britain. On such a spot the Americans and their allies, as the dark invaders, would land with their faces weirdly painted and in jungle-green and brown splotched battle dress so as to obliterate as far as possible the visibility usually silhouetting attackers from the sea. The islands in the naval lane, however, and above all those which the Japanese decided to use and defend were on the whole small. Lt. General Vandegrift said of Batio:

> The atoll is very small. The possessors are able to concentrate their whole defending force with the sole purpose of beating off the incoming troops at the beach. The

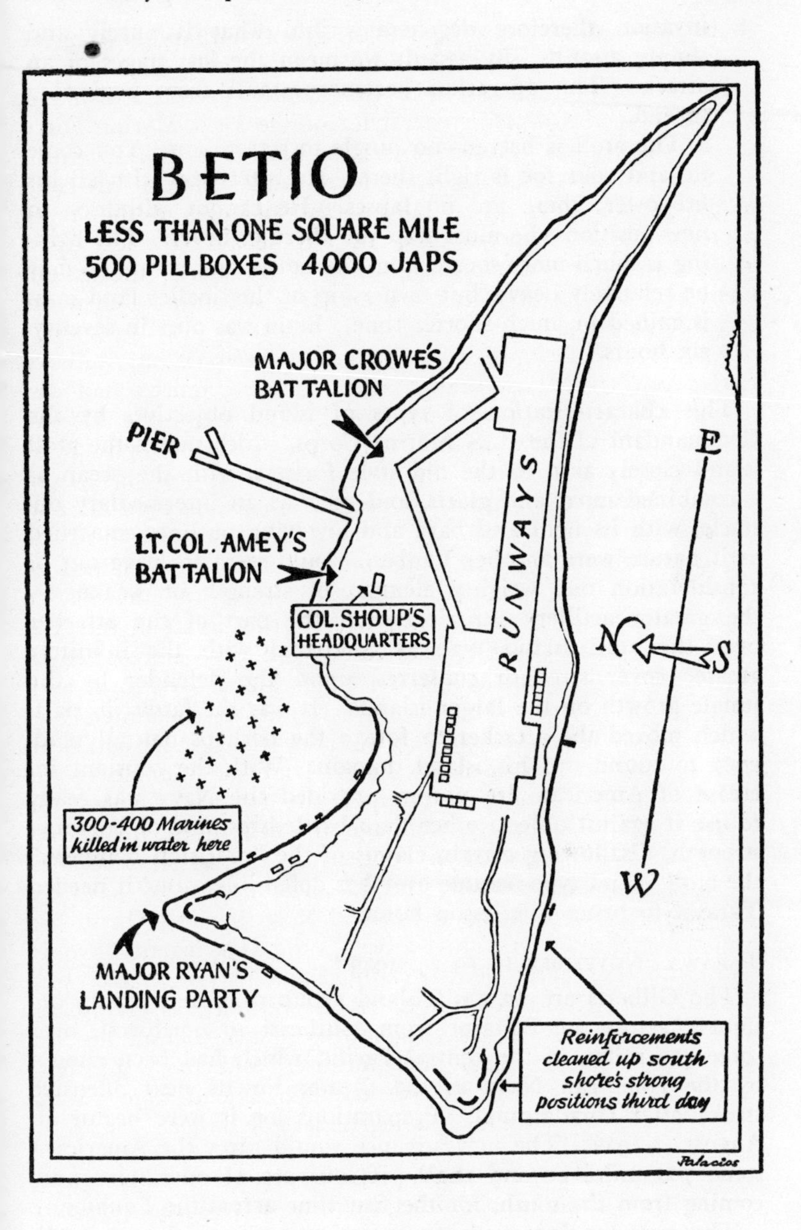
BETIO
LESS THAN ONE SQUARE MILE
500 PILLBOXES 4,000 JAPS
MAJOR CROWE'S BATTALION
PIER
RUNWAYS
E
N
S
W
LT. COL. AMEY'S BATTALION
COL. SHOUP'S HEADQUARTERS
300-400 Marines killed in water here
MAJOR RYAN'S LANDING PARTY
Reinforcements cleaned up south shore's strong positions third day
Palacios

> invasion therefore degenerates into what is purely and simply assault. By assault we mean the last stages of an attack. The operation becomes assault from beginning to end.
>
> The atoll is barren—no jungle to screen you. You come in, and your foe is right there. He meets you with all his firepower; there are no lapses. He cannot withdraw to new positions; he must stop you there or never. The fighting is much more severe from the outset and casualties may be relatively heavy, but ownership of the smaller land mass is gained in much shorter time. Betio was ours in seventy-six hours.[3]

This characterization of types of island objectives by the Commandant of the U. S. Marine Corps, which makes the atoll island closely akin to the old style fortress, with the ocean as a combined moat and glacis, and reefs as an intermediary obstacle, with its bold and bare and low-lying outlines construed as if nature were another Vauban, would seem to leave out of consideration one weighty element of strength or weakness—the greater artillery superiority on the part of the attacker over the atoll island type as compared with the infinitely greater cover strength conferred upon the defender by the jungle growth on the larger islands. It was the latter, in part, which moved the attacker to forego the perhaps natural tendency to round out his island domain. With the constant increase of American fire power, provided the Navy was ready to use it against objects other than battleships, and of bombing strength, disallowing certain claims of the "precision bombers," the atoll island type became ever less defensible. But it needed Tarawa to bring this lesson home.

TARAWA. NOVEMBER 19 TO 23, 1943 [4]

The Gilberts are an atoll island group of sixteen, lying obliquely across the Equator from southeast to northwest, over some 400 miles, in the central Pacific, which had been chosen by the American Navy as the theater for its next offensive stroke after Guadalcanal. Preparations for it were begun on August 1, 1943. The new advance would carry the Americans some 700 miles beyond their previous arc of operations and, coming from the south, for the first time across the Equator.

The heaviest Japanese concentration was on Tarawa atoll,

a group of 25 small islands connected by a coral reef, dry at low tide, thus making it possible to walk from islet to islet along two sides of a triangle enclosing a lagoon, the third side of which is formed by a submerged coral reef; this is broken by a deep water channel which allows boats to enter the lagoon. The main island in the Tarawa atoll is Betio, south of the inlet and forming the southwest corner of the atoll, two and a quarter miles long and half a mile wide at its broadest hipline, with less than a square mile total area. It was to be attacked simultaneously with Makin in the same group, 105 miles to the north of Tarawa, with Abemama to follow a few days later. The reefs around Tarawa were from 2,000 to 5,700 feet off shore, from the high water line. Off Betio they had been strengthened by the Japanese in such ways as to provide boat obstacles and drive the landing boats into the lanes of Japanese fire.

Two of the islands in the Gilberts had been picked by the Japanese to fortify and defend, Tarawa with an airstrip and a harbor of no particular importance, and Makin with a stone pier and potential facilities for hydroplanes. This gave the island objects in many minds "strategic importance" and the American Navy, accepting the challenge, decided to attack. As one Marine major instructed his unit: "We don't intend to neutralize the island. We don't intend to destroy it. We will annihilate it."

The Japanese garrison on Betio, 4,500 men, consisted of 3,500 Imperial Marines, of Japan's amphibious infantry, and the best Tojo had to offer, and 1,500 semi-military labor troops, most of whom seem to have participated in the battle. There were somewhat less than 1,000 Japanese on Makin and less than 200 on Abemama. It has not been stated with how many Japanese the American command had reckoned. According to some reports the garrison had been estimated at its right strength by American intelligence; according to others, which seem to reflect the opinion of Marines in the rank and file, it had been put too low by some 2,000. If this were the case, there would have been underestimation of Japanese garrisons in the small islands as, apparently, overestimation had been made as regards the large Japanese-held islands, from New Guinea down. The Japanese themselves, in medieval fashion that betrays itself in them only slightly oftener than in ourselves, considered Tarawa so strong that one of the Admiralty boasters who intended to dictate peace in the White House, announced that it would not be taken by one million men.

The defenses built up by the Japanese, nearer completion on Tarawa than on Makin, were, if judged by traditional land fortification standards, more elaborate than really strong and included numerous concrete structures, 500 pill boxes, command posts, dugouts, etc. But none were apparently reinforced and none had been dug deep into the ground, on account of the high water table. While a record degree of fortificatory saturation would seem to have been reached, it could by no means be compared in strength with places like Sebastopol, if Sebastopol were to be approached across the water, and assailed by naval guns alone.

Small Caliber Defense Guns

Much has been said about eight foot thick concrete that withstood direct hits from large—from the largest?—naval shells, but still their descriptions do not indicate how these constructions could have withstood 15cm howitzers. Was it the flat trajectory of naval guns that made some so "tough that shell fire and bombing could not reduce them," as Secretary Knox himself insisted? [5] Obviously, the mid-Pacific *Sperrfort* of Tarawa was provided against traditional naval gunnery. But, then, the question of Tarawa was: Could not either the island, or traditional naval gunnery be by-passed?

The defender's artillery was emplaced not so much like traditional coastal artillery, with steel turrets, casemates, cupolas, big caliber guns, for none of his guns was above eight inches and nearly all of them were silenced by pre-invasion fire, but rather with an eye to killing as many Americans as possible. For this reason Betio was fairly studded with small caliber and machine guns, well hidden or protected, and far less easily hit, or silenced by American fire. From some indications noted by Robert Sherrod it would seem that there was a certain difference in opinion respective to island guns between the naval gunnery officers and the Marine officers. To the former it would not make so much difference whether the garrison on Tarawa and the men behind the various guns were 2,000 or 4,000.

According to Marine General Julian Smith's expectations not more than one-third of the garrison could be expected to be killed by pre-landing fire, which is to say, either 667 out of 2,000 or 1,333 out of 4,000. This would make indeed a vast difference in the defender's fire strength, whereas, according to his chief of staff, Colonel Edson, the naval gunnery officers too easily assumed "that land targets are like ships—when you hit a ship it sinks and all is lost, but on land you've got to get direct hits on many installations, and that is impossible, even with three thousand tons of shell and bombs" (Sherrod, pp 41-42). Three thousand tons of steel was almost exactly what gunners and bombardiers put into Tarawa, or as it has been calculated, twenty pounds of missile per square yard.

The troops chosen for the taking of the Gilberts were in part green, but specially trained elements of the Twenty-Seventh Infantry Division, a New York outfit, selected for Makin; and elements of the Second Marine Division who had had experiences in Guadalcanal and who were given the harder nut to crack, Tarawa. As the land and landing force of the Navy, the Marines had been assured that unlike Guadalcanal where the ships had been retired from the scene for their own safety, leaving the Marines to fight it out on land alone and without any support for some time from the sea, the ships including the battle wagons would stay with them to the end and would even take "some chances this time" and assist them in the taking of a well defended beach. "Our Navy will remain with us until our objective is secured and our defenses are established," an order of the day before the battle assured the Marines.

What was called the largest fleet ever assembled in the ocean sailed from an unnamed rendezvous, to which some of the troops had come after a fortnight's trip from New Zealand. The voyage towards the Gilberts was eventless, filled with boredom and rifle cleaning, religious services and movies, and speculations on the part of the officers and men as to whether Tarawa, whose military secrets seemed not to have been satisfactorily penetrated by American aerial reconnaissance, might not prove another Kiska and if not, what naval gunnery could do for the landing men. En route there was a full dress rehearsal at a convenient island. There was cover and screening of the fleet on the water and in the air, although it could not hide the coming of the vast armada from the waiting Japanese and their flyers. There were no ship losses except the carrier

Liscombe Bay, with some 700 navy personnel, which was sunk by an enemy submarine near Makin.

Opening of the Ball

Battle day, November 19, began off Tarawa at 0505 with the opening shot of a battleship against Betio. The bombardment by the warships was followed by that from nearly a thousand planes which added to the damaging effect of the horizontal trajectory that of the vertical. Tiny Betio seemed extinct to the *first assault* waves, after all these sea and air volcanoes had concentrated and vomited their fire against it. More assault waves were getting ready to board their landing craft when unmistakeable signs came that life was still left behind the heaviest Japanese guns, four to five miles away, and transports had to be taken out of the danger zone until at least some of the worst back-talking Japanese guns had been silenced.

Their response upset the original schedules by 31 minutes first, then by 45, and the first wave, carried in Alligators, did not hit the beach before 0919. While the following waves prepared, the fifth sliding down the rope nets at 0635 into the landing craft and reaching the boat rendezvous at nearly nine o'clock, they received word from shore: "We have landed against heavy opposition. Casualties severe."

From then on the waves lost their originally intended sequence. The second, or parts of it, did not go in until much later. Coming under enemy shell fire while moving shoreward, they were told by a control boat to stay back until things had been made safer for them (at 1030). They tried again at 3 and 5 in the afternoon, but either machine gun, or shell fire, sending two more boats down, forced them back. Not before midnight could they try again, under a beautiful, but viciously bright, moon.

Those first to arrive found that Nature had interfered with planning even more than the enemy. An unexpected shift in the wind had so shallowed the water above the reefs that the Higgins landing boats which were to make the passage at high tide, and at high tide only, could not get across, and a shuttle system had to be improvised to get the Marines to the island shore in light-draft Alligators, which could crawl across the reef. Since only limited numbers of these "amphicraft" were available, and negotiation of the reef was a cumbersome process, exposing them to the defender's fire while going over the hump, the whole impetus of the invasion rushing into the chaos prepared by the bombardment, was shattered.

There were no organized waves following the first. These waves, as provided in the Infantry field service regulations, were intended to disperse the enemy's fire, but unlike attackers on land could not lie behind some scant protection—"there are no foxholes in the surf"—to catch a moment's respite while approaching the enemy's firing line. The Marines were brought to, or rather near, the shore in driblets. Since the Alligators had to be spared, these got rid of their passengers at some distance, 700 or 800, some say 400 yards, from the shore and let them wade the rest of their dangerous way to the beach. Many never made the passage, being mowed down while swaying in the water like a partly submerged grainfield by the reaping hooks of Japanese machine gun fire. "There the harvester whose name is Death" . . . . with his sickle keen garnered a rich crop in his strangest field. Still, his heaviest harvest was not in the water for, according to estimates from the spot, more than half of the dead of Tarawa fell in the land fighting against the entrenched Japanese. The same wind that had made the reefs unnavigable had caused the water to recede in front of a four-foot seawall built of coconut logs by the Japanese. Without this wind and the shelter of the seawall the Marines would not have found foothold on even the few yards, or feet of beachhead which was all they were able to gain and hold at first (and which they would not have needed perhaps if they had hit the shores in their Higgins boats). Providence often takes away with the right hand and gives with the left.

On the first day of the landing the beachhead extended to less than fifty yards inland in some places and for a mile along the shore. Japanese resistance had only temporarily been set at defiance. One reads of 300 suicides among them, due to shellshock. But how should these dead be separated from the rest of the later suicides? Resistance intensified most bitterly during the fighting on land, around block houses and pill boxes, or against stretches of trench, or tank traps, fighting in which American flame throwers burned out the enemies' holds like hornets' nests. The combat was like an attack on a trench line system against which one's own artillery had failed to do the expected execution, but from which no retreat was feasible, either for a second try by the attacker, whose supports had the greatest difficulty in reaching him, or by the defender who had no second line to retire to. Conclusions had to be fought out where both stood, isolated as either side was, the Japanese utterly, the Americans increasingly less so.

"Organized resistance"—it is not clear from the reports how far it was truly organized or offered—ended only on the fourth day of the invasion, 75 hours after the first landing. Organized attack, although badly thrown out of gear at the outset, finally won out.

Tarawa Plans Went Wrong

It required only 54 hours of fighting to wrest Makin from the less than 1,000 Japanese there—one newspaper man put them at only 600, 300 Marines and 300 laborers. Things went far more in keeping with plans than at Tarawa. The infantry, assured by their commander beforehand that "more shells per square yard will fall on Makin than ever fell at Verdun, or any other battlefield in history," after a half hour's naval and air bombardment with a total precipitation of 4,250,000 pounds of explosives, were carried from their sea rendezvous, where their LC's were idling waiting the signal to go in, to the shore. Successive waves of craft in V formation rushed to the shore at intervals of five minutes. They crossed the reefs without mishap and landed their men across the ramps of their craft. Shedding their life belts, the infantry dashed at once across the 15 feet of beach into the palm growth and underbrush where the combat developed. Land artillery and tanks followed them in due time, riding down enemy resistance which flared up in the last "crashendo" of a suicide attack.

American air supremacy, based on more than a thousand planes, was overwhelming throughout the Gilberts operation and with the exception of a half dozen Japanese sneak raiders appearing over Tarawa early on the three mornings following the first invasion, it was uncontested. The nearest Japanese air bases, a few hundred miles away in the Marshall Islands and on Nauru, were fully neutralized. American planes were in a position to give the ground forces all the help they could provide. One day after the conquest of Tarawa, labor of the Seabees had restored the usefulness of the airfield on the atoll and American planes were taking off from there, in order to help hold what had been gained against a never likely Japanese counterattack, and to thrust outward towards the next objective of the American advance, the Marshall Islands.

When the dead and the wounded were counted in the Gilberts and it was ascertained that the Marines had "paid the stiffest price in human life per square yard that was ever exacted in the history of the Corps,"[6] many people in the United States found the price excessive. Others, like naval men who were allowed to speak, or an exceptional journalist like Sherrod were inclined to think that this was belittling "the finest victory won by United States troops in this war."[7] In the absence of a valid definition of what constitutes the finest victory, there can be no quarreling with that claim which we take to mean that the action was fought with the great self-sacrifice by men who through nobody's or somebody's fault had not obtained the maximum benefit from planning and from material support.

As announced, the number of killed at Tarawa was 1026 and of wounded 2,557, of a total of the attacking forces not stated except by saying that it was less than a whole division, which would mean less than 15,000 or 19,000 men. In addition 65 were killed and 121 wounded at Makin and one killed and two wounded at Abemama. To these must be added nearly one thousand casualties from among naval personnel including 700 missing after the sinking of the *Liscombe Bay*. Against the American casualties of above 4,000 are set the figures of Japanese losses. Of the 5,700 men garrisoned in the Gilberts few prisoners were taken; nearly all were—had to be, it would seem—killed.

Casting Up the Bill

Official, or officially inspired commentators insisted, to use the terms applied by them, that the price paid in the Gilberts was not excessive, whether compared to the Japanese losses, to the losses to be expected in fighting for similar fortified objects—no one said which earlier enterprise would be quite, or almost comparable—to the casualties suffered by the Americans in Guadalcanal over four months, to the losses suffered in the large Pacific islands where the sickness rate was high, or to "the tremendous value that will accrue to us in possession of these islands." It must remain for history to determine whether it was necessary to take all the Gilberts, whether Makin and Abemama might not have sufficed to furnish the required bases in the style preferred by MacArthur, who had learned to divide an island with the foe and still dominate him and the island object, in a somewhat unexpected, improvised *divide et impera.*

The official post mortem statements, finding it too early to give us a complete and rounded history, even an official one, defended the action against "criticism from armchair strategists" on the whole by insisting that there were not many errors committed and that we had learned a vast amount from it. Secretary Knox, for instance, asked to comment on reports that the attackers had not fully realized the Japanese strength and the extent of their preparations on Tarawa, answered that "the preparations for the attack were the most thoroughgoing that I ever heard of." *(New York Times,* Dec. 1, 1943). He insisted that the Navy "had not been fooled" about the situation on Tarawa before the landing, that since complete data regarding camouflaged island defenses were not always obtainable, "we just assumed that the defenses were there and gave the island a pasting the like of which has not been seen in this war." Doubts whether Tarawa should not have been given a more intensive bombardment he answered by saying that the 2,200 tons of shells plus 700 tons of bombs plus millions of rounds of 50-caliber shells should be compared to the 2,300 tons of bombs dropped on Berlin in the biggest single raid, on an area many times the size of Tarawa. A somewhat less overt complaint was answered by Lt. Gen. Vandegrift declaring that "the warships and aircraft had done everything Marine Corps officers expected them to do." *(New York Times,* Feb. 6, 1944).

NOTES, CHAPTER 52

[1] Frank L. Kluckhohn from Advanced Allied HQ in New Guinea. *N. Y. Times,* Feb. 9, 1943.

[2] A. P. and U. P. despatches from Washington, Sept. 9 and 14, 1944.

[3] Lt. Genl. Vandegrift. *N. Y. Times,* Aug. 6, 1944.

[4] The facts relevant to a description of and judgment on Tarawa are not easily established. While in the cost calculation of economic production losses give rise to a relentless search for causes and faults, losses incurred in a military, destructive undertaking are often clouded up by political considerations. While some, if not most of the reports from Tarawa are emphatic about the unexpectedly low tide at the landing, caused by "a sudden shifting of the wind" and thereby contributing to the high American losses (Secretary Knox, *N. Y. Times,* Dec. 1, 1943), Lt. Gen. Vandegrift, Marine Corps Commandant, declared on one occasion that the tides had nothing to do with the losses. "The landing took place in a rip tide, just as had been anticipated." Ibd., Feb. 6, 1944.

[5] *N. Y. Times,* Dec. 6, 1943.

[6] Richard W. Johnston from Tarawa. *N. Y. Times,* Nov. 28, 1943.

[7] Sherrod, *Tarawa,* 147.

Chapter 53

LANDINGS ON THE ROAD BACK TO THE PHILIPPINES

WITHIN THE BROAD LANE laid out for MacArthur from Australia to the Philippines, paralleling as on a race track the Navy's lane to Tokyo, the main Japanese southward pressure was exerted along the eastern flank, close to the Army-Navy demarkation line, across the Bismarck Sea, New Britain and into the Papua region of New Guinea. In April, 1942, they had taken the Admiralty Islands and New Britain, going ashore against weak Australian forces at Rabaul.

As described in journalistic technicolor, they pounded through the surf, their faces blackened. Green signal flares distorted their features as those of Asiatics on covers of *Fu Manchu* novels. All the while the Australians held their fire. "A Japanese bugler stood on a darkened beach and raised his bugle to his mouth. He got out only three or four notes when the Australians opened fire." Then the tide was reddened with their blood as it had been by the blood of the landing fighters in the Nordic Gudrun Saga. (How far blood letting could be seen in the dark we shall not inquire).

Soon the beach was crowded with Japanese bodies over which the survivors and landing barges thrust forward, grinding the corpses into the sand. "They were squealing like pigs. Hundreds of them had been killed as they tried to get across the wire, and their bodies were slumped there in all sorts of grotesque positions. It was not long before the Japanese realized the value of their dead. They gathered scores of other bodies and threw them across the wire. They clambered over their own dead and came at us."[1]

Further south their offensive carried the Japanese by way of Buna, their advance base in this part of New Guinea, to within thirty miles of Port Moresby (Sept. 12). From there Allied ground and air forces drove them back to the northeastern coast. Early in 1943 this coast as far as Buna was cleared of them in fighting that, towards the end, paralleled the coastline. The land fighters were severely handicapped by want of naval support and of small vessels. Even though according to

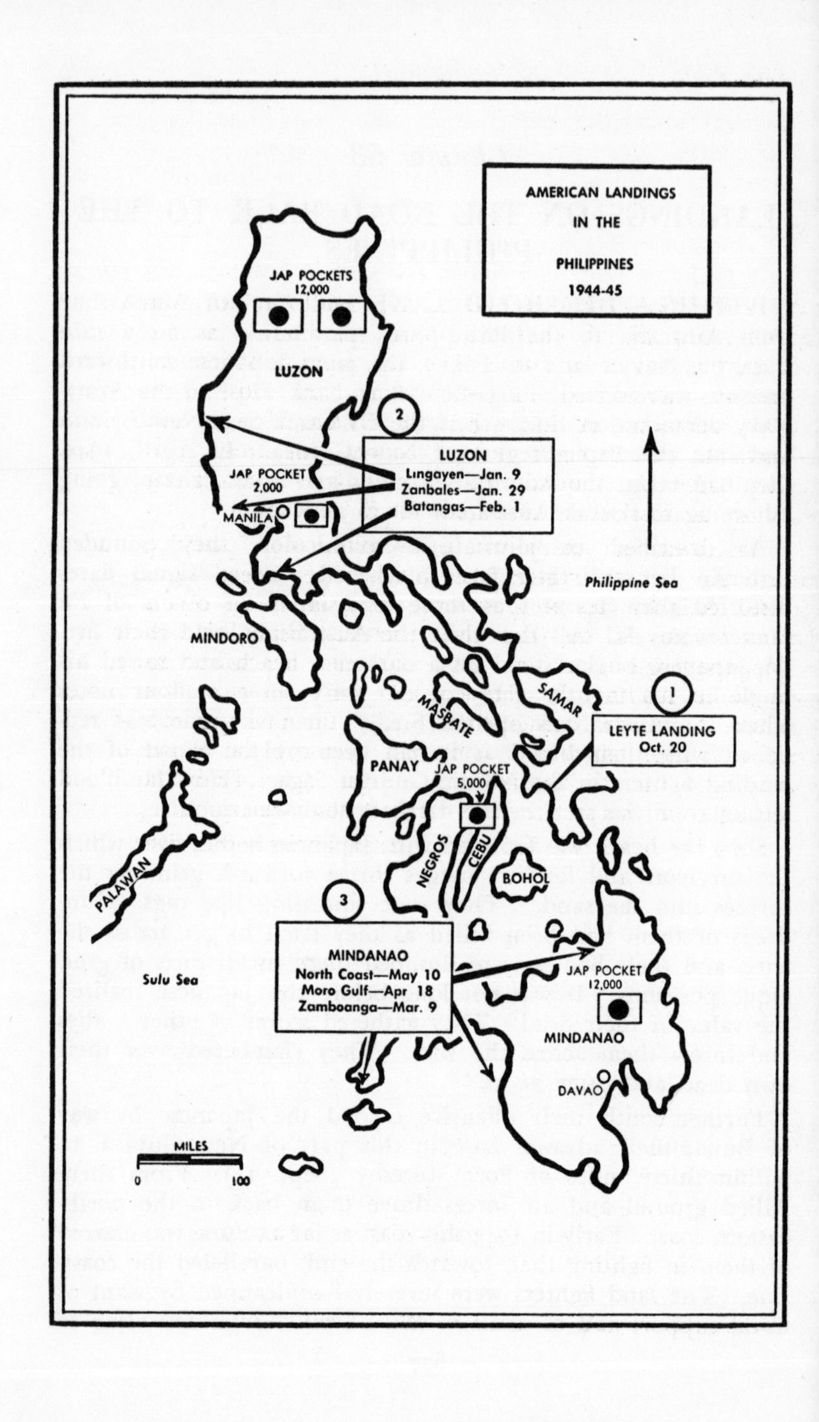
AMERICAN LANDINGS
IN THE
PHILIPPINES
1944-45
JAP POCKETS
12,000
LUZON
2
LUZON
Lingayen—Jan. 9
Zanbales—Jan. 29
Batangas—Feb. 1
JAP POCKET
2,000
MANILA
Philippine Sea
MINDORO
SAMAR
1
LEYTE LANDING
Oct. 20
MASBATE
PANAY
JAP POCKET
5,000
NEGROS
CEBU
BOHOL
PALAWAN
3
MINDANAO
North Coast—May 10
Moro Gulf—Apr 18
Zamboanga—Mar. 9
JAP POCKET
12,000
Sulu Sea
MINDANAO
DAVAO
MILES
0
100

General Marshall's report from about this time, "unified arrangements were welding sea, air and ground forces into efficient fighting teams," the available resources continued to be limited, much too limited as MacArthur's headquarters let the world know.

A conference of commanders and staff officers from the Pacific held in Washington in March, 1943, had become necessary to coordinate the wishes and plans of the local commands, with respect to future investments of forces, with the more general, overall considerations represented by the Washington offices and chiefs of staff. It chose the next objectives, including New Georgia, assigning the required forces and materials. Subsequently, as per June 26, 1943, "the Southwest Pacific and South Pacific were united under General MacArthur's command" (Communiqué from Allied HQ's in Southwest Pacific, March 1, 1944).

In order to reduce the hazards of the drive along the New Guinea coast, it was preceded by inter-island operations in the direction of Rabaul, the center of Japanese aggression in the southwest Pacific which by this time had lost much of its sting. The American offensive in the Solomons, starting from Guadalcanal, was resumed "after months of scraping for men and supplies," as was reported from MacArthur's headquarters on one occasion. All larger units of the Japanese fleet had long retired from the region and no romantic defiance reminiscent of "Joys of old, when knights were bold," by MacArthur, or Secretary Knox—"We will throw everything we have including all my heavy bombers against the Japanese if they try to challenge us" (Ass. Press, Nov. 2, 1943)—would bring them back. But they were still maintaining barge traffic to their advanced positions and an air frontage to the south and southeast of Rabaul, based on airdromes in the New Georgia group and other central Solomons Islands, such as Munda, Vila, Buka, Kieta, Kahili, Ballale and Rekata Bay.

To eliminate and neutralize these fields and turn them wherever possible into nests for American birds, a beginning was made with the taking of Rendova in New Georgia (June 30) and of Munda. Before the occupation of New Georgia was even completed, after a 64-day battle, the American forces moving further north, landed on Vella Lavella (August 15), northwest of Kolombangara Island, with the Vila airdrome on it, thus coming within 400 miles northwest of Rabaul and only 75 miles from the Japanese shipping and air base of Buin, on the southern tip of Bougainville Island. Japanese claims of the successes of their airforces against the Americans approaching Vella Lavella were extravagant, American admissions perhaps a trifle on the "intravagant" side.

Islands Taken and By-Passed

Clearly assured now that a higher degree of manoeuvring among the islands was possible and advisable, and provided with ampler means, American leadership after some more weeks, interrupted by the advance against Lae in New Guinea, resumed landings in the direction of Rabaul, striking at Mono and Stirling Islands of the Treasury groups, south of Bougainville (Oct. 26-27) and in west-central Bougainville (Nov. 1), thus arriving within 260 miles of Rabaul. They by-passed on the way Japanese holdings in the Shortland Islands, 150 miles south of Bougainville, as well as in the southern part of the island itself.

Enough of that island, the largest of the Solomons, 3,500 square miles, was overrun to provide for an airfield and a belt of defenses around it, and wisely the attempt was foregone to complete the conquest of the island, which was thought to be held by 40,000 Japanese, thus resisting the temptation towards rounding out a successful operation, which seems greater in the case of an island—an object so definite and *arrondi*. Islands are too numerous in the Pacific, some 10,000, to make complete possession of most of them feasible, or even important. A strategic choice among them suffices, as for a string of pearls. In fact, the Marine battalion landed on Choiseul Island in a diversionary move while the large landing on Bougainville took place, were withdrawn from there ten days later.

Once established on the northeastern coast of Papua, the advance in New Guinea changed in style, becoming a jump from coastal point to point, usually omitting one or more intermediate stations whose garrisons would subsequently be wiped out, or squeezed into the jungle. Naval vessels were made available to afford this leaping power and as a rule recalled as soon as they could be spared.

The schema for this coastal war was to start from behind A, the foremost point reached on the coast, skip B, the enemy front position, and land behind C, a point further along on the enemy-held coast and then to deprive the enemy of B and C from D, still further up the shoreline. Such an operation was that directed against Lae, a principal Japanese base in New Guinea, which began September 4, 1943. By by-passing Salamaua where, 22 miles southeast from Lae, Americans and Australians were working towards a Japanese airfield, thus making the enemy believe that Salamaua was their first and main or single objective, and by landing east of Lae in the Huon Peninsula it was intended to cut off the enemy's communications to Finchhafen, sixty miles further up the coast.

Judging from reports, the operation went according to plans and better than expectations. Naval units of limited size had been borrowed for the occasion from the South Pacific fleet. At 0613 naval craft laid down a smoke screen half a mile off Lae for the concealment of landing craft, while ships and aircraft started shelling and bombing the place; neither was answered from the shore. At 0650 shelling stopped, while the convoy with the Australians on board in the Huon Bay was closing in on the shore, and landing began at 0702 to the east of Lae. While it was under way, Lae was heavily bombed until beachheads were won while other aircraft racing along the coast held down enemy positions.

Surprise was so complete that no resistance was offered within the first half hour of the operation and Allied casualties in the landing itself were few. Only six hours later did a Japanese air formation, bombers with fighter escort, come to interfere, doing damage to two Allied craft, but the Allied air umbrella repulsed them, downing 21 enemy planes, including 16 out of 35 fighters, themselves losing only two.[2]

Thus began what MacArthur called the "investment of Lae," a new application of a term to a region where there were hardly any lines and where action resulted in driving the Japanese into the largely unsupplied jungle regions of the Markham Valley. A landing at Finchhafen followed less than three weeks later, against stronger, or more alert Japanese resistance on the beach, from where they were not dislodged until after an hour and a half of fighting; counterattacks were made as late as the middle of October.

The next relief granted MacArthur from sheer land fighting through the lending to him of naval forces, "after months of scraping for men and supplies," as was reported from his headquarters, was the landing on New Britain, an inter-island operation beginning December 15, 1943. The purpose of this landing was two-fold, to keep the Japanese from interfering with further movements up the New Guinea coast, particularly with the shipping passing through the Vitiaz Strait from and to Australia and enabling the Allies to extend their air activities from one, or more air bases on New Britain. (Why the Trobriand Islands, further south, had been occupied early in July, 1943, where no enemies had been found although perhaps expected, became soon apparent. From airfields quickly established there support was provided for the ensuing operations, allowing the rapid shifting of fighter aircraft between the Solomons and New Guinea). In MacArthur's language, a hold on New Britain "presaged growing command on Bismarck Sea approaches" and would "shortly bring the Kavieng-Admiralty Islands within decisive reach of our land-based air attack" (Special Communiqué, Dec. 25).

Insular Aircraft Barriers

The landing operations on New Britain began at its western end, at the utmost opposite point from Rabaul, 270 miles away, at Arawe and, a fortnight later, at Cape Gloucester where Marines took the situation in hand, promptly winning an airdrome. At Arawe the various amphibious vehicles evolved by American ingenuity came in for their baptism of fire, but a realistic description of how they stood the test was not forthcoming. Japanese resistance grew after the landings, although they had already frustrated the approach of some boats to the east of Arawe on the 15th. While they had to concede the Allies an airdrome near to the western tip, it was not till early in March that they could be dislodged from the central part of the island and be pressed together in the eastern half, where before Rabaul, in the Gazelle Peninsula, they were still further reduced from April 10 on. In the outcome New Britain, like Bougainville, proved to be a "stationary aircraft carrier," the use of which the two contending parties had to share. There had been an ominously suggestive place name near the point where operations had begun in the island: Sagsag.

Not until further details on strengths, motives, etc., are available will it be possible to judge whether this island campaign was still conceived too much in the land warrior's tradition and whether islands, in the beginning at least, were not viewed as so many water-surrounded "hedgehogs." Japanese resistance showed that such positions can hold out considerably longer than a similar spot defended on a continent. Ringing around enemy-held islands by seizing and holding other islands still farther afield, such as the Green or Nissan Islands, 250 miles southeast of Kavieng, New Ireland, which MacArthur occupied in mid-February, 1944, and which was promptly claimed as cutting off all the ten thousands of Japanese in the Solomons, is still far from equalling the close investment to which a continental encirclement as a rule quickly leads. On the other hand, the attacker can far more easily and safely by-pass island-hedgehogs, not all of which are stationary aircraft carriers.

In order to make doubly sure of this severance of enemy supply lines, reducing him to blockade-running by submarine and single surface craft for the provisioning and evacuating of his "trapped" forces—trap being the term constantly overused in American war journalism—that were estimated at 50,000 in the Bismarck Archipelagus plus another 22,000 in the Solomons, MacArthur landed on February 29, 1944, in the Admiralty Islands, 330 miles northwest of Rabaul and 1300 miles south of the Philippines.

Since Japanese evacuations had been reported with which it might be possible to interfere, troops were carried by fast destroyers rather than transports and by landing craft northward through the Vitiaz Strait. As if to emphasize what the war had done to cavalry, or to make people wonder why troops still bore the name, elements of the U. S. First Cavalry Division (dismounted) who had received amphibian training in Australia, were sent ashore on Los Negros Island where they at once proceeded to take Momote airdrome. Preparation by bombardment from planes and warships had sufficed to drive the weak Japanese forces from the beach into the hills and allowed the cavalrymen who came in only reconnoitering strength, in three waves of four boats and 150 men each, following one another at ten minute intervals, to make the shore without many casualties.

Help From Sea and Air

The fourth wave was fired upon by Japanese heavy machine guns and forced back to their ships until the fire was silenced by naval and aerial barrages. For a time the landed force was under the fire of Japanese 4 in. naval guns which presently woke up, firing from the northern end of Los Negros and the adjacent six other islands of the group, including Manus Island, separated from Los Negros by only a narrow channel. Having won some of these small islands and placed artillery on them, American ground forces crossed over to Manus (March 15), the biggest of the Admiralty Islands, containing Lorengau airdrome.

A week later MacArthur's advance in these waters came closest to the equator and the important pre-war Japanese base at Truk, less than 600 miles away, by taking Emirau Island (March 22), which was declared to be the final stopper in the bottling-up of the Japanese. Points from whence the safety of American surface vessels navigating in the Bismarck archipelago might have been threatened were all taken or neutralized, and they could now move freely in this vast system, shaped like a great oval amphitheater. For the first time since Guadalcanal, and being used also for the first time by MacArthur, American battleships reappeared in these waters during the action against Emirau, submitting Kavieng to bombardment for three and a half hours.

The true continuation of, and improvement beyond, the Lae enterprise was the Hollandia-Humboldt Bay landing, further along the New Guinea coast, in April, 1944. Much had previously been learned about the technical side of such undertakings although not enough concerning the enemy in the chosen locality. This enterprise was preceded by a meeting of MacArthur and Nimitz at which "plans were completely integrated so that a maximum cooperative effort might be exerted against the enemy" (Official statement made simultaneously in Allied HQ in New Guinea and at Pearl Harbor). Whether this was a re-affirmation of older understandings, or a basically new one was not stated, but the zone arrangements as to geography and command of each chief remained as before. A new, more intimate and not merely from-case-to-case understanding was arranged between the two services, still co-existing in spite of the integration forced by war. It was hailed as the ending of the time when "some service-proud officers had taken the stand that one service or another should take the lead in beating Japan" or that "one service could do it alone." [3]

The arrangement provided in practice for the lending of considerable forces from Nimitz's fleet for a landing enterprise of MacArthur's in New Guinea; the General was to command all naval units immediately involved in the landing while the forces screening the enterprise remained under the Admiral's direction, the latter being assured in turn of having at his disposal men, amphibious equipment and planes from MacArthur's command.

The Hollandia operation was a much improved edition of the one against Lae: this time there were available more troops, Americans all, and more ships, more recent types of craft and artillery including rocket firing craft. The margin of safety was still greater, the jump along the coast longer, gaining the Allies some 420 nautical miles on the way to the Philippines, reducing the distance to Davao on Mindanao to 1100 miles. The landing was over a much broader front, some 150 miles, and the expectation was that far more enemies would be cut off from their line of retreat; 60,000 as MacArthur estimated. Taking place astride the Dutch and British New Guinea frontier, the landing would be the first to liberate Dutch territory.

The approach was indirect. From a rendezvous north of the Admiralties the three task forces that were to cover the three separate landings, steamed northwest in the direction of Palau on Friday, April 21, followed by the huge convoy which was to cover altogether 900 miles from its embarkation port, with screening forces of destroyers and escort carriers ahead and astern. At seven P.M. they swung south and were off their assigned landing beaches, Aitape, Hollandia and Tanahmerah Bay, at five A.M. on Saturday. Whether they had adroitly misled the enemy by design, or whether by luck they had remained altogether unobserved, surprise on the New Guinea shore was complete.

After an hour's heavy bombardment from the three big forces of cruisers using up to 8 inch shells, and destroyers, dive bombers and fighters from the carriers came in after gun positions and other shore installations, of which there were not many. Three small offshore islands were bombarded. Rocket boats with their fire brought things for the enemy still nearer the culminating point of horror. Then the first waves of infantry went ashore by landing craft and across a perfectly smooth sea, pushing off the transports at 0645 at Aitape, the easternmost beach. Each craft struck its dot on the beach map. Opposition on the beaches was either non-existent, as amber flares announced, or slight. None of the enemy's planes showed up; a concentration of barges further along the coast reported during mid-morning was sunk.

JAPANESE GARRISONS OVERESTIMATED

For a "bold move cornering fifty to sixty thousand Japanese," as Rear Admiral Barbey, commander of the amphibious forces, called the Hollandia operations,[4] the casualties on either side were surprisingly low. After two weeks and a half MacArthur announced that in the taking of Hollandia including the hinterland 871 Japanese had been killed and 183 taken prisoner, not a high percentage of the 60,000 said to be encircled and in a situation which was Bataan in reverse. It should have been clear enough by that time that there were far fewer Japanese in the island than estimated. The Australians had landed at Madang, which MacArthur had by-passed, on April 24, finding that the enemy's main force there had left a month before. American casualties were only 28 killed and 95 wounded.

The more local objectives of the landed American forces were not so much on the beach, where Hollandia offered good harbor facilities, as the various airdromes which the Japanese had hacked out of jungle farther inland. In the taking of these forest and swamp surrounded air strips the Americans used their Buffaloes, Alligators and Ducks to advantage, pushing with their help across six miles of a freshwater lake. It presented a stimulating image to enthusiasts for the prehistoric, when these fire breathing saurians plunged through the primeval swamps and across Sentani Lake, where they seemed like three new monstrous saltwater fish in addition to the nine kinds already living in that curious lake.[5]

Leaving the Japanese nests of resistance along the northern coast of New Guinea tightly airblockaded and thus shut off from most supplies, to handling by local forces and the "neutralizing" effects of tropical nature, MacArthur pushed on in stages of varying length, these being usually determined by the presence of Japanese airdromes, along the coast and forever Philippine-ward. The land-water-air team for the *reconquista,* provided now with ample ammunition allowing for over-saturation of such points as Wakde, which was plastered with 7,000 tons of bombs during the ten days before the occupation, proved thoroughly coordinated against any form of resistance the Japanese might offer.

Opposition was fairly weak on Wakde, the first étape in the

stride for the Philippines (May 17), but considerably stiffer on Biak, in the Schouten Islands, 200 miles beyond Wakde. Here indeed, it was reported to have been fiercer than anywhere else in the southwest Pacific during the previous eighteen months. The Wakde Islands, 125 miles nearer the Philippines, are close to the coast on which the landing forces, meeting no resistance, first established a beachhead, near Sarmi, before they pushed across the 2½ miles wide sound to the first of the islands and thence to the other which carried the desirable air-strip. So obviously preferable does pushing off from a piece of terra firma than from the ships of a convoy seem even to veteran forces!

The Sarmi airdrome, farther inland, was not taken until six weeks later, after hard jungle fighting. Both on Wakde and Biak one or more airdromes offered as objectives liable to be defended most persistently; hence the obvious choice of landing beaches not too close to the airfields in question. On Biak the Japanese defended the three strips for no less than 25 days against forces who had warships, tanks and superior artillery at their disposal.

Ten days after the completion of the occupation of Biak, another westward jump of 100 miles brought the Americans to Numfor (or Noemfor) Island on July 2. According to one of MacArthur's communiqués, which more often than other generals' announcements try to explain, or emphasize, the landings on Numfor "were made through narrow and difficult coral reefs, generally regarded as impracticable for such a purpose. As a result the location of the attack was completely unexpected by the enemy and his defense preparations were outflanked." On July 6 the last of the three airdromes and all that was worth capture on Numfor was in American hands. The Philippines were now within bombing distance and New Guinea more nearly swept of Japanese, who retained only two major bases, Sorong at the western tip of New Guinea, and Manokwari, at the western outlet of Geelvink Bay.

According to some reckonings, it was the seventeenth major landing of MacArthur's forces which put the Americans down at Sansapor and the islands of Amsterdam and Middleburg, 200 miles beyond Numfor and midway between Sorong and Manokwari, where 15,000 Japanese were said to have been by-passed and isolated (July 30). As there were no airdromes or other installations in the neighborhood, which is suitable however for airfields, opposition was very slight.

At little more than 600 miles distance from the Philippines, the large spider island of Halmahera as the main intervening base, with only a few *nidi* of Japanese resistance left in New Guinea, MacArthur,

> *Like one that stands upon a promontory,*
> *And spies a far-off shore where he would tread,*
> *Wishing his foot were equal with his eye,*

pointed to the Halmahera-Philippines frontage as "the main defense cover for his (the enemy's) conquered empire in the southwest Pacific. Should this line go, all his conquests south of China will be imperiled and in grave danger of flank envelopment" (communiqué of Aug. 1, 1944). It was already flanked farther north, by the advance, somewhat more *pari passu* in the later stages, made in the naval lane to the Marianas in July and August.

NOTES, CHAPTER 53

[1] George H. Johnston, *The Toughest Fighting in the World.* N. Y., 1943.

[2] A. P. and U. P. despatches from Allied HQs in Southwest Pacific. Sept. 6, 1943.

[3] Frank L. Kluckhohn from Allied HQs in New Guinea. *N. Y. Times*, Apr. 28, 1944.

[4] U. P. despatch from Allied HQs in S. W. Pacific, May 10, 1944.

[5] *N. Y. Times*, Apr. 27, 1944.

Chapter 54

UNITED STATES PACIFIC NAVAL CONQUESTS

The Marshall Islands—Kwajalein, Jan. 31-Feb. 8, Eniwetok, Feb. 19-22, 1944, etc.

HAVING FAILED TO REDUCE the Japanese combination concrete-sand-log fortifications at Tarawa even with its big—did it employ the biggest?—guns and bombs, the United States navy, spurred by the heated public criticism of an industrial nation convinced that her gigantic machinery for war production could, or should save lives in the field, began at once experiments with explosives and firing techniques, testing them on concrete blocks of eight inch thickness on its ordnance fields. At the same time new combat methods were devised and tried by land forces at Central Pacific bases. Conditions to be met in the Marshalls were sufficiently similar to those prevailing in the Gilberts to envisage the new operations, ten weeks later, as a similar, although much improved version of substantially the same enterprise.

The next island objective selected was Kwajalein Atoll in the Marshalls, a string of 30 to 50 islets, of which Roi and Namur in the north and Kwajalein at the southern tip are the most important. The islets are ranged in shape of a boomerang around a vast lagoon, 60 miles long, the largest in the world, able to hold the whole American navy. The Marshalls, two parallel lines of atolls, the Ralik (Sunset) and the Radak (Sunrise) chains, running northwest to southwest had been in Japanese hands since the World War I and with good reason were suspected of possessing considerable fortified strength. No one knew, for the Japanese kept visitors from the islands and worked there in secret.

While there were useful small airdromes on Roi and Kwajalein and a radio station, no considerable fortification had been set up on the atoll. Kwajalein is 620 miles northwest of Tarawa and 670 miles south of Wake, 1050 miles east of Truk and more than 2,000 miles from Tokyo and also from Hawaii, the starting point of the armada. It lies rather near the center of the whole group of islands, farther away from American bases than such islands as Mili, Jaluit, and Wotje, which were by-passed as if to demonstrate that this new advance was not island hopping, and also better to insure chances of surprise.

The troops selected for the enterprise, of unannounced strength, were elements of the 7th Infantry Division, who had been at Attu; of the 17th, 184th and 32nd Infantry, and of the Fourth Marines, a yet untried, but intensively trained force. On the way from Pearl Harbor the latter were told, on the eve of their landing on Roi Island on February 1: "You are superior to the enemy in any kind of fighting. Your weapons are superior to those of the enemy. The naval gunfire and air support you are to receive will surpass anything previously provided," [1] in tenor much the same encouragement as Marines wanted to hear and did hear before Tarawa.

The resistance met by the American landing forces proved unexpectedly weak and definitely less than on Tarawa, and American casualties were correspondingly less. The transports were able to enter the lagoon on the first day and thereafter move inside it freely. "Continuous bombardment of beaches by our warships, planes and land-based artillery enabled our forces to make landings on the three principal objectives with little resistance," said one of the first communiqués from Pearl Harbor. Some infantry opposition developed in the final stages which was met systematically, with the help of strong mechanized equipment, rather than impetuously, by the American troops. When, for example, they found themselves face to face with a tank trap actively defended, amphibious tanks took to the water and enfiladed the Japanese line while infantry was attacking frontally. In this particular skirmish 60 to 70 Japanese were said to have been killed without the loss of one man on the American side.

The announced American losses for the whole operation were 286 killed and 82 missing, a relatively high number which might be explained by the sinking of some landing craft; and 1148 wounded, as against known Japanese losses of 8,122 dead and 264 prisoners. The Japanese communiqué, not admitting prisoners, declared that 4,500 men of army and navy units and 2,000 civilians serving with them had died in the fall of Kwajalein.[2]

Ordnance Technique Change

Kwajalein again marked the removal of a blocking fort in the ocean, but effected this time by the artillery that destroys in advance of the infantry that lands and occupies, rather than lands and storms as at Tarawa. Naturally, the changes in ordnance technique, including possibly those in projectiles or fuses have been sparsely publicized except through announcements of the effects achieved. But it has been stated that since penetrating shells, used in intership combat, failed to reduce the log-sand-concrete defenses in the Gilberts, at Kwajalein navy guns at short range, only a few thousand yards from shore, hurled thousands of tons of high explosive point blank at their objectives. This combined with the bombing from the airplanes could "tear loose the very surface of the islets" in the Marshalls,[3] where, however, the defense structures were much weaker than on Tarawa, despite the prolonged occupation by the Japanese.

During the 53 hours preceding the main landings 5,000 tons of shells and 200 tons of bombs were hurled and dropped into Roi-Namur alone and altogether 15,000 tons of shells at the Kwajalein atoll, fifteen times the amount in airplane bombs which were used at the same time.[4] In fact, the Japanese on Kwajalein were fairly smothered by American men and materiel. The armada estimated by some at 2,000,000 tons with 250,000 men, included some of the newest battleships which were no longer considered too important to be employed on undertakings of this kind. No wonder that these islets received "the heaviest naval bombardment in history." "Maybe we had too many men on too many ships for this job," Admiral Turner, in command of amphibious operations in the Marshalls, said in a despatch from there. "But I prefer to do things that way. It meant many lives saved for us, and it should be a discouragement to the Japanese everywhere to know that when we hit we really hit hard and for keeps." [5]

Pin-Point Bombing

Bombing by carrier-based planes was reported to have been "better here than it ever had been before. Pilots were ordered 'to bomb deliberately and accurately' and they did." Preliminary bombardment, started by the planes, based on a score of carriers in the armada, was kept up for three days preceding the main landings. It was complemented by land-based artillery, placed on smaller islands in advance of the landings against the larger ones; islets which the Japanese had either left unoccupied, or from which they were expelled by American troops that landed on them in commando style on January 31.

The first communiqué on operations against Kwajalein mentioned as the objective "the capture of the Marshall Islands." This seems never to have meant domination of all the thirty main atolls, but only of enough of these "unsinkable aircraft carriers" to divide and control the island group and its vicinity by the scepter of air power.

Less than three weeks afterward Eniwotok Atoll, nearly the westernmost of the Marshalls, 300 miles beyond Kwajalein and 760 miles from Truk, was taken by the same tri-elemental combination of overwhelming naval and air force and moderately sized landing forces, elements of a Marine and an infantry regiment, and the same measures of neutralizing more distant Japanese air bases, such as Truk, and occupying close-by islets surrounding, or flanking the main objective, a day or so ahead of the assault itself. The speed-up was so considerable that Engebi Island, one of those in the atoll, carrying an air strip, was in American hands in a little more than six hours after the first landing.

Following the taking of Kwajalein, Eniwotok and Majeru Atolls, in February, and of Wotho in March, ten more of the Marshalls were occupied by American forces early in April, after meeting only light resistance, or none. This brought American holdings to 14 atolls, all except two in the western chain, leaving the Japanese in uncontested possession of most of the older and longer developed plane and seaplane bases and anchorages, such as Mili, Malodelap, Jaluit and Wotje.

Many Island Garrisons By-Passed

Installations in these islands, still held by a garrison of some 10,000 men, as was estimated in June, 1944, were more or less regularly kept in unusable condition by American planes. By-passing was adhered to in spite of all temptation to take such places, including the capital Jaluit, which must have come into the military minds on many occasions; it saved lives and time that would otherwise have been consumed and still gave the Americans all that they really needed—air bases in the direction of their further advance. The occupation of Ujelang Atoll on April 22 and 23, against light opposition "quickly overcome," as well as some earlier occupations, pointed to intentions in the direction of the central Carolines, with Ponape and other enemy bases in that group already under constant air attacks, and of the Marianas.

The basic concept underlying Japanese resistance to the American westward drive across the Pacific was to cause their better equipped enemy such heavy losses through the determined, "suicidal" resistance of the island garrisons that the American people at home, wearied of the sacrifices for "useless" objectives, would clamor for peace. They hoped that the Americans would sit down before every island obstacle in their way, like Louis XIV before the Dutch fortresses, and systematically reduce it, at expense of time, patience, men and materiel.

To meet this threat, or temptation, the American command, alternatively warning the people at home that heavy casualties had to be expected and that the objectives which had to be bought with them were well worth the price exacted, continued to enlarge the margin of material superiority, thus reducing the casualties absolutely and relatively, and created, so to speak, counter-islands in juxtaposition to the Japanese holdings. These were of two kinds—carefully selected islands to be wrested from the enemy and floating islands, consisting of warships in ever increasing numbers, which became ever more like artillery platforms, made fairly safe against island-based fire and aircraft attack; and of aircraft carriers, whose numbers were proportionately increased even more and of which the Americans hoped to possess one hundred before the end of 1944. The warships of more traditional design had the less traditional task of helping to bring certain chosen islands into American possession, while the carriers were to neutralize intervening and by-passed islands as well as those lying certain distances ahead of the intended advance. Although both had complimentary tasks, when warships would participate in wide-flung raids and carrier-based planes in bombardments of invaded islands. These had become main features of American strategy in the Pacific in 1944.

In the advance to the Marianas early in June considerable, but now altogether passive holdings of the Japanese, in the Marshalls and the Carolines, with garrisons estimated at from 20,000 to 25,000 were by-passed. No interference with the movements of the armada seems to have been attempted from these, or other islands through which the American fleet had steamed before, in February, 1944, without meeting resistance save from a few Japanese aircraft. Neither did any frontal attack from Japanese surface forces develop to delay the American armada; the Japanese fleet remained "in being" until after the battle for Saipan had started.

WINNING ARCHIPELAGO FORTS

Then, large enemy naval units were sighted by the Americans, the largest collection of Japanese battleships seen since Guadalcanal, and moving between the Marianas and the Philippines on the 19th. Somewhat mysteriously the Pacific Fleet communiqué announced that "we may have given them some hurt." On the ninth day of the battle for Saipan, an air-sea battle of considerable size, in which Japanese losses are put at 700 planes and at least one carrier, was waged off the Marianas between American carrier-borne planes and Japanese partly carrier-based aircraft which used near-by shore bases as shuttle points. With that half-hearted attempt, the Japanese left the defenders of Saipan, in whom they seem to have aroused hopes for relief, to their doom.

For the reducing of the Marianas the American technique of battery and assault evolved against the atoll type of islands had to undergo certain changes. For the Marianas, volcanic peak-shaped in the north, are in their southern half of considerably larger area. Saipan is 12 miles long and 5 miles wide, and of much higher elevation. The southern islands, Rota, Guam, Aguijan, Tinian and Saipan, are of coralline limestone formation, with steep cliffs rising from the ocean in many places, and fringed for miles by coral reefs. They are densely wooded, which required a return to the jungle style of Pacific warfare.

Saipan, with Garapan its seat of administration, and its various installations including three airfields, a small naval base at Tanapag harbor and a seaplane base, was the most important of the islands, militarily speaking. It is only 1,500 statute miles from the Japanese homeland, four flying hours. "It is now clear" said Pacific Fleet communiqué 81 after four weeks of fighting on Saipan, that it "was built up by the Japanese as the principal fortress guarding the southern approaches to Japan and as a major supply base for Japan's temporary holding in the South Seas areas. The topography of the island lent itself well to defense, and elaborate fortifications manned by picked Japanese troops testified to the importance which the enemy attached to the island. The seizure of Saipan constitutes a major breach in the Japanese line of inner defenses, and it is our intention to capitalize upon this breach with all means available." The other islands contained one or more airfields as well as a few good harbors, such as Apra on Guam.

The American armada began its action against the Marianas while yet hundreds of miles away, on June 10, when the advance carrier-based planes opened a two days' air bombardment of Saipan and others of the island group, with the double purpose of destroying installations and of misleading the enemy as to the first objective among the islands. Landings were begun on Saipan by battalions of Marines, of the Second and Fourth divisions who had been at Tarawa and Roi-Namur respectively, followed by elements of the 27th Infantry Division, "the principal components of the expeditionary force."

Previously a heavy four days' bombardment of surface craft went on, continuing through the night preceding the landings, and knocked out practically all heavy coastal and AA batteries, on the 14th. No detailed descriptions of the landing proper have been forthcoming from the American side. Japanese communiqués insisted that the first two attempts failed and that only the third, at noon, succeeded. American casualties in the ground fighting tend to confirm the suspicion that there were difficulties in landing, for they included no less than 878 missing in action, during the first two weeks fighting on Saipan.

The American forces landed across reefs in the southwestern corner of Saipan, near Agingan Point, advancing from there slowly east, in the direction of Aslito airfield, the best in the island, and north towards the town of Garapan. They soon realized that the Japanese defenses were well-manned, by an estimated 25,000 to 30,000 men, and well organized. Also that they included massed artillery, mortars and machine guns as well as tanks for counter-attack, and that they were even prepared for an amphibious counter-attack when on the 17th a group of troop-carrying barges attempted a landing south of Garapan, behind the American lines, but were repulsed by American armed landing craft, which sank 13 enemy barges, clearly less well armored and armed. On the 18th Aslito airfield was taken and at once put in condition by the Seabees and made usable from the 21st on, although Japanese infiltrations, from a pocket of resistance maintained at Nafutan Point until the 28th, and inter-island artillery fire from Tinian made its use for some time precarious.

Heavy Artillery Landed

While the ships and planes rendered considerable support, the landed forces early had heavy land artillery, laboriously dragged across the reefs, at their disposal. This enabled them to cope on their own with such difficult problems offered as Japanese guns hidden in caves, brought out to fire and then taken back to cover, and with the Japanese forces generally after they had withdrawn to the interior from the coastal plains. Navy fire pursued them into the tropical mountains, rugged but foliage covered, with Mt. Tapotchau (1554 ft.) as its highest elevation, which was taken on the 11th day of the fighting.

Advance was slowed up, due to Japanese opposition, usually in small bodies and not easily found; by climate and terrain, which made it difficult for naval and air forces to support the land forces although their coordination in the "biggest amphibious operation in the Pacific to date" was called "almost perfect" by an admiral returning to Washington; [6] and also perhaps by the prolonged front line service required of the American infantry, lasting beyond what until then had become customary in the Pacific.

It was probably during this phase that the war between the Smiths erupted; Smith, the Marine favoring sharp attacks, if necessary without systematic artillery preparations such as were favored by Smith the Army Commander, whose forces had been pinned down by enemy fire while the Marines were advancing on either flank. In a way the difference in Marine and Army casualties tells the story of this basic disagreement, for these were announced after a fortnight to have been 8482 Marines and 1259 Army soldiers. Unfortunately we do not know the strength of the two forces, or the breakdown between the Marines and soldiers in the final casualty statistics.[7]

Compressed in the last northern third of Saipan, the enemy tried to break out in a final desperate counter-attack staged by several thousand men against the American left wing (July

6). This "Banzai attack" penetrated up to a depth of 2,000 yards and was stopped only after severe fighting and heavy casualties on both sides. The last eleven days of fighting, coming to a close on July 8, continued to take a considerable toll of the American ground forces—885 killed, 4081 wounded and 335 missing. Altogether, Saipan, termed "a prize of the first magnitude" by the new Secretary of the Navy Forrestal, had cost the attacker in the final summary 3,049 killed, 13,049 wounded and 365 missing; of the wounded, however, more than 5,000 were returned to duty by July 25, and participated in the taking of Tinian which the Marines from Saipan tackled after a short rest period. The Japanese dead buried by the Americans were counted at 20,720, to whom should be added the dead buried by the Japanese; 1,700 troops were taken prisoner, a larger body of Japanese than had heretofore surrendered in the Pacific, and 14,192 civilians were interned.

Among the Ladrones (Thieves) Islands, as the Marianas are also called, Guam had been lost early in the Pacific war to the Japanese island robbers by the Americans, who were particularly eager to regain it; for it had acquired a political, in addition to its military importance. There was a somewhat similar feeling among the Japanese, who considered their own realm invaded when the Americans turned to the Marianas. Deep resentment against the leadership that had been unable to halt the Americans led to the fall of Premier Tojo who, less than a year before, had told the Diet that Japan had built an "invincible fortification against any invasion," [8] during the interval between the fall of Saipan and that of Guam.

The troops who had taken Saipan were still in the islands, with additional ones on their way, and also the fleet, including the plane carriers. One thousand two hundred miles from its nearest advanced base in the Marshalls and 3,250 miles from the major base at Pearl Harbor, the fleet showed a degree of self-sustenance and prolonged emancipation from permanent bases which must have surprised the Japanese, if not also Navy old-timers, remembering days when it was held to be axiomatic that any fleet, American or Japanese, that would steam beyond a given longitude in the Pacific would be automatically beaten.

American ships and planes kept the remaining Marianas under fire. For sixteen successive days preceding the invasion Guam had to take heavy bombardment from the air; the weight of bombs was raised to 627 tons plus 147 rockets on the day before the new landing. On the fifth straight day of bombardment from the surface forces the infantry landed (July 20).

Guam, 125 miles to the south of Saipan, with 225 square miles of land and thrice as large as the latter, thus was subjected to considerable softening up, although its rugged and mountainous character probably made it difficult to spot and hit objectives beyond the coast line and called for the employment of land artillery as quickly as possible. The preparatory bombardment allowed the boarding of the island, another, "unsinkable aircraft carrier," in schedule time, after which "moderate ground opposition" evolved.

The American troops who went ashore were again Marines, troops of the Third Marine Division in the north, leading with their first wave at 0828; and of the First Provisional Marine Brigade, formed of ranger elements, as assault forces, rolling to the shore in the south, in nineteen waves. By midday tanks were ashore. They were followed by elements of the 77th Infantry Division as supports. Strength was about equal to the force employed in Saipan and the Americans expected to find on Guam about the same number of defenders as on the other island. Two beachheads were established on either side of Port Apra and the Orote Peninsula, which contained an airstrip, and near the middle of the west coast, about five miles apart. Instead of retiring to the interior, the Japanese preferred to make their main stand in the Orote Peninsula where naval guns including some of 8-inch caliber had been installed and a few dozen tanks were concentrated.

Expanding their beachheads, the Americans began to seal off the peninsula from the rest of the island, while the surface vessels, including battleships, were pounding down on the promontory as in a mortar. It was an inchmeal process, taking until July 28 to overcome resistance in the peninsula and to August 9 until organized defense in the island as a whole was overcome. Again, it was no cheap victory. An interim balance which throws a certain light on the losses incurred in the landing, put the casualties for the first three days at 348 killed. 110 missing and 1,500 wounded, or nearly the Saipan rate. Casualties through August 1 amounted to 1,022 killed, 4,946 wounded and 305 missing, as against enemy dead counted up to that time of 7,419 and of 10,000 up to August 7.

Time, lives and other costs of reducing the Marianas would seem to stand somewhat in direct relationship to their size. Tinian, on which the Marines from Saipan landed on July 23, shortly after fighting had started on Guam, with its 48 square miles claimed American casualties of 208 killed, 1,121 wounded and 32 missing (to August 1), by which time some 5,000 enemy troops had been killed.

The taking of the Marianas was not unlike the boarding and capture of a ship by Marines during the earlier ages of naval warfare—not many of the crew of the boarded vessels would survive, but passengers would usually escape death. Tinian had been under more or less constant fire of American planes, warships and land artillery based on Saipan for nearly six weeks, with the fire immediately preceding the landing stepped up from the 21st on. The Marines landed on beaches directly across from Saipan over a 2½ mile channel and under an umbrella of inter-island artillery fire.

In their progress they received the constant support of fire from warships including battleships brought "nearly to the gates of the town's harbor," as was reported, and of bombing by planes, many of which were now based on Saipan. The campaign for the Marianas was thus an archipelagic war, intended to conquer the necessary military minimum of an entire island group and thereby to learn on a small scale and by actual experience, reducing still larger island groups to the west and the north of this newly won base.

NOTES, CHAPTER 54

[1] Robert Trumbull from Kwajalein. *N. Y. Times,* Feb. 6, 1944.

[2] For tactical minutiae of the fight on the Kwajalein Is. see series of articles by Lt. Col. S. L. A. Marshall in *Infantry Journal,* Aug. to Oct., 1944.

[3] *N. Y. Times,* Feb. 12 and June 12, 1944.

[4] Brodie, *Guide,* 230.

[5] U. P. despatch from Pearl Harbor, Feb. 4, 1944.

[6] Admiral De Lancy, quoted *N. Y. Times,* July 13, 1944.

[7] U. P. despatch from Washington, Sept. 14, 1944.

[8] *N. Y. Times,* Oct. 30, 1943.

Chapter 55

DIEPPE, AUGUST 19, 1942

BY THE SPRING OF 1942 the British Isles were well garrisoned—after the war we shall know how well, or how much over-garrisoned and shall be able to compare this strength, or surplus of strength with that existing from 1914 to 1918. They were well provided with coastal and even inland defenses and fairly crawling with planes of all the United Nations. The urge to re-attack the Fortress Europa was growing stronger, leading to the largely non-military cry for opening *the* second front, although various second fronts had already been opened. These demands generally meant breaching the Atlantic Wall which Germany was constantly strengthening. London and Washington staffs were clearly against this all too-direct approach and, since early in 1942, in favor of indirect assault, by way of the Fortress Africa. But for various reasons, of which the soundest ones seem still unknown to the public, they still agreed to the raid against Dieppe.

The suggestion seems to have originated with the Mountbatten staff. Was it accepted by the higher staffs for the reason that has so often moved superior staffs to agree to the proposals of lower ones: in order not to discourage initiative, to keep men from "deteriorating through inactivity," to give some formations their first battle experience? The Canadian regiments used at Dieppe had been mobilized for three years without seeing action.

The preparation for the raid involved several changes in earlier raid plans. One was quantitative, by employing landing troops by thousands instead of hundreds, including 5,000 Canadians, not complete formations but "large elements of two brigades of the Second Canadian Division" and a battalion of a Canadian tank brigade with heavy tanks. These men, although given a special training, were, of course, less well schooled than the Commandos proper who, besides, had seen action in the past. The second change was in the hour—"We Landed At Dawn," and not in the dark; and the third was in the objective chosen, for instead of open beaches for the landing, a fortified stronghold directly on the water's edge was selected.

It must have been known in London that the town of Dieppe, together with its inhabitants, was nothing but a camouflaged, or reconverted German waterfront fort, with numerous buildings turned into blockhouses and emplacements for machine and other guns. However, for the sampling of the strength of German seaboard defenses it did well enough, besides providing occasion to try out an opposed landing, itself perhaps the most difficult operation in war, to be followed up at once by the one next in difficulty, storming a fortified place and street fighting.

Evaluation of intelligence about Dieppe, planning for cooperation of the forces involved, began early in April, 1942, when the large-scale raid came under discussion. It was an elaborate task, laid down in time tables of Bradshaw size, and in other tables, we are told, and complicated by the variety of speeds with which the various participating units were calculated to move. More airforce than in any previous landing operation was provided. Hope was voiced that the seeming threat of a full dress Allied invasion would bring out the Luftwaffe in strength and enable the Allied airforces to inflict damaging losses on it, which it could ill afford, considering the heavy demands made upon it by the fighting in the East.

While technical cooperation between the forces seems to have functioned well, unity in morale did not prevail to the same extent. While the Canadian government published excerpts from letters by participants, giving testimony of good cooperation, one of the correspondents who took part in the raid, the late A. B. Austin, "met men who were to fight in it who had no understanding of what the air part of the battle may mean," and his impression was, after having lived with the task force for some time, that the soldiers were not given opportunity fully to sense the gravity, danger and details of the whole enterprise through instruction which should have formed "part of this combined operation instructions, so that he may know the significance of what he is doing and be proud of it."

Dieppe "Iron Coast"

The *topos* of the selected *plage* is difficult. Dieppe itself lies in the valley, about a mile wide, cut by the estuary of the river Arques out of a line of steep cliffs, broken by only a few narrow clefts. The beaches at the foot of the cliffs are narrow and rocky, difficult to land on and to debark from. The Germans had done much to strengthen the forbidding character of this "iron coast," including in their coastal defenses in the vicinity of Dieppe two heavy batteries at Berneval and Varengeville-sur-Mer, each less than five miles to the east and west respectively of Dieppe, mounting 5.9 in. guns and howitzers which were to cover the approaches to the town and make it hazardous, if not impossible, for ships to remain off the place within their range during daylight.

If Allied ships were to support the attack and effect and cover the final withdrawal, nothing was more imperative than the silencing of these batteries. This task was consequently confided to the best trained troops, Nos. 3 and 4 Commandos. Bisecting the coast from Berneval to Dieppe and from Varengeville to Dieppe, two further attacks were to be directed, by one Canadian regiment each, against objectives in and about the villages of Puits, near which another heavy gun emplacement was to be eliminated; and Pourville, where a radiolocation station and a flak battery were to be found and taken. Pourville seized, another Canadian regiment was to follow, leap-frogging over the firstcomers, and capture the airfield of St. Aubin-sur-Scie, more than three miles from the sea.

Two other Canadian regiments were to undertake the main attack against Dieppe proper, to be followed so soon as they held the beach by tanks brought to the shore by LCT. It was thought that these tanks would be of great help in street fighting. A further regiment at sea, ready in their landing craft, and the Royal Marine Commando were to wait for further, favorable or unfavorable, developments as a "floating reserve." Fire preparation was to consist of a short, violent bombardment by destroyers and gunboats, to be succeeded by an attack of cannon-firing Spitfires and Hurricanes, directed close behind the Dieppe beachline at the moment when the landing craft was going ashore. No bombing. One does not see that this preparation could have accomplished much more than to alarm the Germans, instead of disorganizing, or softening them as was hoped. As the more outspoken participant in the raid, Austin thought that a heavier and more concentrated air bombardment in the hour before landing would have been a great help, but that perhaps regard for the lives of French civilians had forbidden that.

After prolonged special training, which included carrying of heavy weight equipment, shore marching and which was also

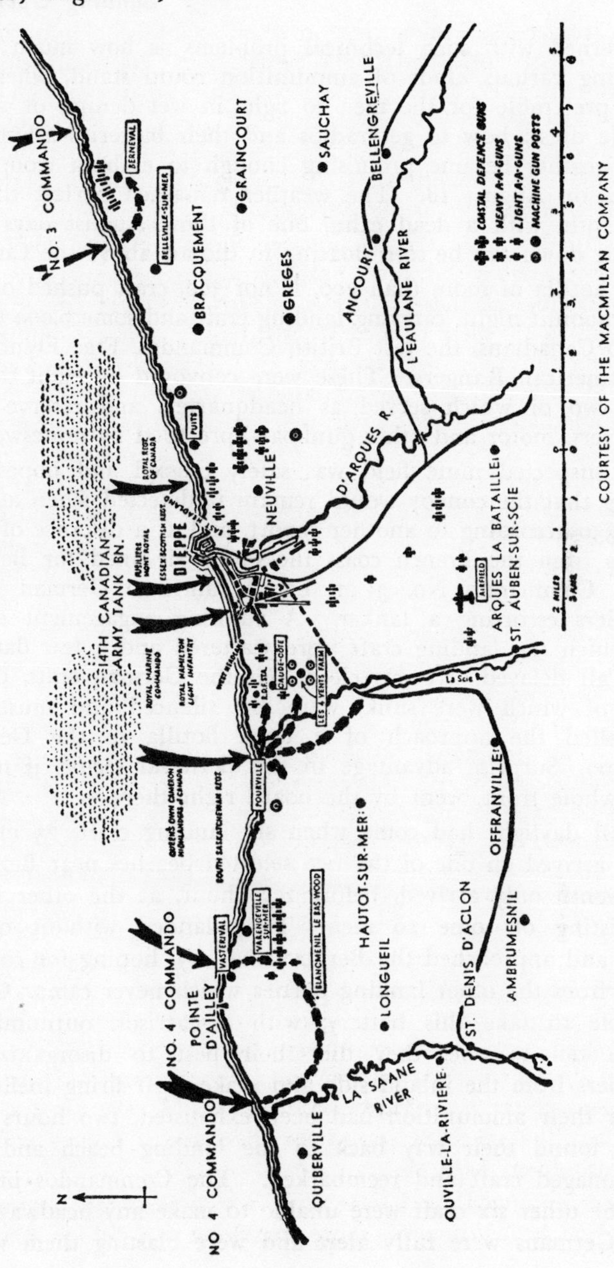

Dieppe Attack, 1942.

concerned with such technical problems as how much water-soaking various kinds of ammunition could stand, whether it was preferable for the men to fight in wet denim, or regular battle dress; how to get radios and their batteries ashore dry, the weather became promising enough to embark troops and stores on August 18. The weather remained perfect through the 19th, with a dead calm, one of those August days when thistle down can be seen floating in the air above the Channel.

A flotilla of more than 200, if not 300, craft pushed off into the moonlit night, carrying landing craft and some 6,000 troops, 5,000 Canadians, the rest British Commandos, Free French and 40 American Rangers. These were convoyed by eight destroyers, two of which served as headquarters and reserve headquarters, motor and other gunboats, preceded by minesweepers.

A suspected mine field was safely crossed and hopes were rising that the convoy would remain undetected when at 0347, or 0330 according to another report, and at a distance of seven miles from the French coast the group destined for Berneval with Commando No. 3 on board ran into German armed trawlers escorting a tanker. A running engagement ensued in which the landing craft were scattered and a few damaged and all delayed. It was clear that the German craft, one or two of which were sunk, were not silenced and must have signalled the approach of a large flotilla to the Germans ashore. Surprise advantage in the Berneval sector, if not on the whole front, went by the board right there.

Full daylight had come when six landing craft, 25 minutes late, arrived on one of the two selected beaches near Berneval; a seventh only arrived, before zero hour, at the other beach, consisting of some 20 men. They landed without opposition and approached the Berneval battery, hoping for cooperation from the other landing parties which never came. Clearly unable to take this battery, with a garrison outnumbering them ten to one, they did their best to disorganize the gunners from the inland side and make their firing ineffective. After their ammunition had been exhausted, two hours later, they found their way back to the landing beach and their undamaged craft and reembarked. The Commandos brought by the other six craft were unable to make any headway since the Germans were fully alert and were blasting them with a withering fire from prepared positions. The survivors eventually surrendered, most of them wounded.

Sharp Work by Commandos

The attack carried by the right wing of the force, against the battery at Varengeville, undertaken by some 250 men of No. 4 Commando, was far more fortunate, on sea and land. Landed in two sections to the right and left of the battery, these two like the unevenly moving claws of a pincer came up behind the battery, the smaller group firing into the munition dump and setting it on fire by a lucky mortar shell and sniping at the German gunners as they attempted to extinguish the fire, while the other group was moving into an assault position. From there, after low flying Spitfires had discharged their cannons into the battery, very much on schedule time, they carried the battery, all of its garrison being killed, except four who were carried off as prisoners when No. 4 Commando re-embarked at half past seven, after having blown up the battery's six guns and fulfilled their task. Their losses were two officers and nine other ranks killed and three officers and 32 others wounded. The two outer wings had done well enough for the forces still on the water, enabling them to hover off the coast throughout the day.

The two landings by the Canadians nearer to Dieppe, at Pourville to the west and at Puits to the east, where in addition to other objectives they were to secure the two headlands dominating Dieppe and thus keep the enemy from interfering with the landings on the beach in front of Dieppe itself, were on the whole unfortunate. Deflected from its course by the encounter with the German trawlers, the unit scheduled to land at Puits at 0450 arrived about 20 minutes late, in broad daylight, and with the enemy ready to repel them from prepared defences. They brought the Canadians under fire even before they had "touched down" and increased their rifle and mortar fire after the invaders had sprung ashore. Casualties were heavy, particularly among officers.

In spite of these heroic endeavors the headland was not gained and from it the Germans dominated both the landing beach at Puits and the more distant one at Dieppe. The Canadian regiment at Puits remained pinned down by German fire. Some were evacuated after their failure had become apparent, others fought on until late in the afternoon when they had to surrender. At Pourville surprise allowed the Canadians to get ashore against very little initial resistance. Their first objectives were reached, including the headland which was, however, soon after lost to a German counter-attack; a few prisoners were made, but enemy resistance stiffened soon and held them up until the time of the withdrawal signal. Through the Pourville beachhead the other Canadian regiment scheduled to come in on a second wave penetrated inland, but not far enough to reach its main objective, the St. Aubin airfield, by the time of reembarkation. Their retreat was costly, most so over the open 350 yards stretch from the cliffs to the landing craft which were constantly forced seaward by the ebbing tide.

The timing of the landings seems to have been based on the landing of the tanks which was to take place after daybreak. Hence the timing of the first landings, those of the Commandos, shortly before dawn. They were so planned as not prematurely to alarm the enemy in Dieppe. This meant a very close and inelastic sequence and left little, or no allowance for the contingencies which were met.

The landing at Dieppe, at about half past five, was too late to profit from the protection of the dawn twilight and the obstacles encountered by the immediately supporting attacks kept the latter from giving any noteworthy support to the Dieppe attack. The Dieppe beach, some 1,700 yards long, measured from the breakwater, was separated from the seaside esplanade, behind which houses and hotels stood, fortified by the Germans, by a seawall paralleling the shore line. Strengthened by the defenders with fences and clusters of barbed wire, it came to play as fatal a rôle as an intermediary obstacle as the reef ring at Tarawa. This two or more of the Canadian regiments were to seize, enabling the tanks to land immediately after, and to push into the town and hold it while the engineers carried out their demolitions.

"Although the Dieppe defenses had been heavily engaged before the assault," as the Canadian report characterizes the preparatory fire of destroyers and the 60 cannon-firing planes, German fire showed no weakening when the regiments ran from the landing craft. Most of the men, it seemed, were pinned down along the sea wall where they met cross fire as well as frontal fire, particularly after the invaders' smoke screen had cleared away and allowed better vision to the guns which the Germans to the eastward had installed in caves dug in the cliffs.

Hard Going for Tanks

The first wave of tanks followed shortly after the infantry in six LCT's, which brought also demolition groups to blast tank obstacles in the streets of Dieppe. The landing craft received their share of fire and all were hit; but while one sank and a second ran aground, all but two tanks of the first wave were landed. On one LCT the ramp chains were severed the moment a tank was going through the door; the ramp fell off and the tank which carried the tank unit commander's pennant, rolled off and sank in eight feet of water. The second wave, thirty minutes later, was hit even harder, one LCT being sunk off the beach. In all 28 tanks were landed, none of which came back, it being understood that in reembarkation personnel was to be given strict priority.

A few tanks got over the sea wall and into the boulevard; the majority, it would seem, were either unable to cross the wall, or were disabled by damage to their treads. One or the other of these immobilized tanks opened fire against the Germans. Owing to the heavy losses among the sappers during the landing, the tank obstacles in the streets could not be removed, and this "altogether unexpected strength of blocks at the ends of streets" (Churchill) kept the tanks from penetrating into the town with the exception of one which crashed through a non-fortified house. One landing party, on the right, was able to take the heavily fortified Casino, the building closest to the beach, but everywhere else the way was blocked by German obstacles and fire.

By 0630, about an hour after the first landing at Dieppe, it appeared to the Military Force Commander, General Roberts, who had received no report of the unfortunate happenings at Puits, or Berneval, but to whom the situation at Pourville and Varengeville seemed not unpropitious, and who had just had word that the tanks had landed, that the situation either permitted, or demanded—his reasoning has not been disclosed—the landing of part of his floating reserves. Besides, enemy fire had slackened temporarily.

A French-Canadian regiment was ordered to land and establish itself on the beach at the edge of Dieppe and possibly take the eastern headland. They reached shore soon after seven, enroute suffering the loss of two landing craft, and heavy casualties so soon as they touched land, when the German fire seems to have been intensified. More than half of this regiment in landing were tugged by a strong tide to the west, beyond the main Dieppe beach, where they found themselves on a narrow stretch of shingle and rock at the foot of unscalable cliffs. Enemy fire, partly from above by mortars, made it impossible for them to deploy either to the right, or the left. Thus imprisoned, after suffering in excess of a hundred casualties, they surrendered by midday. While two parties of the Quebec regiment with more luck worked their way into the town and in the direction of the docks, their main objective, the eastern headland, remained as firmly as ever in enemy hands.

Stiff Resistance at Beach

Shortly after seven, the Military Force Commander, who now knew that the heavy battery at Varengeville had been destroyed, that his troops had penetrated beyond Pourville, that others held the Casino, and who thought that Dieppe might yet be taken, but whose immediate observation was extremely limited and who was unaware how far the eastern headland was out of his control, decided to call in more of his immediately available reserves. These consisted of the Royal Marine Commando, which was specifically ordered to assist in taking the eastern headland.

After the necessary landing craft, lying offshore and at intervals wrapped in frequently renewed smoke screens, had been assembled, they moved to the shore by 0830. At first, they progressed under cover of the smoke off shore which was at times as thick as in an old-fashioned battle and under fire support from a river gunboat and support craft. But once this zone of protection was passed, they were caught by the full force of German fire, against which their own Bren guns firing over the gunwales could avail little. Not many of the firstcomers reached shore unhurt. There was no hope that they could do better than those already landed, and the situation was clearly out of line with what the MFC had imagined it to be. Realizing this, in a heroic resolve to disobey orders, the commander of Marine Commando, donning a pair of white gloves and jumping to the highest deck of his landing craft, waved to the following craft, with some 200 more men on board, to turn around and desist from the futile attack. They grasped the meaning of their commander who fell mortally wounded as he redirected this traffic to retreat from certain disaster.

The engulfing hell into which the Dieppe enterprise ran, luckily for the attackers, did not extend into the air above them. For the Allies, who flew some 2,000 sorties, retained control of the air during most of the day. Air-cooperation with the ground forces is reported as "faultless," the latter most often requesting and receiving smoke curtains. The data of the Allies on German plane losses gave 93 craft destroyed, about half of them bombers, 44 probables and 148 damaged, as against Allied losses of 98. In the British debit score the accidental sinking of one British destroyer by a German plane that had been forced to jettison its bomb load, it is said, should be included. Much has been made of this "triumph in the air" which forced the Germans to come out of hiding and, according to the Canadian statement, suffer losses "amounting to a very serious strategic reverse" at the hands of the RAF and other Allied air forces. With all due allowance made to the hazy usage of the term "strategic" in air operations, this seems definitely more of a claim than a non-political exposé of the raid would reveal.

The most exuberant German claim after Dieppe was that the British had come with the intention to stay, that this was indicated by the presence in their rear in the Channel of "a reserve of six transport ships and three merchantmen and, further in the north, a group of 26 transport ships as an active force, probably a large mass of landing forces. The latter were to get into action as soon as the first landing wave had succeeded in forming a bridgehead around the Dieppe port. It did not come to that."

This communiqué embodied the *Leitmotiv* of the German propaganda: That the British were unable to maintain themselves on the continent for more than nine hours; the spearhead of a big invasion attempt had been smashed by the German local forces alone; after that had become clear, the waiting forces had steamed away; the Festung Europa was impregnable. The Allies never deigned to discuss these German claims about their more extensive designs on August 19 which, the Germans said, were evident from papers found on the prisoners. They have never stated when, after how many hours' stay, they had originally intended to retire from Dieppe. About the decision of their commander on the 19th to withdraw we learn officially that "the withdrawal from the main beaches was timed to begin at 11 A.M." This would mean that he must have realized by ten o'clock that further fighting was hopeless and that orders for reembarkation must be given.

This withdrawal in the face of an unshaken, if not triumphant, enemy was necessarily a dangerous and difficult operation. The whole beach of Dieppe and beyond was under concentrated fire with which Allied air cover would interfere but slightly. Most merciful and effective protection proved to be a smoke curtain laid from headland to headland on either side of the town—"Hung be the heavens with black, yield day to night." Behind it the landing craft came in; behind it and through it the troops endeavored to retire.

Reports for Public Benefit

By 1220 most of those who could be reached on the shore had been taken off by the landing craft "who showed complete disregard of danger in their efforts to take off the troops" (Official Report). A further dash to take off men if more could be found, was made by the Navy shortly before one o'clock. But no larger groups of men could be discovered and at 1308 hours a brigadier general signalled from Dieppe that he was forced to surrender. While the last smoke curtain blew away, the curtain of war politics came down over the historical scene, and behind it the lights were manipulated to make the various publics concerned see things as officialdom wanted them to be seen.

The first thing done behind the lowered curtain was for either party to count losses, draft communiques and give the enterprise a "sense" most acceptable to the English, Canadian and German audiences. On the Allied side this did not pass off without hitches. While Churchill announced on September 8 that most of the attackers had returned, meaning the military, naval and air force, as a whole, the Canadians a week later gave out a statement that out of 5,000 Canadian troops engaged, 3,350 were killed, wounded or missing, a total which was later raised to 3,372. Of these 593 were killed or died of wounds, 1,901 were prisoners and 287 missing, with 591 returned wounded to England. A casualty rate of 65 per cent in a fighting force—it had been 73 per cent at St. Nazaire—is indeed a high rate for modern combat, and the Dieppe experience, as did the Tarawa casualties, made the people at home perhaps unduly apprehensive of disasters to be expected from landing operations. There is no typical expectancy of casualties in such operations; they are too dissimilar to cause averages to mean a great deal.

Some of the German claims which emphasized the high casualties among Allied staff officers—including among the 105 officer prisoners a Canadian brigadier, two colonels, 13 staff officers—are not incompatible with data from Allied sources. The number of 2,195 prisoners claimed by the Germans is not too far from the Allied estimate, but there was a wide gap between the German casualties as reported by the Oberkommando and the London estimates of these losses, 4,000. From the nature of the fighting, with the Germans in prepared positions almost everywhere, and the limited opportunities for the invaders to make and report reliable observations, this would seem a dubious figure indeed, even if the casualties admitted by the Germans should be vastly understated. These were as follows:

Army: 115 killed, 14 missing, 187 wounded.
Navy: 87 killed and missing, 35 wounded.
Airforce: (incl. Flak) 104 killed and missing, 58 wounded.
Total: 320 killed and missing, 280 wounded.

Rather more fantastic were the German claims as to shipping losses suffered by the Allies, who admitted that they included a "fairly large number of landing craft."

"Reconnaissance in Force"

Official statements from the Allied side stressed "that as a combined operation the raid was a successful demonstration of coordination of all three services" (Joint London bulletin of Aug. 20). It was clearly agreed upon to classify it as a "reconnaissance in force," Churchill telling the Commons (Sept. 8) that it "must be considered" as such, and the Canadian official statement doing likewise.

"It was a hard, savage clash such as is likely to become increasingly numerous as the war deepens," said Churchill who was to see, or to see to it, that clashes of this kind were not actually repeated. "We had to get all the information necessary before landing operations on a much larger scale. This raid, apart from its reconnaissance value, brought about an extremely satisfactory air battle in the west, which Fighter Command wish they could repeat every week."

Shocked by the high casualties, Dieppe reminded the peoples of the Dominions and the other English-speaking nations too closely of Gallipoli to enable them to rationalize or find much consolation in these accounts, even if such enterprises were insisted upon as necessary preliminaries to the invasion of the continent. Among their experts who were in position to express themselves most freely, the amphibious veteran Lord Keyes called Dieppe "ill-conceived and ill-fated" and "rather an expensive form of experiment."

To call it a reconnaissance in force, "conceived with the important object of obtaining information and experience vital to the general offensive program," set in motion when "facts essential to the successful prosecution of offensive operations can only be gained by fighting for them," is taking recourse to a somewhat suspected word in the military dictionary. Reconnaissance, says one of the most recent military encyclopedias, "demands in an equal measure cunning, daring and high tactical understanding. Its purpose is to see much and report quickly. Fighting must only be contemplated if the purpose cannot be achieved in any other way. Often adroit evasion and flanking will lead better to the objective in question than attack. If it becomes necessary to thrust through a veil standing in the way of reconnaissance, quick local concentration of forces and surprising push is most apt to achieve results. Right proportioning in the investment of forces shows as everywhere else the art of leadership in reconnaissance; overinvestment means premature consumption of strength." [1]

It means various other things as well: doubt of the sense of the enterprise, doubt as to whether its dearly-bought results could not have been obtained by spying and questioning, doubts whether what was learned could not have been ascertained earlier than 10 A.M. on the day of Dieppe.

It was also asserted that it was most important for the Canadian forces that they "should have an opportunity for practical experience in the landing on an enemy-occupied coast of a large military force, and in particular in the problems arising out of the employment in such a force of heavy armored fighting vehicles" (Canadian statement). But is not such learning in a way invalidated if more than half the learners do not return to make use of their lessons, including many staff officers who were presumed to learn most on such occasions?

Aside from such "strategic" considerations as they have been called, there were also tactical or local objectives, involving the hoped-for destruction of enemy installations in the Dieppe area, of a harbor useful to the enemy in his coastwise shipping; marshalling yards, gas and power plants, some factories. None of the latter suffered much or long lasting damage except a tobacco factory, situated close to the beach, which was burned in the street fighting. The damage to German war economy was therefore slight, less than might have been wrought by a single flight of bombers.

Results Were Negative

The conclusions and obvious lessons to be drawn from Dieppe are on the whole negative. No Allied enterprise of any consequence followed Dieppe on the Atlantic front, where there were only a few Commando raids (as against Sark, Oct. 3, 1942 and December 28, 1943), until two years later, and this was on a totally different scale. No Commando raid involving 6,000 men was again undertaken, which shows that the Allied commanders did not consider it an optimum size for such operations. Was Dieppe intended, by those who conceived the enterprise as a demonstration of what casualties in cases of opposed landings would be likely? While it showed that a small, or medium size force unable to capitalize surprise is apt to pay an extremely high price for a failure, or for whatever it might have as its objective, from this does not follow, as from a physical or chemical experiment or test, that a very large force in a similar enterprise would have to pay the same ratio of losses. As a project for sampling methods and operational expense, Dieppe was ill conceived. Even as a reconnaissance in force it was not given all the force that could have been made available, such as heavy fire from naval guns and air bombardment of the beach. Parachutist and glider landings in the rear of Dieppe would seem to have been hardly advisable, considering the density of German military occupation in the vicinity.

This density proved fatal in still another respect. A town such as Dieppe, with harbor facilities, will attract in like manner defenders and attackers. The latter, should they come with far-reaching intentions, will be guided by the hope of obtaining at the outset the harbor facilities which will eventually be needed. Almost as unavoidably, the defender will center his defense installations and forces in and near such a spot. Realizing this likely concentration of strength, either before or after Dieppe, the Allies everywhere in their subsequent amphibious operations avoided landing near such centers, however great the temptation to win port facilities at once.

The motives behind the Dieppe raid and its consequences, once they are fully disclosed, promise to be more interesting in military history than the enterprise itself. Was it also, with its not unlikely failure, intended to silence the vociferous cries for the "second front" that for the time being could not be opened in the west of Europe? As to consequences, for a time ominous reminiscences of Gallipoli were revived, particularly in the Dominion that had not participated in that earlier disastrous expedition, even though at Dieppe everything was done in a much more accelerated, breathless way, as if to keep that memory suppressed. Among the altogether discreet military critics Admiral Keyes left no doubt that he considered Dieppe a failure of the "colorful" Mountbatten style of amphibious warfare.

In a way the sixth war conference of Roosevelt and Churchill in Quebec in August, 1943, confirmed this when they appointed the young Lord as Supreme Allied Commander in southeast Asia, a theater of war which seemed to call for amphibious warfare, but where, however, such enterprises were bound to proceed in a conservatively offensive way and actually advanced on land rather than in the amphibious style which the sending out of Mountbatten appeared to presage. It was another sign of disapproval when even before him the military commander at Dieppe was retired.

The Dieppe raid, however limited its objectives, provided a test of the spearhead style of landing, of landing on a narrow front. This style might be required, or advisable in landings on limited and isolated objects such as islands, whereas it seemed definitely inadvisable in the case of large assaults designed to force a break-in, or penetration and eventually a break-through against an extensive enemy position. Large-scale landings have to take place on a broad front. Dieppe was a middle-size attempt against large-scale holdings. It proved an unsatisfactory hybrid in this war, in which medium-size actions, except in strictly isolated cases, have proved disappointing and ineffective as compared with large-scale operations or, on the other hand, Commando raids.[2]

NOTES, CHAPTER 55

[1] Franke (ed.), *Handbuch der neuzeitlichen Wehrwissenschaften.* II, 75, art. *Aufklärung.*

[2] This chapter is based on the communiqués issued about Dieppe, statements by Churchill in the House of Commons on Sept. 8 and by the Canadian Minister of Defence of Sept. 18, 1942; *Combined Operations,* 110-146; A. B. Austin, *We Landed at Dawn. The Story of the Dieppe Raid.* N. Y., 1943; Quentin Reynolds, *Dress Rehearsal. The Story of Dieppe.* N. Y., 1943.

Chapter 56

NORTH AFRICA LANDINGS, NOVEMBER, 1942

TWICE IN WHAT AMOUNTS TO this Thirty Years' War against Germany have the Americans regained the initiative in the western theater of war for the enemies of the Reich—in 1918 in France, when the offensive had been made feasible through the arrival in large numbers of the AEF, and again in 1942 in North Africa. Both times their troops were largely new to war as it was being fought on the day of their intervention. The necessary schooling in 1917-18 began in quiet sectors of the western front. The training through combat in 1942 started outside the doors of Europe in Northwest Africa on November 8, 1942. It was a backdoors approach to the Festung Europa, a more direct approach, the "opening of the second front" being considered unfeasible even by the Russians when Stalin in August, 1942, told Churchill that this conclusion "was militarily correct."

Still, in order to be able to create pressure in favor of the Russians across the Channel, where the final blow was to be delivered, but not before really enough men and shipping were available, the Allies agreed to an emergency plan for a diversionary attack on the French coast, in case Red Army resistance to the German onslaught should threaten to collapse. This was to be undertaken by at least six divisions, if the diversion was to be of any value to the Russians, though the problem of gathering enough landing craft for this purpose was for a time serious enough.

The small beginnings of great plans are not easily dated. There are indications aplenty that the North African plans preceded by a year or more the more formal decisions made during Churchill's visit in Washington in January, 1942. The continuance of relations with the Pétain government, the spying through American consulates in North Africa, the conspiring between American agents and Frenchmen not so overly *Pétainiste,* the buying of a certain amount of goodwill in North Africa by maintaining a minimum of commerce with Morocco and Algiers point to the conclusion that the campaign in these regions was American in inception and that therefore it should be, as Churchill agreed, "an American enterprise . . . . under American command."

Politics, rather than purely military considerations, which can only rarely prevail, made it so. During the Prime Minister's next visit, in June, 1942, when more American forces had become available, the North African campaign was again considered and finally resolved upon in July. It was to take place on, or around May 1, 1943, to coincide if possible with an insurrection in France proper. Still later, the plans were "conjuncted" with the planned offensive of the British Eighth Army from the El Alamein line. As to the authorship a *loyal serviteur* like Sumner Welles (in *The Time for Decision*) insists that the original idea was President Roosevelt's, so confirming Churchill's compliment in his Guildhall speech shortly after the landings: "The President is the author of that mighty undertaking. In all of it, I have been his active and ardent lieutenant."

While the North African enterprise was "in point of numbers the largest military movement over the largest number of miles to landings under fire" (Roosevelt's Message to the Congress, September 17, 1943), it was also among the more recent landings the one most under the impact of politics, not only the politics of coalition warfare, which proved fairly simple in the case, but also the politics and diplomacy of winning over France, a tired-out neutral and ex-combatant, to the Allied side, instead of making it an open enemy and driving it still deeper into the camp of Hitler's New Europe. The statement made in the first American communiqué on the North African landings, that "the operation was made necessary by the increasing Axis menace to this territory," that therefore a preventive action on the part of the Allies had become necessary, was only true to the extent that the Nazis had discovered something about the Allied intentions and on their part were getting ready to be beforehand.

It seemed almost as if the Allies had read the one page in Clausewitz dedicated to landings, in which he emphasized that a landing force will be much favored in its attempt if it finds aid and support from elements hostile to the invaded power. The latter in this case, was Germany which had her Armistice Commissions and Gestapo in these regions. In an age when everyone connected with war making and everything intended for war must be organized and often camouflaged into the bargain, aides of the Allies from the inside in North Africa were assembled in and behind a Vichy youth organization which was to imitate the Hitler Youth and the German Labor Service, the *Chantiers de la Jeunesse* in its North African branch.

A Hurry-Up Operation

Through the spying activities of this organization it was ascertained that the Germans intended to forestall Allied plans and themselves take over French North Africa around January 1, 1943. Preparations such as the assembling of water and air transport in Sicily pointed to this. Determined to be beforehand, the Allies accelerated their preparations, which meant among other things foregoing the use of various landing craft which would not be ready before the spring of 1943. As it was, the wish to start the operation in the very early autumn of 1942 proved impracticable, there being as yet not enough landing craft in readiness, since their crews had yet to be trained.

This enforced acceleration left detailed understandings with the anti-Vichyites in North Africa far from completed. To reach these as far as possible, General Mark Clark was sent into Algeria by submarine, during the third week in October, to meet and hold secret council of war with the inside workers. While they furnished him with much information about the French forces, their attitude and loyalties, much of which would remain uncertain until the acid test came; and on equipment, which was largely outmoded and almost devoid of heavy armament, supplies, airfields and more important still, knowledge of the state of mind among French groups, the patriots declared themselves far from ready to cooperate in an early landing.

Originally set for the night of November 27-28, the date was advanced by three weeks by Allied Headquarters where suddenly doubts had arisen whether their confederates in the French possessions had not allowed the secret to leak out and become known to the Germans. If the enemy should intend to come in before the Allies, the latter must not lose a day. Hence the decision, imparted to the patriots on November 1, that the landing would take place on the 7-8th of the same month, which they thought was premature as far as the political preparation for the intended invasion was concerned.

The *Pétainiste* sentiments among the French troops in North West Africa were believed to be so strongly anti-British—after Mers-el-Kebir, Dakar and Madagascar—that all emphasis was on the employment of American troops and American commanders to whom it was hoped the French would show less, if any, hostility. Hence the arrangement that American troops were to make the first landings at Oran under an American general from whom the British General Anderson would take over once the Americans had made good their landing. British forces actually followed the Americans who served as political pacemakers and assault forces at Algiers, and later at Bougie (Nov. 10) and at Philippeville and Bone still later.

Further to overcome French aversions to foreign troops acting on their soil, it was hoped by the leaders of the resistance movement that General Giraud, who had been approached already by American emissaries in France and was to be brought over to Africa, would be invited to become supreme commander of the operations following the landing. Giraud himself had come to hope for this, imagining himself as the future commander of both French and American forces. The Allies remained non-committal on this point for some time and only after Giraud had been taken out of unoccupied France to Eisenhower's headquarters at Gibraltar on November 7 did he learn that no such vast rôle was reserved for him. Happily he agreed, allowing no *amour propre* to stand in the way of unity and thereby inviting more gratitude on the part of the Allies than proved fortunate for them in the long run, when he and de Gaulle began to compete for the supremacy.

Three Task Forces Planned

The plans for the North African campaign as worked out in detail and considerable haste under General Eisenhower as the designated commander of the Allies forces, by an inter-Allied staff in London, since July, 1942, provided for the employment of three task forces who were to strike simultaneously at points in the Mediterranean and the Atlantic coasts of Morocco and Algeria. One American force, consisting of the Third Infantry, the Second Armored and the larger part of the Ninth Infantry Divisions under Major General Patton, was to proceed directly from American ports, Norfolk and vicinity, to the attack points near Casablanca. A second American force under Major General Fredendall, to consist of the First Infantry

Division and one-half of the First Armored Division, coming from England and escorted by the British Navy, was to land at various points around Algiers. Air-borne forces including elements of the 82nd United States Air-borne Division and British elements were to be sent from Britain to the vicinity of Oran in the, up to then, longest distance operation of an air-borne character; and additional elements were water-borne to Oran, to be under the command of General K. N. A. Anderson of the British Army.

The forces entering through the Straits of Gibraltar, including the land-based airforces in that fortress, had to pass a bottle-neck of great potential danger which happily did not develop. The armada coming from America departed on October 24; that from Britain one day later.

The bolder thought of landing still further east, possibly as far inside the Mediterranean as Bizerte, was abandoned nearly from the outset, due according to General Marshall to the lack of shipping of all kinds, transports, landing craft and aircraft carriers. Some other reasons would seem to have weighed also, such as the threat of a German move through Spain and Spanish Morocco, which for several weeks kept the whole American Fifth Army nailed down near Oran and three French divisions as well; and the uncertainty about what hostile Vichy-French might do to the long rearward communications of the Allies, from Bizerte.

On the other hand, the almost traditional hesitation of sailors to proceed against such a strongly fortified and intact naval base at Bizerte and risk ships in the attack, seems not to have entered. As the British commander-in-chief of the Allied North African fleet, Admiral Cunningham, told Ward Price, this "would probably have entailed the sacrifice of at least one battle ship, several cruisers and very likely 25 per cent of the transports . . . . The Navy was prepared to risk it . . . . But on consideration it was decided to take the less hazardous course."

This want of boldness, much discussed since, was to cost the Allies months—and all that months mean in war—for they allowed the Germans to get into Tunisia and fight there for six months. This caution would seem part and parcel of what Nazi General von Kluge, almost plaintively, called the Allies' habit of "conducting the war with methods of certainty." [1]

Whether this is due to the cautious methodical habit of Allied heads of states, chiefs-of-staff and commanders primarily, or to the willingness of the Anglo-Saxon peoples, to which they have to pay close attention, to spend time, rather than lives on a bold throw, the thought remained uppermost in the making of the plans that if the Allies tried and failed in an original push against Bizerte, the whole North African campaign might founder.

The purely military obstacles in the way of the Allies' forces proved far less difficult than the political ones of landing among and against troops who were not yet persuaded that for whatever reasons were presented to them, they should desert from what they regarded as the legal authority of Pétain. The Allied forces ordered not to shoot first, naturally "proceeded on the assumption that determined resistance must be expected. But how were they to deal with undetermined resistance? A code signal 'Play Ball' was to be broadcast to the entire force at the first hostile act on the part of the French in any sector, as a warning to initiate vigorous offensive action." [2]

French Resistance and Fumbling

The French were not at once ready to play ball, and least so in Morocco. Here the main landing was to take place at Fedala and secondary ones at Fort Lyautey (Mehdia) and Safi, all to the north and south of Casablanca. These landings, although made with "comparatively little difficulty" (Admiral King's report), betrayed not a little fumbling on the part of the newcomers to the war—craft missing, or mistaking their assigned beaches, lighters upset by the surf, not a few drownings. It was a favorite subject of talk among American troops "about how far which assault waves landed from which beaches, and how they found their way back." [3] Some had touched down as far as twelve miles off their destination, but none erred farther afield than some parachute battalions after their 1500 mile hop directly from England. Their navigators failed them utterly at the end of their long flight to the vicinity of Oran,[4] demonstrating the greater difficulties of air-to-land navigation as compared with water-to-land orientation. "Junior officers too often failed to show initiative and the troops in general betrayed their greenness. Had we been meeting well-prepared Germans instead of French troops, whose heart was not in the business, the result almost certainly would have been a bloody repulse." [5]

After American ground forces had been landed at Fedala in the very early hours of November 8, French shore batteries opened fire against the American naval forces supporting these landings. While the troops proceeded against objectives, the ships returned the fire of the batteries, which were taken in the early afternoon, and attacked a number of French war ships making sorties from Casablanca which were practically all sunk, or beached.

Around Algiers, largely due to the preparatory labors of French helpers, resistance was weakest, but at Oran an Allied naval contingent of four small ships which tried to break the boom blocking the harbor and disembark Anglo-American troops, was caught by the searchlights and the cross-fire of the defenders and suffered heavily. On the whole, there was more obstruction from the Vichy Navy than from the military force.

Resistance and interference from the "true enemy" came late and in weaker strength than might have been expected. None of the incoming convoys suffered losses outside Gibraltar, but on the 7th a transport on the way to Algiers and 160 miles from there was torpedoed. The troops had to be put into landing craft, some of which were subsequently lost. German aircraft sank another transport on the evening of the 8th and on the 11th and 12th a destroyer, an oiler and three transports, the latter empty and on the homebound trip, were torpedoed. Altogether, the Allies stated their shipping losses as less than three per cent of the near one thousand ships employed.

After the Allies had made good their landings, they found themselves wading knee-deep in the mire of politics and diplomacy rather than in the blood of battle. Their basic assumption had been that they would make allies, and not merely neutrals, of the 50,000 to 70,000, if not 100,000, French troops who momentarily outnumbered them, as well as of the French administrative officials. They were much in need of such aid, for keeping the Arabs peaceful, for maintaining order in the long communications, including a 1,200 mile supply line from Casablanca to the Tunisian front and the rear zones, and for armed help in the struggle as well.

But who was to be the Leader to bring these French organizations into the Allied camp? In the judgment of the Allies, de Gaulle would prove altogether so inacceptable to the Pétainistes in North Africa that he was not even informed about their intention of going there. Giraud, talked out of his hopes for a supreme command over all Allied and French forces and contented by designation as head of French civil and military affairs, once brought over to Algiers, had to confess that he could not control the French forces, least those in Morocco where fighting around Casablanca still went on. The military, that is to say the officers, preferred to stick to the only supposedly legal power, that of Vichy, and the Allies and their conservativism seem never to have thought of appealing to the soldiers over the heads of their officers. They preferred, for some sound military reasons, to have the transfer from Vichy authority take place without any upsetting of existing French military and administrative hierarchy, gradualism rather than anything that smacked of the revolutionary.

Happily for this military legalism, the representative of *Pétainisme* in the person of Admiral Darlan, commander-in-chief of all French forces, was in North Africa at the time on private business. That legally armed, although not otherwise especially commendable, personage was prevailed upon on the 10th to sign an order for general cessation of hostilities. The order, followed by one making Giraud military chieftain, caught on, although Pétain, who was acting under duress, so Darlan declared, repudiated the latter at once.

Darlan's Aid Utilized

The transition of the North Africa French to ally status was thus engineered, with Darlan as a very undesirable wielder and transformer of authority and thaumaturge in the handling of military and other loyalties and oaths. But not before the 14th was Darlan, who momentarily seems to have considered an earlier Axis counterlanding possible, in which case he would be found neutral, brought around to the side of the Allies. This was too late to extend his own and the Allies' influence into Tunisia where some 150 Axis airplanes arrived by midday of November 10, not interfered with by *Pétainiste* commanders who had orders from Vichy not to oppose them, as they could easily have done.

The arrangements with Darlan, called a "temporary military experiment" by Washington, in many eyes crowned a glorious beginning of the liberation of Europe with ignominy. But the military necessities were considered too weighty to yield to such considerations. As General Eisenhower explained them two months afterwards:

> The whole political situation here arose from my attempt to take Tunisia with a rush. If I had been content to sit down and occupy Algeria, I could have maintained control over its local political affairs, though that would have required a lot of troops spread about the country. By pushing on rapidly into Tunisia, so as to seize as much ground there as possible before the Germans got it, I created a sort of vacuum between my front in Tunisia and my base in Algeria.
>
> In this way I lost strength to impose my will on the French of North Africa as I could have done if I had been content to hold Algeria. My 600-mile-long line of communications, with many vulnerable bridges, is my weakness . . . . From the military point of view, in fact, I am not strong enough to impose a political solution. That is why I don't interfere with the French."[6]

When the first contact with the enemy inside Tunisia was established, on the 16th, French forces were already in action together with the Allies.

Like everything else in these landings, even the casualties on the French side were tinged with politics. Resistance during the two days of fighting was spotty and often no more than of token character. It was determined in a few localities, as is indicated by the high ratio of dead to wounded. Besides, figures point to the fact that the French officers were far more determined to resist, were more *Pétainiste* than the rank and file. For during the two days of fighting these casualties were as follows:

	Killed	*Wounded*
Officers	59	72
All others	431	897
	490	969[7]

The American casualties during nearly the same time, to which would have to be added those of the British, were as follows:

U. S. Army .... 960 killed and wounded—350 missing
U. S. Navy.... 10 killed, 150 wounded, 150 missing

The five hundred missing must be considered losses by drowning, as the Washington communiqué stated.[8]

NOTES, CHAPTER 56

[1] Cit. *N. Y. Herald-Tribune,* July 18, 1944.
[2] General Marshall's *Report,* 126.
[3] Ralph Ingersoll, *The Battle is the Pay-Off.* Inf. Journal ed., 62-3.
[4] For details see Col. Edson Raff, *We Jumped to Fight.* N. Y., 1944.
[5] Hanson Baldwin in *N. Y. Times,* May 12, 1943.
[6] G. Ward Price, *Giraud and the African Scene.* N. Y., 1944, 160-1.
[7] A. P. despatch of Nov. 23, 1942.
[8] *N. Y. Times,* Nov. 24, 1942.

This chapter is based on General Marshall's *Report on the Army;* G. Ward Price, *Giraud;* John A. Parris and Ned Russell, in collaboration with Leo Disher and Phil Ault, *Springboard to Berlin.* N. Y., 1943; a series of articles by Hanson Baldwin in *N. Y. Times,* May 11 to 22, 1943; René Gosset, *Le coup d'Alger.* Montreal, 1944.

Since this chapter was written, Winston Churchill published (*Life,* February 4, 1946) his speech in the secret session of Parliament, December 10, 1942, dealing with the North African landings and, in particular, with the transactions with Admiral Darlan.

Chapter 57

THE NTH CONQUEST OF SICILY: "OPERATION HUSKY"

OF THE VARIOUS LARGE-SCALE disengagement manoeuvres of recent military history where the lines of retreat lay across the seas, Dunkirk, Greece, Crete, Tunisia and Sicily, Tunisia was the most disastrous for the retreaters. Blocked by Allied command of sea and air, the Axis enemy was unable to remove any considerable number of men, or equipment or even his high command. These various affairs have been tabulated by Hanson Baldwin as follows: [1]

	Dunkirk	*Greece*	*Crete*	*Tunisia*	*Sicily*
Evacuated safely	335,000	44,865	16-17,000	638[1]	65,000
Casualties	50,000	11,500	11,000[2]	324,000	165,000
Equipment evacuated ..	virtually none	ditto	ditto	none	Considerable

Its complete defeat in Tunisia meant that the Axis had to hold their new lines with new formations drawn from a dwindling man power reservoir, whereas the Allies could advance against them with a victorious army steeled by experience, inspired by success and provided with an abundance of material which came now in free flow from the factories of America. But still, as in the case of the Germans after Dunkirk, there was no immediate thrust across the waters after the retreating enemy.

Long before the battle of Tunisia was over, President Roosevelt announced that "the consequences of Allied victory in Tunisia are actual invasions of the continent of Europe. We do not disguise our intention to make these invasions" (February 12). The Casablanca conference in January had agreed on operations against Sicily and "the logistical arrangements" for them were immediately started. These preparations of amphibious operations which promised to be "of peculiar complexity and hazard on a large scale" (Churchill in House of Commons, June 8, 1943), and the need of the Allies' African armies for a rest caused an interval of two months between the German-Italian surrender at Cape Bon and the landing in Sicily, the most powerful that island had yet seen and the largest combined operation up to that stage of the war. The interval was filled out by some minor air and amphibious operations of the Allies including those against the small Italian islands around Sicily, such as Lampedusa and Pantellaria, involving trials and errors, the latter including the interpretations put on the fall of Pantellaria.

This island, water and air blockaded, though not very closely, had been under fire for three weeks; it had received bombardment from British cruisers and destroyers whose fire, undeterred by coastal artillery, had proved at least as precise as that of the planes. Enemy resistance from the ground and in the air from Sicily, 60 miles away as against the 45 miles from Cape Bon, had been weak throughout. During the last 13 days of the water-and-air siege a total bomb load of from 7,500 to 8,500 tons landed on the target of 45 square miles. Following a severe bombardment, in which 1,700 sorties were flown and a total weight of 1,500 to 2,000 tons was loosed on the island during the last 24 hours, the garrison of 11,000 to 12,000 surrendered to a British division, which had come to assault the island, on June 11, telling it by signals that they were ready to be shipped into captivity.

Spokesmen of airpower in the Allied forces were jubilant. To them the fall of Pantellaria was "definitely a landmark in the history of military aviation." To Major General James H. Doolittle, commanding the Northwest African Strategic Air Force, "it was merely a proposition of steadily increasing the Pantellaria bombardment to a point at which it was physically impossible for them to stand up under it . . . The capitulation proved conclusively that no agency can stand up under the prolonged concentrated bombardment of properly selected objectives." Forgotten were all the disappointments resulting from the bombings of Pacific Islands, or the contrary example of Malta. An island finally had been won primarily by air power, with sea and land power merely later occupants. (It was not the first island conquest by airpower, incidentally; the occupation of Runö Island in the Gulf of Riga on October 13, 1917, had been more substantially that.)

Douhet, it seemed in the clear air of the Mediterranean summer, was proved right after all. The air power enthusiasts, looking down from very great heights and therefore not penetrating to the hearts of the conquered, were thus unable to see the ironic circumstance that Douhet was right only with reference to the Italians, whereas this Mahan of air power had implicitly insisted that his ideas would prove particularly true against democratic nations. The irony of the thing was that air power, air power alone, would take effect only against Italians—there were less than 100 Germans in the island—perhaps for the reason that they had studied Douhet most thoroughly and thus made the exercise of air power against themselves a reason, or pretext for capitulation.

No All-Out Resistance

It is quite safe to say that no Japanese, or German garrison would have surrendered under similar circumstances, with a fourth of the island artillery still intact, rock shelters and caves offering ample refuge and casualties low among both troops and civilians. The island was fully provisioned and while water supply and communications had suffered, neither were by any means destroyed. Unless the political pressure was in full reverse of the usual system in Italy, where it was all downward, it must have been the commanding officers, as they did soon after in case of the removal of Mussolini, who came to the conclusion that surrender had to take place, rather than the rank and file. These, being largely under-age and over-age, displayed low morale even before the bombardment had begun or the civilian inhabitants, of whom there were rather too many for comfort inside the island fortress, as in Corregidor. So far as the civilians were concerned, nothing would indicate that it was they who had decided that the critical moment had arrived when, according to Douhet, "the population, seemingly an unprotected prey to the enemy air fleet, under the common pressure of the self-preservation instinct, will insist on terminating the *à tout prix* fight." Perhaps, the Italian commanders remembered that Douhet's frightfulness was claimed to be more humane in the last analysis of warfare and that they should for that additional reason surrender to convincingly demonstrated air power. "And thus the fortress falls before the image of the weapon."

The air power enthusiasts, searching for support for their doctrine, were in effect insisting that given more planes, "enough" planes, the same effect as in Pantellaria could be achieved against much larger objectives. They conveniently forgot to stipulate how much "enough" would amount to, or ignored the older claims raised on behalf of this, or the other weapon that given "enough" of this or of that weapon the war would be won most easily, or the circumstance that a small fortress and a narrow island fortress in particular make impossible that tactical "evasive" under air and other bombardment which a large scale fortress allows its garrison.

The circumstances under which air power had proceeded against Pantellaria could not have been more favorable—the weather was ideal throughout, in fact, halcyon for the kingfishers of air power, who could bomb from the highest blue above the Mediterranean. The flying distances for the Allies were short, 25% less than those of the Axis, and thus so favorable that bomb weight did not have to be in the least reduced in favor of fuel.

By a simple switching process Allied airforce, immediately following the occupation of Pantellaria, was turned against Lampedusa, which fell on June 11, apparently after an earlier British commando raid had failed on the 7th.[2]

THREE-POINT ATTACK

Sicily is probably the island of the Western world most often invaded. "Her history exists mainly in its relation to the history of other lands" *(Encyclopaedia Britannica)* and her military history consequently comprises almost innumerable landings. It was that more than slightly histrionic personality, General George S. Patton, who seems to have felt most strongly his connection with the line of conquerors when "he leaped into the surf from a landing barge and waded ashore to take personal command of the bitter fighting against German tank units opposing the landing."[3] Journalists, even less than history, can hardly be restrained from producing war parallels.

The Allies of 1943 approached the island from three starting points simultaneously, not only from Africa, whence had come many of her earlier conquerors, but also from England and the United States, the latter the destination of much Sicilian emigration. The three forces, progressing through definite stages of time and space every 24 hours, laid out in a way to puzzle, or mislead Axis observation, unavoidable in the last stages of the approach, were to take a direct course, from waters south of Sicily and in sight of Malta, to their ultimate destination at 1630 on July 9. They arrived off their beaches without any serious interference by the enemy.

The landing beach for the Americans, usually called the Gela beach, ran from six miles west of Licata to Cape Scalambri on the southwestern side of the island-triangle, 38 miles of shoreline. That for the British comprised 37 miles of shoreline, from seven miles west of Cape Correnti to Cape Murro di Porco, a few miles south of Syracuse on the eastern coast. The Canadian sector was laid out against Pachino and around Cape Passero, the southernmost point of the island, and the British to the north of it, that is, in the direction of Messina, the ultimate winding-up point of the whole Sicilian campaign and obviously the point of greatest resistance in the whole triangle. It was a pivotal position quite similar to the one taken by the British on the left wing of the Normandy beachhead around Caen. Not all their Allies were favorably impressed by the slow British pressure northward, General Terry Allen of the American First Division feeling that "they were at the same old game, butting head on. He felt they should have done more manoeuvring. 'I believe that shoe-leather is cheaper than men's lives.' "[4]

The coastal and beach formation in the American sector with a long gentle slope into deep water called for debarkations much further out to the sea than in the British sector, where closer approach of big ships to the coast was possible. The armadas were under the supreme command of Admiral Sir Andrew Browne Cunningham. He was Naval Commander-in-Chief in the Mediterranean, presumably from the moment when the Allies entered that sea with three American and three British admirals as subordinate commanders in their respective battle sectors. Together they disposed for the landings from nearly 2,000 vessels, warships little and big, including four battleships; and merchantmen. Altogether 3,266 surface craft of all types were employed in the Sicilian campaign. Their "primary duty" was, as Admiral Cunningham reminded them on the eve of the invasion,

> . . . to place this vast expedition ashore in minimum time and subsequently to maintain our military and air forces as they drive relentlessly forward into enemy territory. In the light of this great duty, great risks must and are to be accepted. The safety of our own ships and all distracting considerations are to be relegated to second

place or disregarded as the accomplishment of our primary duty may require.[5]

Military supreme commander in the Sicilian campaign was, as in North Africa, General Eisenhower. How far, and from what moment, military command extended over naval and air command, and whether during the approach naval command was supreme over all then on board the Armadas, seems not to have been stated very clearly to the outside world. However, and rather at the insistence of the Americans as Churchill once stated, there was unified command, as in Northwest Africa, a fact which, as Eisenhower's chief-of-staff at the end of the Sicilian campaign explained, "assured that, even though air-land or sea-air impasse were threatened at any stage of a campaign, there was always a man on the spot who could resolve it with a command decision. After that, it was all a matter of good communications and good-will."[6]

Before the arrival of the armadas, the Allied airforces operating from Africa had submitted Sicily's airfields and vital points to such a thorough bombardment, particularly from July 3 on, that no strategic surprise could be expected. Bombers of the Ninth United States Air Force claimed to have smashed the headquarters of the Axis defense at Taormina. Only a few of the Axis airdromes remained usable on the morning of the 10th and for a few days after. Allied air supremacy was but little contested, although enemy resistance was slightly increasing during the day before the landing.

The actual invasion of Sicily was begun by air-borne troops at 2210 on Friday the 9th, when American glider formations came down on the Sicilian plains. By tragic error they had been fired upon by Allied warships during the crossing and had lost several hundred men. Whether this mischance upset the glider pilots or not, their navigational abilities and in particular their night orientation proved to be far from efficient. They were followed 70 minutes later by American paratroopers, jumping from transport planes into the strong wind which threatened havoc to the success of the whole expedition and widely dispersed both the parachutists and the gliders. Some landed as far as thirty miles from their objectives. Too few of the men could be gathered to hold the assigned objectives from which they were soon after driven. Only to the extent that they drew the attention of the enemy away from the sea-borne troops could the activities of the paratroops be called "remarkably successful."

Cavalry Against Paratroops

For the rest, the investment of 6,000 air-borne troops of the United States Army displays rather dubious use of forces, very valuable *per se*. One air-borne detachment is said to have encountered Italian horse cavalry and something of a Mark Twain meeting of mediaevalism and Yankee technicians using tommy guns seems to have ensued. British air-borne units paralleled these activities in the eastern part of the island somewhat more successfully. Due to light flak fire and the absence of enemy night fighters, the formations remained compact in their flight and plane losses were light. But losses among the troops whom they had carried were reported as high, some casualties caused by Allied warships marring the picture of perfect cooperation in the Sicilian campaign which the officials and officers had originally essayed to present for public contemplation.

The inital landing forces of the Allies comprised 160,000 troops, considerably less than the 300,000 with whom the Axis was then supposed to have garrisoned the island. These, however, were widely distributed over the whole area of nearly 10,000 square miles, details about their disposition being unfortunately wanting. The impedimenta of the Allied forces comprised 14,000 vehicles, 1,800 guns and 600 tanks.

They included the British Eighth Army, with Canadians in considerable numbers who as a regional force now saw their first large scale action in this war, and the American Seventh Army under the command of Lt. Gen. Patton, consisting of the First, Third and Forty-Fifth Divisions, the Second Armored and the Eighty-Second Air-Borne Divisions, all battle tested in North Africa except the 45th. This was an Oklahoma National Guard outfit, carried directly across from the Chesapeake area where it had been in training. The total of the Allied forces used in Sicily has never been disclosed, but a strength of considerably above 300,000 seems likely.

At 0100, July 10th, the first troops went down the sides of the transports into landing craft, bouncing and lurching on the high swell which made many men, perhaps as many as half, seasick. This swell continued in various places until the 13th. Air and naval bombardment opened up at 0300 against targets selected and assigned long beforehand. Destroyers, being close to shore, took care of subdivisions of the beaches with barrages to destroy machine gun nests, cut barbed wire, set off mines on shore, destroy or drive off enemy troops. Later they shifted their rapid fire to ascertained strong points, gun emplacements, pill boxes, while the landing troops made their way to the shore.

This fire curtain was swung farther inland, more or less in the style of land artillery supporting an attack on trench lines in World War I, with the usual hit-and-miss results of barrages and curtain fire. An hour before daylight all waves in the American sector, with Commandos, or rather Rangers in the first wave, had obtained their beachheads against slight opposition from what seemed mostly second line Italian coastal defense divisions, and from defenses which were mostly mere barbed wire fencing and machine gun posts. The fire of coastal batteries is usually described as erratic, although other news reports, each covering but a narrow section of the vast beaches, describe it as withering barrage. Anything approaching an over-all history of the landing is still lacking. Losses on the water and the beach were slight. Beach organization was at once set up and, directed through it, supporting troops by the thousands and materiel by hundreds of tons immediately followed.

As General Eisenhower reported on the 17th: "All the initial invasion moves were carried out smoothly, and an astonishing lack of resistance was encountered on the shoreline. Captured Italian generals say we secured complete surprise. The airborne operations, which were executed about three hours ahead of the landing, were apparently the first real notice the defenders had of what was coming."

Within the first 48 hours alone of the invasion 80,000 men, 7,000 vehicles, 300 tanks and 700 guns were landed. In naval official language: "The Navy's primary duty of getting the Army safely on shore continued without intermission . . . ." And a little later: "The Navy's main task of supporting the Army by commanding the sea and disembarking troops and their supplies continues." (July 14). Next to this duty in time and importance came the services of the navies as mobile heavy artillery, the newly discovered, or admitted ability of warships which during World War I was discerned only by perverse civilians, like Lloyd George, unimpressed by fictions as the child in the fairytale who saw the nakedness where all others thought they saw the Emperor's new clothes.

NAVAL GUNS USEFUL

Now it was admitted by all hands that ship artillery "had uses not a few," including those for which it had not been designed, against land targets. According to a statement by Churchill, February 22, 1944, the British Navy alone had bombarded enemy-held coasts on no less than 716 occasions since January 1, 1943. Such bombardment was directed against various targets in Sicily, including Catania which was assailed in full daylight by one of the British battleships, lying from five to seven miles off shore. Enemy heavy fire opened up near the end of that half hour when the capital ship was moving away and destroyers were sent in to deal with these impudent land-guns whose rôle, in the reports as far as we have them, seems almost to be reversed from Nelson-Napoleonic times.

The first Axis counterattack came from bombers during the 10th, causing some casualties among both British landing craft and personnel and American landing forces. Guided by flares, their night bombs were directed against troops on, and ships near, the beach, causing the loss of one American destroyer. Resistance by German troops became fierce when Americans took Gela, the only town of any size in their original beach-head; after the Germans had been driven out twice and the Americans thrown back to the beach as often, the personal intervention of General Patton would seem to have been required to secure the place finally.

On the Licata beach, further to the northwest, an enemy counter-attack with tanks reached a position "from which they could fire on the beaches and at the ships standing by. When this tank attack developed, our cruisers and destroyers moved inshore and opened fire on them, pending the establishment of anti-tank fire on the beach. So effective was naval gunfire on this occasion that the tanks were successfully repulsed at a most opportune time, Had there been no naval gunfire support, or had it been less effective, our landing force in all probability would have been driven into the sea" (Admiral King's Report).

The Axis plans for the defense of Sicily are not easy to reconstruct. Their strength was put by Churchill in a broadcast from Quebec at the unbelievably high figure of 400,000.

but from their distribution, their counter-attacks of varied strength and other opposition, no definite purpose emerges beyond that of preserving by all means the lines of retreat centering on Messina and the narrow Straits between the island and Italy proper. Germany obviously was not willing to invest and lose much of her own forces, which amounted to some 50,000 at the utmost, in the island defense. The Germans were clearly unable to stir the Italians into any determined action either on land, sea, or in the air. With no more than two divisions, the Germans did most of the fighting against the 14 Anglo-American divisions. Numerical superiority was part of the Allies' *science* of war; fighting superiority was part of the Germans' *art* of war, with art increasingly diminishing.

Once the Allies had effected their landing, insular conditions favored them rather than the defender, whose main communications with the exception of the regions along the Straits of Messina were ever more dominated by Allied sea and air power, naval guns covering the roads and railroads that for the most part paralleled the coast. Most of the circumference of the island, so to speak was wrapped in the gun cotton of Allied fire power. Warships and airplanes bombarded day and night and impeded enemy communications through all the three elements, naval gunfire proving particularly helpful in the advance along the coasts whereas close, or tactical support of the ground forces by planes seems not to have been of the best.

As in other places, the Allied airforces seemed inclined to select their own broad, rather than localized, objectives for which they were wanted in Sicily, appearing instead above new targets for the first time, or making whatever else constitutes "records" in the air. Increasing their hydraulic and air pressure within the triangle towards the apex at Messina, the Allies were able at Enna to cut (July 20) the transversal of rail and highway which parallels the triangle's base and runs from the center of the north coast to the center of the east coast where the British had taken Syracuse and Augusta (on July 13).

Outstanding features of the combined operations in Sicily were the various secondary, or accessory landings following the main and original one, behind the enemy flanks, seemingly safely resting on the northern coast. The units used were of varying strength, from patrols that would wade for miles through the surf at night and at some distance from the beach and thus get in the enemy's rear, to that of a regiment combined with a tank detachment which were brought ashore during the night of August 10-11 near Cape Orlando where an American force landed 8½ miles behind the front and then fought its way westward towards the enemy's rear.

Allied and German communiqués on such ventures—three of them within the ten days to August 16—varied considerably, the Allies claiming that "this combined American operation resulted in an immediate withdrawal of enemy forces to new lines and the taking of 1,500 prisoners" and the Germans that "the attempt at enveloping our northern wing by small forces from the sea were frustrated by our troops through counterattacks." [7]

Flight Over Messina Strait

The best and clearest claim that could be raised by the Germans at the end of the 38 day Sicilian campaign was that they had achieved a fairly successful "delanding" of the remnants of their own and their ally's strength across the two-mile wide Straits of Messina. Their ferrying operation, ending at daybreak of August 17, salvaged 65,000 men, including 35,000 Germans, considerable materiel and several thousand Allied prisoners. The Allies' counter-bill disallowed some of the German claims, such as the latter having fought "in a hard and bitter struggle against fourfold or fivefold enemy superiority," pointing to the Allies' 132,000 prisoners of war, of whom, however, very few were Germans. The German claim can only be sustained if the Italians did not put up any fight whatsoever. In any case, the latter contributed the majority of casualties, 130,000 out of a total of 167,000.

Total casualties among the Axis partners were put at 167,000 by the Allies whose own losses in killed, missing and wounded were 31,158, with 7,455 as the American contribution, out of a total never made explicit by the Allies themselves except by President Roosevelt's statement that the initial landing force of 160,000 was "followed every day and every night by thousands of reinforcements," which taken somewhat literally would put the final strength of the Allies at well above 230,000.

Considering the low fighting value of most Italian units, this gave them a fairly comfortable superiority soon after the first landings, not to mention the advantage in materiel. Whether it was this or, as Roosevelt claimed, "the meticulous care with which the operation in Sicily was planned" that paid dividends, later historians may be able to say. According to the President, the casualties in men, ships and materiel "have been low, in fact, far below our estimate," the latter needless to say not being stated, thus denying us an insight into the planners' considerations.

Against the fantastic German claims of having sunk, in the month from July 10 to August 10, 325 American and British ships totalling 1,200,000 tons in the Mediterranean alone, besides damaging 58 transports of 278,000 tons so heavily that their loss could be assumed, the admitted Allied shipping losses were only slightly below 85,000 tons of shipping sunk by enemy action. The losses among warships included one American destroyer, the *Maddox* (1,700 tons) sunk by enemy bombs; three submarines, two British and one American, and eight small war craft.

Large political effects do not always depend on the commensurate size of battles which may inaugurate them. Though of limited military importance or size, the Sicilian campaign, being fought on "sacred" homegrounds, before it was half over brought the Fascist system of government to its fall. On July 25th Mussolini was dismissed by other assistant *duces* who had been in power with him and who felt strongly the shattering effect which an invasion from the sea exercised on the totality of a totalitarian system. Early in September, as President Roosevelt put it in an analysis that was not necessarily complete, "the relentless application of overwhelming Allied power—particularly air and sea power [a concession of potential dangerousness to air power and its not easily measured effects]—convinced the leaders of Italy that it could not continue an active part in the war" (Message to the Congress, Sept. 17, 1943).

Negotiations for an armistice were opened, about which rumors had been heard since early in August, and it was signed in Sicily on September 3. Unfortunately, the Allies did not feel ready to make use of it and of its possible effects on the Germans by breaking into the Festung Europa at once. As Roosevelt phrased it, the armistice "could not be put into effect until September 8, when we were ready to make landings in force in the Naples area. We had planned these landings some time before and were determined to go through with them, armistice or no armistice."

This leaves open the question whether the Allies were not perhaps prisoners of their own plans and timetables and whether that kept them from making bolder use of the opportunity for surprise which the Armistice would have given them, had they pushed at once into the vicinity of Rome. The proposed Allied coup of taking Rome with air-borne troops, worked out in some detail, was given up in the last, or penultimate, hour for reasons as yet not fully explained, but which the troops fighting later at Monte Cassino might have liked to know.

NOTES, CHAPTER 57

[1] *N. Y. Times,* Aug. 19, 1943. The casualties for Dunkirk are British losses for the whole French campaign, not including French and Belgian losses; the casualties for Greece and Crete are only British and Imperial, not including Greek. The casualties for Tunisia refer to Axis losses during the entire campaign, incl. 252,415 prisoners of war taken in Tunisia.

[2] *London Times,* June 9, 1943. Only the Italians and Germans reported this earlier attempt while the Allies, according to their policy of not reporting failures of the Commandos, neither mentioned nor denied such an attempt.

[3] Noel Monks in *London Daily Mail;* A. P. despatch, July 13, 1943.

[4] George Biddle, *Artist at War.* N. Y., 1944, 82.

[5] A. P. despatch from Allied HQs in North Africa, Aug. 6, 1943.

[6] *N. Y. Times,* Aug. 16, 1943.

[7] An eyewitness report of the landing near Cape Orlando is in Jack Belden's *Still Time to Die,* ch. VII.

Chapter 58

LANDINGS ON THE ROAD TO ROME AND GREECE

Calabria, Salerno, Anzio-Nettuno; Kos, Leros, Samos

WHENEVER, OR WHATEVER the Germans learned about the impending Italian armistice, they were in a military way somewhat prepared against it. During the 17 days' interval following the conquest of Sicily, "aware of the widespread Allied belief that Italy will fall apart when the first American or British soldier sets foot on its soil," they took over Italy more completely than before, making sure of their hold on the harbors of Genoa, La Spezia and Livorno; the inland transportation centers of Pisa, Florence and others, and of Rome itself, the country's political center. By the end of August the estimated strength of the Germans in the Peninsula was put at twelve divisions or even 350,000 men.

In one of the few military prognoses of this war that came true, this German mastery of Italy and its likely consequences were depicted as follows: On the Allied side a methodical campaign along the lines of the Sicilian; on the part of the Germans stubborn delaying actions from the toe to the knee of Italy, over mountainous terrain on which up to then the Allies had done least well, a long costly fight lasting beyond the spring of 1944 which even continuous bombing and simultaneous landings on both the Tyrrhenian and Adriatic coasts could not shorten much.[1]

The Allied operations against the Italian mainland as far as they fell before the publication of the Armistice on September 8, began with Commando landings, with units upward to 400 men, across the Strait of Messina near Reggio Calabria on and before August 29. The Germans who knew about the collecting of Allied warships and transports and other details of the assembly, which went on over some ten days, reported that they had annihilated these units, some carrying elaborate communications equipment, except a few men who had taken to the Calabrian hills. According to later statements of the Allies, these had been mere nuisance raids, without any intention of establishing beachheads. Equally unmistakable augury of the invasion was the heavy fire of Allied land based guns around Messina across the Strait and the naval gunfire from battleships like the Rodney and Nelson against objectives on the Calabrian shore, intended to reduce enemy artillery. Allied air activity enlarged the margin of their air supremacy—4:1 as Churchill declared on Sept. 21—opening an era when the Germans had to learn how to fight without planes for days.

As if to emphasize the military return of the British to the continent, the battle for Italy was opened by the British Eighth Army on September 3, at 0430, when British and Canadians landed across a quiet sea on various beaches of the Calabrian coast around Reggio. They were carried across in landing craft of the British Navy and under escort and with the support of cruisers and destroyers, gunboats and monitors. Overhead was a shield of artillery and machine gun fire.

They quickly established various beachheads from near San Giovanni in the north to near Melito in the extreme southern tip of Calabria. A Commando party landed at and captured the town of Bagnara on the coast road leading north, where they were joined by troops of the main body on the second day. Little or no initial resistance was offered, although the German communiqué claimed that one landing attempt made against the rear of their advanced lines had been frustrated with heavy losses to the attacker. (Unless the chronicler of landing operations writing during wartime mentions the enemy versions about landings in addition to the either optimistic, or reticent versions of the Allies, landing operations are apt to appear much easier than several of them have been).

The first unloadings took place well ahead of schedule time. By early afternoon, a "second flight of craft" were moving across the Strait with reinforcements and supplies. Only some barbed wire and a few mines were found on the beaches, which in places were removed by the sappers while the landing craft waited some fifty yards off shore. Behind such obstacles, including demolitions which became more troublesome while the Eighth Army advanced, little resistance was offered. Italians surrendered almost before they were expected to, but there were few Germans among the prisoners.

Capping the Italian Boot

The nature of the terrain and the fact that the only road

in the tip of the Italian toe leads along the shore forced the British to advance to the right and left of the landing beaches, thus laying a kind of metal cap about the extremity of the boot, reaching at the end of the second day for 38 miles from Bagnara to Melito. They were also stabbing inland over nearly roadless terrain in the direction of the Aspremonte range, and eventually secured the first of the few roads leading across the Calabrian peninsula from the Tyrrhenian to the Ionian Sea. They found the Germans wary and unwilling to send greater numbers of troops into this remote part of Italy, which if need be they could always and easily seal off on either the line between the Gulfs of S. Eufemia and Squillace, or that between the Gulfs of Policastro and Tarento. Except perhaps in the psychological field, where it might tempt their *Führer* to battle an enemy who had reentered the continent of Europe, Calabria did not prove strong bait to the German command. That the Allies had entertained such hopes we have been told, though not how much they had to do with the postponement of the Armistice.

Some of their next moves were indeed coupled with the announcement of Italy's retirement from the war. Clearly in the hope of exploiting the confusion to be expected from it, and also perhaps of finding some worthwhile Italian support in the fight against the Germans, the Allies proceeded to land at various points nearer Rome, although not at Rome itself, for the plan to send an air-borne division to the capital was not acted upon, because the Germans would have seized the necessary airfields in time to frustrate it. And no effective Italian help could be offered. The Badoglio Italians had begged postponement for a few days of the announcement of the armistice, not of the landing near Salerno itself, however, about which they knew beforehand, until the Allies were nearer Rome and until they themselves could get ready to assist them more positively. This General Eisenhower declined to do.[2] Aside from the main landing of the American Fifth Army under General Mark Clark in the Gulf of Salerno, some 35 miles below Naples and near the maximum reach of Sicily-based fighter planes, on September 9, there were accessory landings at Taranto by British troops, who took this naval base without opposition; and on the islands of Ventoteme, forty miles west of Naples, and Ischia.

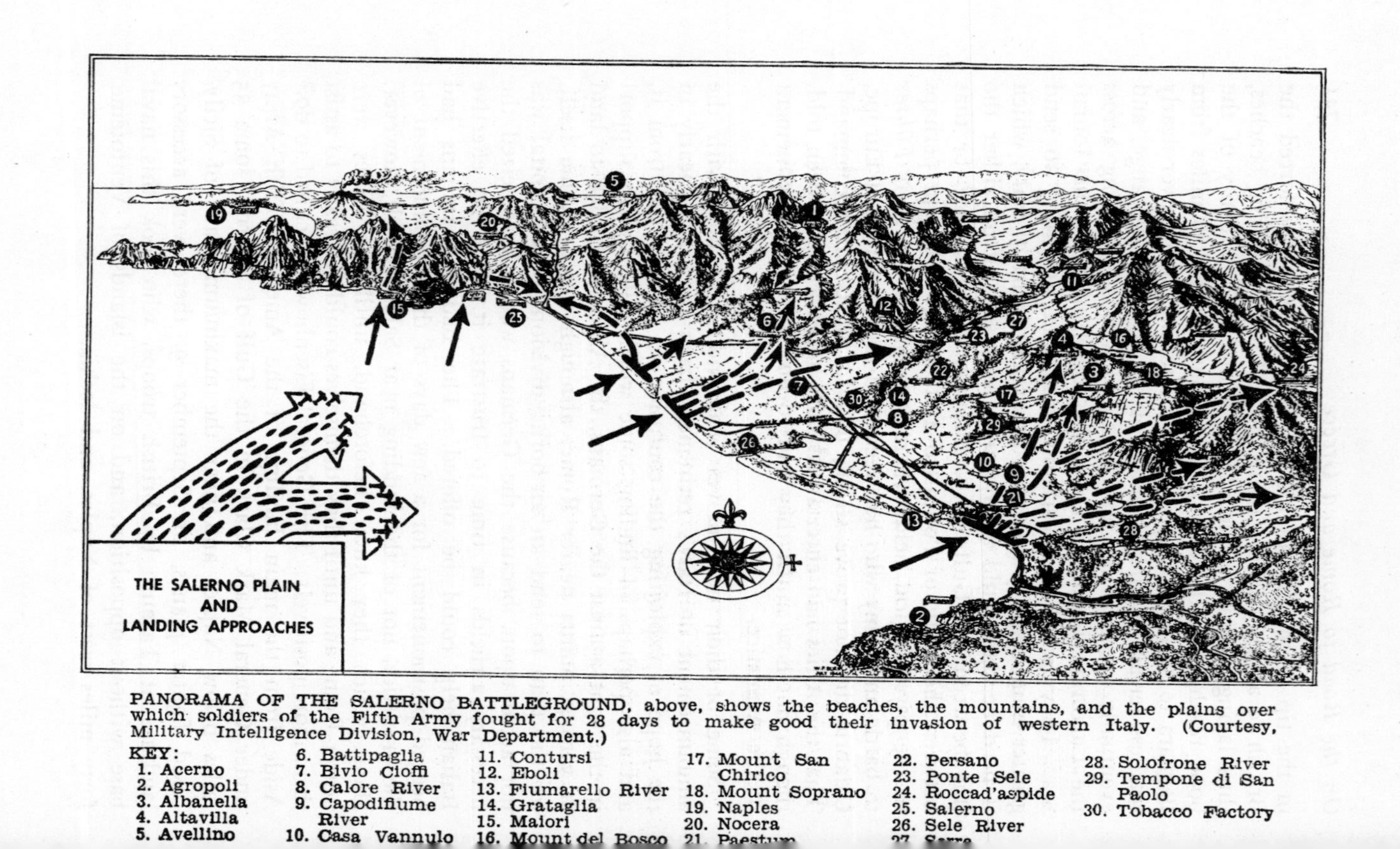

PANORAMA OF THE SALERNO BATTLEGROUND, above, shows the beaches, the mountains, and the plains over which soldiers of the Fifth Army fought for 28 days to make good their invasion of western Italy. (Courtesy, Military Intelligence Division, War Department.)

KEY:
1. Acerno
2. Agropoli
3. Albanella
4. Altavilla
5. Avellino
6. Battipaglia
7. Bivio Cioffi
8. Calore River
9. Capodifiume River
10. Casa Vannulo
11. Contursi
12. Eboli
13. Fiumarello River
14. Grataglia
15. Maiori
16. Mount del Bosco
17. Mount San Chirico
18. Mount Soprano
19. Naples
20. Nocera
21. Paestum
22. Persano
23. Ponte Sele
24. Rocca d'aspide
25. Salerno
26. Sele River
27. Serre
28. Solofrone River
29. Tempone di San Paolo
30. Tobacco Factory

It was around Salerno that the first "real battle for Italy" was fought, and not in remote Calabria. The Allied armada of some 500 vessels came in smaller part from the north coast ports of Sicily where the American 45th Division embarked, and with its main part from North Africa, carrying two British divisions and one American, the 36th, a newcomer to the war. The sea was smooth, the weather fine, that is to say, favorable to the Allies, and it continued so throughout the Salerno fighting. The Sicilian convoy, leaving by dawn of September 8, received its first bombing at 1410 and knew by then that the Germans were alert. At 1500 the two convoys joined.

Screened by cruisers and destroyers against submarines and under constant attack, or observation from German planes which sank one LCT in the North African convoy, the convoy steamed towards the Gulf of Salerno, which they entered by 2100 hours. By this time the news of the Armistice had come over the radio. The selected landing beaches, practically uninhabited in their whole length, were on both sides of the mouth of the Sele river and near the ruins of ancient Paestum, long deserted on account of Saracen raids and malaria. The shore north of the Sele to Salerno, eight miles long, was assigned to a British force—100,000 in all—the one to the South, 4½ miles, to a mixed American-British force of 69,000 men. Each part was subdivided into four beaches, called the Red, Blue, Green and Yellow, later provided with markers for the benefit of troops and landing craft skippers. Commandos and Rangers worked in the Sorrento peninsula, above Salerno, for diversionary purposes.

At 0100 hours the moon, that had silhouetted the ships for the German aviators during their several trips, went down and the ships moved to within five or six miles of the shore, their rendezvous deployment line hardly ever free of German observation. So far as there was hope to surprise them, that could only be looked for in the politics of the situation. Zero hour was set for 0300 hours of September 9 and at 0245 the ships of the convoy laid down a barrage along the shore whence some batteries replied. There was little heavy bombardment of the shore by the naval forces, some in the British sector and none in the American, and it seems that the army commanders had not wanted it, relying greatly on the psychological surprise of the armistice while they realized that there could be no tactical surprise. This was only the first disappointment in expectations of receiving Italian help. As General Eisenhower told newsmen at the end of his Italian campaign, the Italians did not render as much aid to the Allies as had been awaited and that for this reason the Italian campaign had gone so slowly. The absence of an opening-up left most of the German artillery and machine guns intact, and in many cases these were emplaced close to the shore.

Germans on the Alert

The assault troops, three divisions including the 36th, were on their way behind and through a heavy smoke curtain, encountering not a few of the hazards that beset the path of landing forces, among them mines which had broken loose from their moorings and had thus escaped the mine sweepers which had cleared lanes to the shore. The flow through the bottlenecks of the beachheads was slow; some of them were temporarily made unavailable or narrowed down by enemy resistance. The reserve division hovered out a few miles at sea and longer than had been intended. The enemy had "hit the beachhead before the Allies were completely set for him" (General Alexander). In fact, some of the American forces had been greeted by a loudspeaker when touching down at Paestum, calling in English: "Come on in and give up. We have you covered."

The rank and file of the assault divisions while coming in were under the impression that they had landed exactly where the Germans had expected them, since their guns reached them even while the landing craft were a mile off the shore. As John Steinbeck was told by someone who had come in with the second wave: "I swear that is the longest trip I ever took, that mile to the beach. I thought we'd never get there. I figured that if I was only on the beach I could dig down and get out of the way. There was too damn many there in that LCI. I wanted to spread out. That one that hit the mine was still burning when we went on by it. Then we bumped the beach and the ramps went down and I hit the water up to my waist. The minute I was on the beach I felt better." [3]

There were five furious German counter-attacks within the first 24 hours and elements of an American reserve division, the 45th, had to be sent ashore on the second day after the landing to help hold the beachhead, although apparently they had been intended to leapfrog over the first landing force, with the rest and airborne troops brought over from Sicily following soon after. But the first two days were not the most precarious. The battle hung in the balance in the later days, those "very dangerous and extremely critical few days" (General Alexander) in the beachhead where there was more *Todesraum* —death space—than *Lebensraum* for the Allies. Casualties were heavy, one British battalion losing five officers and 100 men in fifteen minutes and 50 per cent of its effectives, around 500 men, to September 29. While no Allied casualties for the Salerno operations as such have been given, their losses in Italy up to October 23 were 8,000 killed, wounded or missing for the British in the Fifth Army and 6,000 for the Americans plus 1,000 for the Eighth Army since its landing in Calabria.[4]

The Germans held prepared positions on ground steadily rising from the sea, "looking down our throat," as put by General Clark, who had to tell one regimental commander as late as the 14th: "If you go back any more, we won't have a beachhead." The situation in this, the worst reported of all Allied battles was, according to Clark, "never desperate," whereas according to Alexander the German counter-blow came "within 1000 yards" of an Allied disaster.[5] It would appear that victory could not have been won but for the overwhelming fire support lent the ground forces by the naval units from destroyers up to battleships, who knew now that their corresponding numbers of the Italian fleet were safely out of the way under the Armistice conditions and that there were no commensurate enemy guns opposite the beachhead. On the sixth day, during the critical period, they were just in time to smother a dangerous German bid to wipe out the beachhead through a tank attack; on the 14th and 15th the situation was so precarious that British news despatches mentioned and the censors passed the ominous comparison with Gallipoli.

Days afterwards Allied naval units were "standing by in readiness to lend support with their gunfire to ground operations" (Naval communiqué of Sept. 28). And it would seem that no landing operation in its later as well as earlier stages had so much depended on naval gunfire for its support as Salerno, if the partisans of airpower had not raised a somewhat similar claim for the air arm. The latter provided fighter plane cover from five escort carriers which were in turn being covered either from Sicily or by the planes from two fleet carriers farther out to the sea and being in a formation with two battleships.

During the night of September 17-18, American patrols found no more limits to no-man's-land. The Germans had gone. On the 20th the latter announced that they had been able to withdraw their forces from farther south, in Calabria and Apulia, and join them with the Sixteenth Panzer Division originally alone at Salerno beach. With these, fanning out in retreat, they were soon to form a front across the peninsula, from the Tyrrhenian to the Adriatic Sea, keeping the Allies out of Naples for 22 days after the first landing.

The original intention or hope of the Allies was to land four divisions, push quickly across to the Adriatic and cut off the two or three German divisions in the deep south of Italy facing the Eighth Army. This somewhat ambitious plan to make use of the Badoglio armistice, had failed, if it had not in fact led to such errors of judgment as sending an untried division into such a delicate operation as an opposed landing.

THE SLOW ROAD TO ROME

General Eisenhower during these days reported that "we are very much in the 'touch and go' stage of this operation . . . Our hold on Taranto and Brindisi is precarious, but we are striving mightily to reinforce. Our worse problem is Avalanche itself (the Salerno enterprise). We have been unable to advance and the enemy is preparing a major counterattack . . . I am using everything we have bigger than a rowboat to get the 3rd Division in to Clark quickly. In the present situation our great hope is the Air Force. They are working flat out and assuming, which I do, that our hold on southern Italy will finally be solidified. We are going to prove once again that the greatest value of any of the three services is ordinarily realized only when it is utilized in close coordination with the other two." (Marshall's *Biennial Report, 1943-45*).

Psychologically, the announcement of the armistice, while causing some perturbation among the Germans and Fascist Italians, seems to have made the Allied troops so confident that, believing the campaign in Italy would thence forth be a walk-over, a good deal of the grim caution required under trying circumstances was thrown to the winds.[6] "The Germans were too quick for us," as General Alexander, commander of the Allied ground forces, later admitted. "Had we cut the Germans off, we would have raced to Rome." [7]

Going over the roads, direct or indirect, from Naples to Rome, proved exceedingly slow. In order to break the impasse, the halt in the mountains and the mud, "to get away from two handicaps, weather and the terrain," as General Sir Henry Maitland Wilson defined the purpose of the Anzio-Nettuno operation, or to overcome their weakness in the mountains by using their strength on or across the sea, the Allies made another major amphibious thrust. It had been preceded by a few minor ones, one by landing craft of the Royal Navy north of the Volturno River in the early hours of October 13 and another on the Adriatic coast on October 3 when elements of the Eighth Army were carried for seventeen miles from the mouth of the Fortore River to Termoli, almost directly east of Rome, with which it is connected by a trunk highway across the Abruzzi. There, the landed force held out until joined by the main forces. On October 6 two British destroyers helped by their fire to dislocate enemy attacks in the Termoli area.

After that there was a long pause in the application of Allied sea power supremacy in the Mediterranean in the form of landings. It was not to carry the Allies to Rome before Christmas. There was even for a time practically no tactical assistance from naval units in the form of fire support given to the troops in the two coastal regions at either end of the front which ran for months across the narrowest part of the Peninsula, from Ortona by Monte Cassino to the Gulf of Gaeta, near Minturno.

Some hesitancy now prevailed in Allied councils, due in part to over-sanguine expectations from the results of the Italian armistice. Assignments of forces and materiel to the various theaters of war had to be changed. All the landing ships and craft originally allotted to Mountbatten's Indian command had to be withdrawn, or withheld from him for what was now considered "the more urgent operations in the West." They were to be employed in the attack on the Anzio beaches and later in the invasion of France,[8] while the amphibious features were cut out of the Indian campaign.

"If not for Christmas, then perhaps for Easter," seemed the

prospect that prompted the Allied landing on the Anzio-Nettuno beachhead, thirty miles south of Rome and over fifty miles behind the front, "deep in the rear of the present enemy front line positions," as the first communiqué said. The beach was some 13 to 15 miles away from the Via Appia, the first of the great military roads of ancient Rome, which the Germans blocked at Minturno, and farther from the Via Casilina, over which the Germans supplied their Monte Cassino lines.

British and American troops of the Fifth Army carried by a mere 243 ships as compared with the altogether 3,000 employed at Salerno, were landed on January 22 at two in the morning on a thirty miles front and in bright winter weather favorable for such an operation. Although the loading of the invasion fleet took place less than 100 miles away, at Naples, the enemy's air reconnaissance had clearly missed the preparations, prevented from peeping by the superior Allied airforce. The surprise was in fact so complete that he did not learn about the landing, it is claimed, until 5½ hours afterwards. In order to make it even more complete, strong Allied attacks on the inland front were timed to coincide with the landing, a *junctim* which it was hoped would consternate, if not frighten, the enemy.

Gunfire That Was Wasted

There were so few Germans in the beachhead that Allied naval gunfire which poured down an average of 200 shells a minute over all the beaches, was practically wasted. At first no obstacles except a few mines and no opposition stood in the way of the Allies. For hours it was like a peace time manoeuvre in which the men who were to play the enemy had not showed up, and from some reports it almost seems as if the absence of resistance had upset schedules. At any rate, the first day objectives were obtained within four hours time. Then there was teatime, as on Gallipoli, and the writing of communiqués, the second one of the Allies speaking of the thrust "seriously threatening the enemy lines of communications leading south and east to the main battle line," a value judgment that should be foreign to communiqués, while the first German communiqué about Anzio-Nettuno told of counter-measures in progress and of attacks of the *Luftwaffe* on the landing fleet. Some of these claims were substantiated later by the announcement of

Allied ship losses which admitted the loss of two light cruisers and two destroyers of the Royal Navy, but said nothing about other shipping losses, which included at least two LCT's and two LCI's.

The prolonged absence of German counter-attacks on the ground, which took practically a full week to develop, gave the Allies unlimited opportunity to get their support weapons, or anything else they wished ashore although clearly there was either not enough boldness, or mechanized equipment landed to thrust deeply into the German communication lines, not to mention reaching Rome. The weather turned against them, robbing them of air support for days at a time and thus foiling their hopes that airpower would make up for shortcomings in ground forces—essentially a fair weather hope. They were never really sufficiently strong to undertake any big thrusts, whatever false expectations the landing might have raised in the Anglo-Saxon nations, for these would have required up to eight or ten divisions, which were not procurable in Italy. All was going into the pool for the invasion of France.

Manoeuvring with insufficient forces failed so soon as the enemy saw through the screening employed by the Allies. The battle for Rome once more became one of attrition when General Mark Clark, explaining the stalemate, told his troops in a message of the middle of February that those in the beachhead should welcome the enemy's assaults, "for it gives you additional opportunity to kill your hated enemy in large numbers. It is an open season in the Anzio beachhead, and there is no limit to the numbers of Germans you can kill." This sounded probably too much like Ypres and Paschendaele and 1915-18 attrition strategy for comfort.[9]

The Germans never took the threat to their lines as seriously as the Allied command had hoped. They allowed themselves enough time to draw troops for their counter-attacks from the north, eight or nine divisions according to Allied statements, rather than weaken the front in the south. With these they considerably reduced the Allied beachhead, in weeks rather than days of battling, recalling Salerno and its pattern. The landing force at Anzio had "to fight for its very existence" suffering "very heavy losses," 20,000 men against 25,000 of the Germans, until a standstill was reached.[10] In this much drawn out repetition of the earlier landing, naval gunfire proved again the strong support of the Allies, although this time the Germans stayed more cautiously outside its range.

Islands Were Easy Prey

Other by-products of the Badoglio armistice were the evacuation and taking of Sardinia, whence the Germans retired to Corsica after some minor clashes with Italian troops. There, the armistice news had set off a resurrection among local patriots. In order to help them Free French troops were brought over from Africa on, or around September 15, who twenty days after their first landings completed the liberation of Corsica. Or would it be more correct to say that it took the Germans and some of the Italians who remained their allies that long to evacuate the island? Of this the Axis had made very slight use as air base, even though the Italian garrison amounted to some 40,000 men, an indication that there was more imperialism in the occupation of Corsica than strategical sense.

The Allied attempts at making good use of the armistice confusion in the eastern Mediterranean, on their right wing in that sea, were far less fortunate. Small British forces on and around September 22 landed on a number of Aegean islands, such as Samos, Stampalia, Castellorizzo, Levita, Patmos, Leros, of importance as a naval base with submarine facilities, and Kos, the second largest of the Dodecanese, with a good airbase. Kos was occupied by parts of an RAF regiment, transported there complete with guns by air. On none of these islands were there any German garrisons while the Italians were either apathetic, treacherous or "delighted to cooperate with the Allies."

Protecting the Balkans

The Germans were concentrated on Rhodes, which they had secured in the moment of our Italian wobbling and indecision, and Crete as well as the Greek mainland. Before the rather vast speculations about the strategic significance of these landings, including a re-invasion of the Balkan peninsula, had gone very far, the Germans returned the blow. They could not well afford to have their Balkan position endangered from the east at a time when the Allies were beginning to threaten it from the west as well. They landed some of their Alpine troops on Corfu where the Italian garrison had established contact with the Allies. They took over the harbor town of Split (Spalato) where Badoglio troops had admitted Yugoslav guerrillas (on or around Sept. 27) and still later the Ionian island of Cephalonia.

This cut off the insurgents from regular waterborne communications with the Allies in Italy who were then moving up the rear side of the boot, having taken Foggia late in September. The few Allied landings on the Dalmatian coast and islands from October on to the time of liberation, so far as news of them made the communiqués, were in the nature of commando raids and blockade breaking undertaken by either Tito partisans, or Allied units acting in combination with them.

The Allies had evidently, in their speculation on the imponderables of the Armistice situation, staked on some amount of Italian cooperation, but in places like Samos they met sheer treachery, such as the delivery of Allied armistice commissioners to the Germans, rather than any safe system of communications and supplies. For they were considerably farther away from their Near Eastern bases—Leros-Cyprus 400 miles as the plane flies, with German-held Rhodes and neutral Turkey intervening—than the Germans were from theirs, with a 175 mile stretch from Candia or Athens to Leros. On October 31 the Germans started their counter-attack, water-and-air-borne, from such bases as the Calato and Maritza airdromes on Rhodes, which the Allied airforces bombed that same day, with a landing on Kos.

Sea-borne troops and parachutists, supported by dive bombers, landed at the same time from the water and the air, undetected at first. While Allied air forces, for most of whose fighter planes Kos was out of reach, tried to interfere with German ship movements, the Nazis brought in air-borne reinforcements as soon as they had established their toehold. In two days of fighting the 600 men of the British garrison and their half-hearted Italian co-belligerents were overwhelmed, the Germans having taken good care to be superior at least in all numerical aspects, with Allied estimates of their strength running as high as 8,000. Most of the garrison were captured while a few were evacuated, those who had taken to the hills as late as three weeks afterwards. Considerable materiel was won which had been brought in by the British, who had also begun to enlarge the airfield on the island during their short stay.

Following a lull in fighting and a failure to surprise Leros on November 4 and 5, the Germans on November 12 turned seriously to the attack on that Dodecanese island which had become rather a thorn in their eastern flank. Using craft large enough for 100 to 200 men, they began their landings at 0630 hours and obtained four footholds, even though the Italian-manned coastal artillery, "fighting side by side with the British," and somewhat better than elsewhere, sank three landing craft. The invaders were able, after a day or two, to bisect the island and separate the defending force. After some fluctuating fighting in the narrow island, in which the Royal Navy interfered "in spite of the enemy's local air superiority," the Germans took Leros and with it 3,000 British and 5,350 Badoglio troops as well as considerable ordnance. Samos to which some of the British defenders had retired was captured on November 22.

The Aegean gamble of the Allies had been lost. Why it should have been played to the bitter end, that is to say, why the troops from untenable Leros, for instance, should not have been evacuated in time, is less easy to understand than why it should have been undertaken in the first instance. The reason given by Sir Henry Maitland Wilson, Allied Commander-in-Chief in the Near East, for the non-evacuation, that it was "also a question of honor," [11] sounds rather antiquated in an era of highly mobile warfare. The usual consolation that this enterprise drew away German forces from places where they were urgently needed, seems not very plausible since the Germans would seem to have undertaken the various island conquests with the occupation forces in and along the Aegean.

NOTES, CHAPTER 58

[1] *N. Y. Times,* Aug. 29, 1943, despatch from London.

[2] For details see Rome despatch by Herbert L. Matthews in *N. Y. Times,* Oct. 11, 1944.

[3] *N. Y. Herald-Tribune,* Oct. 9, 1943.

[4] A. P. despatch of Oct. 23, 1943.

[5] *N. Y. Times,* Nov. 3, 1944.

[6] Milton Bracker from Allied HQs Africa. Ibd., Oct. 10, 1943.

[7] Ibd., Oct. 25, 1943.

[8] Statement of Mountbatten in London, Aug. 25, 1944.

[9] *London Times Weekly,* Feb. 16, 1944.

[10] Churchill in House of Commons, June 6 and Aug. 2, 1944.

[11] *N. Y. Times,* Feb. 17, 1944.

Chapter 59

LANDINGS IN NORMANDY, JUNE 6, 1944, AND AFTER: OPERATION "OVERLORD"

"I promise if the Allies land again on the French coast they will stay exactly nine hours."

Hitler to the Reichstag after Dieppe.

THE STRENGTH OF A FORTRESS never lies in its topography, its garrison, walls, ditches, casemates and armaments alone. It also is built and rests on a credit basis, on a combination of beliefs in its strength, or invincibility, which final analysis proves has probably failed more often than demonstrated its soundness under test.

Following Germany's non-invasion of Britain and other setbacks, and also the promulgation of the Atlantic Charter of August, 1941, (a hazy enough image and promise), German propaganda, beginning in November, 1941, concentrated largely upon building up the concept of the *Festung Europa,* the fortress of Europe, with its 3,600 miles frontage on the Atlantic, and, as the Nazis had it, threatened by engulfment in the red sea of Bolshevism.

So late as the spring of 1943 it was, in the language of German propagandists on occasions, still the impregnable base from which Germany could launch attacks at any moment she chose, although by then the strategic initiative was actually no longer hers. Henceforth defensive aspects, the strength of the outer crust of the fortress, the advantage of interior lines were stressed; the fortress declared to be unassailable, impenetrable; it would be supreme folly for the western allies to allow themselves to be hurled against it by Stalin; American soldiers would never be able to set foot inside that rampart.

Dieppe, declared by the Germans to be an invasion attempt, had proved that, showing in addition, as General Dittmar, the radio spokesman of the Wehrmacht, remarked, that the British fleet could no longer protect a landing force. The landings in North Africa, Sicily, Italy–Calabria, as the propagandists put it–were events merely in the *Vorfeld,* in the *glacis* of the fortress, which was strengthened along its weakest part when all of

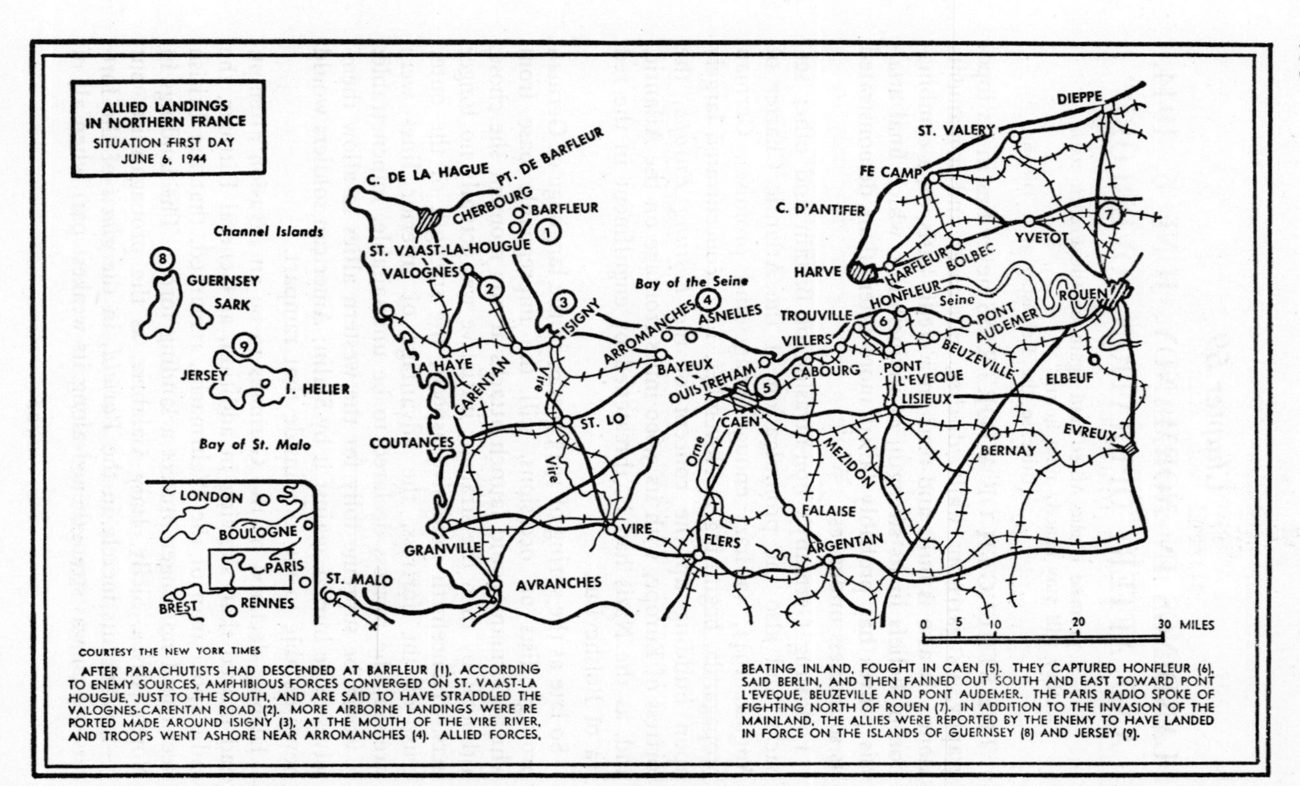

COURTESY THE NEW YORK TIMES

AFTER PARACHUTISTS HAD DESCENDED AT BARFLEUR (1), ACCORDING TO ENEMY SOURCES, AMPHIBIOUS FORCES CONVERGED ON ST. VAAST-LA HOUGUE, JUST TO THE SOUTH, AND ARE SAID TO HAVE STRADDLED THE VALOGNES-CARENTAN ROAD (2). MORE AIRBORNE LANDINGS WERE REPORTED MADE AROUND ISIGNY (3), AT THE MOUTH OF THE VIRE RIVER, AND TROOPS WENT ASHORE NEAR ARROMANCHES (4). ALLIED FORCES, BEATING INLAND, FOUGHT IN CAEN (5). THEY CAPTURED HONFLEUR (6), SAID BERLIN, AND THEN FANNED OUT SOUTH AND EAST TOWARD PONT L'EVEQUE, BEUZEVILLE AND PONT AUDEMER. THE PARIS RADIO SPOKE OF FIGHTING NORTH OF ROUEN (7). IN ADDITION TO THE INVASION OF THE MAINLAND, THE ALLIES WERE REPORTED BY THE ENEMY TO HAVE LANDED IN FORCE ON THE ISLANDS OF GUERNSEY (8) AND JERSEY (9).

unoccupied France was taken after the North African Landings.

The construction activities of the Organization Todt along the Atlantic were featured in word and image, and news reels showed "Field Marshal von Rundstedt's hard-cut features merged with the photographs of the Atlantic Wall." During the hard year, for Germany, of 1943 the propagandists of the Wall vociferated their conclusion that once Britain and the United States had recognized its unbreakable strength, then the time would arrive when they would be forced to admit that German will could not be broken by air raids, or threats of invasion from the south, that the inner contradiction of their alliance with Russia could no longer be ignored or disregarded and that the hour for a peace of understanding would strike.[1]

Should the Allies still insist on the attempt and win beachheads, they would be unable to hold them; to this end German measures were directed. While a landing on the "holy soil" of Germany proper would have great propaganda effect, such attempts, even if they should overcome nearly insuperable navigational difficulties, could "certainly be of short duration only."[2] But from early 1944 the directives of the Propaganda Ministry had to concede that the Allies would certainly make the invasion attempt, and newspapers were allowed, or directed, towards the end of January to print observations in tenor of this, from the *Münchener Neueste Nachrichten,* January 2:

> We have no reason to conceal the fact that we are calmly awaiting the invasion attempt. We are even eager for it to be made, for having confidence in our preparations, we consider it a means of shortening the war. The enemy's gigantic concentration between Iceland and the Azores will not bring him success. Germany, too, has planned every move very carefully.

These soothing utterances proved less reassuring than expected, as is indicated by a rumor which circulated in the Ruhr region immediately afterwards. According to Goering's own *National-Zeitung,* published in Essen:

> Some time ago the rumor spread like wildfire through Oberhausen that strong formations of Anglo-Americans had landed on the Channel coast. It can scarcely be imagined how much time and nervous energy were lost through the

> excitement caused by this rumor. The number of working hours lost cannot be calculated. Of course, the news was discussed endlessly in the factories.

The paper said that the source of this rumor had been discovered, which is more than can be said of most rumors, that a worker had invented it to scare a fellow-worker. "By noon the story was known not only to everyone in the factory, but to everyone in Oberhausen. The next day it spread to the neighboring towns." [3] While there were evidences of pre-invasion jitters among the lower classes in the Reich, brought into the open by the propaganda masters, the higher classes kept their premonitions to themselves until later, although their plotting against Hitler may well have begun at this time.

Second Front Propaganda

War propaganda in a way is a battle of the spirits in the air above the warriors, like that said to have taken place *after* the Battle of Châlons in 451. But in modern times it largely precedes the conflict. One objective that became a subject of sharp controversy among the Allies in their confused pre-battle struggle was the Second Front and when it should be opened. Against all clamorings, against "the request" on Russia's part of early June, 1942, that it take place during that year, the Allied governments, advised by their Combined Chiefs of Staff Board, following General Marshall's visit to London in July, 1942, when North Africa was chosen, insisted that the attempt could not take place where the advocates of the Second Front wanted, in western Europe, neither in 1942, nor in 1943 unless the North Africa-Italy invasion should be put off.

Even at that, the opening of the second front in western Europe would still then have been impossible due to the lack of men and equipment, or to the as yet insufficient depletion of Germany's manpower reserves. During the Roosevelt-Churchill meetings in Washington in June, 1942, the invasion of France in 1944 was indeed decided upon, and accorded priority over the invasion of Japan. Tentative dates for this invasion, late May or early June, depending upon the weather, were set half a year before the Teheran Conference in December, 1943, when Stalin was informed and declared himself satisfied.

These decisions would be more readily understood by peoples accustomed to thinking about the war in military terms. For

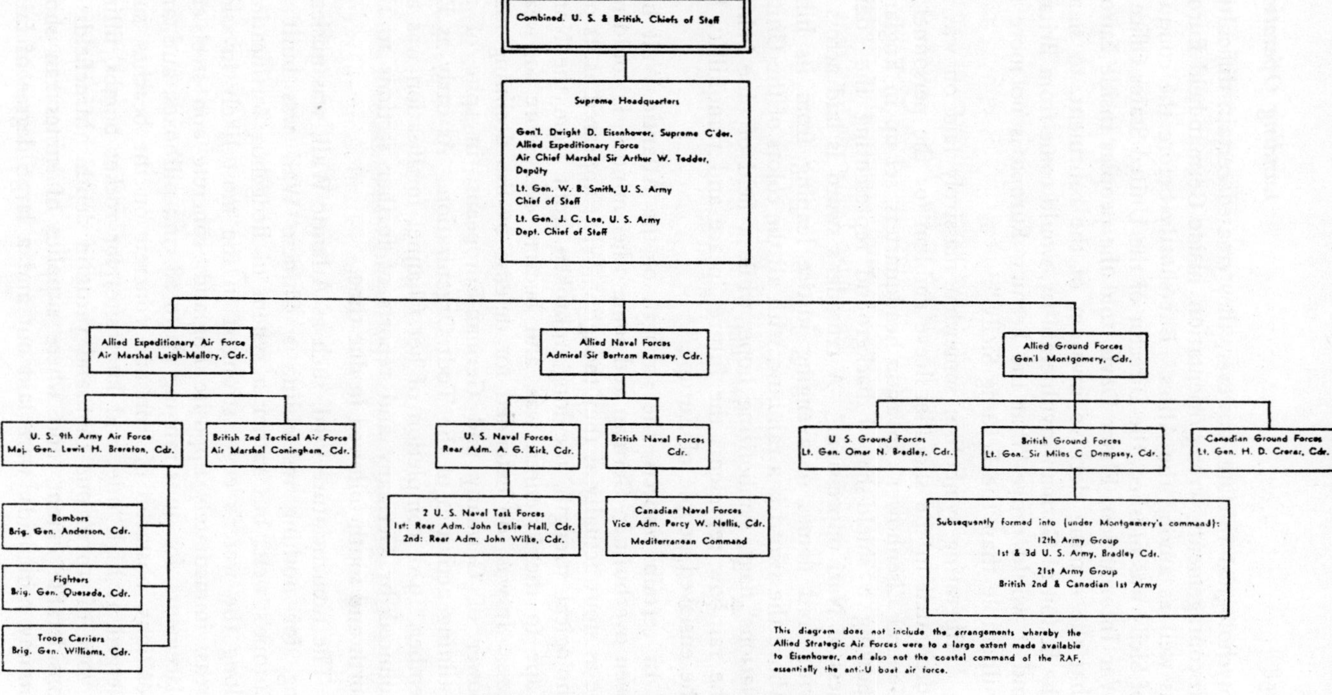

United States-British Command Set-Up

psychological warfare, however, they created certain difficulties, discouragements, wrong orientation, inside German-held Europe as well as among the Allies. But shortly before the conquest of Sicily was finished, the Director of the United States Office of War Information, Elmer Davis, told the peoples inside Europe that the coming decisive invasion of the continent, to smash the fiction of German invulnerability, would come from Britain, and it would prove "that the Festung Europa is no more invulnerable than the Festung Sicily." [4]

A liberating symbolism somewhat massively laid on was incorporated in the shoulder-sleeve emblem for the personnel of General Eisenhower's invasion headquarters, set up in England, showing a shield-shaped background representing the "darkness of Nazi oppression." A crusader's sword is laid across it, with "red flames of avenging justice leaping from its hilt." Above the sword is a rainbow, with all the colors of the United Nations' flags, symbolizing hope, while a field of blue behind the rainbow represents the future "peace and tranquillity for the enslaved people of Europe." [5]

In certain respects, the strength of the Atlantic Wall had been overbuilt in German and some other minds. It could not be as stout actually as they thought. What appeared as its psychological strength, the long unbroken front from the North Cape to the Pyrenees, was, and became ever more so, weakness. It was too extended for defense by the declining manpower of Germany and German-Europeans, in spite of all building activities of the Todt Organization. As early as December, 1940, completion of her Channel fortification was announced by Germany and reports of further sections to the north and south followed in due time.

The nomenclature used, such as Atlantic Wall, was misleading, for nothing resembling a Chinese Wall was built in Europe's west, but rather a system of hedgehog fortifications along the water's edge, strongest in the most likely invasion areas, immediately opposite England; concrete and steel emplacements, for guns ranging up to 28 cm.; pillboxes, anti-tank positions, obstacles of iron and concrete on the beaches, submerged at high tide, and later sites for rocket bombs, filling a zone of up to one thousand yards in depth. Minefields of considerable extent, and where supplies of mines ran short dummy minefields, were laid out and a large degree of faith put in them. For as an order by Rommel of May 22, 1944 put it: "Both the English and Americans dislike to enter areas recognized as being mined. Usually, they will first get specialists who have to clear these areas of mines." Obviously the time-consuming effect in mine clearing fields was considered as more valuable than the explosive effect of mines.

SECOND-LINE TROOPS FACED ALLIES

The land positions became more or less permanently manned, not everywhere as the Allies found out after their landings, with the best soldier material still at Germany's disposal, but, as in Sicily, with second-line troops, in part recruited from the conquered countries, who in the field sometimes proved willing to surrender early rather than late, or never.

Behind the zone of the Wall, reaching from the minefields through the submerged obstacles, the reinforced structures on, and in part above, the beaches to the coastal artillery positions and inundations, were scattered strong points and the tactical reserves. It was into the assembly space of the latter that the Allied air-borne infantry would be dropped. German concentrations of mobile reserves, tactical and strategical, were, as Dieppe had shown, back of the shore line, at or near points where communications centered. They were held in readiness for the delayed counter-attack rather than for immediate resistance against Allied landings. Much depended on where they were located, horizontally, or vertically to the invasion front at the time of the landings; and on the possible means of getting troops behind such fronts, along highways and railroads exposed to the interference of an overwhelming Allied air strength, while en route, or before, when communications centers and bridges, etc., might be smashed, considerably reducing the mobility of the forces.

The German naval personnel distributed at the various Atlantic ports in addition to their regular assignments were organized and trained for coastal defense and anti-invasion fighting under an arrangement of "general service," set up in July, 1943, under the former naval chief, Grand Admiral Raeder. Most of the heavier coastal batteries seem to have been manned by naval personnel. The value of this organization was shown in the vigorous defense which various ports like Brest and Le Havre put up later in spite of the seemingly incongruous elements forming the garrisons. The installation of the German rocket and flying bomb sites came too late for effective use against their logical anti-invasion objectives, the British embarkation ports such as Plymouth and Portsmouth, if concentration on such areas as targets had been possible for such projectiles with their uncertainty of aim.

In their hearts and seemingly from the outset, the German military knew, of course, that the Allies would land at various points and bring ashore considerable forces, some of whom might, if luck favored the Germans, be thrown back into the sea. Thereby one, or the other of the various teeth of the Allied jaw might be knocked out and their biting powers weakened. After that there was a likelihood that they might be contained in beachheads in Anzio style. Then a battle of attrition would set in, during which the Germans hoped to inflict the greater losses and suffer the lesser ones. This battle would be kept as close to the sea as possible. "The coast and its deeply echeloned fortifications must be defended to the last," as Field Marshal von Rundstedt declared several months before the invasion, during the battle of opinions that preceded it.[6]

While Rundstedt was in supreme command in western Europe, Rommel was appointed Commander-in-Chief of the Invasion Front, a division of authority and labor not clearly revealed yet, but perhaps intended as one between the strategical and tactical disposition of the invasion forces. That this meant conflicting counsels in the high command, that there was, as has been presumed, a divergence of views between the two marshals, is not necessarily revealed by a captured order of Rommel, dated May 22, 1944, shortly after his final inspection trip.

Entrusted with the defense of the immediate coastal zone, much that Rommel had to say had a different tone from his earlier pronouncements in favor of highly mobile warfare, but then his tasks were different from those in North Africa. Therefore his order reads as if the desert fox had become almost a coastal fortifications engineer, caught by the "statics" of the situation when he insisted to his subordinates upon the high value of the beach defenses:

> "Again I have to emphasize the purpose of these defenses," he wrote. "The enemy most likely will try to land at night and by fog after a tremendous shelling by artillery and bombers. They will employ hundreds of boats and

> ships unloading amphibious vehicles and waterproofed submergible tanks. We must stop him in the water, not only delaying him but destroying all enemy equipment while still afloat. Some units do not seem to have realized the value of this type of defense."

With the coast made nearly impregnable, the threat from the air, the landing of airborne forces by the enemy, who had such ample reserves of men and machines, seemed to impend more imminently. The enemy might "employ very numerous airborne troops to solve the problem from the land. It may be that shortly the enemy forces will start a mass attack by simultaneous employment of all their strength and put into action the mass of their air-borne troops on the first day," when they might be able to land "three divisions from the air in a chosen area within three minutes."

To meet this danger, Rommel issued instruction on how to watch for such air-borne attacks, which unfortunately have not been published.[7] While emphasizing once more that "the enemy must be annihilated before he reached our main battlefield," Rommel knew and admitted at the time that there was to be, or at least might be, a "main battlefield." No modern coastal defense can be thought of without it, no modern coastal defense consists of a "wall" alone.

Complications Were Many

This main battlefield Rundstedt, as Commander-in-Chief West, envisaged as of considerable depth, and he consequently wished to hold his armored forces for the counter-offensive in a grouping around Paris and east of it, while Rommel wanted to keep them more forward, in positions of readiness close to the coast. In the conflict of opinions, presumably adjudged by Hitler himself, Rommel won out and Rundstedt was replaced by Kluge for a time.

For the Allies the landing in Western Europe was doubly complicated, technically and politically. Both these considerations played their role in the arrangements relating to command and the commanders and the choice of participating forces. With the Chief-of-Staff of the United States Army, General Marshall, found unavailable, and "as a result of conversations between the President of the United States and the Prime Minister," the outlines of a scheme of cooperation emerged by the end of 1943 (See page 757). It would have regard for the political necessities of coalition warfare, which might later call for a severance of the American ground forces from Montgomery's over-all ground forces command, as well as provide the required talents for military leadership. Speculations began as to the ratio of troops in the invasion forces, with Americans, British and Canadians as the main contributors to the great pool and the smaller nations adding their mite, some early estimates putting the American share as high as 73 per cent. Discussions of this sort were discouraged by the Allied Chief-of-Staff, accompanied by a reminder that, after all, the British population, only one-third of that of the United States, had already supplied the larger part of the troops in the Mediterranean theater.

Various hints, not all agreeing, have been given as to how early planning for the invasion of France was begun and when the choice of specific landing beaches was settled. It has been stated, for example, that British staffs worked on the problem of capturing the Cotentin as early as December, 1941, as something to be undertaken by the Commandos after their return from Norway, but that it had been decided that although one division could do it, there was little hope of carrying on the action. [8] More specific planning by a British-American board presided over by General Morgan of the British Army began in April, 1943, surveying "the whole project by decision of the Combined Chiefs-of-Staff Committee." Their plan covering the landing beaches and the main features of the whole scheme, was taken by Churchill to Quebec later in the year, where it received in principle complete agreement. [9]

In any case, time was sufficient for assembling information, to judge enemy and natural strength and obstacles, settle on the required strength of the invasion force, estimate, requisition and ship to, and store in, the British Isles the vast amount of materiel required. With an enemy strength estimated at from 60 to 65 divisions in France and the Low Countries it would seem that the Allies thought necessary and gathered in the Isle from 80 to 100 divisions, including perhaps ten air-borne ones. These figures would not in the least contradict the statement from Allied Headquarters that in the first 20 days of the invasion 1,000,000 men were landed.

It was hoped that means would be found to keep the units

always up to strength, refreshed and fully equipped, something which the enemy was no longer in a position to do with his tired troops and depleted resources. Both man power, with an Allied superiority of at least 3:2 for the whole western theater of war and sometimes more in local theaters, and material power were designed to crush German resistance eventually, although but few expectancy dates were given out by the Allies.

On the water and in the air the Allies were certain of their superiority, which had kept the Germans away from England and left American forces on the way to, and within, Britain, as well as their materials, practically unscathed by German submarines and bombings, although the preparation and assembly of Allied forces were on too gigantic a scale to escape the enemy's observation. Still, his *Luftwaffe* did not interfere, Goering clearly being convinced that fighter protection by day or night over England was too strong to attempt bombing of harbors and other concentration points.

The strategic aim of the landing operations was to force the re-entry of western Europe by the Anglo-Saxon powers. Where should it be attempted? Omitting the factors that argued against other invasion projects in Norway, Denmark, Germany and the Low Countries, or by the Pas de Calais and in the vicinity of the Channel in France, a number of considerations obviously enough stood out in the Allies' choice of their landing beach in Normandy. They preferred an open landing shore, with a minimum of towns on the sea shore, to one close to, or at, one of the French harbor-cities—the result of the experiment of Dieppe which, as a city, proved a valuable adjunct to the Atlantic Wall.

Why Cherbourg Was Chosen

The far-reaching emancipation from dependence upon harbor facilities through the introduction of landing craft enabled them to make this choice, which was coupled with the hope of sooner or later obtaining possession of a regular port through which troops and supplies could be landed in all weathers. Cherbourg was to be this port. They chose a region fairly close to English embarkation points whence many of the landing craft could proceed in shore-to-shore moves and from which supplies and reinforcements could be taken over moderately long, economical steaming distances. At the same time they had to be close enough to England and English airfields to obtain constant fighter air support which, as might be foreseen, could not at once be based in the newly won beachhead, and whose strength was in direct ratio to the distance between base and battle zone.

The more adjacent Pas de Calais region was most strongly held by the Germans and studded with their heaviest artillery. The more distant Brittany coast was at the end of an approach route along which considerable enemy interference from such points as the German-held Channel Islands might be expected, whereas the route into the Bay of the Seine was free of such hindrances. To clinch the choice of Normandy, its coasts along the Bay of the Seine were on the whole sandy and flat and not fronted by offshore islands, sands and reefs, or rock-bound like Brittany and large parts of the Pas de Calais.

Much of the later training of the Allied troops was in preparation for the Normandy objective, though apparently more for what was in store on the seaboard rather than what was to follow inland. Numerous amphibious rehearsals, on small and large scales, for which nine training centers were formed, provided them with a realistic *imago belli* which at least on one occasion was too vivid. In April, 1944, German E-boats piercing through the screen to the sea training furnished by Allied warships, sank two LST's, killing 442 American troops anad wounding 41, in addition to the losses among naval personnel. [10]

FIRST INVASION WAVE STRONG

Meanwhile the deliberations and decisions of Allied HQ's were much under "materialistic" dictates. In fact, industrial production set the date for the opening of battle. When would there at last be enough landing craft "to provide a sufficient margin of safety for the hazardous amphibious assault"? This margin was enlarged rather than narrowed as planning went on. Shortly after his arrival in Britain in mid-January, 1944, Eisenhower found one of the most important questions yet unsettled—that of "increasing the strength of the initial assault wave" in the Normandy operation. The decision was clearly: Make the first wave strong. To make this possible, to obtain more of the still scarce landing craft, the landing was postponed by a full month, and besides it was completely severed from the simultaneously intended landing in southern France, originally an integral part of the Operation Overlord. All available landing craft was to be assembled for the Normandy operation.

Plans, men and equipment, including 200 francs in notes for every man, were ready and invasion was set for the 5th, 6th, or 7th of June. This choice was dictated by tidal conditions. The 5th offered the most favorable set of conditions, with a tide nearly midway between ebb and flood at H-hour, carrying the craft beyond the mud belt on the French shore to a sandy beach, but halting their progress short of the belt of expected beach obstacles, a concatenation which would not again occur until two weeks later.

Ships and craft were on the move, or at assembly points for D-day on the 5th, some with assault troops on board since the last day of May. But on the move as well was a meteorological low pressure area coming eastward across the Atlantic, heading towards the Channel instead of turning north as was expected for a time. On the 3rd, consequently, at 1800 hours, the invasion date was postponed by a minimum of 24 hours, with forward movements of ships to be reversed, or slowed down. Shifts were made in the time tables, with H-hour for the 6th put back half an hour, and on Monday morning early, under a cloudy, cold sky, General Eisenhower decided to undertake the landings on that day.

Those concerned with the historical (or *réclame*) problem of the "beginnings" of a specific tri-elemental operation, in which land power comes in last in point of time, will find it difficult to say which of the two, sea power or air power, started it. In a way it was Allied sea power, but not acting alone, which reduced, if not in places eliminated, the submarine threat that would otherwise have impeded the initial assembly for the landings in Normandy. In still another way it was Allied air power which had steadily reduced German power in the air, first above Britain and then over the continent and German war industries, with the bombing of German airdromes in France constantly pushing back the frontier of German air power. To advance their claims to priority, the extremist advocates of air power went so far, in the autumn of 1943, as to insist that by the following February, or March, thanks to their attacks in the meantime, the armies would be able to march virtually unopposed into the continent.[11] Perhaps for the educational purpose of putting such extremists in their place, or to re-assure the non-flying world, an airman who was an exponent of combined operations, Air Marshal Leigh-Mallory, was appointed commander of the whole tactical air-arm of the Allies.

The thought directing the employment of air power in the attack across the Channel was this, according to General Eisenhower: that applied in overwhelming strength to a particular area it "could paralyze traffic, could immobilize the enemy, could soften up his defenses, could make possible operations that would otherwise be and remain in the realm of the fantastic. We had in the Mediterranean some chance to prove the point that air could prepare the way and could sustain you after you got there in a very, very effective fashion. Pantellaria was not a good test because the defenders wanted to quit. But Salerno was . . . "[12]

Destroying Enemy's Ground Mobility

As invasion day approached, emphasis in strategic bombing was shifted to targets nearer the invasion beaches, from the enemy's industries to his airdromes and communications, ending up in the region of the beach defense before and on invasion day. Its effect was to deprive the enemy from the air, of a large part of his mobility on the ground. According to Churchill's announcement of June 6, there were 11,000 first-line aircraft to be "drawn upon as may be needed for the purposes of the battle".

On the eve of the invasion Allied air superiority over the lands to be invaded was complete to the ratio of at least three to one. Naval preponderance was even greater. Both were no more seriously endangered than is an industrial monopoly by the activities of an independent outsider competitor. There might be minor losses due to German countermeasures, among which those from naval forces were likely to be the least, but no serious interference needed to be awaited by either of the two main activities of the Allied naval forces: (1) Transporting, safeguarding and landing the invasion forces in pre-arranged order, keeping them supplied and reinforced and evacuating their wounded; and (2) serving as the breaching artillery of this force, a rôle it would share with the air forces bombing the coastal region and as supporting artillery until artillery of the ground forces could be landed and take over this task; or until fighting had moved too far inland to be given support from the sea. More than 4,000 ships, not counting the smaller craft, were assigned to these various tasks.

In the tactical sphere there was primacy of air power in point of time. Its action began shortly before midnight Monday with bombing of the invasion coast by more than 1,000 British night bombers, followed, from one a.m. until dawn by the dropping and landing of four divisions, of air-borne troops,

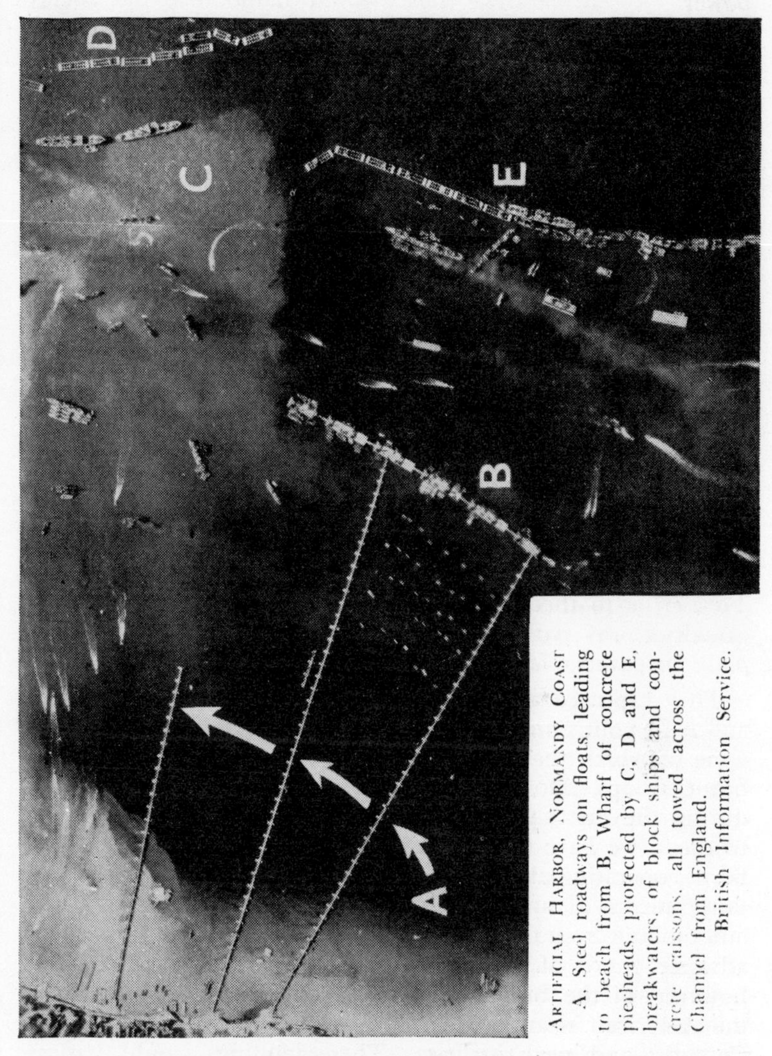

ARTIFICIAL HARBOR, NORMANDY COAST
A. Steel roadways on floats, leading to beach from B, Wharf of concrete pierheads, protected by C, D and E, breakwaters, of block ships and concrete caissons, all towed across the Channel from England.
British Information Service.

two American and two British. As the first Allied communiqué on air activities had it: "Troops carriers and gliders of the U.

S. Ninth Air Force and the RAF flew paratroops and airborne infantry into the zone of operations," reminding one a bit of Sir John Fisher's dictum that troops were the projectile of sea power.

Like bombs at any earlier time, men were now to interrupt, or disturb enemy communications, possibly in conjunction with resistance forces. The arrangements for cooperation with the latter, in spite of all romantic anticipation, were likely to be difficult, however, and it was more probable that they would get into action and prove helpful in the later stages of the operation.

The air-borne troops, carried by more than 1,000 planes, the largest force that had been landed from the air up to then, were two British and two American divisions—the 82nd and 101st, usually two-thirds parachute and one-third glider borne troops. They reached their "drop zones," which they were able to find despite cloud conditions, thanks to new location devices, and against little resistance. Only 2 per cent of the planes were lost. The account of what happened to the troops on the ground, or before they struck ground, is more confused. According to the Germans, they were badly mauled and credit given these highly "expendible" troops from the Allied side has been generous, but general in nature rather than specific. To ascribe to them a maximum of havoc and a minimum of casualties was naturally part of reports which are intended *pour encourager les autres.*

They captured a number of bridges, silenced some batteries and held points on the route to Caen, 8½ miles from the coast, some of which they were able to hold until the arrival of seaborne troops. How far the failure of the plan to win Caen during the first 24 hours of the operation must be attributed to them, we have not yet been told. Still, the British 6th Airborne, operating along the roads to Caen, was said to have been fairly successful and likewise the U. S. 82nd, dropped several miles inland from the east coast of the Cotentin Peninsula, as the advance guard of the VIIth Corps. This could not be said, however, of the British landings south and east of Caen, and the American near Coutances, close to the west coast of the Cotentin and near its base. These landings would indicate what were the outermost perimeter of the Allies' hopes for the beachhead.

The air bombardment of the beach defenses was resumed near dawn by 1,300 American Flying Fortresses and other bombers, "directed at gun emplacements and defensive works in support of landing operations," with fighter-bombers attacking gun batteries and communications in and behind the assault front. Altogether 10,000 tons of bombs and shells were expended during the first day on targets like coastal batteries with whom 640 heavy naval guns fought a rather one-sided duel before, during and after the arrival of the landing forces. During the first day Allied air supremacy remained overwhelming, as 150:1 if measured by the sorties flown by the two sides, 7,500 against the 50 of the *Luftwaffe.*

DAYLIGHT ATTACK A SURPRISE

To one reporter who had participated in other landings, "the big surprise in that attack was that it went in during daylight," [13] the latter being partly obscured however by the use of smoke. H-hour was 0630, an hour after sunrise. Considering the magnitude and complicated character of the operation as well as the vast air and naval superiority of the Allies, predawn hour protection could well be renounced by them, except for some Rangers and Commandos who were landed under cover of darkness to climb cliffs topped by gun positions. The early light, uncertain though it was, made somewhat easier the complicated task of the engineers in removing obstacles such as the "element C," a combination of grill work and cable *chevaux de frise.* This had to be done during an interval of twenty to thirty minutes before the tide could cover them and make them even more formidable. While the light of dawn also helped to direct what remained of German artillery and machine gun fire, it was merely spotty.

In days of vast service réclame the reminder by the American Secretary of the Navy Forrestal that the invasion operations "have depended from their inception upon sea power," [14] was modest enough, if not a sign, among others, that sea power was on the defensive in the inter-service competition. In the tactical sphere the "beginners" on the part of the navies were the mine sweepers. They were out a day and a half before D-day as originally set, sweeping clear their own way and that of the armada into the Bay of the Seine, when they were recalled. only to be told on returning to go back to the same job of clearing and marking the paths of the ships big and small.

They went in as far as 1,500 yards off the coastal batteries. Their task was by no means over with D-day. The sea lanes, ten for the original approach, strewn anew with mines by the Germans night after night from plane and surface craft, had to be swept again and again. The mine bag off the French beaches during the three months from the invasion day equalled one-tenth of all mines garnered, or destroyed by the Allies in five years of war in all theaters. A few minesweepers only were lost on this arduous duty, including the USS *Tide* and *Osprey*.

They were among the smallest in the group of the 800 fighting craft, running all the way up to battleships with 16 inch guns that prepared the way for, and escorted, the 3,200 transports and landing ships, starting from British ports close to, and farther away from, the Channel to their various rendezvous. In this, the greatest of all armadas, provided by the coalition of the sea powers, there were about three British to every two American craft, the latter including 1,300 ships of all types, such as three battleships, three cruisers and more than 30 destroyers, plus contributions from the smaller Allied nations.

A number of ships was detached for a demonstration on D-day against the Calais-Boulogne region, where they "made a lot of noise" and clearly tied down a considerable number of German troops, if not leading the German command to the belief held into July that a still greater Allied landing was intended for that region. The American task force, on the right, was under Rear Admiral Kirk; the British under Rear Admiral Vian.

At 0527 in some places, at 0535 in others, the ships opened up, to contribute their share in the battering of the Atlantic Wall. In numbers and calibers the naval guns were far superior to the enemy's guns, and the novel boldness of old battleships going in as close as 6,000 yards from the shore, as in the case of the nearly ancient USS *Arkansas* of 1912 vintage, with her twelve inch guns, seemed to obliterate the old legend concerning the inferiority of cannon *sur mer* as against cannon *sur terre,* if it did not actually show that the mobile guns on the water possess great advantages over static defenses including artillery in permanent emplacements. The latter returned the fire but weakly, although hitting at least one destroyer in the invasion armada. Twenty minutes after the opening of the bombardment the landing craft, circling around the transports as if obeying a policeman's *"Circulez-Vous!"*, with their first wave broke from their circles into the straight line to the smoke-obscured shore. At 0725 the first landing craft touched down on some of the beaches.

Along the 10,000 yards of the American beaches "every 75 yards a landing craft with assault infantry touched down at H-hour. Assault veterans charged down the ramps, picked their way through the band of obstacles and immediately provided cover for the work of naval and engineer demolition crews which followed close behind. Each crew had a specific task to perform in clearing lanes for subsequent waves of craft carrying infantry, artillery, vehicles and supplies." (Marshall's *Biennial Report*).

RESISTANCE BY SEA NEGLIGIBLE

Only minor German counter-action from the sea marked the ouverture of the invasion. According to the Allies' version of events only three torpedo boats accompanied by armed trawlers tried to interfere with their advance on the 6th. No activity of submarines was reported by the Allies, although a hundred, according to Churchill, lurked around, but it would have been hard for them to operate in the fairly shallow waters of the Channel. However, land-based opposition against the naval craft remained active in spots and was so strong that certain beaches within the strip of the Vth American Corps had to be closed temporarily while at others smaller landing craft offering poorer targets had to be used to take in troops. Nor was this opposition devoid of cleverness. The Germans' dummy gun emplacements along the coast seem to have attracted not a little naval gunfire. Splashes in the water were produced by them to misguide Allied artillery observation and, for the same purpose, although with less success, aircraft spotter signals were faked.

Allied naval losses as announced after five weeks were minor. They included three American destroyers (from 1,700 to 2,200 tons each) and three British (of 1,300 tons and less), eight other small warships and one American transport of 8,100 tons which was struck in the Channel fully laden with troops, practically all of whom, however, were saved by a highly organized rescue service. Two of the three American destroyers and one LST were lost before noon of the 6th. Merchant shipping and landing craft losses included 5 LST's, 24 LCT's and 8 LCI's, lost off Northern France.

The western naval task force, almost exclusively American, carried the First American Army, comprising the Vth and VIIth Corps, into western Europe, facing in part west, against the eastern side of the Cotentin, with its VIIth Corps (14th, 9th and 79th Divisions), and in part south, from Grand Camp to St. Laurent, with the Vth Corps. The British and Canadians, suffering the worst of the weather, were landed on the eastern part of the north shore of the Bay of the Seine. General Montgomery had told commanders to make their men "imbued with one idea—to penetrate quickly and deeply into enemy country and peg out claims in land. The conception that you land on the beaches and get a little bridgehead and then dig in—that is no good. You must penetrate quickly and deeply."

How far his expectations were fulfilled, how far they remained unsatisfied, as measured by the original plans, cannot yet be said. But this much has been admitted: that these plans called for taking Caen on D-day, which would probably have forced the German to forsake the Cotentin, and for a much quicker advance of the Americans who were considerably slowed up in the subsequent "hedgerow war." Due to the determined Allied policy never to admit failures in their own landing attempts, there is only silence about any possible additional attempts of theirs, such as on the Channel Islands, at the Bay of St. Martin, near the northwestern tip of the Cotentin, or at Trouville, all of which were reported by the Germans, some as late as the 9th and 10th. According to their reports, the original invasion front of the Allies, not a continuous one at the outset, reached from St. Vaast-la Hougue in the west to opposite Le Havre in the east, with air-borne troops forming the wings which were to a certain extent clipped by the Germans.

While Prime Minister Churchill announced to Parliament in the afternoon of the 6th that "the invasion goes according to plan, and what a plan! There already are hopes that actual tactical surprise has been attained," the Germans declared that "the long awaited attack on Western Europe was expected by us." Tactical surprise was actually limited. The American Vth Corps, east of the Carentan estuary, ran into the German 352nd Division, in large part composed of Poles, it was said, which had been holding anti-invasion exercises on D-day minus one, and thus into some of the readiest opposition. But, perhaps, this proved fortunate for the invaders after all, as it took the German division out of its assigned defense zone.

The first two divisions landed, the 1st and 29th, whose landing craft had already suffered from the as yet unsilenced German naval guns on the St. Marcouf Islands, were pinned down on beaches not more than 100 yards deep, suffering the heaviest initial casualties. Only after more naval fire and fighter-bomber support had been given and another division, the 2nd, was landed on the 7th were they able to advance southwest, towards Isigny and Carentan. There they hoped to join with the VIIth Corps for a thrust across the base of the Cotentin. But not before early in the 10th was it announced that troops of the VIIth Corps had crossed the Carentan-Valognes road and cut the main railway line into Cherbourg, and later in the day that they had taken Isigny.

The right wing of the British-Canadian army worked its way into Bayeux on the 7th against slight resistance of German observation forces. But the left wing, pressed together on the east bank of the Orne River, was unable to take its main objective, Caen, which was, and remained, strongly defended, ever since the first armored counterattack of the enemy in the Caen region came on Tuesday evening, to be met and repulsed by landed tanks.

Shipping Main Targets

On the 7th all beaches were cleared of the enemy and in some cases contact between them established. A few coastal batteries, however, were still firing and these and some strong points, such as that at Douvres which was manned by some 150 Germans and held out until the 18th, had to be by-passed. Inland targets were kept under naval gunfire, directed by aircraft. Allied reinforcements came in throughout the 7th, partly airborne, because the seas were high and played havoc with the unloading program. The enemy's opposition in the air increased, he losing 176 planes in the air against 289 of the Allies who could much better afford the higher losses, during the period from June 6 at dawn to midday of the 8th. The defensive measures of the weak German naval forces and of the *Luftwaffe,* wisely enough, were, and remained, on the whole directed against Allied shipping which offered the most concentrated targets.

On Friday the 9th the English and American lines seem to have joined sufficiently, at Sully, 16 miles east of Isigny, for the Germans to attempt to split them here, a juncture point, always highly vulnerable in coalition warfare. A British cruiser, miles away, intervened with a barrage, snuffing out the German tank-supported thrust. On the 12th the fusion was complete of all the Allied beachheads, enclosing a crescent-shaped coastal strip of some 60 miles length and from 3 to 15 miles deep, containing some three or more air strips.

With continued Allied landings and concentrations of German reserves "the weight of armor on both sides was increasing," particularly from the 9th on, and a four days battle for Caen developed, fought west of that ancient town to Tilly-sur-Seulles, with the British 7th and 9th Armored Divisions and the German 21st and soon after the 12th SS Panzer Divisions in the fray. The terrain over which the battle fluctuated, centering around Tilly-sur-Seulles, was difficult for the invaders, with woods and rolling hills concealing the enemy's movements. The Germans still threw in only local reserves, with the Allies waiting for the main counter-attack which never materialized.

By the 9th, according to German estimates, Eisenhower had put 18 divisions, or about a quarter of a million men, into the beachhead—actually that many had been landed within the first 24 hours—but not enough to convince the Germans, who had perhaps ten divisions in the battle area, that this was the main endeavor of the Allies who, they thought, had at their disposal some additional fifty divisions, possibly to be thrown into a second beachhead soon to be opened. Practically from its first announcement, the German High Command clung to the view that "new enemy operations must be expected, but have not yet taken place" (June 7).

Horns of German Dilemma

The choice facing Rundstedt and Rommel was between an attempt to throw the Allies back into the sea by employing all, or nearly all they had, or sealing them off in the Cotentin, denying them the use of Cherbourg so long as possible. Very uncertain that the Allies might not land elsewhere while they made an all-out attempt to push the Normandy forces to their boats, aware that their own effort could hardly succeed altogether in view of the strong support to be rendered from Allied naval forces, giving "deep supporting fire" up to 12½ miles inland, and from overwhelming air power, the Germans inescapably concluded that they must try to seal off the Cotentin at its base, to hem the Allies in at another Anzio.

This grew more imperative every day after they had refrained from full-scale counter-attacks during the first days following the 6th of June. In waiting around for additional Allied landings, with their reserves concentrated rather close to the Atlantic, the Germans exposed themselves and their communications far too much to the impact of Allied air power. According to General Guderian's critique, that of a tank expert, of Rommel's handling of the situation "the Wehrmacht could have smashed the Allied invasion of Normandy if its panzer reserves had been distributed deep in France rather than near the coast." [15]

By the 17th, according to General Bradley, the Allied hold on the beachheads was "absolutely secure" and the Germans had lost "their last chance to drive us back into the sea." On the 18th American forces reached the far coast of the peninsula at Berneville-sur-Mer and on the 26th Cherbourg fell to them. By the 20th day of the invasion, one million Allied soldiers were brought ashore, through "symbolic harbors" as Churchill called the landing places. Meanwhile the sealing-off process across the peninsula went on, through hot and fluctuating fighting of no great depth. Behind this lively front, through the port of Cherbourg, the Allies were charging their batteries for the break-through and the break-out in depth beginning late in July.

The logistic importance of the "symbolic harbors" must have been considerable despite the fact that the hopes based on them were only in part fulfilled. There were two of them, one American near St. Laurent-sur-Mer, and one British, near Arromanches, and their combined capacity was expected to be larger than that of the Cherbourg port. Unfortunately, the American auxiliary port, receiving the brunt of the worst weather in forty years, was largely destroyed before completion during the three and a half days' gale from the 19th to the 22nd of June, whereas the British, placed in more sheltered waters, survived. These artificial ports, complete with breakwaters and piers, were built from some sixty blockships, including four old battleships and 32 American merchant vessels, or 300,000 tons shipping in all, concrete caissons and piers brought over from England by tugs in what was considered "perhaps the greatest towing job in history." [16]

Compared with Dieppe and Tarawa, which cost 3,350 and 3,583 casualties, the price paid for this continental springboard must be considered moderate. It was 3,283 American dead and 12,600 wounded during the first eleven days, with casualties on the central beachheads running higher and on the peninsula lower than expected; to these would have to be added the missing and those few thousands taken prisoners by the Germans, as well as the casualties of the other Allies which have not yet been announced. The German losses in prisoners alone to the 18th were 15,000 to whom must be added, considering the tenacity of the fighting, probably at least twice this number in dead and wounded.

POPULAR EFFECT OF NORMANDY LANDING

The Allied landing in Normandy gave a powerful gratifying uplift to Allied peoples and brought commensurate disappointment to German- (or jerry)-built Europe. Planned or not, its beginning coincided with the taking of Rome by the Allies who, however, were unable to persuade the Russians to make their moves from the east in that perfect synchronization that the western peoples (or their newspapers and journalistic military experts) expected. On the American home front industrial production rose and absenteeism dropped, as much as 50 per cent, due to the "new invasion spirit, as the tension of wondering, or waiting has been lifted"; workers felt "that their products are now in the invasion and are proud of it."

Such were the findings of a nation-wide survey—or wish fulfillment—by the War Production Board.[17] Another form of participation, perhaps still more personal or sacrificial, was the increase, at least temporarily, in blood-offerings for plasma from D-day on.

Inside invaded France hopes of resistance forces, whom the Allies had begun months before to provide with arms, on the whole in localities rather distant from the beachheads, took on additional strength, although their aid to the Allies in Normandy itself remained minor. Allied propaganda, about which their own countries on the whole learned less than about German propaganda, reminded the Germans of Hitler's pronouncement after Dieppe that if the Allies should land on the French coast, they would stay just nine hours; this recorded fanfaronade was sent over the radio continuously.

The shock of the Normandy landings was more than the

German propaganda machinery could absorb, or even well cushion. It tried to discount the bad news with data on the vastly overstated, not then yet officially announced, losses caused to Allied shipping in the Channel by German bombers, artillery or surface craft. The Atlantic Wall was now said to have fulfilled its function as a breakwater, never having been intended as a dyke. Beginning June 12-3 the German flying bombs directed at London and vicinity came to Goebbel's help in boosting morale, although some people might have been struck by the parallel between this belated fire on London and the equally belated Big Bertha bombardment of Paris in 1918, and the "too late" factors in this and other parallels between then and now.

Then came the fall of Cherbourg, announced by the Wehrmacht command before the Propaganda Ministry had prepared the people for the now complete break in the Atlantic Wall, the officers having decided that the man in the street should have a more realistic outlook upon the actual situation than Goebbels' ministry was providing. Many of them had long begun to view the situation as serious, too serious to be retrieved by any last hour resort to any "secret weapons."

Conspiracy against Hitler and his system became rife. The landing in Normandy set off dissensions inside the Reich's governing circles. Their effect, if not also their nature, were similar to the train of developments inside Italy after the landings in Sicily. A group of officers, mostly with the noble *von* to their names, "lost courage," as General Guderian, a tank expert and of Armenian descent, accused them in a broadcast, and one of them threw a bomb at Hitler (July 20).

The suppression of this putsch was Hitler's last victory, won on the home front where he had gained his first triumphs. No last hour *levée en masse,* declared July 25, such as Walter Rathenau had proposed in vain a few weeks before the 1918 armistice, could stop the Allies.[18]

NOTES, CHAPTER 59

[1] The data for the above are in Kries-Speier, *German Radio Propaganda.*
[2] *N. Y. Times,* Nov. 23, 1943.
[3] Cit. in *The Nation,* Feb. 26, 1944.
[4] U. P. despatch from Allied HQs North Africa, Aug. 2, 1943.
[5] A. P. despatch from London, March 25, 1944.
[6] *N. Y. Times,* Feb. 15, 1944.
[7] *N. Y. Herald-Tribune,* July 8, 1944.

[8] A. P. despatch from Supreme HQs Allied Exp. Force, June 24, 1944.

[9] Churchill in House of Commons, Aug. 21, 1944.

[10] A. P. despatch from Supreme HQs Allied Exp. Force, Aug. 7, 1944.

[11] *N. Y. Times,* Dec. 5, 1943, despatch from Washington; Hanson Baldwin, ibd., May 1, 1944.

[12] A. P. despatch of June 16, 1945.

[13] Ross Munro in *N. Y. Times,* June 8, 1944.

[14] Ibd., June 15, 1944.

[15] U. P. despatch from Berchtesgaden, May 13, 1945.

[16] For more details see *N. Y. Times,* Oct. 16 and 18, 1944.

[17] A. P. despatches from Washington, June 14, 1944.

[18] The above is largely based on the communiqués and statements from both sides, the reports of Hanson Baldwin to the *N. Y. Times* from the Channel; Charles Christian Wertenbaker, *Invasion.* N. Y., 1944; Col. Conrad H. Lanza, *The Invasion of Normandy. Military Review,* Sept., 1944. Col. Paul W. Thompson's article *D-Day on Omaha Beach (Infantry Journal,* June, 1945) came too late for inclusion.

Chapter 60

RIVIERA LANDING, AUGUST 1944

"Again their ravening eagle rose
In anger, wheel'd on Europe-shadowing wings . . ."
Tennyson: Ode on the Death of the Duke of Wellington, 1852.

THE LANDING OF THE ALLIED SEVENTH ARMY in southern France, beginning August 15, was the clearest proof that the Fortress Europa had become too large for defense and holding by the man power at Germany's disposal; and also of the fact that Hitler's strategy in clinging to far flung territorial gains and holdings was too medieval, or too Napoleonic, or simply too imperialistic. One after the other, the outposts of his temporary jerry-built empire proved untenable. Elba was lost by the middle of June, the Aegean islands later in the year. There were not enough German troops, although they were fattened with non-Germans, like those divisions in southern France who were up to two-fifths Slavic, Russian prisoners of war, Czechs and Poles, to interfere effectually with the Allied landings. Or to hem in those near the shore, Anzio-style, or to slow up their inland advance along the Rhone valley enough to allow the German command to retire all the German forces in the southwest of France in the direction of the Belfort gap.

All better troops that might have been stationed there at some earlier time, had been transferred to the fight in northern France, 400 miles away. Only in the direction of the Italian Riviera did the Germans succeed in blocking the Allies' advance. Kesselring's lines of communications and eventual retreat were too precious to be jeopardized.

The landing in southern France, the opening of the fourth front in Europe, was a branch enterprise of the Mediterranean command and the armies in Italy. Some of the participants, such as the American VIth Army Corps—3d, 36th and 45th Divisions, all sharers of the Anzio enterprise—were withdrawn for the operation from the Arno and other fronts; others, like the French troops, came from North Africa. The invasion troops, under the command of Major General Alexander M. Patch, who had held command in Guadalcanal in the last phase of that operation, received several weeks of intensive training in amphibious operations, while the planning started in the middle of June was receiving finishing touches.

As between the Italian and French Rivieras the choice of the objective for the next Allied drive was fairly obvious. They knew that the forces along the French shore were becoming weaker all the time, that the coast and immediate hinterland were far more favorable to them than in Italy where the Duce long before 1939 had studded the water front with heavy ordnance against a French landing that was never attempted in 1939 or 1940 and where the shore was thickly settled and offered the defender better opportunities for the initial defense in house to house fighting. Hence the choice, among other reasons, of the relatively most uninhabited part of the French Riviera, which was the most inaccessible from the hinterland.

The French interior was also thought to be more favorable politically. After the rather extravagant hopes of aid once placed in the Italians behind the German lines by the Allies had produced such disappointments, the French forces of the Interior gave better promise that in them the landing corps would find what Clausewitz considered the first condition for the success of such an operation—a population in insurrection against its government, which in this case was either Berlin or Vichy.

The selected landing beaches were between Toulon and Cannes, from Le Levandou to St. Tropez, each some twenty miles distant from these cities, behind the Hyères Islands, not far from Fréjus, where Napoleon in 1815 had landed from Elba for the Hundred Days. Along the Napoleonic highway leading north—"from steeple to steeple"—the FFI were to join the American-French invading force. From these resistance groups the Allies had received valuable advance information as to the German forces which, as was later acknowledged, "was borne out in all respects, no units being encountered unexpectedly." It even included such data as the boundary line, against which a special assault was to be directed, between the two German divisions placed nearest the coast.

The landing forces were carried and supported by above 800 ships, which would make the operation about one-fourth or fifth of the size of the initial enterprise in Normandy. The armada included ships of all kinds, "with many mixed vessels," as a naval communiqué stated, with 450 ships contributed to the pool by the Americans, 300 by the British and the rest by the various smaller nations. Among the combat ships were seven British aircraft carriers and three anti-aircraft cruisers. Most of these naval forces had been assembled for some time, for the original D-day had been August 8, which had subsequently been changed to the 14th for "military reasons," and then to the 15th on account of the weather which on that day and afterwards was calm and clear. Their final rendezvous was behind Corsica.

The operation was almost a peace-time general's or admiral's dream (if they had ever dreamed of combined operations). There were in its course and schedules no considerable hitches and just enough enemy forces and resistance to make operations really warlike. All previous experiences in Mediterranean landings were put into, or rejected from, the planning and execution. The unfortunate Sicilian experience of co-national aircraft being shot down by co-national artillery, for example, had led to much improved identification signalling.

AIRBORNE OBJECTIVES GAINED

The heavy bombardment in the days before the 15th had told the Germans that (and where) the Allies were to be expected. Early in the night parachutists and airborne infantry were landed at various points, on the wings of the invasion front and farther inland, even beyond the range of the Monts des Maures, the name a reminder of the Saracen incursions in this land a thousand years earlier. The majority of them did land at their objectives as HQ's stated somewhat emphatically and many were able to join up later with the troops coming in through the beachheads. Pre-dawn landings on Port Cros and Levant in the Hyères group and at Cap Nègre on the continent by specially trained troops were intended to silence coastal batteries.

The naval guns opened up with a heavy bombardment of enemy positions on the shore, firing 15,900 shells of 5 inches and above, of which 12,500 were of at least 12 inch caliber. The main landings, taking place in full daylight, proceeded from the transports at 7, with an hour allowed for the landing craft to cover the seven miles to the shore. All the assault waves were under way by 0740 hours. H-hour was at 8, and the first wave was "exactly on time," as a naval communiqué reported, or "just two minutes late" as omniscient *Time* stated.

Assault and landing craft were preceded by minesweepers which found less to do than was anticipated, and were covered by naval gunfire from ships great and small, from battleship to rocket boat. The latter as well as the landing craft gunboats maintained a running fire until nearly the moment when the first assault wave touched down. By 1000 artillery and tanks could be brought ashore.

Enemy resistance throughout was described as weak or "sporadic," flak fire as feeble, the *Luftwaffe* unrepresented for hours; only twenty of its planes were seen on the first day when the Allies flew 4,285 sorties, losing only seven craft. The local *topos,* in itself favoring the defenders, proved too spacious for the "people without space" and the two divisions they had in the immediate vicinity of the beach. These divisions, taken largely from the conquered east to defend the conquered south of the Third Empire, commanded by German officers and non-com's, were of low combat efficiency and morale. The German naval units trying to interfere from Toulon, comprised nothing above corvettes.

Beach obstacles and mines were few and far between. Coastal artillery was nowhere so strong, or as numerous as one side, or the other had at various times described it. There was thus a good deal of anticlimatic experience in store for the landing troops. Some of the supposed heavy guns on the Hyères Islands were found to be of wood, stove pipes and paper. Another powerful artillery site proved to be good for only 20 millimeter guns.

By the end of the second day Allied penetration had in places reached eight miles in depth, and nearly 700 prisoners had been taken. By the end of the third they held a beachhead with a coastal length of 40 miles and a depth ranging up to 20 miles, and had more than 3,000 prisoners in their hands. On this day the various original beachheads were joined in one embracing 500 square miles, "the largest created in less than three days during this war," as HQ's announced its own record performance. On the 23d Marseilles fell, before Toulon, which was nearer. Possession of the greatest of all French seaports, gave to the Seventh Army the assurance of a steady flow of supplies and reinforcements to carry it far into the interior.

Allied casualties were termed "exceptionally light," less than

300 killed, wounded or prisoners through Tuesday noon the 15th, "as a result of perfect coordination of all arms and services," as a communiqué stated, a tacit admission that on other occasions faulty cooperation had as a matter of fact, if not as a rule, added to the unavoidable losses.

The invasion and reconquest of German-held France by the Anglo-Saxon Allies from the west and the south was a sound military undertaking and performance. It was not attempted before the necessary forces were available, and was devoid of many, if not most, of the frictions natural to a war of coalition. It was in keeping with the classical canons of a Clausewitz to reduce the main actions to a few while keeping all minor ones subordinated; to act with concentrated force, to proceed as fast as possible, but not prematurely. In this, they vindicated the feasible boldness of large style landings, in full contrast to the dispersive landing practices of the Pitts and their predecessors who, as Clausewitz put it at the end of his work, attempted to surround France from Dunkirk to Genoa "by a girdle of armies, while 50 different small objects were aimed at, not one strong enough to overcome the inertia, trammeling and extraneous influences which are produced and reproduced anew everywhere, but most particularly among allied armies." [1]

NOTES, CHAPTER 60

[1] *Of War,* Book VIII, ch. 9.

Chapter 61

THE PACIFIC, 1944-45; BACK TO THE PHILIPPINES; CLOSING IN ON JAPAN

BY MIDSUMMER OF 1945, control of the Pacific was being pushed ever closer to the Japanese mainland. This made the Allies, to use Viking terms, the "sea kings," the "ness kings," i.e. those who hold promontories and from there exercise their sway; and, too modern for precedence, the "air kings" of that ocean. They maintained a sea and air blockade tight enough to endanger increasingly Japanese life lines, while safely insuring their own long communication lines over which unexampled quantities of supplies had to be carried. Eniwetok atoll in the Marshalls was turned into a repair base for the United States Navy and Saipan in five months after the first landing there into the base for the Superfortresses which were to carry destruction against war-industrial Japan.

Pacific *reconquista,* as continued since the fall of 1944, left by-passed for later garnering, numerous islands, parts of intervening islands, or even island groups such as the Carolines, with the exception of the atoll surrounding Ulithi lagoon which was occupied on September 20 and 21, 1944, by Americans in unopposed landings. Some like Truk in the Carolines had once been on the list of desiderata, but on second thought it had been decided that it would be cheaper and easier to neutralize them. Like the best peas in an opened pod, only such islands in the various island groups were selected and taken which were necessary to provide the attackers with way stations for their sea and air craft and to keep Japanese-held islands in the area under control.

This procedure left large numbers of Japanese in the rear of the Allies' advance, estimated by some at half a million early in 1945,[1] but officially at only half that. American naval figures put 83,000 of the enemy on the Carolines, 50,000 on Truk alone, 12 to 14,000 on the Marshall Islands (Wotje, Maloelap, Mili and Jaluit); 4,000 on Ocean and Nauru Islands, 5,000 each on Marcus and Wake Islands, 5,000 on the Marianas (Rota, Pagan, Agrigan); 100,000 to 120,000 in the Solomons and New Guinea region, including Rabaul on New Britain and Kavieng on New Ireland, and 30,000 on Babelthuap in the Palau group.[2]

These Japanese gave evidence of being quite determined to stick it out, those on Bougainville in November, 1944, rejecting an offer of surrendering on honorable terms, whatever these may have implied.[3] The neutralization applied to the remaining Japanese in that portion of the Pacific theatre was in effect a combined land, naval and air blockade, with the air arm the most active, and which, except in the case of Truk, was usually able to keep the Japanese pinned to the ground. Not all aviators found this activity altogether worthwhile, some of the usually outspoken Australians considering their targets as having "an impossible value in the prosecution of the war," involving an appalling waste of planes, bombs and ammunition.[4] Ground forces for the purpose of containing these by-passed enemies were now supplied by the Australians, as the former, or prospective owners of the islands in question, who since November, 1944, took over from the Americans in New Guinea, New Britain, where Rabaul still held out as the prize of the comeback for forces ejected early in 1942; and Bougainville.

Under the cumulative impact of these measures, it was officially hoped, the Japanese would gradually wither on the vine. The latter, however, managed to make the physical vine take root in the islands, proceeding to cultivate tropical subsistence gardens, which under the influence of the optimistic radio communications from the homeland they may have called their victory gardens for all we know. Living off the land during war had found a new theatre in the tropics. Submarines carrying freight provided garrisons with the bare necessities of martial survival, excepting, of course, heavy equipment for which they would have to rely on materiel previously stocked in anticipation of their own invasion of Australia and elsewhere.

These Allied "sea kings," "ness kings," "air kings" were not always in agreement on methods, or objectives or the essential unity of the Pacific war. Simultaneous landings (September 15 and following days) of MacArthur's forces on Morotai, the northermost island of the Dutch Halmaheras or Moluccas, 300 miles beyond New Guinea, a step which carried MacArthur once more north of the Equator; and those of Nimitz and Halsey on Peleliu, 600 miles westward beyond Guam, where the First Division Marines touched down, and on Angaur, target of the 81st Infantry Division, both in the Palau group, both objectives lying on the American road to the Philippines, were at the outset greeted as "the first twin offensive by the commands of General MacArthur and Admiral Nimitz."

RIVAL VIEWPOINTS DISCERNABLE

But it was hard not to detect in either the actual fighting, or the reports about it the old, by now stereotypic, differences of view held by the rival services. With what MacArthur had won at small expense, he had, according to his first communiqué, "penetrated the Halmahera-Philippine line; the enemy conquests to the south are imperiled by threat of envelopment. This would cut off and isolate the enemy garrison in the East Indies, about 200,000 men, and would sever the vital supplies to the Japanese mainland of oil and other war essentials." That seemed as though the Indies had been cheaply and easily brought under control. Again, the communiqué in its tactical part emphasized that "under cover of naval and air bombardment" the General's troops on Morotai readily seized their beachheads. "The point of landing was unexpected, the enemy having expected it in the lower parts of the islands, where he had accumulated very strong forces in heavily defended positions."

On Morotai the Americans ran into a Japanese force consisting of a mere thousand who were quickly routed. They were at first too much demoralized to interfere with the rapid construction of American fields for the air force which was to strike at the Philippines. Gradually, however, the Japanese, given a new commander, built up a force for the counter-attack by first reorganizing the remnants of the original garrison which had been left unpursued in the jungle by the invader, and later by bringing over reinforcements from close-by islands. The American garrison guarding the perimeter around the aviation and naval establishments by the middle of December was so seriously threatened by the imminent Japanese counter-attack that reinforcements had to be shipped in from New Guinea which subsequently routed the enemy in jungle fighting.[5]

If one judges only by the high price paid for them, the Palau Islands were the more valuable win. From there, from air-bases on Peleliu, Angaur, Ngesebus, not only could Babelthuap and the other Palaus, with an estimated garrison of 25,000, be neutralized but also Truk, Ponape and the rest of the Carolines with their 100,000 garrison (these estimates naturally vary with time and purpose). In case of an attack against the central Philippines—Leyte, etc.—rather than the southern archipelago the air fields in the southern Palaus would be nearer the invasion coast than Morotai, some 650 miles from Leyte. It took nearly six weeks before Peleliu was officially "secured," in spite of prolonged air bombardment since September 6 and the expenditure of increasingly larger amounts of aerial bombs and naval shells, of yet further improved methods of "call-fire and close-support fire from warships off shore." The enemy proved to be still better entrenched and dug in in caves than on Guam, and dispensed even with the usual suicidal banzai last hour attacks in final exposures to superior American fire power; whether he evacuated part of his forces to close-by islands of the Palau group, from whence he retook a small island near Peleliu on November 7, has not been stated.

The casualty list for the two divisions landed in the Palaus after three weeks of the hardest fighting included 1,022 dead, 6,106 wounded and 280 missing—the heaviest casualty rate, it was reported, up to then for a single division in the Pacific war as far as the First Marines were concerned with 771 killed, 4,650 wounded and 267 missing. The losses of the Japanese by American count to October 7 were put at 11,083 killed on Peleliu and 1,128 on Angaur and 224 prisoners. In addition, the navy reported that above and around the islands invaded and to be invaded (Palaus, Halmaheras, Philippines) the Japanese from August 20 to September 23 lost 376 aircraft shot down, 592 destroyed on the ground, 137 ships sunk, 182 ships probably sunk, or damaged as against 51 American aircraft lost in combat.[6]

LEYTE, LUZON, MINDANAO

The taking of Peleliu and Morotai indicated positively—as did in a negative way the omission of further island conquests beyond the Marianas—that the main weight of the American forces in the Pacific, growing only slowly while the European fighting approached its decision, was to be thrown against the Philippines. That archipelago emerged ever more clearly as the first sizeable object of the Pacific war rather than the mainland of Japan. Neither Guam, nor Saipan appeared "big enough or near enough to the scene of future operations," against Japan directly; and even the Palaus, if taken in toto, seemed in October still "too far away for probable future operations." [7] Nor were the available forces strong enough to be invested against Japan proper, and the islands along the route to the Japanese mainland could well wait until 1945. Yap in the Carolines was so seriously considered on the very eve of the landing in the Philippines that an American army corps had been put to sea against it from Honolulu on September 15. Reports from the Navy about the heavy fleet and air losses suffered by the Japanese under its foraging blows caused a complete change in plans, resulting in the abandonment of the Yap expedition and the acceleration of the enterprise against the more important Philippines.[8]

Morotai, which the Japanese tried in vain to retake in December, and the Palaus provided flank cover for the approach of MacArthur's flotillas of 600 vessels of all kinds which seem mainly to have come from Dutch New Guinea, some 1,000 miles and eight days from Leyte, with the rest setting out from the Palaus. These island airplane carriers and the far ranging naval forces, which had struck at Marcus Island on October 9 with surface forces, at Okinawa with a carrier task force on the 10th, at the northern part of Luzon on the 11th with airplanes of a carrier group and which operated in Formosan waters from the 12th to the 14th, provided all necessary safety for the transport flotilla.

Discussion of the strategic choice of the Philippines on the whole is of greater interest than the details of the landings on Leyte on and after October 18. These went according to schedules, if not a little ahead of them, except on one beach, and plenty of personnel was available for them, the command allowing it to be reported home that MacArthur had 250,000 men at his disposal. This can only have meant the total of military and naval personnel, two to three sailors now being needed to put one soldier, or marine ashore.[9] The ground forces consisted of the Tenth and Twenty-Fourth Corps, the Third Engineers Brigade, Amphibious, First Battery Division plus supporting and supplying troops. While there was an estimated total of 225,000 Japanese in the archipelago, only a little more than one division was at first on Leyte to meet the American onslaught.

MacArthur presumed that the enemy had expected this elsewhere, in Mindanao, thus causing him to be "caught unawares in Leyte, and beachheads in the Tacoblan area were secured with small casualties." Ever faithful to his communiqué style the General, who as one paper put it "believes in his own clichés" (*Newsweek*, Oct. 30) and likes to refer not only to past achievements, but to future eventualities as well in the daily communiqués, went on to say: "The strategic results of attacking [conquering?] the Philippines will be decisive. The enemy's so-called Greater East Asia co-prosperity sphere will be cut in two . . . A half million men will be cut off without hope of support . . . In broad strategical conception, the defense line of the Japanese, which extends along the coast of Asia from the Japanese islands, through Formosa, the Philippines, the Dutch East Indies to Singapore and Burma, will be pierced in the center, permitting an envelopment to the south and north. Either flank will be vulnerable and can be rolled up at will."

While this may give us the large views which MacArthur held at the outset of the Philippines campaign, their realization took far more time than the formulation of the martial metaphors in which generals and admirals so generously indulged on the occasion, MacArthur calling the Leyte campaign "the key to the Philippines" and Admiral Pratt "the Navy really the key to the first phases of the Philippines invasion."

MacArthur's linear concept of strategy in all essentials presumes that as in earlier times, let us say the 18th century, the enemy holds the same convictions about these so-called lifelines as does the attacker. This, obviously, proved as wrong as the linear concept of the Japanese who expected that the Americans in their advance would attempt to throw them from their island holdings one after the other, and grow tired of this sort of fighting before they finished a long succession of these insular sieges. Even under threat of having their lines to Singapore cut from the vantage point of the reconquered Philippines the Japanese chose not to shorten them, but rather to fight it out with what they had earlier accumulated in the conquered lands of materiel and men, added to what they might be able to bring there later, as they did in the Philippines where some divisions had recently arrived from the Asiatic mainland.

During eras when conditional surrender and maneuvering

and outmaneuvering were prevalent and valid concepts in warfare, the argument presented by threatened lines might have availed. But once America had proclaimed unconditional surrender as the price of peace it meant that American war industrial production and American lives had to be expended till the enemy was smothered beneath their weight. Here the question poses itself: When a civilization decides that everything it has is expendible, does it not come too close to spending its substance? Certain forms of the "art of war" function only when there is agreement on both sides about what constitutes that art. After warfare has become plain destruction, engineering, plus transportation and other logistic performance, most of what art there is in war, the old conventions, goes out.

Debatable Questions In Washington

The question before the American strategists at Washington who had been given "unconditional surrender" as their *mot d'ordre,* as the demand of their constitutional commander-in-chief, was always whether they should not strike at Tokyo, which due to Japanese arrangements of authority remained the center of resistance against unconditional surrender, more immediately and directly; whether landings on and near Formosa would not have been preferable to those in the Philippines, even granting the latter to be an objective highly political in nature. Time certainly could not be gained by going first into the Philippines where, as MacArthur declared at the outset, a protracted campaign had to be expected. On the other hand, the time required to wait for more forces to be become available after the close of the European fighting, later than had been expected, might better be spent in the Philippines than on the nearer approaches to Tokyo where the Japanese would have been able to oppose more troops to the Americans than in the Philippines. This "better spending" of time is again largely a political consideration. A general, a plan, merely considering military aspects, ought to be able to afford to wait. For such reasons it might well be that the preference given to the Philippines was the outcome of a choice between two political objectives, in brief Manila and Tokyo, a decision involving politics as most strategical decisions do, and upon which the tactical-technical considerations were superimposed.

To the special problem involving the technical side of landing operations the Philippines operations, judging by available reports, appear to have contributed little that is novel, although equipment included all that was newest. "The magnificent coordination displayed by the services was as marked as the special tactical efficiency of the various branches," a field order of MacArthur's of October 31 proclaimed, which also duly emphasized the "timely intervention at a critical moment in the decisive naval action following the initial landing," that is to say, against the three-pronged Japanese naval attack attempted along the passages through the archipelago against the dispersed American naval forces and the transport and landing craft anchored in Leyte Gulf (October 23-27).

"The tactical skill with which the troops have been maneuvered, has not only outwitted the enemy, but has resulted in a relatively low casualty rate that is unsurpassed in the history of war." It was reported that at the end of the first week of fighting the Japanese had suffered 14,045 casualties, "practically all of their forces on Leyte," as against American ground casualties of 518 killed, 139 missing and 1,503 wounded, or practically 7:1, plus, however, "considerable casualties" on board the six vessels lost in repulsing the Japanese naval attack. (More exact cost calculation of such combined operations as landings is made impossible by the American navy's policy of not publishing casualties with reference to specific actions, or complementary to announced casualties of the ground forces for specific occasions. This was first given up in the struggle for Okinawa.)

The Leyte campaign, in which MacArthur detected on October 23 among the Japanese "already signs of lack of maneuvering cohesion in the face of the skilful tactics of our local commanders" and the end of which he declared "in sight" as early as November 2, is characterized by what proved the strongest Japanese attempts at counter-landings, on a scale apparently even larger than in the Solomons. This prolonged the campaign much beyond American expectations. Not only were troops brought over by ship from other islands, such as Cebu, to the west coast of Leyte, between the first and 10th of November, but the Japanese also made parachute landings, on the whole futile, near the east coast of Leyte for the purpose of damaging American installations, such as air fields. Most of the first of these Japanese troops seem to have shipped across safely at night time undetected by the American sea and air blockaders. Activity of the latter was handicapped by the unexpectedly slow construction of bases on Leyte due to the heavy rains—23½ inches for November, a record precipitation—and typhoons. Later Japanese reinforcements were far less fortunate, with large numbers being claimed as sunk by American action en route to Leyte.

MUD THREATENED STALEMATE

By the beginning of December "a halt in the mud"—real, not metaphorical—had come about on Leyte. The tri-dimensional chess game of island warfare seemed to approach stalemate. In order to get the ground forces out of the mire, naval craft carried them around the southern tip of the island and disembarked them on the west coast, in the rear of the Japanese who were holding up both the northward and the southward thrust of the Americans in the direction of Ormoc. This was done to break the backbone of Japanese resistance, starting December 7.[10] But not until Christmas day, 1944, after another accessory landing had been performed, on northwest Leyte, could MacArthur regard the Leyte campaign "as closed except for minor mopping-up operations." By the end of the year he put the total Japanese losses on Leyte at 116,770, of whom half were considered killed, with the Japanese dead outnumbering the American dead by at least twenty-four to one.

Before the conquest of Leyte had been completed, American forces were on the move again, within the archipelago, with Manila as their ultimate destination. National capitals were as much set as ultimate objectives in World War II as in the wars of the 19th century, a fact which groups all these wars as those of nationalism. It was on the way to Manila when, after a tortuous voyage through the islands of some 600 miles from Leyte, lasting two days and three nights, a 150 ship convoy landed troops in the southwestern part of Mindoro, the seventh largest of the Philippines, 155 miles south of Manila. The trip although made constantly in sight of land, passing through the Surigao Strait, by Mindanao, Bohol, Cebu, Negros, Panay and Palawan, was only sporadically under attack from enemy aircraft, including suicide pilots of the Kamikaze Special Attack Corps.

This presented a new "first" in amphibious operations—a force pushing through an enemy-held archipelago and under constant threat of air attacks, if not also from submarines and torpedo craft, while en route. But while some losses were suffered by the convoy, the Japanese were unable to block even the narrow waters of the archipelago. This time there had been none of the weeks of preparatory pounding of airfields along the chosen route. Carrier-based planes and army planes from Leyte sufficed to keep the fields in the southern and central Philippines neutralized while the flotilla was en route.

The landing on Mindoro, December 15, although preceded by the usual local and naval and air bombardment, was practically unopposed, this giving MacArthur occasion once more to announce that it was "a complete surprise and met only minor opposition." Dealing with the strategic aspects of the operation his first communiqué declared that it had "driven a corridor from east to west through the Philippine archipelago which is now definitely cut in two and will enable us to dominate the sea and air routes which reach to the China coast. Conquests of Japan to the South are rapidly being isolated, destroying the legendary myth of the Greater East Asia co-prosperity sphere and imperiling the so-called 'Imperial life-line.'" More concretely speaking, Mindoro was to give advanced air bases for the *reconquista* of the heart of the Philippines, Luzon, only half an hour's flight from Mindoro. Besides, this landing clearly foreshadowed that the attack on Luzon would not start from the east coast which was rugged and where the mountains approached closely to the sea and the seasonal winds were unfavorable, but from the south and against the west coast.

Mindoro was quickly put into shape as an air base, the Japanese not interfering at all by ground action and but intermittently from the air, merely pecking at these new holdings. Another Philippine island, the seventh on which American forces came to land, was added when on January 3 Marinduque, only ten miles off the lower coast of Luzon and thirty miles from the nearest point on Mindoro, was occupied without fighting. It does not appear that this was more than a feint.

Japanese Beginnings Followed

American forces for the reconquest of Luzon were already under way while Marinduque was taken over, en route to the best landing beaches on that island, along the Lingayen Gulf, the same region where the main force of the Japanese had begun their invasion a little more than three years before. Far ahead of the convoy, in the north, carrier-based planes were at the same time attacking Formosa and Okinawa, acting like a cavalry screen of old. From the 5th of January on American vessels, finally numbering some 600, showed themselves in the vicinity of Lingayen Gulf and beyond to the north, whither some of the convoy steamed only to turn around and sail south during the night to be ready for the day of landing, the 9th. The troops who went ashore on four beachheads after a prolonged shelling and bombing of the land line, found the Gulf waters unmined and the coastline largely unmanned. The Americans took this as an indication that they had once more surprised the Japanese, who might conceivably abandon the beaches in order to escape the irresistible combination of naval gunfire and air bombardment. It quickly became clear that the enemy had decided not to fight any part of the battle of Luzon on the beaches where, after the full development of American industrial ingenuity and mass production for the purposes of amphibian warfare, American superiority was now evident, but rather in the more difficult terrain of the interior. Hence, to the surprise of many, the battle promptly lost much of its litoral character. "It was not long before Vice Admiral Thomas C. Kinkaid—commander while the convoy was en route—was able to hand over to Lt. General Walter Krueger, Sixth Army commander." [11] The Sixth and the Eighth American armies were to fight for the possession of Luzon against what the Japanese had there, variously estimated at from 100,000 to 150,000, or perhaps 250,000 to 500,000 "fighting men." [12]

There were several more landings on Luzon, the second on January 29 in Zambales province, northeast of Subig Bay, which threatened the Japanese in Bataan, another objective in the backward rolling movie of the Japanese invasion of 1941-2. The third on February 2 at Nasugbu, 20 miles south of Corregidor and 40 miles southwest of Luzon, where the Japanese had landed on December 24, 1941. The fourth touching down on Bataan peninsula itself, near Mariveles, on February 15. But there was so very little resistance on the beaches and the absence of opposition was so obvious that in the landings on January 29 the ships could even dispense with the customary, not to say mechanical, pre-landing bombardment.

The landing near Mariveles took place opposite Corregidor and in defiance of the guns of that island fortress, seven miles away, which previously had been neutralized with bombs and naval gunfire for three days. On February 16 parachute troops came down on the hill crest of Corregidor from the air while a flotilla of landing craft, covered by destroyers, gunboats and rocket craft landed at the same time on its south shore, joining the air borne infantry. These American forces, totaling 3,038, in twelve days of close-in fighting with the help of flame throwers, burning gasoline, bulldozers, and sealing the Japanese defenders in the casemates, reduced an enemy of some 6,000, of whom only 18 were taken alive. American casualties were 136 killed, 531 wounded and 8 missing.

Manila Japanese Main Stand

So much of the American movements on and around Luzon followed the pattern of the Japanese conquest of three years before that it appeared as though in revenge we were dosing the enemy with his own medicine, on the part of the re-invaders, or procceeding in obedience to the obvious dictates of topography, or attempting to surprise the Japanese by employing old tactics instead of new. But nothing could shake the Japanese determination to make the main stand for Luzon around and in Manila.

There organized resistance lasted until February 23 when the walled city was breached. Manila Bay was open early in March, after less than two months' fighting. It had taken a six months' campaign for the Japanese to win it. War had developed downhill speed.

Palawan, where American invasion forces landed on February 28, the westernmost and fifth largest in the archipelago, was the 17th on the string of the retaken Philippines. The communiqué-declared purpose of taking it was much the same as in the case of the preceding sixteen islands—holding that Palawan would "condemn all the enemy's conquests to the south to recapture." Close vicinity to the Dutch East Indies and bordering on the South China Sea gave it indeed more of a controlling position than would many others of the islands. But this control exercised from the Philippines, or elsewhere over the petroleum and other supplies from the Dutch East Indies was "not yet airtight" as Admiral Nimitz reminded the more enthusiastic speculators about "control," for "submarine operations near the China coast were difficult because of the shoals, and blockade runners could hug the coast. Increasing air operations out of the Philippines should tighten this, however." [13]

The reports of the Palawan landing as of those which followed—on Ticao and Burias Islands, south of Luzon, March 3; on Mindanao, beginning March 9, where the veterans of Palawan landed to join up with guerillas who were best organized on this particular island; Panay, March 18; Tawitawi at the southern end of the Sulu group early in April, followed by the landing close to Jolo, the capital of the Sulus, April 10—expressed throughout the smooth functioning of an operational mechanism evolved over a still recent period.

A composite of the reports about the U. S. Eighth Army, which operated largely south of Luzon, making 51 landings on some 25 islands since the close of the Leyte campaign,[14] would read like this: "After a thorough naval and air bombardment" the troops landed "against light opposition," or "went ashore with little loss," or sometimes found the enemy "completely surprised." Only occasionally had mines to be cleared out of the way of the troops, and underwater demolition teams employed at Saipan, Tinian and Peleliu to go in after the mine sweepers, do not seem to have been necessary in the Philippines. The troops quickly seized various inhabited localities, including air strips, while the Japanese "fled to the hills in disorder." "In each case the landings were effected most skillfully with a minimum of resistance, but stubborn and prolonged fighting usually followed in the hills" (Marshall's *Biennial Report*).

Americans Dictated Operations

Their superior landing technique, their complete naval and air supremacy allowed the Americans to dictate, more and more as time went on, where operations would commence, when and where an island was to be taken, or left alone. The Japanese command had practically no opportunity left for counter-concentrating their still available, but widely dispersed troops. They could only hope to inflict on their adversary loss of time and men. The reduction of the islands took months, with the enemy "virtually eliminated" from Luzon only by the end of June, 1945, after 113,593 of his dead had been counted on that island and several thousand prisoners had been taken, at a cost to the Americans of 15,718 casualties, of whom 3,793 were killed.

This counting and comparing of casualties, as part of the

book-keeping system of victory in the Pacific where the conquest of islands made such itemization more feasible than in other theaters, impressive and persuasive as it was, whenever MacArthur seemed to open the book of the dead, still called for cautioning against optimism. Admiral Nimitz, "commander of the 'Pacific Ocean areas,'" early in March, 1945, reminded the American people, while the Philippines and Iwo Jima were yet unreduced, that hitherto only an estimated 10 per cent of the Japanese army forces had been committed in these various operations.[15] The existence of the remaining 90 per cent raised anew the question of what next should be tackled among the big objectives in the Pacific once the Philippines were reduced and where the Japanese had perhaps invested less strength than MacArthur had hoped.

With the reconquest of the southwest Pacific area and MacArthur's original assignment nearly completed, the general eager for new laurels, by early March had "publicly invited a new assignment leading to Tokyo," without saying "specifically that he wanted the over-all command in the Pacific."[16] To the Navy people who had to bear the burden and perhaps criticism of expensive successes in their own taking of islands, it appeared that MacArthur's statements about his low casualties and high Japanese losses constituted an open bid for public favor. The American public, on the whole satisfied with the command arrangements in the European theater, seemed to be invited to back a MacArthurian concept of ending the Pacific war and the General's claim to, or hope for, supreme command in that war's final phase.

In order to resolve public doubts and service rivalry, the joint Chiefs-of-Staff in Washington—i.e. American, and not the combined Chiefs-of-Staff (Anglo-American)—"with the approval of the President" early in April announced that they had "modified the command organization for the war against Japan with a view to giving full effect to the application of our forces against the Japanese, including the large forces to be redeployed from Europe, taking into account the changed conditions resulting from our progress in both the Southwest Pacific and the Pacific Ocean areas. The rapid advances made in both areas, which have brought us into close proximity with the Japanese homeland and the China coast, and the corresponding change in the character of the operations to be conducted, are the considerations which dictated the new directives." Under its terms MacArthur would

> . . . be given command of all Army forces and resources in the Pacific theater. General Arnold will continue in command of the Twentieth Air Force [comprising the B-29 Superfortresses employed against Japan.] The Joint Chiefs-of-Staff will continue to exercise strategic direction of the entire Pacific theater and will charge either General MacArthur or Admiral Nimitz with the over-all responsibility for conducting specific operations or campaigns. Normally General MacArthur will be charged with the conduct of land campaigns and Admiral Nimitz with the conduct of sea campaigns. Each commander will furnish the forces and resources of his service for the joint forces which are required for the conduct of the operation or campaign which has been duly directed by the Joint Chiefs-of-Staff. Essentially the new arrangement permits either Commander-in-Chief to conduct operations or campaigns in any part of the entire theater as directed by the Joint Chiefs-of-Staff, and the choice as to which shall be charged with the responsibility in each case will be dependent on the nature of the operation or campaign which is to be undertaken.[17]

Only so long as the Pacific war had had a largely protractive character, with the priority on the European war, did compartmental arrangements seem permissible. What these new arrangements did was to unify operations in the Pacific theater, although not the services, somewhat more, but not completely. Actually, there remained five commands in the Pacific, MacArthur's, Nimitz', Arnold's, Mountbatten's, Chiang Kai-shek's, with no Eisenhower over them all; no joint command, or joint staff. In a long-range view of combined operations this arrangement appears hardly progressive. In fact, the continued separate status granted to the 20th Air Force as a "global" force, went far in pointing towards a future separate United States air force. The new order seemed also to imply that amphibious operations in time to come might not amount to much as compared with the great land operations. However, one had to hope that modern communications would ensure that a unity even with its center so far away as Washington, well over 7,000 miles, might be preserved, through "suggestion" rather than command.

This new arrangement did not actually relieve MacArthur, now commander-in-chief of army forces in Pacific, from his old command as Allied commander-in-chief of the South Pacific Area, or Nimitz, commander-in-chief of the Pacific fleet, from his command-in-chief of Pacific Ocean areas, except perhaps in part. The continued issuance of separate communiqués and the maintenance of separate HQ's, 1,600 miles apart, would indicate the survival of the old division. Only sporadically was the principle of area replaced by the principle of function.

When for example General Buckner was killed on Okinawa, Nimitz' preserve supposedly, it was MacArthur who appointed his successor, General Stilwell. Still later, the army air forces on Okinawa and the Ryukyus generally were transferred to MacArthur who put them under one of his own men, General Kenney, but the other air forces based on Okinawa remained under Nimitz and Arnold respectively while Army air forces based on Iwo Jima to escort B-29's, remained under Navy command.[18] This seemed to leave Washington in the role of the coordinator and unifier that was perhaps needed in the Pacific, but not necessarily wanted by all the trinitarian services there.

Closer to Japan; Iwo Jima

While both the landings on Iwo Jima and Okinawa carried American power closer to the Japanese mainland, they still left in suspense the ultimate American decision about the finish of the Pacific war. Should it be fought on the Asiatic mainland where the way would presumably lead via Formosa, or should the war be carried at once to the Japanese archipelago itself where it would have to go eventually in any case? The one involved a continental and largely military strategy, with the American Army as the main or commanding instrument; the other a maritime, or amphibious strategy, possibly under naval command. Either concept called for the employment of air power on a large scale, and air power required bases. Both Iwo Jima in the Volcanoes, about 750 miles south of Tokyo and slightly less distant north of Saipan where B-29's were now based, and Okinawa in the Ryukyus were to provide such

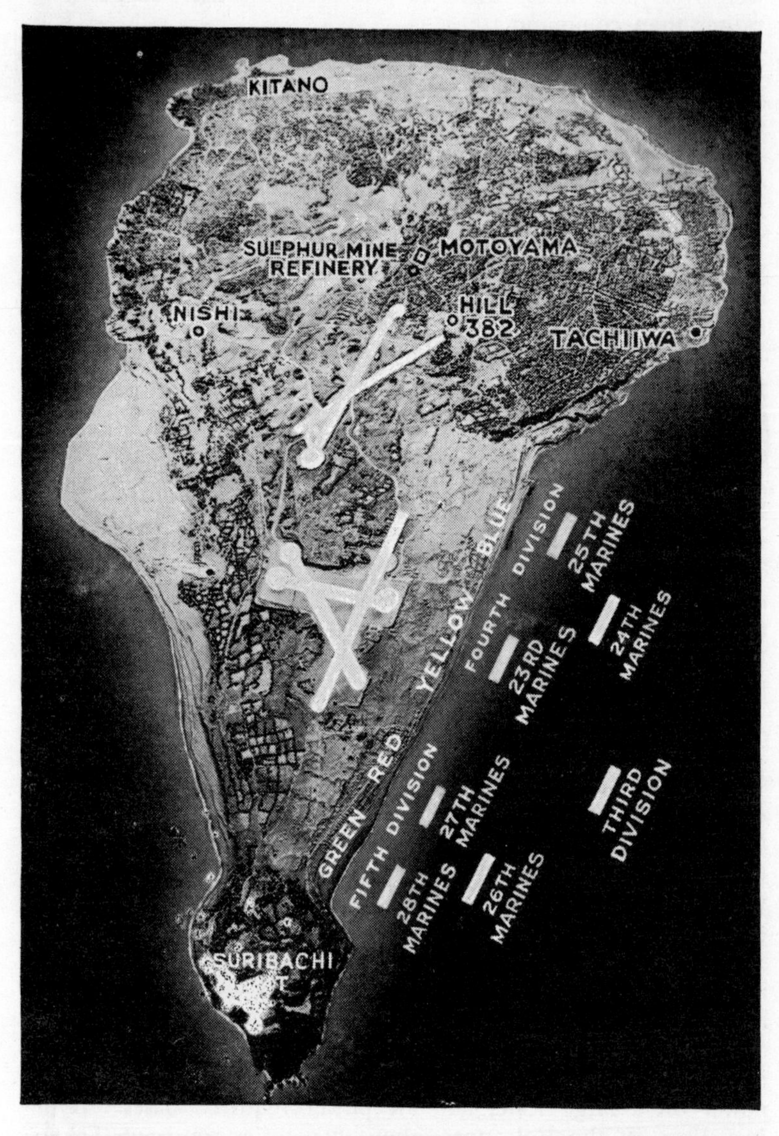

U. S. Signal Corps

Iwo Jima.

bases, from whence the Japanese mainland could be attacked, for purposes either of strategical, or tactical bombing.

In addition, Iwo Jima, situated within the prefecture of Tokyo itself, which was attacked by a carrier task force during February 16 and 17, was the first part of administratively metropolitan Japan to be invaded. From here the Japanese had interfered with the American operations in and from the Marianas whence super-fortresses were striking out, "quite a few" of which, as the U. S. Navy admitted, had been destroyed by Japanese planes operating out of Iwo. From here, the Japanese had been able to warn Tokyo of B-29 raiders. Its possession would bring Tokyo within range of American land-based medium bombers and fighters, the latter to provide cover for the B-29's flying from the Marianas to Japan.

For the choice of a Japan-approaching base there offered in addition, or as alternative to Iwo, Hachijo in the Izu groups, 150-200 miles from Tokyo, hard to assault, with very precipitous beaches—it was bombed by carrier planes at the same time as the attack on Iwo opened—and Chichi in the Bonins which the Navy suspected to be as strongly defended as Iwo. It contained, however, only one airfield as against the two already built on Iwo and one more under construction there. Considering advantages and disadvantages, the choice of the Joint Chiefs-of-Staff in Washington had fallen on Iwo. Japanese resistance was expected to be strong in any case and the expectancy of losses high—some 12,000 casualties perhaps.[19] "But, in the case of a base that is needed," a Navy spokesman explained, "there is only one thing to do—to take it. That is what we are doing at Iwo, but it is a head-on collision."[20]

The American preparations for taking Iwo were meticulous as well as massive. It was submitted to an air bombardment during 72, or 74 successive days—counts vary—and ringed by naval gunfire for three days preceding the landing, not to mention intermittent naval bombardment on some ten earlier occasions. While that had softened up the defense and had eliminated the island as a Japanese air base, the defense was still powerful enough on the day preceding the invasion to damage one of the bombarding warships by shore gunfire.

Most of the heaviest American naval gunfire was supplied by six veteran battleships of which the oldest was nearly 35 years old. They proved that these mobile sea surface fire platforms, even if considerably slower than modern battleships, which were at nearly the same time employed in combination with the aircraft carrier forces, possess great superiority over stationary gunfire such as that based on Iwo, provided one dared to use and risk such ships which the Americans did. The range of caliber in the bombarding guns ran all the way down to the LCI gunboats with their 40 mm. and 20 mm. pieces and their rocket launches, the final beach clearers.

To a higher degree than all the other islands encountered in the Pacific Iwo was a fortress, its strength lying in its topography as much as in its man-made installations, with craters and hundreds of caves linked into a network of ferro-concrete blockhouses and pillboxes; with gun emplacements which somewhat to the surprise of the attackers not always faced the sea and therefore were not exposed to naval gunfire. These were intended for use in land fighting. While not aware of the details of these fortifications, the invader still showered on them a pre-landing bombardment of length and intensity unparalleled in the Pacific war. It seems to have cleared away most of the immediate beach defenses, for the first waves arrived without losses. The Japanese endeavored no longer to cling tenaciously to this line of greatest exposure for the defender, but rather tried to make the beaches and the immediate approaches the zone of greatest danger for the invader, who was forced to crowd them with men and materiel, by laying barrages from time to time on the only beaches available for landings on Iwo.

With a land mass of some eight square miles, the Japanese could to a certain extent "go on the evasive," vertically, by hiding in caves and other shelter under American gunfire and air bombardment and popping back into firing positions in time to meet the attack. But for the Marines there "was no maneuver" on Iwo, no horizontal enveloping, no accessory landing except near the end of the fighting when some rocky islets off Iwo were taken, from whence the Japanese had molested the attackers. More than 80,000 fighting men were finally crowded into these 8 square miles, a maximum—and definitely not an optimum for tactical moves—of saturation without precedence in military history.

Escort Armada of 800 Ships

The Fourth and Fifth Marine Divisions, the former a veteran

force, the latter as yet untried, brought to Iwo and its landing beaches on February 19, 1945 (Feb. 18, U. S. time) by an armada of over 800 ships of all sorts, encountered at first only sporadic artillery and mortar fire from, and on, the beaches. Opposition, however, "increased markedly after the drive inland began" (Admiral Nimitz's second communiqué), calling for constant fire support from the fleet, particularly during the first 24 hours when the Marines' own artillery could not get to the shore or beyond it.

Soft volcanic sands halted numerous vehicles of the landing forces before they had gone ten feet and made supply movements difficult, all of which had to be manhandled for the first two days until steel mats had been spread over the clogging sand. The southern Iwo airdrome, the prize possession in the island, was captured on the second morning of the invasion and a cut across the southern end of the island achieved, leaving Mt. Suribachi (564 feet), a volcano, on the left flank and two thirds of the island to the north on the right flank of the attack in the hands of the Japanese, estimated at some 20,000.

For the first time the enemy used rockets, a secret imparted to them by the Germans. They hid behind extensive mine fields, with holes for the land mines prepared long in advance, on which a number of American tanks were knocked out, and were making general use of the advantages of holding high ground. These holdings were "as fanatically defended as any yet encountered in the war in the Pacific" (Pacific Fleet communiqué No. 267). On the 21st, when the original two divisions had been halted altogether in their northward push, elements of the Third Marine Division were thrown in, bringing the total of the investing force to some 60,000, minus the casualties for two days which were approaching 4,000—more than the total of Tarawa—and passed 4,500 on February 21. Not all casualties were suffered ashore. The U. S. Navy's biggest and oldest carrier, the *Saratoga*, on February 21, was struck by Japanese aircraft. Although she was able to keep afloat and eventually was repaired, her casualties totalled 123 killed, or missing and 192 wounded. Offshore air protection and air observation—with Marine observation planes beginning to operate from the first Iwo airdrome on February 26—and artillery support were unavoidably costly.

The advance across the island continued to be slow. While Mt. Suribachi was taken on the 23d, the northward push became almost stalled for days when gains were measured by feet and yards. Japanese resistance, deep rooted, had to be disposed of little by little by artillery and flame throwers. Only in the evening of March 16 could "organized resistance" be declared as overcome.

"Victory was never in doubt. Its cost was." The cost of taking Iwo was excessively heavy, as the commanders on the spot had been trying to make the American people understand almost from the outset; so high that they had to put the responsibility for it squarely up to Washington, or on the strategy employed, if one prefers. "When the capture of an enemy position is necessary to win a war it is not within our province to evaluate the cost in money, time, equipment or, most of all, in human life. We are told what our objective is to be and we prepare to do the job, knowing that all evaluations have been considered by those who gave us our orders." Thus the highest commanding Marine officer, Lt. Gen. Holland M. Smith.[21]

The twenty-six days of fighting cost the 61,000 Marines in action on Iwo, 4,189 officers and men killed, 441 missing and 15,308 wounded, or 32.6 per cent; and the Japanese their whole garrison, 23,244 dead and 1,038 prisoners. It was the bloodiest fighting in the Pacific to-date, considerably worse in numbers and in percentages than on Tarawa, Saipan and Peleliu, according to an official Navy Department analysis which incidentally revealed what ground forces had been employed in the early conquests (percentages in totals of strength).[22]

	Wounded	*Killed*	*Total*
Tarawa	2,191 (12.8%)	984 (5.8%)	3,175 (8.6%)
Saipan	8,910 (18.7%)	2,337 (4.9%)	11,247 (23.6%)
Peleliu	4,974 (20.1%)	1,198 (4.8%)	6,172 (24.9%)
Iwo Jima ...	15,308 (25.8%)	4,189 (6.8%)	19,938 (32.6%)

A democratic people will always be inclined to ask whether too many of these died, or suffered on Iwo; or if they did truly for the greatest purpose—that others may live. In order to quell such doubts it was revealed that from March 4 to the end of June, 1945, Iwo served as an island of refuge where no less than 821 B-29's, with 9,361 trained aviators on board, driven off their home-bound course, their fuel exhausted, safely made emergency landings, most of whom would have been lost without this aeronautical live-saving station.[23]

The next position from, and around, which the war of attrition against Japan, and its air force in particular, was to be fought, pending the fall of the Festung Europa and the redeployment of superior American forces released there to the Far East, was Okinawa, main island of the Ryukyus. Situated halfway between Formosa and the Japanese mainland, it brought American forces within 350 miles' striking distance of the latter.

With this advance the expectancy rate of losses in men and materiel was bound to rise. "As we approach the main islands of the enemy the damage to our ships and the loss of our men are becoming more severe. In the future we shall have to expect more damage rather than less," President Truman told the Congress while Okinawa was being reduced (Message of June 1, 1945). This was bound to come, not so much on account of the long approaches and communications for the American fleet—with only a few Japanese submarines left to be active against it—but because of the ever stronger enemy positions encountered, the better troops manning these, the greater exposure of American ships to enemy air attacks from bases still in Japanese hands. The necessity for a fleet operating over such wide distances, (distances figuring large in enemy calculations), to take its own vast supply train along from the outset and thus maintain a mobile base rather than fixed ones, added considerably to the objects exposed to enemy air attacks, such as ammunition and repair ships, floating dry docks, oilers, water tankers, hospital and barrack craft—the entire vast armada of the Pacific Fleet Service Force.[24] Enemy resistance, growing in proportion with the size of the contested island, was bound to keep larger numbers of American ships at work close to such an objective for purposes such as giving fire support, providing screens on the water and in the air and distributing supplies.

Intended as a kind of partial substitute for battle ships, "which form the principle gunfire support," the attack on Okinawa was preceded by the capture of the Kerama island group west of the southern tip of Okinawa, commenced on March 26 by the 77th Inf. Division, from the Philippines. "It was completed prior to the main landing on Okinawa and heavy artillery is now emplaced there and in support of the

Okinawa attack" (Pacific Fleet communiqué No. 317, of April 1). Thus heavy land-based artillery was ready to participate in the naval gunfire and air bombarding in their later stages, the former having opened fire on Okinawa nine days ahead of the invasion and the latter having begun their pounding on March 18, when the maritime phase of the Okinawa operation had begun.

Following this, then the heaviest bombardment in the Pacific, American forces went ashore on April 1, Easter morning. This worked still another stepping up in the crescendo of the Pacific war; the landing itself was the largest amphibious operation in that theater to date, with 1400 vessels of all types participating, carrying and protecting the largest ground forces yet employed in the central and western Pacific. Commands and responsibilities were arranged as follows: Admiral R. A. Spruance commander of the Fifth Fleet, in over-all tactical command of the operation; the amphibious phase under the command of Vice Admiral Richmond Kelly Turner, Commander Amphibious Forces of the Pacific Fleet, with Lieutenant General Simon Bolivar Buckner, U.S.A., who was to find his death on Okinawa, in command of the invading troops, the Tenth Army, which included the 24th Army Corps and the Marine Third Amphibious Corps.

Not to be outdone longer by MacArthur and his flamboyant forecasting communiqués, Admiral Nimitz announced at the outset of the operations that Okinawa "will give us bases 325 nautical miles from Japan, which will greatly intensify the attacks of our fleet and air forces against Japanese communications and against Japan itself. As our sea and air blockade cuts the enemy off from the world and as our bombing increases strength and proficiency our final decisive victory is assured" (Communiqué 317).

The landing took place on the west coast of the island, which is sixty miles long and ten in breadth at its widest, at the northern end of the southernmost third, at 8:30 A.M. and over an eight-mile beach. To the surprise, it seems, of nearly everybody, initial resistance was again practically lacking. Within 75 minutes after the first assault, landing tanks rolled ashore, followed by marine artillery in the early afternoon. By 6 P.M. the beachhead had reached a depth of three miles at some points. Again, the enemy, at first estimated at from 60,000 to 75,000,

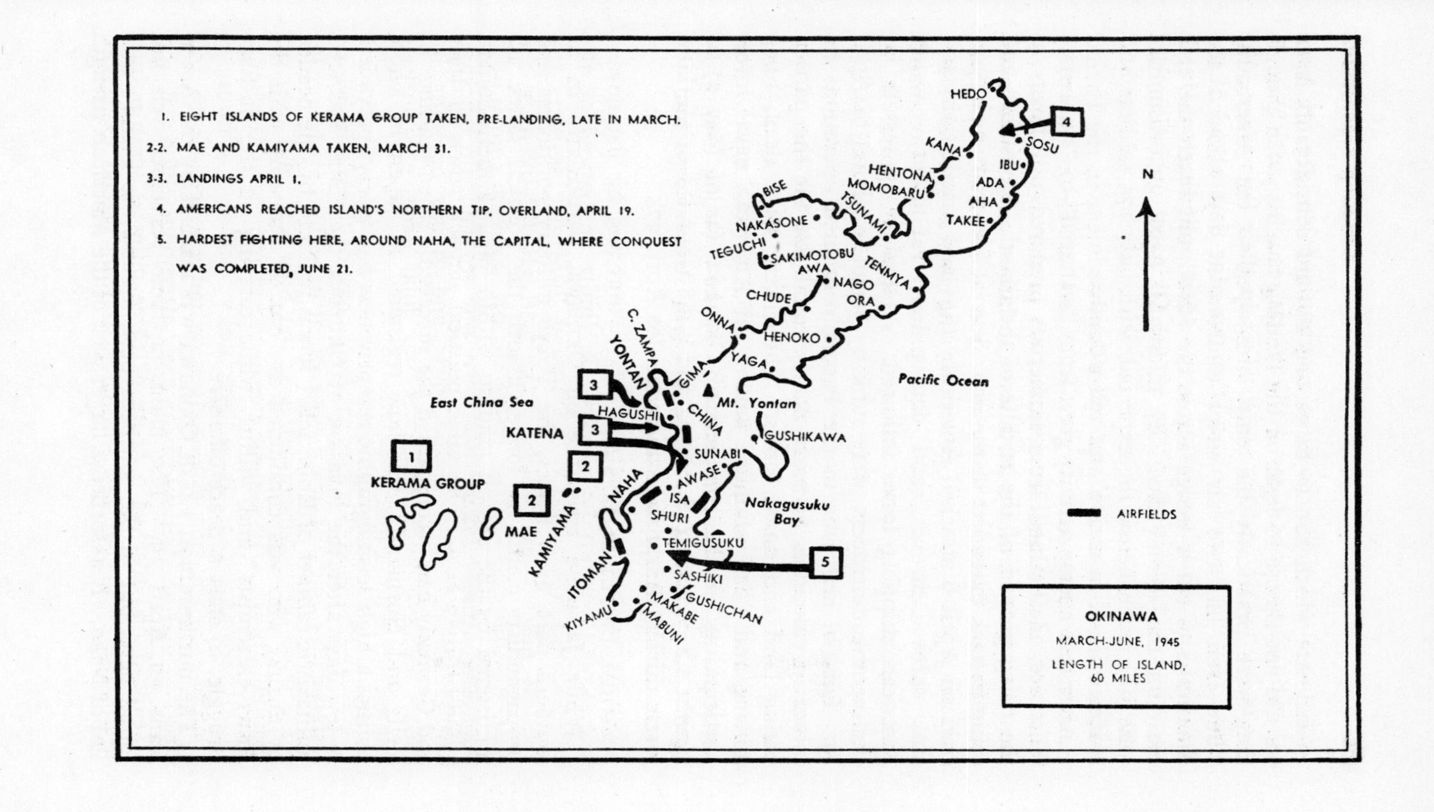
1. EIGHT ISLANDS OF KERAMA GROUP TAKEN, PRE-LANDING, LATE IN MARCH.
2-2. MAE AND KAMIYAMA TAKEN, MARCH 31.
3-3. LANDINGS APRIL 1.
4. AMERICANS REACHED ISLAND'S NORTHERN TIP, OVERLAND, APRIL 19.
5. HARDEST FIGHTING HERE, AROUND NAHA, THE CAPITAL, WHERE CONQUEST WAS COMPLETED, JUNE 21.
HEDO
KANA
SOSU
IBU
ADA
AHA
TAKEE
HENTONA
MOMOBARU
TSUNAMI
BISE
NAKASONE
TEGUCHI
SAKIMOTOBU
AWA
NAGO
ORA
TENMYA
CHUDE
ONNA
HENOKO
YAGA
C. ZAMPA
YONTAN
GIMA
Mt. Yontan
HAGUSHI
CHINA
GUSHIKAWA
KATENA
SUNABI
AWASE
ISA
SHURI
NAHA
Nakagusuku Bay
TEMIGUSUKU
SASHIKI
GUSHICHAN
MAKABE
MABUNI
KIYAMU
ITOMAN
KAMIYAMA
MAE
KERAMA GROUP
East China Sea
Pacific Ocean
N
AIRFIELDS
OKINAWA
MARCH-JUNE, 1945
LENGTH OF ISLAND, 60 MILES

an estimate which had to be revised upward considerably later on, did not choose to fight on the landing beaches, even though they were practically the only ones open to the Americans. They even left two air strips undefended and allowed the Americans to cut a wedge across the island and reach the east coast on the second day. By midnight, April 4-5, casualties were still much below the expected, with only 175 soldiers and Marines killed in action and 198 wounded.

After that enemy activity grew from "scattered" to "stubborn" resistance, maintained from numerous prepared pill boxes in the rising ground of the middle of the island. Resistance and counterattack extended to the air. Several hundred enemy aircraft on April 6 attacked American ships and installations and sank three destroyers and damaged several others. Other American shipping losses followed. A total of 15 vessels, including five destroyers, were sunk by Japanese planes to April 19. For the first time in the Pacific, combined operations resulted in casualties of more nearly equal size for the participating land and naval forces. There were 1,131 dead, 1,604 missing and 2,816 wounded among the American naval forces participating in the conquest of Okinawa (as of May 2) as against 1,847 dead, 418 missing and 9,148 wounded of the land forces during nearly the same time (to April 27).

Political Effect in Japan

While Japanese resistance on Okinawa, particularly in its southern part, grew fiercer, the Tokyo military politicians and propagandists could no longer absorb the political shock administered by the second landing within Japan's metropolitan administrative area. The same political impact which in Italy and Germany had accompanied the successful Allied landings in Sicily and Normandy, was now registered in Japan in whose politics a high seismographic receptiveness had always prevailed. A few days after the invasion of Okinawa the "Sure Victory," militaristic cabinet of Koiso fell. It was replaced by one headed by Suzuki who was considered as "out of sympathy with the Army extremists" and willing, while carrying on the military struggle, to start a peace offensive.

The northernmost tip of Okinawa was reached by the Americans on April 19. The main Japanese resistance was not found here, but in the southern part, around the island's capital of Naha. It absorbed a larger part of the American strength than seems to have been originally assigned to Okinawa, including B-29's which were used for tactical bombing of Okinawa objectives instead of strategical objectives elsewhere. At the end of the fighting here it was stated that escort carrier planes had flown 35,000 sorties in support of the Tenth Army, an altogether unprecedented number.

Naval units, including carriers, were forced to stay on the job much longer than was intended, or considered desirable by many aboard these ships. Critics, mainly from among these, began to ascribe the ensuing greater risks which the fleet units incurred, to the slow and cautious ground action of the Army which, as compared with the Marines, seemed rather inclined to spare soldiers on land than sailors on board the supporting ships. This was opposite of what the more rapid procedures and *tempi* of the Marines were expected, or intended, to do for the fleet. Naturally, while the Army employed these naval units, considering them expendible, some of the latter were lost to the enemy's suicide bombers, or even his land-based guns which sank one destroyer stranded on a reef, one mile off Naha, while giving close-in fire support.

It was in the middle phase of the Okinawa fighting that the majority of the American casualties were suffered. They were so impressive that more ancient concepts of military economy—belonging to ages preceding the late President Roosevelt's demand for unconditional surrender everywhere—reemerged and near the middle of June moved General Buckner to offer the Japanese means and ways "in which an orderly and honorable cessation of hostilities may be arranged." But the Japanese commander did not answer; neither did Tokyo avail itself of the opportunity of opening negotiations which possibly might have extended beyond the capitulation of Okinawa.

During the early days of the Okinawa operation there was not a little press reporting, from official sources, of course, about progress ahead of schedules. But of this there was none since May when fighting became stationary for days, when even setbacks occurred and progress was often measured by the yard rather than the mile. This led to outbreaks of criticism at home, obviously following the receipt of Pacific mail, about "the military fiasco" on Okinawa. It is always interesting to trace tactical, or strategical concepts back to specific officer groups, their temperament, their outlook on things, their interest, even. The views in opposition about the termination of the fighting on Okinawa clearly went back to junior officers of the Navy. They suggested that instead of the continued frontal attacks on the Naha fortified area the Marine corps which had speedily cleaned up the northern end of the island, should after that have been employed in new landings behind the Japanese lines in the south. But "our tactics were ultra-conservative. Instead of an end run we persisted in frontal attacks," failing to take air bases fast enough to relieve the naval units.[25]

A certain malaise was spread by comments like these, and after a fortnight Admiral Nimitz found it necessary to answer them personally. "Fully aware that delays ashore would increase the losses afloat," he had flown to Okinawa and conferred with General Buckner only to concur with his views and those of the senior naval commanders concerned. He now came to the defense of the Okinawa operations, attributing the slow-down to heavy rains and mud and the absence of good material such as reef coral for the building of roads and airstrips. Additional landings to the south, "against an alerted enemy defense," "would have involved heavy casualties and would have created unacceptable supply problems."

Staff officers on Okinawa denied that such landings, considered by them several times in detail, were at all feasible on account of the reefs and the terrain of the proposed landing beach. Besides, they insisted, there would have been the greatest difficulty in keeping the men along the newly opened front properly supplied, on account of the limited shipping available; and the more so were these troops to be stronger than one division, as they would have to be if they were to make an impression at all on the enemy with his behind-the-front strength.[26]

NEARLY THREE MONTHS' FIGHTING

When the commanders on Okinawa resigned themselves to these conclusions, there remained nothing to do but crush Japanese resistance peacemeal at the smallest possible cost to their own forces. The end came on June 21, after 82 days of fighting as against 26 days at Iwo Jima; and 12 days later than originally estimated[27] and at higher cost than experts, and obviously the commanders as well, had anticipated. American casualties from March 18 up to and through June 20 totalled 46,319—11,897 killed and 34,422 wounded. Japanese losses according to American counting amounted to 110,549 of which 8,698 were prisoners and the rest dead.

American ground casualties were 4,417 dead or missing and 17,033 wounded in the four Army divisions; and 2,573 dead or missing and 12,565 wounded among the Marines (two divisions plus one regimental combat team). Of these 1,658 killed and 8,255 wounded were lost during the last 26 days (from May 24). Since the strength of these various units has not been given, it is again impossible to state which did fight more economically, Army or Marine units. Of the total American casualties 4,907 killed or missing and 4,824 wounded were those of naval personnel, the largest share, it would appear, that that partner had yet paid in combined operations, for its shipping losses off Okinawa were 33 light ships sunk—"light" meaning vessels up to destroyers—and 54 others damaged. Losses of 1,290 out of the total of 9,731 occurred after May 23.[28]

That naval support of land operations proved such a "costly and serious business" around Okinawa (Secretary of Navy Forrestal, May 16), was in large part due to Japanese air attacks, kamikaze and otherwise, which struck first and hardest those ships which formed the outer screen, 25 to 50 miles seaward beyond the main anchorage. On this outpost work alone, where they did warning and anti-aircraft duty, shooting down 490 planes in the 82 days' siege, and where they were always the first objects of Japanese air attacks, these little ships suffered 1,000 casualties.[29]

Slightly ahead of the final accounting of this, for both sides the longest and costliest of the various campaigns in the central and western Pacific, the following data about American and Japanese casualties in these island conquests had been given out at Guam Headquarters (for American casualties through May 29, for Japanese through June 19). Since they include losses among American naval personnel, they brought out, by inference at least, naval losses, not heretofore published.

	Japanese		*American*	
	Killed	*Captured*	*Killed*	*Wounded*
Okinawa	87,343	2,565	11,260	33,769
Iwo	23,244	1,038	4,630	15,308
Saipan	27,586	2,161	3,426	13,099
Guam	17,442	524	1,437	5,648
Palau	13,354	435	1,302	6,115
Tarawa	5,000	150	913	2,037
	173,969	6,873	22,968	75,976

(Figures for Americans killed include missing.)

Loss and gain calculations, estimating gains in miles at the expense of men's lives, are always grim. Leaving aside the question of cost of casualties per square mile of Okinawa, possession of this island reduced by more than 1,000 miles the earlier B-29 route from bases in the Marianas to the industrial centers in the Japanese mainland and back, once Superfortresses could operate from there. These shorter flying distances would enable them to carry greater bomb loads and run more missions per plane.[30] Possession of Okinawa made cooperation with Chinese forces easier and brought Japanese industries in Manchuria within easy reach of bombers. Such bombing was to form a large part of the siege warfare which would fill out the several months until the arrival of many more men and machines from Europe which were to carry the war against Japan into the final phase.

With requirements such as 16 tons of shipping for every American fighting man in the first 30 days of a campaign and at least six tons every month thereafter—these were the standard necessities as estimated in autumn, 1944—a burden which grew with the extension of distances from home ports, the logistic tasks involved in the initial assembly for the final attack in the Far East were stupendous. It was to ease these burdens that MacArthur on May 1, 1945, sent Australian troops together with Dutch contingents into Tarakan, off the northeast coast of Borneo, a rich oil field and refining center. The Japanese had overrun this place in 1942, when the Dutch, among other defenses, had dug a moat and filled it with oil to be set on fire in case of a landing in the region where they expected it. But the enemy had come from a different direction.[31]

Although much of the installations on Tarakan, rebuilt by the Japanese, had been destroyed by Allied bombing, it was hoped that the field, yielding an exceptionally pure oil, could be speedily redeveloped and that Tarakan alone would furnish some ten per cent of all the oil needed in the last phase of the Far Eastern war by the Allies who, indeed, by early June were already tapping oil from the shallower wells on Tarakan, even though the island was not yet then fully reduced by the invaders. The landing was followed by attacks on Brunei (June 10) whereby, according to MacArthur, the enemy "definitely lost the war of strategy in the Southwest Pacific"; and Balik Papan, an even richer Borneo oil center. At Balik Papan a

Japanese plan "to flood the beaches with burning oil misfired,"[32] an indication of the difficulties to be expected in creating a Nibelungen zone of flames. Once in production again, these oil fields were to spare the Allies thousands of shipping miles over which heretofore the vast amounts of oil products—15,000,000 barrels of fuel oil for the seven weeks of the Marianas campaign alone—had to be brought from either the United States, or the Near East.

NOTES, CHAPTER 60

[1] *Liberty Magazine,* Jan. 20, 1945.
[2] *N. Y. Times,* Feb. 14, 1945.
[3] *Ibd.,* Nov. 12, 1944.
[4] U. P. despatch from Sydney, May 13, 1945, concerning the protest of Australian aces based on Morotai in the Halmaheras.
[5] For details see *Infantry Journal,* July 11-15, 1945.
[6] *N. Y. Times,* Oct. 19, 1944.
[7] Admiral William V. Pratt, ret., in *Newsweek,* Oct. 30, 1944.
[8] A. P. despatch from Washington, Nov. 13, 1944.
[9] President Truman's message to Congress, June 1, 1945.
[10] For details see "The Fall of Ormoc on Leyte." *Military Review,* Aug., 1945.
[11] U. P. despatch from Lingayen Gulf, Jan. 10, 1945. "In any amphibious operation, command of all forces engaged rests in the hands of the naval command until the troops have been put ashore and have established their command organization. At this point, the landing force commander advises the naval commander that he has assumed command of his troops ashore." Admiral King's Annual Report, March 27, 1945.
[12] U. P. despatch from Leyte, Jan. 10 and *N. Y. Times,* Jan. 10, 1945.
[13] *N. Y. Times,* March 9, 1945.
[14] *Time,* July 16, 1945.
[15] *N. Y. Times,* March 9, 1945.
[16] Ernest K. Lindley in *Newsweek,* March 5, 1945; *Army and Navy Journal,* Feb. 10, 1945.
[17] A. P. despatch from Washington, April 5, 1945.
[18] *Time,* July 2 and 23, 1945.
[19] Hanson W. Baldwin in *N. Y. Times,* March 5, 1945.
[20] *Ibd.,* Feb. 23, 1945.
[21] *Ibd.,* March 16, 1945.
[22] *Ibd.,* March 17, 1945.
[23] *Ibd.,* June 26, 1945. A first little history of the taking of Iwo Jima, largely pictorial in character, is *Iwo Jima. Springboard to Victory.* Text by Capt. Raymond Henri. N. Y., 1945.
[24] For details of the origin and working of this mobile supply base which goes back to the invasion of the Marshalls in February, 1944, see *N. Y. Times,* Oct. 24, 1944 and April 2, 1945.
[25] For these criticisms see David Lawrence's column "Washington Slant" of May 30, 1945.

[26] For these controversies see U. P. despatch from Guam, June 17, *N. Y. Times* despatch from 10th Army Hq's on Okinawa, June 14, and Admiral Pratt in *Newsweek,* July 2, 1945. The casualty date given in General Marshall's report are somewhat at variance.

[27] *Time,* July 2, 1945.

[28] For these casualty data see *N. Y. Times,* June 28, and A. P. despatches from Guam, June 21 and 22, 1945.

[29] *Time,* July 9, 1945.

[30] General Henry H. Arnold. U. P. despatch from Honolulu, June 24, 1945.

[31] *Yank,* July 27, 1945.

[32] *Time,* July 16, 1945.

CONCLUSION

Said Brasidas, "it was not right to be chary of timbers, and put up with the enemy's having built a fort in their country. Perish the ships and force a landing."

THUCYDIDES IV, 11.

IN UNDERTAKING THEIR numerous landings the combatants of World War II realized and performed something that might have been done in the preceding war, which in so many respects and viewed in retrospect was a "war of missed opportunities." What might not have been done! Instead of inordinately sacrificing the Allied infantry, allowing them to be massacred, "one might have debarked secretly on a suitably chosen point of the litoral behind the German front troops whose appearance, combined with a frontal attack, might have provoked the enemy to fall back and might have resulted in his defeat . . . ."[1] What for various reasons and with sundry excuses could be shunned from 1914 to 1918, could no longer be put off in the world's second and still more total holocaust. There were no longer any alternatives, or evasions such as the traditionally separated services were devising before and after 1914.

It was like a history-conscious reminder of what might have been effected at that earlier time and in a specific theater of war when on November 1, 1944, Royal Marine commandos landed at Flanders shores, on the island of Walcheren. They were covered by the bombardment from warships including one battleship and various monitors and from land-based artillery firing across the Schelde estuary.

These ground forces were to wrest control of the approaches to Antwerp from the Germans whose big coastal guns (250 mm and less) survived the pre-landing bombardments much better than in other places and than they had been expected to do here. Instead of the hoped-for 90% elimination, only 60% of these guns had been knocked out prior to the landing and some of the damaged ones were repaired while the fighting was going on. They sank or seriously damaged twenty out of twenty-five of the ships of an inshore support squadron, of the LCI and LCT type. "These at close range drew the enemy's fire while the assault went through" as a British Admiralty communiqué stated. Consequently casualties, while not given in detail, were as heavy as on Tarawa and at Dieppe, that is to say, well over 2,000.

From 1939, or more correctly from 1940 on, "the being master of the sea" left the great Allied maritime power no longer as in times gone by "at great liberty to act and to take as much, or as little of the war as it pleases" (Francis Bacon). It was now a case of *Hic Rhodus hic salta.* If Rhodes or any other island were wanted, there had to be jumping to the land held by the enemy. To be able to do this the traditional sea powers had to become superior in land power as well—partly with the help of the Red Army—and bring their strength to land, through the instrumentality of sea power and air power, on the enemy-held shores. Such was the categorical imperative of World War II, resulting in landing strategy, landing tactics and techniques, landing machinery, landing operations, the risking of certain components of sea power not hitherto considered expendible in such operations, partly to avoid another "massacre of the infantry," like that in favor of "sea power" in the World War I. How successfully this turned out is shown by the reduction of British Empire casualties, from 1,357,800 dead in World War I to 216,287 (to February 28, 1945), in World War II.

When the last of these landings on hostile shores was accomplished, the victors of this war were left with two major military problems on their hands that concern landings, problems that called for consideration during, or before, or after the peace negotiations. First the problem of the flying bomb as the vehicle of manless invasion across the seas; and second, the question of the unification of the two, or three existing armed services, as the best preparation and organization for landing operations. Both problems are fraught with political implications, the former with foreign policies; the latter with home, including service, politics. Both demand consideration of the experiences of the past, recent if not also remote, and the contingencies of the future.

Penalty For Flying Bombs

The flying bomb invasion at the stage at which the invention was left when the war in Europe ended presents primarily a problem to England. There have been indications that its appearance has deeply upset the people at the Archimedic point from which the balance of power in Europe has been manipulated for so long, the very policy that was meant to make any invasion of England impossible; that it has driven them into a determination to dictate an even harder peace to the Germans, the inventors, or first appliers, of this instrument designed to destroy the last remnants of their isolation.

While this intention to inflict punishments upon those who have brought a new and lethal scourge into this militarily evil-cursed world, is understandable enough, it will alter the progress of military technique just as little as the blowing of early artillerymen from their own guns by the various "last knights" who resented the advent and use of gunpowder. England seems to be seeking for the solution, in a political way, of the problem of the flying bomb launched from the continent, largely by leaving Europe to the influence of a friendly Russia. This seems to her, in her present state of mind or strength, preferable to any unification of Europe, to be put between herself and Russia, an idea which first Napoleon and later Hitler have utterly discredited, from her viewpoint.

After such military experience it apparently seemed preferable to open all doors in Western Europe to the influence of Russia, in the hope that her ambitions will be sated by continental surfeit, that she will be satisfied with these new spheres of dominance and refrain from disturbing the British Isles.

So far as there is military calculation in this, it would seem that Britain reckons on some such political arrangements accomplishing more than any others to keep the continent from becoming a vast site for launching flying bombs, into the secrets of which Russia tried to probe when she occupied the Danish island of Bornholm, a German rocket laboratory, and Peenemünde, another such center. France, England's nearest neighbor, in spite of its proximity figured little in these post-war considerations on England's part following the dropping of the V-1's and V-2's on her soil, although one parliamentarian, the member for Cambridge University, declared during the troubles in the Levant in May and June, 1945, that "we all knew with V-1 and V-2 we could not sleep in our beds ever again if France were not on our side."

Besides the bomb, or manless, invasion there remains the danger for England of the sea and air borne invasion. These threats Churchill had in mind when, after his return from the

Normandy beaches, where he had observed the activities of amphibious and other vehicles, he told Parliament that "a new light is playing on the possibility of invading across the Channel, light which I hope will not be altogether lost upon our own people in the days when we have handed over our burdens to others" (August 2, 1944).

Farsighted, if not hypermetropic, men in the United States have found the flying bomb disturbing even to the far western calm of their land. Murray, the Democratic Senator from Montana, warning New York rather than his own constituents, declared that he had been "reliably informed that in the progress of aeronautical science it may soon be possible to produce an electronic-controlled robot bomb which could fly the Atlantic from any section of Europe, select the island of Manhattan as the general target, and then select the Empire State Building as a specific target. The implications of this fact, from the viewpoint of national security and international relations, are obvious." [2]

How obvious these implications are, others will doubt, wondering also how they will be considered—probably not very openly—in the future peace arrangements.

Experiences in combined or landing operations have provided one of the several incitements for officers and civilians to consider, if not to demand, the unification of services in the United States. Much less is heard of such demand in supposedly more tradition-bound Britain, although there too the experiences and exigencies of total war have given rise to proposals for a merger. Such former high ranking officers as Lord Chatfield, First Sea Lord from 1933 to 1938; Field Marshal Lord Milne, Chief of the Imperial General Staff from 1926 to 1933, and Air Marshal Sir John Salmon, Chief of the Air Staff from 1930 to 1933, as well as Lord Hankey, for more than 25 years Secretary of the Committee of Imperial Defense, declared as early as 1943 that:

> It is not possible in modern war to judge the strength of a nation by the individual strengths of its armed forces. The capacity for making war is greater than the power of all three. Sea power is the carrying power necessary for the maintenance of the whole war effort. Land power is the seizing and holding power. Air power is the spearhead of belligerent power able to strike swiftly and sometimes

> even decisively in pursuit of a common strategy of total war. Civil power supplies and sustains the means of sustaining war in every sphere and form. Each of these constituent powers as components of a belligerent power is inevitably dependent on the others and fluctuates in relative effectiveness. The extent and importance of the roles they fill in policy, strategy and tactics are continually changing with varying efficacy of offensive and defensive technique.
>
> The three fighting services comprise in reality one service. Only when the true meaning of total war is realized and all allegiances merged into a larger patriotism can the full strength of a nation's belligerent power be developed . . . . To think of the separate services as separate entities, each working its own element, and to reckon their strength in terms only of their respective members is to be guilty of a grave mistake. Even in their respective spheres the fighting services have always been incomplete and now they are often individually impotent. Today all operations are combined operations." [3]

In the United States the demands for the unification of the services under one Secretary of the Armed Forces go back to the stunning experiences of Pearl Harbor when the invader struck at and profited by, the very cleavage existing there between the two services, or their commanders. Whether or not the conclusion of the Roberts Commission that the absence of Army-Navy cooperation on the spot was apparently the chief cause of the success of the attack remains tenable under a closer re-examination, this was the basic shock as felt by the American politico-military psyche. Until this was altered, the then Senator Truman declared, "our scrambled military setup" would remain "an open invitation to catastrophe."

Unification Pro and Con

The reasons for a merger have been adduced from the fields of command, strategy, economy and, perhaps least where they may be found most often, in that of tactical experience; while esprit de corps and tradition,[4] or progress as already achieved without the merger were made to speak against it.[5] While nothing like a full story of tactical experiences, successes, mistakes and shortcomings in tactical cooperation can be given during war time, or immediately after, it is clear that combat lessons have greatly contributed to the unification move.

After an inspection trip to the various American-held fronts in the summer of 1943 Senator Lodge found "senior officers of both the Army and the Navy deeply impressed with the need for unity of the services when our new military policy is framed."[6] Some of the mistakes, or wrong arrangements in combined operations were clearly repetitions of ancient errors. German analyses, for example, of the Oesel operation of 1917 brought out that the different signal systems of the participating armies and navies had produced unnecessary misunderstandings during the action. The same hitch was reported by Congressman Maas as having occurred in some Pacific operations; when a naval officer declared that this was being remedied, the Congressman, himself a naval reserve-officer, shot back: "As a temporary expedient during this war; but isn't unity one of the things we are fighting for?"[7]

This unity had not been achieved when Allied naval units, though informed beforehand of the intention of bringing in troop-carrying planes two nights after the original landing in Sicily, shot down twenty-three such co-national planes. This tragic occurrence convinced Generals Eisenhower and Spaatz "that we are going to have to do some very earnest basic training in both ground and naval forces. Otherwise, we will finally get our air forces to the point where they will simply refuse to come over when we want them. Generally speaking, we are on the strategic offensive, which means we *must* have air superiority. Therefore, we should teach our people *not* to fire at a plane unless it definitely shows hostile intent."[7a])

On the whole, the United States Navy, for reasons not all disclosed yet, has tended to oppose unity, rather than the Army, which has become its champion, with General Eisenhower, as the supreme director of the landings most heavily fraught with consequences, proposing "to make one unit of the Army, Navy, and Air Forces and put them all into the same uniform."[8] The American Navy, through Admiral King's Report of April, 1944, and the stand its officers have taken before Congressional committees and in public, had declared itself satisfied that Army-Navy teamwork had been firmly established without unification and that both were already welded into "one national military force"; that the principle of "unified command" in amphibious operations was successfully operating in Anglo-American military and naval forces as well as in

American forces acting by themselves; that "our Army and Navy forces have learned how to fight as one team." [9] Sublime politician that he was, and the Navy's firm friend, the late President Roosevelt "made it plain that he thought the question a war was being fought." [10] President Truman later favored it, after V-E day.

As far as this work is a comprehensive history of landings, as the supreme operational amalgam of services, and a history directed not so much by reference to such a present-day problem as the unification of services as it is motivated by the desire of presenting this great problem in its ever-changing time-setting, and regardless of whether it contributes anything to this discussion or to the final decision, it must speak out in favor of unification.

The past does not have to be accepted as a heritage, but later military generations have too often entered upon its supposed heritage in their attempt to master contemporary problems by reference to traditions of service self-sufficiency. Too often in this specific past have landings either been hindered by a want of adequate teamwork, or, and which should prove the more startling fact, have simply been left undone because the services could not get together for such operations—they remained "unthinkable" under conditions of departmentalization and "departmentality." However, super-departmentality is to be evolved, or even maintained after the present conflict. That which the stern god of war has wedded,[11] in a shot-gun marriage, to the accompaniment of a number of painful incidents, softer-minded indulgent peace must not be allowed to divorce.[12]

The end of the Second World War produced the atomic bomb. This in turn might be said to have brought on the end of that war, or, in the minds of the hopeful and the fearful, the end of war generally. As regards the attack and defense of islands, the appearance of the bomb finished a complete pendulum swing from a high degree of insular invulnerability to an equally high degree of vulnerability. In the history of war islands once were, or seemed untouchable. This is no longer the case. (Was perhaps the result of the British elections of 1945 an indication of the strong sense of vulnerability, politically expressed?) Henceforth, so far as they can be reached by air power—and which island could not?—they offer, the smaller in size the more clearly, the perfect target, the most concentrated target for the most concentrated explosive. So much so, that while the atomic bomb and its carriers prevail, a large continental land mass will prove less vulnerable than any island.

The atomic bomb, to date the greatest single piece of "destruction engineering," which now has become the dominant aspect of war-making, was not the product of any of the traditional departments organized and maintained for the making of war. It was extra- (or supra-) departmental in its conception, its technical work-out, its application. Even before it had arrived, experiences in combined and landing operations had provided one of the several incitements for officers and civilians to consider, if not to demand, the unification of services in the United States.

NOTES, CONCLUSION

[1] General Percin, *Le massacre de notre infanterie.* Paris, n. d., 159.

[2] In the Senate, August 15, 1944. For German intentions of bombing the U. S. with V-2 rocket projectiles, see *N. Y. Times,* June 14, 1945.

[3] *The Times,* October 1, 1943.

[4] It might have been due to the fear of a more or less complete return to pre-war conditions and set-ups in post-war times that General MacArthur recommended to the U. S. War Department the retention as part of the permanent post-war army of a corps of what for short might be called landing engineers. For details see *N. Y. Times,* April 24, 1945.

[5] Many of these reasons were brought out during the Congressional hearings on unification of the services beginning in April 1944, others in an article by the then Senator Truman in *Collier's Weekly,* August 21, 1944.

[6] A. P. despatch from Washington, September 30, 1944.

[7] *N. Y. Times,* May 11, 1945.

[8] To the West Point cadets. *Newsweek,* July 2, 1945.

[9] For a summing up of the stand taken on unification by the services see Hanson W. Baldwin in *N. Y. Times,* June 8, 1945.

[10] *Ibd.,* April 21, 1945.

[11] "A Military Wedding." Combined Operations. By Comdr. Thomas W. Jones. *U. S. Naval Institute Proceedings,* December, 1944.

[12] "Now there is one thing you have in war that you do not have in peace. You have unification, compelling by a very threatening danger. In other words, Franklin's old saying, 'If we don't hang together, we'll hang separately,' applies in war more definitely than it does in peace." General Eisenhower A. P. despatch from Paris, June 15, 1945.

INDEX: GEOGRAPHICAL

INDEX: TOPICAL

INDEX: PERSONS

To T. V., with the understanding that we share the beach head!

-JET.